Advanced Applications in Heat Exchanger Technologies

Advanced Applications in Heat Exchanger Technologies presents the most recent developments in enhancing heat exchanger performance, reliability, and resilience, including the implementation of artificial intelligence, machine learning, and additive manufacturing.

Covering the essential parts of many commercial endeavors, ranging from aerospace to marine applications to oil and gas, the book discusses various heat exchanger types and interdisciplinary industry applications. It encompasses several different techniques, such as nanofluids, microchannel heat exchangers, computer modeling, advanced manufacturing, and optimization. The book addresses real-world concerns that impact long-term heat exchanger performance and dependability, such as fouling, corrosion prevention, and maintenance measures.

This book is intended for researchers and graduate students who are interested in heat exchanger R&D and the diverse range of industrial applications of heat exchanger technologies in contemporary practice.

Advanced Applications in Heat Exchanger Technologies

AI, Machine Learning, and Additive Manufacturing

Edited by
Sunil Kumar, Kavita Rathore, and Debjyoti Banerjee

CRC Press
Taylor & Francis Group
Boca Raton London New York

CRC Press is an imprint of the
Taylor & Francis Group, an **informa** business

First edition published 2026
by CRC Press
2385 NW Executive Center Drive, Suite 320, Boca Raton FL 33431

and by CRC Press
4 Park Square, Milton Park, Abingdon, Oxon, OX14 4RN

CRC Press is an imprint of Taylor & Francis Group, LLC

ISBN: 978-1-032-83544-0 (hbk)
ISBN: 978-1-032-83545-7 (pbk)
ISBN: 978-1-003-50983-7 (ebk)

DOI: 10.1201/9781003509837

Typeset in Times
by Deanta Global Publishing Services, Chennai, India

Contents

Preface

Heat exchangers are critical components in power generation, chemical processing, refrigeration, and heating, ventilation, and air conditioning (HVAC) systems, and they facilitate energy transmission. It is critical to maximize their thermal performance since it has a direct influence on energy efficiency, operational costs, and environmental sustainability. In light of this, we are pleased to provide the edited book *Advanced Applications in Heat Exchanger Technologies: AI, Machine Learning, and Additive Manufacturing*. The goal of this book is to offer a complete review of the most recent breakthroughs in heat exchanger thermal performance enhancement strategies. Our contributing authors have meticulously assembled current information and focused on the latest technical advancements designed to increase heat exchanger thermal efficiency. This compilation attempts to bridge the gap between fundamental heat transfer concepts and their practical applications, appealing to a varied audience of thermal engineering researchers, engineers, practitioners, and students. This book's material covers a wide range of classic and contemporary strategies for improving heat exchanger performance. Novel extended surface configurations, surface treatment methods, fluid flow manipulation, nanofluids, phase-change materials, microchannel heat exchangers, additive manufacturing, computational modeling, and optimization are all thoroughly investigated. The presented chapters delve into providing useful insights for readers wishing to improve their grasp of thermal performance optimization. In addition to these advanced subjects, the book offers real-world case studies and experimental findings demonstrating the usefulness of various heat exchanger layouts. Furthermore, it discusses the difficulties encountered in attaining thermal performance optimization while considering practical features and concerns such as fouling, corrosion, cleaning, and fouling avoidance measures. Readers will get useful insights into heat exchanger maintenance, long-term performance, and dependability issues because of in-depth discussions. Our major objective is to encourage readers' inventive thinking by promoting the ideation and investigation of unique strategies that will drive the future improvement of thermal performance enhancement techniques for heat exchangers. We hope to motivate readers to be active contributors to innovation in this sector by supporting this holistic approach that includes modeling, simulations, design, development, manufacturing, assembly, and testing. The book also serves as a useful rapid reference for heat transfer improvement strategies in a variety of industrial applications that include analytical, computational, and experimental methodologies. We would like to express our deepest appreciation to all the contributing authors for their remarkable knowledge and devotion in crafting the content of this book. We would also like to thank CRC Press for providing us with a venue to display this extraordinary array of modern thermal performance-enhancing strategies. We are convinced that this book will be an invaluable resource for thermal engineering academics, engineers, practitioners, and students. Whether improving heat exchangers in power generation, chemical processing, refrigeration, or HVAC systems, the useful insights included within these pages will guide experts toward more efficient, sustainable, and innovative heat exchanger solutions.

Editors

Editors

Sunil Kumar is an Associate Research Scientist in the Department of Biological and Agricultural Engineering at Texas A&M University (TAMU). He earned his PhD in Nuclear Engineering, specializing in nuclear reactor thermal hydraulics. His international experience includes developing radiation detectors at the European Organization for Nuclear Research (CERN), Switzerland, and working on high-energy ion cleaning at the Max-Planck Institute for Nuclear Physics (MPIK), Germany. Dr. Kumar possesses expertise in experimental and computational heat and mass transfer, including heat exchangers, aerosol technologies, HVAC systems, and materials drying. His contributions span a range of impactful projects, such as developing passive safety systems for the Advanced Heavy Water Reactor (AHWR), enhancing heat exchanger performance, improving electronic cooling, modeling indoor environments, drying materials, and studying particulate emissions. A recognized expert in his field, Dr. Kumar has delivered numerous conference talks, served as an invited speaker, and contributed as an editorial panel member for the *International Journal of Thermofluid Science and Technology.* He has also chaired sessions at the Thermal and Fluids Engineering Conference (TFEC) organized by the American Society of Thermal and Fluids Engineers (ASTFE) and has reviewed for prestigious journals, including those published by ASME, Elsevier, MDPI, AIP, and others. Dr. Kumar has published several peer-reviewed journal articles, conference papers, and book chapters. He also teaches a graduate-level course at TAMU, further demonstrating his commitment to advancing education and research in his field.

Kavita Rathore is a Senior Postdoctoral Researcher and Stanton Nuclear Security Fellow at Texas A&M University (TAMU). She earned her PhD in Mechanical Engineering from the Indian Institute of Technology Kanpur, India, and is the first Indian woman to receive the prestigious Stanton Fellowship at Texas A&M University, awarded by the Stanton Foundation. Dr. Rathore's expertise encompasses thermal fluid systems, nuclear reactor thermal hydraulics, heat exchangers, nuclear nonproliferation, and the application of plasma in energy generation, decontamination, and heavy oil processing. Her research contributions are reflected in numerous peer-reviewed journal publications and presentations at national and international conferences, where she has been recognized with many awards. In addition to her academic accomplishments, Dr. Rathore holds US and Indian patents, further demonstrating her innovative contributions to the field. She has served as a peer reviewer for research journals, organized conferences, delivered numerous invited talks, and presented at prominent conferences worldwide. Dr. Rathore's work exemplifies her dedication to advancing knowledge and innovation in energy, nuclear security, and applications of plasma.

Debjyoti Banerjee is a Professor of Mechanical Engineering at Texas A&M University (TAMU). He was elected as a Fellow of the American Society of Mechanical Engineers (ASME) in 2016 and as the Fellow/Dean's Fellow of EnMED (the Inter-Collegiate School of Engineering Medicine) in 2019/2021. He was invited to the membership of the Scientific Council (SC) of the International Committee on Heat and Mass Transfer (ICHMT) in 2023. He is currently an Associate Editor (AE) of the *ASME Open Journal of Engineering.* He served as an AE for the *ASME Journal of Energy Resources and Technology* as well as the *International Journal of Electronics.* He also served as a Guest Editor (GE) of special issues on machine learning and artificial intelligence. He is an invited Member of the Editorial Board of the *International Journal of Interfacial Phenomena and Heat Transfer.* He has served multiple times as AE and GE (for special issues) of the *ASME Journal* *of Heat Transfer/Journal of Heat and Mass Transfer* and the *ASME Journal of Nanotechnology in Engineering and Medicine.* He has received 17 US patents and more than 200 peer-reviewed publications from his research work involving bio-/micro-/nanotechnologies, phase change heat transfer (multiphase flows), and engineering ethics, at Applied Biosystems Inc./ABI (Applera/ Life Technologies), NanoInk Inc., Coventor Inc., Tata (India), and TAMU.

Contributors

Debjyoti Banerjee
Texas A&M University
Vikas Bargah
C.V. Raman Global University
India

Tapasvi Bhatt
BITS Pilani, Pilani Campus
India

Suvanjan Bhattacharya
BITS Pilani, Pilani Campus
India

Amit Choudhari
Cleveland State University
Niloofar Fatemipayam
Oklahoma State University

Gaurab Kumar Ghosh
Adarsha College of Engineering
India

Kaustav Moni Goswami
Centre of Rajiv Gandhi Institute of Petroleum Technology
India

Mayank Goswami
Indian Institute of Technology Roorkee
India

Ashish Kumar Gupta
Oklahoma State University

Shashikant Gupta
GD Goenka University
India

Mohd Naqueeb Shaad Jagirdar
Centre of Rajiv Gandhi Institute of Petroleum Technology
India

Naman Jain
BITS Pilani, Pilani Campus
India

Subrata Jana
Mody University of Science and Technology
India

Hilol Joldar
C.V. Raman Global University
India

Prathamesh P. Khandekar
Institute of Chemical Technology
Mumbai, India

Sumit Khatri
BITS Pilani, Pilani Campus
India

Sunita Khod
Indian Institute of Technology Roorkee
India

Abhishek Kumar
Texas A&M University

Sunil Kumar
Texas A&M University

Chaitanya R. Mali
CSIR-National Chemical Laboratory
India

Nancy Maurya
BITS Pilani, Pilani Campus
India

Kuber Nash Mishra
OP Jindal University
India

Sudhansu S. Mishra
Government College of Engineering
India

Taraprasad Mohapatra
C.V. Raman Global University
India

Vaibhav More
Institute of Chemical Technology, Mumbai
India

Hakeem Niyas
Centre of Rajiv Gandhi Institute of Petroleum Technology
India

Sikta Panda
Adarsha College of Engineering
India

Aditya Rane
Oklahoma State University

Shreesha Yogish Rao
Lam Research Corporation
Fremont, California

Kavita Rathore
Texas A&M University

Sudhansu S. Sahoo
Odisha University of Technology and Research
India

Mithilesh Kumar Sahu
OP Jindal University
India

Md Sazzadul Islam Sajol
Oklahoma State University

Prathamesh Sapale
Oklahoma State University

Arya Shah
Institute of Chemical Technology
Mumbai, India
India

Karan Sharma
Institute of Chemical Technology
Mumbai, India

Mohit Sharma
Medi-Caps University
India

Vaishali Sharma
Indian Institute of Technology Roorkee
India

Prashant Singh
University of Tennessee

Ravi Teja A. Tirumala
Oklahoma State University

Abhishek Tiwari
Indian Institute of Technology
Hyderabad, India

Dinkar Verma
GD Goenka University
India

Manish D. Yadav
Institute of Chemical Technology
Mumbai, India

Acknowledgments

We extend our gratitude to CRC Press (Taylor & Francis Group) and their esteemed editorial team for their invaluable suggestions and unwavering support throughout the completion of this book. We are also deeply appreciative of CRC Press for providing a platform to showcase this exceptional compilation of modern thermal performance enhancement strategies.

Our sincere thanks go to all the contributing authors and reviewers for their insightful perspectives on each chapter of the book. Their remarkable expertise and dedicated efforts have been instrumental in shaping the content of this book: *Advanced Applications in Heat Exchanger Technologies: AI, Machine Learning, and Additive Manufacturing.* Their contributions have greatly improved the quality and depth of this work, and we are profoundly grateful for their commitment and excellence.

This book is dedicated to all engineers, researchers, and academicians.

Editors

1 Heat Transfer Fundamentals and Introduction to Heat Exchangers

Sunil Kumar, Kavita Rathore, and Debjyoti Banerjee

1.1 INTRODUCTION

Heat transfer is a cornerstone of thermal engineering and a fundamental process governing energy exchange across various systems and applications. This chapter delves into the essential principles of heat transfer—conduction, convection, and radiation—providing a comprehensive foundation for understanding their mechanisms, mathematical formulations, and interdependence in relevance to heat exchangers. These principles not only dictate the movement of thermal energy but also enable the design and optimization of systems for efficient thermal management across industries. Heat transfer occurs whenever there is a temperature gradient, with energy naturally flowing from regions of higher temperature to lower temperature to achieve thermal equilibrium. The complexity of this process lies in the simultaneous interplay of different heat transfer modes. For instance, the boiling of water on a stove illustrates how conduction transfers heat through the pot, convection circulates thermal energy within the water, and radiation contributes energy from the flame to the pot's surface. Each mode adheres to specific physical laws: Fourier's law for conduction, Newton's law of cooling for convection, and the Stefan–Boltzmann law for radiation offering distinct pathways for energy transfer. Advances in material science, computational modeling, and technological innovations have significantly enhanced the understanding and application of heat transfer. From the atomic-scale interactions driving conduction to the macroscopic fluid flows characterizing convection, and the electromagnetic waves underpinning radiation, the principles of heat transfer provide the basis for designing high-performance systems. Whether optimizing industrial processes, advancing renewable energy technologies, or enhancing everyday appliances, the study of heat transfer remains pivotal to achieving energy efficiency, sustainability, and operational excellence [1]. This chapter systematically explores the core principles of conduction, convection, and radiation, highlighting their unique contributions and mutual dependencies. It also sets the stage for subsequent discussions on heat exchangers, thermal performance enhancement, and applications across diverse industries. By understanding these foundational principles, readers will gain the analytical tools needed to tackle complex thermal challenges and innovate in the ever-evolving field of heat transfer.

1.2 PRINCIPLES OF CONDUCTION, CONVECTION, AND RADIATION

Heat transfer is a fundamental process that occurs whenever there is a temperature gradient between two objects, with energy naturally flowing from the region of higher temperature to the region of lower temperature. This process ensures thermal equilibrium over time and is governed by well-defined physical principles. The direction of heat transfer always aligns with decreasing temperature, moving from a hotter object to a colder one. While the mechanisms of heat transfer conduction, convection, and radiation are governed by distinct laws and principles, they often occur simultaneously

DOI: 10.1201/9781003509837-1

in real-world scenarios, enabling complex thermal interactions [2]. For instance, boiling water on a stove is a classic example where all three modes of heat transfer operate concurrently. Conduction, the transfer of thermal energy through a solid medium, plays a critical role as heat flows from the stove's metal surface into the base of the pot. This process depends on the thermal conductivity of the materials involved, with metals being highly efficient conductors. Boiling models have been developed by researchers for different engineering applications [3]. Simultaneously, radiation contributes to the heat transfer process as thermal energy is emitted from the flame of the gas burner in the form of electromagnetic waves. This energy radiates through space and is absorbed by the bottom of the pot, further increasing its temperature. Finally, convections govern heat transfer within the water itself. As the pot's base heats up, the water in contact with it also warms, becoming less dense and rising to the surface. Cooler, denser water then sinks to replace it, creating a convective current that evenly distributes heat throughout the liquid. This interplay of conduction, radiation, and convection ensures efficient heat transfer, ultimately bringing the water to the boil. This example illustrates how the three modes of heat transfer are interconnected and often operate in tandem, each contributing uniquely to the overall thermal process.

1.2.1 Conduction Heat Transfer

Heat conduction is the transfer of thermal energy through direct molecular or atomic interactions without bulk motion of the material. In fluids, this process occurs via inelastic molecular collisions, while in nonconductive solids, it is facilitated by lattice vibrations or atomic oscillations. In conducting solids, such as metals, conduction is primarily governed by the motion of free electrons, which enables significantly higher rates of heat transfer. Examples of conduction include the transfer of heat from a heated cooking utensil to a hand, from a soldering iron to solder, and from a warm roof surface to snow. In opaque solids, conduction is the sole mechanism for heat transfer. In contrast, transparent solids, such as glass, quartz, and certain plastics, can also transfer heat via radiation due to their optical properties. In moving fluids, conduction is often complemented by convection, and in transparent fluids, by radiation as well. Materials exhibit a wide range of thermal conductivities: metals have the highest values due to their electron-based energy transfer; nonmetals have lower values due to phonon-based conduction, and liquids and gases exhibit progressively lower thermal conductivities due to their molecular structure and density [4].

Materials with low thermal conductivity are classified as thermal insulators, and their effectiveness can be enhanced by altering their physical structure. For example, highly conductive solids can function as effective insulators when processed into dispersed forms such as granules, powders, fibers, or foams. These forms increase the path length for heat transfer while reducing the effective cross-sectional area, significantly impeding the overall heat flow. Common examples of such insulation materials include mineral wool, glass fiber, diatomaceous earth, glass foam, extruded polystyrene foam (e.g., Styrofoam), corkboard, Celotex, and magnesia. In contrast, materials with high thermal conductivity are crucial in applications requiring efficient heat dissipation. For instance, high-conductivity thermal adhesives are utilized to bond fins to microchips, facilitating the rapid transfer of heat from the chip to the fins. The fins, in turn, increase the surface area available for convective heat transfer, enabling effective thermal management in electronic systems. This dual manipulation of thermal conductivity—enhancing it for heat dissipation or reducing it for insulation—underscores the versatility of heat conduction principles in diverse applications.

1.2.2 Convection Heat Transfer

Heat transfer by convection is a fundamental process wherein thermal energy is transferred through fluid motion, combining molecular interactions and bulk transport. This dual mechanism distinguishes convection from conduction, where heat transfer occurs solely through molecular collisions without the transport of matter. Convection is mathematically described by Newton's law of cooling, which relates the heat flux to the temperature difference between a surface and a fluid. A key parameter in this relationship is the heat transfer coefficient, which is influenced by both the thermophysical properties of the fluid such as density, viscosity, specific heat, and thermal conductivity, and the physical characteristics of the system, including fluid velocity, flow geometry, and surface roughness [5, 6]. The heat transfer coefficient encapsulates the complexities of convective processes, bridging the gap between theoretical models and real-world heat exchange systems.

The molecular transport mechanisms in convection involve diffusion and advection, often occurring simultaneously. Diffusion refers to the transport of thermal energy driven by gradients, such as temperature or concentration gradients, within the fluid. This mechanism is localized and relies on molecular interactions to redistribute energy. In contrast, advection involves the bulk motion of fluid particles, such as water flowing in a river or air moving over a heated surface, allowing for more efficient thermal energy transport over larger distances. For instance, in a river where the bottom water is colder than the surface water, heat is initially transferred by conduction from the warmer riverbed to the cooler water at the interface. The bulk motion of the fluid then distributes this energy through advection, with diffusion further smoothing out temperature gradients within the fluid.

Convection at the interface between a fluid and a surface inherently relies on conduction. Heat transfer begins with conduction across the thin boundary layer at the interface, where molecular collisions enable energy exchange between the surface and the fluid. This conduction process serves as the starting point for subsequent convective transport within the fluid, driven by either natural or forced mechanisms. In natural convection (or buoyant convection), fluid motion arises due to density variations caused by temperature differences, such as the upward movement of warm air and the descent of cooler air. In forced convection, external forces such as fans, pumps, or stirrers drive fluid motion, enabling controlled and often more efficient heat transfer.

Convective heat transfer is critical to numerous engineering applications and natural phenomena. For example, in steam or hot-water radiators, most of the heat supplied to a room is transferred through convective currents, with warm air rising and circulating to distribute thermal energy. Similarly, the heating and cooling systems in vehicles rely heavily on forced convection, where air circulation, driven by fans, ensures uniform temperature regulation. In water heaters, heat generated by the combustion of natural gas is transferred to the water through convection, providing efficient and uniform heating. Natural convection also explains why lakes and ponds freeze at the surface rather than the bottom. As water cools to 4°C, it reaches its maximum density and sinks, allowing less dense, colder water to rise and freeze at the surface.

In systems involving phase changes, convection plays an even more significant role. Phase transitions such as evaporation, condensation, sublimation, and solidification are associated with the absorption or release of latent heat, dramatically enhancing the overall heat transfer. Machine learning models are further improving the latent heat storage systems for different engineering applications [7, 8]. For instance, the human body dissipates excess thermal energy through the evaporation of sweat, a process that combines natural convection with the latent heat of vaporization to cool the skin below ambient temperature. Similarly, the sublimation of dry ice (solid carbon dioxide) absorbs substantial amounts of heat, making it a highly effective cooling agent in cryogenic applications. In industrial boilers, heat transfer from combustion gases through metal tube walls converts water into steam, with the latent heat of vaporization significantly contributing to the system's efficiency.

The study of convective heat transfer is vital for the optimization of thermal systems in engineering. The mechanisms of diffusion and advection, coupled with the interdependence of convection and conduction, enable the design of efficient systems for heating, cooling, and energy transfer. Phase-change phenomena further amplify the effectiveness of convective processes, finding applications in diverse fields such as power generation, electronics cooling, environmental management, and biomedical systems [9]. By employing advanced modeling techniques, such as computational fluid dynamics (CFD), and experimental methods, engineers can predict, analyze, and optimize convective heat transfer in increasingly complex systems, meeting the demands of modern technological and environmental challenges [5, 10, 11].

1.2.3 Radiation Heat Transfer

All materials, regardless of their temperature, continuously emit thermal radiation in all directions. Unlike conduction and convection, radiation heat transfer does not require a material medium for the transfer of energy; it can occur in a vacuum. Plasma is one of the best examples of such heat transfer [12, 13]. This unique property underpins fundamental natural processes, such as the transfer of energy from the Sun to the Earth through the vacuum of space. Radiation heat transfer is governed by the Stefan–Boltzmann law, which states that the power radiated by a surface is proportional to the fourth power of its absolute temperature (T^4). The intensity and spectral distribution of this radiation can also be described using Planck's law, while Wien's displacement law explains the shift of peak radiation to shorter wavelengths as temperature increases [14].

The process of radiation heat transfer occurs between surfaces at different temperatures and involves the emission, absorption, reflection, and transmission of electromagnetic waves. The efficiency of these processes is determined by material-specific properties such as emissivity (ε), absorptivity (α), reflectivity (ρ), and transmissivity (τ). Emissivity, for instance, quantifies how effectively a surface emits radiation compared with a perfect blackbody, while absorptivity represents the fraction of incident radiation absorbed by a material. Reflectivity and transmissivity describe how much radiation is reflected or transmitted, respectively. The net energy transferred by radiation is determined by the difference between the radiation emitted by a surface and the radiation absorbed from its surroundings. Surfaces with high emissivity, such as matte black finishes, are efficient radiative emitters and absorbers, whereas surfaces with high reflectivity, like polished metals, emit and absorb less radiation.

Radiation heat transfer is inherently spectral, meaning the intensity of emitted radiation depends on the wavelength of the electromagnetic waves. For solids, the emitted radiation typically spans a continuous spectrum, including wavelengths in the infrared, visible, and sometimes ultraviolet regions. The spectral intensity and peak wavelength are temperature-dependent, as described by Wien's displacement law, which states that the wavelength corresponding to the maximum radiation intensity inversely correlates with the absolute temperature. For example, as an iron bar is heated to around 800 K (approximately 525°C or 1000°F), its radiation shifts from the infrared region, invisible to the human eye, to the visible spectrum, causing the bar to glow dark red. At higher temperatures, the emission shifts further toward shorter wavelengths, intensifying and transitioning to orange or yellow hues. This phenomenon is similarly observed in the filament of an incandescent bulb, where increased voltage raises the filament's temperature, shifting its emission from red to yellow as the temperature increases and radiation becomes more intense at shorter wavelengths.

Liquids and gases exhibit more selective spectral behavior compared with solids. Their molecular structures allow them to interact with specific wavelengths of radiation, making them transparent to certain wavelengths while being highly absorptive at others. For instance, molecular oxygen in the Earth's atmosphere absorbs all vacuum ultraviolet radiation (<200 nm) from the Sun, shielding life from harmful high-energy radiation. Similarly, greenhouse gases like carbon dioxide (CO_2) and methane (CH_4) are highly effective absorbers of infrared radiation, trapping heat in the atmosphere and contributing to global warming. Organic liquids also exhibit selective absorption bands in the infrared and ultraviolet regions, which are exploited in various scientific applications, such as spectroscopy and the development of heat transfer fluids.

The implications of heat transfer radiation extend across a broad spectrum of scientific and engineering disciplines. Many engineering systems and techniques, such as optical emission tomography, have been developed by harnessing the properties of the radiation spectrum [15, 16]. In spacecraft design, materials with high emissivity are used for thermal management in the vacuum of space, ensuring efficient heat dissipation. In building construction, low-emissivity glass is employed to minimize heat loss or gain by reflecting infrared radiation while allowing visible light transmission. Understanding the radiative properties of atmospheric gases is fundamental to developing accurate climate models and strategies for mitigating the effects of global warming. Additionally, the selective spectral properties of liquids, gases, and solids are harnessed in spectroscopy, enabling precise analysis of material compositions and thermal characteristics [14].

Radiation heat transfer plays a pivotal role in energy exchange processes across natural and engineered systems. Its reliance on electromagnetic wave propagation and its spectral dependency highlights the complexity and versatility of this mode of heat transfer. Through advanced modeling techniques, such as computational radiative transfer simulations, and experimental studies, researchers and engineers can deepen their understanding of radiative processes, enabling the development of innovative technologies in fields ranging from energy systems and environmental management to advanced material design and thermal control.

1.3 INTRODUCTION TO HEAT EXCHANGERS

Heat transfer is a critical operation in a wide array of industrial and engineering applications, where devices specifically designed to facilitate thermal energy exchange between two substances are referred to as heat exchangers. These systems are engineered to manage thermal energy efficiently, playing a pivotal role in processes such as power generation, chemical synthesis, heating, ventilation, and air conditioning (HVAC) systems, and industrial manufacturing. The design and operation of heat exchangers are grounded in the principles of thermodynamics and heat transfer, encompassing conduction, convection, and, in some cases, radiation. Heat exchangers function by utilizing physical separation between the working fluids, typically achieved through tubes, plates, or specialized surfaces, which act as the heat-transfer interface [17].

1.3.1 Classification and Types of Heat Exchangers

Heat exchangers are broadly classified based on their configuration, flow arrangement, and application. The two most prevalent designs in industrial applications are the tube-and-shell heat exchanger and the double-pipe heat exchanger, while plate heat exchangers are widely employed in compact and high-efficiency applications [18].

1.3.1.1 Tube-and-Shell Heat Exchangers

This design consists of a bundle of tubes enclosed within a cylindrical shell. One working fluid flows through the tubes, while another flows through the shell side. The tube walls serve as the primary heat-transfer interface, with convection dominating on both sides of the wall and conduction through the tube material facilitating heat exchange. This configuration is well-suited for applications requiring high pressure and temperature differentials, oil processing plants and recovery, and power plants. For example, shell-and-tube heat exchangers play an essential role in microplasma ball reactors for JP-8 hydrocarbon conversion by providing efficient heat transfer, precise thermal control, fouling resistance, and energy optimization. Their integration enhances operational efficiency, scalability, and sustainability, improving the economic and environmental feasibility of plasma-driven processing [19]. Common variants include U-tube, floating-head, and fixed-tube-sheet designs, each tailored to specific operational constraints such as thermal expansion, ease of cleaning, and flow characteristics. [20].

1.3.1.2 Double-Pipe Heat Exchangers

Double-pipe exchangers employ a concentric tube arrangement, with one fluid flowing through the inner tube and the other flowing in the annular space between the inner and outer tubes. This simple design provides flexibility in handling small to moderate heat loads and flow rates. The heat transfer is governed by the overall heat transfer coefficient (U), which depends on the convective heat transfer coefficients of the fluids, as well as the thermal conductivity of the tube material [21, 22].

TABLE 1.1
Comparison of Heat Exchangers with Typical Hydraulic Diameters and Surface Area to Volume Ratio

Heat Exchanger (HX) Type	Hydraulic Diameter D_h (mm)	Surface Area to Volume Ratio (A_S/V) (m^2/m^3)
Plate	10–20	180–350
Plain tube and shell and tube	6–40	60–600
Strip and louvered	0.5–10	350–7100
Automotive radiators	2.5–5	710–1500
Cryogenic	1.7–3.7	100–2500
Matrix, wire, and sphere	0.2–2.5	1500–18,000
Gas turbine rotary	0.5–1.2	3000-7100

1.3.1.3 Plate Heat Exchangers

Composed of thin, corrugated metal plates arranged in a stacked configuration, plate heat exchangers allow fluids to flow alternately between adjacent plates. This design maximizes the heat transfer area while maintaining a compact footprint, making it ideal for applications with limited space or where high thermal efficiency is required [23]. The corrugation enhances turbulence, increasing the Nusselt number (Nu) and improving convective heat transfer. Plate heat exchangers are widely used in food processing, pharmaceuticals, and HVAC systems [24, 25].

A comparison of diverse types of heat exchangers in terms of surface area and volume ratios is shown in Table 1.1.

1.3.2 Effect of Flow Configurations on Heat Exchangers

The arrangement of fluid streams significantly influences the thermal performance of heat exchangers. The three primary configurations are:

- *Parallel flow (concurrent).* Both fluids flow in the same direction. This arrangement provides a high-temperature gradient at the inlet but becomes less efficient as the fluids approach thermal equilibrium.
- *Counter flow.* The fluids flow in opposite directions, maintaining a larger temperature gradient throughout the length of the exchanger. This configuration achieves the highest log mean temperature difference (LMTD) and is thermodynamically superior for heat exchange.
- *Crossflow and mixed flow.* These configurations involve perpendicular or partially mixed flow arrangements. Crossflow exchangers are common in air-cooled heat exchangers and automotive radiators.

Heat transfer within heat exchangers is predominantly steady state, where temperatures remain constant at a given location over time. The temperature profile of the fluids continuously changes along the flow direction, creating a varying temperature gradient (dT/dx) that drives the heat transfer process. Analytical methods, such as the effectiveness-NTU (number of transfer units) method, are employed to evaluate heat exchanger performance under specific operating conditions [18, 26].

1.3.3 Heat Exchanger Applications and Scientific Insights

Heat exchangers are ubiquitous across industries, ranging from large-scale energy systems to everyday applications. Condenser and evaporator units in refrigeration and air conditioning systems utilize phase-change heat transfer, where the latent heat of vaporization significantly enhances the overall heat transfer rate. Similarly, ordinary cooking utensils, such as pots and pans, exemplify simple heat exchangers where conduction, convection, and radiation combine to transfer heat effectively. In industrial applications, heat exchanger performance is often evaluated through dimensionless groups such as the Reynolds number (Re), Prandtl number (Pr), and Péclet number (Pe), which characterize fluid flow and heat transfer regimes. Computational tools, including CFD, are increasingly employed to model complex flow and thermal interactions in heat exchangers, enabling the optimization of performance and energy efficiency [27].

1.3.4 Advanced and Emerging Technologies

Innovations in heat exchange design continue to evolve with advancements in material science and manufacturing techniques. Additive manufacturing (AM), for example, allows for the fabrication of intricate geometries that enhance heat transfer by promoting turbulence and increasing surface area [28, 29]. Materials with tailored thermal conductivity and corrosion resistance are being developed to extend the lifespan and efficiency of heat exchangers in extreme environments. Further, the integration of nanofluid-engineered fluids containing nanoparticles with high thermal conductivity has shown promise in enhancing the heat transfer capabilities of conventional systems. The use of advanced materials such as graphene and metallic foams is also gaining traction in high-performance heat exchanger applications.

1.4 HEAT TRANSFER ANALYSIS METHODS RELEVANT TO HEAT EXCHANGERS

There are several parameters on which heat transfer in heat exchangers is evaluated. Some of which are briefly discussed in this section.

1.4.1 Overall Heat Transfer Coefficient

The overall heat transfer coefficient (U) is a fundamental parameter in heat transfer. It is used for heat exchanger design and analysis, representing the total resistance to heat transfer through a composite medium. It helps in combining the effect of conduction, convection, and wherever required, radiation across multiple layers or interfaces. The "U" is used to quantify the rate of heat transfer per unit area and per unit temperature difference between the two fluids in a heat exchanger [30, 31]. Mathematically, the overall heat transfer coefficient is defined as

$$q = U.A.\Delta T_{eff} \tag{1.1}$$

Where

q is the rate of heat transfer (W),

U is the overall heat transfer coefficient (W/m²·K),

A is the heat transfer area (m²), and

ΔT_{eff} is the effective temperature difference (K), which is generally calculated as the LMTD or using effectiveness-NTU methods.

1.4.1.1 Overall Heat Transfer Coefficient for a Plane Wall

The overall heat transfer coefficient (U) accounts for the combined effects of thermal resistance due to conduction (resistance through the solid material separating the fluids), convection (resistance from the boundary layer of fluid on either side of the heat transfer surface), and fouling (resistance caused by deposits, scaling, or fouling on the heat exchange surfaces). In heat exchangers, the hot and cold fluids are generally separated by a conductive wall (as illustrated in Figure 1.1). This wall may be flat (e.g., in plate heat exchangers and fins) or cylindrical (e.g., in tube-in-tube heat

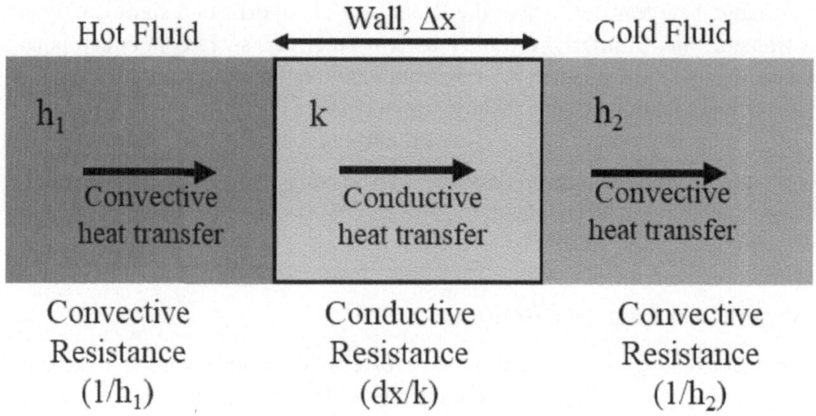

FIGURE 1.1 Heat transfer through a composite medium.

exchangers) [32, 33]. The overall heat transfer coefficient is related to the thermal resistance in the system as follows:

$$U = \frac{1}{A\,.\sum R_{th}} \tag{1.2}$$

Where A is the heat transfer area and $\sum R_{th}$ is total thermal resistance. Therefore, finding the overall heat transfer coefficient reduces the total thermal resistance of the system. In reference to Figure 1.1, the combined thermal resistance for heat transfer is given by:

$$\frac{1}{U} = \frac{1}{h_1} + \frac{\Delta x}{k} + \frac{1}{h_2} + R_{f,1} + R_{f,2} \tag{1.3}$$

Where h_1 and h_2 are convective heat transfer coefficients for hot and cold fluids respectively. Δx is thickness of solid wall, k is thermal conductivity of wall material (W/m · K), $R_{f,1}$ and $R_{f,2}$ are fouling resistance on hot and cold sides.

1.4.1.2 Overall Heat Transfer Coefficient for a Cylindrical Wall

Shell-and-tube-type heat exchangers typically fall under this category. The hot fluid could be flowing in the center tube or the outer tube. In both cases, the heat transfer across the center tube takes place. Figure 1.2 shows the schematic of flow of hot water at the center and cold water in the annulus created between the two pipes [34].

For a cylindrical wall, conductive thermal resistance is defined as

$$R_{th,\,cylinder} = \frac{ln\left(\frac{r_0}{r_i}\right)}{2\pi kl} \tag{1.4}$$

The convective thermal resistance is $(1/hA)$ and resistance is in series. The area to be considered has to be specified since the inner surface area and the outer surface area of the cylinder are different. In general form

$$UA = \frac{1}{\sum R_{th}} \tag{1.5}$$

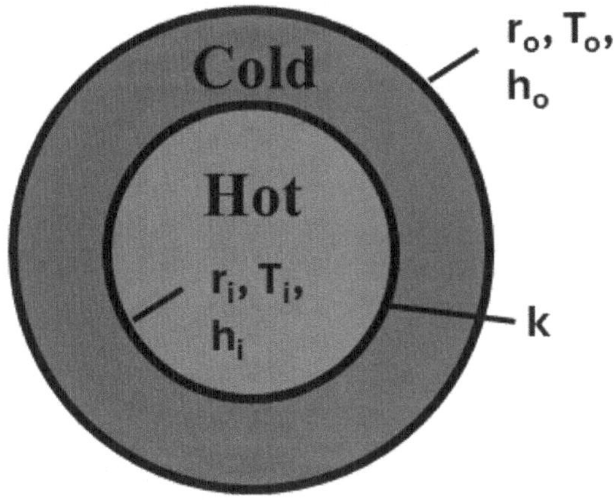

FIGURE 1.2 Heat transfer through a pipe wall.

It can be written as

$$U_oA_o = U_iA_i = \frac{1}{\sum R_{th}} \tag{1.6}$$

Considering the outer surface area, the overall heat transfer coefficient U becomes

$$U_oA_o = \frac{1}{\frac{1}{h_iA_i} + \frac{ln\left(\frac{r_0}{r_i}\right)}{2\pi kL} + \frac{1}{h_oA_o}} \tag{1.7}$$

Here A_o, A_i are outer and inner surface area. Then

$$U_o = \frac{1}{\frac{A_o}{h_iA_i} + \frac{ln\left(\frac{r_0}{r_i}\right)}{2\pi kL}A_0 + \frac{A_o}{h_oA_o}} \tag{1.8}$$

Thus

$$U_o = \frac{1}{\frac{1}{h_i}\left(\frac{r_o}{r_1}\right) + \left(\frac{r_o}{k}\right)ln\left(\frac{r_0}{r_i}\right) + \frac{1}{h_o}} \tag{1.9}$$

Similarly, if inner surface area is considered, the U becomes

$$U_iA_i = \frac{1}{\frac{1}{h_iA_i} + \frac{ln\left(\frac{r_0}{r_i}\right)}{2\pi kL} + \frac{1}{h_oA_o}} \tag{1.10}$$

In this case, U becomes

$$U_i = \frac{1}{\frac{A_i}{h_iA_i} + \frac{ln\left(\frac{r_0}{r_i}\right)}{2\pi kL}A_i + \frac{A_i}{h_oA_o}} \tag{1.11}$$

Or

$$U_i = \frac{1}{\frac{1}{h_i} + \frac{r_i}{k}.\ln\left(\frac{r_o}{ri}\right) + \frac{1}{h_o}\left(\frac{r_i}{ro}\right)} \tag{1.12}$$

However, if the thermal resistance of the wall is negligible (due to high thermal conductivity), compared with fluid or other resistance, the equation for universal heat transfer can be simplified to

$$U = \frac{1}{\frac{1}{h_i} + \frac{1}{h_o}} \tag{1.13}$$

The above equation indicates that "U" is primarily controlled by the smaller heat transfer coefficient. Thus, improvement on a smaller heat transfer coefficient will enhance the overall heat transfer coefficient.

In industries, the condensation of gases (or liquification) is one of the important processes. In a gas-to-liquid heat exchanger, the heat transfer coefficient is smaller on the gas side. Following the above equation, it is recommended to improve the smaller heat transfer coefficient for enhancing the overall heat transfer coefficient. Therefore, designers provide fins on the gas side surface for enhancing "U" [35–37]. In this scenario, the total heat transfer area on the gas side becomes

$$A_{total} = A_{unfinned} + A_{fin} \tag{1.14}$$

Where, $A_{unfinned}$ is area of tube without fins, and A_{fin} is surface area of fins. However, the fins have associated efficiency that depends on size, shape, pitch, and material. Additionally, the total surface area of tube and fins combined is not available for effective condensation. Let us define fin efficiency as η_{fin} and total surface efficiency as η_o. Following which above equation can be rewritten as

$$\eta_o A_{total} = A_{unfinned} + \eta_{fin} A_{fin} \tag{1.15}$$

Substituting the value of $A_{unfinned}$ from Equation 1.15, the above equation can be arranged as follows

$$\eta_o A_{total} = (A_{total} - A_{fin}) + \eta_{fin} A_{fin} \tag{1.16}$$

$$\eta_o = 1 - \frac{A_{fin}}{A_{total}}.(1 - \eta_{fin}) \tag{1.17}$$

Thus, for determining the overall heat transfer coefficient, $\eta_o A_{total}$ needs to be used for the finned surfaces irrespective of inner surface, outer surface, or both. Therefore, if the outer surface area of the tube is finned and thermal resistance of tube material is neglected then

$$U_i A_i = U_o A_o = \frac{1}{\frac{1}{h_i A_i} + \frac{1}{h_o (\eta_o A_{total})_{outer\_surface}}} \tag{1.18}$$

Typical values of the overall heat transfer coefficient for commonly used heat exchanger types are shown in Table 1.2.

1.4.1.3 Fouling Factor

As the time passes, the heat exchanger surfaces get dirty due to scaling, deposits, corrosion, and so on, resulting in reduction in the heat transfer coefficient due to thermal resistance to heat transfer.

TABLE 1.2
Overall Heat Transfer Coefficient for Commonly Used Heat Exchangers

Type of Heat Exchanger	Overall Heat Transfer Coefficient "*U*" (W/m²C)
Steam condenser	1000–6000
Water-to-water	850–1700
Feed water heaters	1000–8500
Water-to-oil	100–350
Steam-to-light fuel oil	200–400
Steam-to-heavy fuel oil	50–200
Freon condenser (water-cooled)	300–1000
Ammonia condenser (water-cooled)	800–1400
Gas-to-gas	10–40
Water-to-air in finned tubes	30–60
Steam-to-air in finned tubes	400–4000

Fouling generally develops due to scaling or precipitation, deposits of finely divided particulates, chemical reactions, corrosion, attachments of algae or other biological materials, and crystallization on surfaces due to subcooling [38]. The effect of fouling is accounted for by a term called the "Fouling factor":

$$R_f = \frac{1}{U_{dirty}} - \frac{1}{U_{clean}} \tag{1.19}$$

Thus, fouling factor R_f is zero for new heat exchangers. By taking into account the effect of fouling, "Fouling resistance" is added in the thermal resistance. Therefore, Equation 1.19 becomes

$$U_iA_i = U_oA_o = \frac{1}{\sum R_{th}} = \frac{1}{\dfrac{1}{h_iA_i} + \dfrac{R_{fi}}{A_i} + \dfrac{ln\left(\dfrac{r_0}{r_i}\right)}{2\pi kL} + \dfrac{1}{h_oA_o} + \dfrac{R_{fo}}{A_o}} \tag{1.20}$$

Here, R_{fi}, and R_{fo} are fouling factors for inside and outside surfaces, respectively. The fouling buildup is affected by flow velocity and operating temperature. Increase in fouling is observed at low fluid velocity and high temperature [39]. Some of the fouling factor values suggested by the Tubular Exchanger Manufacturers' Association (TEMA) are presented in Table 1.3.

1.4.1.4 Factors Affecting the Overall Heat Transfer Coefficient

Critical factors influencing the overall heat transfer coefficient are as follows [30, 41]:

- *Fluid properties.* Fluids with higher thermal conductivity and lower viscosity generally result in higher convective heat transfer coefficients, thereby increasing *U*. Gases have lower convective coefficients compared with liquids, leading to lower *U*
- *Flow regime.* Turbulent flow enhances convective heat transfer by reducing the boundary layer resistance, leading to higher *h* values and consequently a higher *U*. Laminar flow, in contrast, has a thicker boundary layer and lower *h*, resulting in a lower *U*.
- *Wall materials.* Materials with high thermal conductivity (e.g., copper, aluminum) reduce the conductive resistance $\Delta x/k$, which helps increase *U*. Insulating materials with low *k*, decreases *U*.

TABLE 1.3
Fouling Factor Values for Different Cooling Fluid Types in Heat Exchangers [40]

Fluid	Fouling Factor (R_f) (m²C/W)
Water (type: river, sea, distilled, and boiler feed) (if operating temperature below 50°C)	0.0001
Water (type: river, sea, distilled, and boiler feed) (if operating temperature above 50°C)	0.0002
Air	0.000119–0.000238
Steam (sat., oil-free)	0.000119–0.000357
Fuel oil	0.00088
Transformer oil	0.0007
Hydraulic fluid	0.000238
Molten salts	0.000119
Organic heat transfer media	0.000119
Vegetable oil	0.000715
Gasoline, kerosene	0.000238
Light gas oil	0.000476
Heavy gas oil	0.000715
Solvent vapors, acid gas, natural gas	0.000238

- *Surface conditions.* Clean and smooth surfaces are used to optimize U. Fouling, scaling, or corrosion increases the resistance (R_f)
- *Heat exchanger configuration.* The design (e.g., tube-and-shell, plate, or double-pipe) affects the surface area (A) and flow patterns, influencing U. Baffles and fins are often used to increase turbulence and heat transfer area, enhancing U.

1.4.2 The Logarithmic Mean Temperature Difference Method

In a heat exchanger, the transfer of thermal energy occurs between two fluids, typically one hot and one cold, across a separating surface. The temperature difference between the fluids is not constant but varies continuously along the flow path due to heat exchange dynamics. This variation plays a critical role in determining the rate of heat transfer. The LMTD or ΔT_{lm} method addresses this complexity by providing a mathematically derived average temperature difference that accurately represents the nonlinear temperature gradient across the heat exchanger. By incorporating the effects of these variations, the LMTD ensures a precise calculation of the heat transfer rate, making it a pivotal parameter in the design, performance evaluation, and optimization of heat exchanger systems. Its reliability and accuracy under steady-state conditions make it indispensable in engineering applications where efficiency and thermal performance are paramount [42, 43].

Heat exchangers are operated with minimal changes in their operation for a longer period. Therefore, heat exchangers are modeled as steady-state devices. Let us consider a section of a heat exchanger as shown in Figure 1.3. The location of notations T_{hi} and T_{ci} indicate that this section of heat exchanger belongs to a parallel flow heat exchanger.

Applying overall energy balances to the hot and cold fluids.

For hot fluid:

$$q = m_h C_{p,h} \left(T_{h,i} - T_{h,o}\right) \tag{1.21}$$

For cold fluid:

$$q = m_c C_{p,c} \left(T_{c,o} - T_{c,i}\right) \tag{1.22}$$

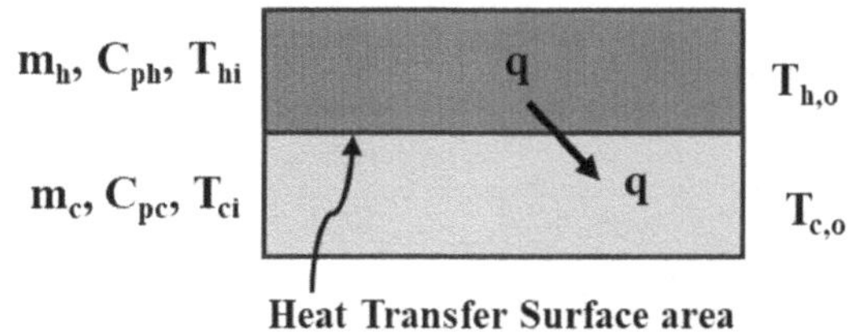

FIGURE 1.3 A section of a heat exchanger.

The following expression relates the total heat transfer rate q to the temperature difference ΔT between the hot and cold fluids, where

$$\Delta T = T_h - T_c \tag{1.23}$$

However, ΔT varies along the length of the heat exchanger and therefore a rate equation of the following form is needed:

$$q = UA\Delta T_m \tag{1.24}$$

Where U is overall heat transfer coefficient, A is surface area for heat transfer, and ΔT_m is the appropriate mean temperature difference. The mean temperature difference is used for the heat transfer analysis for different flow configurations in heat exchangers. For simplification in the following section only parallel and counter flow heat exchangers have been discussed.

1.4.2.1 Parallel Flow Heat Exchangers

In parallel heat exchangers, hot and cold fluid both flow in the same direction or enter from the left (or right) side of the heat exchanger as shown in Figure 1.4a.

In this configuration, the hot fluid keeps transferring heat to the colder fluid (Figure 1.4b, 1.4c) while flowing along the length of heat exchanger. Thus, the temperature of cold fluid keeps increasing. For thermal analysis, applying an energy balance to a differential element of thickness in the hot and cold fluids.

For hot fluid:

$$dq = -\dot{m}_h c_{p,h} dT_h = -C_h dT_h \tag{1.25}$$

For cold fluid:

$$dq = \dot{m}_c c_{p,c} dT_c = C_c dT_c \tag{1.26}$$

In this element, the heat transfer will be

$$dq = U.\Delta T.dA \tag{1.27}$$

$$d(\Delta T) = dT_h - dT_c \tag{1.28}$$

Upon integrating the above equation, the following could be found:

$$\int \frac{d(\Delta T)}{\Delta T} = -U\left(\frac{1}{C_h} + \frac{1}{C_c}\right)\int dA \tag{1.29}$$

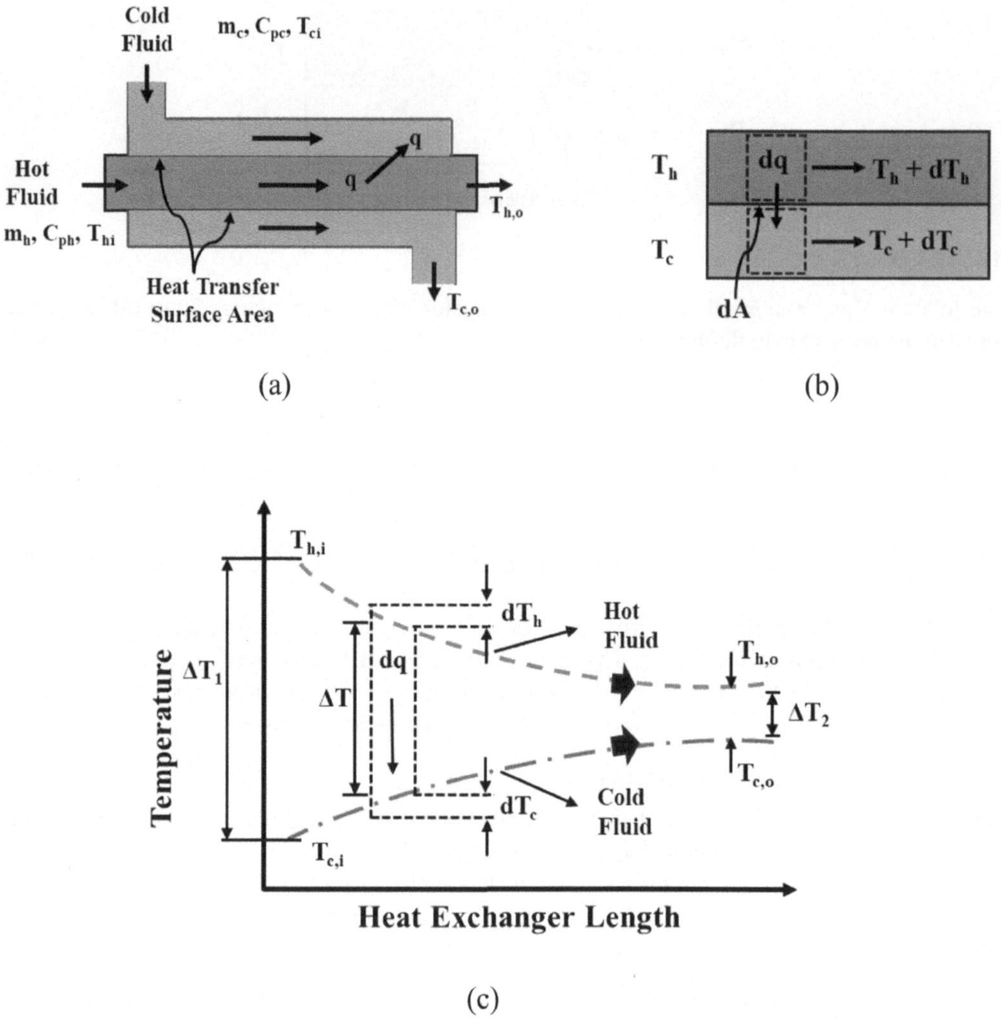

FIGURE 1.4 Illustration of heat exchanger flow and temperature behavior: (a) Parallel flow configuration; (b) a section of a heat exchanger with a differential element of length dx; (c) the temperature variation of hot and cold fluids along the length of a heat exchanger.

Which on taking integral indicates

$$\ln\left(\frac{\Delta T_2}{\Delta T_1}\right) = -UA\left(\frac{1}{C_h} + \frac{1}{C_c}\right) \tag{1.30}$$

Substituting for C_h and C_c:

$$ln\left(\frac{\Delta T_2}{\Delta T_1}\right) = -UA\left(\frac{\left(T_{h,i} - T_{h,o}\right)}{q} + \frac{\left(T_{c,o} - T_{c,i}\right)}{q}\right) \tag{1.31}$$

For a parallel flow heat exchanger:

$$\Delta T_1 = T_{h,i} - T_{c,i} \text{ and } \Delta T_2 = T_{h,o} - T_{c,o} \tag{1.32}$$

Thus, the above Equation 1.31 shows that

$$q = UA\frac{\Delta T_2 - \Delta T_1}{ln\left(\frac{\Delta T_2}{\Delta T_1}\right)} \tag{1.33}$$

Comparing this with the energy analysis Equation 1.24:

$$\Delta T_{lm} = UA\frac{\Delta T_2 - \Delta T_1}{ln\left(\frac{\Delta T_2}{\Delta T_1}\right)} \tag{1.34}$$

Where

$$\Delta T_1 = T_{h,1} - T_{c,1} \equiv T_{h,i} - T_{c,i} \tag{1.35}$$

$$\Delta T_2 = T_{h,2} - T_{c,2} \equiv T_{h,o} - T_{c,o} \tag{1.36}$$

1.4.2.2 Counter Flow Heat Exchangers

In counter flow heat exchangers, hot and cold fluid flows in the opposite direction or enters from the left and right side of the heat exchanger as shown in Figure 1.5.

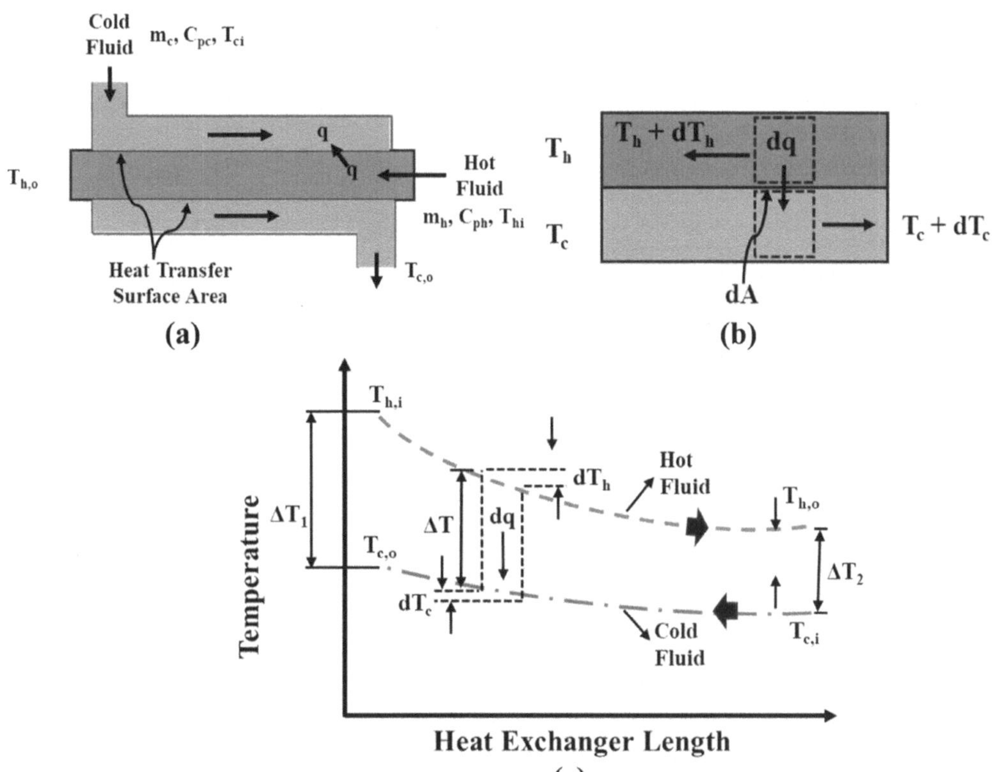

FIGURE 1.5 Illustration of counter flow heat exchanger configuration and associated temperature behavior: (a) counter flow configuration; (b) differential control volume element along the exchanger length; (c) temperature profiles of hot and cold fluids along the exchanger.

The heat transfer analysis performed for the parallel flow heat exchanger is also applicable to the counter flow heat exchanger. The equation for LMTD is thus

$$\Delta T_{lm} = UA\frac{\Delta T_2 - \Delta T_1}{ln\left(\frac{\Delta T_2}{\Delta T_1}\right)} \tag{1.37}$$

However, the temperature difference ΔT_1 and ΔT_2 are calculated as follows.

$$\Delta T_1 = T_{h,1} - T_{c,1} \equiv T_{h,i} - T_{c,o} \tag{1.38}$$

$$\Delta T_2 = T_{h,2} - T_{c,2} \equiv T_{h,o} - T_{c,i} \tag{1.39}$$

From the LMTD equation above and redefinition of temperature for the counter flow heat exchanger, it is found that for the same inlet and outlet temperatures, the log mean temperature difference for a counter flow configuration is higher than that for a parallel flow configuration.

$$\Delta T_{lm,\,CF} > \Delta T_{lm,\,PF} \tag{1.40}$$

For the same amount of heat transfer, counter flow heat exchangers need a smaller surface compared with parallel flow heat exchangers [44].

1.4.2.3 Correction Factors for Heat Exchangers Design

Counter flow heat exchangers are deployed to achieve a higher rating of heat transfer compared with parallel flow and the LMTD relations developed are limited to these two configurations only. Developing LMTD relations for the crossflow or a multipass shell-and-tube-type heat exchanger is complicated due to complex flow conditions involved in the process of heat transfer. Therefore, in these conditions, an equivalent temperature difference can be related to the LMTD for the case of counter flow heat exchanger with the help of a correction factor [45, 46]. So that

$$\Delta T_{lm} = F.\Delta T_{lm,CF} \tag{1.41}$$

$\Delta T_{lm,\,CF}$ will be calculated for a counter flow heat exchanger considering the same inlet and outlet temperature of the heat exchanger under the design or evaluation process. The correction factor depends on the geometry, inlet, and outlet temperature of the hot and cold fluid streams. Generally, $F \leq 1$, where a limiting value of 1 corresponds to the counter flow heat exchanger. In this manner, F provides a deviation of the LMTD from the corresponding value of a counter flow heat exchanger case.

Thus, for a shell-and-tube heat exchanger, where the cooling fluids are passing in a crossflow or multipass flow manner, heat transfer rate can be computed using the following equations:

$$q = F.U.A.\Delta T_{lm,CF} \tag{1.42}$$

Where $\Delta T_{lm,\,CF}$ is represented using Equation 1.37 or Equation 1.34 and shown below for reference:

$$\Delta T_{lm,\,CF} = UA\frac{\Delta T_2 - \Delta T_1}{ln\left(\frac{\Delta T_2}{\Delta T_1}\right)} \tag{1.43}$$

Here

$$\Delta T_1 = T_{h,1} - T_{c,1} \equiv T_{h,i} - T_{c,o} \quad (1.44)$$

$$\Delta T_2 = T_{h,2} - T_{c,2} \equiv T_{h,o} - T_{c,i} \quad (1.45)$$

The value of "F" is obtained using correction factor charts. Below is an example of applying the correction factor approach in a two-shell and four-tube tubular exchanger.

For a two-shell and four-tube TEMA class "E" type heat exchanger, in which stream 1 enters the heat exchanger at 40°C and exits at 80°C, while stream 2 on the shell side enters the heat exchanger at 150°C and exits at 100°C. The calculated heat capacity ratio (R), temperature effectiveness (P), and LMTD values are 1.2500, 0.3636, and 64.8716, respectively. Figure 1.6 shows the correction factor chart and highlighted curve in reference to capacity ratio 1.25. This provides a correction factor of 0.9799. Thus, the corrected LMTD value will be 63.5656.

1.4.3 The Effectiveness-NTU Method for Heat Exchanger Analysis

The LMTD method is a cornerstone in heat exchanger design and analysis, particularly well-suited for scenarios where all four terminal temperatures are known or can be readily determined. By leveraging the LMTD value, the required heat transfer area (A), a key parameter in defining the size of the heat exchanger, can be accurately calculated using the equation $Q = U \cdot A \cdot \text{LMTD}$. This makes the LMTD approach a reliable and efficient choice for sizing heat exchangers when complete temperature data is available. However, its application becomes challenging when the problem involves incomplete information, such as when only the inlet temperatures, flow rates, and the overall heat transfer coefficient are specified, and the outlet temperatures and heat transfer rate need to be determined. Solving such problems using the LMTD method often requires iterative calculations [48].

To address these limitations, the effectiveness-NTU (number of transfer units) method, developed by Kays and London in 1955, offers a robust and systematic alternative. Unlike the LMTD method, the effectiveness-NTU approach eliminates the need for detailed terminal temperature information, making it especially valuable in rating problems and off-design analysis. For example, this method enables engineers to assess the performance of a heat exchanger under varying operational conditions, such as when the flow rate of one fluid is reduced to 75% of the design value. By simplifying the process and avoiding iterative computations, the effectiveness-NTU method has

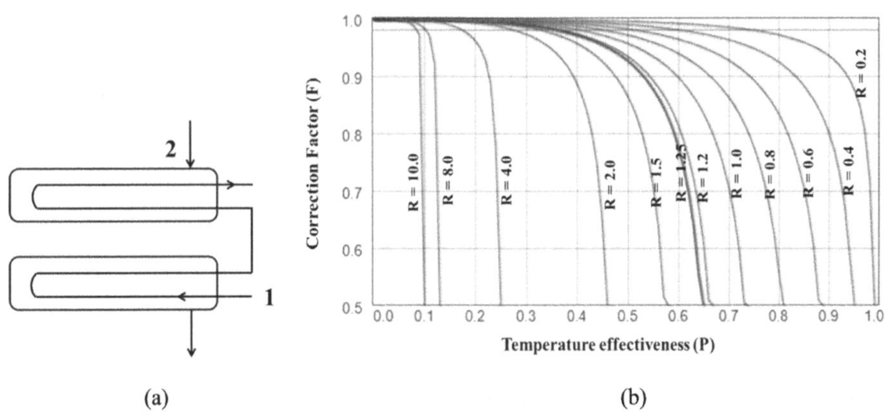

FIGURE 1.6 Representation of a multi-pass shell-and-tube heat exchanger and associated correction factor chart: (a) two-shell and four-tube pass TEMA class "E" type heat exchanger configuration; (b) correction factor (F) chart for use in LMTD calculations (source: https://checalc.com/solved/LMTDChart.html) [47].

become an essential tool for addressing complex heat exchanger challenges and ensuring optimal performance under dynamic operating conditions [48, 49].

This method is primarily based on the dimensionless parameter called the heat transfer effectiveness (ε), which is defined as ration of heat transfer rate and maximum heat transfer rate.

$$\varepsilon \equiv \frac{q}{q_{max}} \tag{1.46}$$

However, the actual heat transfer rate is given by

$$q = C_c\left(T_{c,o} - T_{c,i}\right) = C_h\left(T_{h,i} - T_{h,o}\right) \tag{1.47}$$

and the maximum possible heat transfer rate for any heat exchanger can be obtained using the maximum temperature difference, i.e., $\Delta T_{max} = T_{h,i} - T_{c,i}$

Here C is known as capacity rate, thus, if

$$C_c < C_h \text{ then } q_{max} = C_c\left(T_{h,i} - T_{c,i}\right) \tag{1.48}$$

Or if

$$C_h < C_c \text{ then } q_{max} = C_h\left(T_{h,i} - T_{c,i}\right) \tag{1.49}$$

Therefore

$$q_{max} = C_{min}\left(T_{h,i} - T_{c,i}\right) \tag{1.50}$$

Where

$$C_{min} = min\left(C_h, C_c\right) \tag{1.51}$$

and, in each case, capacity rate of the other fluid is designated as C_{max}. In this manner, the capacity ratio is defined as

$$C = \frac{C_{min}}{C_{max}} \tag{1.52}$$

Substituting these relationships for the effectiveness of the actual heat transfer rate, we get

$$\varepsilon = \frac{C_c\left(T_{c,o} - T_{c,i}\right)}{C_{min}\left(T_{h,i} - T_{c,i}\right)} \tag{1.53}$$

Or

$$\varepsilon = \frac{C_h\left(T_{h,i} - T_{h,o}\right)}{C_{min}\left(T_{h,i} - T_{c,i}\right)} \tag{1.54}$$

Therefore, if the quantities such as ε, $T_{h,i}$, $T_{c,i}$ are known, the calculations for the actual heat transfer rate can be performed using following equation:

$$q = \varepsilon C_{min}(T_{h,i} - T_{c,i}) \tag{1.55}$$

Hence, for any heat exchanger:

$$\varepsilon = f\left(NTU, {}^{C_{min}}\!\big/_{C_{max}}\right) \tag{1.56}$$

Where the Number Transfer Units (NTU), which is a dimensionless parameter, is defined as

$$NTU \equiv {}^{UA}\!\big/_{C_{min}} \tag{1.57}$$

The effectiveness of the heat exchanger typically depends on geometry, and the flow arrangement [49].

1.4.3.1 Effectiveness-NTU Correlations for Parallel Flow Heat Exchanger

For this correlation, let us assume that

$$C_{min} = C_h \tag{1.58}$$

Then using Equation 1.54, the following equation for effectiveness is obtained:

$$\varepsilon = \frac{\left(T_{h,i} - T_{h,o}\right)}{\left(T_{h,i} - T_{c,i}\right)} \tag{1.59}$$

In this parallel flow configuration, the heat transferred by the hot fluid to the cold fluid is

$$q = m_h c_{p,h}\left(T_{h,i} - T_{h,o}\right) \tag{1.60}$$

$$q = m_c c_{p,c}\left(T_{c,o} - T_{c,i}\right) \tag{1.61}$$

The capacity ratio is given by

$$C = \frac{C_{min}}{C_{max}} = \frac{m_h c_{p,h}}{m_c c_{p,c}} = \frac{T_{c,o} - T_{c,i}}{T_{h,i} - T_{h,o}} \tag{1.62}$$

Following the equations of LMTD for parallel flow heat exchanger

$$ln\left(\frac{T_{h,o} - T_{c,o}}{T_{h,i} - T_{c,i}}\right) = -\frac{UA}{C_{min}}\left(1 + \frac{C_{min}}{C_{max}}\right) \tag{1.63}$$

Or

$$\left(\frac{T_{h,o} - T_{c,o}}{T_{h,i} - T_{c,i}}\right) = exp\left[-NTU\left(1 + \frac{C_{min}}{C_{max}}\right)\right] \tag{1.64}$$

With rearrangement and simplification of the above equation, the following equation is obtained for the parallel flow heat exchanger:

$$\varepsilon = \frac{1 - exp\left[-NTU\left(1 + \left\{\frac{C_{min}}{C_{max}}\right\}\right)\right]}{1 + \left(\frac{C_{min}}{C_{max}}\right)} \tag{1.65}$$

$$\varepsilon = \frac{1 - exp\left[-NTU(1+C)\right]}{1+C} \tag{1.66}$$

Similar expressions can be developed for different heat exchanger flow configurations as shown in Table 1.4.

1.5 IMPORTANCE OF THERMAL PERFORMANCE ENHANCEMENT

Heat exchangers are fundamental components in a wide array of industrial applications, ranging from power generation and HVAC systems to automotive engines, chemical processing, and renewable energy systems. The thermal performance of heat exchangers significantly affects the overall efficiency, reliability, and environmental sustainability of these processes. Enhancing their thermal performance is, therefore, a critical focus for engineers and researchers [50, 51].

1.5.1 Significance of Thermal Performance Enhancement

1.5.1.1 Energy Efficiency and Optimization

Improved heat exchanger performance leads to better utilization of thermal energy, minimizing waste and reducing energy consumption. This directly impacts the energy efficiency of systems like power plants, refrigeration units, and industrial reactors. Research in this field highlights that optimizing heat exchangers can reduce energy consumption by up to 30%.

TABLE 1.4
Effectiveness Relationships for Different Flow Configurations of Heat Exchangers

Flow Arrangement in Different Heat Exchangers	Relations
Parallel flow	$\varepsilon = \frac{1 - exp\left[-NTU(1+C)\right]}{1+C}$
Counter flow	$\varepsilon = \frac{1 - exp\left[-NTU(1-C)\right]}{1 - exp\left[-NTU(1-C)\right]}$
Crossflow (single pass)	$\varepsilon = 1 - exp\left[\frac{1}{C}.(NTU)^{0.22}\left\{exp\left[-C\left(NTU^{0.78}\right)\right]-1\right\}\right]$
Shell and tube (one shell pass)	$\varepsilon = 2\left[1 + C + \left(1+C^2\right)^{\frac{1}{2}} \frac{1 + exp\left[-NTU\left(1+C^2\right)^{\frac{1}{2}}\right]}{1 - exp\left[-NTU\left(1+C^2\right)^{\frac{1}{2}}\right]}\right]^{-1}$
Shell and tube n-shell pass	$\varepsilon = \left[\left(\frac{1-\varepsilon C}{1-\varepsilon}\right)^n - 1\right] * \left[\left(\frac{1-\varepsilon C}{1-\varepsilon}\right)^n - C\right]^{-1}$
All exchangers (C = 0)	$\varepsilon = 1 - \exp(-NTU)$

1.5.1.2 Compact and Lightweight Designs

By enhancing thermal performance, heat exchangers can achieve higher heat transfer rates in smaller geometries, resulting in compact and lightweight designs. For example, microchannel and plate-fin heat exchangers have revolutionized industries like aerospace, where minimizing weight without compromising efficiency is critical.

1.5.1.3 Environmental Sustainability

Enhanced thermal performance reduces the demand for fuel in heat-driven systems, significantly cutting greenhouse gas emissions. This aligns with global climate goals, such as those outlined in the Paris Agreement. For instance, advancements in heat exchanger design in HVAC systems have reduced CO_2 emissions in buildings by 20–25%.

1.5.1.4 Economic Benefits

High-efficiency heat exchangers reduce operational costs by lowering energy usage and maintenance requirements. Additionally, the increased durability and reliability of modern heat exchangers reduce downtime, saving industries millions annually.

1.5.1.5 Adaptation to Advanced Applications

Efficient heat exchangers are essential in cutting-edge technologies like supercritical CO_2 cycles, concentrated solar power systems, and next-generation nuclear reactors (e.g., small modular reactors) [52, 53]. These applications require systems capable of handling extreme temperatures and pressures with minimal energy losses.

1.5.2 Advanced Methods for Thermal Performance Enhancement

1.5.2.1 Passive Techniques

Passive techniques rely on design and material innovations to improve heat transfer without external energy input [50, 54].

- *Extended surfaces and fins.* Extended surfaces increase the effective heat transfer area, enhancing thermal performance. Computational studies have shown that louvered fins in automotive heat exchangers can increase heat transfer by 25–35% compared with plain fins.
- *Surface modifications.* Incorporating grooves, dimples, or corrugations on heat transfer surfaces enhances turbulence, disrupting the boundary layer and improving convective heat transfer. Studies report that dimpled surfaces can achieve a 15–20% increase in heat transfer with minimal pressure drop penalties [55].
- *Nanofluids.* Nanofluids, created by dispersing nanoparticles such as A12O3, CuO, or graphene into base fluids, significantly improve thermal conductivity and convective heat transfer coefficients [56]. Studies demonstrate that nanofluids can increase thermal performance by up to 40%.
- *Phase change materials (PCMs).* PCMs utilize latent heat during phase transitions to enhance thermal storage and transfer capabilities [57, 58]. They are increasingly applied in renewable energy systems and electronic cooling including buildings [59, 60]. Composite PCMs with encapsulated nanoparticles further improve thermal conductivity and heat management.

1.5.2.2 Active Techniques

Active techniques use external energy to manipulate flow or heat transfer characteristics [61–63].

- *Mechanical agitation.* Devices such as rotating or vibrating inserts induce turbulence, enhancing mixing and heat transfer. For example, research on rotating tube heat exchangers shows a 30% increase in heat transfer performance.
- *Electrohydrodynamic (EHD) and magnetohydrodynamic (MHD) methods.* These methods utilize electric or magnetic fields to influence fluid flow, enhancing convective heat transfer. Studies indicate that MHD methods can increase heat transfer rates by up to 50% in electrically conductive fluids.
- *Thermally conductive coatings.* Advanced coatings like graphene or carbon nanotubes reduce thermal resistance and increase heat flux density. These coatings are particularly beneficial in high-temperature applications, such as gas turbines or concentrated solar power systems.

1.5.2.3 Hybrid Techniques

Hybrid techniques combine passive and active methods to maximize performance [64, 65].

- *Nanofluids with mechanical vibrations.* The combination of nanofluids and induced vibrations enhances heat transfer through increased turbulence and improved thermal conductivity. This synergy is especially effective in compact heat exchangers.
- *Extended surfaces with PCMs.* Integrating extended surfaces like fins with PCMs allows for efficient heat storage and transfer, particularly in thermal energy storage systems.

1.5.2.4 Optimization and Computational Techniques

Advanced computational methods and material innovations play a significant role in enhancing heat exchange performance [66].

- *Computational fluid dynamics.* CFD simulations enable precise modeling of complex heat transfer processes, allowing for the optimization of heat exchanger geometries and flow configurations. Studies have shown that CFD-informed designs can improve efficiency by 15–20%.
- *Additive manufacturing.* Additive manufacturing (3D printing) enables the creation of intricate geometries and internal structures, such as lattice designs, which enhance turbulence and heat transfer [67, 68]. Research published in *Additive Manufacturing* demonstrates that 3D-printed heat exchangers can achieve up to a 50% increase in efficiency compared with conventional designs.
- *Advanced materials.* Using materials with high thermal conductivity, such as aluminum alloys, copper composites, or advanced ceramics, significantly improves heat exchanger performance [69]. Nano-engineered materials further enhance thermal and mechanical properties.

Enhancing the thermal performance of heat exchangers is critical for improving energy efficiency, reducing costs, and achieving sustainability goals across industries. The integration of passive, active, and hybrid techniques, combined with computational advancements and material innovations, enables significant improvements in heat exchanger performance. Continued research and development in this area helps the adoption of more efficient, compact, and environmentally friendly heat exchangers, meeting the demands of modern technology and global energy systems.

1.6 APPLICATIONS OF HEAT EXCHANGERS IN DIFFERENT INDUSTRIES

Heat exchangers are indispensable devices used to transfer heat between fluids in a wide range of industries. Their design and functionality vary significantly based on the application, operating conditions, and desired outcomes. The following sections are descriptions of their applications, emphasizing the types of heat exchangers used in different industries and their roles in enhancing system efficiency, sustainability, and performance.

1.6.1 Power Generation

In the power generation sector, heat exchangers are central to thermal energy management, ensuring efficient conversion of energy into electricity. Shell-and-tube heat exchangers are extensively used in steam power plants for condensers, where they condense steam exiting turbines back into water for reuse. Similarly, in nuclear power plants, they serve as steam generators, transferring heat from the reactor core to secondary systems [70–72]. Air-cooled heat exchangers are employed in gas turbines for exhaust heat recovery, while compact plate heat exchangers are used in renewable energy systems like concentrated solar power (CSP) and geothermal plants to transfer heat to working fluids. Regenerative heat exchangers, such as rotary regenerators, recover waste heat in combined cycle power plants, enhancing efficiency and reducing energy losses.

1.6.2 Automotive Industry

In the automotive sector, heat exchangers play a pivotal role in thermal management. Radiators, a type of air-to-liquid heat exchanger, cool engine fluids, preventing overheating. Intercoolers are utilized to reduce the temperature of compressed air in turbocharged engines, enhancing combustion efficiency. Oil coolers, typically plate-fin heat exchangers, regulate the temperature of engine and transmission oils to ensure optimal performance. For electric vehicles (EVs), battery thermal management systems (BTMS) incorporate liquid-cooled plate heat exchangers or microchannel heat exchangers, maintaining battery temperature within safe limits to improve lifespan and efficiency [73, 74].

1.6.3 HVAC Systems

In HVAC systems, heat exchangers are essential for maintaining thermal comfort and energy efficiency. Air-to-air heat exchangers, such as energy recovery ventilators (ERVs), transfer heat between outgoing exhaust air and incoming fresh air, reducing heating or cooling loads. Plate heat exchangers are commonly used in water-cooled air conditioning systems, transferring heat from refrigerants to cooling towers. Evaporative condensers and finned-tube heat exchangers are employed in heat pump systems, enabling efficient heat transfer for both heating and cooling applications [75].

1.6.4 Chemical and Petrochemical Industries

The chemical and petrochemical industries rely heavily on robust and high-performance heat exchangers due to the demanding nature of their processes. Shell-and-tube heat exchangers dominate in applications such as distillation, where they function as reboilers and condensers, and in high-pressure or high-temperature chemical reactions. Spiral heat exchangers are used in processes involving viscous fluids, offering high heat transfer efficiency and compact designs. For gas processing, plate-fin heat exchangers efficiently cool and separate hydrocarbon gases, while cryogenic heat exchangers liquefy natural gas for transportation [76].

1.6.5 Oil and Gas Industry

In the oil and gas sector, heat exchangers are critical across the upstream, midstream, and downstream processes. Double-pipe heat exchangers are commonly used for heat recovery from crude oil and refining streams. In liquefied natural gas (LNG) production, cryogenic heat exchangers facilitate gas liquefaction by cooling it to extremely low temperatures. Plate heat exchangers are used in desalination processes within offshore platforms, ensuring a steady supply of potable water for operations [77].

1.6.6 Food and Beverage Industry

In the food and beverage industry, heat exchangers ensure safe and efficient thermal processing. Plate heat exchangers are widely employed for pasteurization, rapidly heating and cooling milk, juices, and other beverages to eliminate pathogens while preserving quality. Tubular heat exchangers handle viscous fluids such as sauces and purees, providing uniform heating or cooling. Scraped surface heat exchangers are used for products prone to fouling, such as chocolate or caramel. In brewing, counter flow heat exchangers maintain precise temperature control during wort cooling, essential for fermentation [78].

1.6.7 Aerospace Industry

The aerospace industry requires lightweight and high-performance heat exchangers for thermal management in extreme conditions. Compact microchannel heat exchangers are used in rocket engines to pre-cool fuel or oxidizers, optimizing combustion efficiency. In aircraft, plate-fin heat exchangers manage cabin air temperature and dissipate heat from engines. For satellites, specialized heat exchangers maintain thermal balance in the vacuum of space, ensuring the reliable operation of onboard systems [79, 80].

1.6.8 Renewable Energy Systems

Renewable energy systems are essential for achieving the United Nations' Sustainable Development Goals (SDGs) [81]. Heat exchangers play a critical role in these systems by enabling efficient energy transfer and storage. Plate heat exchangers are widely used in geothermal power plants to extract heat from underground reservoirs. In solar thermal systems, shell-and-tube heat exchangers transfer solar heat to working fluids, facilitating its conversion into electricity or storage in thermal energy systems. Additionally, PCM-based heat exchangers and an artificial intelligence-driven approach to continuously keep the panels facing the sun to achieve maximum irradiation can enhance thermal storage capabilities, ensuring a steady energy supply even during intermittent solar conditions [82, 83].

1.6.9 Electronics and Data Centers

In the electronics and data center industries, heat exchangers ensure effective thermal management of high-heat-generating components. Liquid-to-liquid heat exchangers cool processors, graphics processing units, and server racks, dissipating heat to maintain performance and reliability. Air-to-liquid heat exchangers are often integrated into cooling loops for high-density server environments, reducing energy costs and improving sustainability [84].

Thus, by employing a diverse range of heat exchanger designs, industries achieve tailored solutions for their specific thermal management challenges. Advances in materials, design, and computational modeling are continuously improving heat exchanger performance, enhancing energy efficiency, reducing costs, and supporting sustainable development across a wide spectrum of applications. This adaptability ensures that heat exchangers remain a cornerstone of engineering systems worldwide.

1.7 CONCLUSION

This chapter has delved into the core principles of heat transfer—conduction, convection, and radiation—underscoring their critical importance in energy exchange across a wide range of applications and systems. By examining the mathematical models and interconnected nature of these mechanisms, it has underscored their relevance in the development and refinement of thermal systems. Heat exchangers, essential to industries like power generation, HVAC, automotive engineering, and renewable energy, have been a focal point, highlighting their dependence on effective thermal management. Recent breakthroughs in materials science, nanotechnology, and computational modeling have reshaped the landscape of thermal design, paving the way for innovative, compact, and energy-efficient solutions. Emerging technologies such as nanofluids, extended surfaces, and hybrid systems exemplify the application of these principles to tackle contemporary engineering challenges. With a blend of theoretical foundations and real-world applications, this chapter prepares readers for navigating complex thermal scenarios and driving advancements in sustainable and efficient energy systems. It also lays a solid groundwork for further exploration into advanced heat transfer technologies.

REFERENCES

1. S. Kumar, K. Rathore, Renewable energy for sustainable development goal of clean and affordable energy, *Int. J. Mater. Manuf. Sustain. Technol.* 2 (2023): 1–15. https://doi.org/10.56896/IJMMST.2023.2.1.001.
2. *What Is Heat Transfer (Heat Flow)? Complete Guide | SimScale*, n.d. https://www.simscale.com/docs/simwiki/heat-transfer-thermal-analysis/what-is-heat-transfer/ (accessed December 15, 2024).
3. C.R. Mali, V. Vinod, A.W. Patwardhan, Comparison of phase interaction models for high pressure subcooled boiling flow in long vertical tubes, *Nucl. Eng. Des.* 324 (2017): 337–359. https://doi.org/10.1016/J.NUCENGDES.2017.09.010.
4. P. Cheng, C.-T. Hsu, Heat conduction, *Transp. Phenom. Porous Media.* (1998): 57–76. https://doi.org/10.1016/B978-008042843-7/50003-9.
5. W.H. Giedt, Heat convection, *AccessScience.* (2024). https://doi.org/10.1036/1097-8542.160000.
6. S. Kumar, M.D. King, Numerical investigation on indoor environment decontamination after sneezing, *Environ. Res.* 213 (2022): 113665. https://doi.org/10.1016/j.envres.2022.113665.
7. P. Anand, P.K.S. Tejes, B.K. Naik, H. Niyas, Design analysis and performance prediction of packed bed latent heat storage system employing machine learning models, *J. Energy Storage.* 72 (2023): 108690. https://doi.org/10.1016/J.EST.2023.108690.
8. M.N.S. Jagirdar, G. Priyadarshi, B.K. Naik, H. Niyas, Implementation of adaptive neuro-fuzzy inference system in design and process optimization of latent heat storage system, *J. Energy Storage.* 89 (2024): 111810. https://doi.org/10.1016/J.EST.2024.111810.
9. D. Verma, M.S. Kalra, P. Wahi, Dynamics of a BWR with inclusion of boiling nonlinearity, clad temperature and void-dependent core power removal: stability and bifurcation characteristics of advanced heavy water reactor (AHWR), *Nucl. Eng. Des.* 308 (2016): 297–311. https://doi.org/10.1016/J.NUCENGDES.2016.08.045.
10. J.B. Joshi, K. Nandakumar, A.W. Patwardhan, A.K. Nayak, V. Pareek, M. Gumulya, C. Wu, N. Minocha, E. Pal, M. Kumar, V. Bhusare, S. Tiwari, D. Lote, C. Mali, A. Kulkarni, S. Tamhankar, Computational fluid dynamics, *Adv. Comput. Fluid Dyn. Nucl. React. Des. Saf. Assess.* (2019): 21–238. https://doi.org/10.1016/B978-0-08-102337-2.00002-X.
11. S. Kumar, M. King, R.G. Hardin, Modeling of flows through porous media, *Comput. Fluid Flow Heat Transf.* (2024): 133–158. https://doi.org/10.1201/9781003465171-8.
12. K. Rathore, S. Bhattacharjee, P. Munshi, Characterization of microwave plasma in a multicusp using 2D emission based tomography: bessel modes and wave absorption, *Phys. Plasmas.* 24 (2017): 063503. https://doi.org/10.1063/1.4984248.
13. E.R.G. Eckert, E. Pfender, Advances in plasma heat transfer, *Adv. Heat Transf.* 4 (1967): 229–316. https://doi.org/10.1016/S0065-2717(08)70275-2.

14. R.D. Watson, K.S. Chapman, *Radiation Heat Transfer*, McGraw-Hill Education, 2002. https://www.accessengineeringlibrary.com/content/book/9780070684997/chapter/chapter6 (accessed December 18, 2024).
15. K. Rathore, P. Munshi, S. Bhattacharjee, Design of a portable optical emission tomography system for microwave induced compact plasma for visible to near-infrared emission lines, *Rev. Sci. Instrum.* 87 (2016): 033501. https://doi.org/10.1063/1.4942768.
16. K. Rathore, S. Bhattacharjee, D.N. Patel, P. Munshi, Optical emission spectroscopy-based tomography for compact low-pressure microwave plasma in a multicusp, *IEEE Trans. Plasma Sci.* 45 (2017): 2492–2503. https://doi.org/10.1109/TPS.2017.2734963.
17. S. Kakaç, H. Liu, A. Pramuanjaroenkij, *Heat Exchangers: Selection, Rating, and Thermal Design*, CRC Press, 2002.
18. H. Nemati, M.M. Ardekani, J. Mahootchi, J.P. Meyer, Chapter 1 - heat exchanger classifications, in: H. Nemati, M.M. Ardekani, J. Mahootchi, J.P.B.T.-F. of I.H.E. Meyer (Eds.), Elsevier, 2024, 3–5. https://doi.org/10.1016/B978-0-443-13902-4.00021-4.
19. K. Rathore, S.I. Bhuiyan, S.M. Slavens, D. Staack, Microplasma ball reactor for JP-8 liquid hydrocarbon conversion to lighter fuels, *Fuel.* 285 (2021): 118943. https://doi.org/10.1016/J.FUEL.2020.118943.
20. A.A. Abbasian Arani, R. Moradi, Shell and tube heat exchanger optimization using new baffle and tube configuration, *Appl. Therm. Eng.* 157 (2019): 113736. https://doi.org/10.1016/j.applthermaleng.2019.113736.
21. *Metalforms, Hairpin Heat Exchangers - Double Pipe, Multitube | Metalforms Heat Transfer*, n.d. https://www.metalformsheattransfer.com/hairpin-heat-exchangers.
22. M. Medrano, M.O. Yilmaz, M. Nogués, I. Martorell, J. Roca, L.F. Cabeza, Experimental evaluation of commercial heat exchangers for use as PCM thermal storage systems, *Appl. Energy.* 86 (2009): 2047–2055. https://doi.org/10.1016/j.apenergy.2009.01.014.
23. S. Kumar, A. Thyagarajan, D. Banerjee, Experimental investigation of thermal energy storage (TES) platform leveraging phase change materials in a chevron plate heat exchanger, *Int. Mech. Eng. Congr. Expo.* (2022): 1–7. https://doi.org/10.1115/IMECE2022-96226.
24. *GL | Alfa Laval*, n.d. https://www.alfalaval.com/products/heat-transfer/plate-heat-exchangers/gas-to-liquid-plate-heat-exchangers/gl/.
25. Compact and micro-heat exchangers, *Process Intensif.* (2008): 77–101. https://doi.org/10.1016/B978-0-7506-8941-0.00005-5.
26. Y. Wang, K. Liu, W. Chen, J. Chen, K. Wang, G. Zhou, Analysis of stream flow and its impact on thermal performance in shell-side of heat exchangers based on flow dead zone, *Appl. Therm. Eng.* 249 (2024): 123414. https://doi.org/10.1016/j.applthermaleng.2024.123414.
27. C. T'Joen, Y. Park, Q. Wang, A. Sommers, X. Han, A. Jacobi, A review on polymer heat exchangers for HVAC&R applications, *Int. J. Refrig.* 32 (2009): 763–779. https://doi.org/10.1016/j.ijrefrig.2008.11.008.
28. D. Roy, A.K. Gupta, M.S. Alam, S. Srikanth, B.K. Jha, Enhancement of properties of micro-alloyed Low-Carbon Ni-Added steel by thermomechanical treatment, *J. Mater. Eng. Perform.* 29 (2020): 7952–7963. https://doi.org/10.1007/S11665-020-05311-W/FIGURES/10.
29. A.K. Gupta, D. Boruah, N. Suresh, N. Kamal, A.K. Singh, Preparation effect of mould systems on microstructure and mechanical properties of spheroidised graphite iron, *Int. J. Eng. Res. Appl.* 6 (2016): 68–73. https://pureportal.coventry.ac.uk/en/publications/preparation-effect-of-mould-systems-on-microstructure-and-mechani (accessed December 14, 2024).
30. M. Stewart, O.T. Lewis, Heat transfer theory, *Heat Exch. Equip. F. Man.* (2013): 1–91. https://doi.org/10.1016/B978-0-12-397016-9.00001-4.
31. K.J. Bell, Overall heat transfer coefficient, A-to-Z Guid. to thermodyn. *Heat Mass Transf. Fluids Eng.* (2011). https://doi.org/10.1615/ATOZ.O.OVERALL_HEAT_TRANSFER_COEFFICIENT.
32. Q. Wang, B. Xu, X. Huang, Q. Chen, H. Wang, Heat transfer and flow characteristics of straight-type PCHEs with rectangular channels of different widths, *Nucl. Eng. Des.* 391 (2022): 111734. https://doi.org/10.1016/J.NUCENGDES.2022.111734.
33. *Heat Exchanger - Heat Transfer Coefficient - U-Factor | Nuclear-power.com*, n.d. https://www.nuclear-power.com/nuclear-engineering/heat-transfer/heat-exchangers/heat-exchanger-heat-transfer-coefficient-u-factor/ (accessed December 18, 2024).
34. *Overall Heat Transfer Coefficient - U-factor | Definition | Nuclear-power.com*, n.d. https://www.nuclear-power.com/nuclear-engineering/heat-transfer/convection-convective-heat-transfer/overall-heat-transfer-coefficient-u-factor/ (accessed December 18, 2024).
35. *18.2 Heat Transfer From a Fin*, n.d. https://web.mit.edu/16.unified/www/FALL/thermodynamics/notes/node128.html (accessed December 18, 2024).

36. H. Hussein, B. Freegah, Q. Saleh, Investigation the influence of the number and configuration of fins on the hydrothermal behavior of a double-pipe heat exchanger, *J. Eng. Res.* (2023). https://doi.org/10.1016/J.JER.2023.11.006.
37. E.J. Gregory, Plate fin heat exchangers, A-to-Z Guid. to thermodyn. *Heat Mass Transf. Fluids Eng.* (2011). https://doi.org/10.1615/ATOZ.P.PLATE_FIN_HEAT_EXCHANGERS.
38. J.G. Speight, The concept of fouling, *Fouling in Refineries.* (2015): 1–30. https://doi.org/10.1016/B978-0-12-800777-8.00001-2.
39. T. Lestina, Heat exchangers fouling, cleaning and maintenance, *Handb. Therm. Sci. Eng.* (2017): 1–33. https://doi.org/10.1007/978-3-319-32003-8_24-1.
40. *Heat Transfer Typical Fouling Factors Values*, n.d. https://www.hcheattransfer.com/fouling_factors2.html (accessed December 18, 2024).
41. H. Yin, Z. Kong, Y. Qian, C. Yang, Analysis of the factors affecting heat transfer performance and prediction of heat transfer coefficient for MVR evaporator: a case study, *Case Stud. Therm. Eng.* 55 (2024): 104140. https://doi.org/10.1016/J.CSITE.2024.104140.
42. M.I. Stewart, Heat transfer theory, *Surf. Prod. Oper.* (2014): 39–97. https://doi.org/10.1016/B978-0-12-382207-9.00003-2.
43. *Logarithmic Mean Temperature Difference - LMTD | Nuclear-power.com*, n.d. https://www.nuclear-power.com/nuclear-engineering/heat-transfer/heat-exchangers/logarithmic-mean-temperature-difference-lmtd/ (accessed December 18, 2024).
44. S.J.M. Cartaxo, F.A.N. Fernandes, Counterflow logarithmic mean temperature difference is actually the upper bound: a demonstration, *Appl. Therm. Eng.* 31 (2011): 1172–1175. https://doi.org/10.1016/J.APPLTHERMALENG.2010.12.015.
45. L.Y. Chen, V.S.K. Adi, R. Laxmidewi, Shell and tube heat exchanger flexible design strategy for process operability, *Case Stud. Therm. Eng.* 37 (2022): 102163. https://doi.org/10.1016/J.CSITE.2022.102163.
46. J. Hao, Q. Chen, X. Li, T. Zhao, A correction factor-based general thermal resistance formula for heat exchanger design and performance analysis, *J. Therm. Sci.* 30 (2021): 892–901. https://doi.org/10.1007/S11630-021-1369-8/METRICS.
47. *CheCalc - Logarithmic Mean Temperature Difference LMTD Correction Factor Charts*, n.d. https://checalc.com/solved/LMTD_Chart.html (accessed December 25, 2024).
48. Z.Y. Guo, X.B. Liu, W.Q. Tao, R.K. Shah, Effectiveness–thermal resistance method for heat exchanger design and analysis, *Int. J. Heat Mass Transf.* 53 (2010): 2877–2884. https://doi.org/10.1016/J.IJHEATMASSTRANSFER.2010.02.008.
49. Y.H. Cho, H.M. Chang, An effectiveness-NTU method for triple-passage counterflow heat exchangers, *KSME J.* 7 (1993): 282–289. https://doi.org/10.1007/BF02970972/METRICS.
50. S. Bhattacharyya, D.K. Vishwakarma, A. Srinivasan, M.K. Soni, V. Goel, M. Sharifpur, M.H. Ahmadi, A. Issakhov, J. Meyer, Thermal performance enhancement in heat exchangers using active and passive techniques: a detailed review, *J. Therm. Anal. Calorim.* 147 (2022): 9229–9281. https://doi.org/10.1007/S10973-021-11168-5.
51. S. Chtourou, H. Djemal, M. Kaffel, M. Baccar, Thermal efficiency and performance enhancement examination in a new PHE design, *Case Stud. Therm. Eng.* 28 (2021): 101502. https://doi.org/10.1016/J.CSITE.2021.101502.
52. S. Kumar, P.K. Vijayan, U. Kannan, M. Sharma, D.S. Pilkhwal, Experimental and computational simulation of thermal stratification in large pools with immersed condenser by, *Appl. Therm. Eng.* 113 (2017): 345–361. https://doi.org/10.1016/j.applthermaleng.2016.10.175.
53. S. Kumar, R.B. Grover, H. Yadav, P.K. Vijayan, U. Kannan, A. Agrawal, Experimental and numerical investigation on suppression of thermal strati fi cation in a water-pool: PIV measurements and CFD simulations, *Appl. Therm. Eng.* 138 (2018): 686–704. https://doi.org/10.1016/j.applthermaleng.2018.04.070.
54. S. Liu, M. Sakr, A comprehensive review on passive heat transfer enhancements in pipe exchangers, *Renew. Sustain. Energy Rev.* 19 (2013): 64–81. https://doi.org/10.1016/J.RSER.2012.11.021.
55. C.R. Mali, A.W. Patwardhan, G.K. Pandey, I. Banerjee, V. Vinod, CFD study on the effect of various geometrical parameters of honeycomb type orifices on pressure drop and cavitation characteristics, *Nucl. Eng. Des.* 370 (2020): 110880. https://doi.org/10.1016/J.NUCENGDES.2020.110880.
56. S. Mandal, A.K. Gupta, B.H. Beavers, V. Singh, J. Narayan, R. Sachan, Atomic-scale insights on large-misfit heterointerfaces in LSMO/MgO/c-Al2O3, *Cryst.* 11 (2021): 1493. https://doi.org/10.3390/CRYST11121493.

57. H. Niyas, S. Prasad, P. Muthukumar, Performance investigation of a lab–scale latent heat storage prototype – numerical results, *Energy Convers. Manag.* 135 (2017): 188–199. https://doi.org/10.1016/J.ENCONMAN.2016.12.075.
58. H. Niyas, C.R.C. Rao, P. Muthukumar, Performance investigation of a lab-scale latent heat storage prototype – Experimental results, *Sol. Energy.* 155 (2017): 971–984. https://doi.org/10.1016/J.SOLENER.2017.07.044.
59. M.F. Rahi, M. Arıcı, A.M. Abdulateef, H. Niyas, Numerical investigation and optimization of macroencapsulated phase change material capsules in building roof slab, *J. Energy Storage.* 68 (2023): 107806. https://doi.org/10.1016/J.EST.2023.107806.
60. S. Kumar, D. Banerjee, S. Kumar, D. Banerjee, A review on phase change materials for sustainability applications by leveraging machine learning, *Energy Consum. Conversion, Storage, Effic.* (2024). https://doi.org/10.5772/INTECHOPEN.114380.
61. S. Thapa, S. Samir, K. Kumar, S. Singh, A review study on the active methods of heat transfer enhancement in heat exchangers using electroactive and magnetic materials, *Mater. Today Proc.* 45 (2021): 4942–4947. https://doi.org/10.1016/J.MATPR.2021.01.382.
62. N. Wang, S.P. Ghoushchi, K. Sharma, I. Elbadawy, A. Mouldi, H. Loukil, A.F. Deifalla, Thermal performance enhancement in a double tube heat exchanger using combination of bubble injection and helical coiled wire insert, *Case Stud. Therm. Eng.* 52 (2023): 103722. https://doi.org/10.1016/J.CSITE.2023.103722.
63. K. Shank, S. Tiari, A review on active heat transfer enhancement techniques within latent heat thermal energy storage systems, *Energies.* 16 (2023): 4165. https://doi.org/10.3390/EN16104165.
64. K. Nedjem, M. Teggar, T. Hadibi, M. Arıcı, Ç. Yıldız, K.A.R. Ismail, Hybrid thermal performance enhancement of shell and tube latent heat thermal energy storage using nano-additives and metal foam, *J. Energy Storage.* 44 (2021): 103347. https://doi.org/10.1016/J.EST.2021.103347.
65. J.M. Mahdi, S. Lohrasbi, E.C. Nsofor, Hybrid heat transfer enhancement for latent-heat thermal energy storage systems: a review, *Int. J. Heat Mass Transf.* 137 (2019): 630–649. https://doi.org/10.1016/J.IJHEATMASSTRANSFER.2019.03.111.
66. D. Bacellar, V. Aute, Z. Huang, R. Radermacher, Design optimization and validation of high-performance heat exchangers using approximation assisted optimization and additive manufacturing, *Sci. Technol. Built Environ.* 23 (2017): 896–911. https://doi.org/10.1080/23744731.2017.1333877.
67. A.K. Gupta, A. Choudhari, A. Rane, A. Tiwari, P. Sharma, A. Gupta, P. Sapale, R.T.A. Tirumala, R. Muthaiah, A. Kumar, Advances in nickel-containing high-entropy alloys: from fundamentals to additive manufacturing, *Materials.* 17 (2024): 3826. https://doi.org/10.3390/MA17153826.
68. *Materials Response to Laser/Ion Irradiation in Complex Metal Oxides: Atomic-Scale Studies of Defects and Structural Rearrangements - ProQuest*, n.d. https://www.proquest.com/openview/bc50b760cda00f4335b039834359effd/1?pq-origsite=gscholar&cbl=18750&diss=y (accessed December 14, 2024).
69. S. Chakraborty, A.K. Gupta, D. Roy, A. Basumallick, Studies on nano-metal dispersed Cu-Cr matrix composite, *Mater. Lett.* 257 (2019): 126739. https://doi.org/10.1016/J.MATLET.2019.126739.
70. D. Verma, K.N. Iyer, Dynamics and bifurcation characteristics of a boiling channel with forced circulation, *Sadhana - Acad. Proc. Eng. Sci.* 48 (2023): 1–12. https://doi.org/10.1007/S12046-023-02088-1/FIGURES/17.
71. C.R. Mali, V. Vinod, A.W. Patwardhan, Effect of wall temperature profile inputs on thermal hydraulic characteristics in long vertical steam generator tube, *SSRN Electron. J.* (2020). https://doi.org/10.2139/SSRN.3679529.
72. C.R. Mali, V. Vinod, A.W. Patwardhan, New methodology for modeling pressure drop and thermal hydraulic characteristics in long vertical boiler tubes at high pressure, *Prog. Nucl. Energy.* 113 (2019): 215–229. https://doi.org/10.1016/J.PNUCENE.2019.01.026.
73. D. Luo, Z. Wu, Y. Yan, D. Ji, Z. Cheng, R. Wang, Y. Li, X. Yang, Optimal design of a heat exchanger for automotive thermoelectric generator systems applied to a passenger car, *Appl. Therm. Eng.* 227 (2023): 120360. https://doi.org/10.1016/J.APPLTHERMALENG.2023.120360.
74. M. Jhariya, Ashok K. Dewangan, Syed Q. Moinuddin, S. Kumar, A. Ahmad, K. Yadav, A. Kumar, A.K. Yadav, Research progress on efficient battery thermal management system (BTMs) for electric vehicles using composite phase change materials with liquid cooling and nanoadditives, *J. Therm. Anal. Calorim.* 149 (2024): 13653–13680. https://doi.org/10.1007/S10973-024-13752-X.
75. L. Vanbaelinghem, A. Costantino, F. Grassauer, N. Pelletier, Alternative heating, ventilation, and air conditioning (HVAC) system considerations for reducing energy use and emissions in egg industries in temperate and continental climates: a systematic review of current systems, insights, and future directions, *Sustainability.* 16 (2024): 4895. https://doi.org/10.3390/SU16124895/S1.

76. K.J. Bell, Heat exchangers, *Encycl. Phys. Sci. Technol.* (2003): 251–264. https://doi.org/10.1016/B0-12-227410-5/00310-0.
77. *Types of Heat Exchangers in O&G - Applications & How They Work | IFS*, n.d. https://ifsolutions.com/types-of-heat-exchangers-in-oil-gas-applications-how-heat-exchangers-work/ (accessed December 20, 2024).
78. G. Tonyali Karsli, D. Cekmecelioglu, Design and simulation of heat exchangers for the food industry, *Therm. Process. Food Prod. by Steam Hot Water Unit Oper. Process. Equip. Food Ind.* (2023): 67–108. https://doi.org/10.1016/B978-0-12-818616-9.00012-2.
79. F. Careri, R.H.U. Khan, C. Todd, M.M. Attallah, Additive manufacturing of heat exchangers in aerospace applications: a review, *Appl. Therm. Eng.* 235 (2023): 121387. https://doi.org/10.1016/J.APPLTHERMALENG.2023.121387.
80. A. Carozza, A. Carozza, Heat exchangers in the aviation engineering, *Heat Exch. - Adv. Featur. Appl.* (2017). https://doi.org/10.5772/67486.
81. K. Rathore, S. Kumar, Net zero emissions for our future generations through renewables: a brief review, *Int. J. Energy Resour. Appl.* 2 (2023): 39–56. https://doi.org/10.56896/IJERA.2023.2.1.004.
82. *Heat Exchanger Technologies for Sustainable Renewable Energy Systems*, CRC Press, 2025. https://www.routledge.com/Heat-Exchanger-Technologies-for-Sustainable-Renewable-Energy-Systems/Awasthi-Kumar-Dutt-Thangavel/p/book/9781032878508 (accessed December 20, 2024).
83. A. Dawahdeh, H. Sharadga, S. Kumar, *Novel MPPT Controller Augmented with Neural Network for Use with Photovoltaic Systems Experiencing Rapid Solar Radiation Changes*, 2024.
84. M. Azarifar, M. Arik, J.Y. Chang, Liquid cooling of data centers: a necessity facing challenges, *Appl. Therm. Eng.* 247 (2024): 123112. https://doi.org/10.1016/J.APPLTHERMALENG.2024.123112.

2 Heat Exchanger Classifications and Their Areas of Application

Sunil Kumar, Kavita Rathore, Mithilesh Kumar Sahu, Mohit Pramod Sharma, Kuber Nath Mishra, and Debjyoti Banerjee

2.1 INTRODUCTION

Heat exchangers serve as pivotal components in thermal management systems, facilitating the precise and efficient transfer of thermal energy between diverse media—liquids, gases, or solids—and by utilizing natural temperature differentials. This method of heat exchange is independent of external power sources and promotes both energy efficiency and reliability, making heat exchangers essential in applications demanding stringent temperature regulation, including heating, cooling, and phase transition processes such as condensation and evaporation. The versatility of heat exchangers renders them indispensable across various industries, from power generation and chemical processing to food production, electronics cooling, waste heat recovery, HVAC, aerospace, and so on [1–3]. Their role in enhancing system sustainability and operational efficiency is evident through applications that encompass waste heat reclamation, fluid sterilization, distillation, and the optimization of chemical reactions, contributing to reduced energy consumption and heightened process safety. A well-known example is the automotive radiator, which protects engine integrity by channeling heat away from the engine and dissipating it into the surrounding air using a water and antifreeze solution. This setup transfers heat from the engine coolant to the air, thereby preventing overheating and ensuring sustained engine performance. Another example is a steam generation system, such as a boiler, where heat exchangers configured as tubular bundles efficiently transfer heat from combustion gases or electrical elements to water, producing steam or hot water essential for industrial processes. Through their critical function in regulating thermal energy, heat exchangers support operational excellence, efficiency, and safety across various industrial and engineering applications, establishing themselves as foundational elements in modern thermal systems [4].

2.2 CLASSIFICATION APPROACH OF HEAT EXCHANGERS

The selection of heat exchange equipment is a strategic and highly consequential process, integral to achieving the desired operational efficiency and thermal performance in technological and manufacturing environments. This selection requires a rigorous analysis of the thermal application to establish essential process parameters, such as the type of heat exchange media, thermophysical properties, flow rates, and the specific temperature and pressure conditions of the streams involved. A clear understanding of these factors is critical in aligning the heat exchanger's design and operational characteristics with the unique demands of the application. Different heat exchanger configurations and construction techniques provide distinct performance benefits, enabling end users to meet precise thermal requirements and optimize system efficiency. Primarily, heat exchangers can be classified to address a broad spectrum of different kinds of needs while enhancing process effectiveness, sustainability, and safety as shown in Figure 2.1 [5–7].

DOI: 10.1201/9781003509837-2

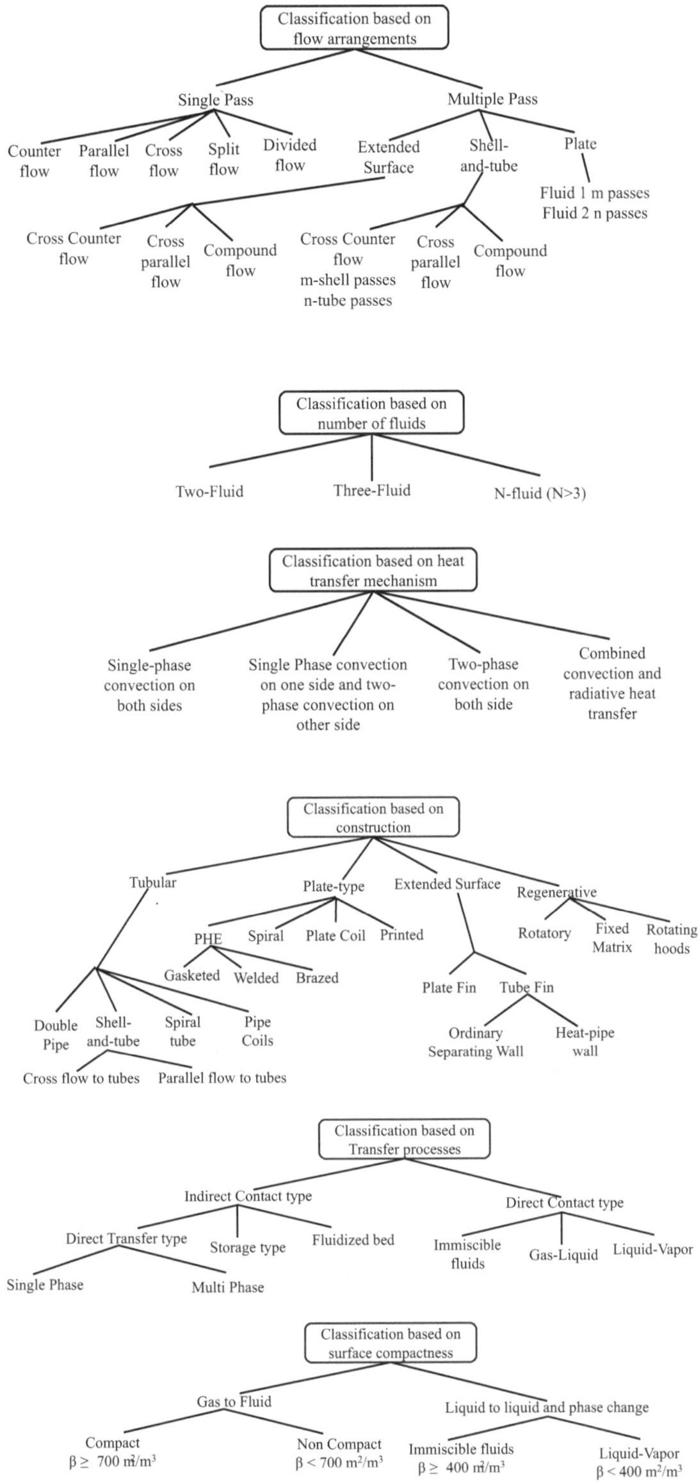

FIGURE 2.1 Classification of heat exchangers [6].

The classification of heat exchangers presented in Figure 2.1 is comprehensive. As per the practical characterization of the heat exchangers, they may be combined into specific classification groups such as:

1. Flow arrangements
2. Number of fluids
3. Heat transfer mechanisms
4. Geometry of construction
5. Transfer processes
6. Surface compactness

These classifications and some of the subclassifications of heat exchangers are shown in the following sections.

2.2.1 Classification Based on Flow Arrangements

Heat exchangers can generally be categorized into several fundamental types. The four most used types based on the configuration of their flow paths are shown in Figure 2.2.

Figure 2.2 shows that heat exchangers can be configured in several distinct flow arrangements, each offering unique thermal advantages such as:

- In a parallel-flow setup, also known as concurrent flow, often used in car radiators and air-cooled condensers in refrigeration systems, both fluid streams enter from the same end and travel side by side in the same direction, ultimately exiting together on the other side.

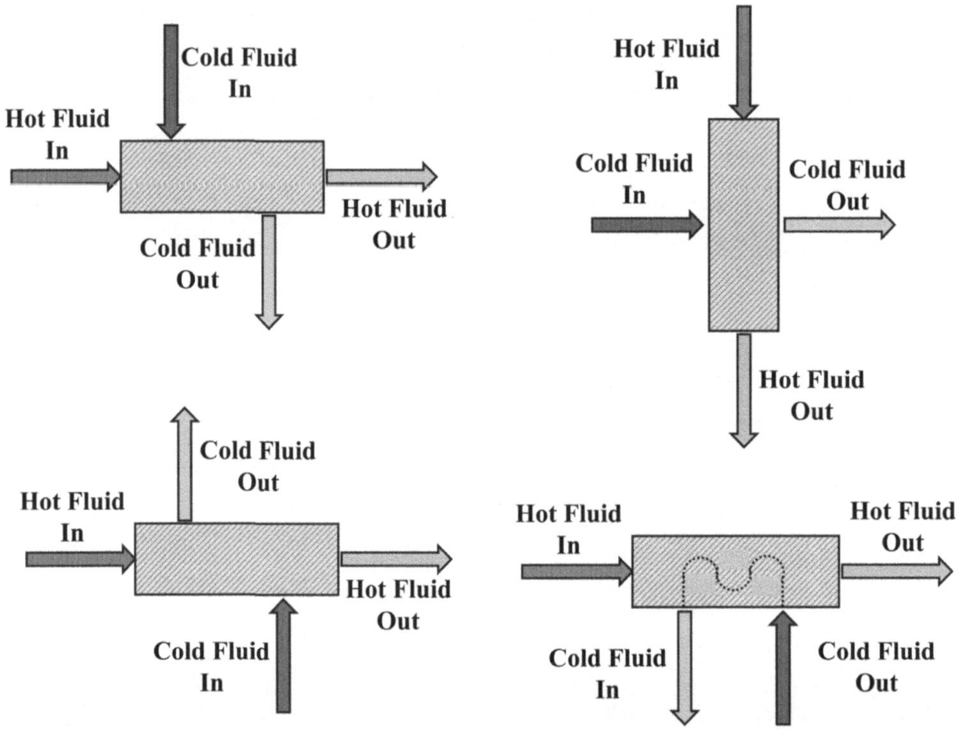

FIGURE 2.2 Flow path configuration for heat exchangers: (a) parallel flow, (b) counter flow, (c) single-pass crossflow, and (d) multipass crossflow.

This configuration allows for initial rapid heat transfer but may limit the temperature range that can be achieved.

- A counterflow design used in a heat recovery ventilation system, where fluids move in opposite directions, often achieves greater overall heat transfer by maximizing the temperature difference between the fluids throughout the length of the exchanger.
- Single-pass crossflow systems used in air conditioners (AC), where one fluid moves through the heat transfer surface perpendicular to the other, allow for effective cooling or heating in applications where complete temperature exchange is not necessary. Single-pass crossflow heat exchangers can be further subdivided into: (1) crossflow with both fluids unmixed and (2) crossflow with fluid 1 mixed and fluid 2 unmixed.
    - In the case of crossflow with both fluids unmixed, neither fluid undergoes self-mixing. Both fluids flow through predetermined channels, maintaining separate paths as they pass through the exchanger. During this process, heat is transferred from the hot fluid to the colder fluid across a shared heat transfer surface.
    - In the case of crossflow with fluid 1 mixed and fluid 2 unmixed, fluid 2 flows through predetermined channels, while fluid 1 passes over and around these channels within a larger enclosure or housing. This setup allows fluid 1 to mix as it moves across the bank of channels, while fluid 2 remains unmixed, confined within its designated flow path.
- In a multipass crossflow heat exchanger used in the HVAC system, the fluid repeatedly crosses back and forth across the opposing fluid's path, creating an alternating pattern that provides a crossflow similar to counterflow, enhancing heat exchange across the system [].

2.2.2 Classification Based on Number of Fluids

In thermal engineering and industrial applications, the processes of heating, cooling, heat recovery, and heat rejection predominantly involve the transfer of heat between two fluid streams. As a result, two-fluid heat exchangers constitute the most common and widely implemented category within the heat exchanger domain. These systems facilitate energy transfer between two distinct fluids, optimizing the thermal efficiency while minimizing the exergy loss across a shared heat transfer surface. Typical configurations include counterflow, parallel flow, and crossflow arrangements, which are integral to applications in HVAC systems, power generation, refrigeration cycles, and various industrial processes. In these scenarios, fluids such as water, air, oil, or refrigerants are directed through designated channels to achieve efficient thermal interaction with minimal energy loss [10].

Three-fluid heat exchangers, although less common, are essential in specialized applications where precise thermal regulation is critical, particularly in cryogenic systems and specific chemical processes. Within cryogenics, three-fluid configurations are vital for achieving ultra-low temperatures required in air separation units, helium-air separation systems, and hydrogen liquefaction processes. In the latter, three-fluid heat exchangers enable effective thermal stratification and energy management, essential for hydrogen purification, liquefaction, and ammonia synthesis. By incorporating a third fluid stream, these systems can achieve intricate thermal control across multi-stage cooling processes, significantly enhancing energy efficiency and process reliability. According to studies on cryogenic processes, these systems effectively manage energy transfer under extreme conditions, leveraging unique thermal gradients and enthalpic control to reach temperatures as low as −253°C in hydrogen liquefaction and −196°C in air separation [11].

In even more complex applications, such as advanced chemical processes, multifluid heat exchangers with as many as 12 fluid streams have been designed to handle simultaneous heat exchange between multiple fluids. These highly specialized configurations enable precise manipulation of temperature and flow conditions for each fluid, critical for the multi-stream requirements of specific chemical reactions and separations. Multi-stream heat exchangers are particularly suited for processes involving multiple exothermic and endothermic reactions, where simultaneous heating

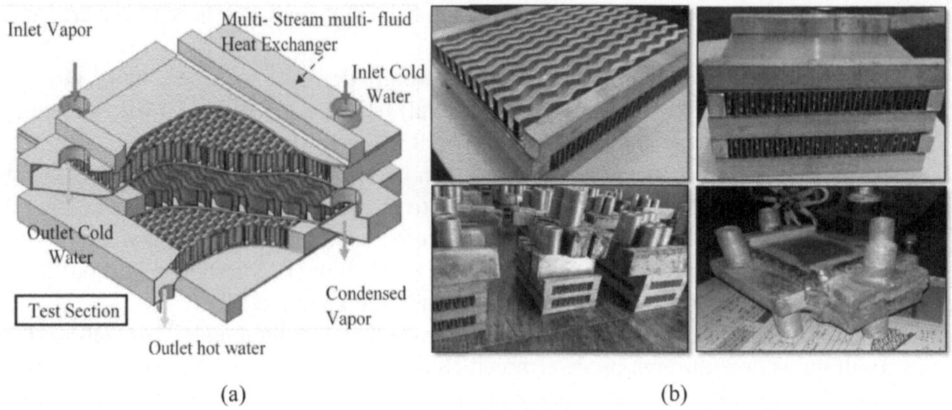

FIGURE 2.3 Multifluid compact plate-fin heat exchanger: (a) schematic with fins and (b) geometrical structure [14].

and cooling of different streams are required to maintain reaction equilibrium and optimize yield. Advanced designs such as plate-and-frame or shell-and-tube exchangers, when adapted for multistream flow, provide the necessary surface area and temperature control required for such applications [12].

The design and theoretical modeling of three-fluid and multifluid heat exchangers are shown in Figure 2.3. The design poses considerable algebraic complexity due to the intricate interaction of multiple fluid streams and the need for rigorous thermal balance equations. Unlike simpler two-fluid systems, multifluid configurations involve complex energy exchange matrices and necessitate advanced computational methods for accurate prediction of temperature and heat flux distributions. These calculations often require specialized software and modeling techniques to ensure reliable operation within the stringent thermal parameters dictated by industrial and chemical processes [13].

2.2.3 Classification Based on Heat Transfer Mechanism

Heat exchangers can be classified based on the mechanisms of heat transfer as shown in Figure 2.4, each influencing the thermodynamic performance, design intricacies, and applications. The key classifications are shown in the following sections.

2.2.3.1 Single-Phase Convection on Both Sides

In heat exchangers utilizing single-phase convection on both sides, heat is transferred between two fluids without any phase change, with both fluids remaining in the same phase (liquid or gas) throughout the process. This mode of heat transfer relies solely on convective currents to transport thermal energy, making it prevalent in applications where high flow stability and predictable thermal gradients are required. Common implementations include economizers, air heaters in boiler systems, compressor intercoolers, automotive radiators, regenerators, oil coolers, and space heaters.

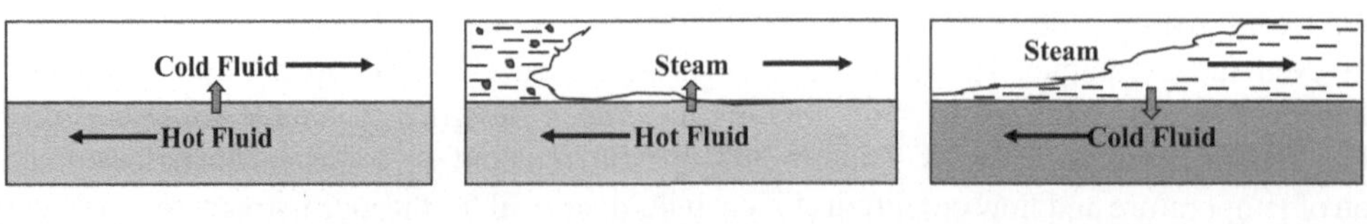

FIGURE 2.4 Heat transfer mechanism: (a) single-phase convection on both sides and (b and c) single-phase convection on one side and two-phase convection on the other (evaporation and condensation process).

The rate of heat transfer in these systems can be calculated using the Nusselt number (Nu), which correlates convective heat transfer to conductive heat transfer and is heavily influenced by parameters such as Reynolds number (Re) and Prandtl number (Pr) of the fluids. The thermal performance in single-phase convection exchangers largely depends on fluid properties, including thermal conductivity, viscosity, density, and specific heat, as well as flow characteristics (laminar or turbulent). To enhance thermal transfer, these exchangers often incorporate fins or turbulators to increase surface area and promote mixing. Compact heat exchangers with high surface-area-to-volume ratios, such as plate-fin and micro-channel heat exchangers, are examples where single-phase convection on both sides is optimized to achieve high heat transfer efficiencies in a compact footprint [15].

2.2.3.2 Single-Phase Convection on One Side and Two-Phase Convection on the Other

This category of heat exchangers involves single-phase convection in one fluid stream while the other undergoes a phase change, such as evaporation or condensation. Phase-change heat transfer is highly efficient due to the latent heat of phase change—a significant amount of energy is transferred with little or no change in temperature. This mechanism is pivotal in condensers, boilers, and steam generators, such as those used in pressurized water reactors (PWRs) and power plants [16, 17]. It also features prominently in evaporators and radiators within HVAC systems. In this configuration, the heat transfer rate is governed by the latent heat of vaporization or condensation and is significantly impacted by parameters such as vapor quality, mass flux, and heat flux at the phase-change boundary. For instance, in a condensation process, the heat transfer rate is influenced by film-wise or drop-wise condensation modes, with drop-wise condensation offering a higher heat transfer coefficient due to minimal thermal resistance at the interface. In boiling processes, factors like nucleate boiling and critical heat flux (CHF) determine thermal efficiency, with CHF marking the transition to film boiling where heat transfer decreases sharply [18, 19].

2.2.3.3 Two-Phase Convection on Both Sides

This configuration incorporates phase changes in both fluid streams, often with one stream undergoing boiling and the other condensing. Such designs maximize heat transfer efficiency by exploiting latent heat on both sides, thereby allowing for substantial energy transfer with relatively compact equipment sizes. These exchangers are particularly advantageous in high-performance applications such as cryogenics, refrigeration, and advanced thermal management systems where stringent heat loads and precise thermal control are required. Two-phase heat exchangers with both boiling and condensation modes demand rigorous design considerations, as dynamic factors like interfacial instabilities, pressure drop, and flow regime transitions (e.g., bubbly, slug, or annular flow) [3, 4] significantly affect performance. The complexity of modeling two-phase convection on both sides often involves multiphase flow equations, computational fluid dynamics (CFD), and empirical correlations tailored to specific flow configurations [20]. CFD is used to solve various engineering problems including power generation, biological material drying studies, aerosol dispersion, indoor environment management, and so on [21, 22].

2.2.4 Classification Based on the Geometry of Construction

Direct-transfer-type heat exchangers, also known as transmural heat exchangers, facilitate heat transfer directly through a wall separating two fluids without any intermediate phase or storage. These heat exchangers are defined by their construction, with the main types being tubular, plate, and extended-surface heat exchangers. Each construction type offers distinct advantages and is suited for specific applications based on factors like heat transfer efficiency, ease of maintenance, pressure tolerance, and space requirements.

2.2.4.1 Tubular Heat Exchangers

Tubular heat exchangers are constructed using circular tubes, facilitating heat transfer between two fluids by allowing one fluid to flow inside the tubes (the tube-side fluid) and the other fluid to flow outside the tubes (the shell-side fluid). This arrangement creates a robust and versatile design capable of handling high-pressure and high-temperature applications due to the structural integrity of the tubes. The flexibility in tubular heat exchanger design is achieved through variations in tube diameter, number of tubes, tube length, tube pitch (spacing between tubes), and tube arrangement (such as straight, U-tube, or helical patterns). These design modifications allow tubular heat exchangers to be precisely tailored to meet specific thermal and hydraulic requirements, optimizing heat transfer efficiency, pressure drop, and maintenance needs. Tubular heat exchangers are commonly classified into three main types [23, 24]:

2.2.4.1.1 Double-Pipe Heat Exchangers

A double-pipe heat exchanger (DPHE) (refer to Figure 2.5) is a basic yet effective heat exchanger design, characterized by its concentric tube configuration where one fluid flows through an inner tube while the other flows in the annular space between the inner and outer tubes. This design is especially advantageous in counterflow configurations, which maximize the temperature gradient, enhancing heat transfer rates. DPHEs are ideal for small to medium applications, especially where fluids exhibit significant temperature differentials, and are widely used in industries like chemical processing, oil and gas, and HVAC. Key design parameters include tube diameters, length, and material selection based on fluid properties and operating conditions, with materials like stainless steel or copper chosen for their durability under high pressures and temperatures. Modular by design, DPHEs can be arranged in series or parallel configurations to accommodate process needs, while enhancements like finned inner tubes can improve heat transfer efficiency in the annular region. Although DPHEs are easy to fabricate, maintain, and clean, their limited heat transfer area restricts their use in large-scale operations, and the cost per unit of the heat transfer area is relatively high. Applications generally involve sensible heating or cooling of fluids, such as in interstage heating or cooling in chemical processes or heat recovery in refineries and HVAC systems. Scientific evaluations of DPHE performance utilize methods like log mean temperature difference (LMTD) and effectiveness-NTU (number of transfer units) to optimize heat exchange, with recent innovations including nanofluid additives and fouling-resistant coatings to enhance efficiency. Although simple, the DPHE's adaptability and efficiency in high-pressure, high-temperature environments ensure its continued relevance, with ongoing research promising further advancements in performance and applicability across a range of thermal management needs [25].

FIGURE 2.5 Double-pipe and multitube heat exchangers [26].

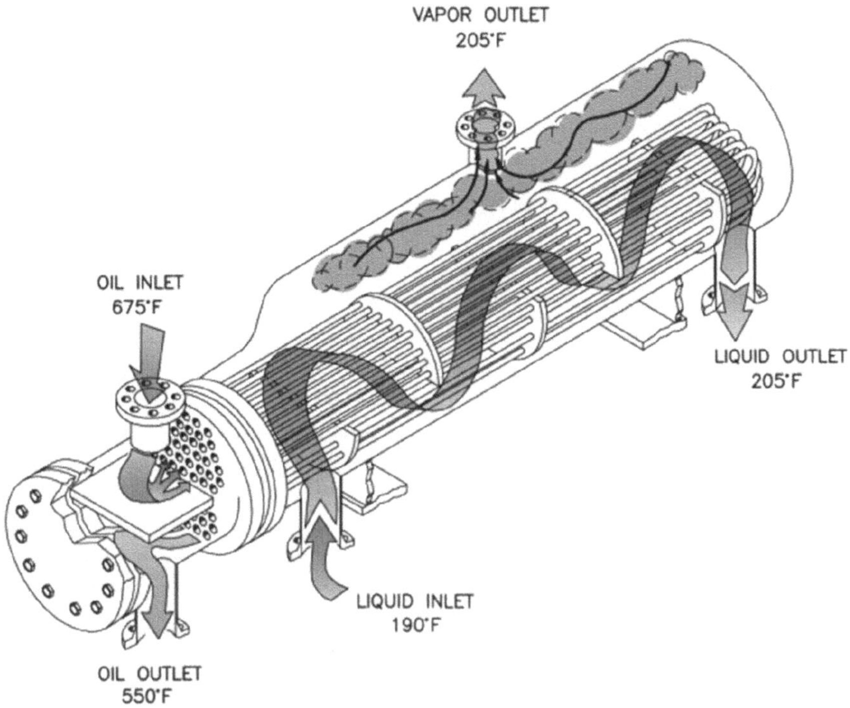

FIGURE 2.6 Internal views of a kettle-type reboiler (a type of shell-and-tube heat exchanger with baffle plates) [28].

2.2.4.1.2 Shell-and-Tube Heat Exchangers

Shell-and-tube heat exchangers (STHEs), shown in Figure 2.6, represent one of the most extensively utilized configurations in thermal management and industrial heat transfer applications, renowned for their structural robustness, thermal efficiency, and adaptability to high-pressure and high-temperature environments. STHEs are composed of a bundle of tubes enclosed within a cylindrical shell, where one fluid flows within the tubes (tube-side flow) and another circulates in the surrounding shell (shell-side flow). This dual-flow arrangement enables efficient energy transfer between the fluids, achieved through conduction and convection processes. STHEs are highly valued in industries such as petrochemical refining, power generation, and chemical processing, where precise thermal control and high-capacity energy exchange are essential [27].

2.2.4.1.2.1 Structural Configuration and Design Variability

STHEs are known for their modular design, which enables the customization of key components such as tube diameter, length, shell diameter, baffle orientation, and the number of passes on the tube side. Tubes are often arranged in triangular, square, or rotated square pitch patterns to optimize space utilization and enhance the convective heat transfer coefficient by increasing the fluid turbulence in both the shell and tube sides [29, 30]. Tubes are typically constructed from materials like stainless steel, copper, titanium alloys, or even high-nickel alloys such as Inconel or Hastelloy, depending on the thermal conductivity, corrosion resistance, and mechanical strength requirements imposed by the operating conditions and chemical composition of the fluids [31, 32]. The multipass configuration in STHEs is a critical design element, allowing tube-side fluid to traverse the tube bundle multiple times, thus extending the fluid's residence time and maximizing thermal contact. This configuration can involve even numbers of passes (e.g., two, four, six), which ensure balanced fluid distribution and reduced pressure drop across the exchanger. The counterflow configuration, where fluids flow in opposite directions, is particularly advantageous as it maximizes the temperature gradient along the length of the exchanger, enhancing the overall LMTD, and thus, the heat transfer rate. Conversely, parallel-flow or crossflow arrangements may be selected for specific process constraints, such as reduced pressure drop or lower flow resistance [33].

2.2.4.1.2.2 Baffle Design and Flow Optimization

Baffles are an integral component of STHEs, serving both to support the tubes and to manipulate the shell-side flow, directing it across the tube bundle in a crossflow or even helical pattern. Conventional segmental baffles create perpendicular flow relative to the tubes, effectively increasing turbulence and promoting higher convective heat transfer coefficients. However, segmental baffles also result in significant pressure drops. Advanced baffle designs, such as helical baffles and disk-and-doughnut baffles, provide smoother flow paths with reduced pressure loss, making them suitable for applications where energy efficiency is critical. Helical baffles, for instance, create a spiraling flow around the tubes, which maintains a more consistent flow velocity and pressure, thus optimizing thermal performance while minimizing energy loss [34].

2.2.4.1.2.3 Specialized Tube Bundle Configurations

Various tube bundle configurations address distinct operational demands in STHEs, each with trade-offs in terms of thermal expansion capabilities, ease of cleaning, and cost. In fixed tube sheet exchangers, the tube sheet is welded directly to the shell, creating a structurally simple and cost-effective solution. However, fixed tube sheets limit the exchanger's ability to accommodate differential thermal expansion between the shell and tubes, a factor that is partially mitigated by expansion joints or bellows. U-tube exchangers, on the other hand, are designed with a single tube sheet, allowing for unrestricted thermal expansion but limiting mechanical cleaning due to the sharp U-bend. This design is advantageous in high-temperature applications where thermal stresses are a concern, though the inability to clean the tube side restricts its use to fluids with minimal fouling tendencies. Floating head exchangers incorporate a floating tube sheet, permitting the tube bundle to "float" within the shell to absorb thermal expansion without imparting stress on the tube sheet welds. The floating head allows the tube bundle to be easily removed for cleaning, making this configuration ideal for heavily fouling fluids. Literature highlights the use of pull-through floating heads, which minimize the need for shell-side disassembly, providing efficient maintenance options at a higher cost. The Tubular Exchanger Manufacturers Association (TEMA) classifies heat exchangers based on their configuration [35]. TEMA defines these configurations with codes such as AEL, BEM, and BEU, representing different combinations of "front end stationary head types," "shell types," and "rear end head types," each tailored to specific thermal and mechanical requirements.

2.2.4.1.2.4 Material Selection and Anti-Fouling Innovations

Material selection in STHEs is driven by factors such as corrosion resistance, thermal conductivity, and mechanical integrity under extreme conditions. Corrosion-resistant alloys like Hastelloy, Monel, and titanium are widely used in highly corrosive environments, as seen in chemical processing and oil refineries. Nanofluids—fluids enhanced with nano-sized particles of metals, metal oxides, or carbon-based materials—have emerged as promising additives for improving the thermal conductivity and heat transfer efficiency of working fluids in STHEs [37]. Research has shown that graphene-based nanofluids can significantly increase thermal conductivity due to graphene's high surface area and thermal properties, while alumina (Al_2O_3) and titanium dioxide (TiO_2) nanoparticles offer notable improvements when dispersed in conventional fluids, such as water or oil. Fouling, a major challenge in STHE operations, reduces heat transfer efficiency and increases pressure drop, necessitating frequent maintenance [38, 39]. Recent advancements in anti-fouling coatings and surface treatments are helping to address this issue. Coatings like silica-based nanocoating and titanium nitride (TiN) are applied to tube surfaces to mitigate fouling by creating smoother, non-stick surfaces that reduce particle adhesion. Additionally, plasma-enhanced chemical vapor deposition

(PECVD) has been used to apply ultra-thin, fouling-resistant coatings that reduce surface roughness and improve heat transfer. Additionally, advanced techniques such as optical emission tomography systems can help optimize surface treatments or coatings for enhanced heat transfer and durability [40]. Electrochemical surface treatments, including anodization and electropolishing, further enhance corrosion resistance and minimize biofouling in applications involving seawater or other aggressive fluids [41].

Thus, shell-and-tube heat exchangers remain a cornerstone in industrial heat transfer applications due to their structural versatility, high thermal efficiency, and adaptability to extreme operating conditions. With advancements in materials science, anti-fouling technologies, nanofluid applications, and computational modeling, STHEs continue to evolve, meeting the increasingly complex demands of modern industries. Through continual refinement of baffle designs, tube materials, and flow configurations, STHEs demonstrate the potential to balance traditional engineering principles with state-of-the-art technology, ensuring their critical role in energy-intensive sectors requiring resilient, customizable, and high-performance thermal management solutions [42]. In specialized applications, such as in a microplasma ball reactor for JP-8 liquid hydrocarbon conversion to lighter fuels, a shell-and-tube heat exchanger ensures efficient heat transfer, precise thermal control, fouling resistance, and energy integration. The integration of a shell-and-tube heat exchanger enables high-performance operation, scalability, and system durability. These configurations can significantly enhance the economic and environmental feasibility of plasma-driven hydrocarbon processing [43].

2.2.4.1.3 Spiral-Tube-Type Heat Exchangers

Spiral-tube heat exchangers, shown in Figure 2.7, are highly efficient, compact heat transfer devices featuring one or more coiled tubes arranged in a spiral or helical structure, which significantly increases the surface area-to-volume ratio, thereby promoting enhanced heat transfer. This design induces secondary flow patterns, such as Dean vortices, that intensify mixing and disrupt thermal boundary layers, leading to superior convective heat transfer. The spiral configuration supports both counterflow and parallel-flow arrangements, with counterflow often preferred for maximizing the temperature gradient, enhancing overall thermal efficiency. Spiral-tube exchangers are particularly advantageous in applications requiring high efficiency in limited space, such as cryogenic processing, pharmaceutical manufacturing, and waste heat recovery, and are known for their centrifugal

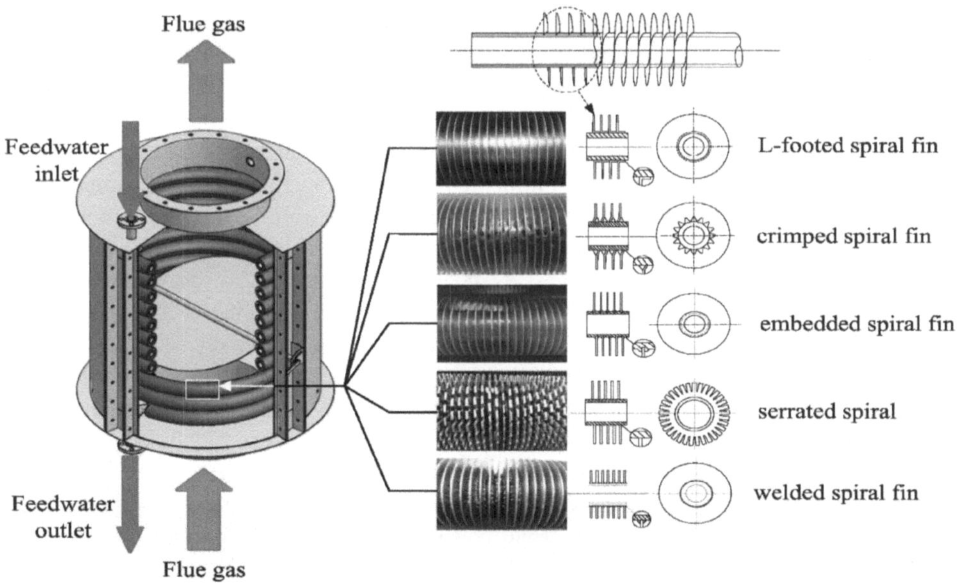

FIGURE 2.7 Spiral-tube-type heat exchanger with the fins [45].

self-cleaning effect, which reduces fouling by pushing particulates toward the outer tube walls for easier flushing. This feature, along with the flexibility to accommodate thermal expansion, makes them ideal for viscous or fouling-prone fluids and environments with space constraints, such as marine and mobile HVAC systems. Advanced materials like stainless steel and high-nickel alloys, combined with anti-fouling nanocoating, extend their durability in aggressive or high-temperature environments. Recent innovations like variable pitch coils and helical inserts have further optimized the thermal and hydraulic performance of spiral-tube exchangers, allowing for enhanced heat transfer efficiency, reduced maintenance, and longer operational lifespans in various industrial settings [44].

2.2.4.2 Plate Heat Exchangers

Plate heat exchangers are constructed using thin metal plates that form flow channels for the fluid streams, providing a large surface area for heat transfer in a compact design. These plates, arranged in a stack, separate the fluid streams, which are either single-phase or two-phase, enabling efficient heat exchange across gas, liquid, and multiphase flows. The design of the plates is versatile: they can be smooth for less resistance or corrugated to increase turbulence, enhancing heat transfer rates and preventing fouling. Plate heat exchangers are commonly used in applications requiring high thermal efficiency, such as HVAC systems, refrigeration, chemical processing, food and beverage production, and pharmaceuticals. The compact design allows for high thermal performance relative to size, making them ideal for situations with space constraints or where rapid temperature changes are required [46, 47].

2.2.4.2.1 Gasketed Plate Heat Exchangers

Gasketed plate heat exchangers, shown in Figure 2.8, represent an innovative approach to efficient heat transfer, utilizing a stack of thin, corrugated plates that both maximize surface area and facilitate seamless flow separation between two fluid streams. The corrugations on each plate enhance turbulence within the flow channels, creating a high degree of mixing at the micro-level. This unique surface geometry not only improves heat transfer efficiency but also minimizes fouling by generating local shear forces that prevent sediment from settling on the plates. With each plate positioned to align specific corner ports, fluids enter alternate channels, allowing one stream to flow through designated passages while the other occupies the adjacent channels. This clever channel arrangement optimizes the exchanger for a counterflow configuration, an efficient setup that maintains high thermal gradients across the exchanger's length, ensuring robust thermal performance [48].

Held together by compression bolts, the plates form a secure assembly that balances durability with the flexibility of easy disassembly. The gaskets, lining each plate's edges, serve as a multipurpose seal, preventing fluid mixing, containing leaks, and channeling the fluids through specific pathways. Made from elastomeric materials, these gaskets are carefully selected to withstand the chemical and thermal conditions of the application, offering secure operation within defined limits. However, the material properties of these gaskets mean that gasketed plate exchangers are generally limited to moderate temperatures (up to 250°C) and pressures (typically under 25 bars). One of the distinct advantages of these exchangers lies in their modular design, which allows plates to be added or removed to adjust capacity as required. This flexibility makes them ideal for industries where scalability is important, such as in food and beverage processing, pharmaceutical manufacturing, and dairy production. The modularity, combined with the ease of disassembly, enables thorough cleaning, maintenance, and gasket replacement, meeting stringent hygiene standards in industries that prioritize cleanliness and regulatory compliance. The compact nature of gasketed plate heat exchangers means they deliver a high thermal capacity within a small physical footprint, often achieving greater efficiency per unit volume than traditional heat exchanger types. The high turbulence and resulting heat transfer coefficients provide effective heat exchange even with minimal space requirements, making these exchangers particularly suitable for installations where space is at

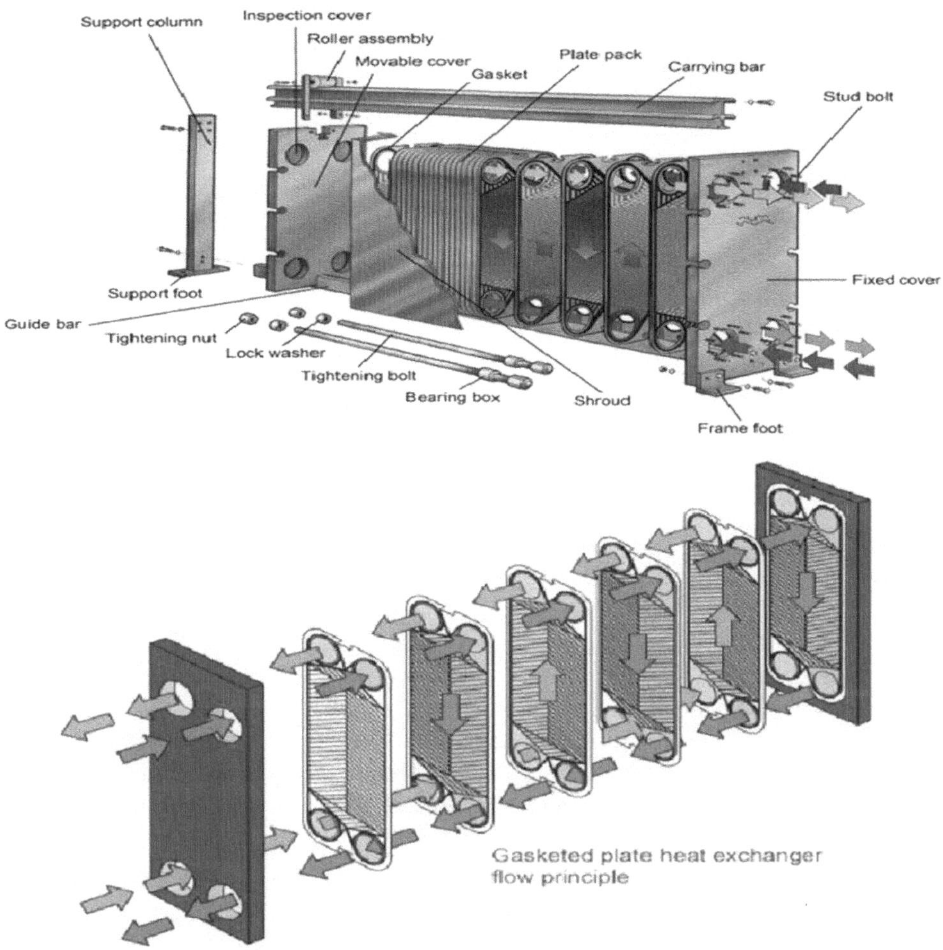

FIGURE 2.8 Gasketed plate heat exchanger [49].

a premium. In short, gasketed plate heat exchangers excel in applications requiring moderate operating conditions, high efficiency, and easy maintenance. Their design strikes an effective balance between efficiency, adaptability, and cleanliness, establishing them as a preferred solution across industries where both high performance and strict hygiene are essential [50].

2.2.4.2.2 Spiral Plate Heat Exchangers

Spiral plate heat exchangers, shown in Figure 2.9, are crafted by winding two elongated, parallel metal plates into a spiral shape around a mandrel, with the edges welded to create distinct flow channels. Spacing pins are welded along the plates to maintain consistent separation between the surfaces of the spiral channels, with adjustable pin lengths (typically between 5 mm and 20 mm), allowing for customized spacing based on desired flow rates. This adaptability enables engineers to achieve optimal flow conditions and minimize the required heat transfer surface area. The dual spiral pathways in this design introduce secondary flow patterns, enhancing heat transfer efficiency and reducing the accumulation of fouling deposits. Spiral plate heat exchangers are compact and space-efficient, but their specialized fabrication makes them relatively expensive. These units can vary in size, with heat transfer surfaces ranging from 0.5 to 500 square meters, and can typically handle pressures up to 15 bar and temperatures up to 500°C. They are particularly effective for

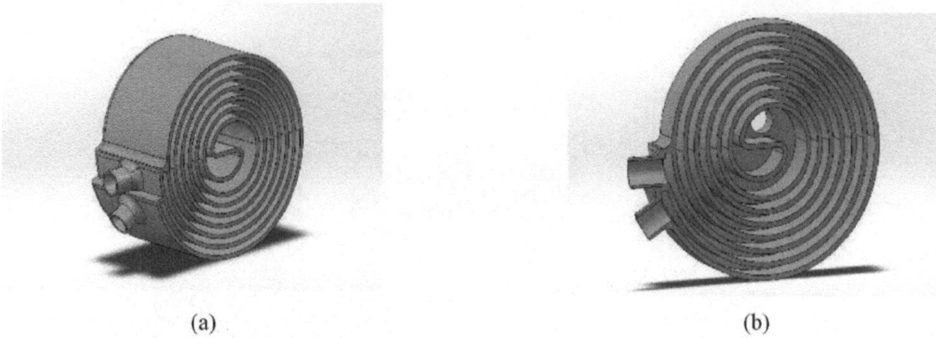

FIGURE 2.9 Spiral plate heat exchanger [52].

applications involving sludge, viscous fluids, or suspensions with solid particles, such as slurries [51].

There are three primary types of spiral heat exchangers, each tailored to distinct flow arrangements and connection designs.

2.2.4.2.2.1 Type I: Counterflow Configuration

Type I spiral exchangers use flat covers over the spiral channels, with each medium flowing counter current to the other through connections at the center and periphery. This configuration is suitable for applications without phase changes, such as liquid-liquid, gas-liquid, or gas-gas exchanges. One medium enters at the center and flows outward, while the other enters from the periphery and flows inward, achieving a true counterflow that maximizes temperature gradients for enhanced heat transfer [53].

2.2.4.2.2.2 Type II: Crossflow Design

Type II exchangers are set up for crossflow, where one channel is fully seal-welded and the other is open along both sheet metal edges, resulting in one medium flowing in a spiral path and the other in crossflow. The spiral channel's edges are sealed, while the open channel allows crossflow. This arrangement is particularly effective as a vaporizer, with two spiral elements often stacked within the same housing for increased capacity and efficiency [54].

2.2.4.2.2.3 Type III: Condensation with Subcooling

Similar to type I but featuring a specially designed top cover, type III exchangers are intended for condensing applications with subcooling and vapor separation. The top cover includes a distribution cone to evenly disperse vapor along the spiral channels, maintaining steady vapor velocity. To enable subcooling, the outer turns of the vapor channel are covered, creating a counterflow with the cooling medium. This design separates the condensate and vapor-gas mixture, with the condensate exiting through a downward outlet and the gas through an upward outlet [55].

In addition, type G spiral exchangers are used specifically as condensers. Here, vapor enters through a central tube, reverses direction in the upper shell, and condenses as it flows downward through the spiral channels in crossflow, making it ideal for controlled condensation processes. Spiral heat exchangers thus provide a compact, efficient solution for specialized applications, particularly where high fouling resistance and effective handling of viscous or particulate-laden fluids are required. Each type offers a tailored approach to specific thermal management needs, from straightforward heat transfer to complex vaporization and condensation tasks [56].

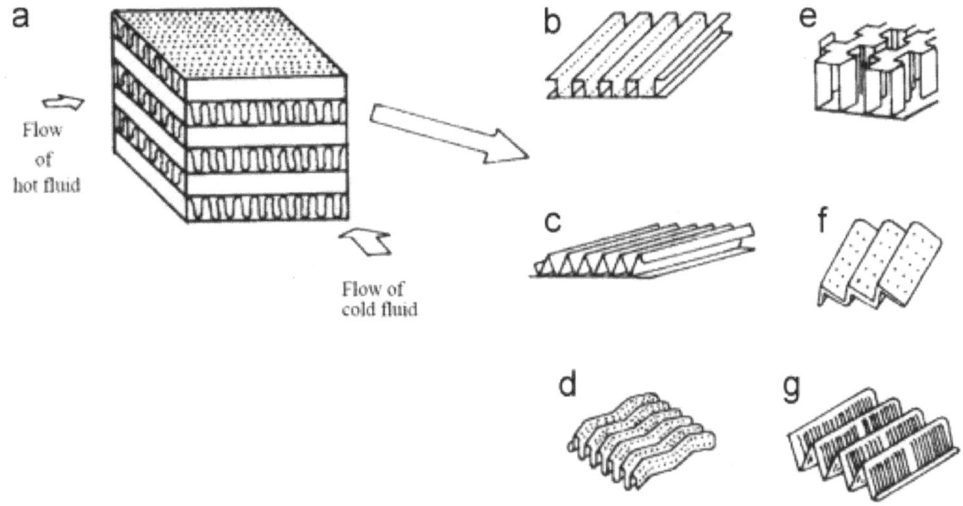

FIGURE 2.10 Plate-fin heat exchanger and its surface geometries: (a) plain rectangular fins, (b) plain triangular fins, (c) wavy fins, (d) offsets strip fins, (e) perforated fins, and (f) louvered fins [64].

2.2.4.2.3 Lamella Heat Exchangers

The lamella (Ramen) type heat exchanger, shown in Figure 2.10, is an advanced adaptation of the traditional shell-and-tube design, characterized by a set of thin, flat channels—referred to as lamellae or flat tubes—stacked in parallel within a cylindrical shell. These lamellae are formed by welding two profiled metal strips together, creating internal flow paths for one fluid while allowing the second fluid to flow around the lamellae within the shell. This unique structure eliminates the need for internal baffles, enabling smooth, longitudinal flow across the channels, which encourages true countercurrent heat exchange—a highly efficient configuration for maximizing temperature gradients and optimizing heat transfer. The lamella is securely fixed at both ends by steel spacer bars, which allow for precise control of the spacing between each lamella, tailoring the channel width to suit specific flow requirements. To maintain a tight seal, the ends of the lamellae are welded to a channel cover, which connects seamlessly to the inlet and outlet nozzles, ensuring leak-proof operation [57]. A flange system with a spigot and recess design provides stability while also accommodating differential thermal expansion, reducing stress on the components. The floating box mechanism further protects against thermal stress, making this design exceptionally reliable in applications with varying temperatures and external forces. One of the standout features of this heat exchanger is its ease of maintenance. With a removable end flange and quick-release shell flanges, the entire lamella bundle can be extracted from the shell, allowing for straightforward cleaning and inspection. The smooth surfaces inside the lamella channels are ideal for chemical cleaning, making it practical to route fouling-prone fluids through the shell side. The high turbulence within the lamella channels prevents sediment buildup, minimizing fouling while promoting efficient heat transfer. Engineered to handle pressures up to 35 bar and temperatures up to 200°C with Teflon gaskets—or up to 500°C with asbestos gaskets—the lamella heat exchanger offers robust performance across a broad range of operating conditions. This innovative design provides a compact, maintenance-friendly solution for industries that demand high heat transfer efficiency, reliability, and reduced fouling in a versatile and resilient format [58, 59].

2.2.4.3 Extended-Surface Heat Exchangers

Extended-surface heat exchangers are specifically designed to improve heat transfer performance by adding fins or protrusions to the primary surfaces of tubular or plate-type exchangers. This configuration addresses the disparity in thermal conductivity between fluids, particularly when one of the fluids—often a gas—has a significantly lower heat transfer coefficient compared to the other, which is typically a liquid. In such cases, the gas flow creates an insulating boundary layer that hinders efficient heat exchange. The inclusion of fins increases the overall surface area, thus compensating for the lower thermal conductivity of gases without necessitating a larger physical unit. The fins, usually made from highly conductive materials like aluminum or copper, enhance heat transfer by intensifying turbulence in the gas flow, which disrupts the boundary layer and boosts the convective heat transfer rate. This design approach is instrumental in applications that require compact yet highly efficient heat exchangers, especially for gas-to-gas or gas-to-liquid systems where maximizing heat dissipation within a limited space is crucial. The fin geometry and layout are meticulously tailored to align with specific flow patterns and thermal demands, ensuring that the extended-surface contributes optimally to the exchanger's performance. By leveraging the increased surface area and induced turbulence, extended-surface heat exchangers achieve significant improvements in thermal efficiency, making them indispensable in scenarios where space is limited. Still, high heat transfer rates are essential [60].

2.2.4.3.1 Plate-Fin Heat Exchangers

Plate-fin heat exchangers, shown in Figure 2.10, are designed to maximize heat transfer within a compact structure by stacking thin plates with fins arranged between them. This setup creates a dense network of flow channels, allowing flexibility in flow configurations, such as crossflow, counterflow, and parallel flow. The various fin types—louvered, wavy, offset strip, or perforated—are specifically chosen to disrupt fluid flow, generating turbulence that enhances heat transfer efficiency. This turbulence effect is especially significant with offset strip fins, which periodically reset the boundary layer along the channel, making plate-fin heat exchangers highly effective for applications requiring efficient heat exchange within a limited space. In gas-to-gas heat transfer, plate-fin exchangers are ideal, especially where minimizing size and weight is critical, such as in automotive, aerospace, and cryogenic systems. Tube-fin exchangers, on the other hand, are more suitable for liquid-to-gas applications. Thanks to their compact design, plate-fin exchangers are increasingly used in settings where mass and volume reduction are essential, such as in energy recovery systems, industrial processes, and refrigeration. These exchangers achieve a remarkable heat transfer area density, approximately 2,000 m^2 per cubic meter, making them one of the most efficient solutions in terms of space utilization. Fluid streams are separated by flat plates, with corrugated fins sandwiched between them to guide the flow, while specialized manifolds at the inlet ensure even distribution, enabling a uniform and optimized heat transfer process across the exchanger [61].

In plate-fin heat exchangers, corrugated sheets are positioned between the flat plates to not only increase the available heat transfer surface but also to provide structural reinforcement to the assembly. These corrugated sheets enable the exchanger to withstand pressure differentials and maintain its compact form. Among the various types of corrugated sheets used, the most commonly employed designs include plain fin, plan-perforated fin, serrated fin, and herringbone or wavy fins. Using fins that are discontinuous in the flow direction can effectively break up boundary layers, enhancing heat transfer by preventing the formation of a stable thermal layer on the surface. Similarly, a wavy surface along the flow path thins or disrupts the boundary layers, leading to increased heat transfer coefficients. However, these design features also result in a higher pressure drop, as the flow encounters more resistance due to the increased turbulence [62].

Plate-fin heat exchangers have narrow flow channels, which require lower mass velocities to minimize pressure drop. However, the compact size of these channels makes them susceptible to fouling when fluids contain impurities. Due to the tightly packed design, these exchangers cannot be mechanically cleaned, so they are typically used with clean fluids to ensure optimal performance and prevent blockages. They are commonly applied in processes like air liquefaction, where fluid purity aligns well with the exchanger's design, allowing for efficient heat transfer in a compact footprint [63].

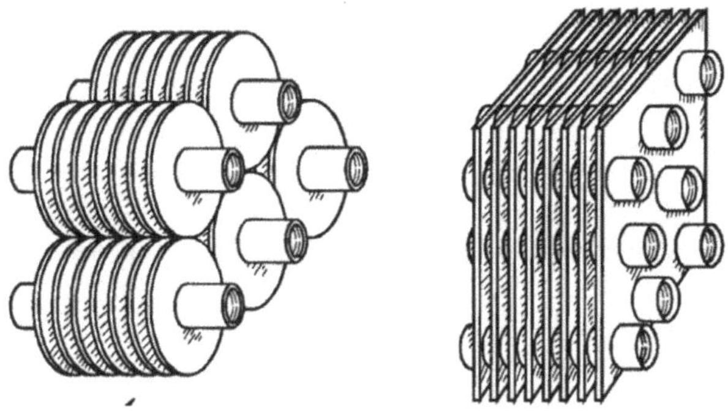

(a) (b)

FIGURE 2.11 (a) Individually and (b) continuous fins in fin-and-tubular heat exchangers [67].

2.2.4.3.2 Tubular-Fin Heat Exchangers

Tubular-fin heat exchangers, shown in Figure 2.11, are specialized extended-surface devices that incorporate fins on the exterior surfaces of tubes to maximize the effective area for heat transfer. This design is particularly beneficial in applications where one fluid, often air or gas, has a much lower heat transfer coefficient compared with the other fluid, which is usually a liquid. By expanding the surface area on the gas side, the fins help improve heat dissipation, balancing the thermal exchange between the fluids even when the gas has low thermal conductivity. Commonly employed in air conditioning systems, automotive radiators, industrial cooling, HVAC, and power generation, tubular-fin heat exchangers are versatile, allowing for efficient heat dissipation in both gas-to-liquid and gas-to-gas configurations. Their construction typically includes a core arrangement of tubes, which can be aligned in either an in-line or staggered pattern, each with distinct effects on fluid flow. In an in-line arrangement, the tubes are set up in rows, providing a smoother flow path that minimizes pressure drop, although this may also result in reduced turbulence and consequently lower heat transfer rates. Conversely, a staggered configuration offsets the tubes, creating more turbulence that enhances the heat transfer coefficient but also increases the pressure drop across the exchanger [65].

The fins themselves, usually made from thermally conductive materials like aluminum or copper, can be tailored in different forms—flat, serrated, louvered, or helical—each serving specific thermal and hydraulic needs. Flat fins provide a straightforward surface area increase without significantly disturbing the flow, making them suitable for steady flows where a low-pressure drop is essential. Serrated or louvered fins, on the other hand, introduce small interruptions along the fin surface, breaking up the boundary layer and inducing localized turbulence, which markedly enhances the heat transfer coefficient. These fins are ideal for high-efficiency applications that require compact designs. Helical fins, wrapped around the tubes in a spiral pattern, generate a swirling effect that improves heat transfer in high-viscosity fluid applications, although this configuration tends to introduce an additional pressure drop [66].

Tubular-fin heat exchangers offer several key benefits, including enhanced thermal efficiency, as the fins provide a greater area for heat exchange on the gas side, where thermal conductivity is typically low. They are also compact in design, achieving a larger heat transfer surface without substantially increasing the physical size of the exchanger, making them particularly suited for space-constrained settings. Furthermore, their adaptability across various configurations and fluid types allows them to meet the demands of diverse industries. However, these advantages come with certain challenges. The added surface area from the fins, especially in highly finned or staggered arrangements, can result in higher pressure drops, which may require adjustments in system design to maintain optimal flow rates. Additionally, the finned surfaces are more prone to fouling in dusty or particulate-laden environments, complicating maintenance and cleaning, as the fins can trap particles that are difficult to remove. This necessitates careful material selection for the fins, as they must balance thermal conductivity with durability and resistance to environmental factors like corrosion to ensure longevity and performance under various conditions. Overall, tubular-fin heat exchangers are highly efficient solutions for heat dissipation, with their design flexibility enabling them to perform effectively in a range of thermal management applications [68].

In heat exchangers, fins are often added to the outside of tubes to enhance heat transfer by increasing the surface area exposed to the fluid. These fins can be oriented in various ways, such as transverse, helical, or longitudinal, depending on the specific thermal requirements. Longitudinal fins are particularly beneficial in double-pipe or shell-and-tube configurations without baffles, making them effective for handling gases or thicker fluids like oils. Another approach is to apply continuous fin sheets across a bank of tubes, which can be round, rectangular, or elliptical, thus optimizing heat transfer. These fins are typically attached through various methods, including soldering, brazing, welding, extrusion, or mechanical fitting, chosen based on durability and thermal conductivity needs. Plate-fin-tube heat exchangers are widely used in applications like HVAC, refrigeration, and ventilation, where efficient heat dissipation in compact systems is essential. Compact heat exchangers, characterized by having a surface area density above 700 m^2/m^3 on one or both sides, are ideal for applications where one side involves gas flow, such as plate-fin, tube-fin, and regenerative systems. These high-density designs enable significant heat transfer rates by creating small hydraulic diameters, which enhance the convective heat transfer efficiency for gases [69].

In refrigeration systems, especially in condensers and evaporators, extended surfaces are commonly integrated inside tubes to improve heat transfer performance. For air-cooled condensers and waste heat boilers, the tube-fin design is often used, where horizontal tube bundles are arranged for air or gas flow across the tubes while condensation or boiling occurs inside. In induced-draft air-cooled condensers, fans positioned above the tubes draw air over them, aiding heat dissipation even with the lower air-side heat transfer rates. These exchangers often include only a few rows of tubes, with the fluid passing through multiple stages to maximize heat exchange. However, in multipass condensers, the redistribution of the two-phase mixture between passes can be challenging. This is often addressed with U-tube configurations or by dedicating separate passes for specific processes, like subcooling or superheating. In these multipass setups, each stage is arranged lower than the last to allow gravity to assist with condensate drainage. Detailed standards and guidelines for air-cooled heat exchangers can be found in resources provided by the American Petroleum Institute, offering design and operational insights for enhanced performance [70].

2.2.5 Classification According to Transfer Processes

In this category, heat exchangers can primarily be classified into indirect and direct-contact types based on the transfer processes.

2.2.5.1 Indirect-Contact Heat Exchangers

In the indirect-contact heat exchanger, the two fluid streams are isolated from each other, with heat passing from one fluid to the other through a solid barrier that remains impermeable. This design prevents any direct mixing of the fluids, ensuring only thermal energy is exchanged across the dividing wall. These exchangers, also known as surface heat exchangers, are structured into different types based on function and application needs, including direct-transfer models, storage-based exchangers, and fluidized-bed systems. Each type provides a unique approach to achieving efficient thermal exchange without direct fluid interaction [71, 72].

2.2.5.1.1 Direct-Transfer Type Heat Exchangers

In this form of heat exchanger, heat is transferred from a hot fluid to a cold fluid across a solid barrier, allowing for continuous thermal exchange without direct fluid interaction. Each fluid flows in its own channel, preventing any mixing, while both streams move simultaneously through the exchanger. This setup, commonly referred to as a recuperative heat exchanger or simply a recuperator, typically has no moving parts, enhancing durability and reducing maintenance needs. Recuperative exchangers cover a range of designs, including tubular, plate, and extended-surface types, although terms like "recuperator" are not always applied to standard shell-and-tube or plate models in the industry. Recuperators can be categorized further into prime surface and extended-surface types. Prime surface recuperators, such as plain tubular and smooth plate exchangers, do not incorporate fins or additional surface enhancements, relying instead on the base material for heat transfer. Recuperative exchangers are widely used across industries, offering a versatile and efficient solution for applications where fluid separation and robust thermal transfer are essential [73].

2.2.5.1.2 Storage-Type Heat Exchangers

Storage-type heat exchangers function by storing and subsequently releasing heat, with each fluid passing through the same channels in an alternating sequence. This setup includes a thermal medium, such as a matrix with a structured cellular design or a permeable packed bed, which temporarily retains the heat from the hot fluid. When hot gas flows through the exchanger, the matrix absorbs and holds the heat, cooling the gas in the process. Later, when the flow switches and cold gas moves through the same channels, the matrix releases the stored energy, warming the incoming cold gas. Unlike continuous direct-transfer heat exchangers, this approach operates through a cycle of energy retention and release, making it ideal for processes requiring phased heating and cooling. This type of heat exchanger, often referred to as a regenerative heat exchanger or regenerator, is particularly effective in gas-to-gas applications [74]. To sustain consistent operation, the gases or matrix components are periodically alternated, ensuring that each channel handles both hot and cold fluid flows in turn. The interval when hot gas passes through the matrix is called the "hot period," while the "cold period" refers to the time when cold gas flows through, absorbing the matrix's stored heat. These intervals may vary in length, depending on the system's thermal requirements. A minor effect of this cyclical process is "carryover leakage," in which a small amount of fluid remains in the channels after the flow switch, briefly mixing with the opposite fluid. Additionally, when fluids are at different pressures, there is potential for "pressure leakage," where fluid may leak from the high-pressure side to the low-pressure side, typically around seals or valves. Due to these characteristics, regenerative heat exchangers are mainly used in gas-to-gas systems focused on sensible heat transfer. In specific cases, they also allow for limited moisture exchange, enabling a small amount of humidity (up to 5%) to transfer from a moist to a dry gas stream [75].

2.2.5.1.3 Fluidized-Bed Heat Exchangers

In a fluidized-bed heat exchanger, shown in Figure 2.12, a collection of fine particles, like sand or coal, envelops one side of the exchanger—typically a set of tubes. The operation of this system depends significantly on the fluid's velocity. At low fluid speeds, the particles stay in place, allowing fluid to navigate through the gaps between them. However, as the fluid velocity increases to a critical threshold, it exerts a force that slightly exceeds the particles' weight, causing them to enter a "fluidized" state. In this state, the particles behave much like a fluid, spreading out and mixing extensively, which leads to a consistent temperature across the bed. This fluidized behavior offers significant benefits for heat transfer, as the continuous mixing within the bed promotes efficient energy exchange between the particles and fluid, effectively increasing the thermal conductivity of the system. Such an arrangement is highly beneficial for processes demanding intense thermal interaction, making fluidized-bed exchangers particularly suitable for applications such as drying, reactor engineering, adsorption, waste heat recovery, and coal combustion. In combustion scenarios, the fluidized particles help ensure even heat distribution and controlled reaction conditions. However, because the temperature differential between the particles and fluid decreases in a fluidized state, these systems may exhibit reduced thermal performance, which necessitates adjustments to traditional heat transfer analysis methods, like the effectiveness-NTU approach. The presence

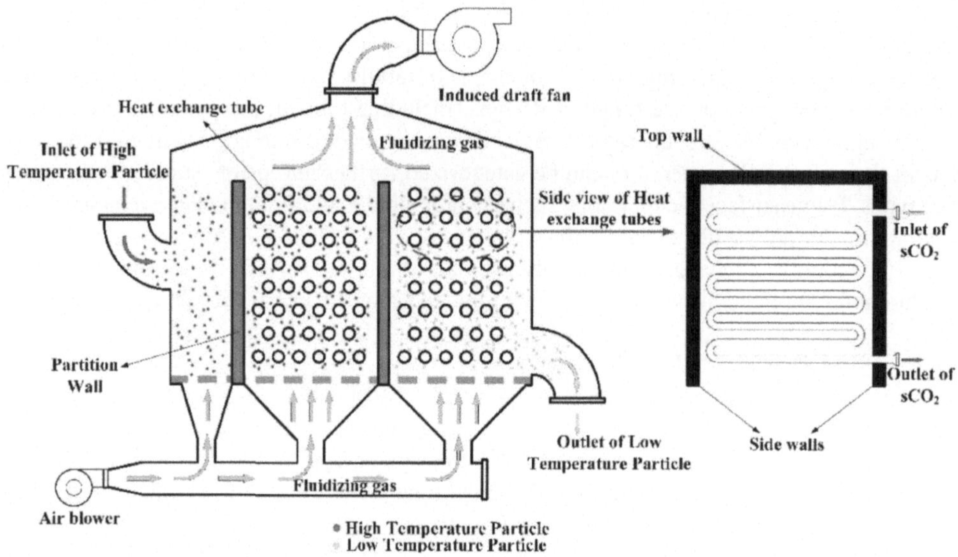

FIGURE 2.12 Schematic diagram of fluidized-bed solid particle/sCO_2 heat exchanger [77].

of chemical reactions or combustion in the bed also adds layers of complexity to designing these exchangers, requiring careful consideration in engineering applications [76].

2.2.5.2 Direct-Transfer Type Heat Exchangers

Direct-contact heat exchangers, enable heat transfer by allowing two fluid streams to come into direct physical contact, after which they are separated. This type of exchanger is especially useful in applications where both heat and mass transfer are needed, such as in evaporative cooling or rectification processes, where the phase change in one or both fluids contributes significantly to the energy transfer. The phase change enhances the rate of heat transfer, making direct-contact exchangers highly effective. Unlike indirect-contact heat exchangers, such as recuperators and regenerators, direct-contact exchangers offer several distinct advantages: they can achieve exceptionally high heat transfer rates, have a simpler and more cost-effective construction, and largely eliminate fouling issues due to the absence of a separating wall between the fluids. However, they are limited to applications where the fluids can be in direct contact without contamination or other issues. Detailed design theories for these exchangers are complex and are not covered here, but they can be classified into various types based on their specific functions and operational needs [78].

2.2.5.2.1 Immiscible Fluid Heat Exchangers

This type of heat exchanger operates by bringing two immiscible fluids into direct contact, enabling thermal exchange without blending the fluids. The process can involve simple heat transfer between fluids in the same phase or may include phase changes, such as when condensation or vaporization occurs. Typical applications involve cooling organic or oil vapors by condensing them with a contact fluid, like water or air, which absorbs the heat and facilitates the phase transition [79].

2.2.5.2.2 Gas-Liquid Heat Exchangers

In this configuration, the heat exchanger facilitates direct contact between a gaseous medium, typically air, and a low-pressure liquid, usually water, allowing for efficient separation post thermal interaction. These systems are predominantly utilized for either cooling the liquid or humidifying the gas. The process involves a partial evaporation of the liquid, with the resultant vapor being entrained in the gas flow. This evaporative mechanism is responsible for more than 90% of the energy transfer in such exchanges, predominantly through mass transfer, with convective heat transfer minimally contributing to the overall process. Prominently, this type of heat exchanger is exemplified by "wet" cooling towers, which may operate under forced or natural draft conditions to optimize airflow and enhance thermal efficiency. Additional implementations include various types of spray systems used in air conditioning, such as spray chambers, as well as in industrial applications like spray dryers, spray towers, and spray ponds. These applications leverage the same core principles of heat and mass transfer to achieve desired temperature regulation and air moisture content [80].

2.2.5.2.3 Liquid-Vapor Heat Exchangers

In this configuration, steam is condensed—either partially or fully—by direct contact with cooling water, or alternatively, water is heated using waste steam. This process yields non-condensable gases along with residual steam and hot water as the outlet streams. Such direct-contact heat exchange is typically employed in applications requiring efficient thermal transfer and gas removal, as seen in desuperheaters and open feedwater heaters, also referred to as deaerators, within power plants. These systems are instrumental in optimizing energy efficiency and ensuring high-quality feedwater by removing dissolved gases [81].

2.2.6 Classification Based on Surface Compactness

Compact heat exchangers are meticulously designed to deliver exceptional thermal efficiency through a significantly high surface area-to-volume ratio, setting them apart from traditional shell-and-tube models. The increased surface area density brings a range of operational benefits, including reduced footprint, lighter weight, minimized structural requirements, and decreased fluid volume, which in turn lower energy demands and cut operational costs. These attributes make compact heat exchangers ideally suited for systems requiring precise thermal control and integrated, space-efficient configurations. Figure 2.13 shows the area densities of a range of compact heat exchangers. For a gas-to-fluid heat exchanger to qualify as compact, it must exhibit a surface area density above 700 m^2/m^3 or a hydraulic diameter of 6 mm or less, providing ideal conditions for heat transfer in gaseous media. In liquid or phase-change applications, a surface area density of 400 m^2/m^3 or greater is the standard for compact classification. Further subcategories enhance this classification: meso-scale heat exchangers achieve surface area densities exceeding 3,000 m^2/m^3 with hydraulic diameters ranging from 1 mm to 100 mm, while micro-scale heat exchangers have surface densities over 15,000 m^2/m^3, with hydraulic diameters ranging from 1 mm to as small as 100 μm. This indicates that such highly compact systems are critical in applications demanding maximum thermal exchange within minimal space [82].

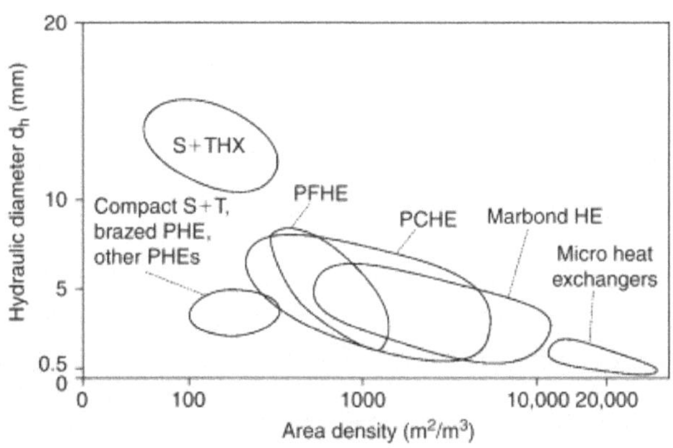

FIGURE 2.13 The area densities of a range of compact heat exchangers [83].

In contrast, conventional shell-and-tube exchangers generally maintain a surface area density under 100 m^2/m^3 on one side when using plain tubes, with minor gains from finned designs. Plate heat exchangers, however, provide roughly double the average heat transfer coefficient or overall heat transfer coefficient (U) of shell-and-tube systems in water-to-water applications due to their inherently higher surface density and optimized flow dynamics. The flow arrangements in compact heat exchangers are specifically designed to maximize heat transfer, commonly employing single-pass crossflow, counterflow, and multipass cross-counterflow configurations. These configurations enhance thermal effectiveness by preserving temperature gradients between fluids. Examples of compact designs include plate-fin and tube-fin exchangers for gas applications, as well as gasketed, welded, and printed-circuit exchangers for liquid or phase-change flows. By achieving high surface density within a compact structure, these heat exchangers enable precise, efficient thermal management solutions crucial for advanced applications in fields such as HVAC, power generation, chemical processing, and aerospace engineering, where high efficiency and minimized footprint are essential [46].

From the standpoint of compactness in heat exchangers, surface area density plays a critical role. For example, a shell-and-tube exchanger with tubes around 5 mm (0.2 in.) in diameter approaches what is considered a compact design due to its relatively high surface area density. In contrast, larger tubes, such as the 25.4 mm (1 in.) diameter tubes commonly used in power plant condensers, have a much higher surface area density, classifying these systems as non-compact. Automotive radiators, particularly those from the 1990s with a configuration of about 790 fins per meter (or 20 fins per inch), achieve a much higher surface area density—approximately 1870 m^2/m^3 (570 ft^2/ft^3) on the air side. This density is similar to what would be achieved with tubes measuring 1.8 mm (0.07 in.) in diameter. Advanced regenerators for gas turbines further enhance compactness, reaching densities close to 6600 m^2/m^3 (2000 ft^2/ft^3), comparable to tubes only 0.5 mm (0.02 in.) in diameter. The human lung represents an extraordinary natural example of compactness, with a surface area density of about 17,500 m^2/m^3 (5330 ft^2/ft^3), similar to tubes with an incredibly small diameter of 0.19 mm (0.0075 in.). Inspired by this biological model, some cutting-edge micro-heat exchangers are being developed to replicate or even surpass the lung's compactness, achieving very high surface area densities to maximize heat transfer within extremely confined spaces. As compact surfaces can achieve structural stability and strength with thinner-gauge material, the gain in a lower exchanger mass is even more pronounced than the gain in a smaller volume [84].

2.2.6.1 Gas-to-Fluid Heat Exchangers

The lower heat transfer coefficient (h) of gases, compared with that of liquids such as water and oil, significantly impacts the design and efficiency of heat exchangers used in applications involving gas-to-liquid (shown in Figure 2.14) or gas-to-gas heat transfer. Because gases transfer heat at much lower rates, they inherently require a much larger surface area to achieve the same thermal conductance as liquids. This mismatch creates a fundamental challenge in designing heat exchangers that can maintain thermal equilibrium across both fluid paths, which is essential for optimal efficiency. In a gas-to-liquid heat exchanger, achieving a similar thermal conductance (hA product, where A is the surface area) on both the gas and liquid sides typically necessitates a larger surface area on the gas side. Traditional shell-and-tube heat exchangers, commonly used in industry, are often inadequate for this purpose due to their structural limitations. Shell-and-tube designs primarily rely on circular tubes arranged within a cylindrical shell, which constrains the available surface area on the gas side due to spatial limitations. Consequently, in applications demanding high heat transfer rates from gases, this design falls short in providing the necessary surface area without substantially increasing the exchanger's overall size and weight [85].

To overcome these challenges, compact heat exchangers with high surface area densities have become a preferred solution, especially for applications requiring efficient gas-to-liquid, gas-to-gas, or gas-to-phase change heat transfer. Compact heat exchangers are designed with extended surfaces, such as fins, that increase the surface area available for heat exchange while minimizing

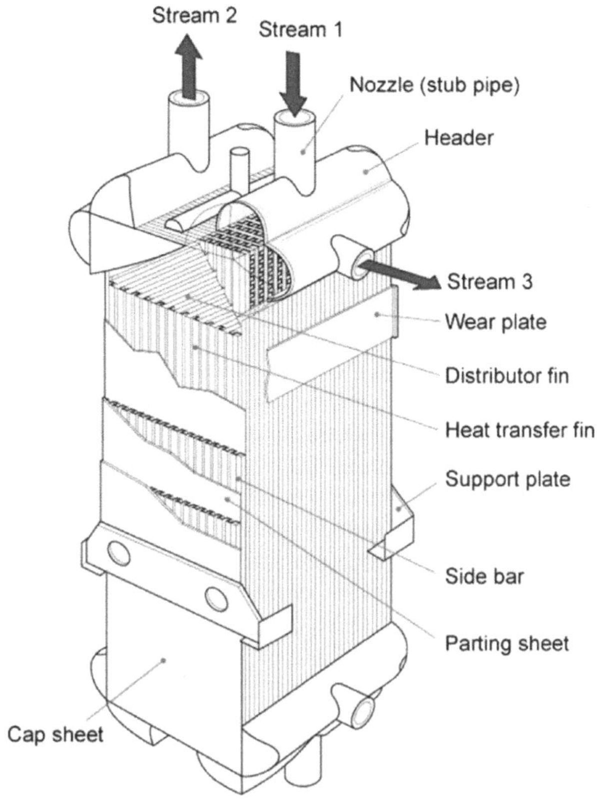

FIGURE 2.14 Gas-to-liquid plate heat exchanger [86].

the volume and weight of the exchanger. Plate-fin and tube-fin types are prominent examples of compact exchangers. They consist of closely spaced plates or tubes equipped with fins, which add surface area without requiring large volumes, thus facilitating heat transfer in a smaller, more efficient package. Compact heat exchangers offer several distinct advantages over traditional shell-and-tube models. First, they allow for a highly flexible arrangement of surface area densities, enabling precise tailoring to the heat transfer requirements on each side of the exchanger. This customization is particularly advantageous in systems where there is a significant discrepancy in the heat transfer coefficients of the fluids involved. Additionally, the compact structure allows designers to distribute the area between the hot and cold sides more effectively, achieving the desired thermal balance with reduced size and weight. This compactness is essential for applications where physical space is limited, such as in automotive, aerospace, and HVAC systems, where minimizing volume and weight is critical to overall system performance [87].

Another key benefit of compact heat exchangers is their potential for cost savings. Due to their smaller size and reduced material requirements, they generally offer a lower initial material cost compared with traditional exchangers. Their lightweight nature also translates to operational savings, as systems using compact exchangers require less support infrastructure and can often reduce pumping power requirements, given their efficient heat transfer capabilities. However, the design and operation of compact heat exchangers come with specific considerations. Since one of the fluids is typically a gas with a low heat transfer coefficient, compact exchangers need clean, non-corrosive fluids to avoid fouling within the narrow flow channels. Fouling, which is the buildup of unwanted materials on the surfaces, can severely impact heat transfer efficiency, especially in exchangers with complex fin geometries and limited clearance. To minimize maintenance and extend operational life, compact exchangers are best suited for clean, low-fouling fluids, such as filtered air, light hydrocarbons, and certain refrigerants.

Pressure drop management is another critical factor in compact heat exchanger design. Due to the narrow flow passages and intricate surface structures, there is a tendency for higher pressure drops, which can require more pumping power to maintain desired flow rates. Therefore, the design of compact exchangers must carefully balance pressure drop against thermal performance, ensuring that energy efficiency is not compromised. Compact heat exchangers are typically assembled through brazing or mechanical expansion, techniques that secure fins and plates in place to create a highly efficient heat transfer path. However, these construction methods often limit the operating pressure and temperature ranges of compact exchangers compared with shell-and-tube designs. While shell-and-tube exchangers can handle high pressures and temperatures due to their robust construction, compact exchangers are generally suited to moderate pressure and temperature applications [84].

One design characteristic of compact exchangers is the large frontal area coupled with shorter flow paths, which helps to optimize heat transfer by exposing more surface area directly to the incoming fluid. This configuration necessitates careful header design to ensure even flow distribution across the exchanger. Uneven flow distribution can lead to performance issues, such as localized overheating or reduced heat transfer efficiency, so proper header design is crucial for uniform thermal performance. Despite their advantages, compact heat exchangers also entail higher initial investment costs, particularly in terms of specialized manufacturing tooling and equipment needed for their intricate designs. Consequently, they are typically most economical in high-demand applications where the cost savings from their operational efficiencies outweigh the initial investment.

Fouling remains a notable limitation of compact heat exchangers, as their intricate fin structures and narrow channels are prone to clogging in the presence of particulates or corrosive fluids. This limitation restricts their use to applications where the fluids are clean and well-maintained, such as filtered air or purified refrigerants. Ensuring clean fluids helps maintain heat exchanger performance and reduces maintenance requirements, making compact exchangers ideal for environments with strict fluid quality control [88].

2.2.6.2 Liquid-to-Liquid and Phase-Change Heat Exchangers

Liquid-to-liquid and phase-change heat exchangers come in various designs tailored for specific applications, including gasketed plate-and-frame, welded plate, spiral plate, and printed-circuit exchangers. These configurations cater to diverse thermal requirements, providing flexibility in handling different fluids and operating conditions. Gasketed plate-and-frame exchangers, for instance, are commonly used in applications that require frequent maintenance or cleaning, as they allow easy disassembly. Welded plate exchangers offer a more robust, leak-proof design suitable for higher pressures and temperatures. Spiral plate exchangers are compact and efficient, making them ideal for handling fouling-prone fluids, as their design promotes self-cleaning through induced turbulent flow. Printed-circuit exchangers, known for their high thermal efficiency and compactness, are often used in applications requiring high-pressure capabilities. Detailed descriptions and specific applications of liquid-to-liquid and phase-change heat exchangers are covered in Section 2.6 providing an in-depth look at their unique design features and operational benefits [89].

2.3 APPLICATIONS OF HEAT EXCHANGERS

Heat exchangers play a vital role across various industries, providing essential thermal management in systems where efficient heat transfer is necessary. The most common configurations are two-fluid systems, where heat transfer occurs between two separate fluids, maximizing thermal energy exchange. However, some specialized fields, such as cryogenics and complex chemical processes, require three-fluid heat exchangers to achieve precise temperature control and meet demanding operational needs. In cryogenics, for example, three-fluid exchangers are used to achieve extremely low temperatures essential for air separation and hydrogen liquefaction, where specific temperature levels allow gases to condense into liquids or solidify. In chemical synthesis, such as ammonia production, multifluid exchangers enable controlled temperature and phase conditions that are essential for efficient reaction processes. Advanced multifluid exchangers can handle more intricate thermal processes, including multiphase heat transfer, often required in hydrocarbon distillation, where mixed vapor streams condense simultaneously within a single exchanger unit [90].

The versatility of heat exchangers is evident across a wide range of applications, from power generation and process engineering to air conditioning, refrigeration, cryogenics, and even waste heat recovery. In power plants, various heat exchanger designs are employed to support high-efficiency steam cycles and cooling processes [91, 92]. Boilers, steam generators, condensers, regenerators, and cooling towers are common components that allow for efficient energy transfer and conversion [93]. For example, modern boilers operate at pressures over 1200 PSI (80 bar), integrating furnace water-wall tubes, superheaters, and economizers to maximize thermal efficiency. Pressurized water reactors (PWRs), widely used in nuclear power, incorporate large U-tube steam generators capable of producing between 300 and 400 MW of electrical power, demonstrating the advanced capability of heat exchangers to handle both high-pressure and large thermal loads [94, 95].

In the process industry, two-phase flow exchangers are widely used to facilitate essential tasks, such as vaporization, condensation, crystallization, and catalytic reactions within fluidized beds. For example, in crystallization, exchangers control temperature and phase conditions to allow dissolved substances to form solids. In air conditioning and refrigeration, condensers and evaporators are key to regulating temperature and humidity levels, supporting climate control in commercial and residential spaces. An energy-efficient application of heat exchangers lies in direct-contact condensation, where vapor condenses directly into a high-pressure liquid. This method allows for effective thermal energy storage, as the condensed liquid can be kept in a reservoir until energy is required. At that point, depressurizing the liquid causes it to flash into vapor, which can then be used for heating or as a working fluid in a power cycle. This storage and release method allows for on-demand energy usage, contributing to overall system efficiency and reducing the reliance on external energy sources [96].

The development of heat exchanger technology has been a significant driver of industrial advancements. One of the early breakthroughs was the introduction of water-tube boilers, which allowed engines to operate at higher pressures and capacities. This evolution enabled the production of boilers capable of withstanding high pressures and temperatures, incorporating advanced elements like economizers, air heaters, and superheaters to capture and reuse energy. As power demands increased, engineers designed larger, more sophisticated boilers and condensers, a progression that has been instrumental in modernizing power plant efficiency. In the chemical industry, heat exchangers are integral to processes involving liquid vaporization and solvent recovery. Evaporators are used extensively to concentrate on solutions or separate solvents from mixtures, with water commonly removed in the process. In many cases, valuable solvents are recovered and reused, showcasing heat exchangers' flexibility and adaptability to specific processing needs. This diversity in heat exchanger applications underscores their essential role in industrial efficiency, supporting everything from large-scale energy production to precise chemical reactions and sustainable solvent recovery [97].

2.4 SELECTION CRITERIA FOR HEAT EXCHANGERS

Selecting a suitable heat exchanger involves a meticulous assessment of numerous parameters to ensure alignment with operational, environmental, and economic objectives. Each criterion influences the system's overall performance, sustainability, and long-term efficiency. The following sections are a detailed breakdown of the key factors that are commonly considered during the selection process:

2.4.1 Meeting Process Requirements

A heat exchanger must align with the specific thermal and fluid dynamics requirements of a given process. The exchanger should accommodate temperature ranges defined by the inlet and outlet conditions, as well as pressure levels dictated by the system's constraints. Its design should account for flow rates, which impact heat transfer coefficients and Reynolds number—a dimensionless parameter indicating the flow regime. The system must maintain thermal equilibrium to ensure stability during operational cycles. Futureproofing with design margins or modular features can address process variability, such as increased throughput or altered thermal loads.

2.4.2 Optimized Thermal Performance

The heat exchanger's effectiveness is governed by the thermal conductivity of materials and heat transfer coefficients achieved through appropriate geometry and design. The Nusselt number, a dimensionless measure of convective heat transfer relative to conductive heat transfer, plays a central role in design evaluation. Materials like copper, aluminum alloys, or advanced ceramics enhance thermal conduction. Compact heat exchanger configurations, such as plate-fin or micro-channel designs, optimize heat flux per unit area, making them ideal for applications where space and weight constraints exist.

2.4.3 Resistance to Service Conditions

The heat exchanger must demonstrate resilience against corrosive and erosive stresses posed by internal and external environments. Corrosion resistance involves selecting materials with low electrochemical activity, such as stainless steel, titanium, or nickel alloys, particularly for applications involving chloride-rich or acidic fluids. Fouling resistance is equally critical; fouling refers to the deposition of unwanted materials on heat transfer surfaces, which diminishes thermal efficiency. Anti-fouling coatings, surface treatments, or self-cleaning designs can mitigate these effects, ensuring operational consistency in environments with high particulate or organic loads.

2.4.4 Ease of Maintenance and Accessibility

Maintenance accessibility ensures system longevity and minimizes downtime. Heat exchangers prone to fouling, such as shell-and-tube types, must allow for mechanical cleaning through removable components. The design should also consider vibration-induced wear, a common issue in turbulent flow systems. Modular and replaceable parts, such as gaskets, seals, or damaged plates, ensure cost-effective repairs. Accessibility is enhanced by spatial layouts that permit tool entry, adhering to ergonomic principles for maintenance personnel.

2.4.5 Economic Feasibility

Economic evaluations encompass the total cost of ownership (TCO), which includes capital expenditure (CapEx), operational expenditure (OpEx), and maintenance costs. CapEx considers fabrication complexity, material selection, and installation logistics, while OpEx involves energy consumption dictated by pressure drops and heat transfer inefficiencies. Maintenance costs are influenced by the mean time to failure (MTTF) and the mean time to repair (MTTR). Cost-benefit analyses utilize net present value (NPV) calculations to assess long-term viability, incorporating factors like operational lifespan and projected energy savings.

2.4.6 Physical and Spatial Constraints

Physical constraints are often dictated by geometric limitations, such as diameter-to-length ratios or surface area density, which affect installation feasibility and performance. Compact designs, like plate heat exchangers, are preferred in confined spaces, while shell-and-tube exchangers suit applications with higher thermal loads but ample space. Weight constraints impact the selection of mounting structures and transport logistics, with larger units requiring reinforced foundations or specialized lifting mechanisms.

2.4.7 Material Compatibility and Durability

Material selection is guided by the principle of chemical and mechanical compatibility with the process fluids and operating conditions. For example, stainless steels exhibit excellent corrosion resistance due to a passive chromium oxide layer, while titanium alloys withstand high temperatures and saline environments. Compatibility extends to seals and gaskets, which must resist chemical attack, thermal degradation, and mechanical wear. Material durability is often quantified by metrics like creep resistance, fracture toughness, and pitting corrosion resistance.

2.4.8 Energy Efficiency and Environmental Impact

Energy-efficient designs reduce operational costs and align with sustainability goals. Heat exchanger efficiency can be quantified by parameters like the effectiveness-NTU method, which assesses the performance relative to theoretical maximum heat transfer. Compact designs with high surface area-to-volume ratios minimize energy consumption. Environmental impacts include carbon footprints associated with material sourcing, fabrication, and disposal, necessitating the selection of materials with lower embodied energy and recyclability.

2.4.9 Fluid Dynamics and Flow Configuration

Flow configuration significantly influences convective heat transfer coefficients and the pressure drop across the exchanger. Counterflow arrangements, where fluids move in opposite directions, maximize the LMTD and are thermally more efficient than parallel flow or crossflow designs. Laminar flow offers lower pressure drops but reduced heat transfer rates, while turbulent flow enhances heat transfer at the cost of higher energy consumption.

2.4.10 Pressure Drop and Pumping Power Optimization

Minimizing pressure drop is crucial for reducing energy costs associated with pumping fluids. The relationship between pressure drops and flow velocity is governed by the Darcy–Weisbach equation, where excessive turbulence or fouling increases flow resistance. Advanced designs, such as streamlined channels or enhanced fins, reduce frictional losses without compromising thermal performance. Optimizing flow channel geometries and surface textures helps balance heat transfer rates and energy requirements.

2.4.11 Safety and Regulatory Compliance

Compliance with international standards ensures safe operation and mitigates legal risks. Standards such as the ASME Boiler and Pressure Vessel Code, API 660 (Shell and Tube Heat Exchangers for General Refinery Services), and TEMA (Tubular Exchanger Manufacturers Association) guide design, fabrication, and testing. Adherence to pressure vessel classifications and weld quality inspections ensures structural integrity and safety under specified operating conditions.

2.4.12 Process Integration and Control

Modern heat exchangers are often integrated with smart technologies, including sensors for real-time monitoring of thermal and hydraulic parameters. Parameters like surface temperature, flow velocity, and pressure differentials are monitored using technologies such as resistance temperature detectors (RTDs) and piezoelectric pressure sensors. Integration with control systems enables adaptive responses to operational variations, enhancing efficiency and safety.

Thus, incorporating these detailed criteria and leveraging advanced scientific principles ensures the selection of a heat exchanger tailored to operational demands, sustainability objectives, and long-term economic feasibility. Such a comprehensive approach guarantees robust performance and adaptability in diverse industrial applications, supporting both immediate and future operational goals [98].

2.5 ROLE OF AI/ML IN HEAT EXCHANGER(S) CLASSIFICATION AND APPLICATION

The integration of artificial intelligence (AI) and machine learning (ML) has revolutionized the field of heat exchanger technology, introducing transformative methodologies for classification, design, and operational optimization [99, 100]. By leveraging AI/ML, traditional processes have been enhanced to achieve greater precision, efficiency, and adaptability across a wide range of industrial applications.

2.5.1 Redefining Heat Exchanger Classification

AI/ML has introduced a groundbreaking shift in heat exchanger classification by automating and enhancing the traditional manual evaluation of design parameters. Intelligent design interpretation utilizes advanced models like convolutional neural networks (CNNs) to analyze CAD schematics and operational images, categorizing heat exchangers with remarkable accuracy into types such as shell-and-tube, plate, or finned-tube exchangers [101]. These systems not only reduce human error but also accelerate the selection process for engineers. Additionally, structural and dimensional insights provided by AI algorithms assess geometric layouts and configurations, introducing a deeper layer of classification that focuses on functional and application-specific categories. Furthermore, dynamic material profiling incorporates machine-learning-based material databases to recommend optimal materials suited to operational and environmental conditions. For instance, corrosion-resistant alloys are suggested for marine environments, while lightweight composites are tailored for aerospace applications [102]. This approach ensures durability and optimal performance in diverse scenarios.

2.5.2 AI/ML as Enablers in Heat Exchanger Applications

AI/ML technologies are instrumental in enhancing the functionality and efficiency of heat exchangers, going beyond classification to enable advanced analytics, predictive modeling, and operational optimization. Intelligent performance analytics integrate real-time sensor data with predictive models, allowing engineers to simulate and forecast the behavior of heat exchangers under varying conditions. This results in actionable insights for improving heat transfer efficiency, reducing pressure drops, and mitigating fouling tendencies. Advanced fault prognostics leverage historical data to detect anomalies like scaling or flow disruptions, preventing potential failures and minimizing downtime through timely corrective actions. Moreover, adaptive systems powered by reinforcement learning adjust dynamically to fluctuating operational conditions, optimizing flow rates and coolant distributions to maintain consistent performance while reducing energy consumption. Proactive maintenance strategies enabled by AI-driven digital twins simulate operational wear and tear, predicting failures and scheduling interventions in advance. This approach not only prevents unexpected breakdowns but also enhances resource allocation, extending the lifespan of heat exchangers.

2.5.3 Enhanced Sector-Specific Applications

Renewable energy is critically important to fight climate change and fulfill the Sustainable Development Goals (SDGs) outlined by the United Nations (UN) [103, 104]. AI/ML-driven advancements have transformed the application of heat exchangers across various industries by addressing unique challenges and improving operational efficiency. In renewable energy, AI optimizes solar thermal systems by dynamically adjusting to diurnal variations, ensuring maximum energy capture, while ML predicts fouling in geothermal systems to sustain long-term efficiency [105]. In the process industry, AI simulations model complex thermal processes for chemical and petrochemical plants, optimizing heat exchanger designs for reactor cooling and distillation processes while adapting to multiphase flows and reactive environments. HVAC systems benefit from AI-driven smart thermal management solutions that balance energy efficiency with indoor air quality, reducing costs and improving occupant comfort. In the automotive and aerospace sectors, AI enhances the performance of radiators by predicting thermal loads in real-time and optimizing lightweight heat exchangers to achieve superior performance-to-weight ratios critical for aviation and space exploration. Additionally, AI optimizes liquid-cooled heat exchangers in data centers, ensuring reliable thermal management for high-density computing loads and uninterrupted operations.

2.5.4 The Technological Core: AI/ML Techniques for Advancing Heat Exchangers

AI/ML technologies utilize a wide array of techniques to innovate heat exchanger design and operation. Neural networks, such as artificial neural networks (ANNs) and recurrent neural networks (RNNs), are employed to capture complex nonlinear relationships between operating parameters and predict performance trends over extended periods. Reinforcement learning enables autonomous decision-making, refining operational strategies to maximize efficiency under variable conditions. Natural language processing (NLP) extracts valuable insights from technical documentation, enhancing maintenance protocols and tailoring models for specific industrial applications. Generative AI tools revolutionize design configurations by iteratively refining heat exchanger layouts based on predefined objectives, such as maximizing the heat transfer area or minimizing energy losses. Clustering techniques, like k-means, group exchangers based on performance metrics and provide data-driven insights to customize designs and improve operational strategies.

2.5.5 Practical Impacts and Industry Implications

The practical benefits of integrating AI/ML into heat exchanger systems are significant and multifaceted. These technologies optimize costs by reducing energy consumption and maintenance expenditure while ensuring consistent operational efficiency under varying conditions. Predictive maintenance and anomaly detection extend the lifespan of heat exchangers by minimizing wear and preventing unexpected failures. Additionally, AI-driven optimizations contribute to sustainability by enhancing energy efficiency and aligning with global environmental goals. The adoption of AI/ML in heat exchanger technology not only addresses current industrial challenges but also paves the way for more resilient and adaptive systems.

2.6 CONCLUSIONS

Heat exchangers are indispensable components of contemporary thermal management systems, playing a pivotal role in advancing energy efficiency, sustainability, and operational reliability. This chapter underscores the critical importance of heat exchangers in addressing diverse operational challenges through innovative designs, advanced material applications, and emerging technologies. By employing sophisticated classification methodologies, heat exchangers are meticulously tailored to meet the distinct demands of industries such as power generation, chemical processing, aerospace, and HVAC systems. Advancements in materials science, nanotechnology, and surface coating technologies have significantly enhanced the thermal and mechanical performance of heat exchangers, enabling superior functionality under stringent operational conditions. Furthermore, compact and modular configurations effectively address spatial constraints while maintaining high efficiency and adaptability. The integration of AI and ML represents a transformative development in the field, facilitating precise classification, real-time operational monitoring, predictive maintenance, and optimized design frameworks. These technologies substantially improve operational efficiency, minimize downtime, and accelerate the development of high-performance, sustainable thermal management solutions. As global demand for energy-efficient and environmentally sustainable systems continues to grow, heat exchangers, leveraging state-of-the-art designs and AI/ML-enabled innovations, are well-positioned to play a central role in shaping the future of industrial and technological progress. This evolution highlights their essential contribution to fostering adaptability, resilience, and excellence in meeting the complex challenges of modern thermal management applications.

REFERENCES

1. S. Kumar, P.K. Vijayan, U. Kannan, M. Sharma, D.S. Pilkhwal, Experimental and computational simulation of thermal stratification in large pools with immersed condenser by, *Appl. Therm. Eng.* 113 (2017): 345–361. https://doi.org/10.1016/j.applthermaleng.2016.10.175.
2. S. Kumar, A. Thyagarajan, D. Banerjee, An investigation on chevron plate heat exchanger filled with organic phase change material, in: *Proc. ASME 2020 Heat Transf. Summer Conf. HT2020 July 13-15, 2020, Virtual, Online, American Society of Mechanical Engineers (ASME)*, 2020, 1–6. https://doi.org/10.1115/HT2020-9122.
3. K. Rathore, S. Kumar, Net zero emissions for our future generations through renewables: a brief review, *Int. J. Energy Resour. Appl.* 2 (2023): 39–56. https://doi.org/10.56896/IJERA.2023.2.1.004.
4. M.A. Dekhil, J.V. Simo Tala, O. Bulliard-Sauret, D. Bougeard, Development of an innovative heat exchanger for sensible heat storage in agro-food industry, *Appl. Therm. Eng.* 177 (2020): 115412. https://doi.org/10.1016/j.applthermaleng.2020.115412.
5. H. Nemati, M.M. Ardekani, J. Mahootchi, J.P. Meyer, Chapter 1 - heat exchanger classifications, in: H. Nemati, M.M. Ardekani, J. Mahootchi, J.P.B.T.-F. of I.H.E. Meyer (Eds.), Elsevier, 2024, 3–5. https://doi.org/10.1016/B978-0-443-13902-4.00021-4.
6. S. (Sadık), A.E. Bergles, F. Mayinger, N.A.S.I.N.A.S.I. on H.E.-T.-H.F.D. Kakaç, Heat exchangers: thermal-hydraulic fundamentals and design, No Title. n.d. https://cir.nii.ac.jp/crid/1130000793697243136 (accessed December 9, 2024).
7. M.P. Sharma, A.K. Nayak, Determination of two-phase turbulent-mixing rate in simulated subchannels of a natural-circulation pressure tube-type BWR, *Nucl. Sci. Eng.* 184 (2016): 280–291. https://doi.org/10.13182/NSE15-112.
8. Y. Wang, K. Liu, W. Chen, J. Chen, K. Wang, G. Zhou, Analysis of stream flow and its impact on thermal performance in shell-side of heat exchangers based on flow dead zone, *Appl. Therm. Eng.* 249 (2024): 123414. https://doi.org/10.1016/j.applthermaleng.2024.123414.
9. M.P. Sharma, A.K. Nayak, Experimental investigation of two phase turbulent mixing rate under bubbly flow regime in simulated subchannels of a natural circulation pressure tube type BWR, *Exp. Therm. Fluid Sci.* 76 (2016): 228–237. https://doi.org/10.1016/J.EXPTHERMFLUSCI.2016.03.016.
10. K. Wang, Y. Peng, J. Chen, P. Wu, Investigating subcooled flow boiling heat transfer and fluid dynamics within a shell-and-plate heat exchanger, *Appl. Therm. Eng.* 258 (2025): 124679. https://doi.org/10.1016/j.applthermaleng.2024.124679.
11. E. Tavousi, N. Perera, D. Flynn, R. Hasan, Heat transfer and fluid flow characteristics of the passive method in double tube heat exchangers: A critical review, *Int. J. Thermofluids.* 17 (2023): 100282. https://doi.org/10.1016/j.ijft.2023.100282.

12. K. Kowalski, P. Błasiak, S. Pietrowicz, A review of scraped surface heat exchangers: parameters describing thermal-fluid processes, influence of their internal geometric parameters and energy consumption, *Int. J. Heat Mass Transf.* 236 (2025): 126336. https://doi.org/10.1016/j.ijheatmasstransfer.2024.126336.
13. Z. Lin, H. Wang, Y. Wang, Y. Cao, X. Lei, S. Bai, X. Xu, J. Sun, Y. Liu, Experimental analysis on drying performance of a condensing clothes dryer with two new plate-fin heat exchangers, *Case Stud. Therm. Eng.* 42 (2023): 102769. https://doi.org/10.1016/j.csite.2023.102769.
14. A. Vaisi, K. Javaherdeh, R. Moosavi, Condensation heat transfer performance in multi-fluid compact heat exchangers with wavy and strip fins, *Int. J. Heat Mass Transf.* 182 (2022): 121968. https://doi.org/10.1016/J.IJHEATMASSTRANSFER.2021.121968.
15. T.S. Khan, M.S. Khan, M.-C. Chyu, Z.H. Ayub, Experimental investigation of single phase convective heat transfer coefficient in a corrugated plate heat exchanger for multiple plate configurations, *Appl. Therm. Eng.* 30 (2010): 1058–1065. https://doi.org/10.1016/j.applthermaleng.2010.01.021.
16. C.R. Mali, V. Vinod, A.W. Patwardhan, Effect of wall temperature profile inputs on thermal hydraulic characteristics in long vertical steam generator tube, *SSRN Electron. J.* (2020). https://doi.org/10.2139/SSRN.3679529.
17. C.R. Mali, V. Vinod, A.W. Patwardhan, New methodology for modeling pressure drop and thermal hydraulic characteristics in long vertical boiler tubes at high pressure, *Prog. Nucl. Energy.* 113 (2019): 215–229. https://doi.org/10.1016/J.PNUCENE.2019.01.026.
18. T. Alam, M.-H. Kim, A comprehensive review on single phase heat transfer enhancement techniques in heat exchanger applications, *Renew. Sustain. Energy Rev.* 81 (2018): 813–839. https://doi.org/10.1016/j.rser.2017.08.060.
19. D. Verma, K.N. Iyer, Dynamics and bifurcation characteristics of a boiling channel with forced circulation, *Sadhana - Acad. Proc. Eng. Sci.* 48 (2023): 1–12. https://doi.org/10.1007/S12046-023-02088-1/FIGURES/17.
20. X. Tao, M.P. Nuijten, C.A. Infante Ferreira, Two-phase vertical downward flow in plate heat exchangers: flow patterns and condensation mechanisms, *Int. J. Refrig.* 85 (2018): 489–510. https://doi.org/10.1016/j.ijrefrig.2017.10.008.
21. S. Kumar, M. King, R.G. Hardin, Modeling of flows through porous media, *Comput. Fluid Flow Heat Transf.* (2024): 133–158. https://doi.org/10.1201/9781003465171-8.
22. M.D. King, M. Klassen, R. Hardin, M.D. King, *Dispersion of Sneeze Droplets in a Meat Facility Indoor Environment – Without Partitions.* n.d.
23. C.R. Mali, A.W. Patwardhan, G.K. Pandey, I. Banerjee, V. Vinod, CFD study on the effect of various geometrical parameters of honeycomb type orifices on pressure drop and cavitation characteristics, *Nucl. Eng. Des.* 370 (2020): 110880. https://doi.org/10.1016/J.NUCENGDES.2020.110880.
24. A. Dabestani, M. Kahani, CFD analysis of rotation effect on flow patterns and heat transfer enhancement in a horizontal spiral tube heat exchanger, *Case Stud. Therm. Eng.* 64 (2024): 105494. https://doi.org/10.1016/j.csite.2024.105494.
25. R.A. Rinik, A.A. Bhuiyan, M.R. Karim, Enhancement of heat transfer using elliptical twisted inner pipe with convergent conical ring turbulator for turbulent flow in double pipe heat exchanger, *Int. J. Therm. Sci.* 210 (2025): 109558. https://doi.org/10.1016/j.ijthermalsci.2024.109558.
26. W. Wang, Y. Shuai, B. Li, B. Li, K.S. Lee, Enhanced heat transfer performance for multi-tube heat exchangers with various tube arrangements, *Int. J. Heat Mass Transf.* 168 (2021): 120905. https://doi.org/10.1016/j.ijheatmasstransfer.2021.120905.
27. Z. Jin, Z. Yang, X. Li, X. Zhao, B. Zhang, Dynamics and heat transfer characteristics on tube side and shell side of micro-tube compact air-to-air heat exchanger, *Appl. Therm. Eng.* 259 (2025): 124895. https://doi.org/10.1016/j.applthermaleng.2024.124895.
28. R.A. Parisher, R.A. Rhea, Mechanical equipment, *Pipe Draft. Des.* (2012): 112–133. https://doi.org/10.1016/B978-0-12-384700-3.00006-2.
29. H. Niyas, S. Prasad, P. Muthukumar, Performance investigation of a lab–scale latent heat storage prototype – numerical results, *Energy Convers. Manag.* 135 (2017): 188–199. https://doi.org/10.1016/J.ENCONMAN.2016.12.075.
30. H. Niyas, C.R.C. Rao, P. Muthukumar, Performance investigation of a lab-scale latent heat storage prototype – experimental results, *Sol. Energy.* 155 (2017): 971–984. https://doi.org/10.1016/J.SOLENER.2017.07.044.
31. D. Roy, A.K. Gupta, M.S. Alam, S. Srikanth, B.K. Jha, Enhancement of properties of micro-alloyed low-carbon Ni-added steel by thermomechanical treatment, *J. Mater. Eng. Perform.* 29 (2020): 7952–7963. https://doi.org/10.1007/S11665-020-05311-W/FIGURES/10.

32. A.K. Gupta, D. Boruah, N. Suresh, N. Kamal, A.K. Singh, Preparation effect of mould systems on microstructure and mechanical properties of spheroidised graphite iron, *Int. J. Eng. Res. Appl.* 6 (2016): 68–73. https://pureportal.coventry.ac.uk/en/publications/preparation-effect-of-mould-systems-on-microstructure-and-mechani (accessed December 14, 2024).
33. A.C. Caputo, A. Federici, P.M. Pelagagge, P. Salini, On the design of shell-and-tube heat exchangers under uncertain operating conditions, *Appl. Therm. Eng.* 212 (2022): 118541. https://doi.org/10.1016/j.applthermaleng.2022.118541.
34. A.A. Abbasian Arani, R. Moradi, Shell and tube heat exchanger optimization using new baffle and tube configuration, *Appl. Therm. Eng.* 157 (2019): 113736. https://doi.org/10.1016/j.applthermaleng.2019.113736.
35. R.C. Byrne, *Standards of The Tubular Exchanger Manufactures Association (TEMA)*, 10th ed., TEMA. Inc., 2019.
36. Z. Yang, Y. Ma, N. Zhang, R. Smith, Design optimization of shell and tube heat exchangers sizing with heat transfer enhancement, *Comput. Chem. Eng.* 137 (2020): 106821. https://doi.org/10.1016/j.compchemeng.2020.106821.
37. A.K. Gupta, A. Choudhari, A. Rane, A. Tiwari, P. Sharma, A. Gupta, P. Sapale, R.T.A. Tirumala, R. Muthaiah, A. Kumar, Advances in nickel-containing high-entropy alloys: from fundamentals to additive manufacturing, *Materials.* 17 (2024): 3826. https://doi.org/10.3390/MA17153826.
38. S. Mandal, A.K. Gupta, B.H. Beavers, V. Singh, J. Narayan, R. Sachan, Atomic-scale insights on large-misfit heterointerfaces in LSMO/MgO/c-Al2O3, *Crystals.* 11 (2021): 1493. https://doi.org/10.3390/CRYST11121493.
39. *Materials Response to Laser/Ion Irradiation in Complex Metal Oxides: Atomic-Scale Studies of Defects and Structural Rearrangements - ProQuest*, n.d. https://www.proquest.com/openview/bc50b760cda00f4335b039834359effd/1?pq-origsite=gscholar&cbl=18750&diss=y (accessed December 14, 2024).
40. K. Rathore, P. Munshi, S. Bhattacharjee, Design of a portable optical emission tomography system for microwave induced compact plasma for visible to near-infrared emission lines, *Rev. Sci. Instrum.* 87 (2016): 033501. https://doi.org/10.1063/1.4942768.
41. G. Yi, X. Tang, L. Du, X. Li, S. Zhao, Hollow fiber polytetrafluoroethylene membrane heat exchanger with anti-corrosion properties, *J. Memb. Sci.* 678 (2023): 121664. https://doi.org/10.1016/j.memsci.2023.121664.
42. C. Françolle de Almeida, M. Saget, G. Delaplace, M. Jimenez, V. Fierro, A. Celzard, Innovative fouling-resistant materials for industrial heat exchangers: a review, 39 (2023): 71–104. https://doi.org/doi:10.1515/revce-2020-0094.
43. K. Rathore, S.I. Bhuiyan, S.M. Slavens, D. Staack, Microplasma ball reactor for JP-8 liquid hydrocarbon conversion to lighter fuels, *Fuel.* 285 (2021): 118943. https://doi.org/10.1016/J.FUEL.2020.118943.
44. Q. Mao, Y. Zhao, Y. Zhang, Study on thermal storage performance of a novel conical spiral tube heat storage system, *Appl. Therm. Eng.* 245 (2024): 122910. https://doi.org/10.1016/j.applthermaleng.2024.122910.
45. P. Pongsoi, S. Pikulkajorn, S. Wongwises, Heat transfer and flow characteristics of spiral fin-and-tube heat exchangers: A review, *Int. J. Heat Mass Transf.* 79 (2014): 417–431. https://doi.org/10.1016/J.IJHEATMASSTRANSFER.2014.07.072.
46. J.E. Hesselgreaves, R. Law, D. Reay, *Compact Heat Exchangers: Selection, Design and Operation*, Butterworth-Heinemann, 2016.
47. S. Kumar, A. Thyagarajan, D. Banerjee, Experimental investigation of thermal energy storage (TES) platform leveraging phase change materials in a chevron plate heat exchanger, *Int. Mech. Eng. Congr. Expo.* (2022): 1–7. https://doi.org/10.1115/IMECE2022-96226.
48. G.S.M. Martins, R.S. Santiago, L.E. Beckedorff, T.S. Possamai, R. Oba, J.L.G. Oliveira, A.A.M. de Oliveira, K. V Paiva, Structural analysis of gasketed plate heat exchangers, *Int. J. Press. Vessel. Pip.* 197 (2022): 104634. https://doi.org/10.1016/j.ijpvp.2022.104634.
49. A.L.M. Nahes, M.J. Bagajewicz, A.L.H. Costa, Simulation of gasketed-plate heat exchangers using a generalized model with variable physical properties, *Appl. Therm. Eng.* 217 (2022): 119197. https://doi.org/10.1016/J.APPLTHERMALENG.2022.119197.
50. H. Hajabdollahi, M. Naderi, S. Adimi, A comparative study on the shell and tube and gasket-plate heat exchangers: The economic viewpoint, *Appl. Therm. Eng.* 92 (2016): 271–282. https://doi.org/10.1016/j.applthermaleng.2015.08.110.
51. M.C. Returi, R. Konijeti, A. Dasore, Heat transfer enhancement using hybrid nanofluids in spiral plate heat exchangers, *Heat Transf. Res.* 48 (2019): 3128–3143. https://doi.org/10.1002/htj.21534.

52. V. Irabatti, Y. Patil, S. Kore, V. Barangule, A. Kothe, Comprehensive review of spiral heat exchanger for diverse applications, *Mater. Today Proc.* 72 (2023): 1328–1334. https://doi.org/10.1016/J.MATPR.2022.09.308.
53. M.A. Boda, S.S. Deshetti, M.A. Gavade, Design and development of parallel-counter flow heat exchanger, *Int. J. Innov. Res. Adv. Eng.* 4 (2017): 29–35.
54. G. Starace, M. Fiorentino, M.P. Longo, E. Carluccio, A hybrid method for the cross flow compact heat exchangers design, *Appl. Therm. Eng.* 111 (2017): 1129–1142. https://doi.org/10.1016/j.applthermaleng.2016.10.018.
55. J.R. García-Cascales, F. Vera-García, J. Gonzálvez-Maciá, J.M. Corberán-Salvador, M.W. Johnson, G.T. Kohler, Compact heat exchangers modeling: condensation, *Int. J. Refrig.* 33 (2010): 135–147. https://doi.org/10.1016/j.ijrefrig.2009.08.013.
56. J. Wu, L. Wang, Y. Liu, Research on film condensation heat transfer of the shell side of the spiral coil heat exchanger, *Int. J. Heat Mass Transf.* 125 (2018): 1349–1355. https://doi.org/10.1016/j.ijheatmasstransfer.2018.05.029.
57. A. Coskun, B. Demir, Comparative analysis of different lamella geometries used in exhaust gas heat exchangers, *Appl. Therm. Eng.* 100 (2016): 1–10. https://doi.org/10.1016/j.applthermaleng.2016.02.007.
58. U. Ikhtiar, A.A. Bin Hairuddin, A. Bin Asarry, K.A.B.M. Rezali, H.M. Ali, Numerical investigation of lamella heat exchanger for engine intake charge air cooling utilizing refrigerant as coolant medium, *Alexandria Eng. J.* 65 (2023): 661–673. https://doi.org/10.1016/j.aej.2022.10.008.
59. U.E. Ola, *Plate Heat Exchanger Design Methodology*, 2011.
60. A.D. Kraus, A. Aziz, J. Welty, D.P. Sekulic, Extended surface heat transfer, *Appl. Mech. Rev.* 54 (2001): B92–B92.
61. T.A. Khan, W. Li, Optimal design of plate-fin heat exchanger by combining multi-objective algorithms, *Int. J. Heat Mass Transf.* 108 (2017): 1560–1572. https://doi.org/10.1016/j.ijheatmasstransfer.2017.01.031.
62. M. Picon-Nuñez, G.T. Polley, E. Torres-Reyes, A. Gallegos-Muñoz, Surface selection and design of plate–fin heat exchangers, *Appl. Therm. Eng.* 19 (1999): 917–931. https://doi.org/10.1016/S1359-4311(98)00098-2.
63. S. Sanaye, H. Hajabdollahi, Thermal-economic multi-objective optimization of plate fin heat exchanger using genetic algorithm, *Appl. Energy.* 87 (2010): 1893–1902. https://doi.org/10.1016/j.apenergy.2009.11.016.
64. H.E. Ahmed, H.A. Mohammed, M.Z. Yusoff, An overview on heat transfer augmentation using vortex generators and nanofluids: approaches and applications, *Renew. Sustain. Energy Rev.* 16 (2012): 5951–5993. http://www.sciencedirect.com/science/article/pii/S1364032112003838 (accessed October 5, 2013).
65. P. Ocłoń, S. Łopata, Study of the effect of fin-and-tube heat exchanger fouling on its structural performance, *Heat Transf. Eng.* 39 (2018): 1139–1155. https://doi.org/10.1080/01457632.2017.1363619.
66. A. Chumpia, K. Hooman, Performance evaluation of single tubular aluminium foam heat exchangers, *Appl. Therm. Eng.* 66 (2014): 266–273. https://doi.org/10.1016/j.applthermaleng.2014.01.071.
67. A. Sadeghianjahromi, C.C. Wang, Heat transfer enhancement in fin-and-tube heat exchangers – a review on different mechanisms, *Renew. Sustain. Energy Rev.* 137 (2021): 110470. https://doi.org/10.1016/J.RSER.2020.110470.
68. L. Han, K. Yang, J. Yang, R. Li, Y. Li, L. Deng, D. Che, A thermal calculation model for tubular condensing heat exchanger, *Appl. Therm. Eng.* 244 (2024): 122701. https://doi.org/10.1016/j.applthermaleng.2024.122701.
69. S. Gururatana, R. Prapainop, S. Chuepeng, S. Skullong, Development of heat transfer performance in tubular heat exchanger with improved NACA0024 vortex generator, *Case Stud. Therm. Eng.* 26 (2021): 101166. https://doi.org/10.1016/j.csite.2021.101166.
70. P. Ocłoń, S. Łopata, T. Stelmach, M. Li, J.-F. Zhang, H. Mzad, W.-Q. Tao, Design optimization of a high-temperature fin-and-tube heat exchanger manifold – a case study, *Energy.* 215 (2021): 119059. https://doi.org/10.1016/j.energy.2020.119059.
71. T. Nomura, M. Tsubota, T. Oya, N. Okinaka, T. Akiyama, Heat storage in direct-contact heat exchanger with phase change material, *Appl. Therm. Eng.* 50 (2013): 26–34. https://doi.org/10.1016/j.applthermaleng.2012.04.062.
72. M.F. Rahi, M. Arıcı, A.M. Abdulateef, H. Niyas, Numerical investigation and optimization of macroencapsulated phase change material capsules in building roof slab, *J. Energy Storage.* 68 (2023): 107806. https://doi.org/10.1016/J.EST.2023.107806.

73. D.L. Dalluge, L.E. Whitmer, J.P. Polin, Y.S. Choi, B.H. Shanks, R.C. Brown, Comparison of direct and indirect contact heat exchange to improve recovery of bio-oil, *Appl. Energy.* 251 (2019): 113346. https://doi.org/10.1016/j.apenergy.2019.113346.
74. M. Medrano, M.O. Yilmaz, M. Nogués, I. Martorell, J. Roca, L.F. Cabeza, Experimental evaluation of commercial heat exchangers for use as PCM thermal storage systems, *Appl. Energy.* 86 (2009): 2047–2055. https://doi.org/10.1016/j.apenergy.2009.01.014.
75. S. Jaisankar, J. Ananth, S. Thulasi, S.T. Jayasuthakar, K.N. Sheeba, A comprehensive review on solar water heaters, *Renew. Sustain. Energy Rev.* 15 (2011): 3045–3050. https://doi.org/http://dx.doi.org/10.1016/j.rser.2011.03.009.
76. A.A.B. Pécora, M.R. Parise, Heat transfer coefficient in a shallow fluidized bed heat exchanger with a continuous flow of solid particles, *J. Brazilian Soc. Mech. Sci. Eng.* 28 (2006): 253–258.
77. Q. Yu, Y. Yang, Z. Wang, H. Zhu, Modeling and parameter sensitivity analysis of fluidized bed solid particle/sCO2 heat exchanger for concentrated solar power plant, *Appl. Therm. Eng.* 197 (2021): 117429. https://doi.org/10.1016/J.APPLTHERMALENG.2021.117429.
78. D.P. Sekulić, R.K. Shah, Thermal design theory of three-fluid heat exchangers, in: J.P. Hartnett, T.F.B.T.-A. in H.T. Irvine (Eds.), Elsevier, 1995, 219–328. https://doi.org/10.1016/S0065-2717(08)70297-1.
79. D.D. Nikodijević, Ž.M. Stamenković, M.M. Jovanović, M.M. Kocić, J.D. Nikodijević, Flow and heat transfer of three immiscible fluids in the presence of uniform magnetic field, *Therm. Sci.* 18 (2014): 1019–1028.
80. S. Ben Saad, C. Gentric, J.-F. Fourmigué, P. Clément, J.-P. Leclerc, CFD and experimental investigation of the gas–liquid flow in the distributor of a compact heat exchanger, *Chem. Eng. Res. Des.* 92 (2014): 2361–2370. https://doi.org/10.1016/j.cherd.2014.02.002.
81. R. Wen, X. Ma, Y.-C. Lee, R. Yang, Liquid-vapor phase-change heat transfer on functionalized nanowired surfaces and beyond, *Joule.* 2 (2018): 2307–2347.
82. R.K. Shah, A.D. Kraus, D. Metzger, Compact heat exchangers, *Handb. Heat Transf. Appl.* 2 (1985): 4–174.
83. Compact and micro-heat exchangers, *Process Intensif.* (2008): 77–101. https://doi.org/10.1016/B978-0-7506-8941-0.00005-5.
84. L.B. Erbay, N. Uğurlubilek, Ö. Altun, B. Doğan, Compact heat exchangers, *Eng. Mach. Mag.* 54 (2013): 37–48.
85. Q. Jiang, M. Zhuang, Q. Zhang, Z. Zhu, M. Geng, L. Sheng, P. Zhu, Experimental study on the thermal hydraulic performance of plate-fin heat exchangers for cryogenic applications, *Cryogenics (Guildf).* 91 (2018): 58–67. https://doi.org/10.1016/j.cryogenics.2018.02.006.
86. D. Popov, K. Fikiin, B. Stankov, G. Alvarez, M. Youbi-Idrissi, A. Damas, J. Evans, T. Brown, Cryogenic heat exchangers for process cooling and renewable energy storage: A review, *Appl. Therm. Eng.* 153 (2019): 275–290. https://doi.org/10.1016/j.applthermaleng.2019.02.106.
87. R.K. Shah, J.M. Robertson, Compact heat exchangers for the process industry, in: *Energy Efficient Process Technology*, Springer, 1993, pp. 565–580.
88. S. Teir, A. Jokivuori, Thermal design of heat exchangers, *Steam Boil. Technol.* Eb. (2002).
89. M. Edalatpour, N.T. Agathocleous, J. Byun, J.G. Tasiopoulos, C.W. Engberg, W. Abate, Two-phase computational modeling of evaporative film boiling in plate heat exchangers with offset-strip fins, *Int. J. Heat Mass Transf.* 236 (2025): 126260. https://doi.org/10.1016/j.ijheatmasstransfer.2024.126260.
90. M.M. Aslam Bhutta, N. Hayat, M.H. Bashir, A.R. Khan, K.N. Ahmad, S. Khan, M. Mahmood, A. Bhutta, N. Hayat, M.H. Bashir, A.R. Khan, K.N. Ahmad, S. Khan, CFD applications in various heat exchangers design: a review, *Appl. Therm. Eng.* 32 (2012): 1–12. https://doi.org/10.1016/j.applthermaleng.2011.09.001.
91. S. Kumar, R.B. Grover, H. Yadav, P.K. Vijayan, U. Kannan, A. Agrawal, Experimental and numerical investigation on suppression of thermal strati fi cation in a water-pool : PIV measurements and CFD simulations, *Appl. Therm. Eng.* 138 (2018): 686–704. https://doi.org/10.1016/j.applthermaleng.2018.04.070.
92. S. Kumar, R.B. Grover, P.K. Vijayan, U. Kannan, Numerical investigation on the effect of shrouds around an immersed isolation condenser on the thermal stratification in large pools, *Ann. Nucl. Energy.* 110 (2017): 109–125. https://doi.org/10.1016/j.anucene.2017.06.022.
93. D. Verma, M.S. Kalra, P. Wahi, Dynamics of a BWR with inclusion of boiling nonlinearity, clad temperature and void-dependent core power removal: Stability and bifurcation characteristics of advanced heavy water reactor (AHWR), *Nucl. Eng. Des.* 308 (2016): 297–311. https://doi.org/10.1016/J.NUCENGDES.2016.08.045.

94. J.B. Joshi, K. Nandakumar, A.W. Patwardhan, A.K. Nayak, V. Pareek, M. Gumulya, C. Wu, N. Minocha, E. Pal, M. Kumar, V. Bhusare, S. Tiwari, D. Lote, C. Mali, A. Kulkarni, S. Tamhankar, Computational fluid dynamics, *Adv. Comput. Fluid Dyn. Nucl. React. Des. Saf. Assess.* (2019): 21–238. https://doi.org/10.1016/B978-0-08-102337-2.00002-X.
95. C.R. Mali, V. Vinod, A.W. Patwardhan, Comparison of phase interaction models for high pressure subcooled boiling flow in long vertical tubes, *Nucl. Eng. Des.* 324 (2017): 337–359. https://doi.org/10.1016/J.NUCENGDES.2017.09.010.
96. C. T'Joen, Y. Park, Q. Wang, A. Sommers, X. Han, A. Jacobi, A review on polymer heat exchangers for HVAC&R applications, *Int. J. Refrig.* 32 (2009): 763–779. https://doi.org/10.1016/j.ijrefrig.2008.11.008.
97. M. Ghalandari, M. Irandoost Shahrestani, A. Maleki, M. Safdari Shadloo, M. El Haj Assad, Applications of intelligent methods in various types of heat exchangers: a review, *J. Therm. Anal. Calorim.* 138 (2021): 1–12.
98. S. Kakaç, H. Liu, A. *Pramuanjaroenkij, Heat Exchangers: Selection, Rating, and Thermal Design*, CRC Press, 2002.
99. P. Anand, P.K.S. Tejes, B.K. Naik, H. Niyas, Design analysis and performance prediction of packed bed latent heat storage system employing machine learning models, *J. Energy Storage.* 72 (2023): 108690. https://doi.org/10.1016/J.EST.2023.108690.
100. S. Kumar, D. Banerjee, S. Kumar, D. Banerjee, A review on phase change materials for sustainability applications by leveraging machine learning, *Energy Consum. Conversion, Storage, Effic.* (2024). https://doi.org/10.5772/INTECHOPEN.114380.
101. M.N.S. Jagirdar, G. Priyadarshi, B.K. Naik, H. Niyas, Implementation of adaptive neuro-fuzzy inference system in design and process optimization of latent heat storage system, *J. Energy Storage.* 89 (2024): 111810. https://doi.org/10.1016/J.EST.2024.111810.
102. S. Chakraborty, A.K. Gupta, D. Roy, A. Basumallick, Studies on nano-metal dispersed Cu-Cr matrix composite, *Mater. Lett.* 257 (2019): 126739. https://doi.org/10.1016/J.MATLET.2019.126739.
103. S. Kumar, K. Rathore, Renewable energy for sustainable development goal of clean and affordable energy, *Int. J. Mater. Manuf. Sustain. Technol.* 2 (2023): 1–15. https://doi.org/10.56896/IJMMST.2023.2.1.001.
104. U.N.D. of E. and S. Affairs, The sustainable development goals report 2019, in: *United Nations Department of Economic and Social Affairs (UNDESA)* (2019), 64. https://unstats.un.org/sdgs/report/2022/%0Ahttps://www.un-ilibrary.org/content/books/9789210018098%0Ahttps://www.un-ilibrary.org/content/books/9789210478878.
105. A. Dawahdeh, H. Sharadga, S. Kumar, *Novel MPPT Controller Augmented with Neural Network for Use with Photovoltaic Systems Experiencing Rapid Solar Radiation Changes*, 2024.

3 Conventional Methods for Enhancing Heat Transfer Relevant to Heat Exchangers

Sumit Khatri, Suvanjan Bhattacharyya, and Sunil Kumar

3.1 INTRODUCTION

Over the past era, urbanization, industrialization, and population growth have boosted energy demand. Thus, researchers are developing new energy sources and energy-saving methods. Traditionally, energy is generated from fossil fuels like coal, natural gas, oil, and so on. Naturally exhaustible, these fuels harm the environment. However, scientists have tried to lessen fossil fuel emissions' environmental impact using renewable energy systems [1, 2]. Efficient and compact thermal solutions are developing rapidly to save energy. Heat exchangers are practical thermal systems used in homes, businesses, and industries. Heat exchangers efficiently transmit thermal or heat energy across two or more fluids without interaction. It improves the performance and efficiency of the system, making it essential in HVAC and industrial applications. Common types include shell and tube, air-cooled, and plate exchangers. Researchers have optimized heat exchanger thermal transfer, design, size, and performance. Heat exchangers come in small to large sizes to fulfill specific needs and optimize thermal performance. More efficient heat exchangers are needed as demand rises [3]. Heat exchanger performance depends on fluid characteristics and flow rate. Heat exchanger performance can be improved via magnetic fields, electric fields, inserts, nanofluids, hybrid fluids, and other methods. Heat transfer (HT) improvement strategies are mixed, passive, and active.

Active heat transfer systems use external energy to increase performance. Extreme temperatures or heat can shorten the lifespan and functionality of some thermal systems. Premature degradation and failure pose serious safety issues. To sustain dependability, performance, and durability, system components must be cooled by dissipating hot spots or transferring heat to a heat sink. In some cases, system energy supply can be difficult. Active techniques can adjust flow modification to system requirements, such as magnetic fields controlling ferrofluids. Vibrating heat transfer surfaces, pulsing flow, and electric fields are other active techniques [4]. However, active technologies require more work to create small, efficient heat exchangers. Passive approaches use a small amount of system energy to increase fluid turbulence and heat transfer. Industrialists like these approaches for their simplicity, cost-effectiveness, and reliability. Surface modification and turbulators promote flow turbulence and heat transmission. Fins increase surface area and heat transfer, as well as increasing heat transmission and flow field turbulence. Passive approaches improve heat transfer; however, they cause a loss in pressure and require higher pumping power from fans or pumps. Low-profile ribs that do not obstruct flow reduce this power requirement. Near the ribs, flow separation and reattachment only affect the thermal boundary layer, improving heat transmission. The passive method uses the system's energy and requires more pumping power [5].

 DOI: 10.1201/9781003509837-3

3.2 PASSIVE METHODS

The term "passive techniques" refers to adjustments that are made to thermal systems to improve heating performance without using external energy sources. A passive decay heat removal system (PDHRS) is an example of the successful implementation of a passive method [6–9]. The use of porous materials inserts like turbulators, swirl flow tools, wire coils, twisted strips, corrugated surfaces, and extended surfaces like dimples, fins, nanofluids, protrusions, and varied tube geometries (such as helical and spiral) are some examples of these [5]. Even though these technologies require more pumping force due to higher friction, they are appealing since they are quick to apply, cost-effective, and require low maintenance [10]. Nanofluids are useful in improving thermal conductivity and making flow more turbulent. Turbulator inserts create flow disturbances, fins increase surface area and turbulence, and changes to geometry can enhance both turbulence and surface area. These are all effective ways to improve performance. These strategies concentrate on lowering the thermal boundary layer by employing disturbance, secondary flow, and increased surface area. They frequently combine different methods to achieve the best outcome in terms of heat transfer enhancement [11].

3.2.1 Fins and Extended Surfaces

Heat exchangers with fins or expanded surfaces improve heat transfer between a preliminary surface and ambient fluid. Industrial needs are covered with rectangular, square, annular, cylindrical, and tapered fins in various geometric configurations [12, 13]. Optimizing these fins can boost heat dissipation or reduce fin weight while maintaining exact heat dissipation. Recent innovations include permeable and perforated fins. This chapter examines types of fin heat transfer enhancement using thermophysical and geometric characteristics. Extended surface fins include variable area straight fins, constant area straight fins, annular fins, and pin fins. To increase surface contact area and convection, improving thermal conductivity and cooling efficiency, annular fins are used. Their industrial and home applications include radiators, engines, compressors, and electronics. The chapter also discusses longitudinal fin research, stressing efficiency and optimization for thermal management. Using perforated concave rectangular winglet pair vortex generators (PCRWP VGs) with plates inside rectangular ducts, Heriyani et al. [14] improved heat exchanger heat transmission. This technology is used in in-line and staggered arrangements to enhance heat transfer efficiency and lower pressure drop. The in-line configuration had a reduced cost-benefit ratio (CBR) of 3.56 and a thermal enhancement factor (TEF) of 1.29 throughout 0.4–2 m/s fluid flow velocity (Reynolds number 2143 to 11,763) compared with the staggered configuration. Nadooshan et al. [15] used a wind tunnel to study how perforated fins affect circular tube heat transfer. Perforated fins increased heat transfer by 9.23% at low external Reynolds numbers but decreased by 10.6% at high Reynolds numbers. Larger cross-sectional areas make internal and external flow rates more noticeable. Ishaq et al. [16] presented a diamond-shaped fin design for double-pipe heat exchangers to improve the transfer of heat. Optimal fin location and height for heat transfer efficiency varied by fin number and annulus width according to experimental data. These configurations must be adjusted if the inner-to-outer pipe radii ratio changes. The summary is shown in Table 3.1.

3.2.2 Surface Roughness

Corrugated, ribbed, and other rough surfaces induce flow collisions and promote flow mixing, improving heat transfer in many systems over smooth surfaces. These surfaces disrupt and rebuild boundary layers, enhancing thermal performance. Optimizing this effect requires geometric characteristics including height, pitch, arrangement, and shape. Research on sharp, sinusoidal, trapezoidal, triangular, and arc-shaped corrugated channels shows that phase shifts, relative waviness, amplitude, and wavelength affect thermal performance. Both numerical simulations and experimental

TABLE 3.1
Overview of Heat Transfer Improvements Using Various Fin and Vortex Generator Configurations

Authors	Extended Surface Type	Reynolds Number	Applications	Outcomes
Heriyani et al. [14]	Perforated concave rectangular winglet pair vortex generators (PCRWP VGs)	2143–11,763	Heat exchangers in rectangular ducts	The in-line configuration had a lower cost-benefit ratio (3.56) and a thermal enhancement factor (1.29)
Nadooshan et al. [15]	Perforated fins on circular tubes	Low and high external Reynolds numbers	Circular tube heat transfer	Heat transfer increased by 9.23% at lower Reynolds numbers but dropped by 10.6% at higher Reynolds numbers
Ishaq t al. [16]	Diamond-shaped fins in double-pipe heat exchangers	Not specified	Double-pipe heat exchanger thermal performance	Optimal heat transfer depends on fin number, height, and annulus width, with adjustments needed for inner-to-outer pipe radius changes

research show that altering these factors improves heat transmission. Studies have also evaluated buoyant force, pulsing fluid flow, and channel designs on heat transfer. Compared with plain channels, arc-shaped corrugated channels with recirculation zones improve heat transfer, whereas sinusoidal channels minimize entropy and maximize thermal performance. Visualizations of airflow via inclined grooves show how deviations swirl longitudinally and secondary flows improve heat transfer. Overall, surface roughness, especially in corrugated channels, improves heat transmission by improving flow mixing and boundary layer interruption. Geometric parameter tuning is crucial for varied applications. Elshafei et al. [17] found that corrugated channels with constant wall temperature and fixed corrugation ratio (Re 3220–9420) improved heat transfer and pressure drop at 3.0 spacing ratio and 90° phase shift. In computational studies on turbulent flow between corrugated plates with buoyancy, Forooghi and Hooman [18] found that heat transfer declines with Grashof number for buoyancy-aided flows until recovery at a critical point and grows monotonically for buoyancy-opposed flows. Corrugated channels need higher Grashof numbers than vertical pipes to affect heat transmission. Dean vortices and chaotic advection improve heat transfer in periodic wavy channels, which outperform straight channels in efficiency and penalties in pressure drops, according to Sui et al. [19]. Wavy surfaces break the thermal boundary layer, making them excellent for heat exchangers, and Paisarn's [20] comprehensive research shows that they improve heat transmission and pressure drop over wavy channels. The summary is shown in Table 3.2.

3.2.3 Swirl Flow Devices

These include helical inserts and twisted tapes that induce secondary recirculation in axial flow, enhancing the transfer of heat in the heat exchangers for single-phase and two-phase flows. Twisted tapes, manufactured from metallic strips twisted into various shapes, help heat transfer and reduce friction. Studies have shown that twisted tapes increase turbulence and swirl, boosting thermal performance. Despite pressure drops, helically twisted tapes, tapes with rods or spacers, and multiple twisted tapes optimize flow patterns and break boundary layers, improving thermal efficiency. Esmaeilzadeh et al. [21] found that transmission of heat and friction in g-Al_2O_3/water nanofluid flow

TABLE 3.2
Summary of Heat Transfer and Pressure Drop Improvements in Different Channel Shapes and Flow Types

Authors	Extended Surface Type	Reynolds Number	Applications	Outcomes
Elshafei et al. [17]	Corrugated channels with constant wall temperature	3220–9420	Heat transfer and pressure drop in corrugated channels	Improved heat transfer and pressure drop with a 3.0 spacing ratio and 90° phase shift
Forooghi and Hooman [18]	Corrugated plates with buoyancy	Not specified	Turbulent flow with buoyancy effects	Heat transfer declines with Grashof number for buoyancy-aided flows until recovery; monotonically increases for buoyancy-opposed flows
Sui et al. [19]	Periodic wavy channels	Not specified	Heat exchangers and turbulent flow applications	Dean vortices and chaotic advection enhance heat transfer, outperforming straight channels in efficiency despite pressure drop penalties
Paisarn [20]	Wavy surfaces	Not specified	Heat exchangers	Breaking the thermal boundary layer improves heat transfer and pressure drop compared with smooth channels

in a tube having twisted tapes of varying thicknesses. Despite higher friction coefficients, thicker twisted tapes improved thermal performance and convective heat transfer in laminar flow (Re 150–1600). New Nusselt number-thermal performance correlations were proposed. Bhattacharyya et al. [22] used ANSYS Fluent 19.2 to calculate magnetic nanofluid (Fe_3O_4) pressure drop and heat transfer in a heat exchanger tube with innovative hybrid ribs. The results explain that hybrid ribs and magnetic nanofluid increased the heat transfer coefficient more than magnetic nanofluid alone, consistent with correlations. Thianpong et al. [23] investigated that airflow in a twisted tape swirl generator with a dimpled tube boosted the friction factor and heat transfer more compared with a plain tube. The study found that lower pitch and twist ratios enhance these effects for Reynolds numbers 12,000 to 44,000. Empirical correlations for Nusselt number (Nu) and friction factor (f) were derived. Sarada et al. [24] found that the turbulent flow heat transferred in a horizontal tube with varied width twisted tape inserts. Tapes that are 10–22 mm wide do a better job of balancing heat transfer improvement with pressure drop reduction compared with 26 mm tapes. Full-width tapes had 36-48% heat transfer increases and reduced-width tapes 33-39%, with novel Nusselt number correlations and friction factor based on a modified twist ratio. The summary is shown in Table 3.3.

3.2.4 Displacer Devices (Baffles, Inserts)

Helical baffles reduce pressure by minimizing the dead zones and improving heat transfer. However, reducing pressure drop boosts pumping power and system efficiency. Heat exchanger flow visualization shows diverging flows converge near baffles, causing longitudinal swirl and secondary flows from upward fluid movement. Entropy generation and thermohydraulic performance were studied numerically in wavy channels with sinusoidal, trapezoidal, and triangular corrugation profiles by Akbarzadeh et al. [25]. Between Reynolds numbers 400 and 1400, triangular channels produced

TABLE 3.3
Overview of Methods to Improve Heat Transfer Using Extended Surfaces and Nanofluids

Authors	Extended Surface Type	Reynolds Number	Applications	Outcomes
Esmaeilzadeh et al. [21]	Twisted tapes in g-Al_2O_3/water nanofluid flow	150–1600	Laminar flow heat transfer in a tube	Thicker twisted tapes improved convective heat transfer despite higher friction coefficients; proposed new *Nu*-thermal performance correlations
Bhattacharyya et al. [22]	Hybrid ribs with magnetic nanofluid (Fe_3O_4)	Not specified	Heat exchanger tube	Hybrid ribs with magnetic nanofluid enhanced heat transfer coefficient more than nanofluid alone, matching theoretical correlations
Thianpong et al. [23]	Dimpled tube with twisted tape swirl generator	12,000–44,000	Airflow heat transfer in dimpled tubes	Lower pitch and twist ratios significantly enhanced heat transfer and friction factor; derived empirical *Nu* and friction factor correlations
Sarada et al. [24]	Varied width twisted tape inserts	Not specified	Turbulent flow heat transfer in horizontal tubes	Full-width tapes enhanced heat transfer by 36–48%; reduced-width tapes by 33–39%, with modified twist ratio correlations for *Nu* and friction factors

the maximum thermal entropy, followed by sinusoidal and trapezoidal channels. Sinusoidal profiles were recommended for better performance and reduced entropy. A circular tube having 45° V-baffles was analyzed for periodic laminar flow and heat transmission using 3D numerical simulations by Jedsadaratanachai et al. [26]. The finite volume approach and SIMPLE algorithm for Reynolds numbers 100 to 2000 showed that V-baffles increase longitudinal vortices, heat transmission, Nusselt number, and friction factor, especially with larger blockage ratios. For V-upstream and V-downstream configurations, baffle heights of 0.20 and 0.25 times the diameter of the tube yielded optimal thermal enhancement factors of 3.20. Wang et al. [27] improved heat exchangers with helical baffles by blocking triangular leakage zones between simple baffles with fold baffles. Numerical simulations revealed reduced helical streamline pitch and increased radial and axial velocity vectors, resulting in significant improvements in shell-side heat transfer coefficient ($\alpha 0$) by 17.7% to 34.2% and heat transfer coefficient (K) by 7.9% to 9.7%, despite increased pressure drop (ΔP), thermal performance factors above 1.0 indicated higher integrated performance, balancing heat transfer efficiency and pressure drop for energy-efficient heat exchanger designs with optimized baffles. Zhou et al. [28] used FLUENT14.0 and the RNG k-ε model to study heat transfer and fluid flow in shell and tube heat exchangers with trefoil-hole baffles. After the first trefoil-hole baffle, fluid flow was fully established with periodic axial heat transfer coefficients and fluctuations in pressure drop. Jet and secondary flow near baffles reduced boundary layer thickness, improving heat transmission. The summary is shown in Table 3.4.

TABLE 3.4
Summary of Heat Transfer and Flow Improvements Using Different Baffle and Channel Configurations

Authors	Extended Surface Type	Reynolds Number	Applications	Outcomes
Akbarzadeh et al. [25]	Wavy channels (sinusoidal, trapezoidal, triangular)	400–1400	Thermohydraulic performance analysis	Triangular channels produced maximum thermal entropy; sinusoidal profiles were recommended for better performance and reduced entropy
Jedsadaratanachai et al. [26]	Circular tube with 45° V-baffles	100–2000	Heat transfer and flow visualization	V-baffles increased longitudinal vortices, Nusselt number, and friction factor; optimal TEF of 3.20 for baffle heights 0.20–0.25 times tube diameter
Wang et al. [27]	Helical baffles with fold baffles	Not specified	Shell-side heat transfer in heat exchangers	Improved α_o by 17.7%–34.2% and K by 7.9%–9.7%; thermal performance factors above 1.0 despite increased pressure drop
Zhou et al. [28]	Trefoil-hole baffles	Not specified	Shell and tube heat exchangers	Established periodic axial heat transfer and pressure drop fluctuations; jet and secondary flows improved heat transfer by reducing boundary layer thickness

3.3 Active Methods

Active heat transfer methods use external energy sources, such as injection, electric or magnetic fields, vibrations, jet impingement, or suction, to improve the rate of heat transfer [23]. These methods necessitate external power input, making them less feasible for many practical applications due to the need for additional equipment. Consequently, active techniques have shown less potential compared with passive methods, as providing the necessary external power can often be challenging [29]

3.3.1 Mechanical Aids (Agitation, Vibration)

To achieve the goal of increasing the heat transfer rate, it becomes important to disrupt the thermal boundary layer that is located near the heated surface. This can be accomplished by mechanically moving or churning the fluid. The rotation of the heat-transferring surface with the application of mechanical power is a technique that is frequently used in this category. Turbulence is produced in the flow as a result of this rotation, which increases the rate of heat transfer. Li et al. [30] experimental investigations revealed that cross-sectional shape and rotation affect heat transfer performance differently, with notable disparities between the trailing and leading sides. The impact of Reynolds and rotation numbers on Nusselt number ratios highlights the intricate relationship between heat transfer and flow dynamics in rotating systems. Qui et al. [31] indicate that the Nusselt number ratio generally increases linearly with the rotation number, while friction ratios may oscillate as the rotation number rises. Research highlights that rotation significantly affects heat transfer, particularly at sharp turns, enhancing impingement effects and influencing downstream passages. Additionally, buoyancy forces have been found to reduce friction resistance in smooth U-ducts, especially under high rotation number conditions. Morris and Abadi's [32] experimental studies on rotating circular pipes with internal ribs have shown that secondary flow generated by Coriolis forces enhances heat transfer on the trailing edge. Research on rotating rectangular ducts equipped with various rib geometries, including W-shaped and V-shaped ribs, revealed that discrete rib configurations performed better in both non-rotating and rotating modes. Moreover, these configurations even exhibited higher friction factors, indicating a trade-off between enhanced heat transfer and increased resistance. Wright et al.'s [33] investigations on rotating ribbed channels with various rib geometries have shown that W-shaped and discrete W-shaped ribs deliver superior heat transfer performance, albeit with higher frictional losses. Discrete angled ribs and discrete V-shaped demonstrated the least pressure drop, balancing heat transfer enhancement as well as pressure penalties. Overall, discrete V-shaped and W-shaped ribs were found to offer the best thermal performance in both non-rotating and rotating channels, outperforming the angled rib configuration. The summary is shown in Table 3.5.

3.3.2 Surface and Fluid Vibration

Surface or fluid vibrations in heat exchangers play an important role in improving the transfer of heat by disrupting the thermal boundary layer and increasing turbulence. This leads to more

TABLE 3.5
Heat Transfer and Flow Behavior in Rotating Systems with Different Surface and Duct Designs

Authors	Extended Surface Type	Reynolds Number	Applications	Results
Li et al. [30]	Rotating surfaces with varying cross-sectional shapes	Not specified	Rotating heat transfer systems	Cross-sectional shape and rotation affect heat transfer differently; leading and trailing sides show disparities
Qui et al. [31]	Rotating systems with smooth U-ducts	Not specified	Rotational heat transfer in ducts	With the rotation number, the Nusselt number ratio increases linearly; friction ratios oscillate; buoyancy reduces friction at high rotation numbers
Morris and Abadi [32]	Rotating circular pipes with internal ribs	Not specified	Rotating systems with secondary flow induced by Coriolis force	Enhanced heat transfer on the trailing edge due to Coriolis forces; higher friction factors observed
Wright et al. [33]	Rotating ribbed channels (W-shaped, V-shaped, discrete ribs)	Not specified	Rotating and non-rotating duct systems	Discrete W-shaped and W-shaped ribs offer better heat transfer but with higher friction losses; discrete V-shaped ribs balance pressure drop and heat transfer

efficient energy exchange between the surfaces and the working fluid. Vibrations may significantly boost heat transfer rates while also impacting the pressure drop. The type and intensity of vibration, such as amplitude and frequency, influence the degree of augmentation. Overall, incorporating vibrations can optimize heat exchanger performance for applications requiring high thermal efficiency. Mohammed et al. [34] investigated the impact of forced mechanical vibrations on a concentric pipe heat exchanger with zigzag-shaped turbulators using air as the working fluid. Results show that vibration enhances both heat transfer and pressure loss, with the highest heat transfer increase at 116% and friction factor at 95%. Frequency has a greater effect on heat transfer improvement than acceleration or signal type changes. Cheng et al.'s [35] experimental and numerical studies show that utilizing these vibrations significantly boosts the convective heat transfer coefficient and reduces fouling resistance. This method results in sustained heat transfer improvement. Hosseinian and Isfahani [36] showed that surface vibration in heat exchangers can effectively enhance heat transfer, as demonstrated in experiments with a PVDF double-pipe heat exchanger using electro-dynamic vibrators. Results indicate that higher vibration levels and increased mass flow rates significantly boost the heat transfer coefficient, with a maximum observed improvement of 97%. This highlights the potential of controlled surface vibration to optimize thermal performance in heat exchanger systems. Pettigrew and Taylor [37] designed guidelines to prevent tube failures in shell and tube heat exchangers due to vibrations induced by the flow of fluid. Vibration analysis typically involves steps such as evaluating flow distribution, dynamic parameters, excitation mechanisms, and damage assessment. These guidelines are crucial for various heat exchangers, including steam generators and condensers, to ensure structural integrity and reliability. The summary is shown in Table 3.6.

TABLE 3.6
Heat Transfer Improvements and the Effects of Vibration in Heat Exchangers

Authors	Extended Surface Type	Reynolds Number	Applications	Outcomes
Mohammed et al. [34]	Concentric pipe heat exchanger with zigzag turbulators and forced mechanical vibration	Not specified	Airflow in heat exchangers	Vibration increased heat transfer by 116% and friction factor by 95%; frequency had the most significant effect
Cheng et al. [35]	Vibrated surfaces in heat exchangers	Not specified	Heat exchangers for sustained heat transfer	Vibrations boosted the convective heat transfer coefficient and reduced fouling resistance, ensuring long-term performance
Hosseinian and Isfahani [36]	PVDF double-pipe heat exchanger with electro-dynamic vibrators	Not specified	Heat transfer in vibrated double-pipe systems	Controlled surface vibration enhanced heat transfer coefficient by up to 97%, with improvements at higher vibration levels and mass flow rates
Pettigrew and Taylor [37]	Shell and tube heat exchangers	Not specified	Design and safety against flow-induced vibration	Developed guidelines to prevent tube failures by analyzing flow distribution, excitation mechanisms, and dynamic parameters

3.3.3 Electrohydrodynamic Techniques

Electrohydrodynamic (EHD) methods enhance heat transfer by applying an electric field to the heat transfer fluid, typically a dielectric fluid that reacts to the field. This process involves using low current and high voltage, which converts electric energy into kinetic energy as the fluid interacts with the electric field. This interaction increases fluid motion, effectively obstructing the boundary layer in the radial direction and boosting heat transfer performance. Laohalertdecha et al. [38] showed that heat transfer in heat exchangers can also be enhanced using active techniques (e.g., surface or acoustic vibrations, electric field) or passive methods (e.g., special surface designs, fluid additives). EHD techniques, classified as active, were recognized for their potential to improve heat transfer. The review provides an overview of EHD-based heat transfer research, serving as a foundational guide for researchers interested in utilizing EHD techniques for enhancement. Lin and Jang [39] examined how fluid flow and heat transfer work in a heat exchanger with a special setup of wire electrodes that use electric fields to improve performance. They look at how different electrode placements (square or diagonal), tube arrangements (lined up or staggered), and applied voltages (from 0 to 16 kV) affect the system. The study focuses on flow conditions where the fluid moves smoothly, with Reynolds numbers ranging from 100 to 1000. The results indicate that EHD enhancement is most efficient at lower Reynolds numbers and higher voltages, with the staggered tube pitch and square wire electrode configuration achieving optimal heat transfer. At 16 kV and Re = 100, this configuration yields a maximum 218% increase in the average Nusselt number and a 56% in fin area reduction compared with cases without EHD. Thapa et al. [29] reviewed techniques for minimizing energy consumption by enhancing heat transfer in heat exchangers utilized across various applications, including solar heaters and power plant units. Active methods, involving external power sources, and passive methods, utilizing surface modifications, were compared, noting that both can increase friction and pumping power requirements. Notably, electrohydrodynamic methods can boost heat transfer by up to 20 times with only a three-fold friction factor increase, while magnetic materials combined with passive techniques show a 10% Nusselt number enhancement. The summary is shown in Table 3.7.

TABLE 3.7
Summary of Studies on Improving Heat Transfer with Electrohydrodynamic (EHD) Techniques and Surface Designs

Authors	Extended Surface Type	Reynolds Number	Applications	Outcomes
Laohalertdecha et al. [38]	EHD techniques	Not specified	General heat exchangers	Active EHD techniques reviewed for enhancing heat transfer performance using electric fields
Lin and Jang [39]	Tube heat exchanger and plate-fin with EHD wire electrodes	100–1000	Heat exchangers with dielectric fluids	Square wire electrode configuration and staggered tube pitch and at 16 kV, Re = 100 increased Nusselt number by 218% and reduced fin area by 56%
Thapa et al. [29]	Electrohydrodynamic methods and magnetic materials	Not specified	Solar heaters, power plants, and general systems	EHD methods enhance heat transfer up to 20 times with a three-fold friction factor increase; magnetic materials boost the Nusselt number by 10%

3.4 COMPOUND METHODS

3.4.1 Combination of Passive Techniques

Passive techniques are often used independently, but when combined, they form what is known as compound enhancement. The interaction of multiple enhancement techniques working together results in higher heat transfer coefficients compared with the sum of their individual effects. Examples of passive compound techniques include integral axial corrugation roughness combined with twisted tapes with oblique teeth and integral helical rib roughness paired with center-cleared twisted tapes. Babu et al. [40] reviewed passive compound heat transfer augmentation techniques (PCHTAT), which combine multiple methods, such as helical-ribbed tubes with twisted tape inserts or dimpled tubes with swirl generators, to achieve greater enhancement than single techniques. Such combinations, applicable in both laminar and turbulent flows, can significantly boost the overall heat transfer coefficient. Promvonge et al. [41] experimentally investigated the heat transfer and friction loss characteristics of V-nozzle turbulator inserts combined with a snail entry in a circular tube, covering Reynolds numbers from 8000 to 18,000. The results indicate that using both V-nozzles and a snail entry significantly enhances the Nusselt number and friction factor compared with using either a V-nozzle alone or a plain tube, with the PR = 2.0 configuration showing the highest performance. Empirical correlations for Nusselt number, friction factor, and performance evaluation criteria are developed based on the Reynolds number and pitch ratio. Saha [42] studied the turbulent airflow in rectangular and square ducts with combined internal axial corrugations and twisted tape inserts (with and without oblique teeth), which showed enhanced heat transfer and pressure drop characteristics. Twisted tapes with oblique teeth outperform those without when combined with axial corrugations, achieving up to a 55% increase in heat duty at constant pumping power and up to a 47% reduction in pumping power at constant heat duty. Developed correlations for Nusselt number and friction factor accurately predict experimental outcomes, though only marginal improvements are observed with full-length and short-length twisted tapes compared with those without oblique teeth. Bhattacharyya et al. [43] experimentally investigated the effects of short-length (SL) and full-length (FL) swirl generators on heat transfer and flow in a solar air heater tube and revealed that FL swirl generators achieved higher thermal energy transport coefficients but incurred greater pressure penalties. The study showed that using full-length spring and twisted tape inserts increases heat duty by up to 22%–27% at constant pumping power compared with short-length configurations while pumping power increases by 27%–33% at constant heat duty. This extensive dataset supports the design of efficient solar thermal heaters and heat exchangers. The summary is shown in Table 3.8.

3.4.2 Combination of Active Techniques

Active methods for heat transfer enhancement integrate external power sources with additional techniques to boost performance. While they can achieve significant heat transfer improvements, their design complexity and reliance on external power limit their widespread application [44, 45]. These methods are often employed selectively due to the higher energy input and system sophistication required. Chamkha and Aly [45] studied compound active heat transfer methods combining external energy sources with other enhancement techniques to optimize thermal performance. This approach, which may involve integrating magnetic fields, flow suction/injection, or thermoelectric elements, can significantly enhance heat transfer but often adds design complexity and operational costs. Such methods are applied in specialized scenarios where high-efficiency heat transfer is paramount despite the need for additional energy input. Pour and Esmaeilzadeh [46] experimented using EHD actuators with wire electrodes and showed that significant heat transfer improvements can be achieved, particularly at low Reynolds numbers, while optimizing power consumption. Such methods highlight the synergy between electrical fields and conventional convective techniques to

TABLE 3.8
Comparison of Extended Surface and Swirl-Generation Techniques for Enhancing Heat Transfer

Authors	Extended Surface Type	Reynolds Number	Applications	Result
Babu et al. [40]	Helical-ribbed tubes with twisted tape inserts, dimpled tubes with swirl generators	Not specified	General heat exchangers	Compound techniques significantly boost heat transfer coefficient over single methods in both laminar and turbulent flows
Promvonge et al. [41]	V-nozzle turbulator inserts with snail entry	8000–18,000	Circular tube	PR = 2.0 configuration shows the highest enhancement; developed empirical correlations for *Nu*, friction factor, and performance evaluation
Saha [42]	Internal axial corrugations with twisted tapes (with/without oblique teeth)	Turbulent flow	Rectangular and square ducts	Twisted tapes with oblique teeth improve heat duty by up to 55% and reduce pumping power by 47%
Bhattacharyya et al. [43]	Short-length (SL) and full-length (FL) swirl generators	Not specified	Solar air heater tubes	FL swirl generators increase heat duty by 22–27% at constant pumping power but with 27–33% increased pressure penalty

enhance overall heat transfer. Bhattacharyya et al. [47] explored the enhancement of heat transfer at low Reynolds numbers using a magnetic field and Fe_3O_4 nanofluid. The application of magnetic forces creates turbulence, leading to heat transfer improvements up to 135% at high field strengths, though with increased pressure drops. The results demonstrate that magnetic fields effectively boost heat transfer performance while balancing flow resistance, especially at lower magnetic intensities. The summary is shown in Table 3.9.

3.4.3 Hybrid Approaches

A hybrid approach to enhancing heat transfer in heat exchangers blends passive and active techniques to achieve optimal efficiency. Passive methods, like surface alterations and inserts, improve heat transfer without needing extra energy. Meanwhile, active techniques, such as using external power sources or applying magnetic and electrohydrodynamic fields, create controlled flow disturbances to further enhance heat transfer. Combining these methods results in significantly better performance compared with using a single technique, striking a balance between improved thermal output, energy use, and design feasibility. This approach is especially valuable for applications where space and energy are limited but high-efficiency heat management is crucial. Agrawal et al. [48] tackled the challenge of managing optimal power flow (OPF) in hybrid energy systems that included various energy sources, focusing on the unpredictability of renewable energy. It introduces a multi-objective thermal exchange optimization (MOTEO) approach, which leverages Newton's law of

TABLE 3.9
Advanced Active Methods for Heat Transfer Enhancement in Specialized Applications

Authors	Extended Surface Type	Reynolds Number	Applications	Result
Chamkha and Aly [45]	Compound active methods (e.g., magnetic fields, suction/injection, thermoelectric elements)	Not specified	Specialized heat exchangers	Significant heat transfer enhancement but adds design complexity and operational costs
Pour and Esmaeilzadeh [46]	Electrohydrodynamic actuators with wire electrodes	Low Reynolds numbers	Systems requiring enhanced convection	Improved heat transfer at low Re while optimizing power consumption; synergy between electrical fields and convection
Bhattacharyya et al. [47]	Magnetic fields with Fe_3O_4 nanofluid	Low Reynolds numbers	Heat exchangers for magnetic nanofluid flow	Heat transfer improvements up to 135% at high field strengths; increased pressure drop balanced at lower intensities

cooling alongside advanced non-dominated sorting and crowding distance techniques for effective multi-objective problem-solving. Testing on a modified IEEE 30-Bus system showed that MOTEO provides efficient solutions while adhering to strict system constraints, enhancing OPF management reliability in hybrid systems. Patel [49] explored optimizing energy efficiency by integrating a solar heater with a shell-tube heat exchanger using qualitative research and SOLIDWORKS modeling. The design features a sturdy iron frame supporting the glazing, absorber, and water pipes, enhancing durability, heat transfer, and cost-effectiveness through strategic material choices. Results indicate effective solar energy collection and conversion, proving the model's practical application for real-world heating systems and emphasizing the importance of material selection for peak performance. Starace et al. [50] introduced an innovative design method combining numerical and analytical approaches, where heat transfer and fluid dynamics were modeled and predictor functions developed using regression. By dividing the heat exchanger into control volumes and employing an iterative analytic method, the approach improves accuracy and enables in-depth analysis of fin effects on flow, heat transfer, and pressure losses. Al-Bakhit [51] proposed a hybrid approach where nonlinear momentum equations for one or two channels are solved using commercial CFD software, and the resulting velocity fields are applied in a custom code to solve the linear energy equation across the entire heat exchanger. This method provides an efficient way to simulate parallel-flow heat exchangers effectively. The summary is shown in Table 3.10.

3.4 SPECIFIC TECHNIQUES FOR VARIOUS HEAT EXCHANGERS

3.4.1 Shell and Tube Heat Exchangers

Shell and tube heat exchangers are widely used for efficient heat transfer between two fluids, where one flows through the tubes and the other around the outside within a shell. Their design is versatile, making them suitable for high-pressure and high-temperature applications in industries such as power plants, chemical processing, and HVAC systems. The structure can accommodate various tube arrangements to enhance heat exchange performance. Maintenance and cleaning are relatively

TABLE 3.10
Optimization and Simulation Techniques for Enhanced Heat Exchanger Performance

Authors	Extended Surface Type	Reynolds Number	Applications	Result
Agrawal et al. [48]	Multi-objective thermal exchange optimization (MOTEO) approach	Not applicable	Hybrid energy systems, optimal power flow (OPF)	Improved OPF management, enhanced reliability under renewable energy variability, adhering to strict constraints
Patel [49]	Solar heater integrated with shell-tube heat exchanger	Not applicable	Solar heating systems	Effective energy collection and conversion with cost-efficient, durable designs optimized through material selection
Starace et al. [50]	Control-volume-based design with numerical/analytical modeling	Not specified	Heat exchanger design optimization	Accurate modeling of fin effects, enabling enhanced analysis of heat transfer, flow dynamics, and pressure loss
Al-Bakhit [51]	Hybrid CFD and custom code approach for parallel-flow exchangers	Not applicable	Simulation of parallel-flow heat exchangers	Efficient simulation combining nonlinear momentum and linear energy equations, improving modeling precision

straightforward, which adds to their reliability and long service life. This type of exchanger is favored for its robust construction and capacity to handle large temperature differentials. Darbandi and Abdollahpour [52] introduced a novel approach that combined numerical simulations and analytical methods to accurately predict STHE thermal performance, bypassing costly experiments and reducing computational expenses. By integrating numerical-based correlations into an analytical P-NTU model, designers can efficiently assess and optimize STHEs for improved thermal outcomes across different operating scenarios. Prajapati et al. [53] analyzed a segmental baffled, water-to-water shell and tube heat exchanger designed using the Kern method. Exergy and advanced exergy analyses revealed significant potential for reducing exergy destruction, with 97.5% of it being endogenous and avoidable, suggesting that improvements in design, configuration, and operational parameters could enhance system performance. Al-darraji et al. [55] explored the novel impact of air bubble injection in the shell with disc and ring baffles, including those with circular holes (DRCHB). Experimental and numerical results revealed that DRCHB with maximum air injection significantly improved effectiveness by 202% to 231%, achieving superior heat transfer performance and a lower pressure drop compared with other baffle configurations. The summary is shown in Table 3.11.

3.4.2 Plate Heat Exchangers and Air-Cooled Heat Exchangers

Plate heat exchangers are compact and efficient devices used for transferring heat between fluids, known for their high heat transfer rates due to large surface areas and minimal space requirements. They are widely used in HVAC, refrigeration, and industrial processes for effective heat exchange.

TABLE 3.11
Innovative Techniques and Analyses for Shell and Tube Heat Exchanger Performance Enhancement

Authors	Application	Result
Darbandi and Abdollahpour [52]	Shell and tube heat exchanger thermal performance optimization	Novel approach combining numerical simulations and analytical methods for predicting thermal performance and reducing computational expenses
Prajapati et al. [53]	Water-to-water shell and tube heat exchanger design	Exergy and advanced exergy analyses revealed potential for reducing exergy destruction by 97.5%, suggesting improvements in design and operational parameters for enhanced performance
Fetuga et al. [54]	Segmental baffled water-to-water shell and tube heat exchanger	Similar to Prajapati et al., advanced exergy analysis indicated significant potential for reducing exergy destruction and improving system performance
Al-darraji et al. [55]	Air bubble injection in shell with disc and ring baffles	Maximum air injection with disc and ring baffles improved effectiveness by 202% to 231%, achieving superior heat transfer performance and lower pressure drop

Air-cooled heat exchangers use ambient air to cool process fluids, eliminating the need for water. They are ideal for areas with water scarcity and are commonly used in power plants, petrochemical industries, and HVAC systems for their cost-effectiveness and energy efficiency. Jadhav et al. [56] investigated the use of partially filled, high porosity-graded aluminum and copper foams to optimize heat transfer and reduce pressure drop in heat exchangers. Three configurations (M1, M2, and M3) with varying porous layer thicknesses and orientations (positive, negative, and compound) were evaluated using DEF flow and LTNE models. Results show that the decreasing graded foams in models M1 and M2 enhance heat transfer by up to 23.32% while significantly lowering pumping power by up to 84.02% compared with configurations without graded foams. Ryms et al. [57] explored the effect of spacing between vertical plates on free convective heat transfer and aimed to optimize this distance. Experimental tests were conducted on a heat exchanger model with varying channel widths, using temperature and velocity field data to calculate heat flux and efficiency. Numerical simulations complemented the experiments, leading to an optimal single-channel width of 0.035 m and multi-channel spacing of 0.0123 m for improved heat transfer. Jin et al. [58] studied heat dissipation in fuel cell vehicles using a plate-fin heat exchanger. Experimental findings revealed a 20% to 80% improvement in cooling performance under various conditions with a consistent spray coverage of 45%. An empirical correlation was developed, showing that 92.73% of data points fell within a 20% error margin, and the average deviation was 9.3%. Dizaji et al. [59] used a mesh-net insert to boost the efficiency of air-to-air plate heat exchangers, particularly for building heat recovery units. Testing showed significant improvements, with the overall heat transfer coefficient increasing from 24 to 36 W/m^2K and the Nusselt number rising by up to 75% based on airflow rates. Despite the additional fan power due to higher pressure drop, economic analysis indicated positive profitability, especially in areas with high electricity prices and when using systems like resistance heaters or low-performance heat pumps. The summary is shown in Table 3.12.

3.4.3 Microchannel Heat Exchangers

Microchannel heat exchangers are compact and highly efficient devices designed for applications requiring high heat transfer rates in a small volume. Their structure features numerous tiny

TABLE 3.12
Advanced Materials and Designs for Enhanced Heat Exchanger Efficiency

Authors	Application	Result
Jadhav et al. [56]	Optimizing heat transfer in heat exchangers with graded aluminum and copper foams	Decreasing graded foams in models M1 and M2 enhanced heat transfer by up to 23.32% and reduced pumping power by up to 84.02% compared with configurations without graded foams
Ryms et al. [57]	Free convective heat transfer in heat exchangers with varying plate spacing	Optimal channel width of 0.035 m and multi-channel spacing of 0.0123 m improved heat transfer
Jin et al. [58]	Heat dissipation in fuel cell vehicles using plate-fin heat exchangers	Cooling performance improved by 20% to 80%, with empirical correlation showing 92.73% data points within a 20% error margin and an average deviation of 9.30%
Dizaji et al. [59]	Air-to-air plate heat exchangers for building heat recovery	Mesh-net insert improved heat transfer coefficient from 24 to 36 W/m^2K and Nusselt number by up to 75%, with positive profitability despite higher pressure drop

channels that increase the surface area-to-volume ratio, promoting superior heat transfer performance. They are commonly used in advanced cooling systems, such as in electronics and automotive industries, due to their lightweight and space-saving design. Microchannels also reduce the refrigerant charge needed, making them environmentally friendly. Their intricate design, however, can pose manufacturing and maintenance challenges. Al-Fatlawi and Niazmand [60] explored how nanoparticle shapes impact thermal conductivity in base fluids, using mono and hybrid nanofluids like TiO_2, Al_2O_3, and TiO_2–Al_2O_3 in water and ethylene glycol. Simulations indicate that a 4% hybrid nanofluid (50% Al_2O_3, 50% TiO_2) enhances thermal conductivity by 18.6%, outperforming mono-nanofluids. The results offer insights for optimizing nanofluid-based cooling in microelectronics. Liang Du and Wenbo Hu [61] reviewed recent advances in microchannel designs, internal reinforcements, surface treatments, and material types, highlighting diamond microchannel heat sinks and proposing future research to balance heat transfer efficiency with pressure drop. Kuruva et al. [62] numerically explored the thermohydraulic performance of microchannel heat exchangers (MCHEs) with microchambers and cylindrical inserts using finite volume analysis. Results showed that MCHEs with both microchambers and double inserts (SMCHE-MCDCI) outperformed other configurations in heat transfer, achieving up to 51.28% higher heat transfer coefficients. The addition of inserts enhances fluid mixing and reduces pressure drop, boosting overall performance by 14.57% compared with MCHEs without inserts. Dwivedi et al. [63] highlight that MCHEs are effective for managing high heat flux in compact spaces. Research indicates that nanofluids enhance thermal performance more than conventional fluids and incorporating flow disrupters like ribs or baffles boosts heat transfer. Ceramic microchannels, especially with Al_2O_3-water nanofluids and wavy designs, show superior thermal behavior compared with traditional metallic and rectangular microchannels. The summary is shown in Table 3.13.

3.5 PERFORMANCE EVALUATION EXPERIMENTAL AND NUMERICAL METHODS

Evaluating the performance of a heat exchanger involves examining its effectiveness in transferring heat under different operating conditions. The main factors to consider are the heat transfer rate, thermal efficiency, pressure drop, and heat transfer coefficient. The heat transfer rate is calculated based on the temperature difference between the inlet and outlet fluids, while thermal efficiency measures how well the heat exchanger utilizes the potential for heat transfer. Pressure drop indicates

TABLE 3.13
Advancements in Microchannel Heat Exchangers and Nanofluid Applications for Thermal Management

Authors	Application	Result
Al-Fatlawi and Niazmand [60]	Nanoparticle shape impact on thermal conductivity in nanofluids for microelectronics cooling	4% hybrid nanofluid (50% Al_2O_3, 50% TiO_2) enhances thermal conductivity by 18.6%, outperforming mono-nanofluids
Liang Du and Wenbo Hu [61]	Review of microchannel designs, materials, and surface treatments for heat sinks	Highlights diamond microchannel heat sinks and suggests future research to balance heat transfer efficiency with pressure drop
Kuruva et al. [62]	Thermohydraulic performance of MCHEs with microchambers and cylindrical inserts	MCHEs with microchambers and double inserts achieve 51.28% higher heat transfer coefficients and 14.57% better overall performance compared with MCHEs without inserts
Dwivedi et al. [63]	High heat flux management in compact spaces with microchannel heat exchangers	Nanofluids and flow disruptors like ribs or baffles significantly enhance thermal performance; ceramic microchannels with Al_2O_3-water nanofluids and wavy designs outperform traditional designs

the resistance to fluid flow, with higher drops often signaling inefficiencies. Common methods for performance assessment include the NTU-effectiveness method, log mean temperature difference (LMTD), and exergy analysis, all of which help optimize design and improve energy efficiency. Vishwakarma et al. [64] experimentally investigated the heat transfer and pressure drop characteristics of a circular tube with wavy tape inserts in the transition flow regime. Tests were conducted on a tube with a 20 mm internal diameter and 2000 mm length, with Reynolds numbers ranging from 533 to 7002. Nine wavy tape inserts with different wave and width ratios were used under three constant heat fluxes (1, 2, and 3 kW/m²). The research found that the transition flow boundaries shifted, with the onset of transition occurring earlier when tapes were used. Additionally, the transition limits were influenced by the heat flux, wave ratio, and width ratio. The study also developed correlations to predict the Nusselt number and friction factor for laminar and turbulent flows. Bhattacharyya et al. [65] experimentally examined the effect of a novel ribbed wave tape on heat transfer enhancement in a solar air heater operating in the transitional flow regime. The experiment used three different wave ratios, with the tube uniformly heated and Reynolds numbers varying from 521 to 10,194. For a plain tube, transition begins at a Reynolds number of 2810 and ends at 3724, while for the tube with ribbed wavy tapes, transition starts earlier and ends at higher Reynolds numbers. For a heat flux of 2 kW/m², the transition for the wavy tape with a wave ratio of 3 begins at a Reynolds number of 2589, and for a heat flux of 4 kW/m², it begins at 2745. The transition ends at Reynolds numbers 3969 and 4139 for wave ratio 3 at heat fluxes of 2 and 4 kW/m², respectively. As the wave ratio increases, the transition length decreases at a uniform heat flux. The average Nusselt number enhancement for ribbed wavy tapes with wave ratio 3 at 4 kW/m² heat flux was 262% in the laminar flow regime and 85% in the turbulent flow regime. Two empirical correlations were developed to predict the Nusselt number and friction factor across all flow regimes. Bhattacharyya et al. [66] experimentally investigated the impact of a perforated circular disc swirl generator on heat transfer (HT) and flow characteristics in a solar air heater with a helical corrugated tube. The thermal energy transport coefficient is analyzed for different values of the corrugation angle (θ), corrugation pitch ratio (y), perforation ratio (k), and perforation disc pitch ratio (s), with Reynolds numbers (Re) ranging from 10,000 to 52,000. Isothermal pressure drop and heat transfer experiments under uniform heat flux conditions were conducted. Results show that the presence of a perforated circular disc inside the helically corrugated tube increases heat transfer by 50–60%. Entropy generation and exergy efficiency were also evaluated. The combination of the corrugated tube and swirl generator enhances the thermal energy transport coefficient with minimal pressure penalty. A predictive correlation for the Nusselt number and friction factor was developed, and a comprehensive data set was created, which is valuable for designing solar thermal air heaters and heat exchangers. Everts and Meyer [67] experimentally examined and compared the pressure drop and heat transfer characteristics in smooth horizontal circular tubes across various flow regimes, including laminar, transitional, quasi-turbulent, and turbulent. A custom experimental setup was designed, built, and validated against existing literature. The smooth circular test section had an inner diameter of 11.5 mm and a maximum length-to-diameter ratio of 872. Pressure drop and heat transfer data were collected at Reynolds numbers ranging from 500 to 10,000 with varying heat fluxes using water as the test fluid. A total of 317 mass flow rate measurements, 34,553 temperature measurements, and 2536 pressure drop readings were recorded. The results showed that the Reynolds numbers at which transition began and ended were consistent for both pressure drop and heat transfer. Based on the findings, correlations were developed to describe the relationship between heat transfer and pressure drop, along with average Nusselt numbers for both developing and fully developed flows under mixed convection conditions. The study concluded that the pressure drop-heat transfer relationship could serve as an additional criterion for distinguishing between flow regimes. Bhattacharyya et al. [69] explored the use of broken twisted tape (BTT) inserts in triangular axial corrugated tubes to enhance heat transfer performance, both experimentally and numerically. The goal was to improve thermohydraulic performance by incorporating these inserts into a corrugated tube design. The analysis covers Reynolds numbers ranging from 10,000 to 80,000 in the turbulent flow regime, with the shear stress transport model applied as the turbulence model. Air is used as the working fluid, and various twisted tape inserts, including full-length tapes with different twist ratios and broken tapes with different twist ratios and gap ratios, are tested. The experiments maintain a constant angular pitch of 120° between neighboring corrugations and a 60° corrugation angle while investigating three different corrugation heights. The results demonstrate a strong correlation between measurements and existing published data. For all cases examined, the performance factors consistently exceeded unity, indicating improved thermohydraulic performance. Correlations for predicting friction factor and Nusselt number were also developed [68, 70, 71]. The summary is shown in Table 3.14.

3.6 SPECIFIC TECHNIQUES FOR VARIOUS HEAT TRANSFER DEVICES

Heat transfer devices are fundamental to the efficient conversion and recovery of thermal energy across a broad spectrum of industrial and domestic applications. These include critical processes such as liquid boiling and steam condensation in power generation, thermal operations in the pharmaceutical and chemical industries, precise temperature control in dairy production, heating of fluids in concentrated solar power systems, and cooling mechanisms for electrical machinery and electronic components. Enhancing the performance of these systems is vital for achieving significant energy savings, optimizing material usage, and reducing operational costs. Heat transfer enhancement techniques are developed to minimize thermal resistance, either by expanding the effective surface area for heat exchange or by inducing turbulence within the fluid flow. Approaches such as surface roughening or the incorporation of extended surfaces aim to increase the available heat transfer area, while devices such as inserts, winglets, and turbulators are employed to generate turbulence, thereby improving overall heat transfer efficiency. However, these enhancements are often associated with an increase in pumping power, which can result in elevated operational expenses.

The efficacy of these heat transfer enhancement techniques is typically evaluated using the thermal performance factor (TPF), denoted as η. The TPF provides a measure of the relative improvement in heat transfer rate in relation to the corresponding change in friction factor. It is calculated using the following equation.

TABLE 3.14
Extended Surface and Swirl-Inducing Techniques for Heat Transfer and Flow Optimization

Authors	Extended Surface Type	Reynolds Number	Applications	Result
Vishwakarma et al. [64]	Wavy tape inserts	533–7002	Heat transfer and pressure drop in a circular tube	Transition flow boundaries shift earlier with tapes; developed correlations for Nusselt number and friction factor
Bhattacharyya et al. [65]	Ribbed wavy tape	521–10,194	Heat transfer enhancement in solar air heaters	262% Nusselt number enhancement in laminar flow and 85% in turbulent flow with wavy tape inserts
Bhattacharyya et al. [66]	Perforated circular disc swirl generator	10,000–52,000	Solar air heater with helical corrugated tube	Heat transfer increased by 50–60% with minimal pressure penalty; developed predictive correlations
Everts and Meyer [67]	Smooth circular tube	500–10,000	Pressure drop and heat transfer in smooth tubes	Correlations developed for Nusselt number and pressure drop for mixed convection conditions
Abolarin et al. [68]	Smooth circular tube	500–10,000	Pressure drop and heat transfer in smooth tubes	Developed correlations to describe pressure drop-heat transfer relationship for distinguishing flow regimes
Bhattacharyya et al. [69]	Broken twisted tape inserts in triangular axial corrugated tubes	10,000–80,000	Heat transfer performance enhancement in corrugated tubes	Performance factors consistently exceed unity, indicating improved thermohydraulic performance

$$\eta = \frac{Nu / Nu_0}{\left(f / f_0\right)^{1/3}}$$

Where $\left(Nu = hl / k\right)$ is the Nusselt number and $f = \dfrac{\Delta P}{\left(\rho u^2 / 2\right)\left(l / D_h\right)}$ is known as the friction factor.

There are multiple types of heat exchangers that are involved. Widely used heat exchangers are shell and tube heat exchangers, plate heat exchangers, air-cooled heat exchangers, and microchannel heat exchangers. The conventional methods for heat transfer enhancement specific to these four types of heat exchangers are in the following sections.

3.6.1 Shell and Tube Heat Exchangers

A wide range of studies has been conducted to increase the heat transfer rate and reduce the size and cost of shell and tube heat exchangers (STHE). Conventional methods for heat transfer enhancement in STHE are predominantly focused on optimizing fluid flow patterns and increasing the effective surface area without the reliance on external energy sources. These approaches are generally categorized as passive techniques and are extensively utilized across a wide range of industries, including power generation, HVAC systems, nuclear power plants, chemical processing, and petrochemical operations. Below are some of the generally employed conventional methods.

3.6.1.1 Extended Surfaces (Fins)

Fins are employed to increase the surface area of contact with fluids. They also induce turbulence and secondary flows to further enhance heat transfer. Fins can be added to the tube side or the shell side to increase the heat transfer surface area without needing to increase the size of the heat exchanger. The longitudinal fins are typically used on the shell side to reduce thermal resistance, while transverse fins are used on the tube side for similar effects.

3.6.1.2 Baffles

Baffles are placed on the shell side to direct the flow of fluid. They help in creating a crossflow pattern or turbulent flow rather than allowing the fluid to flow parallel to the tubes. By forcing the shell-side fluid to flow over the tubes multiple times (crossflow), baffles increase fluid velocity and improve heat transfer efficiency. Segmental baffles are the most common type—they improve fluid velocity and increase turbulence. Disk and doughnut baffles offer less pressure drop and a smoother flow pattern.

3.6.1.3 Tube Inserts

Inserts like twisted tapes or wire coils are placed inside the tubes to induce swirl or turbulence in the fluid flow. Inserts disturb the laminar boundary layer inside the tube, which increases turbulence and improves the heat transfer coefficient. Helical strips are placed inside tubes to promote turbulent flow. They have a similar effect to twisted tapes and are used to disturb the flow and enhance heat transfer.

3.6.1.4 Tube Layout

Varying the arrangement of tubes inside the heat exchanger, such as using triangular or square layouts, provides more tube surface area in contact with the shell-side fluid, which increases turbulence and enhances heat transfer. This also encourages better mixing of the fluid, which leads to more effective heat exchange.

3.6.1.5 Tube Corrugation or Ribbed Tubes

Tubes with corrugated or ribbed surfaces are used to increase turbulence and heat transfer area. The ribbed or corrugated profile increases the internal and external surface area and disturbs the flow, enhancing the heat transfer process.

3.6.1.6 Multi-Pass Flow

Shell and tube heat exchangers can be designed with multiple passes on the shell or tube sides. By forcing the fluid to pass through the heat exchanger more than once, the overall heat transfer rate is increased because of better fluid mixing and extended exposure to heat transfer surfaces.

With increased heat transfer efficiency, shell and tube exchangers can be made more compact while maintaining their heat transfer capacity. By improving heat exchange efficiency, less energy is required to maintain the desired process temperatures. Some enhancement methods, particularly those that increase fluid turbulence (e.g., baffles, tube inserts), may lead to higher pressure drops, requiring more pumping power. Certain techniques, like surface roughening or finned surfaces, can increase fouling rates due to their more complex surface geometry.

3.6.2 Plate Heat Exchangers

Conventional methods for heat transfer enhancement in plate heat exchangers (PHEs) focus on improving the efficiency of heat transfer by altering surface design, optimizing flow conditions, and increasing the effective heat transfer area. Below are some key conventional methods used to enhance heat transfer in PHEs.

3.6.2.1 Corrugated Plates

The heat transfer plates in PHEs are often manufactured with corrugated or wavy patterns. The most commonly used pattern is the chevron (herringbone) design, which involves a series of V-shaped grooves across the surface of the plates. Corrugated plates increase the effective surface area and promote turbulence in the fluid, disrupting laminar flow and enhancing convective heat transfer. The turbulence improves the overall heat transfer coefficient by preventing the formation of stagnant boundary layers of fluid near the surface. Corrugations also help mix the fluid more effectively as it flows through the narrow channels between plates, leading to a higher heat transfer rate. However, the increased turbulence may cause a higher pressure drop across the heat exchanger. High-angle chevron plates increase turbulence and improve heat transfer rates but with a higher pressure drop. Low-angle chevron plates offer a lower pressure drop but slightly reduced heat transfer performance.

3.6.2.2 Plate Thickness Reduction

The thickness of the plates used in PHEs can be minimized to reduce the resistance to heat transfer. Thinner plates reduce the distance that heat must travel between the hot and cold fluids, thereby lowering thermal resistance and improving overall heat transfer efficiency. Thinner plates also allow for faster heat exchange between fluids. However, the plate material must still be strong enough to withstand the mechanical stresses and pressures associated with the application, particularly in high-pressure industrial processes.

3.6.2.3 Enhanced Surface Textures

Surface modifications such as dimples, grooves, or other micro-textured designs are applied to the heat transfer surfaces of the plates. These surface textures further enhance turbulence and improve fluid mixing, which leads to better heat transfer. Textures disrupt the boundary layer of fluid, encouraging more effective heat exchange and reducing the thermal resistance at the surface. Surface roughening is especially beneficial when dealing with viscous fluids or fluids with a tendency to develop laminar flow. Despite the improvement in heat transfer, textured surfaces may increase the risk of fouling in certain fluid environments.

3.6.2.4 Increasing Plate Surface Area

By increasing the number of plates or adjusting their design, the overall surface area available for heat transfer can be increased. More plates or larger surface areas provide additional paths for heat exchange, allowing for more efficient heat transfer between the fluids. This is particularly effective in applications where a high thermal load must be managed within a compact space. However, increasing the number of plates also raises the pressure drop and can complicate maintenance or cleaning.

3.6.2.5 Optimized Plate Spacing

The spacing between plates is adjusted to strike a balance between fluid velocity, heat transfer, and pressure drop. Reducing the distance between plates increases the fluid velocity, creating more turbulence and enhancing convective heat transfer. At the same time, it increases the overall surface area exposed to the fluids. However, narrowing the gaps can increase the likelihood of fouling and cause a higher pressure drop, which may require more pumping power. Optimizing plate spacing is crucial for achieving high heat transfer efficiency without excessive energy consumption.

3.6.2.6 Counter-Flow Arrangement

A counter-flow configuration is used where the hot and cold fluids flow in opposite directions across the plates. A counter-flow arrangement maximizes the temperature difference between the two fluids along the entire length of the heat exchanger. This allows for more efficient heat exchange compared with a parallel-flow arrangement, where the temperature gradient decreases as the fluids move through the exchanger. A counter-flow configuration can achieve higher heat transfer efficiency, resulting in greater energy savings.

3.6.2.7 Flow Distribution Optimization

The inlet and outlet ports and channels of the PHE are designed to distribute the fluid evenly across the plates. Proper flow distribution prevents areas of low flow or stagnation, ensuring that the entire heat transfer surface is used efficiently. Uneven distribution can reduce the effectiveness of the heat exchanger, as parts of the plates may not participate in the heat transfer process. Ensuring uniform flow distribution improves both heat transfer performance and overall system reliability.

3.6.2.8 Use of Multi-Pass Flow

In a multi-pass arrangement, the fluid is directed through multiple passes across the plates on one or both sides of the heat exchanger. Multi-pass configurations allow the fluids to traverse the heat exchanger several times, increasing the contact time with the heat transfer surfaces and promoting greater heat exchange. This approach is particularly useful in applications with a large temperature difference between the fluids, or where space constraints require a compact design. However, multi-pass designs can increase pressure drop, so careful design is necessary to balance heat transfer efficiency and pumping requirements.

3.6.3 Air-Cooled Heat Exchangers

Conventional methods for heat transfer enhancement in air-cooled heat exchangers (ACHEs) aim to improve the performance of heat transfer systems by optimizing the surface area, airflow patterns, and thermal contact between the heat exchanger and the air. ACHEs are commonly used in industries such as power generation, petrochemical processing, refineries, HVAC systems, and automotive cooling systems, where air is used as the cooling medium. A detailed description of the most widely used conventional methods for enhancing heat transfer in air-cooled heat exchangers is presented in the following sections.

3.6.3.1 Extended Surfaces

Fins are typically attached to the external surface of the tubes in ACHEs. These fins increase the surface area in contact with the air, allowing for more efficient heat dissipation from the fluid inside the tubes to the surrounding air. Fins increase the surface area available for heat transfer without expanding the overall size of the heat exchanger. By increasing the contact surface, more heat can be transferred from the fluid inside the tubes to the ambient air. Finned surfaces are particularly useful in low-density fluids like air, where the natural heat transfer rates are lower than in liquids. Different fin geometries are used based on the application and desired performance such as the following:

- Plain fins: simple, flat fins that extend perpendicularly from the tubes, providing additional surface area.
- Louvered fins: these fins have small cuts or louvers that create airflow disturbances, increasing turbulence and promoting better heat transfer.

- Corrugated or wavy fins: the fins have an undulating pattern, which improves the mixing of air as it flows over the fins, leading to higher turbulence and better heat transfer.
- Spiral fins: these fins are helically wound around the tubes, increasing surface area and improving air interaction, which enhances convective heat transfer.

While fins enhance heat transfer by increasing the surface area, they can also introduce additional airflow resistance, increasing the pressure drop across the heat exchanger. This means that more fan power may be required to maintain the desired airflow, which could lead to higher operational costs.

However, the shape, density, and arrangement of fins can be modified to optimize heat transfer while balancing pressure drop and fan power requirements. Changing the fin density (number of fins per inch), fin height, or fin thickness allows for better control of heat transfer efficiency and pressure drop. For example, higher fin density increases the heat transfer surface area but also increases airflow resistance, which can lead to higher fan power consumption. On the other hand, lower-density fins reduce pressure drop but may not provide as much surface area for heat dissipation. Optimized geometries such as serrated fins (fins with small cuts) and louvered fins create more turbulence in the airflow, which helps to break up the thermal boundary layer on the fin surface, improving convective heat transfer.

3.6.3.2 Tube Surface Modifications

In addition to modifying the fins, the surface of the tubes themselves can be altered to improve heat transfer. Techniques include roughening or corrugating the outer surface of the tubes. Tube surface modifications such as roughened or corrugated surfaces increase turbulence in the air flowing over the tubes. This reduces the thickness of the thermal boundary layer and improves the rate of convective heat transfer. Corrugated tubes create small vortices in the airflow, which disrupts laminar flow and enhances heat transfer by promoting better fluid mixing. These modifications are especially useful in enhancing the heat transfer performance of the air side, which typically has lower heat transfer coefficients compared with the fluid side inside the tubes. However, surface modifications can slightly increase the manufacturing complexity and cost of the heat exchanger. Additionally, rough or corrugated surfaces may be more prone to fouling or dirt accumulation, especially in dusty environments, which could reduce their efficiency over time.

3.6.3.3 Surface Modifications for Tubes

Instead of using conventional round tubes, different tube shapes like elliptical, flat, or oval tubes can be used. Additionally, tubes can be arranged in staggered or in-line configurations to optimize airflow. The shape of the tubes significantly impacts both the heat transfer and the pressure drop characteristics. Elliptical or oval tubes reduce airflow resistance compared with round tubes, which leads to lower pressure drops. These shapes also provide a larger surface area for heat transfer, enhancing the efficiency of the heat exchanger. A staggered tube arrangement forces the air to weave around the tubes, increasing turbulence and promoting better heat transfer. On the other hand, an in-line tube arrangement provides a smoother airflow with less turbulence, resulting in a lower pressure drop but potentially less heat transfer efficiency. While elliptical or staggered tubes improve heat transfer, they may require more space or careful design considerations to ensure that the overall heat exchanger remains compact and cost-effective. Additionally, increased turbulence from staggered tubes may lead to higher fan power requirements.

3.6.3.4 Multi-Pass Flow Configuration

Air-cooled heat exchangers can be designed to allow for multiple passes of the fluid inside the tubes. The fluid flows through the heat exchanger more than once before exiting, increasing its contact with the air. A multi-pass configuration enhances the heat transfer by increasing the fluid's exposure to the air-cooled surface. With each pass, the temperature gradient between the fluid and the air becomes smaller, leading to more efficient heat exchange. This configuration is particularly effective when there are large temperature differentials between the incoming and outgoing fluids. For example, in a two-pass configuration, the fluid flows through half the tubes in one direction and then reverses through the remaining tubes, effectively doubling its exposure to the air. Although multi-pass configurations improve heat transfer, they can also increase the pressure drop on the fluid side, requiring more pumping power. Careful design is needed to ensure that the benefits of enhanced heat transfer outweigh the additional energy costs.

3.6.3.5 Surface Coatings

Coatings such as hydrophilic, anti-corrosive, or thermal conductivity-enhancing coatings can be applied to the fins and tubes. Hydrophilic coatings prevent water droplets from accumulating on the fins in humid environments, which can block airflow and reduce heat transfer. By promoting better water drainage, these coatings allow air to flow more freely over the fin surfaces. Anti-corrosive coatings protect the metal surfaces from corrosion caused by exposure to moisture, chemicals, or pollutants, ensuring the longevity and efficiency of the heat exchanger. Thermal conductivity-enhancing coatings improve heat transfer by facilitating faster heat conduction between the surface of the tubes and fins and the surrounding air. While coatings can improve performance and extend the lifespan of the heat exchanger, they add to the manufacturing cost. Furthermore, some coatings may wear off over time, especially in harsh operating environments, requiring periodic maintenance or reapplication.

3.6.3.6 Baffles

Baffles are used to direct and redistribute air through the heat exchanger, ensuring that airflow is more uniform across the finned surfaces. Baffles can help eliminate dead zones, where airflow is inadequate or stagnates, leading to poor heat transfer. By directing the air more evenly across the heat exchanger, baffles ensure that all parts of the surface contribute to heat dissipation. Baffles can also be used to increase air velocity in certain regions of the exchanger, which enhances heat transfer efficiency by promoting more turbulence and better fluid mixing. Higher air velocity can improve heat transfer rates, but it also increases pressure drop and may lead to higher fan power requirements. Furthermore, increased airflow can generate more noise, and in some cases, higher wind speeds might lead to dust accumulation or fouling on the heat exchanger surface, necessitating more frequent cleaning or maintenance.

3.6.3.7 Flow Distribution Optimization

The design of the heat exchanger's air and fluid distribution system can be optimized to ensure uniform distribution across the finned surface and inside the tubes. Ensuring that both the air and the fluid are evenly distributed across the heat exchanger's surfaces is critical for maximizing heat transfer efficiency. Poor distribution can lead to regions of the heat exchanger that are underutilized or overused, reducing overall efficiency. Optimizing the inlet and outlet design for the air and fluid ensures that all parts of the heat exchanger are used effectively. This improves thermal performance without requiring additional equipment or energy inputs. While flow distribution optimization can enhance performance, the design process may become more complex, particularly in large-scale systems. Additionally, changes in fluid distribution might require more precise control mechanisms or flow management devices, which could increase upfront costs.

3.6.4 Microchannel Heat Exchangers

Conventional methods for heat transfer enhancement in MCHEs focus on maximizing the efficiency of heat exchange within highly confined spaces by leveraging the benefits of small channel dimensions. MCHEs are known for their compact design, high heat transfer rates, and low fluid inventory, making them ideal for applications in refrigeration, automotive systems, aerospace, electronics cooling, and other areas where space and weight constraints are critical. In such systems, the fluid flow often operates in laminar regimes due to the small dimensions, so various enhancement techniques are employed to improve heat transfer while mitigating the pressure drop. A detailed description of the most widely used conventional methods for enhancing heat transfer in MCHEs is presented in the following sections.

3.6.4.1 Micro Pin Fins

Micro pin fins, or small cylindrical or rectangular projections, are embedded within the microchannels to increase surface area and promote flow disturbance. Micro pin fins dramatically increase the effective surface area for heat transfer by providing additional contact points between the fluid and the solid surfaces. These fins break up the flow, creating localized turbulence and mixing. This reduces the thermal boundary layer near the channel walls, resulting in more efficient convective heat transfer. In addition, the increased surface area provided by the fins allows for a higher heat flux to be managed in a compact space, making MCHEs with pin fins especially effective for applications with high thermal loads, such as electronics cooling. The inclusion of pin fins adds flow resistance, increasing the pressure drop within the channels. This can require higher pumping power to maintain fluid flow. Pin fin structures also add complexity to the manufacturing process, as precise fabrication is required to ensure the fins are uniformly distributed and aligned.

3.6.4.2 Surface Roughening

Internal surfaces of the microchannels are roughened or textured with micro- or nano-scale features. Surface roughening disrupts the laminar boundary layer, which tends to develop in microchannels due to the small flow dimensions. By introducing controlled surface roughness, the flow becomes more turbulent even at low Reynolds numbers, improving the mixing of the fluid and increasing convective heat transfer. The rough surface causes vortices and eddies in the flow, which enhances heat exchange between the fluid and the channel walls. While surface roughening significantly improves heat transfer, it can also lead to an increase in frictional losses, resulting in higher pressure drops. Additionally, roughened surfaces may promote fouling or clogging, especially in fluids that contain particulates or dissolved solids. Careful design is needed to strike a balance between heat transfer enhancement and pressure drop.

3.6.4.3 Wavy or Curved Channels

Instead of straight microchannels, the channels are designed with a wavy, sinusoidal, or serpentine pattern. Wavy or curved channels induce secondary flows and increase the interaction between the fluid and the channel walls. The curvature of the channels causes fluid to continuously change direction, disrupting the formation of stable boundary layers. This leads to enhanced mixing of the fluid and a higher heat transfer coefficient. Wavy channels also provide longer effective paths for the fluid, increasing the time the fluid remains in contact with the channel walls, which further improves heat transfer. These designs are particularly beneficial in applications where space is constrained but high heat removal is required, such as in microelectronics. The use of wavy or curved channels increases the overall pressure drop due to the additional flow resistance from the continuously changing flow direction. The more complex geometry also adds to the manufacturing difficulty and cost, particularly when producing channels with precise and consistent curvature.

3.6.4.4 Multi-Pass and Counter-Flow Configurations

Microchannel heat exchangers can be designed to use multi-pass flow configurations or counter-flow arrangements to enhance thermal performance. In a multi-pass configuration, the fluid passes through the microchannels multiple times, increasing its exposure to the heat transfer surface. Each pass allows additional heat to be exchanged, improving the overall efficiency of the system. Multi-pass systems are particularly beneficial in applications with large temperature gradients, as they help ensure that more heat is transferred from the fluid to the surrounding environment or vice versa. In counter-flow configurations, the two fluids in the heat exchanger flow in opposite directions. This maximizes the temperature difference between the fluids along the length of the exchanger, resulting in higher heat transfer efficiency compared with parallel-flow systems. Multi-pass configurations increase the length of the flow path, which can lead to higher pressure drops on the fluid side. Counter-flow arrangements, while highly effective in maximizing heat transfer, may require more complex routing of the channels and careful control of fluid flow rates to maintain efficient heat transfer without significant pressure loss.

3.6.4.5 Hydrophilic and Hydrophobic Surface Coatings

The internal surfaces of the microchannels can be coated with hydrophilic or hydrophobic materials to optimize heat transfer, especially in systems involving phase change, such as condensation or boiling. Hydrophilic coatings improve the wettability of the surface, promoting the formation of a thin liquid film during condensation processes. This reduces thermal resistance by allowing the condensate to spread uniformly across the surface, facilitating efficient heat transfer. Conversely, hydrophobic coatings repel water, encouraging the formation and quick shedding of droplets. This is particularly beneficial in two-phase heat transfer processes, where the rapid removal of condensate droplets or vapor bubbles can improve heat transfer efficiency by maintaining a clean and effective heat transfer surface. While surface coatings can greatly enhance phase-change heat transfer, they can degrade over time, particularly in harsh operating environments. Coating applications and maintenance add to the overall cost and complexity of the system. Additionally, coating uniformity is critical for achieving consistent performance across all channels.

3.6.4.6 Nanofluids and Advanced Coolants

Nanofluids, which are engineered fluids containing suspended nanoparticles, are used to improve the thermal conductivity and heat transfer capabilities of the working fluid in microchannel systems. Nanofluids exhibit significantly higher thermal conductivity than conventional coolants such as water, ethylene glycol, or refrigerants. The nanoparticles within the fluid enhance the heat transfer properties by improving the conduction of heat through the fluid and increasing the thermal energy transport within the microchannels. This allows for higher heat transfer rates, even with small channel sizes. Nanofluids are particularly advantageous in applications that require the management of high heat fluxes, such as power electronics, semiconductor cooling, or aerospace thermal systems. While nanofluids can significantly enhance thermal performance, they can also increase the risk of clogging or fouling within microchannels, especially if the nanoparticles aggregate. The long-term stability of nanofluids can be a concern, as particle settling may occur over time. Additionally, nanofluids can be more expensive than traditional coolants, and their use may require specialized handling and maintenance procedures.

3.6.4.7 Microchannel Shape and Size Optimizations

The cross-sectional shape and size of microchannels are optimized to achieve the best possible heat transfer and flow characteristics for a given application. The shape of the microchannels (e.g., circular, rectangular, triangular, or trapezoidal) affects the heat transfer surface area and the flow characteristics. Optimizing the channel shape can enhance fluid interaction with the walls, improving convective heat transfer. Smaller microchannel sizes increase the surface area-to-volume ratio, allowing for more efficient heat exchange, but they also result in higher pressure drops. The choice of channel shape and size is often dictated by the specific heat transfer requirements of the application, balancing heat transfer efficiency with acceptable pressure loss. Smaller microchannels offer higher heat transfer rates but come with increased pressure drop, which can necessitate higher pumping power. Furthermore, microchannels with very small dimensions are more prone to clogging or fouling, particularly in applications with particulate-laden fluids. The fabrication of non-standard shapes or very small channels can also add to the manufacturing complexity and cost.

3.6.4.8 Phase-Change Materials

Phase-change materials (PCMs) are incorporated into microchannel heat exchangers to enhance thermal performance in systems that experience rapid temperature fluctuations or require efficient thermal management during phase transitions (e.g., boiling or condensation). PCMs absorb or release large amounts of latent heat during phase transitions, which significantly enhances the heat transfer capacity of the system. For example, during boiling, the PCM absorbs heat from the fluid as it changes from solid to liquid, stabilizing the temperature within the system. Conversely, during condensation, the PCM releases stored heat as it transitions back to a solid state. The use of PCMs in microchannels allows for more effective temperature control and heat dissipation, particularly in applications where thermal spikes or fluctuations must be managed. While PCMs provide significant benefits in terms of thermal management, they can add complexity to the design and operation of the heat exchanger [72–74]. PCMs typically have lower thermal conductivity than other materials, so they may need to be combined with thermal conductivity enhancers such as metal foams or expanded graphite. Additionally, the long-term stability of PCMs and their ability to withstand repeated phase changes without degradation are key considerations.

3.6.4.9 Parallel-Flow Optimization

In microchannel heat exchangers, ensuring uniform distribution of fluid across all channels is critical for maximizing heat transfer. Flow distribution can be optimized through the design of manifolds and inlets. Uniform flow distribution ensures that all microchannels are fully utilized for heat transfer. Poor flow distribution can lead to the underutilization of certain channels, reducing the overall thermal performance of the exchanger. Optimized inlet and outlet manifold design helps balance the flow across all channels, ensuring that each channel carries an equal proportion of fluid. This leads to more efficient heat transfer and reduces the likelihood of hot spots or areas with insufficient cooling. Achieving uniform flow distribution requires precise design and manufacturing of channel geometry and manifold systems, which can increase the cost of the heat exchanger. Moreover, ensuring uniform flow in highly compact systems can be challenging, especially when operating under variable flow conditions.

3.6.4.9 Vortex Generators

Vortex generators are small structural elements, such as fins or tabs, introduced into the microchannels to create localized turbulence and vortices. Vortex generators disrupt the smooth flow of fluid through the microchannels, creating small vortices that enhance fluid mixing and reduce the thermal boundary layer near the channel walls. This localized turbulence increases the heat transfer rate by improving the convective heat transfer coefficient. Vortex generators are especially effective in laminar flow regimes, where natural turbulence is minimal, making them a key tool for enhancing heat transfer in microchannel systems. While vortex generators are highly effective at enhancing heat transfer, they also increase flow resistance and pressure drop, which may require more pumping power. Additionally, their presence complicates the design and fabrication of the microchannels, adding to the overall system cost.

3.7 CHALLENGES AND LIMITATIONS

The design and production of heat exchangers involve several material and manufacturing constraints that must be carefully considered, especially when incorporating conventional methods for enhancing heat transfer. These constraints significantly influence the feasibility, cost, durability, and performance of the heat exchanger, and balancing them is critical to achieving optimal thermal efficiency while maintaining reliability, ease of manufacturing, and long-term operational stability. The following sections show a comprehensive examination of the key material and manufacturing constraints relevant to conventional heat transfer enhancement techniques used in heat exchangers.

3.7.1 Material Constraints

The choice of materials used in heat exchangers directly impacts the effectiveness of heat transfer, resistance to environmental conditions, mechanical durability, and overall lifecycle costs. The materials must meet stringent thermal, chemical, and mechanical performance requirements to enable efficient heat transfer while withstanding operational stresses.

3.7.1.1 Thermal Conductivity

High thermal conductivity is essential for efficiently transferring heat between the fluids and the heat exchanger's surface. Materials with poor thermal conductivity will limit heat transfer rates, reducing the overall efficiency of the exchanger. Materials such as copper and aluminum are highly valued for their excellent thermal conductivity, allowing for rapid heat transfer. These materials are widely used in heat exchangers for applications like HVAC, refrigeration, and electronics cooling. Despite their high thermal conductivity, copper and aluminum may not be suitable for all operating environments, such as in corrosive settings or applications requiring high structural strength. Stainless steel, titanium, or nickel-based alloys are often chosen for corrosive or high-temperature environments, though these materials have lower thermal conductivity, which may reduce heat transfer performance.

3.7.1.2 Corrosion Resistance

Heat exchangers often operate in environments where the working fluids are corrosive, leading to material degradation over time if proper materials are not used. Corrosion resistance is critical to maintaining the integrity and efficiency of the heat exchanger over its operational life. Stainless steel, titanium, and nickel alloys are commonly chosen for applications that involve corrosive fluids, such as chemical processing, petrochemical plants, or marine environments. These materials resist corrosion but often have lower thermal conductivity, which can negatively impact the heat transfer rate. Although corrosion-resistant materials ensure long-term durability and reduce maintenance costs, they are often more expensive and less thermally efficient than alternatives like copper or aluminum. The designer must carefully evaluate the operating environment to select the right balance between heat transfer efficiency and material durability.

3.7.1.3 Corrosion Resistance

Heat exchangers must endure various mechanical stresses, such as high pressures, thermal cycling, and vibrations, depending on the application. The material must maintain its structural integrity under these conditions without failing due to stress or fatigue. Materials like stainless steel and titanium offer excellent mechanical strength and resistance to fatigue, making them suitable for high-pressure or high-temperature applications such as in nuclear reactors, steam generation systems, and chemical plants. While materials with superior mechanical properties are necessary for certain applications, they can have lower thermal conductivity, reducing heat transfer performance. Furthermore, materials like titanium are more expensive and harder to machine, increasing manufacturing costs.

3.7.1.4 Compatibility with Working Fluids

The material used in the heat exchanger must be chemically compatible with the working fluids to avoid corrosion, scaling, or material degradation over time. For example, copper is commonly used in HVAC and refrigeration systems but is susceptible to corrosion when exposed to ammonia-based refrigerants, necessitating the use of alternative materials such as aluminum or stainless steel. Material compatibility may require choosing a material with lower thermal conductivity to ensure resistance to chemical attack, which can reduce heat transfer performance. Material selection must balance the need for chemical durability with efficient heat transfer.

3.7.2 Manufacturing Constraints

Manufacturing processes play a crucial role in determining the feasibility of implementing heat transfer enhancement techniques. Many of these techniques involve complex geometries, surface modifications, and micro-structuring, all of which can pose significant challenges in terms of manufacturability and cost.

3.7.2.1 Fabrication of Complex Geometries

Techniques such as the inclusion of micro fins, corrugated surfaces, or baffle designs enhance heat transfer by increasing surface area or creating turbulence, but they also add significant complexity to the manufacturing process. Manufacturing intricate geometries often requires specialized processes such as precision casting, laser cutting, 3D printing, or micromachining. These methods are more expensive and time-consuming compared with standard fabrication techniques. The complexity of these geometries also increases the likelihood of manufacturing defects such as warping, misalignment, or residual stresses, which can compromise the performance of the heat exchanger. While these advanced geometries can significantly improve heat transfer rates, they come at the cost of higher production complexity, increased lead times, and potentially reduced scalability for mass production. Designers must consider whether the performance benefits justify the additional manufacturing costs.

3.7.2.2 Surface Treatments and Coatings

Many heat transfer enhancement techniques involve applying specialized surface coatings or treatments, such as hydrophilic/hydrophobic coatings, anti-corrosion layers, or thermal conductivity-enhancing coatings. These surface treatments can improve the heat exchanger's performance and durability, but they require additional manufacturing steps. Applying coatings requires processes such as electroplating, anodizing, or chemical vapor deposition (CVD), all of which increase manufacturing costs and complexity. Additionally, maintaining coating uniformity is critical to ensuring consistent heat transfer and durability across the heat exchanger. Inconsistent coatings can lead to performance degradation, such as localized corrosion or poor heat transfer. Coatings improve heat transfer performance and extend the life of the heat exchanger, but they add cost and manufacturing complexity. The durability of these coatings must also be considered, as some may degrade over time, particularly in harsh operating environments, requiring reapplication and increased maintenance costs.

3.7.2.3 Miniaturization and Precision Manufacturing

Enhancements such as microchannel designs, micro pin fins, and other microscale features require extremely high precision in fabrication, which adds to manufacturing complexity. Manufacturing microscale features necessitates advanced techniques such as laser cutting, micromachining, micro-milling, or 3D microfabrication. These processes are expensive, require specialized equipment, and have tight tolerances that must be maintained to ensure uniformity and optimal performance. While miniaturized features significantly improve heat transfer by increasing surface area and promoting turbulence, they also raise production costs, increase the likelihood of defects, and can result in more challenging quality control. Additionally, smaller channels are more prone to fouling and clogging, which can reduce heat exchanger performance over time, requiring more frequent maintenance.

3.7.2.4 Welding and Joining Techniques

Many heat exchangers involve the joining of multiple components, such as tubes, fins, or plates. The joining process must maintain the structural integrity and thermal performance of the materials. Advanced joining techniques such as laser welding, brazing, or diffusion bonding are commonly used to join complex geometries in heat exchangers. These techniques must be executed with precision to avoid leaks, maintain mechanical strength, and ensure optimal thermal contact between joined components. Poorly executed joints can lead to reduced mechanical durability or thermal inefficiencies. Advanced joining methods improve the performance and longevity of heat exchangers but increase manufacturing complexity and cost. The precision required in these processes may necessitate specialized equipment and skilled labor, raising production costs further.

3.7.2.5 Mass Production Limitations

Some heat transfer enhancement methods, particularly those involving intricate geometries or advanced materials, are not easily scalable for mass production. Processes such as additive manufacturing (3D printing) or micromilling, which are well-suited for creating complex features in small-batch or custom heat exchangers, are not easily adapted to large-scale production. Scaling these manufacturing processes can result in higher costs and longer production times. Custom heat exchangers with advanced features are more efficient but are limited in terms of large-scale production due to cost and complexity. These manufacturing limitations may restrict the widespread use of enhanced heat transfer designs in high-volume industries, where cost constraints are a major factor.

3.7.3 Manufacturing Constraints

In addition to material and manufacturing constraints, operational durability and ease of maintenance are critical factors to consider when implementing heat transfer enhancement techniques. Certain design enhancements may improve heat transfer but can also lead to increased fouling, higher maintenance requirements, and reduced operational life.

3.7.3.1 Fouling Susceptibility

Heat exchangers with small channels, roughened surfaces, or intricate fin structures are more prone to fouling from debris, scale, or biological growth. Fouling increases thermal resistance and reduces the efficiency of the heat exchanger. Fouling on enhanced surfaces can significantly reduce heat transfer efficiency over time by blocking fluid pathways and insulating heat transfer surfaces. Antifouling coatings or self-cleaning designs can help mitigate fouling, but they add to manufacturing complexity and cost. Enhanced surfaces improve initial heat transfer performance but can degrade quickly if fouling occurs. Designing heat exchangers to be easily cleanable or incorporating fouling-resistant materials can help mitigate this issue but may increase costs or reduce design flexibility.

3.7.3.2 Fouling Susceptibility

Heat exchangers are often exposed to significant temperature variations, which lead to thermal expansion and contraction of materials. Materials with different coefficients of thermal expansion (CTE) can experience stress at joints or between components. Materials must be carefully selected to handle thermal cycling without compromising the structural integrity of the heat exchanger. For example, mismatched CTEs between the tubes and the fins can cause mechanical stress, leading to cracking or joint failure over time. Using materials with compatible CTEs improves durability, but this may require selecting materials with lower thermal conductivity, reducing heat transfer efficiency. The designer must balance the need for thermal and mechanical durability with optimal heat transfer performance.

3.7.4 Cost and Economic Constraints

The overall cost of materials and manufacturing processes plays a significant role in the design and implementation of heat transfer enhancement techniques. Even the most efficient heat exchanger design must remain economically viable to be adopted widely in industry.

3.7.4.1 Cost of Advanced Materials

High-performance materials like titanium, nickel alloys, or specialized coatings are often significantly more expensive than conventional materials like aluminum or steel. While these materials provide excellent corrosion resistance, thermal performance, or durability, their high cost may be prohibitive for certain applications, particularly in cost-sensitive industries such as automotive or consumer HVAC systems. While advanced materials enhance performance, they raise the overall cost of the heat exchanger. This necessitates a cost-benefit analysis to determine whether the performance gains justify the additional material costs.

3.7.4.2 High-Cost Manufacturing Processes

Manufacturing processes that involve complex geometries, high-precision machining, or specialized surface treatments add significant cost to the production of heat exchangers. Processes like micromachining, 3D printing, or advanced coating applications are expensive and require sophisticated equipment and skilled labor, which drives up production costs. These processes may also increase production time, further contributing to higher costs. Although advanced manufacturing processes enable highly efficient heat exchanger designs, the increased costs may limit their use in certain markets or industries. Careful consideration of the market application, production volume, and lifecycle costs is essential to justify the higher manufacturing expenses.

In conclusion, material and manufacturing constraints play a critical role in the successful design and implementation of heat transfer enhancement techniques for heat exchangers. A careful balance must be struck between optimizing heat transfer performance, ensuring long-term durability, managing manufacturing complexity, and maintaining cost-effectiveness. Designers must consider specific application requirements, including operating environment, mechanical stresses, and fluid compatibility, to select appropriate materials and manufacturing processes that can deliver efficient, reliable, and economically viable heat exchanger solutions.

REFERENCES

1. K. Rathore, S. Kumar, Net zero emissions for our future generations through renewables: a brief review, *International Journal of Energy Resources Applications* 2 (2023): 39–56. https://doi.org/10.56896/IJERA.2023.2.1.004.
2. S. Kumar, K. Rathore, Renewable energy for sustainable development goal of clean and affordable energy, *International Journal of Materials Manufacturing and Sustainable Technologies* 2 (2023): 1–15. https://doi.org/10.56896/ijmmst.2023.2.1.001.
3. R.K. Shah, B. Thonon, D.M. Benforado, *Opportunities for Heat Exchanger Applications in Environmental Systems*, n.d. www.elsevier.com/locate/apthermeng.
4. 33880-final, (n.d.).
5. T. Alam, M.H. Kim, A comprehensive review on single phase heat transfer enhancement techniques in heat exchanger applications, *Renewable and Sustainable Energy Reviews* 81 (2018): 813–839. https://doi.org/10.1016/j.rser.2017.08.060.
6. S. Kumar, R.B. Grover, H. Yadav, P.K. Vijayan, U. Kannan, A. Agrawal, Experimental and numerical investigation on suppression of thermal stratification in a water-pool: PIV measurements and CFD simulations, *Applied Thermal Engineering* 138 (2018): 686–704. https://doi.org/10.1016/j.applthermaleng.2018.04.070.
7. S. Kumar, P.K. Vijayan, U. Kannan, M. Sharma, D.S. Pilkhwal, Experimental and computational simulation of thermal stratification in large pools with immersed condenser, *Applied Thermal Engineering* 113 (2017): 345–361. https://doi.org/10.1016/j.applthermaleng.2016.10.175.
8. S. Kumar, R.B. Grover, P.K. Vijayan, U. Kannan, Numerical investigation on the effect of shrouds around an immersed isolation condenser on the thermal stratification in large pools, *Annals of Nuclear Energy* 110 (2017): 109–125. https://doi.org/10.1016/j.anucene.2017.06.022.

9. S. Kumar, R.B. Grover, H. Yadav, P.K. Vijayan, U. Kannan, A. Agrawal, *Experimental Investigations on Thermal Stratification in a Large Pool of Water with Immersed Isolation Condenser*, 2020. http://asmedigitalcollection.asme.org/ICONE/proceedings-pdf/ICONE2020/83785/V003T13A033/6580030/v003t13a033-icone2020-16647.pdf.
10. S.A. Marzouk, M.M. Abou Al-Sood, E.M.S. El-Said, M.M. Younes, M.K. El-Fakharany, A comprehensive review of methods of heat transfer enhancement in shell and tube heat exchangers, *Journal of Thermal Analysis and Calorimetry* 148 (2023): 7539–7578. https://doi.org/10.1007/s10973-023-12265-3.
11. E. Tavousi, N. Perera, D. Flynn, R. Hasan, Heat transfer and fluid flow characteristics of the passive method in double tube heat exchangers: A critical review, *International Journal of Thermofluids* 17 (2023). https://doi.org/10.1016/j.ijft.2023.100282.
12. M.L.G. Ho, C.S. Oon, L.L. Tan, Y. Wang, Y.M. Hung, A review on nanofluids coupled with extended surfaces for heat transfer enhancement, *Results in Engineering* 17 (2023). https://doi.org/10.1016/j.rineng.2023.100957.
13. N. Nagarani, K. Mayilsamy, A. Murugesan, G.S. Kumar, Review of utilization of extended surfaces in heat transfer problems, *Renewable and Sustainable Energy Reviews* 29 (2014): 604–613. https://doi.org/10.1016/j.rser.2013.08.068.
14. O. Heriyani, M. Djaeni, Syaiful, A.K. Putri, Perforated concave rectangular winglet pair vortex generators enhance the heat transfer of air flowing through heated tubes inside a channel, *Results in Engineering* 16 (2022). https://doi.org/10.1016/j.rineng.2022.100705.
15. A. Ahmadi Nadooshan, R. Kalbasi, M. Afrand, Perforated fins effect on the heat transfer rate from a circular tube by using wind tunnel: An experimental view, *Heat and Mass Transfer/Waerme- Und Stoffuebertragung* 54 (2018): 3047–3057. https://doi.org/10.1007/s00231-018-2333-3.
16. M. Ishaq, A. Ali, M. Amjad, K.S. Syed, Z. Iqbal, Diamond-shaped extended fins for heat transfer enhancement in a double-pipe heat exchanger: an innovative design, *Applied Sciences (Switzerland)* 11 (2021). https://doi.org/10.3390/app11135954.
17. E.A.M. Elshafei, M.M. Awad, E. El-Negiry, A.G. Ali, Heat transfer and pressure drop in corrugated channels, *Energy* 35 (2010): 101–110. https://doi.org/10.1016/j.energy.2009.08.031.
18. P. Forooghi, K. Hooman, Effect of buoyancy on turbulent convection heat transfer in corrugated channels - A numerical study, *International Journal of Heat and Mass Transfer* 64 (2013): 850–862. https://doi.org/10.1016/j.ijheatmasstransfer.2013.05.028.
19. Y. Sui, C.J. Teo, P.S. Lee, Direct numerical simulation of fluid flow and heat transfer in periodic wavy channels with rectangular cross-sections, *International Journal of Heat and Mass Transfer* 55 (2012): 73–88. https://doi.org/10.1016/j.ijheatmasstransfer.2011.08.041.
20. N. Paisarn, *Study on the Heat-Transfer Characteristics and Pressure Drop in Channels with Arc-Shaped Wavy Plates*, n.d.
21. E. Esmaeilzadeh, H. Almohammadi, A. Nokhosteen, A. Motezaker, A.N. Omrani, Study on heat transfer and friction factor characteristics of γ-Al2O3/water through circular tube with twisted tape inserts with different thicknesses, *International Journal of Thermal Sciences* 82 (2014): 72–83. https://doi.org/10.1016/j.ijthermalsci.2014.03.005.
22. S. Bhattacharyya, A.K. Sharma, D.K. Vishwakarma, K. Saini, A.R. Paul, Z. Huan, Thermo-hydraulic performance of magnetic baffles for cooling using magnetic nanofluid in a mini channel, *Sustainable Energy Technologies and Assessments* 57 (2023). https://doi.org/10.1016/j.seta.2023.103194.
23. C. Thianpong, P. Eiamsa-ard, K. Wongcharee, S. Eiamsa-ard, Compound heat transfer enhancement of a dimpled tube with a twisted tape swirl generator, *International Communications in Heat and Mass Transfer* 36 (2009): 698–704. https://doi.org/10.1016/j.icheatmasstransfer.2009.03.026.
24. S.N. Sarada, A. V Sita, R. Raju, K. Kalyani Radha, L. Shyam Sunder, *Enhancement of Heat Transfer Using Varying Width Twisted Tape Inserts*, 2010. www.ijest-ng.com.
25. M. Akbarzadeh, S. Rashidi, J.A. Esfahani, Influences of corrugation profiles on entropy generation, heat transfer, pressure drop, and performance in a wavy channel, *Applied Thermal Engineering* 116 (2017): 278–291. https://doi.org/10.1016/j.applthermaleng.2017.01.076.
26. W. Jedsadaratanachai, N. Jayranaiwachira, P. Promvonge, 3D numerical study on flow structure and heat transfer in a circular tube with V-baffles, *Chinese Journal of Chemical Engineering* 23 (2015): 342–349. https://doi.org/10.1016/j.cjche.2014.11.006.
27. S. Wang, J. Wen, H. Yang, Y. Xue, H. Tuo, Experimental investigation on heat transfer enhancement of a heat exchanger with helical baffles through blockage of triangle leakage zones, *Applied Thermal Engineering* 67 (2014): 122–130. https://doi.org/10.1016/j.applthermaleng.2014.03.017.

28. G.Y. Zhou, J. Xiao, L. Zhu, J. Wang, S.T. Tu, A numerical study on the shell-side turbulent heat transfer enhancement of shell-and-tube heat exchanger with trefoil-hole baffles, in: *Energy Procedia*, Elsevier Ltd, 2015, 3174–3179. https://doi.org/10.1016/j.egypro.2015.07.656.
29. S. Thapa, S. Samir, K. Kumar, S. Singh, A review study on the active methods of heat transfer enhancement in heat exchangers using electroactive and magnetic materials, in: *Materials Today: Proceedings*, Elsevier Ltd, 2021, 4942–4947. https://doi.org/10.1016/j.matpr.2021.01.382.
30. H. Li, R. You, H. Deng, Z. Tao, J. Zhu, Heat transfer investigation in a rotating U-turn smooth channel with irregular cross-section, *International Journal of Heat and Mass Transfer* 96 (2016): 267–277. https://doi.org/10.1016/j.ijheatmasstransfer.2015.12.071.
31. L. Qiu, H. Deng, J. Sun, Z. Tao, S. Tian, Pressure drop and heat transfer in rotating smooth square U-duct under high rotation numbers, *International Journal of Heat and Mass Transfer* 66 (2013): 543–552. https://doi.org/10.1016/j.ijheatmasstransfer.2013.07.055.
32. W.D. Morris, K. Farhadi Rahmat-Abadis, Convective heat transfer in rotating ribbed tubes, *International Journal of Heat and Mass Transfer* 39 (1996): 2253–2266.
33. L.M. Wright, W.-L. Fu, J.-C. Han, *Thermal Performance of Angled, V-Shaped, and W-Shaped Rib Turbulators In Rotating Rectangular Cooling Channels (AR = 4:1)*, n.d. http://www.asme.org/about-asme/terms-of-use.
34. A.M. Mohammed, S. Kapan, M. Sen, N. Celik, Effect of vibration on heat transfer and pressure drop in a heat exchanger with turbulator, *Case Studies in Thermal Engineering* 28 (2021). https://doi.org/10.1016/j.csite.2021.101680.
35. L. Cheng, T. Luan, W. Du, M. Xu, Heat transfer enhancement by flow-induced vibration in heat exchangers, *International Journal of Heat and Mass Transfer* 52 (2009): 1053–1057. https://doi.org/10.1016/j.ijheatmasstransfer.2008.05.037.
36. A. Hosseinian, A.H. Meghdadi Isfahani, Experimental study of heat transfer enhancement due to the surface vibrations in a flexible double pipe heat exchanger, *Heat and Mass Transfer/Waerme- Und Stoffuebertragung* 54 (2018): 1113–1120. https://doi.org/10.1007/s00231-017-2213-2.
37. M.J. Pettigrew, C.E. Taylor, Vibration analysis of shell-and-tube heat exchangers: an overview - Part 1: Flow, damping, fluidelastic instability, *Journal of Fluids and Structures* 18 (2003): 469–483. https://doi.org/10.1016/j.jfluidstructs.2003.08.007.
38. S. Laohalertdecha, P. Naphon, S. Wongwises, A review of electrohydrodynamic enhancement of heat transfer, *Renewable and Sustainable Energy Reviews* 11 (2007): 858–876. https://doi.org/10.1016/j.rser.2005.07.002.
39. C.W. Lin, J.Y. Jang, 3D Numerical heat transfer and fluid flow analysis in plate-fin and tube heat exchangers with electrohydrodynamic enhancement, *Heat and Mass Transfer/Waerme- Und Stoffuebertragung* 41 (2005): 583–593. https://doi.org/10.1007/s00231-004-0540-6.
40. R.B. C., P. Kumar, S. Roy, R. Ganesan, A comprehensive review on compound heat transfer enhancement using passive techniques in a heat exchanger, *Materials Today: Proceedings* 54 (2022): 428–436. https://doi.org/10.1016/j.matpr.2021.09.541.
41. P. Promvonge, S. Eiamsa-ard, Heat transfer augmentation in a circular tube using V-nozzle turbulator inserts and snail entry, *Experimental Thermal and Fluid Science* 32 (2007): 332–340. https://doi.org/10.1016/j.expthermflusci.2007.04.010.
42. S.K. Saha, Thermohydraulics of turbulent flow through rectangular and square ducts with axial corrugation roughness and twisted-tapes with and without oblique teeth, *Experimental Thermal and Fluid Science* 34 (2010): 744–752. https://doi.org/10.1016/j.expthermflusci.2010.01.003.
43. S. Bhattacharyya, The effects of short length and full length swirl generators on heat transfer and flow fields in a solar air heater tube, *Journal of Thermal Analysis and Calorimetry* 140 (2020): 1355–1369. https://doi.org/10.1007/s10973-019-08764-x.
44. K.K.I. Al-Chlaihawi, M.R. Hasan, A.L. Ekaid, Compound passive heat transfer augmentation techniques: A comprehensive review, *Heat Transfer* 53 (2024): 363–421. https://doi.org/10.1002/htj.22955.
45. A.J. Chamkha, A.M. Aly, MHD free convection flow of a nanofluid past a vertical plate in the presence of heat generation or absorption effects, *Chemical Engineering Communications* 198 (2011): 425–441. https://doi.org/10.1080/00986445.2010.520232.
46. M. Shakouri Pour, E. Esmaeilzadeh, Experimental investigation of convective heat transfer enhancement from 3D-shape heat sources by EHD actuator in duct flow, *Experimental Thermal and Fluid Science* 35 (2011): 1383–1391. https://doi.org/10.1016/j.expthermflusci.2011.05.006.
47. S. Bhattacharyya, A.K. Sharma, D.K. Vishwakarma, A.R. Paul, Thermo-hydraulic performance of magnetic baffles for removal of concentrated heat fluxes in a heated mini channel, *Applied Thermal Engineering* 216 (2022). https://doi.org/10.1016/j.applthermaleng.2022.118992.

48. S. Agrawal, S. Pandya, P. Jangir, K. Kalita, S. Chakraborty, A multi-objective thermal exchange optimization model for solving optimal power flow problems in hybrid power systems, *Decision Analytics Journal* 8 (2023). https://doi.org/10.1016/j.dajour.2023.100299.
49. A. Patel, *Optimizing the Efficiency of Solar Heater and Heat Exchanger Integration in Hybrid System*, 2023. https://www.researchgate.net/publication/373458388.
50. G. Starace, M. Fiorentino, M.P. Longo, E. Carluccio, A hybrid method for the cross flow compact heat exchangers design, *Applied Thermal Engineering* 111 (2017): 1129–1142. https://doi.org/10.1016/j.applthermaleng.2016.10.018.
51. H. Al-Bakhit Research Engineer, A. Fakheri Professor, *A Hybrid Approach for Full Numerical Simulation of Heat Exchangers*, n.d. http://www.asme.org/about-asme/terms-of-use.
52. M. Darbandi, M.S. Abdollahpour, An innovative cost-effective numerical-analytical procedure to derive heat correlations for shell and tube heat exchangers equipped with smooth and dimple tubes, *Numerical Heat Transfer, Part A: Applications* (2024). https://doi.org/10.1080/10407782.2023.2301521.
53. P. Prajapati, B.D. Raja, H. Savaliya, V. Patel, H. Jouhara, Thermodynamic evaluation of shell and tube heat exchanger through advanced exergy analysis, *Energy* 292 (2024). https://doi.org/10.1016/j.energy.2024.130421.
54. I.A. Fetuga, O.T. Olakoyejo, S.M. Abolarin, J.K. Gbegudu, A. Onwuegbusi, A.O. Adelaja, Numerical analysis of thermal performance of waste heat recovery shell and tube heat exchangers on counter-flow with different tube configurations, *Alexandria Engineering Journal* 64 (2023): 859–875. https://doi.org/10.1016/j.aej.2022.09.017.
55. A.R. Al-darraji, S.A. Marzouk, A. Aljabr, F.A. Almehmadi, S. Alqaed, A. Kaood, Enhancement of heat transfer in a vertical shell and tube heat exchanger using air injection and new baffles: experimental and numerical approach, *Applied Thermal Engineering* 236 (2024). https://doi.org/10.1016/j.applthermaleng.2023.121493.
56. P.H. Jadhav, N. Gnanasekaran, M. Mobedi, Analysis of functionally graded metal foams for the accomplishment of heat transfer enhancement under partially filled condition in a heat exchanger, *Energy* 263 (2023). https://doi.org/10.1016/j.energy.2022.125691.
57. M. Ryms, K. Tesch, W.M. Lewandowski, Optimization of the distance between the vertical plates in the convective air heat exchanger, *International Journal of Thermal Sciences* 185 (2023). https://doi.org/10.1016/j.ijthermalsci.2022.108064.
58. Q. Jin, Y. Yu, Y. Xia, Experimental study on heat transfer characteristics of spray cooling system with plate-fin heat exchanger, *Experimental Heat Transfer* 37 (2024): 767–779. https://doi.org/10.1080/08916152.2023.2204314.
59. H.S. Dizaji, S. Pourhedayat, H. Moria, S. Alqahtani, S. Alshehery, A.E. Anqi, Performance boost of a commercial air-to-air plate heat recovery unit by mesh-net insert; thermal-frictional, economic, and effectiveness-NTU analysis, *Energy* 290 (2024). https://doi.org/10.1016/j.energy.2023.130037.
60. A.W. Al-Fatlawi, H. Niazmand, Thermal analysis of hybrid nanofluids inside a microchannel heat exchanger for electronic cooling, *Journal of Thermal Analysis and Calorimetry* 149 (2024): 4119–4131. https://doi.org/10.1007/s10973-024-12991-2.
61. L. Du, W. Hu, An overview of heat transfer enhancement methods in microchannel heat sinks, *Chemical Engineering Science* 280 (2023). https://doi.org/10.1016/j.ces.2023.119081.
62. K.M. Kuruva, K.V.K. Reddy, I.A. Badruddin, A.S. Shaik, M.A. Ali, Numerical study of microchamber and cylindrical inserts on enhancement of heat transfer in microchannel heat exchanger, *Journal of Thermal Analysis and Calorimetry* 149 (2024): 2385–2401. https://doi.org/10.1007/s10973-023-12809-7.
63. A. Dwivedi, M.M. Khan, H.S. Pali, A comprehensive review of thermal enhancement techniques in microchannel heat exchangers and heat sinks, *Journal of Thermal Analysis and Calorimetry* 148 (2023): 13189–13231. https://doi.org/10.1007/s10973-023-12451-3.
64. D.K. Vishwakarma, S. Bhattacharyya, M.K. Soni, V. Goel, J.P. Meyer, Evaluating the heat transfer and pressure drop in the transitional flow regime for a horizontal circular tube fitted with wavy-tape inserts, *International Journal of Thermal Sciences* 196 (2024). https://doi.org/10.1016/j.ijthermalsci.2023.108677.
65. S. Bhattacharyya, D.K. Vishwakarma, M.K. Soni, The influence of a novel ribbed wave tape on the enhancement of heat transfer in a solar air heater operating in a transitional flow regime: an experimental study, *Experimental Heat Transfer* (2023). https://doi.org/10.1080/08916152.2023.2265208.
66. S. Bhattacharyya, M. Pathak, M. Sharifpur, S. Chamoli, D.R.E. Ewim, Heat transfer and exergy analysis of solar air heater tube with helical corrugation and perforated circular disc inserts, *Journal of Thermal Analysis and Calorimetry* 145 (2021): 1019–1034. https://doi.org/10.1007/s10973-020-10215-x.

67. M. Everts, J.P. Meyer, Heat transfer of developing and fully developed flow in smooth horizontal tubes in the transitional flow regime, *International Journal of Heat and Mass Transfer* 117 (2018): 1331–1351. https://doi.org/10.1016/j.ijheatmasstransfer.2017.10.071.
68. S.M. Abolarin, M. Everts, J.P. Meyer, Heat transfer and pressure drop characteristics of alternating clockwise and counter clockwise twisted tape inserts in the transitional flow regime, *International Journal of Heat and Mass Transfer* 133 (2019): 203–217. https://doi.org/10.1016/j.ijheatmasstransfer.2018.12.107.
69. S. Bhattacharyya, A.C. Benim, R. Bennacer, K. Dey, Influence of broken twisted tape on heat transfer performance in novel axial corrugated tubes: experimental and numerical study, *Heat Transfer Engineering* 43 (2021): 437–462. https://doi.org/10.1080/01457632.2021.1875168.
70. S. Bhattacharyya, D. Sarkar, R. Roy, S. Chakraborty, V. Goel, E. Almatrafi, Application of new artificial neural network to predict heat transfer and thermal performance of a solar air-heater tube, *Sustainability (Switzerland)* 13 (2021). https://doi.org/10.3390/su13137477.
71. M. Everts, J.P. Meyer, Relationship between pressure drop and heat transfer of developing and fully developed flow in smooth horizontal circular tubes in the laminar, transitional, quasi-turbulent and turbulent flow regimes, *International Journal of Heat and Mass Transfer* 117 (2018): 1231–1250. https://doi.org/10.1016/j.ijheatmasstransfer.2017.10.072.
72. S. Kumar, A. Thyagarajan, D. Banerjee, *Experimental Investigation of Thermal Energy Storage (TES) Platform Leveraging Phase Change Materials in a Chevron Plate Heat Exchanger*, n.d. http://asmedigitalcollection.asme.org/IMECE/proceedings-pdf/IMECE2022/86700/V008T11A057/6982377/v008t11a057-imece2022-96226.pdf.
73. S. Kumar, A. Thyagarajan, D. Banerjee, *An Investigation on Chevron Plate Heat Exchanger Filled with Organic Phase Change Material*, 2020. http://asmedigitalcollection.asme.org/HT/proceedings-pdf/HT2020/83709/V001T11A016/6568988/v001t11a016-ht2020-9122.pdf.
74. S. Kumar, D. Banerjee, A review on phase change materials for sustainability applications by leveraging machine learning, in: *Energy Consumption, Conversion, Storage, and Efficiency*, IntechOpen, 2024. https://doi.org/10.5772/intechopen.114380.

4 Improved Heat Transfer Using Nanoparticles, Coatings, and Surface Engineering

Prathamesh Sapale, Niloofar Fatemipayam, Md Sazzadul Islam Sajol, Ravi Teja A. Tirumala, Amit Choudhari, and Ashish Kumar Gupta

4.1 INTRODUCTION

Nanofluids significantly enhance heat transfer due to their superior thermal properties, making them essential for various industrial applications (1, 2). Figure 4.1 illustrates different ways to improve heat transfer efficiency. These methods are categorized into two main categories: passive and active. The passive approach modifies the system's design without external energy input. On the other hand, active techniques require external energy sources to boost heat transfer. Improving thermal conductivity boosts system performance, energy efficiency, and safety, which helps reduce energy consumption and prevent overheating in electronic devices and machinery. Effective heat transfer is crucial not only for product quality and reducing operating costs but also for ensuring reliability and extending the lifespan of critical components. Additionally, nanofluids' advanced heat transfer techniques contribute to sustainability by lowering greenhouse gas emissions and reducing the carbon footprint (3–6). Thus, nanofluids offer a sustainable solution for optimal performance, safety, and environmental impact. This chapter focuses on passive methods, such as nanofluids, nanoparticle/colloidal paints, coatings, and surface treatments. Based on specific system requirements, we will focus on the mechanisms involved in significantly improving thermal management and performance.

4.2 NANOFLUIDS

Nanofluids are advanced heat transfer solutions with nanometer-sized particles (1–100 nm), like SiO_2, TiO_2, and CuO, dispersed in base fluids such as water oil or ethylene glycol (1, 2, 7, 8). These systems offer enhanced thermal properties and increased heat transfer efficiency, benefiting various industries, i.e., transportation to electronics (8, 9). Their high surface area and effective dispersion due to Brownian motion ensure thermal stability and outperform conventional fluids. Nanofluids minimize clogging, reduce pumping power, and improve heat management with superior thermal conductivity and specific heat transfer.

4.3 Nanofluid Synthesis Techniques

Nanofluids can improve thermal and other pertinent properties to facilitate efficient heat transfer. However, synthesizing nanofluids is crucial, with agglomeration being a primary challenge. These can be synthesized via single-stage and two-stage synthesis processes (10). In the single-stage method, nanofluid synthesis and nanoparticle dispersion will be done in one step (11). Single-stage

 DOI: 10.1201/9781003509837-4

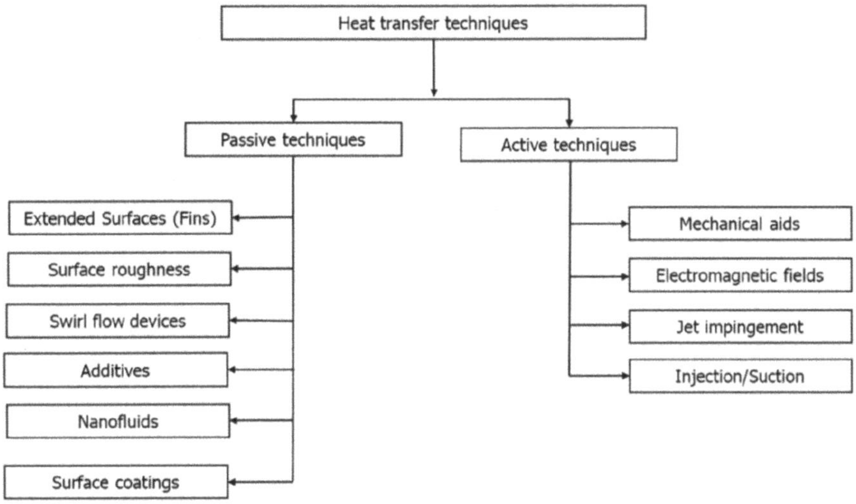

FIGURE 4.1 Schematic representation of heat transfer techniques.

methods include physical vapor deposition (12, 13), chemical vapor deposition (14, 15), laser ablation (16, 17), and submerged arc (18). The two-stage method of nanofluid preparation comprises two distinct stages for dispersing nanoparticles in a base fluid (8). During the initial dispersion phase, nanoparticles undergo synthesis via chemical precipitation, sol-gel synthesis, or physical vapor deposition and are then dispersed in the base fluid. Surfactants, homogenization, ultrasonication, and stirring play vital roles in achieving a uniform distribution of nanoparticles throughout the base fluid (19). The base fluid is initially separated to meet the specified desired properties and purity standards. Depending on the specific requirements of nanofluids, this may involve using deionized water, specific types of oils, or ethylene glycol. In the second stage, the prepared base fluid is blended with nanoparticles, often incorporating an appropriate surfactant or stabilizer. Techniques such as ultrasonication or mechanical stirring ensure uniform dispersion (20). Various two-stage methods exist for preparing nanofluids: sol-gel method (15, 21–23), ball milling (24–26), and ultrasonic agitation (27, 28).

The single-stage method is an integrated approach where nanoparticles are synthesized and dispersed simultaneously, leading to improved dispersion. This process also enhances the stability of nanofluids by minimizing aggregation since the nanoparticles are created directly in the base fluid. Additionally, it allows the production of nanoparticles with specific properties, such as size and shape, tailored to the desired applications. This method reduces the risk of contamination by eliminating the need to handle nanoparticles separately. Moreover, it is cost-effective as it eliminates the need for additional dispersing agents and complex post-processing steps. However, the process is limited in terms of scalability. Furthermore, it offers limited flexibility in choosing the appropriate base fluid and nanoparticles (10).

On the other hand, the two-stage method is more suitable for scalability, making it ideal for industrial applications. It allows for better control over nanoparticle shape, size, and distribution within the base fluid and offers greater flexibility in selecting nanoparticles tailored to specific applications. A vital advantage of this method is its compatibility with a wide range of base fluids, making it more versatile than single-stage methods. However, it also has drawbacks, such as the tendency for nanoparticles to agglomerate, particularly during nanofluid storage. The process is more complex due to the need for additional steps like storage, drying, and redispersion, which also increases costs. Moreover, separate dispersing agents may be required to stabilize nanoparticles in the base fluid, potentially affecting rheology and thermal properties. Another significant disadvantage is the risk of contamination, as handling and processing nanoparticles separately can lead to quality issues (19).

4.3.1 Effect of Physical Properties for Thermal Conductivity Enhancement

Different nanoparticles play an important role in increasing a nanofluid's thermal conductivity, which depends on certain factors, including nanoparticle material, size, shape, concentration, base fluid properties, temperature influences, and stability. Consequently, these aspects directly affect the efficiency and suitability of nanofluids in various thermal systems. In the following section, each factor's influence will be discussed.

The nanoparticles used in the nanofluid have an essential role in determining the nanofluid's thermal conductivity. Metal-based nanoparticles such as copper or silver tend to show better heat conductivities than other metals or alloys since they are known for their high heat transfer properties. Silica and alumina are among some metal oxides that exhibit good enhancement, although not as high as carbon-based materials (29). The shape of the nanoparticles also changes thermal conductivity since it has been shown that there is a clear correlation between temperature and the shape factor. A recent study argues that platelet-shaped nanoparticles have the highest retention of heat in comparison with cylinders and block nanoparticles, showing a strong association between nanoparticle morphology and enhanced thermal conductivity (30). It has also been shown that some shapes, like nanorods or nanosheets, are much more effective at increasing the heat transfer rate (31 –33) than spherical particles due to higher surface area, effective orientation within the fluid, and lower aggregation (34). The nanoparticle size is also an important aspect of their thermal performance. Small-sized nanoparticles provide a larger surface area-to-volume ratio, leading to better fluid interaction and improved heat transfer characteristics. Also, smaller nanoparticles can increase Brownian motion, raising their thermal conductivity. Therefore, optimal nanoparticle sizes range from a few nanometers to tens of nanometers to balance surface area and stability (34).

Another influential factor is the base fluid's concentration, i.e., nanoparticle volume fraction. A higher concentration of particles makes nanofluid more conductive. That is because increasing particle numbers leads to more pathways for heat conduction. However, thermal conductivity enhancement diminishes or drops after reaching a certain concentration level. This decline results from the aggregation of particles, which leads to a decrease in effective surface area for heat transfer and increases viscosity due to particle interaction, thus lowering thermal performance. The properties of the base fluid, including the viscosity and internal thermal conductivity changes, determine how well the nanofluid will work, owing to some specific fluid features. When fluids have lower viscosities, there is maximum heat transfer since they allow the free movement of nanoparticles, thereby enhancing conduction and convection within such liquid substances. Temperature has complex implications on thermal conductivity as well. Brownian motion usually increases at higher temperatures, leading to a rise in thermal conductivity (1).

4.3.2 Brownian Motion and Thermophoresis for Thermal Efficiency Optimization

Brownian motion and thermophoresis have a significant impact on the nanofluid's characteristics. The former is the term for the unpredictable movement of nanoparticles in the base fluid following collisions with the fluid molecules. This form of motion results in enhanced thermal conductivity by promoting energy transfer through increased particle interactions. It has been shown that the heat transfer rate increases as particles move more randomly. This effect becomes more pronounced when nanoparticles are smaller. However, excessive Brownian motion may lead to instability and sedimentation issues at tiny particle sizes (35). Thermophoresis, conversely, refers to a nanoparticle's movement in a fluid due to temperature gradients. This phenomenon has been considered positive when particle movement is from a hot to a cold region and negative when the reverse happens. The effect of thermophoresis is different from Brownian motion because the particle's movement coordinates with the direction of the temperature gradient. Also, contrary to Brownian motion, the relation of thermophoresis and thermal conductivity is independent of particle size. Therefore, the thermal characteristics of nanofluids are more influenced by nanoparticles' Brownian motion than thermophoresis (36).

4.3.3 Key Characteristics of Nanofluids

Incorporating nanoparticles into base fluids significantly improves their heat transfer effectiveness, surpassing both colloidal solutions and the base fluids alone. Nanosized particles boost thermophysical properties, facilitating the efficient transfer of heat. This improved capability can be crucial for industrial applications like heating and cooling systems and automotive engines, where effective thermal management is vital for optimal performance. This section will also cover nanofluid density, viscosity, convective heat transfer, and rheological properties.

4.3.3.1 Thermal Conductivity

Thermal conductivity refers to the capacity of a given material to transfer heat from one point to another. The clustering or networking of nanoparticles aids in efficient heat transfer, and the liquid layering around the nanoparticles can further modify the thermal charge (37). The Maxwell model, developed by James Clerk Maxwell, is one of the earliest models. It describes how poorly dispersed nanoparticles can alter conductivity. However, its effectiveness relies on each component's thermal conductivity and volumetric ratios. However, the model underestimates nanofluids' evolution and is based on several assumptions: negligible particle-particle interactions, spherical particle shape, homogeneous dispersion, and zero interfacial thermal resistance.

4.3.3.2 Specific Heat Capacity

Specific heat refers to the energy needed to raise the temperature of a substance's unit mass or volume by one degree Celsius or Kelvin (38). Nanofluids exhibit non-linear behavior at higher concentrations due to interactions between particles and the base fluid, influencing specific heat transfer. Equation 4.1 represents a mathematical model correlating specific heat transfer based on mixing theory.

$$\mathrm{Cp}_{,nf} = \left(1-\Phi\right)\mathrm{Cp}_{,bf} + \Phi \mathrm{Cp}_{np} \tag{4.1}$$

$$\rho_{nf}\mathrm{Cp}_{,nf} = \left(1-\Phi\right)\rho_{bf}\mathrm{Cp}_{,bf} + \Phi \rho_{np}\mathrm{Cp}_{np} \tag{4.2}$$

Equation 4.2 assumes that nanoparticles and the base fluid are in thermal equilibrium, incorporating classical and statistical principles. Here, Cp_{np}, Cp_{bf}, and Cp_{nf} represent the specific heat capacities of nanoparticles, base fluid, and nanofluids. Similarly, ρ_{np}, ρ_{bf}, and ρ_{nf} denote the densities of nanoparticles, base fluid, and nanofluid. Additionally, Φ indicates the nanoparticle concentration as a volume percentage. The temperature range for model 1 ranges from 25–65°C, and model 2 ranges from 25–40°C (39). Specific heat varies with temperature and particle concentration. It increases at elevated temperatures but decreases with higher particle concentrations. As a result, the base fluid has a higher specific heat than nanofluids.

4.3.3.3 Convective Heat Transfer

Nanofluids have been introduced as potential media for enhancing convectional heat transfer. It has been shown that nanofluids generally possess higher convective heat transfer coefficients under turbulent flow conditions than their base fluids (1). The improvement depends on nanoparticle types, size, concentration, etc. In addition, some vital parameters change different aspects of the "heat transfer" processes. For example, higher values of thermal conductivity and improved convective abilities improve the Mouromtseff number. Moreover, nanoparticles influence the Prandtl and Peclet numbers, signifying a more efficient heat transfer process.

Density is a fundamental property of nanofluids. It is defined as the mass per unit volume of a substance, typically measured in kilograms per cubic meter (kg/m³). Including nanoparticles in the base fluid alters the overall density. To calculate the density of a nanofluid ρ_{nf}, you take the weighted average of the densities of the base fluid ρ_{bf}, and the nanoparticles ρ_{np}, Shown in Equation 4.3. In this equation, Φ represents the nanoparticle volume fraction, typically less than 5% (40).

$$\rho_{nf} = (1-\Phi)\rho_{bf} + \Phi\rho_{np} \tag{4.3}$$

As discussed earlier, density significantly influences a nanofluid's thermal conductivity, heat capacity, and thermal diffusivity, which are crucial for heat transfer applications. As per Equation 4.4, the thermal diffusivity, α is the ratio of thermal conductivity, k, to the product of density, ρ, and specific heat capacity, C_p. Nanoparticle concentration and the uneven distribution led to variations in density and impact thermal and fluid dynamics properties (41–42). Choosing the correct nanoparticles and base fluid is crucial for applications; if the nanoparticles are denser than the base fluid, there will be sediment-to-stability concerns.

$$\alpha = \frac{k}{\rho C_p} \tag{4.4}$$

The stability of nanofluids is essential for effectively implementing heat transfer applications. Stable nanofluids maintain a uniform dispersion of nanoparticles within base fluids, ensuring consistent thermal conductivity and specific heat capacity, both essential for effective heat absorption and transfer. Furthermore, well-dispersed nanoparticles prevent clogging in flow channels and deposition on the heat exchanger surfaces (fouling effect), thereby ensuring smooth operational conditions (42). This also enhances convective heat transfer and promotes effective Brownian motion. The stability of nanofluids relies on the characteristics of the nanoparticles and the base fluids, as well as the ability of the particles to remain stable within the base fluid. Depending on the forces in the base fluids, these particles can stay suspended for weeks or even months. Several methods are available to assess the stability of nanofluids, including dynamic light scattering (DLS), transmission electron microscopy (TEM), scanning electron microscopy (SEM), zeta potential analysis, the sedimentation method, and UV-Vis spectrophotometry. Various factors influence the stability of nanoparticles, including their shape and size; smaller particles exhibit better stability because of their lower sedimentation rate and reduced aggregation tendency. The pH and ionic strength help maintain electrostatic repulsion via surface charge. In this section, we discuss DLS, morphological analysis, measurement of zeta potential, the sedimentation balance method, and UV-Vis spectrophotometry.

4.3.4 Practical Applications

These novel fluids produce much enhanced convective heat transfer characteristics manifested by considerable performance enhancements in heat exchangers and cooling systems. These can result in smaller, more efficient heat exchangers and better cooling in electronics and automobile industries to prevent overheating and increase system reliability. Nanofluids have been extensively used in heat transfer devices such as heat exchangers, chillers, and solar panels, which aim to improve heat transport efficiency and optimize thermal systems (43).

In comparison with traditional fluids, nanofluids display high and temperature-dependent thermal conductivity. The quality can, therefore, be the reason for better heat transfer performance in diverse applications. There is a potential for substituting conventional coolants with nanofluids applicable in various areas, including engine cooling, solar water heating, cooling of electronics,

TABLE 4.1
Diverse Applications of Nanofluids Across Different Areas

Application Area	Specific Use	Nanofluid Composition	Performance Improvement	Challenges
Renewable Energy	Solar water heating	Water-TiO_2	Increased thermal efficiency by 6.8%	Stability of nanofluid, cost
HVAC	Air conditioning systems	Water-CuO	Increased heat transfer coefficient by 23%	Potential clogging, stability
Industrial Cooling	Heat exchangers (45, 46)	Ethylene glycol-Al_2O_3	Improved cooling efficiency by 15%	Long-term stability, dispersion
Automotive	Engine coolant	Water-alumina	Increased thermal conductivity by 20%	Agglomeration, cost considerations
Electronics	Cooling electronic devices	Water-copper	Enhanced heat dissipation, lower temperatures	Nanoparticle stability, cost
Geothermal energy	Drilling fluids	Water-alumina	Enhanced cooling, reduced equipment wear	Cost, scaling issues
Power generation	Coolant in power plants	Water-Cu	Improved heat transfer efficiency	Stability, cost

and improving the efficiency of diesel generators. Nanofluids are practical materials for use as coolant systems' additives mainly because this group has a broad range of industries, including materials, chemicals, food and beverages, oil and gas extraction, paper, printing, and textiles. As to these benefits, they may be used in combined heat power stations or any other air conditioning systems where they can contribute to efficient energy savings alongside environmental advantages like low CO_2 emissions (44). Table 4.1 shows various applications, performance enhancements, and other nanofluid-related issues in different fields.

4.3.5 Critical Environmental Risks and Mitigation Strategies

The manufacturing and disposal of nanofluids pose significant environmental challenges, with nanoparticles potentially disrupting ecosystems and posing health risks. Issues such as nanoparticle aggregation, oxide formation in metallic particles, and inherent toxicity underscore the urgency of addressing these environmental and health hazards. If released, nanoparticles' large surface area-to-volume ratio causes agglomeration during storage and environmental pollution. Oxidized nanoparticles can penetrate biological membranes, interacting with cells and potentially causing toxic effects, raising concerns about long-term exposure for humans and wildlife to industrial nanoparticle emissions (47). When these environmental issues are addressed, nanofluids in energy and heat transfer applications will become more sustainable. These can be addressed by implementing rigorous filtration systems to capture nanoparticles before discharge, developing closed-loop manufacturing processes (40, 48–50) to prevent environmental release, conduct comprehensive environmental impact assessments to monitor nanoparticle dispersion and effects on ecosystems, deploy sensors, encapsulate nanoparticles, and develop reclamation strategies.

4.4 COLLOIDAL PAINTS

Heat transfer efficiency is crucial in various systems, from industrial processes to consumer electronics. Advanced techniques using nanoparticle-infused colloidal paints and innovative surface treatments have demonstrated significant potential in enhancing heat transfer capabilities. Incorporating materials such as metal nanoparticles (e.g., Ag, Cu) and metal oxides like Cu_2O, Al_2O_3, Fe_2O_3, and TiO_2 into colloidal paints significantly boosts thermal conductivity, resulting in improved heat dissipation and performance in thermal management applications (51–53). Cu_2O nanoparticles, in particular, stand out for their excellent scattering and charge carrier recombination properties and versatility in applications ranging from photocatalysis to advanced thermal coatings (54–58). Nanofluids, suspensions of nanoparticles in base fluids, have been extensively researched for their potential to enhance thermal conductivity and heat transfer properties. However, colloidal solutions offer distinct advantages over typical nanofluids in heat transfer applications, such as improved stability and superior thermal properties. This section delves into these benefits, emphasizing these advanced colloidal solutions' stability, thermal enhancements, and practical applications.

4.4.1 Strategies for Enhancing Stability

One of the significant challenges with nanofluids is maintaining stability over time. Nanoparticles tend to agglomerate due to van der Waals forces, leading to sedimentation and uneven distribution within the fluid, thereby compromising heat transfer efficiency (59). In colloidal solutions, nanoparticles exhibit Brownian motion, characterized by random zigzag movements due to collisions with the dispersion medium's molecules (60). The stability of colloidal solutions, including nanofluids, is crucial for their applications and is well explained by the Deryaguin–Landau–Verwey–Overbeek (DLVO) theory (61). The DLVO theory, crucial for understanding colloidal solution stability, balances van der Waals attraction and electrostatic repulsion. This interplay dictates whether particles stay dispersed or aggregated, which is essential for maintaining consistency and functionality in various applications. Colloidal solutions employ organic polymer dispersants, surfactants, surface-modified nanoparticles, and other mechanical methods to mitigate agglomeration and ensure stability. These additives prevent particle aggregation and sedimentation, maintaining a stable dispersion crucial for sustained and efficient heat transfer performance. This stability is essential for maximizing the enhancements in thermal conductivity observed in colloidal nanofluid applications, driven by mechanisms such as Brownian motion and the formation of conductive and convective pathways through nanoparticle interactions (34, 62).

4.4.2 Heat Transfer Efficiency: The Interplay of Size, Concentration, and Key Mechanisms

Several vital mechanisms primarily explain improvements in thermal conductivity in colloidal nanofluid applications. These include the Brownian motion of metal oxide nanoparticles dispersed within the fluid, which enhances heat transfer through constant particle movement (63). Additionally, conductive coatings form around the nanoparticles due to the ordered arrangement of fluid molecules, further facilitating efficient thermal conduction. The aggregate of nanoparticles can create conductive bridges that contribute to enhanced heat transfer pathways within the nanofluids, as shown in Figure 4.2. Moreover, the propagation of thermal energy through phonons plays a significant role in the overall thermal conductivity improvement observed in colloidal nanofluids containing metal oxide nanoparticles (64–66). This type of heat transport is particularly substantial in nanomaterials and low-dimensional systems, where the mean free path of phonons can be longer than the material's characteristic dimensions, leading to enhanced thermal conductivity.

To gain deep quantitative and qualitative insights into particle dynamics, which is essential for understanding these mechanisms, the Stokes–Einstein (SE) equation, a fundamental relation in physical chemistry that describes the diffusion of colloidal particles suspended in nanofluids, can be used. The SE equation is a fundamental relation in physical chemistry that describes the diffusion of particles suspended in a fluid (67). The SE equation, as shown in Equation 4.5, helps us understand

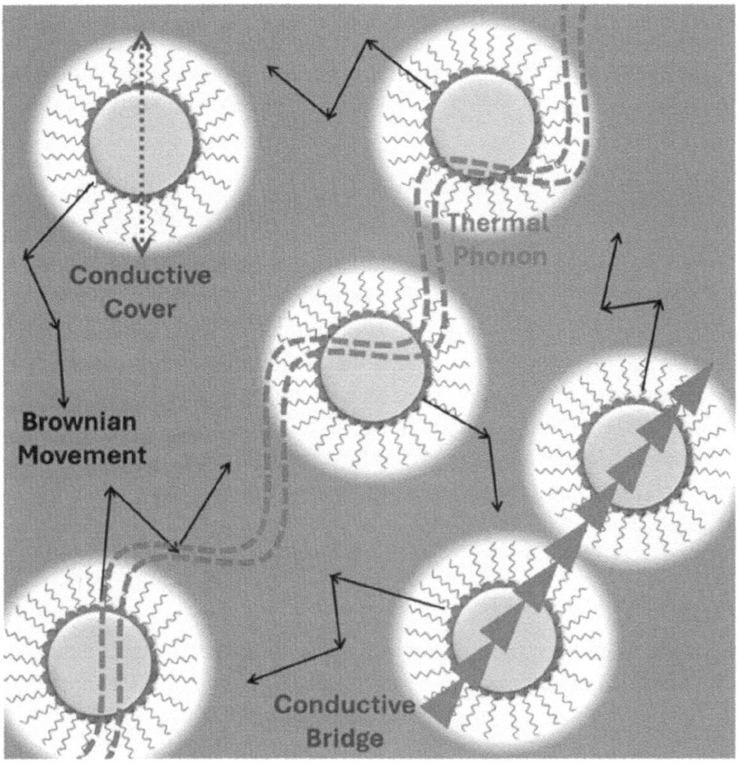

FIGURE 4.2 Schematic illustration of conductive and convective mechanisms in metal oxide nanofluids.

how the diffusion of colloidal particles affects thermal conductivity. For example, for a spherical colloidal nanoparticle with radius R located at a large distance from other nanoparticles or a wall, in motion in a fluid with dynamic viscosity η and temperature of the solution T, the diffusion coefficient D can be calculated using the SE equation:

$$D = \frac{kT}{f}, \quad f = 6\pi\eta R \tag{4.5}$$

The SE equation has two main characteristics: thermal diffusion kT and hydrodynamic behavior of the fluid, derived from the Einstein equation and Stokes friction factor f, respectively. Here, k is defined as the Boltzmann constant. By optimizing the size, geometry, and concentration of colloidal nanoparticles, we can significantly enhance heat transfer efficiency in various applications, such as cooling systems and electronic devices. Smaller particles diffuse faster due to their smaller radius R, as indicated by the SE equation (as shown in Equation 4.5). Fast-diffusing particles can enhance thermal conductivity by facilitating more efficient heat transfer through collisions and interactions with the fluid molecules. However, if the particles are too small, they may agglomerate due to van der Waals forces, reducing the suspension's stability and negatively affecting heat transfer. Thus, an optimal particle size must be found to balance diffusion rate and stability. Higher concentrations of colloidal particles are desired to increase the number of interactions between particles and the fluid, leading to better heat transfer. Excessive concentration can lead to particle clustering or increased fluid viscosity, reducing thermal conductivity (68). Therefore, finding optimal concentration is crucial for maximizing heat transfer efficiency. Colloidal solutions have demonstrated superior thermal properties compared with typical nanofluids. The thermal conductivity of colloidal solutions can be significantly enhanced by incorporating high-conductivity materials such as metal oxides or carbon-based nanoparticles. For example, studies have shown that incorporating graphene or carbon nanotubes in colloidal solutions can drastically improve thermal conductivity, enhancing heat transfer efficiency (68, 69). This is particularly advantageous in applications where rapid and efficient heat dissipation is critical, such as in electronic cooling or solar thermal systems.

4.4.3 Functionalized Colloidal Solutions

Using encapsulated phase change materials (PCMs) in colloidal solutions is another innovative approach offering significant advantages over traditional nanofluids (70). PCMs can absorb and release large amounts of latent heat during phase transitions (e.g., solid to liquid), providing enhanced thermal management. This property is beneficial in applications requiring temperature regulation and thermal storage. Integrating PCMs into colloidal solutions makes it possible to create fluids that conduct heat efficiently and provide thermal buffering, making them ideal for dynamic thermal environments (71). Nanofluids with encapsulated PCMs further boost heat capacity, making them highly effective for thermal regulation. However, challenges include maintaining colloidal stability, preventing particle agglomeration, and ensuring uniform dispersion—critical for consistent thermal performance.

4.4.4 Practical Applications and Versatility

Colloidal solutions are more versatile in practical applications due to their enhanced stability and superior thermal properties (23). They can be used in various heat transfer applications, including cooling systems for electronics, automotive radiators, and HVAC systems (72–74). Additionally, using colloidal solutions in paints and coatings offers innovative applications in building and construction, where enhanced thermal conductivity can improve energy efficiency and reduce heating and cooling costs.

4.4.5 Phase Change Heat Transfer

Due to its high latent heat transfer capability, phase change heat transfer (PCHT) is crucial for efficient thermal management. PCHT transforms substances between phases, such as liquid to vapor (boiling) and vapor to liquid (condensation), each characterized by substantial energy exchange and high heat transfer coefficients. Boiling, where liquid absorbs latent heat to form vapor, and condensation, where vapor releases latent heat upon liquefying, are essential for power generation and cooling (40). Boiling can occur as pool boiling in open spaces or flow boiling in confined channels, both pivotal for efficient heat transfer in various systems. Flow boiling offers more heat transfer enhancement than pool boiling by increasing the surface area for vapor bubble formation, thus improving thermal efficiency. The continuous liquid movement through the channel renews the liquid at the heated surface, maintaining effective heat transfer rates. Increased flow velocity aids in removing vapor bubbles, enhancing cooling efficiency, and delaying critical heat flux (CHF) by preventing vapor film formation (75). Pool boiling, relying on weaker natural convection currents, results in slower liquid circulation due to buoyancy differences (76). The pool boiling process includes natural convection, nucleate, transition, and film boiling. In nucleate boiling, vapor bubbles form, grow, and detach at heated surface sites, optimizing heat transfer. Transition boiling leads to an unstable vapor film, causing temperature and heat transfer rate fluctuations. During film boiling, a stable vapor film insulates the surface, significantly reducing heat transfer efficiency, known as the Leidenfrost point (77).

4.4.6 Surface Wettability

Surface wettability significantly impacts the efficiency of condensation and boiling heat transfer processes. Super hydrophilic surfaces (contact angle <5°) enhance heat transfer by promoting rapid liquid spreading and small bubble formation, which quickly merges and detaches, maintaining efficient heat exchange (78, 79). Hydrophilic surfaces (5°–90°) support moderate wetting and fluid spreading, while hydrophobic surfaces (90°–150°) facilitate larger bubble formation and detachment, increasing turbulence and heat removal efficiency. Superhydrophobic surfaces (contact angle >150°) exhibit minimal wetting and form an insulating vapor film at high heat fluxes, reducing heat transfer efficiency (78). This dynamic is crucial for optimizing surface properties in various thermal management applications, as wettability influences bubble dynamics and overall heat transfer rates (80, 81).

4.5 COATINGS

Various coating techniques modify surfaces by applying functional micro- or nanoparticle layers, creating hydrophilic or hydrophobic properties. These methods result in complex surface structures with randomly distributed shapes and sizes, as illustrated in Figure 4.3.

4.5.1 Chemical Processing

4.5.1.1 Chemical Vapor Deposition

Chemical vapor deposition (CVD) involves creating a thin layer over a heated surface by decomposing chemical vapors. With the presence of those gases, there will be a new chemical reaction on the surface of the substrate, which will be turned into a new substance. These products will be deposited on the surface as a thin layer, which requires activation energy (82). Thin films can be generated using techniques such as the following: (a) thermal decomposition, breaking down a compound into simpler substances through the application of heat; (b) oxidation, the substance loses electrons, often involving the addition of oxygen or the removal of hydrogen; and (c) hydrolysis, the

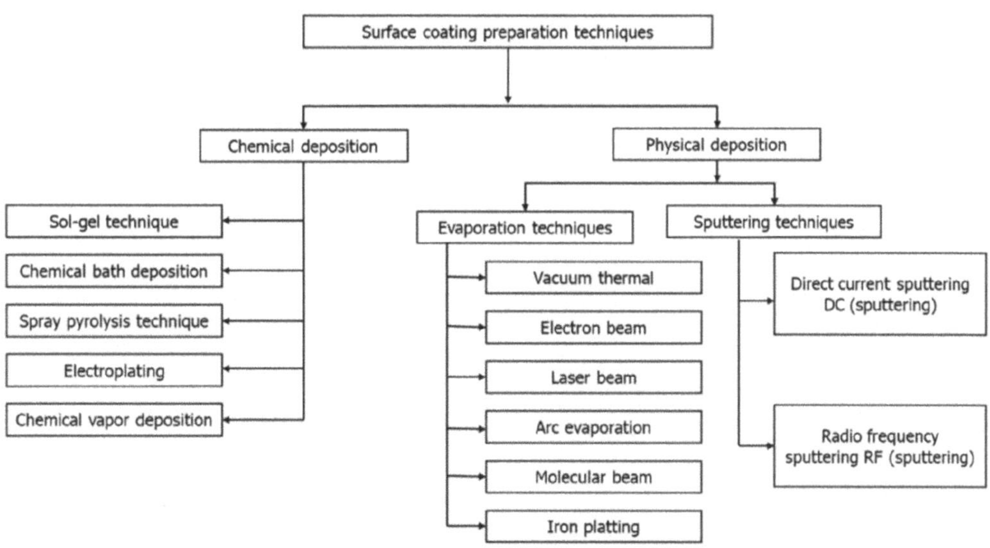

FIGURE 4.3 Overview of various surface coating preparation techniques (82).

compound is broken down by reacting with water. The CVD coatings are versatile and have a broad range of applications. For instance, the CVD has significantly improved condensation heat transfer in graphene coatings on copper substrates using low pressure and atmospheric pressure. However, the major drawback is that this method requires prolonged experiments with several trials to get the needed result (83).

4.5.1.2 Electrochemical Method

Electrochemical methods enhance heat transfer coefficient and critical heat flux, create microporous structures on surfaces, and improve stability, waterproofing, and high-temperature resistance. Electroplating at the cathode, using elements like Cu^{2+} and H^+, forms porous coatings by controlling electroplating time, as demonstrated with copper sulfate and sulfuric acid solutions. Anodic electrochemical techniques, suitable for materials like aluminum and titanium, involve using vital electrolytes to corrode and dissolve the anode, creating porous structures (84). This process includes oxide layer formation, ion migration, and oxide layer dissolution, as shown in the creation of TiO_2 nanotube arrays, which enhance surface roughness and wettability. Surface preparation typically involves ultrasonic cleaning, drying, and polishing to ensure cleanliness and porosity before coating.

4.5.1.3 Sol-Gel Method

Figure 4.4 represents the sol-gel method, widely used for inorganic coatings, which operates at lower thermal conditions with precise control, producing diverse films with distinct morphologies and chemical compositions. This technique extensively modifies TiO_2 films, with studies demonstrating that varying proportions of silver, adjusting pH levels, and calcination temperatures significantly influence film properties (85). Enhanced wettability and durability are achieved through such modifications. Research highlights include using nano-TiO_2 for improved hydrophilicity and TiO_2-polydopamine hybrid films for excellent wear resistance. These studies underscore the sol-gel method's versatility and effectiveness in film modification (86).

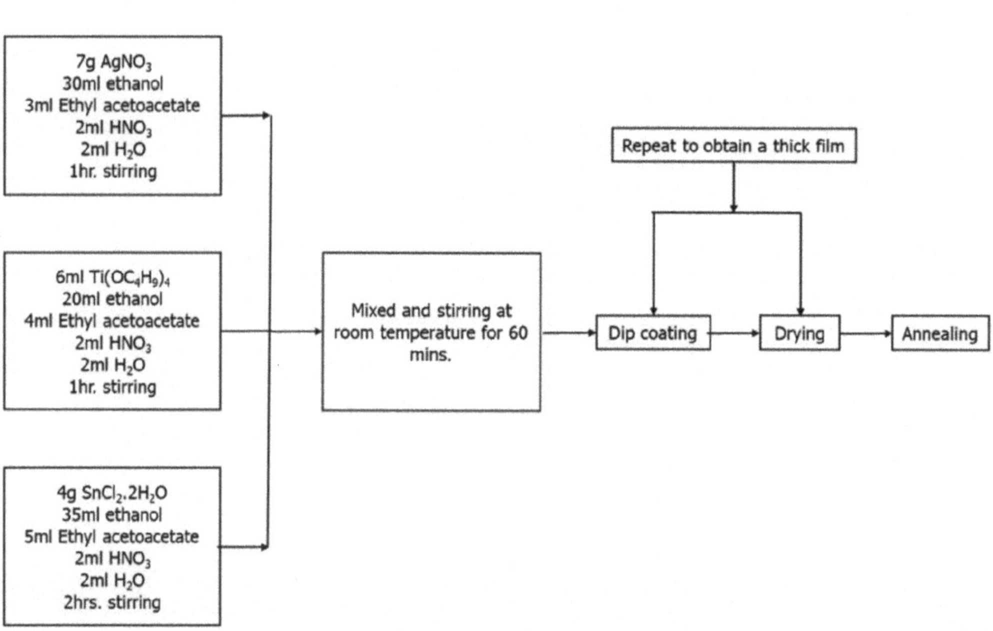

FIGURE 4.4 Procedure for the synthesis of tin and silver-doped titanium dioxide via the sol-gel technique (86).

4.5.1.4 Self-Assembled Monolayers

Self-assembled monolayers (SAMs) consist of a single layer of organic compounds on a surface, significantly enhancing heat transfer by minimizing heat resistance, especially on hydrophobic coatings. Research indicates that SAMs significantly improve thermal performance (87). The solution-based self-assembly method involves immersing polished and plasma-cleaned nickel substrates in benzene solutions of decanethiol (DT), octadecanethiol (ODT), and a mixed DT-ODT for 24 hours, followed by rinsing, drying under nitrogen, and heating at 100°C to enhance bonding. This process ensures the formation of durable SAMs on nickel substrates (88).

4.5.2 Physical Processing

4.5.2.1 Magnetron Sputtering

This flexible technique applies coatings to various industrial materials, such as metals and oxides, on surfaces of different shapes. It operates using either non-reactive (e.g., argon) or reactive gases (e.g., nitrogen, oxygen) (89). The process involves high-energy ions bombarding a target under an electric field, causing target atoms to sputter and deposit onto a substrate. A magnetic field (magnetic cage) near the target surface enhances secondary electrons' movement and collision rate, improving deposition efficiency. This method is widely used, including in applications like lithium-ion battery electrodes, due to its effective particle deposition and coating capabilities.

4.5.2.2 Dip Coating

Figure 4.5 illustrates the dip coating technique, an economical and reliable method where a substrate is submerged in hydrolyzable metal compounds and then steadily withdrawn, creating a liquid layer. After drying, the solvent evaporates, creating a thin film that may require heat treatment for hardening. Adding nanoparticles or volatile components like hydrogen peroxide can create mesoporosity, leading to layers with pore sizes ranging from 2 nm to 50 nm (90). This technique allows control over coating thickness, influenced by solution viscosity, evaporation rate, and substrate withdrawal angle. Although faster than other wet chemical methods, it may lead to non-uniform coatings and require high-temperature sintering. Despite its drawbacks, it is widely used in industrial and laboratory settings to produce high-quality films.

4.5.2.3 Spin Coating

One of the most utilized methods for fabricating ultra-thin films with uniform thickness and precise coating properties. This process involves depositing a fluid onto the center of a substrate, which is

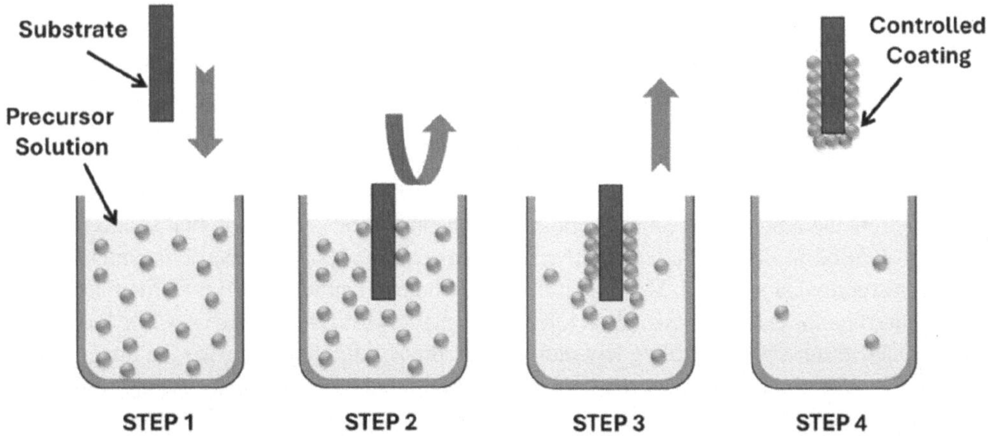

FIGURE 4.5 Schematic representation outlining the steps involved in the dip coating technique. (91)

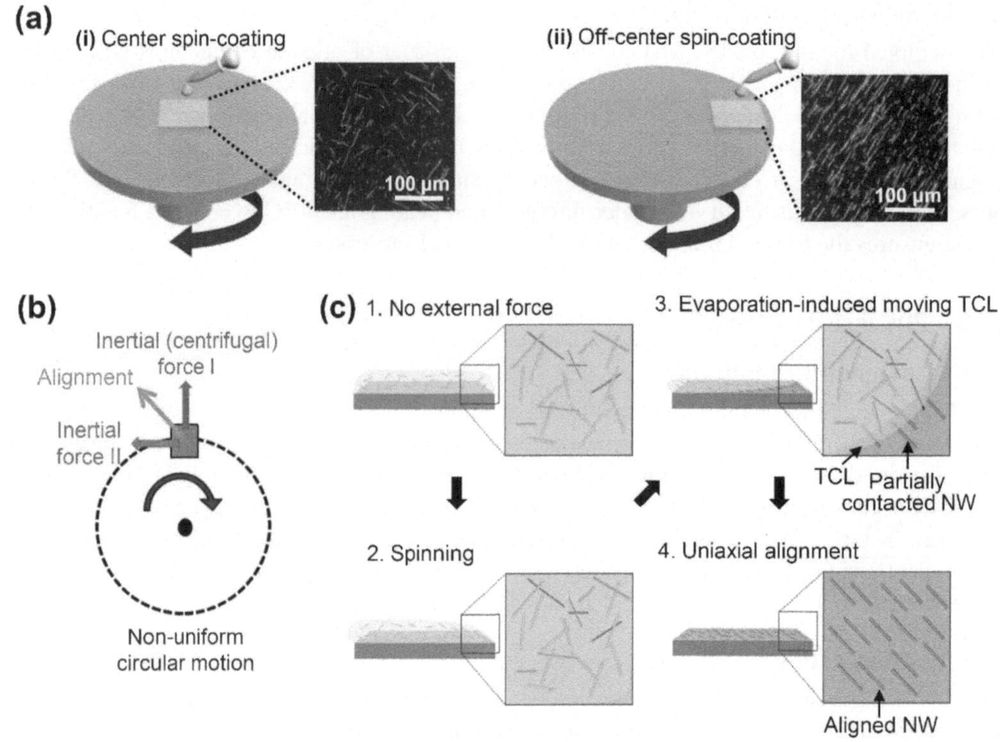

FIGURE 4.6 Diagrams illustrating the complete spin-coating process and its influence on the alignment of nanowires (NWs) (93).

then spun at high speeds. The centrifugal force disperses those fluids evenly from the center to the edges of the substrate, creating a thin and consistent coating. The film's thickness and uniformity are affected by the fluid's viscosity, the speed and span of the spinning, and the solvent's evaporation rate (94). Figure 4.6 shows the spin-coating process, which is highly regarded for its ability to generate consistent and homogeneous coatings, making it ideal for applications that require precise film characteristics. This technique is suitable for a wide range of materials and substrates, combining aspects of both physical and chemical coating processes to enhance surface properties like wettability and adhesion. For instance, combining sol-gel technology with spin coating has improved silicon substrates' optical properties and coatings' effectiveness in thermal applications (95). The primary advantage of spin coating lies in its ability to produce films with high uniformity, which is crucial for various industrial and research applications (96–98).

4.5.2.4 Sintering

Sintering plays a crucial role in materials science by forming solid structures from powdered materials by heating them below their melting point, which causes the particles to bond. This technique is extensively used in developing conductive and wear-resistant coatings. Various sintering methods, such as thermal sintering, UV-Vis sintering, and chemical sintering, offer distinct advantages. UV-Vis sintering, for example, uses ultraviolet-visible light to sinter nanoparticles like nickel-silver on heat-sensitive substrates, enabling low-temperature processing that preserves the integrity of the substrate. This method has shown promising results in producing conductive coatings with significant electrical properties (91). Thermal sintering, while effective, requires high temperatures, which can damage sensitive materials. Chemical sintering, using agents like oxalic acid, facilitates particle bonding at lower temperatures through chemical reactions, offering another viable alternative (92).

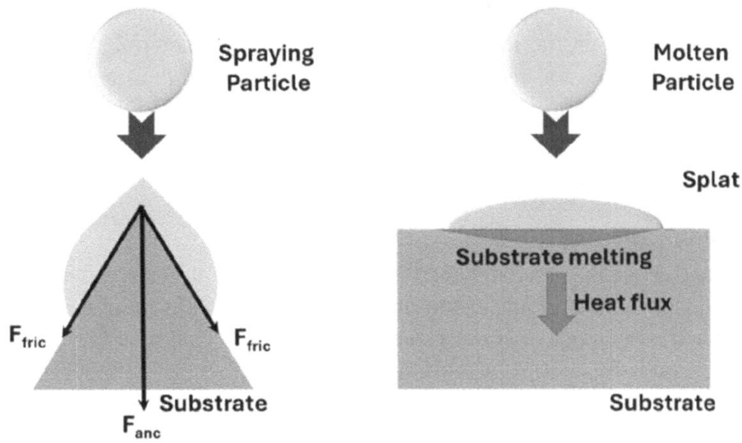

FIGURE 4.7 Bonding mechanisms of plasma spray coating (103).

These sintering techniques are pivotal in creating advanced coatings tailored for electronics and wear-resistant applications.

4.5.2.5 Plasma Spray Coating

Figure 4.7 represents plasma spray coating, a coating technique commonly used in the biomedical sector, especially for hydroxyapatite (HA). The process involves heating ceramic or metal precursor solutions using an ionized gas and applying the melted compounds onto metal or alloy surfaces to create a protective layer (99). This layer has the potential to act as a shield against rust, erosion, and elevated temperatures (99). The primary benefits of this spraying are the cheap cost and fast deposition rate, and its ability to reduce the thermal degradation risk compared with other moderate and high-temperature operations due to the chemically inert nature of the plasma flame and precise temperature control during deposition (100). Despite these advantages, plasma-sprayed HA coatings (32, 101) often face challenges such as inadequate adhesion to the substrate and potential microstructural alterations in the coating material. However, recent research indicates that plasma-sprayed HA coatings on high melting point metals like titanium are effective in orthopedic applications, promoting strong adhesion and bone bonding, which enhances the performance of implants by facilitating direct bone integration and achieving robust adhesion between the new bone and the HA coating (102).

4.5.3 Surface Treatments

Surface Treatments are vital in heat transfer. They enhance efficiency and performance, enabling better heat exchange and fluid distribution in thermal management systems. Based on scale, they can be categorized as macroscale, microscale, nanoscale, or hybrid.

4.5.4 Macrostructure

Macroscale geometrical fins significantly enhance boiling heat transfer by improving bubble nucleation and increasing surface area. Techniques such as laser melting create 3D-thin wall grid structures on steel, achieving high CHF values. Textured copper pillars, hollow conical structures, and honeycomb designs also promote nucleation and improve heat transfer coefficients (HTC) and CHF (104). Modifying surface structures to alter instability wavelengths has shown further improvements. Studies with fluids like HFE-7100 and FC-72 reveal that fins with greater height and less spacing enhance nucleate boiling by providing more nucleation sites. Additionally, pressing fins against heating surfaces and placing fine particles on these surfaces have proven effective (105). These advancements highlight the importance of geometric and material parameters in optimizing macro-structured surfaces for heat transfer applications, though their benefits are increasingly supplemented by micro/nanofabrication techniques (105).

4.5.5 Microstructure

4.5.5.1 Surface Roughening

The most significant advantage of a roughened surface over a smooth surface is its ability to increase the contact area and the number of nucleation sites, resulting in excellent phase change heat transfer. There are many popular methods of surface roughening, such as sandblasting, shown in Figure 4.8, etching, machining, and so on (106). Sand or abrasive blasting involves propelling abrasive materials at high speed against a material's surface. According to Poongavanam et al., this technique effectively removes surface contaminants and creates a uniformly rough texture, enhancing the adhesion properties of coatings or treatments. (107). Shot blasting can form V-corrugations on an aluminum absorber plate by etching precise grooves and ridges with high-velocity abrasive media. These corrugations increase surface area and turbulence, significantly enhancing heat transfer coefficients. Shot peening creates rough surfaces by bombarding a surface with small, spherical media at high velocities, causing plastic deformation. It uses the same principles as sandblasting. This process generates a textured surface with microscopic indentations and protrusions (108, 109).

The rough sodium surface was used to find the enhancement of the heat transfer coefficient and the onset of boiling. Using a surface roughness between 0.15 and 5 μm, Marto and Rohsenow (109) enhanced the critical heat flux for very low surface tension liquids such as hexane and pentane.

4.5.5.2 Lithography

Unlike surface roughening and coating methods, which often create random surface structures, lithographic techniques can produce precise surface textures. These patterns, made at the micro level using photolithography and at the nano level using electron-beam lithography, involve etching processes guided by lithographic masks. These masks can be formed via photoresist coatings or soft lithography and then used to etch the substrate, creating detailed textures. Though lithographic

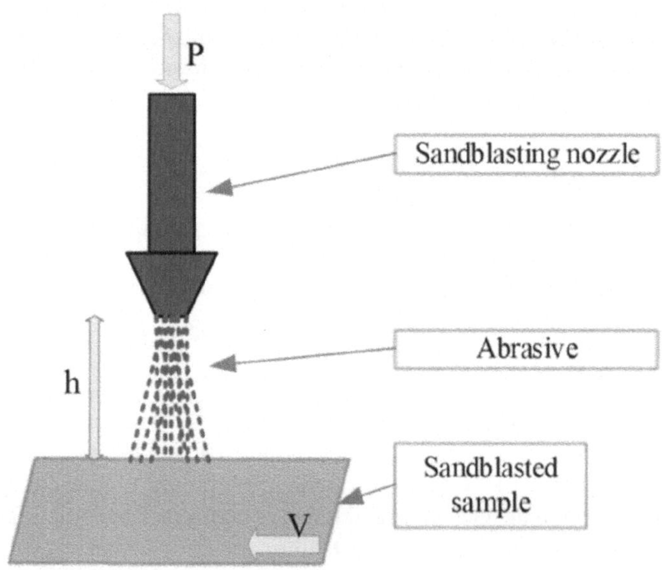

FIGURE 4.8 The sandblasting process creates roughness (110).

methods require clean rooms and are costly, making them impractical for large areas, they uniquely enable the study of phase change on precisely textured surfaces. These techniques are essential in micro-electro-mechanical systems (MEMS) and integrated circuit (IC) manufacturing, offering capabilities to create exact designs (78). Despite their limitations, such as high costs and small fabrication areas, alternatives like soft lithography can mitigate these issues, and materials other than silicon, including glass and polymers, can also be used with some limitations.

4.5.5.3 Micro-fin Surface

Patterning micro-fins on heating surfaces significantly enhances heat transfer by increasing surface area and inducing capillary flow, which helps separate vapor and liquid pathways. These benefits depend on optimizing fin size, shape, and spacing. Figure 4.9 shows various micro-fin configurations, such as vertical, inclined, square, and cylindrical fins, have been studied, showing that CHF increases with fin height and width for FC-72 (111). New micro-fin structures using electro-coating have demonstrated consistent wall superheat with R-141b. Separating vapor and liquid paths using a parallel-tube pumpless loop has proven effective for path separation (112).

4.5.5.4 Microgrooved Surface

Microgroove surfaces, engineered with tiny, precisely controlled grooves or channels, enhance heat transfer by increasing surface area and promoting efficient fluid flow and evaporation. These surfaces are typically created using laser ablation, electric spark wire cutting, or photolithography. Experiments on copper substrates with grooves ranging from 0.2 to 1.3 mm in width and 0.5 to 1.4 mm in depth showed significant improvements in pool boiling performance, with deeper grooves having a more pronounced effect. Similarly, laser-ablated textures on silicon substrates for pool boiling applications demonstrated a lower onset temperature for nucleate boiling. They produced more active nucleation sites, leading to smaller bubble departure diameters (115). This resulted in a 50% maximum HTC enhancement over rough silicon surfaces and a 235% enhancement over polished surfaces.

4.5.5.5 Microporous Coating

Microporous coatings, created through techniques like the sol-gel process, welding, electrospinning, and CVD, allow precise control over porosity, enhancing surface area for applications such as catalysis and filtration (116). These coatings improve fluid flow efficiency and thermal stability in cooling systems by offering increased surface area, active nucleation sites, enhanced permeability, and efficient vapor release pathways. Optimal improvements in nucleate boiling HTC and CHF are achieved with particle sizes of 10 to 25 μm, while gains with larger sizes are less significant (117). Additionally, innovative coatings with non-uniform thickness can separate liquid from vapor phases, reducing counterflow resistance and significantly improving CHF, as demonstrated with

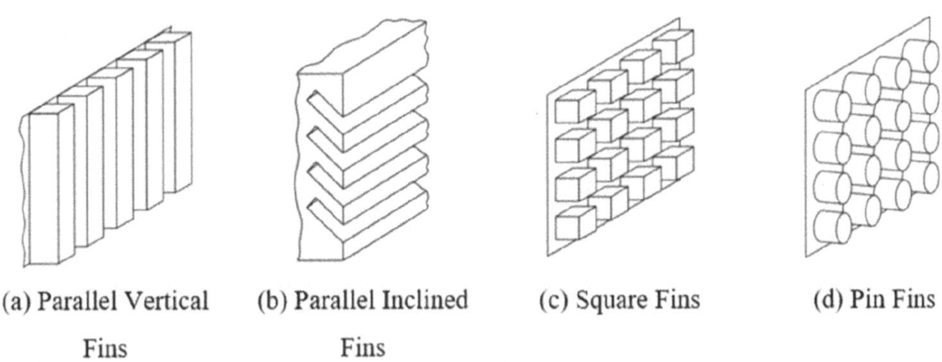

FIGURE 4.9 Different shapes of micro-fin to enhance boiling efficiency (113, 114).

sintered copper coatings that enhanced CHF for pentane nearly threefold compared with plain surfaces (118).

4.5.6 Nanostructure

4.5.6.1 Nanotubes

Figure 4.10 shows boiling curves of light carbon nanotubes (CNTs), cylindrical nanostructures of carbon atoms arranged in a hexagonal lattice, known for their exceptional mechanical, electrical, and thermal properties. They typically have diameters in the nanometer range and can be several micrometers long. CNTs are fabricated using CVD, arc discharge, and laser ablation methods. In CVD, hydrocarbon gases decompose on a metal catalyst to grow CNTs at high temperatures, while arc discharge and laser ablation vaporize carbon sources under specific conditions. Plasma-enhanced CVD (PECVD) produces aligned, size-controlled CNTs that can wet liquids with moderate surface tension (<180 mN/m). By adjusting CNT arrangement and surface chemistry, the wettability of CNT-coated surfaces can be tailored (119). Studies have shown that CNT-coated surfaces on various substrates, such as silicone, stainless steel, and titanium, significantly enhance the HTC and CHF during boiling processes, improving thermal performance in cooling applications (120).

4.5.6.2 Nanowire

A nanowire is a nanostructure with a diameter typically less than 100 nanometers and an extended length. Due to its high aspect ratio and quantum confinement effects, it often exhibits unique electrical, thermal, and mechanical properties (121). Nanowires are used in electronics, sensors, and energy devices; they are composed of various materials such as metals, semiconductors, and oxides (121). Studies have shown that silicon surfaces coated with silicon nanowires significantly enhance water's HTC and CHF. Enhancing the hydrophilicity of silicon surfaces with silicon nanowires has been shown to increase CHF. In contrast, copper nanowires on silicon substrates have achieved significant CHF enhancement with increasing nanowire height up to 2 μm (121).

4.5.6.3 Nanofiber

A nanofiber is a fiber with a diameter in the nanometer range, typically less than 100 nanometers, produced from polymers and characterized by a high aspect ratio and unique physical properties. Nanofibers enhance boiling efficiency by increasing the surface area for heat transfer and promoting

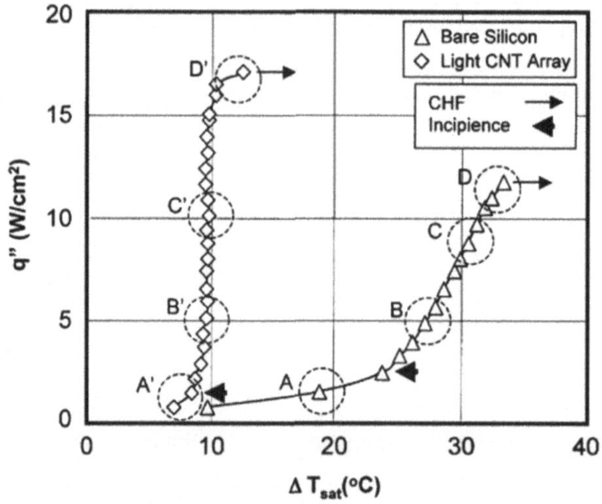

FIGURE 4.10 Boiling curves of FC-72 for light CNT array and bare silicon (121).

nucleate boiling, which leads to higher heat transfer coefficients and critical heat flux. Additionally, they improve wettability and vapor removal, reducing the risk of surface dry-out. Jun et al. (122) utilized copper nanofibers in their boiling experiments and observed a three- to eightfold increase in the HTC for both water and ethanol compared with the bare surface. Sahu et al. (123) Employed copper surfaces coated with copper-plated polymer nanofibers and found enhancements in the HTC and CHF for Novec 7300.

4.5.6.4 Etching

Etching methods to create nanostructures include reactive ion etching, which uses chemically reactive plasma for precise material removal, and wet chemical etching, which involves immersing the substrate in a chemical solution that selectively dissolves the material. These techniques allow for the creation of highly controlled and defined nanoscale features. Laser-based etching techniques are also increasingly popular for creating high-precision and resolution nanostructures. For example, the laser etching method has been used to develop superhydrophilic surfaces with a contact angle of 155° and superhydrophobic glasses utilizing femtosecond laser etching, which exhibit both abrasion resistance and thermal stability. Researchers have explored enhancing HTC in stainless steel heat exchangers through electrochemical etching, achieving HTC improvements ranging from 0.6% to 17.7% for 316 and 304 stainless steels (75).

4.5.7 Nano Coating

4.5.7.1 Nanoparticle Coating

Nanoparticle coating involves applying a thin layer of nanoparticles onto a surface to enhance HTC and CHF. This technique improves thermal performance by increasing surface wettability and thermal conductivity, making it ideal for applications requiring efficient heat dissipation and management. For instance, a hydrophilic titanium oxide surface coated with silicon oxide nanoparticles has significantly enhanced HTC and CHF. A copper substrate coated with SiO_2 nanoparticles achieved an 80% enhancement in HTC for water (124). Additionally, heterostructures by coating copper surfaces with graphene and carbon nanotubes resulted in a 150% increase in HTC and a 40% enhancement in CHF for FC-72 compared with bare surfaces. (125)

4.5.7.2 Nanoporous Coating

Nanoporous coating involves applying a layer with nanoscale pores onto a substrate, significantly enhancing its surface properties. This advanced coating technique increases surface area, improves thermal conductivity, and promotes efficient fluid dynamics, making it ideal for applications requiring superior heat transfer performance and effective fluid management. For example, aluminum substrates coated with nanoporous alumina achieved a 30% reduction in incipience superheat for FC-72. Similarly, nanoporous coatings on copper surfaces resulted in approximately 173% enhancement in the HTC for water (126). Figure 4.11 represents studies on aluminum alloy surfaces coated with Al_2O_3 nanoparticle nanoporous structures that initially showed consistent boiling performance (127). Still, a decline was observed after 900 hours of testing due to nanostructure changes. Despite these challenges, nanoporous coatings continue to present significant advancements in heat transfer efficiency.

4.6 CONCLUSION

Nanofluids represent a promising advancement in heat transfer technology, offering significant improvements in thermal conductivity, convective heat transfer, and overall energy efficiency. By incorporating nanoparticles such as SiO_2, TiO_2, and CuO into base fluids, nanofluids can outperform conventional fluids in various industrial applications, from electronics to transportation. The

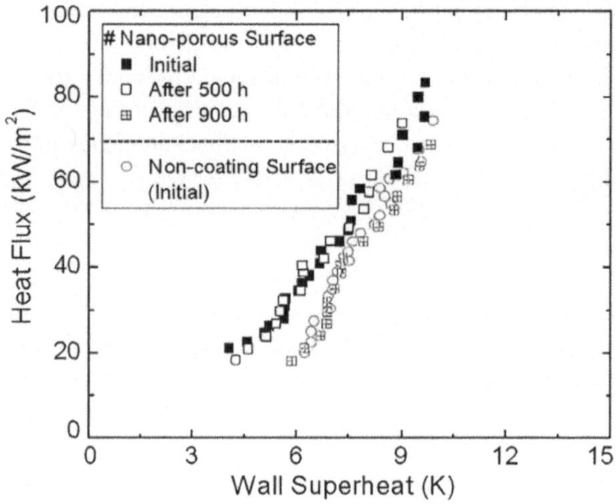

FIGURE 4.11 Comparison of boiling performance of water on an aluminum surface coated with Al_2O_3 nanoporous structures (127).

synthesis methods, whether single-stage or two-stage, play a crucial role in determining the properties and stability of nanofluids. While the single-stage method provides ease of dispersion and better nanoparticle stability, the two-stage method offers scalability and versatility for industrial applications.

Various factors, including the nanoparticle material, size, shape, concentration, and base fluid properties, influence the thermal efficiency of nanofluids. Fundamental mechanisms, such as Brownian motion and nanofluids, represent a promising advancement in heat transfer technology, significantly enhancing thermal conductivity, convective heat transfer, and overall energy efficiency. By incorporating nanoparticles such as SiO_2, TiO_2, and CuO into base fluids, nanofluids can outperform conventional fluids in various industrial applications, including electronics and transportation. The synthesis methods used—either single-stage or two-stage—play a crucial role in determining the properties and stability of these nanofluids. While the single-stage method offers ease of dispersion and better nanoparticle stability, the two-stage method provides scalability and versatility for industrial applications. The thermal efficiency of nanofluids is influenced by several factors, including the type of nanoparticle, its size, shape, concentration, and the characteristics of the base fluid. Fundamental mechanisms, such as Brownian motion and thermophoresis, contribute to enhanced heat transfer. However, excessive particle aggregation and instability can pose challenges. Optimizing the nanoparticle volume fraction and the properties of the base fluid and temperature conditions is vital for achieving optimal performance.

Moreover, the stability of nanofluids is crucial for ensuring long-term efficiency, preventing clogging, and avoiding deposition issues that can impair heat transfer processes. Techniques such as DLS, zeta potential analysis, and TEM are essential for assessing and maintaining the stability of nanofluids. Nanofluids offer a sustainable and efficient solution for improving thermal management systems, contributing to better performance, enhanced energy efficiency, and reduced environmental impact. However, continued research and development are necessary to address their synthesis, stability, and scalability challenges, ensuring that nanofluids reach their full potential in real-world applications.

Nanofluids that contain metal-based nanoparticles with platelet shapes and larger surface area-to-volume ratios are more effective at enhancing thermal conductivity. Additionally, Brownian motion and thermophoresis are essential factors that cause nanoparticles to develop a slip velocity relative to the base fluid. These dynamics are crucial for designing effective two-component nanofluid models. Applying nanoparticles in base fluids can significantly improve heat transfer efficiency and offer environmental benefits when managed according to standard protocols. This underscores the importance of responsible practices when utilizing nanofluids in industrial applications, such as heat exchangers, chillers, and solar panels, to optimize the thermal performance of these devices.

Colloidal nanofluids offer significant advancements in thermal conductivity due to mechanisms such as Brownian motion, phonon propagation, and optimal nanoparticle dispersion. Including high-conductivity materials like metal oxide nanoparticles further enhances heat transfer efficiency due to their excellent scattering and charge carrier recombination properties. Functionalized colloidal solutions, especially those with encapsulated PCMs, demonstrate superior thermal management, making them ideal for dynamic applications. With their versatility and efficiency in applications ranging from photocatalysis to thermal coatings, metal oxide nanoparticles highlight their potential to significantly enhance thermal properties in diverse practical settings.

Advanced surface engineering techniques enhance heat transfer efficiency, including thermal coatings and treatments. Modifying surface properties such as thermal conductivity, wettability, and texture optimizes key processes like boiling and condensation, which are crucial for systems operating under high thermal loads. These innovations improve energy efficiency and ensure the reliability of industrial heat exchangers, electronic cooling systems, and other critical applications. As research progresses, the continued refinement of these techniques will drive further advancements in thermal management, contributing to more sustainable and efficient energy solutions.

REFERENCES

1. Yu W, France DM, Timofeeva EV, Singh D, Routbort JL. Comparative review of turbulent heat transfer of nanofluids. *International Journal of Heat and Mass Transfer.* 2012 Oct 1;55(21):5380–96.
2. Xuan Y, Li Q. Heat transfer enhancement of nanofluids. *International Journal of Heat and Fluid Flow.* 2000 Feb 1;21(1):58–64.
3. Ghalambaz M, Doostani A, Izadpanahi E, Chamkha AJ. Phase-change heat transfer in a cavity heated from below: the effect of utilizing single or hybrid nanoparticles as additives. *Journal of the Taiwan Institute of Chemical Engineers.* 2017 Mar 1;72:104–15.
4. Hamada Y, Ohtsu W, Fukai J. Thermal response in thermal energy storage material around heat transfer tubes: effect of additives on heat transfer rates. *Solar Energy.* 2003 Oct 1;75(4):317–28.
5. Sheikholeslami M. Effect of uniform suction on nanofluid flow and heat transfer over a cylinder. *Journal of the Brazilian Society of Mechanical Sciences and Engineering.* 2015 Nov 1;37(6):1623–33.
6. Zuckerman N, Lior N. Jet Impingement heat transfer: physics, correlations, and numerical modeling. In: Greene GA, Hartnett† JP, Bar-Cohen A, Cho YI, editors. *Advances in Heat Transfer [Internet].* Elsevier; 2006 [cited 2024 Jun 19]. 565–631. Available from: https://www.sciencedirect.com/science/article/pii/S0065271706390065.
7. Wei X, Zhu H, Tiantian K, Wang L. Synthesis and thermal conductivity of Cu2O nanofluids. *International Journal of Heat and Mass Transfer.* 2009 Sep 1;52:4371–4.
8. Regulation of natural convection heat transfer for SiO2–solar salt nanofluids by optimizing rectangular vessels design - Hu – 2020. *Asia-Pacific Journal of Chemical Engineering - Wiley Online Library [Internet].* [cited 2024 Oct 29]. Available from: https://onlinelibrary.wiley.com/doi/abs/10.1002/apj.2409.
9. Chetan, Behera BC, Ghosh S, Rao PV. Application of nanofluids during minimum quantity lubrication: a case study in turning process. *Tribology International.* 2016 Sep;101:234–46.
10. Wang XQ, Mujumdar AS. A review on nanofluids - part I: theoretical and numerical investigations. *Brazilian Journal of Chemical Engineering.* 2008 Dec;25:613–30.
11. Yasmin H, Giwa SO, Noor S, Aybar HŞ. Influence of preparation characteristics on stability, properties, and performance of mono- and hybrid nanofluids: current and future perspective. *Machines.* 2023 Jan;11(1):112.
12. Hwang Y, Lee JK, Lee JK, Jeong YM, Cheong S ir, Ahn YC, et al. Production and dispersion stability of nanoparticles in nanofluids. *Powder Technology.* 2008 Aug 11;186(2):145–53.

13. Rane AV, Kanny K, Abitha VK, Thomas S. Chapter 5 - methods for synthesis of nanoparticles and fabrication of nanocomposites. In: Mohan Bhagyaraj S, Oluwafemi OS, Kalarikkal N, Thomas S, editors. *Synthesis of Inorganic Nanomaterials [Internet].* Woodhead Publishing; 2018 [cited 2024 May 24]. 121–39. (Micro and Nano Technologies). Available from: https://www.sciencedirect.com/science/article/pii/B9780081019757000051.
14. Chemical vapor deposition and its application in surface modification of nanoparticles. *Chemical Papers [Internet].* [cited 2024 May 24]. Available from: https://link.springer.com/article/10.1007/s11696-019-00963-y.
15. Sidik NAC, Mohammed HA, Alawi OA, Samion S. A review on preparation methods and challenges of nanofluids. *International Communications in Heat and Mass Transfer.* 2014 May 1;54:115–25.
16. Ramakrishnan SB, Khatri N, Addanki Tirumala RT, Mohammadparast F, Karuppasamy K, Kalkan AK, et al. Cupric oxide Mie resonators. *Journal of Physical Chemistry C.* 2022 Sep 29;126(38):16272–9.
17. Semaltianos NG. Nanoparticles by laser ablation. *Critical Reviews in Solid State and Materials Sciences.* 2010 Apr;35(2):105–24.
18. Lo CH, Tsung TT, Chen LC. Shape-controlled synthesis of Cu-based nanofluid using submerged arc nanoparticle synthesis system (SANSS). *Journal of Crystal Growth.* 2005 Apr 15;277(1):636–42.
19. Shrimal P, Jadeja G, Patel S. A review on novel methodologies for drug nanoparticle preparation: microfluidic approach. *Chemical Engineering Research and Design.* 2020 Jan 1;153:728–56.
20. Synthesis of hybrid nanofluid with two-step method. *E3S Web of Conferences [Internet].* [cited 2024 May 24]. Available from: https://www.e3s-conferences.org/articles/e3sconf/abs/2018/42/e3sconf_i-trec2018_03057/e3sconf_i-trec2018_03057.html.
21. Danks A, Hall S, Schnepp Z. The evolution of 'sol–gel' chemistry as a technique for materials synthesis. *Materials Horizons.* 2016;3(2):91–112.
22. Applied Sciences | Free Full-Text | Preparation Techniques of TiO2 Nanofluids and Challenges: A Review [Internet]. [cited 2024 May 25]. Available from: https://www.mdpi.com/2076-3417/8/4/587
23. Dhinesh Kumar D, Valan Arasu A. A comprehensive review of preparation, characterization, properties and stability of hybrid nanofluids. *Renewable and Sustainable Energy Reviews.* 2018 Jan 1;81:1669–89.
24. Tulinski M, Jurczyk M. Nanomaterials synthesis methods. In: *Metrology and Standardization of Nanotechnology [Internet].* John Wiley & Sons, Ltd; 2017 [cited 2024 May 25]. 75–98. Available from: https://onlinelibrary.wiley.com/doi/abs/10.1002/9783527800308.ch4.
25. Erdemir F. Study on particle size and X-ray peak area ratios in high energy ball milling and optimization of the milling parameters using response surface method. *Measurement.* 2017 Dec 1;112:53–60.
26. Cryomilling as environmentally friendly synthesis route to prepare nanomaterials - Nirmal Kumar Katiyar, Krishanu Biswas, C. S. Tiwary, 2021 [Internet]. [cited 2024 May 25]. Available from: https://journals.sagepub.com/doi/full/10.1080/09506608.2020.1825175?casa_token=94kD5BUulEIAAAAA%3AfUtJ9HSpoGJ_fDxS_eUa9C79Y1HqoMZFdWY1f_Pb2NBTPbxnu4okMwx-WssF1PXZcRGzvgNrnKqi.
27. Habibi R, Neild A. Sound wave activated nano-sieve (SWANS) for enrichment of nanoparticles. *Lab on a Chip.* 2019;19(18):3032–44.
28. Gkikas G, Barkoula NM, Paipetis AS. Effect of dispersion conditions on the thermo-mechanical and toughness properties of multi walled carbon nanotubes-reinforced epoxy. *Composites Part B: Engineering.* 2012 Sep 1;43(6):2697–705.
29. Shima PD, Philip J. Role of thermal conductivity of dispersed nanoparticles on heat transfer properties of nanofluid. *Industrial & Engineering Chemistry Research.* 2014 Jan 15;53(2):980–8.
30. Significance of nanoparticles shape for enhanced heat transfer over on elastic sheet - Abdul Hamid Ganie, Basharat Ullah, Noreen, Nisar Ahmad Koka, Umar Khan, 2024 [Internet]. [cited 2024 Aug 11]. Available from: https://journals.sagepub.com/doi/abs/10.1177/23977914231224487.
31. Electrohydrodynamic capillary instability of Rivlin–Ericksen viscoelastic fluid film with mass and heat transfer - Awasthi - 2024 - Heat Transfer - Wiley Online Library [Internet]. [cited 2024 Oct 29]. Available from: https://onlinelibrary.wiley.com/doi/abs/10.1002/htj.22944.
32. Kumar S, Vijayan PK, Kannan U, Sharma M, Pilkhwal DS. Experimental and computational simulation of thermal stratification in large pools with immersed condenser. *Applied Thermal Engineering.* 2017 Feb 25;113:345–61.
33. Kumar S, Vijayan PK, Grover RB, Kannan U, Yadav H, Agrawal A. Experimental Investigations on Thermal Stratification in a Large Pool of Water With Immersed Isolation Condenser. In [cited 2024 Oct 29]. Available from: https://dx.doi.org/10.1115/ICONE2020-16647.

34. Alami AH, Ramadan M, Tawalbeh M, Haridy S, Al Abdulla S, Aljaghoub H, et al. A critical insight on nanofluids for heat transfer enhancement. *Scientific Reports* 2023 Sep 15;13(1):15303.
35. Rayleigh–Bénard instability in nanofluids: a comprehensive review. *Micro and Nano Systems Letters [Internet].* [cited 2024 May 30]. Available from: https://link.springer.com/article/10.1186/s40486-020-00123-y.
36. Thermal conductivity of nanofluids-A comprehensive review. *International Journal of Thermofluid Science and Technology [Internet].* 2020 [cited 2024 Aug 11]. Available from: http://ijtf.org/2020/thermal-conductivity-of-nanofluids-a-comprehensive-review/.
37. Molecular dynamics simulation of effect of liquid layering around the nanoparticle on the enhanced thermal conductivity of nanofluids. *Journal of Nanoparticle Research [Internet].* [cited 2024 May 27]. Available from: https://link.springer.com/article/10.1007/s11051-009-9728-5.
38. On the Specific Heat Capacity of CuO Nanofluid - Le-Ping Zhou, Bu-Xuan Wang, Xiao-Feng Peng, Xiao-Ze Du, Yong-Ping Yang, 2010 [Internet]. [cited 2024 May 27]. Available from: https://journals.sagepub.com/doi/full/10.1155/2010/172085.
39. Riazi H, Murphy T, Webber GB, Atkin R, Tehrani SSM, Taylor RA. Specific heat control of nanofluids: a critical review. *International Journal of Thermal Sciences.* 2016 Sep 1;107:25–38.
40. Pak BC, Cho YI. Hydrodynamic and heat transfer study of dispersed fluids with submicron metallic oxide particles. *Experimental Heat Transfer.* 1998 Apr 1;11(2):151–70.
41. Chon CH, Paik S, Tipton JB, Kihm KD. Effect of nanoparticle sizes and number densities on the evaporation and dryout characteristics for strongly pinned nanofluid droplets. *Langmuir.* 2007 Mar 1;23(6):2953–60.
42. Dey D, Sahu DS. Nanofluid in the multiphase flow field and heat transfer: a review. *Heat Transfer.* 2021;50(4):3722–75.
43. Fan L, Khodadadi JM. Thermal conductivity enhancement of phase change materials for thermal energy storage: a review. *Renewable and Sustainable Energy Reviews.* 2011 Jan 1;15(1):24–46.
44. Saidur R, Leong KY, Mohammed HA. A review on applications and challenges of nanofluids. *Renewable and Sustainable Energy Reviews.* 2011 Apr 1;15(3):1646–68.
45. Talkar S, Choudhari A, Rayar P. Building envelope optimization and cost-effective approach in hvac to support smart manufacturing. In: Vasudevan H, Kottur VKN, Raina AA, editors. *Proceedings of International Conference on Intelligent Manufacturing and Automation.* Singapore: Springer; 2020. 299–308.
46. Choudhari A, Rane A, Talkar S, Rayar P, Shukla D. Designing and prototyping for conservation and effective utilization of waste heat from air conditioner. *IOP Conference Series: Materials Science and Engineering.* 2021 Mar;1104(1):012007.
47. Nanofluids Research: Key Issues | Discover Nano [Internet]. [cited 2024 May 30]. Available from: https://link.springer.com/article/10.1007/s11671-010-9638-6.
48. Choudhari A, Rayar P, Shimpi S, Pawar N, Ambetkar S. Design and development of vacuum frying machine for the production of high-quality fried products. In: Vasudevan H, Kottur VKN, Raina AA, editors. *Proceedings of International Conference on Intelligent Manufacturing and Automation.* Singapore: Springer Nature; 2023. 521–32.
49. Gupta A, Choudhari A, Kadaka T, Rayar P. Design and analysis of vertical vacuum fryer. In: Vasudevan H, Kottur VKN, Raina AA, editors. *Proceedings of International Conference on Intelligent Manufacturing and Automation.* Singapore: Springer; 2019. 133–49.
50. Gupta* AK, Choudhari A, Kumar A, Gupta A. Custom implants and beyond: the biomedical potential of additive manufacturing. *Archives of Case Reports.* 2024 May 17;8(2):050–5.
51. Saleh R, Putra N, Wibowo RE, Septiadi WN, Prakoso SP. Titanium dioxide nanofluids for heat transfer applications. *Experimental Thermal and Fluid Science.* 2014 Jan 1;52:19–29.
52. Atiganyanun S, Plumley JB, Han SJ, Hsu K, Cytrynbaum J, Peng TL, et al. Effective radiative cooling by paint-format microsphere-based photonic random media. *ACS Photonics.* 2018 Apr 18;5(4):1181–7.
53. Maaza M, Khamliche T, Akbari M, Kana N, Tandjigora N, Beukes P, et al. A novel approach for engineering efficient nanofluids by radiolysis. *Scientific Reports.* 2022 Jun 24;12(1):10767.
54. Addanki Tirumala RT, Ramakrishnan SB, Mohammadparast F, Khatri N, Arumugam SM, Tan S, et al. Structure–property–performance relationships of dielectric Cu2O nanoparticles for mie resonance-enhanced dye sensitization. *ACS Applied Nano Materials [Internet].* 2022 May 11 [cited 2022 May 20]; Available from: https://doi.org/10.1021/acsanm.2c00730.
55. Rajashekar Reddy N, Abhilash P, Rahul M. Enhancement of heat transfer in radiator using copper oxide nano fluids. *Materials Today: Proceedings.* 2021 Jan 1;39:643–8.

56. Gyawali S, Tirumala RTA, Loh H, Andiappan M, Bristow AD. Photocarrier recombination dynamics in highly scattering Cu2O nanocatalyst clusters. *Journal of Physical Chemistry C*. 2024 Feb 8;128(5):2003–11.
57. Mohammadparast F, Ramakrishnan SB, Khatri N, Tirumala RTA, Tan S, Kalkan AK, et al. Cuprous oxide cubic particles with strong and tunable mie resonances for use as nanoantennas. *ACS Applied Nano Materials*. 2020 Jul 24;3(7):6806–15.
58. Advances in Nickel-Containing High-Entropy Alloys: From Fundamentals to Additive Manufacturing [Internet]. [cited 2024 Oct 29]. Available from: https://www.mdpi.com/1996-1944/17/15/3826.
59. Tamadondar MR, Zarghami R, Boutou K, Tahmasebpoor M, Mostoufi N. Size of nanoparticle agglomerates in fluidization. *The Canadian Journal of Chemical Engineering*. 2016;94(3):476–84.
60. Yang B. Thermal conductivity equations based on Brownian motion in suspensions of nanoparticles (Nanofluids). *Journal of Heat Transfer [Internet]*. 2008 Mar 18 [cited 2024 Jun 19];130(042408). Available from: https://doi.org/10.1115/1.2789721.
61. Chakraborty S, Kumar P, Chakraborty S. Nanofluids Long-term Stability Challenges and Guidelines. [cited 2024 Jun 19]; Available from: https://books.rsc.org/books/edited-volume/2039/chapter/4660184/Nanofluids-Long-term-Stability-Challenges-and.
62. Roy D, Pal S, Tiwary CS, Gupta AK, Babu PN, Mitra R. Stable nanocrystalline structure attainment and strength enhancement of Cu base alloy using bi-modal distributed tungsten dispersoids. *Philosophical Magazine*. 2022 Feb 1;102(3):189–209.
63. Zhu D, Wang L, Yu W, Xie H. Intriguingly high thermal conductivity increment for CuO nanowires contained nanofluids with low viscosity. *Scientific Reports*. 2018 Mar 27;8(1):5282.
64. Wan X, Pan D, Zong Z, Qin Y, Lü JT, Volz S, et al. Modulating thermal conductivity via targeted phonon excitation. *Nano Letters*. 2024 Jun 12;24(23):6889–96.
65. Maire J, Anufriev R, Yanagisawa R, Ramiere A, Volz S, Nomura M. Heat conduction tuning by wave nature of phonons. *Science Advances*. 2017 Aug 4;3(8):e1700027.
66. Kothari K, Maldovan M. phonon surface scattering and thermal energy distribution in superlattices. *Scientific Reports*. 2017 Jul 17;7(1):5625.
67. Kholodenko AL, Douglas JF. Generalized Stokes-Einstein equation for spherical particle suspensions. *Physical Review E*. 1995 Feb 1;51(2):1081–90.
68. Younes H, Mao M, Sohel Murshed SM, Lou D, Hong H, Peterson GP. Nanofluids: key parameters to enhance thermal conductivity and its applications. *Applied Thermal Engineering*. 2022 May 5;207:118202.
69. Mohd Zubir MN, Badarudin A, Kazi SN, Huang NM, Misran M, Sadeghinezhad E, et al. Highly dispersed reduced graphene oxide and its hybrid complexes as effective additives for improving thermophysical property of heat transfer fluid. *International Journal of Heat and Mass Transfer*. 2015 Aug 1;87:284–94.
70. Shchukina EM, Graham M, Zheng Z, Shchukin DG. Nanoencapsulation of phase change materials for advanced thermal energy storage systems. *Chemical Society Reviews*. 2018 Jun 5;47(11):4156–75.
71. Pereira J, Souza R, Moreira A, Moita A. A Review on the Nanofluids-PCMs Integrated Solutions for Solar Thermal Heat Transfer Enhancement Purposes. *Technologies*. 2023 Dec;11(6):166.
72. Bacha HB, Ullah N, Hamid A, Shah NA. A comprehensive review on nanofluids: Synthesis, cutting-edge applications, and future prospects. *International Journal of Thermofluids*. 2024 May 1;22:100595.
73. Okonkwo EC, Al-Ansari T. Parametric investigation of a chilled water district cooling unit using mono and hybrid nanofluids. *Scientific Reports*. 2021 Sep 28;11(1):19227.
74. Zhu Q, Chua MH, Ong PJ, Cheng Lee JJ, Le Osmund Chin K, Wang S, et al. Recent advances in nanotechnology-based functional coatings for the built environment. *Materials Today Advances*. 2022 Aug 1;15:100270.
75. Qin S, Ji R, Miao C, Jin L, Yang C, Meng X. Review of enhancing boiling and condensation heat transfer: Surface modification. *Renewable and Sustainable Energy Reviews*. 2024 Jan;189.
76. Optimization of Enhanced Surfaces for High Flux Chip Cooling by Pool B H1B oiling [Internet]. 1993. Available from: http://asme.org/terms.
77. Liang G, Mudawar I. Review of pool boiling enhancement by surface modification. *International Journal of Heat and Mass Transfer*. 2019;128: 892–933.
78. Hu H, Xu C, Zhao Y, Ziegler KJ, Chung JN. Boiling and quenching heat transfer advancement by nanoscale surface modification. *Scientific Reports*. 2017 Dec;7(1). doi:10.1038/s41598-017-06050-0.
79. Mandal S, Kumar Gupta A, Echeverria E, McIlroy DN, Poplawsky JD, Sachan R. Laser-assisted nanofabrication of multielement complex oxide core–shell nanoparticles. *Materials & Design*. 2022 Aug 1;220:110882.

80. Attinger D, Frankiewicz C, Betz AR, Schutzius TM, Ganguly R, Das A, et al. Surface engineering for phase change heat transfer: a review. Vol. 1, In: *MRS Energy and Sustainability*. Springer Nature; 2014.
81. Mandal S, Gupta AK, Konečná A, Shirato N, Hachtel JA, Sachan R. Creation of multi-principal element alloy NiCoCr nanostructures via nanosecond laser-induced dewetting. *Small*. 2024 Jul;20(28):e2309574.
82. Katarkar AS, Majumder B, Pingale AD, Belgamwar SU, Bhaumik S. A review on the effects of porous coating surfaces on boiling heat transfer. In: *Materials Today: Proceedings*. Elsevier Ltd; 2021. 362–7.
83. Madhuri KV. Thermal protection coatings of metal oxide powders. In: *Metal Oxide Powder Technologies: Fundamentals, Processing Methods and Applications*. Elsevier; 2020. 209–31.
84. Khatib SJ, Oyama ST. Silica membranes for hydrogen separation prepared by chemical vapor deposition (CVD). Vol. 111. In: *Separation and Purification Technology*. 2013. 20–42.
85. Hong S, Jiang S, Hu Y, Dang C, Wang S. Visualization investigation of the effects of nanocavity structure on pool boiling enhancement. *International Journal of Heat and Mass Transfer*. 2019 Jun;136:235–45.
86. Ilkhechi NN, Ghorbani M, Mozammel M, Khajeh M. The optical, photo catalytic behavior and hydrophilic properties of silver and tin co doped TiO2 thin films using sol–gel method. *Journal of Materials Science: Materials in Electronics*. 2017 Feb;28(4):3571–80.
87. Jiang Y, Shi K, Tang H, Wang Y. Enhanced wettability and wear resistance on TiO2/PDA thin films prepared by sol-gel dip coating. *Surface and Coatings Technology*. 2019 Oct;375:334–40.
88. Chen L, Liang S, Yan R, Cheng Y, Huai X, Chen S. N-Octadecanethiol self-assembled monolayer coating with microscopic roughness for dropwise condensation of steam. *Journal of Thermal Science*. 2009 Jun;18(2):160–5.
89. Yang J, Zhou M, Liu J, Wang H, Weng C. Fabrication and tribological properties of self-assembled monolayers of alkanethiols on nickel substrates. *Applied Surface Science*. 2021 Sep;559.
90. Slomski EM, Fischer S, Scheerer H, Oechsner M, Stephan P. Textured CrN thin coatings enhancing heat transfer in nucleate boiling processes. *Surface and Coatings Technology*. 2013 Jan;215:465–71.
91. Neacşu IA, Nicoară AI, Vasile OR, Vasile BŞ. Inorganic micro- and nanostructured implants for tissue engineering. In: *Nanobiomaterials in Hard Tissue Engineering: Applications of Nanobiomaterials*. Elsevier Inc.; 2016. 271–95.
92. Pajor-Świerzy A, Szyk-Warszyńska L, Duraczyńska D, Szczepanowicz K. UV-Vis Sintering process for fabrication of conductive coatings based on Ni-Ag core–shell nanoparticles. *Materials*. 2023 Nov;16(22).
93. Lee G, Kim H, Lee SB, Kim D, Lee E, Lee SK, et al. Tailored uniaxial alignment of nanowires based on off-center spin-coating for flexible and transparent field-effect transistors. *Nanomaterials*. 2022 Jan;12(7):1116.
94. Magdassi S, Grouchko M, Berezin O, Kamyshny A. Triggering the sintering of silver nanoparticles at room temperature. *ACS Nano*. 2010 Apr;4(4):1943–8.
95. Xu S, Wang ZL. One-dimensional ZnO nanostructures: Solution growth and functional properties. Vol. 4. In: *Nano Research*. Tsinghua University Press; 2011. 1013–98.
96. Satvekar RK, Phadatare MR, Karande VA, Patil RN, Tiwale BM, Pawar SH. Influence of silane content on the optical properties of Sol Gel derived spin coated silica thin films [Internet]. *International Journal of Basic and Applied Sciences*. 2012;1:468–76. Available from: www.sciencepubco.com/index.php/IJBAS.
97. Roy D, Chakraborty S, Gupta AK, BasuMallick A, Scattergood RO, Koch CC. Synergistic effect of Nb and Zr additions on the structure-property relationships of nanocrystalline Cu processed by mechanical alloying and hot pressing. *Journal of Alloys and Compounds*. 2021 Feb 15;854:157174.
98. Chakraborty S, Gupta AK, Roy D, Basu Mallick A. Nanomechanical properties of mechanically alloyed and spark plasma sintered W-nanoparticulate dispersed Cu-Nb alloys. *Materials Letters*. 2020 Sep 1;274:128004.
99. Hijón N, Cabañas MV, Peña J, Vallet-Regí M. Dip coated silicon-substituted hydroxyapatite films. *Acta Biomaterialia*. 2006;2(5):567–74.
100. Chu H, Xu N, Yu X, Jiang H, Ma W, Qiao F. Review of surface modification in pool boiling application: Coating manufacturing process and heat transfer enhancement mechanism, Vol. 215. In: *Applied Thermal Engineering*. Elsevier Ltd; 2022.
101. Kumar S, King M, Hardin RG. Modeling of flows through porous media. In: *Computational Fluid Flow and Heat Transfer*. CRC Press; 2024.
102. Herman H. Plasma-sprayed coatings. *Scientific American*. 1988 Sep;259(3):112–17.
103. Song Z, Li H. Plasma spraying with wire feeding: A facile route to enhance the coating/substrate interfacial metallurgical bonding. *Coatings*. 2022 May;12(5):615.

104. Mohseni E, Zalnezhad E, Bushroa AR. Comparative investigation on the adhesion of hydroxyapatite coating on Ti-6Al-4V implant: a review paper. *International Journal of Adhesion and Adhesives.* 2014;48:238–57.
105. Nasersharifi Y, Kaviany M, Hwang G. Pool-boiling enhancement using multilevel modulated wick. *Applied Thermal Engineering.* 2018 Jun;137:268–76.
106. Yu CK, Lu DC. Pool boiling heat transfer on horizontal rectangular fin array in saturated FC-72. *International Journal of Heat and Mass Transfer.* 2007 Aug;50(17–18):3624–37.
107. Kurihara HM, Myers JE. The effects of superheat and surface roughness on boiling coefficients. *AIChE Journal.* 1960 Mar;6(1):83–91.
108. Poongavanam GK, Panchabikesan K, Leo AJD, Ramalingam V. Experimental investigation on heat transfer augmentation of solar air heater using shot blasted V-corrugated absorber plate. *Renewable Energy.* 2018 Nov;127:213–29.
109. Marto PJ, Rohsenow WM. *Effects of Surface Conditions on Nucleate Pool Boiling of Sodium [Internet].* 1966. Available from: http://heattransfer.asmedigitalcollection.asme.org/.
110. Miturska-Barańska I, Rudawska A, Doluk E. The influence of sandblasting process parameters of aerospace aluminium alloy sheets on adhesive joints strength. *Materials.* 2021 Jan;14(21):6626.
111. Yuda AW, Supriadi S, Saragih AS. Surface modification of Ti-alloy based bone implant by sandblasting. In: *AIP Conference Proceedings.* American Institute of Physics Inc.; 2019.
112. Wei JJ, Honda H. Effects of fin geometry on boiling heat transfer from silicon chips with micro-pin-fins immersed in FC-72. *International Journal of Heat and Mass Transfer.* 2003;46(21):4059–70.
113. Assistant GR. *Parametric Investigation into the Effects of Pressure, Subcooling Surface Augmentation and Choice of Coolant on Pool Boiling in the Design of Cooling Systems for High-Power-Density Electronic Chips [Internet].* 1990. Available from: http://www.asme.org/about-asme/terms-of-use.
114. Anderson TM, Mudawar I. microelectronic cooling by enhanced pool boiling of a dielectric fluorocarbon liquid. *Journal of Heat Transfer.* 1989 Aug 1;111(3):752–9.
115. Professor A, ASME DAM. I. Mudawar Microelectronic Cooling by Enhanced Pool Boiling of a Dielectric Fluorocarbon Liquid [Internet]. 1989. Available from: http://heattransfer.asmedigitalcollection.asme.org/
116. Singh SK, Sharma D. Review of pool and flow boiling heat transfer enhancement through surface modification, Vol. 181. In: *International Journal of Heat and Mass Transfer.* Elsevier Ltd; 2021.
117. Assistant MccR. *Characteristics of Nucleate Pool Boiling from Porous Metallic Coatings [Internet].* 1982. Available from: http://heattransfer.asmedigitalcollection.asme.org/.
118. Jun S, Kim J, Son D, Kim HY, You SM. enhancement of pool boiling heat transfer in water using sintered copper microporous coatings. *Nuclear Engineering and Technology.* 2016 Aug;48(4):932–40.
119. Austin AJ, Echeverria E, Wagle P, Mainali P, Meyers D, Gupta AK, et al. High-temperature atomic layer deposition of GaN on 1D nanostructures. *Nanomaterials.* 2020 Dec;10(12):2434.
120. Liu H, Zhai J, Jiang L. Wetting and anti-wetting on aligned carbon nanotube films. *Soft Matter.* 2006;2(10):811–21.
121. Ujereh S, Fisher T, Mudawar I. Effects of carbon nanotube arrays on nucleate pool boiling. *International Journal of Heat and Mass Transfer.* 2007 Sep;50(19–20):4023–38.
122. Jun S, Sinha-Ray S, Yarin AL. Pool boiling on nano-textured surfaces. *International Journal of Heat and Mass Transfer.* 2013;62(1):99–111.
123. Sahu RP, Sinha-Ray S, Sinha-Ray S, Yarin AL. Pool boiling of Novec 7300 and self-rewetting fluids on electrically-assisted supersonically solution-blown, copper-plated nanofibers. *International Journal of Heat and Mass Transfer.* 2016 Apr;95:83–93.
124. Nguyen DH, Kim KM, Shim GH, Kim JH, Lee CH, Lim ST, et al. Experimental study on the thermal-hydraulic performance of modified chevron plate heat exchanger by electrochemical etching method. *International Journal of Heat and Mass Transfer.* 2020 Jul;155:119857.
125. Wu W, Bostanci H, Chow LC, Hong Y, Su M, Kizito JP. Nucleate boiling heat transfer enhancement for water and FC-72 on titanium oxide and silicon oxide surfaces. *International Journal of Heat and Mass Transfer.* 2010 Apr;53(9–10):1773–7.
126. Kumar GU, Soni K, Suresh S, Ghosh K, Thansekhar MR, Babu PD. Modified surfaces using seamless graphene/carbon nanotubes based nanostructures for enhancing pool boiling heat transfer. *Experimental Thermal and Fluid Science.* 2018 Sep;96:493–506.
127. Lee CY, Bhuiya MMH, Kim KJ. Pool boiling heat transfer with nano-porous surface. *International Journal of Heat and Mass Transfer.* 2010 Sep;53(19–20):4274–9.

5 Advanced Methods for Enhancing Heat Transfer Using Nanofluids

Sunita Khod, Vaishali Sharma, and Mayank Goswami

5.1 INTRODUCTION TO NANOFLUIDS

Heat transfer refers to thermal energy transfer from one system to another due to temperature differences. The three heat transfer methods include conduction, convection, and radiation. The convective heat transfer, which refers to heat transfer through fluid motion, is important for several industries, including automotive, electronics, power generation, aerospace and aviation, chemical, pharmaceutical, and so on [1]. Since energy demand is rising, enhancing heat transfer has become essential to save energy. Heat transfer enhancement increases the efficiency and performance of the systems, provides safety and reliability, and lowers maintenance costs and environmental impact [2].

Heat transfer rates can be improved by using active and passive techniques. Active methods require external power in the form of electric and magnetic fields, whereas passive methods do not require any external power input. These include using porous media such as nanofluids, nanoparticles, and so on, and surface treatments using colloidal paints and surface coatings. Heat transfer enhancement using nanofluids improves the heat transfer rate by increasing the effective surface area, thermal conductivity, and heat transfer coefficient [3, 4].

Nanofluids are fluids containing nano-meter-sized particles known as nanoparticles. These are synthesized by suspending nanoparticles in the conventional base fluid such as water, ethylene glycol, or oil (as shown in Figure 5.1). The nanoparticles can be metallic (Cu, Ag, Au), metallic oxide (Al_2O_3, TiO_2, SiO_2), carbon-based (carbon nanotubes, CNTs, graphene, diamond), polymeric (nylon, polyvinyl chloride, polysulfide, etc.), and so on. These nanoparticles are synthesized by physical and chemical methods. Physical methods of nanoparticle preparation include evaporation, condensation, physical vapor deposition, electrospraying, laser pyrolysis, and laser-ablation. Sol-gel method, microemulsion, hydrothermal and polyol synthesis, chemical vapor synthesis, and the plasma-enhanced chemical vapor deposition technique are chemical methods [5, 6]. The dimension of nanoparticles ranges from 1–100 nm. Better long-term stability, little pressure drop, and higher thermal conductivity are benefits of nanofluids compared with micro-meter-sized particles. The utilization of nanofluids solves the problem of particle clogging, sedimentation, and erosion while they flow in a tube or micro-channel.

The nanofluid can be unitary or hybrid. A unitary nanofluid is when only a single type of nanoparticle is added to the base fluid. The hybrid nanofluid consists of more than one type of nanoparticle suspended in the base fluid. The nanoparticles for unitary nanofluid can be metallic (Ag, Cu), metal oxide (Al_2O_3, TiO_2), carbon-based nanoparticles (CNT, graphene), polymers, and so on. The nanoparticles for hybrid nanofluid consist of metallic and metal oxide combination (Cu/Al_2O_3 and Ag/TiO_2), metallic and carbon-based combination (Cu/CNT and Ag/graphene), and metal oxide and carbon-based combination (Al_2O_3/CNT and TiO_2/graphene). The unitary nanofluids are less

DOI: 10.1201/9781003509837-5

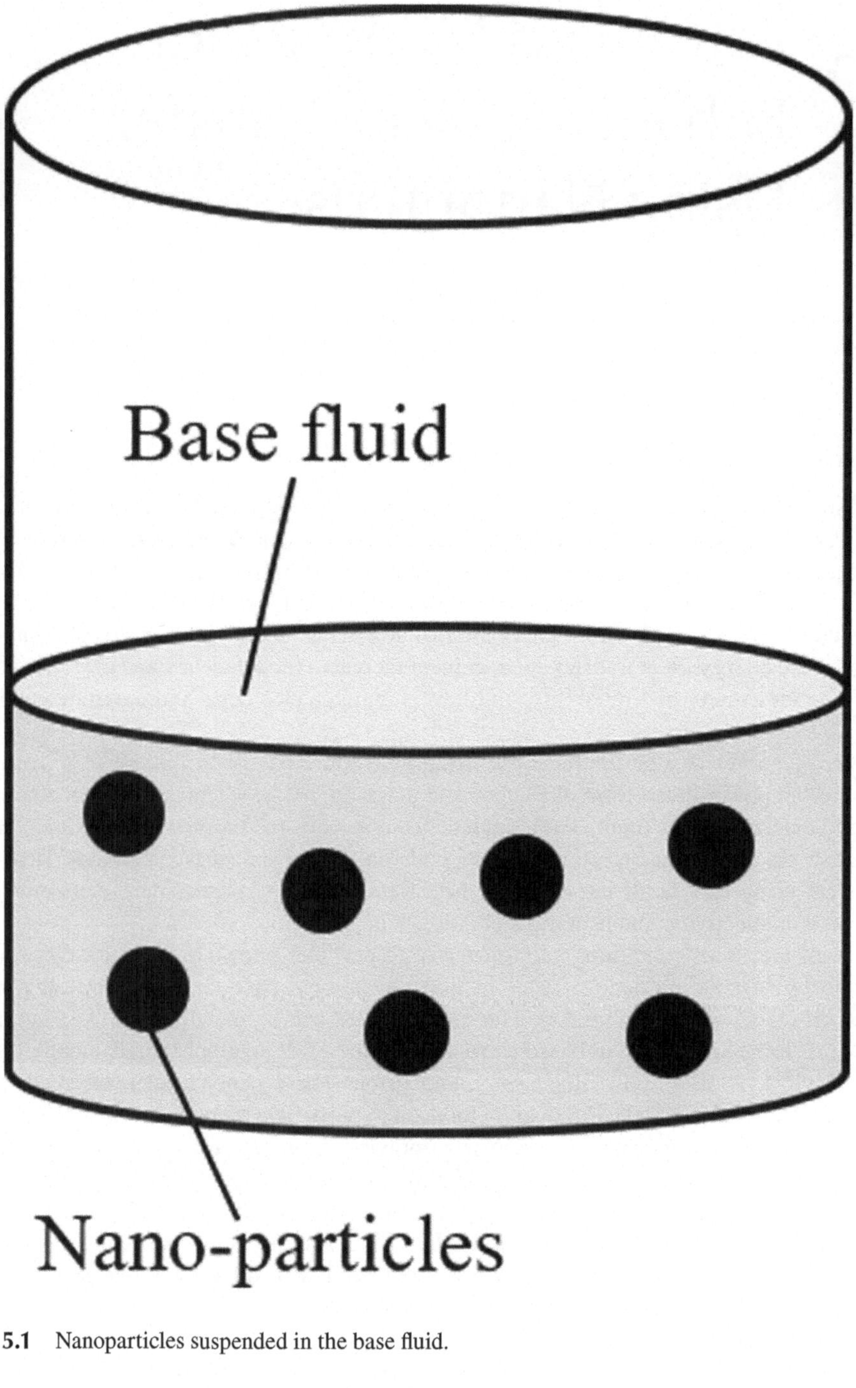

FIGURE 5.1 Nanoparticles suspended in the base fluid.

expensive and simpler to prepare, while hybrid nanofluids possess superior thermal properties and are costly.

5.1.1 Preparation of Nanofluids

The nano-sized particles should be well dispersed to produce stable nanofluids. The nanofluids are prepared using two methods: (1) the one-step method and (2) the two-step method.

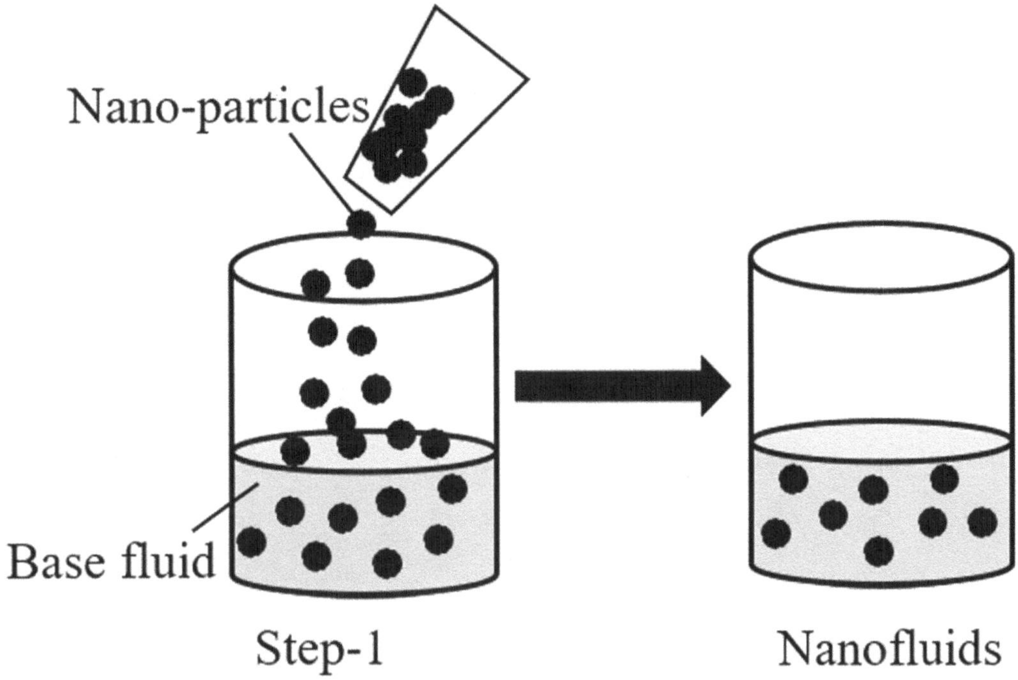

FIGURE 5.2 One-step method for nanofluid preparation.

5.1.1.1 One-Step Method

The nanofluids are prepared in one single step (Figure 5.2). The one-step method consists of directly making and dispersing the nanoparticles in the base fluids simultaneously. The intermediate steps, such as drying, storage, dispersion, and so on, are not essential. The nanofluids synthesized using a one-step method are more stable because of less nanoparticle aggregation in the base fluid. The control over the shape and size of the nanoparticle is the main advantage of this method. However, small-scale production and high cost are the limitations. The submerged arc nanoparticle synthesis and laser-ablation method are the commonly used one-step methods for nanofluid preparation [7, 8].

5.1.1.2 Two-Step Method

The nanofluids are prepared in two steps (as shown in Figure 5.3). First, the nanoparticles are prepared using physical and chemical techniques such as physical vapor deposition, chemical vapor deposition, pulsed laser deposition, sputtering, ion implantation, and so on. Second, the nanoparticles are dispersed in the base fluid using intensive magnetic force agitation, ultrasonic agitation, high-shear mixing, homogenizing, sol-gel vapor phase, or ball milling.

The instability of the nanofluid is the main limitation of this method. The nanoparticles are prone to aggregation due to high surface area and surface activity. Aggregation adversely affects nanofluid stability. Sedimentation and agglomeration of nanoparticles are the main problems. Sedimentation can be decreased by ultrasonication and stirring. Particle aggregations can be controlled using surfactants, including gum Arabic, SDBS, chitosan, polyvinylpyrrolidone, and cetyltrimethylammonium bromide. The large-scale and cost-effective production is the main advantage of this method. The method is suitable for preparing oxide-based nanofluids. Direct synthesis, sol-gel, and chemical precipitation are the commonly used two-step methods for nanofluid preparation [7, 8].

5.1.1.3 Mathematical Expressions

The mass of nanoparticles used for suspension is given by:

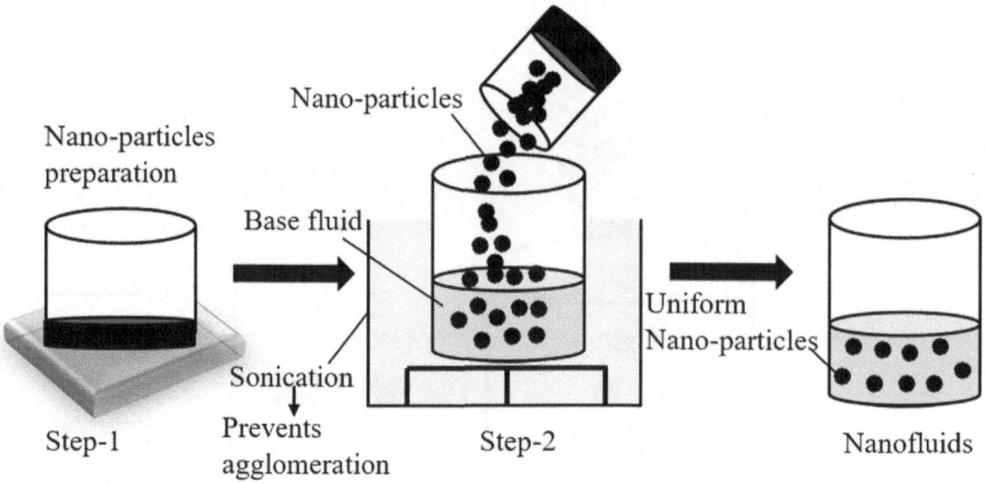

FIGURE 5.3 Two-step method for nanofluid preparation.

$$M_{np} = V_{total} \times C_{np} \times \rho_{np} \tag{5.1}$$

Where M_{np} is the mass of the nanoparticle used, V_{total} is the total volume of the fluid, C_{np} is the concentration of nanoparticles, and ρ_{np} is the density of the nanoparticle. M_{np} is measured in units of kg, V_{total} is in m^3, ρ_{np} is in kg/m^3, and C_{np} is unitless.

The nanoparticle concentration (C_{np}) is equivalent to the volume fraction (Φ_{np}), and the mathematical expression for the percentage volume fraction of nanoparticles is given by:

$$\Phi_{np} = \frac{V_{np}}{V_{np} + V_{bf}} \times 100 \tag{5.2}$$

Where Φ_{np} is the % volume fraction of the nanoparticles, V_{np} is the nanoparticle volume, and V_{bf} is the volume of the base fluid.

Nanoparticle concentration affects thermal conductivity, viscosity, specific heat capacity, and stability. An increase in nanoparticle concentration increases thermal conductivity and viscosity. The concentration should be sufficient to achieve the desired enhancement in thermal conductivity without causing adverse effects. Higher viscosity increases the requirement of pumping power. An excessively high concentration might lower the specific heat capacity and stability of the nanofluid. Therefore, nanoparticle concentration should be optimized to balance enhanced thermal conductivity and specific heat with manageable viscosity levels. The concentration should also ensure the stability of nanoparticles in the base fluid. Hence, the nanoparticle concentration should be optimized depending on the specific requirement.

The nanoparticle concentration for heat transfer applications varies depending on the system requirement, including thermal conductivity enhancement, viscosity constraints, stability, and overall system performance. The typical range of nanoparticle concentrations for heat exchanger applications generally falls between 0.01% and 10% by volume. A low concentration (0.01– 0.5%) is suitable for applications where minor enhancements in thermal conductivity are sufficient and where maintaining low viscosity is critical. A moderate concentration (0.5–2%) is used in heat exchanger applications where a balance between thermal enhancement and manageable viscosity is needed. A high concentration (2–10%) is required for high-performance cooling systems or advanced industrial heat exchanger applications.

5.1.2 Stability of Nanofluids

The performance of the nanofluid depends on its stability. Nanofluid stability should be investigated prior to the application for high heat transfer. Nanoparticle collision during the suspension causes agglomeration. The phenomenon of precipitation is an obstacle to nanofluid production. The clustering of particles leads to the clogging of the micro-channels. The operational conditions, such as temperature, concentration, shearing rate, and so on, also adversely affect nanofluid stability. Nanofluid instability for hybrid nanofluids is a big challenge due to the different types of nanoparticles, base fluids, and surfactants. Nanofluid instability influences thermo-mechanical performance and results in inefficient heat transfer. The production of stable nanofluids is essential for heat transfer applications. The use of surfactants and surface modifications improves nanofluid stability [9, 10]. This section briefly explains the methods to evaluate and enhance the stability of nanofluids.

5.1.2.1 Methods for Stability Evaluation of Nanofluids

Nanofluid stability can be evaluated by various methods presented in Figure 5.4. A short description of these methods is presented in the following subsections.

5.1.2.1.1 Zeta Potential Analysis

Zeta potential is an important measure for evaluating the stability of nanofluids. It represents the electric potential at the slipping plane, the interface between the dispersion medium and the stationary layer of fluid adhering to the dispersed nanoparticles. This potential arises due to the electric surface charges around the nanoparticles, creating an electrostatic field that influences the ions in the solution. The thermal motion of ions and the electrostatic field together screen the electric surface charge, balancing the net surface charge with the resulting electric charge in magnitude but with opposite polarity.

It is a measure of the electrostatic repulsion between particles in a colloidal system. The zeta potential is typically measured in millivolts (mV). A higher absolute value of zeta potential (either positive or negative) indicates greater electrostatic repulsion, which helps to keep the particles apart

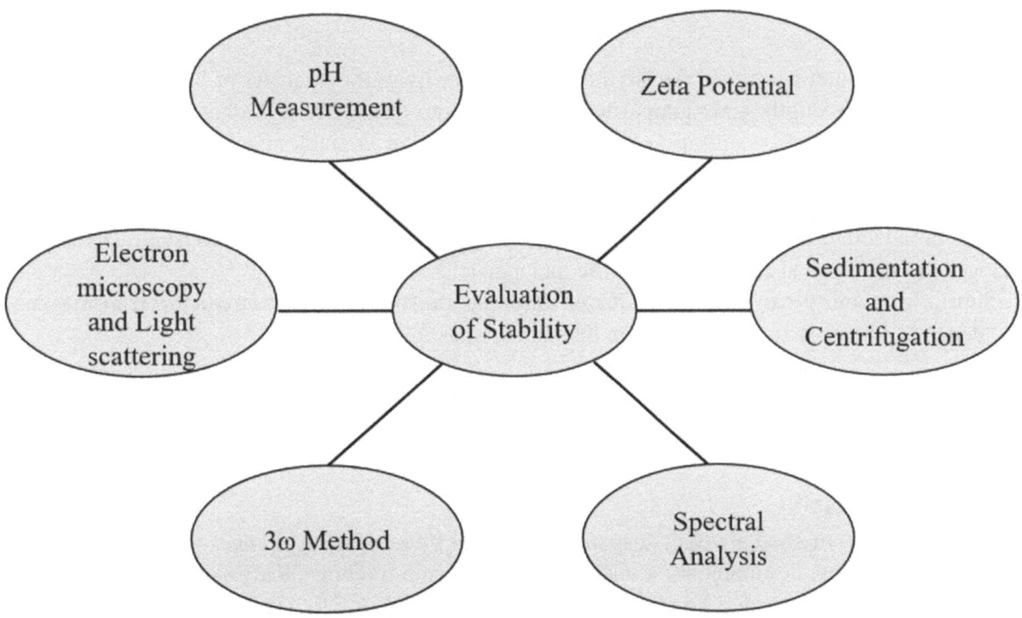

FIGURE 5.4 Methods for the evaluation of nanofluid stability.

and thus increases stability. Typically, a zeta potential above ±30 mV suggests good stability, while values closer to 0 mV suggest potential instability and a tendency for particles to aggregate [11, 12].

5.1.2.1.2 Sedimentation and Centrifugation Method

The sedimentation method, also referred to as the photographic technique, is a straightforward and direct way to assess the stability of nanofluids. This method involves observing the settling behavior of nanoparticles over time under the influence of gravity. A sample of the nanofluid is left undisturbed over time, and any sedimentation is observed and recorded, often visually or using imaging techniques. The rate at which nanoparticles settle can indicate the stability of the nanofluid.

Sedimentation is a time-consuming method. To expedite this process, the centrifugation method is used as an alternative. It accelerates the evaluation of colloidal stability by subjecting the nanofluid to high centrifugal forces, causing the particles to sediment based on their size and density. The amount of sediment formed and the rate of sedimentation can provide insights into the stability and tendency of particles to agglomerate. Little to no sediment formation after centrifugation indicates high stability. Significant sediment formation and a clear supernatant suggest that the nanofluid is prone to aggregation and settling [13].

5.1.2.1.3 Spectral Analysis

The spectral analysis method refers to the use of spectroscopy techniques to analyze the optical properties of nanofluids and assess their stability. The most commonly used spectral analysis methods include UV-visible spectroscopy (UV-Vis) and Fourier transform infrared spectroscopy (FTIR). These techniques provide information about the dispersion state of nanoparticles, the presence of agglomerates, and chemical interactions within the nanofluid.

UV-Vis spectrophotometer measurement is a reliable technique for spectral analysis that can quantitatively measure the concentration of nanofluids. This method is applicable for extinction bands ranging from 190 to 1100 nm wavelength. To evaluate nanofluid stability, the absorption peak characteristics are tested over time. The relationship between absorbance (A_λ) and concentration (c) is given as [14]:

$$A_\lambda = log_{10}\left(\frac{I_0}{I}\right) = \alpha \times l \times c \tag{5.3}$$

Where I_0, I, α, l, and c are the intensity of incident laser light, the intensity of laser beam, absorptivity, optical path length, and nanoparticle concentration, respectively. A decrease in absorbance signifies a reduction in concentration caused by sedimentation. A stable suspension typically shows a consistent absorbance spectrum over time, with sharp and well-defined peaks corresponding to the specific wavelengths absorbed by the nanoparticles. Changes in the absorbance spectrum, such as peak broadening, shifting, or a decrease in intensity, can indicate aggregation, sedimentation, or changes in the chemical environment of the nanoparticles.

Similarly, stability can also be evaluated through transmittance measurement. Transmittance (T_λ) depends on absorbance (A_λ), and can be expressed as follows [14]:

$$T_\lambda = \frac{I}{I_0} \tag{5.4}$$

5.1.2.1.4. 3ω Method

The 3-omega (3ω) method assesses suspension stability by measuring changes in thermal conductivity. This technique is suitable for a wider range of volume fractions. Karthikeyan et al. observed an 87.5% drop in the thermal conductivity of CuO nanofluid due to agglomeration and nanoparticle sedimentation. Particle clustering reduces both the effective thermal interaction area and the surface-to-volume ratio [15]. Conversely, Oh et al. [35] analyzed the increase in thermal conductivity in nanofluids based on ethylene glycol (EG) and water. This increase was attributed to moderate particle agglomeration. They discovered a peak value for particle agglomeration, beyond which thermal conductivity decreases as clustering increases [16]. However, the application of the 3ω method for evaluating colloidal stability is still quite limited.

5.1.2.1.5 Electron Microscopy and Light Scattering

Electron microscopy and light scattering techniques are useful for measuring particle size distributions and analyzing particle aggregation. In electron microscopy, two main methods are used to visualize high-resolution images of particle clusters in dried nanofluid samples: scanning electron microscopy (SEM) and transmission electron microscopy (TEM). Both SEM and TEM can detect and analyze the microstructure and distribution of nanoparticles. Uniform particle distribution indicates well-dispersed nanofluids, while clusters and aggregations in the test sample suggest unstable nanofluids.

5.1.2.1.6 pH Measurement

pH measurement is an effective and simple technique for assessing the stability of nanofluids, as the pH can significantly influence the surface charge and electrostatic interactions between nanoparticles, thereby affecting their dispersion and stability. The stability of a nanofluid is often enhanced at a certain pH level where the electrostatic repulsion between particles is maximized, preventing aggregation and sedimentation. Qi et al. employed transmittance to study the relationship between pH values and nanofluid stability. They analyzed four samples of Al_2O_3-water and TiO_2-water at pH levels of 7, 8, 9, and 10. Their findings indicated that nanofluids showed better stability at a pH of 8, as this was the condition under which the samples exhibited the lowest transmittance over time [17].

5.1.2.2 Methods for Stability Enhancement of Nanofluids

The enhancement of nanofluid stability is crucial to prevent the aggregation and sedimentation of nanoparticles, which can negatively impact their thermal properties and overall performance in applications. To enhance nanofluid stability, both physical and chemical methods are employed. Physical methods typically involve applying high energy to disintegrate nanoparticle clusters. In contrast, chemical methods stabilize nanoparticles through various approaches such as pH adjustments, surface modifications, and other additive techniques. The various methods for stability enhancement of nanofluids are shown in Figure 5.5. The following subsections provide the details of each of the methods.

5.1.2.2.1 Physical Methods

5.1.2.2.1.1 Mechanical Stirring

Mechanical stirring is a straightforward and effective method for enhancing the stability of nanofluids by improving the dispersion of nanoparticles within the base fluid. It involves the use of a stirring device or mixer to induce fluid motion to mix nanoparticles uniformly within the base fluid. The agitation provided by mechanical stirring helps to overcome the natural tendency of nanoparticles to agglomerate due to van der Waals forces, ensuring a stable suspension. Mechanical stirring ensures that nanoparticles are evenly distributed throughout the nanofluid, improving consistency in thermal properties and performance. It is a simple and economic method to enhance nanofluid stability.

5.1.2.2.1.2 Ultrasonic Vibration

Ultrasonic vibration is a more effective mechanical method to enhance the stability and uniform dispersion of nanoparticles in nanofluids. Ultrasonic vibration involves the application of high-frequency sound waves (20 kHz to several MHz) to a liquid medium. The ultrasonic waves generate high-intensity acoustic cavitation, which creates bubbles in the nanofluid. These bubbles rapidly

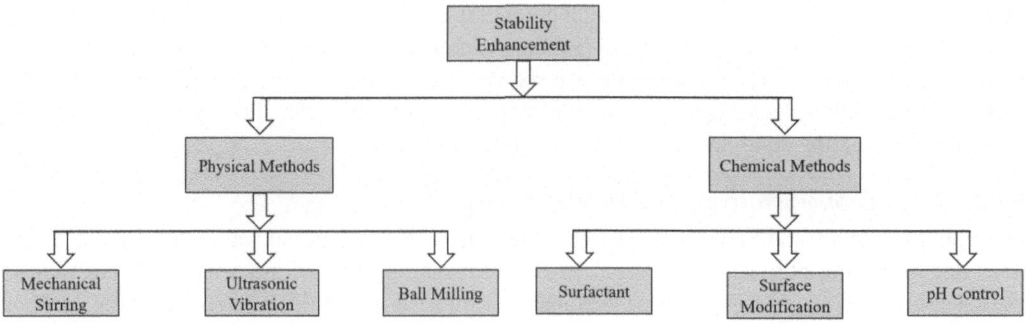

FIGURE 5.5 Methods for stability enhancement of nanofluids.

collapse, producing intense localized shock waves and shear forces that help to break apart agglomerates of nanoparticles and disperse them more uniformly. It provides a high level of control over the dispersion process and can be adapted for various types of nanofluids.

5.1.2.2.1.3 Ball Milling

Ball milling is a mechanical technique used to reduce the particle size and enhance the dispersion of nanoparticles in a nanofluid. It involves the use of a rotating cylinder filled with grinding media (balls) to mill or grind materials into finer particles. As the cylinder rotates, the balls collide with the nanoparticles and the container walls, applying mechanical forces that break down larger nanoparticles into smaller ones and enhance their dispersion. Ball milling generates significant heat, which can affect temperature-sensitive nanoparticles or base fluids. Consequently, this technique is infrequently reported and is often used in conjunction with other stabilization methods to achieve well-dispersed nanofluids.

5.1.2.2.2 Chemical Methods

Chemical methods for enhancing stability involve adding surfactants, modifying surfaces, and adjusting pH values. These techniques are effective in improving the stability of nanofluids. Detailed descriptions of these various methods are provided in the following subsections.

5.1.2.2.2.1 Surfactant

Surfactants play a crucial role in enhancing the stability of nanofluids by preventing the aggregation and sedimentation of nanoparticles. Surfactants are chemical agents that adhere to nanoparticle surfaces to reduce the surface tension of the liquid. By doing so, surfactants help to slow down the aggregation of nanoparticles. Various types of surfactants are used in nanofluids, such as SDS, oleic acid, and PVP. The primary purpose of using surfactants is to alter the properties of the nanofluid, thereby enhancing the dispersion of nanoparticles [18, 19]. Both electrostatic repulsive forces and hydrophobic surface forces are involved in breaking down nanoparticle agglomerations.

Adding surfactants is a practical approach to optimizing nanoparticle dispersion, but it does come with certain limitations. For one, surfactants are prone to forming foams at elevated temperatures. Additionally, the attachment of surfactant molecules can increase thermal resistance between the base fluid and the nanoparticles, which may negatively affect thermal conductivity.

5.1.2.2.2.2. Surface Modification

Surface modification is a highly effective technique that does not rely on surfactants and involves functionalizing the surfaces of nanoparticles. This method holds promise for achieving long-term stability of nanofluids when functionalized nanoparticles are introduced into base liquids. Yang et al. developed a nanofluid by applying silane modifications to the surface of SiO_2 nanoparticles within a solution. Their research revealed that, following a pool boiling process, no deposition layer formed on the heated surface [14]. Similarly, Tang et al. utilized poly-methacrylic acid (PMAA) to modify the surface of ZnO nanoparticles. PMAA interacts with hydroxyl groups on the ZnO surface, leading to the formation of zinc methacrylate, which enhances the stability of ZnO nanofluids [20]. This technique requires meticulous control over reaction conditions to achieve optimal results.

5.1.2.2.2.3. pH Control

pH control directly impacts the electrokinetic properties of nanofluids, which are crucial indicators of nanofluid stability. High surface charge density resulting from appropriate pH levels stabilizes well-dispersed nanosuspensions through repulsive forces. Li et al. identified that water-based copper nanofluid exhibits optimal stability at a pH of 9.5. However, their study also indicated that maintaining the pH of stable nanofluids around neutral (pH 7) is preferable. This is because acidic or alkaline solutions can potentially damage heat transfer devices, especially at elevated temperatures. Consequently, nanofluids demonstrating the best stability at neutral pH levels are considered promising candidates for industrial applications [21].

5.2 PHYSICAL AND THERMAL PROPERTIES OF NANOFLUIDS

The efficiency of the thermal system for heat transfer is determined by the thermophysical properties of the working fluid. Adding the nanoparticles to the base fluid enhances the thermophysical properties that affect the convective heat transfer. The various thermophysical properties of the nanofluid that influence heat transfer include thermal conductivity, viscosity, specific heat capacity, and density as shown in Figure 5.6. Surface tension, thermal diffusivity, electrical conductivity, and rheological properties are other properties of nanofluids. Purity level, concentration, shape, and size of the nanoparticles alter the nanofluid's thermophysical properties.

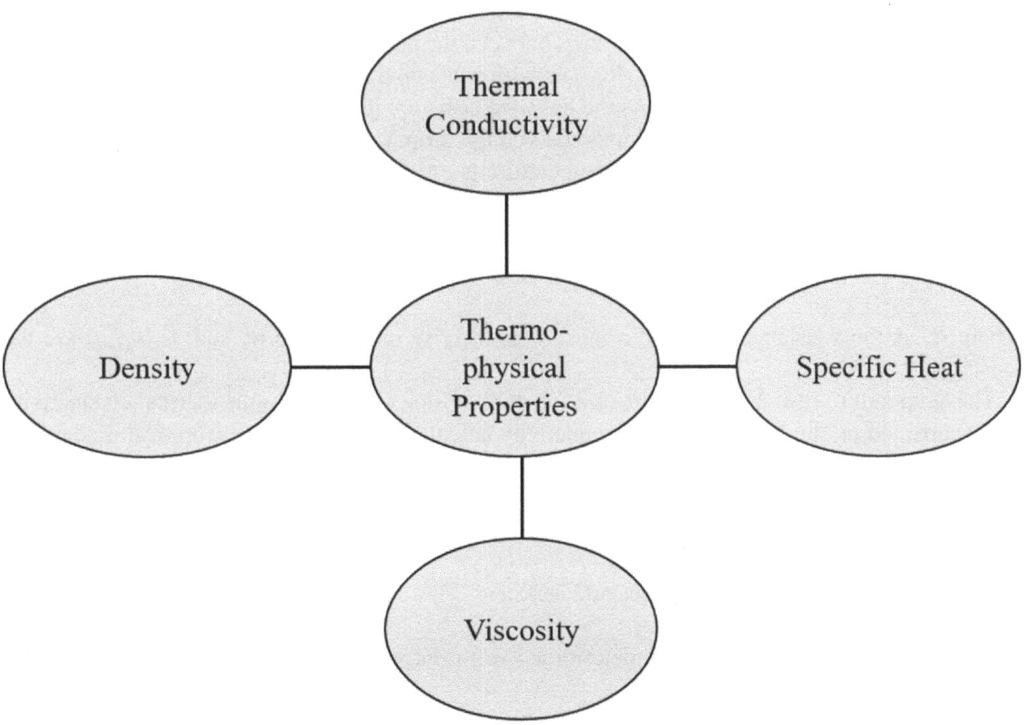

FIGURE 5.6 Thermophysical properties of nanofluids.

The nanofluid's higher thermal conductivity and specific heat increase the convective heat transfer coefficients and improve energy storage and heat transfer capability. The increased viscosity of the nanofluid can enhance the thermal boundary layer but leads to a lower Reynolds number, higher pressure, and increased pumping power requirements. Higher nanofluid density can improve natural convection, increase volumetric heat capacity, and potentially improve forced convection by raising the Reynolds number. However, it increases the pressure drop and pumping power requirement as well. The thermophysical property of the nanofluid has a significant impact on the enhancement of heat transfer; therefore, optimizing the thermophysical properties is crucial.

5.2.1 Thermal Conductivity

The thermal conductivity of the nanofluid is the measure of the heat energy transport by conduction. The thermal conductivity of the nanofluid is higher than that of the base fluid. This is because of the Brownian motion of the nanoparticles and the presence of an interfacial layer between the nanoparticle and bulk liquid. The different techniques used for its measurement include the transient hot-wired method, the 3ω method, the thermal constant analyzer technique, and the cylindrical cell method. This section discusses the various techniques for thermal conductivity measurement and the measured thermal conductivity for different nanofluids reported in the literature.

5.2.1.1 Measurement Techniques

5.2.1.1.1 Transient Hot-Wired Method

This method is widely used to measure the thermal conductivity of a fluid and is based on the transient heat conduction principle. The setup of the transient hot-wired method is shown in Figure 5.7. A long and thin wire is immersed along the axis of the cylindrical vessel containing nanofluid, with one end connected to a power source and the other to the set of resistances (R_1, R_2, and R_3).

The electrical circuit forms a Wheatstone bridge between the three known resistances and the wire. The electrical circuit provides a constant current through the wire by adjusting the resistance of potentiometer R_3 such that no current flows between the points 1 and 2. At the balance condition, the resistance of the wire is $R_w = R_1R_3/R_2$. The wire heats up due to Joule's law of heating when an electric current is passed through it. As the hot wire heats up, its electrical resistance changes with temperature. The electric circuit measures the voltage drop V across the wire, which is related to the wire's resistance R_w ($V = IR_w$). The temperature is estimated by using the quadratic resistance between resistance and temperature as given by:

$$R_w = a_0 + a_1T + a_2T^2 \tag{5.5a}$$

Where R_w is the resistance of the thin wire, T is the rise in temperature, and a_0, a_1, a_2 are the constants.

The temperature rise, ΔT of the wire as a function of time t is related to the thermal conductivity of the surrounding fluid. The thermal conductivity calculation requires the theoretical model, and is calculated by assuming [22]:

a) The wire should be infinitely long, have infinite thermal conductivity, and have zero specific heat.
b) The fluid should be homogeneous and isotropic.

The thermal conductivity, K can be determined from the slope of the temperature vs. time plot using the relationship given by [8]:

$$\Delta T(t) = \frac{q}{4\pi K}\ln(t) + C \tag{5.5b}$$

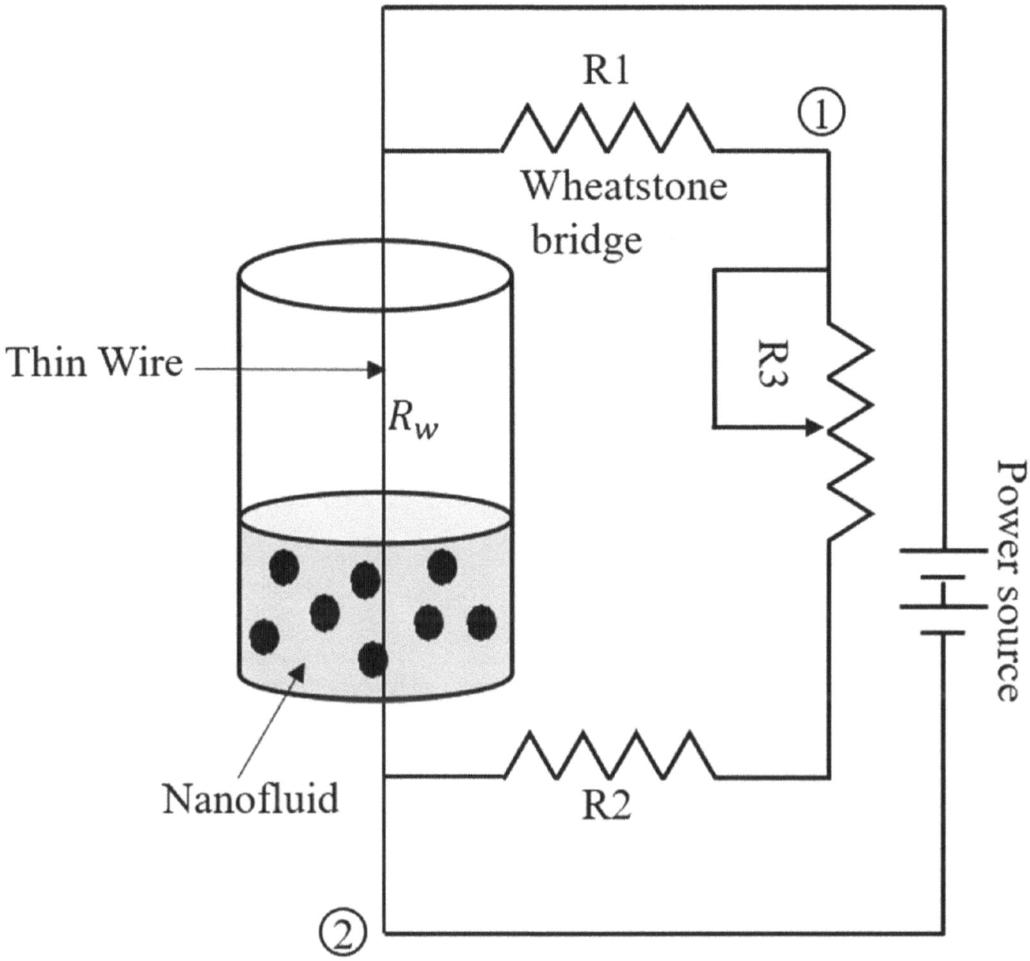

FIGURE 5.7 Schematic of the transient hot-wired method.

Where $\Delta T(t)$ is the temperature rise of the wire as a function of time t, q is the heat generated per unit length, K is the thermal conductivity of nanofluid, and C is the constant that is equivalent to the intercept of $\Delta T(t)$ vs. t plot. The quantity $\Delta T(t)$ is measured in units of kelvin (K), q in watts per meter (W/m), K in watts per meter kelvin (W/m.K), t in seconds (s), and C is a constant with units of K.

The thermal conductivity of nanofluid, K is given by [8]:

$$K = \frac{q}{4\pi} \frac{\Delta(\ln t}{\Delta T}) \tag{5.6}$$

Where K is the thermal conductivity (W/m.K), q is the heat per unit length (W/m), T is the temperature (K), and t is the time (s).

5.2.1.1.2 3ω Method

The 3ω method is an advanced method to measure the lower thermal conductivity of the materials. A thin metal wire in the form of a heater is deposited around the material (as shown in Figure 5.8). A sinusoidal AC of angular frequency ω is passed through the heater and causes the periodic

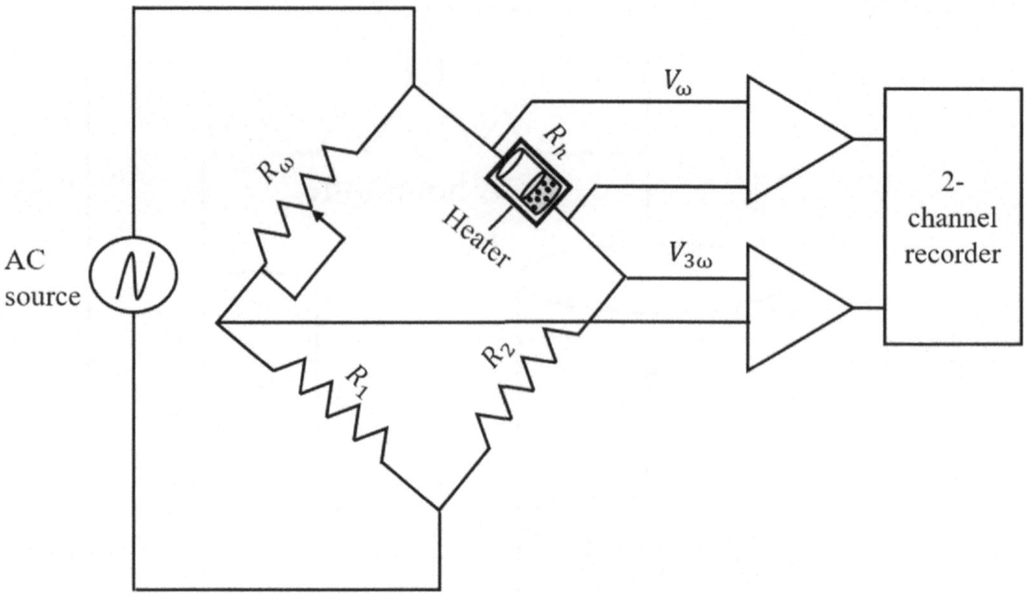

FIGURE 5.8 Schematic of the 3ω method.

heating of the wire. The twice amplification of input frequency leads the metal temperature to oscillate at twice the angular frequency (2ω). The periodic heating causes temperature oscillation in the material. The resistance of the metal heater is a linear function of temperature, so the temperature oscillation can be calculated indirectly by measuring the 3ω voltage of the metallic heater. The thermal conductivity of the nanofluid can be measured by the amplitude and phase of the temperature oscillation. The relation between the thermal conductivity of the nanofluid, temperature oscillation, and heat rate is given by [8]:

$$\Delta T = \frac{P}{l\pi K}\int_0^{\infty} \frac{sin^2 kb}{(kb)^2 (k^2 + q^2)^{1/2}} dk \tag{5.7}$$

Where ΔT is the temperature rise of the metal wire, P is the input power, K is the thermal conductivity of nanofluid, l is the length of the hot wire, b is the radius of the wire, k represents wave number, and q is a constant. The units of ΔT is kelvin (K), P is watts (W), K is watts per meter kelvin (W/m.K), l is meters (m), b is meters (m), k is meter inverse (m^{-1}), and q is a constant with units of meter inverse (m^{-1}).

5.2.1.1.3 Thermal Constant Analyzer Technique

The thermal constant analyzer technique is similar to the transient hot wire method and is based on the Fourier heat conduction law. The technique measures the thermal conductivity, diffusivity, and specific heat of nanofluids. The technique is fast, no sample preparation is required, and a wide range of nanofluid thermal conductivity (0.02 to 200 W/m.K) can be measured.

The experimental setup comprises a thermometer, a vessel, a constant temperature bath, and a constant thermal analyzer as shown in Figure 5.9. The nanofluid is filled in the container with a thermal constant analyzer probe that is vertically submerged. The container is placed in a constant temperature bath. The temperature of the nanofluid is measured by dipping the thermometer into the container. Nanofluid thermal conductivity is calculated by measuring probe resistance [23]. The correlation used to calculate thermal conductivity is given by [8]:

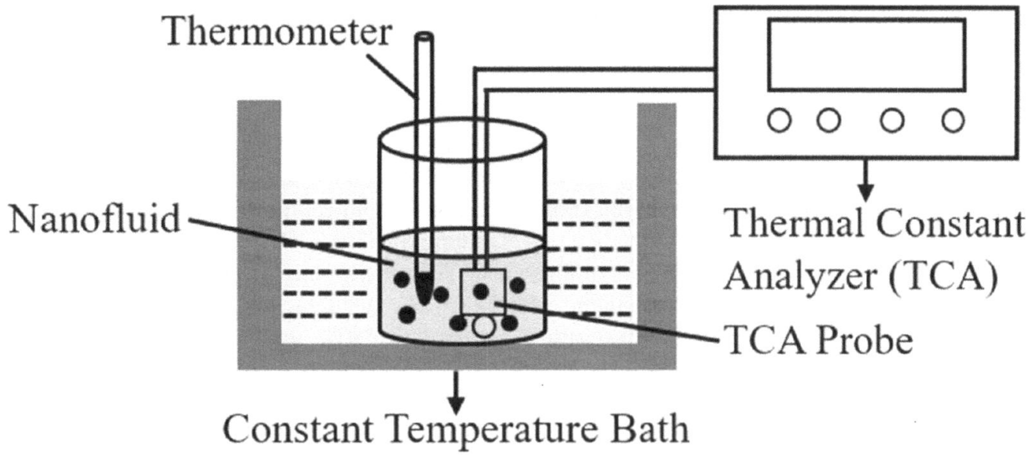

FIGURE 5.9 Schematic of the thermal constant analyzer technique.

$$\Delta T(\tau) = \frac{1}{\alpha}\left[\frac{R_P(\tau)}{R_0} - 1\right] \tag{5.8}$$

Where α is the temperature coefficient of the electric resistance, $R_P(\tau)$ is the probe resistance at time τ, R_0 is the electric resistance at $\tau=0$, and τ is the electrification time.

$$\tau = \sqrt{\frac{tR}{r_p^2}} \tag{5.9}$$

Where t is the measuring time, R is the nanofluid thermal diffusivity, and r_p is the probe radius [8].

5.2.1.2 Factors Affecting Thermal Conductivity

The thermal conductivity of nanofluids is affected by various factors, which include nanoparticle size, shape, material, base fluid material, temperature, concentration, and so on (shown in Figure 5.10). The effect of different parameters on nanofluid thermal conductivity is described in the following sections [24].

5.2.1.2.1 Temperature

The temperature of the base fluid and nanoparticles affects the thermal conductivity of the nanofluid, and it is found to increase with increasing temperature. The increased Brownian motion at higher temperatures is responsible for this. The higher temperature reduces the thermal resistance at the interface between nanoparticles and the base fluid, facilitating better heat transfer. The agglomeration of nanoparticles also reduces at higher temperatures, leading to better dispersion of

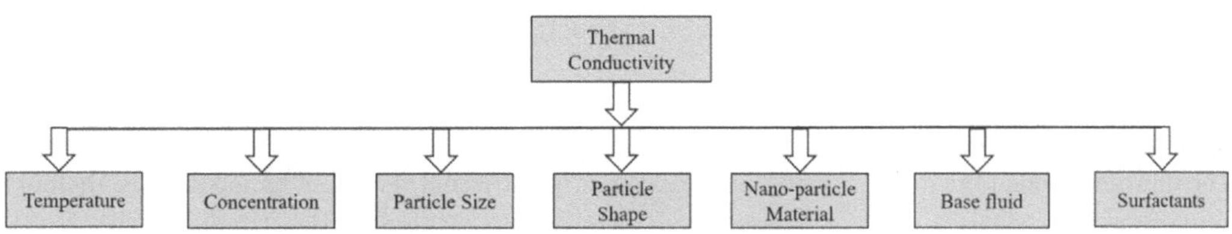

FIGURE 5.10 Factors affecting the thermal conductivity of nanofluids.

nanoparticles and improving thermal conductivity. Yu et al. examined the thermal conductivity of nanofluid containing ZnO nanoparticles [25]. The results show that higher temperature enhances thermal conductivity.

Mintsa et al. reported 30% enhancement of thermal conductivity at 51°C compared with thermal conductivity at 21°C for 0.4% of CuO-water nanofluid [26]. The 1% of SiO_2 nanoparticles dispersed in ethylene glycol shows a 15% increase in thermal conductivity at 60°C compared with 25°C [27]. Kole and Dey et al. investigated the thermal conductivity enhancement for CuO-gear oil nanofluids. An enhancement of 11.9% is reported with a 0.025 volume fraction of CuO nanoparticles at a temperature of 80°C [28]. The nanofluid containing 3% of TiO_2 in ethylene glycol shows 25% thermal conductivity enhancement at 65°C compared with 25°C [29]. Quantitatively, thermal conductivity increases between 10% and 30% for temperatures greater than 25°C.

5.2.1.2.2 Concentration

The higher nanoparticle concentration augments the thermal conductivity of the nanofluids. This is because of the increases in the intensity of the Brownian motion at higher concentrations, the availability of a larger surface area for efficient heat transfer, and aggregation of nanoparticles, which create pathways for higher thermal conductivity.

Moldoveanu et al. reported the enhanced thermal conductivity of Al_2O_3-SiO_2/water and Al_2O_3-TiO_2/water nanofluids for increasing total particle concentration [30]. The Al_2O_3-SiO_2/water hybrid nanofluid is more sensitive to increasing concentration. At higher constant concentrations, the thermal conductivity curve becomes steeper. The thermophysical properties of ZnO-MgO nanoparticles suspended in water-ethylene glycol base fluid are also reported for 0.025–0.1% nanoparticles concentration and 30–50°C temperature [31]. An enhancement of thermal conductivity up to 17.5% is observed due to reduced clustering and increased Brownian motion of the nanoparticles. Ilyas et al. investigated the f-ZnO/oil nanofluid for increasing concentration, and nonlinear behavior is observed due to the inhomogeneous distribution of the nanoparticles [32].

5.2.1.2.3 Particle Size

The size of the nanoparticles significantly influences the nanofluid's thermal conductivity. Nanoparticles in the size range of 5–100 nm can be fabricated. Various experimental studies by Paul et al. [33], Teng et al. [34], and Das et al. [35] regarding the effect of nanoparticle size on thermal conductivity have been reported. The results show that the decreased particle size enhances thermal conductivity. This is because the surface area-to-volume ratio is higher for smaller-sized nanoparticles. The larger surface area provides more interface for heat transfer between the nanoparticles and the base fluid, facilitating more efficient energy exchange. The smaller-sized nanoparticles exhibit higher Brownian motion due to smaller size and mass, therefore, the increased micro-convection increases overall thermal conductivity. Smaller nanoparticles are less likely to agglomerate as compared with larger particles and so remain well dispersed within the fluid. The heat transfer path length is reduced due to the small size of the particles, allowing quicker thermal energy transfer and augmenting thermal conductivity [36, 37].

Quantitatively, smaller nanoparticles of sizes less than 10 nm can lead to thermal conductivity enhancements up to 40–50%. The moderate-sized nanoparticles (10–20 nm) increase the thermal conductivity up to 20–30% and the larger-sized nanoparticles (20–50 nm) generally show 10–20% improvement in thermal conductivity. Hence, the increase in thermal conductivity is more pronounced for nanoparticles of size less than 10 nm [38–40].

5.2.1.2.4 Particle Shape

The thermal conductivity of the nanofluid is different for different shapes of nanoparticles. Usually, cylindrical and spherical-shaped nanoparticles are used to synthesize the nanofluids. Xie et al. investigated spherical and cylindrical-shaped nanoparticles for SiC-water nanofluids. The cylindrical-shaped colloidal dispersion shows higher thermal conductivity enhancement [41]. The decrease in the sphericity increases the thermal conductivity [42]. This is because cylindrical nanoparticles have a higher aspect ratio (length to diameter) as compared with spherical nanoparticles, and therefore can create percolation networks at lower volume fractions, providing continuous pathways for heat conduction. These pathways facilitate more efficient heat transfer through the fluid. The larger surface area per unit volume of cylindrical particles enhances interaction with the base fluid, improving heat transfer. Cylindrical nanoparticles can align themselves along the direction of the heat flux under certain conditions and create continuous thermal pathways, enhancing the effective thermal conductivity of the nanofluid. The shape and contact area of cylindrical nanoparticles with the base fluid can reduce the interfacial thermal resistance compared with spherical particles and increase the nanofluid's thermal conductivity [38, 39].

Quantitatively, Choi et al. reported an enhancement of thermal conductivity up to 150% for a nanofluid containing 1% volume fraction of MWCNTs with an aspect ratio of ~1000 in water [39]. Xie et al. found that with a 0.6% volume fraction of MWCNTs in water, the thermal conductivity increased by 39% for an aspect ratio of 1000 [43]. A nanofluid with 4% volume fraction of spherical Al_2O_3 in water showed a thermal conductivity enhancement of 10–20% for aspect ratio around ~1–3 [38]. Li and Xuan reported a 40% higher thermal conductivity for a 0.1% volume fraction of Cu nanoparticles in water (aspect ratio ~1) [40]. Nanofluids with a 3% volume fraction of SiO_2 nanoparticles in water showed an enhancement of about 12% (aspect ratio ~1–5) [44].

5.2.1.2.5 Particle Material and Base Fluid

The nanofluid is prepared by dispersing nanoparticles in the base fluid. The thermal conductivity of the nanofluid depends on the material of the nanoparticles and base fluids used to prepare the nanofluid. Nanoparticles with higher intrinsic thermal conductivity will generally enhance the thermal conductivity of the nanofluid more effectively. Metallic nanoparticles such as Ag and Cu enhance the thermal conductivity of the nanofluid more effectively compared with non-metallic nanoparticles such as Al_2O_3 and SiO_2. This is because metallic nanoparticles possess higher thermal conductivity than non-metallic nanoparticles. Quantitatively, 40% and 70% thermal conductivity enhancement of 0.1% and 1% volume concentrations of Cu nanoparticles in water is reported [40]. With the 5% volume concentration of Al_2O_3 and SiO_2 in water, an enhancement of 30% and 20% for thermal conductivity is obtained [38, 44].

The base fluid's properties determine how well it can complement the thermal conductivity of the nanoparticles. The compatibility of the base fluid with nanoparticles affects interfacial thermal resistance. The high thermal conductivity base fluids such as water (thermal conductivity ~0.6 W/m.K) show more pronounced enhancements in the thermal conductivity of nanofluids as compared with ethylene glycol (thermal conductivity ~0.25 W/m.K) based nanofluids. For example, adding 1% Cu nanoparticles to water can increase thermal conductivity by 25–40%, while adding 2% Al_2O_3 to ethylene glycol can increase thermal conductivity by about 20%.

5.2.1.2.6 Surfactants

Nanoparticle suspension causes agglomeration and decreases the thermal conductivity of the nanofluids. The use of surfactants prevents clustering and enhances thermal conductivity. Eastman et al. estimated the thermal conductivity of Cu-ethylene glycol nanofluid with and without the addition of surfactants [45]. The results show that thermal conductivity improved significantly by adding surfactants. Kole and Dey studied Cu-gear oil nanofluids at 0.11–2% volume concentration and an oleic acid surfactant. An enhancement of 24% for 2% volume concentration at room temperature is observed. Li et al. reported the thermal conductivity of Cu-H_2O nanofluids using sodium dodecyl benzene sulfonate as a surfactant [46].

5.2.2 Specific Heat Capacity

The specific heat capacity is one of the important thermophysical properties of nanofluids affecting the heat transfer rate. The specific heat capacity of the nanofluid is defined as the amount of heat required to increase the temperature of one gram of nanofluid by one degree centigrade. The specific heat of the nanofluid can increase or decrease depending on the nanoparticle and base fluid material. It is measured using the differential scanning calorimeter (DSC) method [47]. This section introduces the specific heat measurement method and the measured specific heat of various nanofluids.

5.2.2.1 Measurement of Specific Heat

The DSC technique is used to measure the specific heat of nanofluids. The rise in the temperature of the sample material is measured with respect to the reference material (water) due to heating the material (as shown in Figure 5.11). The measurements reported in the literature generally employ the DSC technique due to the high accuracy of the method.

It is also measured by the different equations given in the literature. The first model was proposed by Pak et al. [48] and is given by:

$$C_{p,nf} = \Phi C_{p,np} + (1-\Phi) C_{p,bf} \tag{5.10}$$

Where $C_{p,nf}$ is the specific heat of the nanofluid, Φ is the volume fraction of the nanoparticle, and $C_{p,np}$ and $C_{p,bf}$ is the specific heat of the nanoparticle and base fluid.

The second model was given by Xuan et al. [49] based on nanoparticle and base fluid densities. Mathematically, it is given by:

$$C_{p,nf} = \frac{\Phi C_{p,np}\rho_{np} + (1-\Phi) C_{p,bf}\rho_{bf}}{\Phi \rho_{np} + (1-\Phi)\rho_{bf}} \tag{5.11}$$

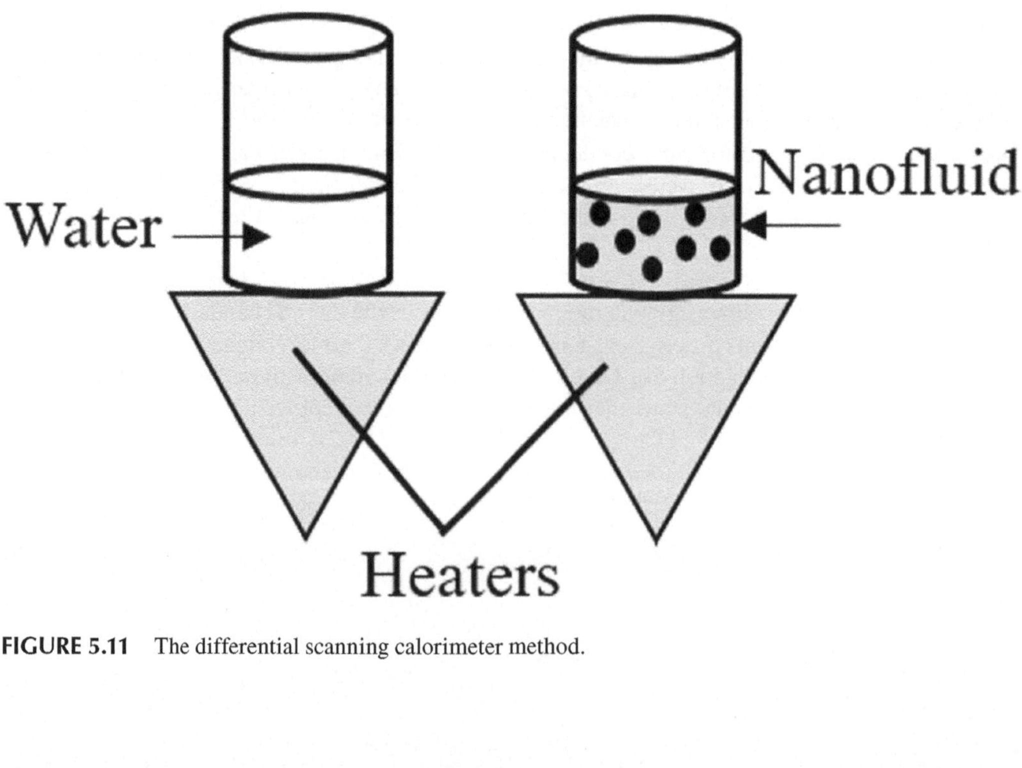

FIGURE 5.11 The differential scanning calorimeter method.

Where $C_{p,nf}$, $C_{p,np}$, and $C_{p,bf}$ are the specific heat of the nanofluid, nanoparticle, and base fluid. Φ is the volume fraction of the nanoparticle, ρ_{np} is the nanoparticle density, and ρ_{bf} is the density of the base fluid.

These equations are used to estimate the specific heat of unitary nanofluids. Sundar et al. [50] proposed a new method to estimate the specific heat capacity of hybrid nanofluids by determining the average heat capacity of nanoparticles. The expression is given by:

$$C_{p,hnp} = \frac{w_1\rho_{np,1} + w_2\rho_{np,2}}{w_1 + w_2} \tag{5.12}$$

Where $C_{p,hnp}$ is the average specific heat capacity of the hybrid nanoparticles. $\rho_{np,1}$ and $\rho_{np,2}$ is the density of nanoparticle 1 and nanoparticle 2, w_1 and w_2 are the weight percentages of nanoparticle 1 and nanoparticle 2. The average value of specific heat capacity, $C_{p,hnp}$ should be used in the model given by Pak et al. and Xuan et al.

5.2.2.2 Factors Affecting the Specific Heat of Nanofluids

The specific heat of nanofluids is affected by several factors, including temperature, concentration, nanoparticle material, and base fluid material. The nanoparticle size and chirality do not affect the specific heat of the nanofluids.

Yang et al. investigated specific heat for super carbon nanotubes (ST) [28]. The analysis shows that the specific heat of ST is independent of the length, diameter, and chirality at a particular temperature. Roberties et al. [29] reported the specific heat determination for Cu-ethylene glycol nanofluids using the DSC method. The results show that the addition of copper nanoparticles reduces the specific heat of the nanofluids in the whole temperature range.

The experimental study for the specific heat of CuO-base oil nanofluids is reported by Saeedinia et al. [30] for weight fractions of 0.2–2%. The specific heat of the nanofluid is lower than the base fluid. The value decreased with the increased nanofluid concentration. A decrease of 23% from the base fluid is observed at a 2% weight fraction and 40°C temperature. Tiznobaik et al. [51] examined SiO_2-salt eutectic nanofluids with four different-sized (5, 10, 30, and 60 nm) nanoparticles at high temperatures. An enhancement of 25% in the specific heat value of the nanofluid is observed over the base fluid, regardless of nanoparticle size.

Wole-Osho et al. [52] reported the specific heat of Al_2O_3-ZnO/H_2O hybrid nanofluids for a 25–65°C temperature range, and volume concentration varies from 0.33–1.67%. The analysis shows that the specific heat of nanofluids increases with the temperature, and an equal concentration of Al_2O_3 and ZnO nanoparticles gives the highest specific heat. Moldoveanu et al. [53] developed an experimental study to investigate the specific heat of unitary and hybrid nanofluids with Al_2O_3, SiO_2, and TiO_2 nanoparticles. The volume concentration varies between 0.01% and 0.03% in room temperature conditions. The highest specific heat is estimated for Al_2O_3-SiO_2/H_2O hybrid nanofluids.

Çolak et al. [54] conducted an experimental study to determine the specific heat of Al_2O_3/H_2O, Cu/H_2O, and Al_2O_3-Cu/H_2O nanofluids. The results show that the specific heat increases with the temperature while decreasing with the nanoparticle volume concentration. Also, the lowest specific heat is estimated for the Cu/H_2O nanofluid. Satti et al. [55] reported the heat capacity of Al_2O_3, ZnO, CuO, TiO_2, and SiO_2 nanoparticles dispersed in propylene glycol and H_2O. The variation of specific heat is observed for the 30–90°C temperature range and 0.5–6% volume concentration. The specific heat value is lowest for the highest volume concentration, irrespective of the nanoparticle size.

Similarly, in the experimental study reported by Sekhar et al. for Al_2O_3-H_2O nanofluids, the decrease in specific heat with the rise in volume concentration is observed because of the thermal diffusivity improvement [56]. Popa et al. [57] investigated the specific heat of CuO and Al_2O_3 nanoparticles in ethylene glycol and water with temperature and volume concentration variations. The ethylene glycol-based nanofluid has lower specific heat than water-based nanofluid.

The various experimental studies investigating the effect of various parameters on the specific heat of nanofluids can be summarized as follows:

- The specific heat of a nanofluid increases with the rise in temperature.
- The increase in volume concentration lowers the specific heat of nanofluids.
- The nanoparticle size and chirality do not affect the nanofluid's specific heat capacity.
- The specific heat of a nanofluid depends on the nanoparticle and the base fluid's material. So, the specific heat of the nanofluid can be higher or lower than the base fluid.
- Hybrid nanofluids have higher specific heat than unitary nanofluids.

5.2.3 Viscosity

The viscosity of a nanofluid is an important thermophysical property that considerably affects heat transfer performance, pressure drop, and flow characteristics, and determines the rheological properties. It is a measure of resistance offered by the fluid particles to deformation by shear. The viscosity of the nanofluids depends on the temperature, volume concentration, particle diameter, base fluid, and nanoparticle loading. This section briefly discusses viscosity measurement methods and the various parameters that affect the nanofluid's viscosity.

5.2.3.1 Measurement of Viscosity

The viscosity of the nanofluids is a crucial transport property. It is categorized as kinematic and dynamic viscosity. Nanofluids can show Newtonian or non-Newtonian behavior. A nanofluid behaves as a Newtonian fluid when the shear stress is proportional to the deformation rate, whereas, in the non-Newtonian fluid, viscosity traces the change of shear rate. For example, nanofluids with low and high viscosity fluids, such as water and ethylene glycol, show Newtonian behavior at lower nanoparticle concentrations, whereas for higher nanoparticle concentrations, water-based nanofluids show non-Newtonian behavior. The viscosity of the nanofluids is measured by using viscometers (as shown in Figure 5.12). The different types of viscometers include:

- Capillary viscometer
- Rotational viscometer
- Oscillating viscometer
- Brookfield viscometer

The viscosity of nanofluids is estimated by using various model equations. Einstein's equation for dilute nanoparticle suspension is given by [58]:

$$\mu_{nf} = \mu_{bf}\left(1 + 2.5\Phi\right) \tag{5.13}$$

Where μ_{nf} is the viscosity of the nanofluid, μ_{bf} is the viscosity of the base fluid, and Φ is the volume fraction of nanoparticles.

At higher nanoparticle concentrations, the particles interact, so the viscosity increases. Batchelor modified Einstein's model for higher concentrations and is given by:

$$\mu_{nf} = \mu_{bf}\left(1 + 2.5\Phi + 6.2\Phi^2\right) \tag{5.14}$$

Brinkman's model equation for Newtonian behavior of nanofluids is given by [58]:

$$\mu_{nf} = \frac{\mu_{bf}}{\left(1 - \Phi\right)^{2.5}} \tag{5.15}$$

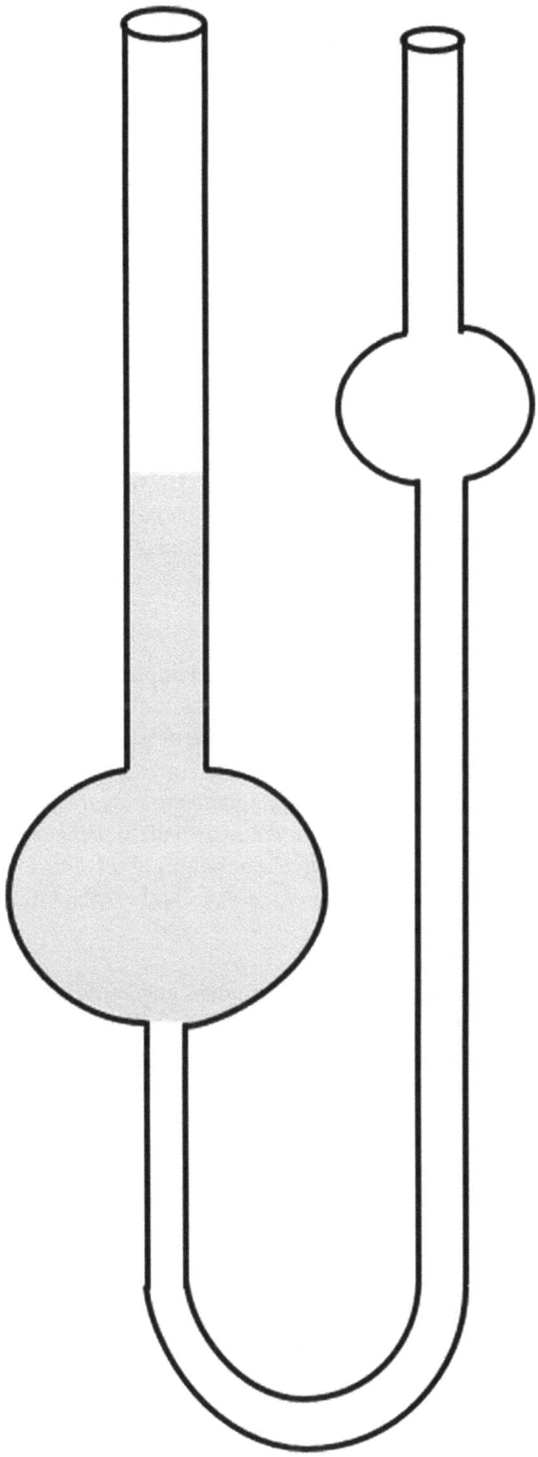

FIGURE 5.12 Capillary viscometer.

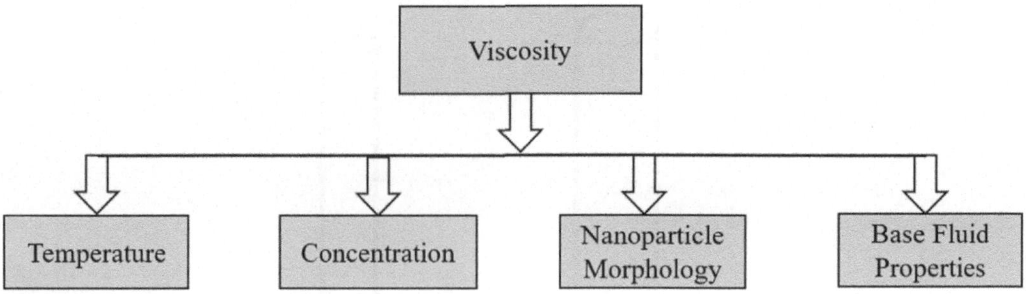

FIGURE 5.13 Factors affecting the viscosity of nanofluids.

5.2.3.2 Factors Affecting the Viscosity of Nanofluids

The various proposed viscosity models show that the viscosity of the nanofluid is a function of base fluid viscosity and nanoparticle volume fractions. However, the different parameters, such as temperature, nanoparticle concentration, particle morphology, base fluid material, particle-fluid interaction, shear rate, and so on, affect the viscosity of nanofluids (as shown in Figure 5.13). The effect of various parameters is described in the following sections.

5.2.3.2.1 Temperature

The viscosity of nanofluids decreases with the rise in temperature because Brownian motion is enhanced and intermolecular adhesion weakens. Nguyen et al. [59] investigated the dynamic viscosity of Al_2O_3-H_2O nanofluid with 36 nm and 47 nm particle sizes. The viscosity of the nanofluid decreases with a rise in temperature. Mena et al. [60] also measured the viscosity in the temperature range of 5–20°C. Suganthi et al. [61] studied ZnO nanoparticles dispersed in H_2O, and viscosity decreases with the temperature at 35–55°C. Manikandan et al. [62] experimentally reported the viscosity of sand-propylene glycol nanofluids with temperatures of 29–140°C and nanoparticle loading of 0–2% volume concentration. The viscosity of the nanofluid is found to be 46% lower at 29°C.

5.2.3.2.2 Nanoparticle Concentration

The viscosity of the nanofluid increases as the volume concentration of nanoparticles increases. Hung et al. [63] examined the viscosity of CNTs suspended in water. The concentration of CNTs affected the viscosity. Rashin et al. [64] found that the viscosity of CuO-coconut oil nanofluid increased with varying nanoparticle concentrations when manufactured in two steps. Namburu et al. [65] observed that viscosity increased with volume concentration. Saeedinia et al. [66] studied the viscosity of stable CuO-base oil nanofluids with particle weight fractions ranging from 0.2% to 2% at various temperatures. These nanofluids exhibit Newtonian behavior at all particle loadings and temperatures.

5.2.3.2.3 Nanoparticle Morphology

Nanoparticle morphology, which includes shape and size, influences the viscosity. The smaller-sized nanoparticles increase the viscosity more significantly than larger-sized nanoparticles. This is because smaller-sized particles have a larger surface area-to-volume ratio. Non-spherical nanoparticles have a higher viscosity due to the high aspect ratio compared with spherical-shaped nanoparticles. The viscosity of SiO_2-H_2O nanofluid at 0.5% volume concentration and three different particle sizes (15, 30, and 50 nm) is reported [63]. Viscosity drops with an increase in nanoparticle size. In an experimental study of Al_2O_3-H_2O nanofluid, a 250 nm-sized nanoparticle has a higher viscosity than a 200 nm-sized particle. Chevalier et al. reported increased viscosity for SiO_2-water and SiO_2-ethylene glycol nanofluids with 35, 94, and 190 nm-sized nanoparticles [64].

5.2.3.2.4 *Base Fluid Properties*

Viscosity varies depending on the basic nanofluid. The intrinsic viscosity of the base fluid determines the viscosity of nanofluids. A greater viscosity base fluid will produce higher viscosity nanofluids. Nanofluids based on ethylene glycol (EG) will have a higher viscosity than water-based ones. Water-based nanofluids display Newtonian behavior, which may be approximated using the well-known Einstein viscosity equation. An experimental investigation examined the rheological effects of Al_2O_3, TiO_2, and SiO_2 nanoparticles on water as a base fluid. The viscosity of all nanofluids decreases as temperature increases, with SiO_2 having the highest viscosity and Al_2O_3 having the lowest [67].

5.2.4 Density

The density of nanofluids influences the Reynolds number, pressure drop, stability, friction factor, Nusselt number, and heat transfer rate. It is defined as the mass per unit volume. The density of nanofluids is higher than the base fluids. Determining the nanofluid's density is essential for determining the heat transfer performance.

5.2.4.1 Measurement of Density

The density of the nanofluids is measured by using the density meter. The different density meter types include float, static, vibrating, and so on. The density of nanofluid is calculated by using the following correlation:

$$\rho_{nf} = \frac{m_{np} + m_{bf}}{V_{np} + V_{bf} + V_v} \tag{5.16}$$

Where ρ_{nf} is the density of the nanofluid, m_{np} is the mass of the nanoparticles, m_{bf} is the mass of the base fluid, V_{np} is the volume of the nanoparticles, V_{bf} is the volume of the base fluid, and V_v is the volume of the voids.

The density of the nanofluids can also be calculated using the equations proposed in the literature. The equation proposed by Pak et al. [68] for a nanofluid's density calculation is given by:

$$\rho_{nf} = \Phi\rho_{np} + (1-\Phi)\rho_{bf} \tag{5.17}$$

Where ρ_{nf} is the density of the nanofluid, ρ_{np} is the density of the nanoparticles, ρ_{bf} is the density of the base fluid, and Φ is the nanoparticle volume fraction.

The following equation can be used to calculate the density of the hybrid nanofluids:

$$\rho_{hnf} = \Phi_1\rho_{np,1} + \Phi_2\rho_{np,2} + (1-(\Phi_1+\Phi_2))\rho_{bf} \tag{5.18}$$

Where ρ_{hnf}, $\rho_{np,1}$, $\rho_{np,2}$, and ρ_{bf} are the density of the hybrid nanofluid, nanoparticle 1, nanoparticle 2, and the density of the base fluid. Φ_1 and Φ_2 are the particle fraction of nanoparticle 1 and nanoparticle 2.

5.2.4.2 Factors Affecting Nanofluid Density

The factors that affect the nanofluid's density include temperature, the density of the nanoparticle and base fluid, and nanoparticle concentration. The density of the nanofluid increases with the nanoparticle concentration because of the higher density of nanoparticles while decreasing with the rise in temperature due to increased molecular motion. Vidhya et al. [69] investigated the ZnO-MgO hybrid nanofluid using an H_2O-ethylene glycol mixture of 40:60% as a base fluid. The temperature varies from 30°C to 80°C for a volume concentration range of 0.0125–0.1%. The highest density of the nanofluid is calculated for the lowest temperature and highest volume concentration. The experimental study conducted by Yarmand et al. [70] to investigate the thermophysical properties of graphene nanoplatelets (GNP)-Ag/H_2O hybrid nanofluids also shows the negative trend of density with temperature and a positive trend with concentration of nanoparticles.

Wanatasanappen et al. [71] reported the thermophysical properties of Al_2O_3-CuO/H_2O-ethylene glycol hybrid nanofluid at 1% nanoparticle concentration and 30°C temperature. An enhancement of 4.8% in the nanofluid density is calculated compared with the base fluid. Hozien et al. [72] also reported heat transfer analysis for TiO_2/H_2O, ZnO/H_2O, and Ag/H_2O nanofluids at 0.25% volume concentration and 30–60°C temperature range. The Ag/H_2O nanofluid exhibits the highest density (24%) compared with water. Asokan et al. [73] investigated the thermophysical properties of two unitary nanofluids (Al_2O_3/DI H_2O-EG and CuO/DI H_2O-EG) and one hybrid nanofluid (Al_2O_3-CuO/DI H_2O-EG) with nanoparticle concentration between 0.02% and 0.06% at 30°C. The unitary nanofluid (Al_2O_3/DI H_2O-EG) possesses a slightly higher density than the hybrid nanofluid (Al_2O_3-CuO/ DI H_2O-EG).

The effect of various parameters on the density of the nanofluids can be summarized as follows:

- The nanofluid's density falls as the temperature rises.
- The density of the nanofluid increases with increasing nanoparticle concentration.
- The density of the unitary nanofluid is slightly higher than hybrid nanofluids.
- The density is different for different nanoparticles and base fluid materials.

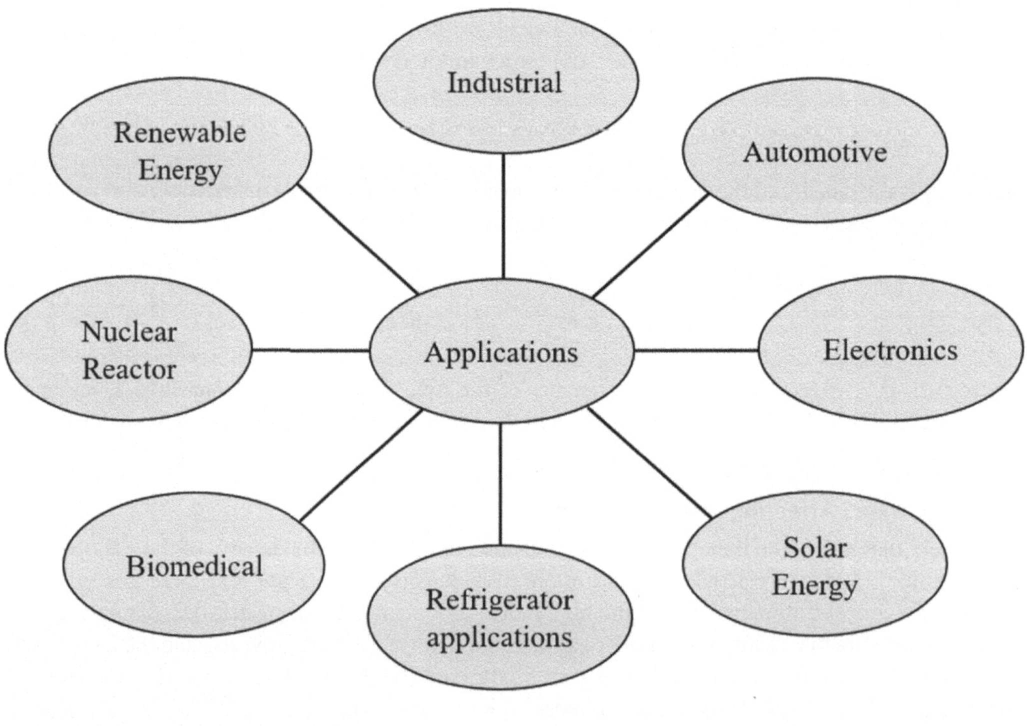

FIGURE 5.14 Applications of nanofluids.

5.3 APPLICATIONS IN HEAT EXCHANGERS

5.3.1 Applications of Nanofluids

Industries are expanding, and energy demand is continuously rising. Sustainable development requires efficient heat transfer; therefore, compact heat exchangers are needed for heat transfer applications. Nanofluids with augmented thermal conductivity and specific heat capacity are optimistic solutions for efficient heat transfer applications (as shown in Figure 5.14). The various heat exchanger applications using nanofluids are given in the following sections [74, 75].

5.3.1.1 Industrial Heat Exchangers

Nanofluids are used in the heating and cooling systems of power plants, such as nuclear and thermal power plants. The heat transfer rate increases, and therefore, energy consumption is reduced. Using nanofluids in heat exchangers for petrochemical plants enhances heat transfer during chemical reactions and results in better control over the process.

5.3.1.2 Automotive Applications

Nanofluids can be used in the radiator and heat engine for automotive applications. The use of nanofluids improves heat dissipation and increases cooling efficiency, preventing the radiator and engine from overheating. Enhanced cooling allows engines to operate at optimal temperatures, reducing wear and emissions. This improvement in thermal management translates to more consistent and reliable engine operation, extending the lifespan of the engine components.

5.3.1.3 Electronics Cooling

In high-performance electronics, efficient thermal management is critical to ensure reliable operation and longevity of electronic components. Nanofluids can significantly enhance the performance of cooling systems, such as micro-channel heat sinks, by improving heat transfer rates. This enhancement lowers thermal resistance, which is crucial for preventing overheating and thermal damage.

5.3.1.4 Solar Energy Collectors

Using nanofluids in solar energy collectors increases the absorption of solar energy and improves efficiency. Carbon nanotubes are used in solar water heating systems to increase solar energy conversion efficiency.

5.3.1.5 Refrigeration Systems

The use of nanofluids as refrigerants improves the heat transfer performance of refrigeration systems. For example, using nanofluids with TiO_2 or Al_2O_3 nanoparticles improves the thermal conductivity of refrigerants used in air conditioners and refrigerators.

5.3.1.6 Biomedical Applications

Nanofluids are used in various biomedical applications. Magnetic nanofluids with nanoparticles like Fe_2O_3 are used in hyperthermia devices to transfer precise heat to cancerous tissues and targets.

5.3.1.7 Nuclear Reactor

Nanofluids containing Cu or Al_2O_3 nanoparticles are used as coolants in nuclear reactors to improve heat transfer and reactor efficiency.

5.3.1.8 Renewable Energy Systems

In geothermal systems, improved heat exchange efficiency facilitated by nanofluids can make these systems more practical and economically feasible. The use of nanofluids in renewable energy systems underscores their potential to contribute to sustainable energy solutions by optimizing thermal management and improving overall system efficiency.

5.3.2 Nanofluids for Heat Exchangers

Unlike conventional fluids, heat exchangers are employed with nanofluids to meet rising energy demands. Heat exchanger devices with nanofluids perform better. The process of cooling in heat exchangers is described by Newton's law given by:

$$Q = h \times A \times \Delta T \tag{5.19}$$

Where Q is the heat transfer rate, h is the convective heat transfer coefficient, A is the surface area through which heat is transferred, and ΔT is the temperature difference between the source and nanofluid.

The effective temperature difference, ΔT, is calculated by using log mean temperature difference (LMTD) given by:

$$\Delta T_{lm} = \frac{\Delta T_1 - \Delta T_2}{\ln\left(\frac{\Delta T_1}{\Delta T_2}\right)} \tag{5.20}$$

Where ΔT_{lm} is the LMTD, ΔT_1 is the temperature difference between the flowing fluids at one end, and ΔT_2 is the temperature difference between the flowing fluids at the other end.

The convective heat transfer coefficient, h is given by:

$$h = \frac{Nu \times k}{L} \tag{5.21}$$

Where Nu is the Nusselt number, k is the thermal conductivity of the used fluid, and L is the dimension of the heat exchanger device. The Nusselt number is determined using the empirical relations between the Reynolds number, Re, and the Prandtl number, Pr. The Nusselt number for turbulent flow inside a circular pipe is given by [74]:

$$Nu = 0.021 \times Re^{0.8} \times Pr^{0.5} \tag{5.22}$$

The Reynolds number, Re, [76] and the Prandtl number, Pr [77] can be calculated by:

$$Re = \frac{\rho \times v \times D}{\mu} \tag{5.23}$$

$$Pr = \frac{C_p \times \mu}{k} \tag{5.24}$$

Where ρ is the density of the flowing fluid, v is the velocity, D is the pipe diameter, μ is the dynamic viscosity, C_p is the specific heat capacity, and k is the thermal conductivity of the flowing fluid.

The heat transfer rate (Q) can be enhanced by improving the convective heat transfer coefficient (h), which can be increased by improving the Nusselt number (Nu), enhancing the thermal conductivity (k) of the fluid, and miniaturization of heat exchanger dimension (L). The current section introduces the enhancement of heat transfer by augmenting the thermal conductivity (k) of the flowing fluids using Nanofluids. The applications of nanofluids for different types of heat exchangers are explained in the following sections.

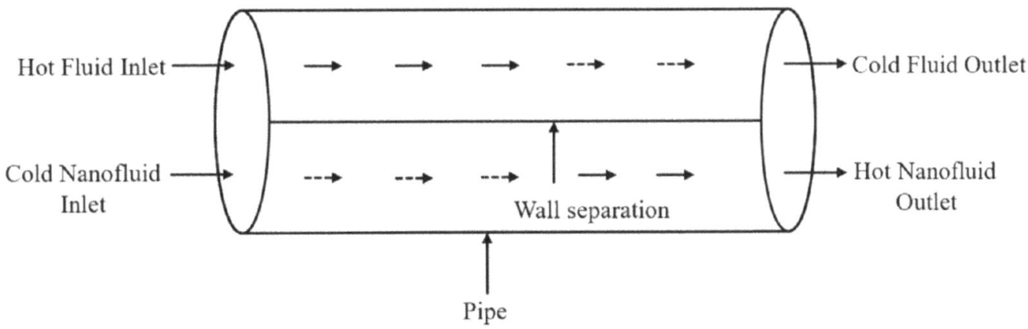

FIGURE 5.15 Single-tube heat exchanger.

5.3.2.1 Single-Tube Heat Exchangers

A single-tube heat exchanger (HX) consists of one tube to transfer heat between two fluids. That is, the hot and cold fluids flow in the same tube as shown in Figure 5.15. It is a simple and effective device for heat transfer applications. Depending on the relative direction of the fluid flow, a single-tube HX operates in parallel, counter, or crossflow configurations.

The Nusselt number (Nu) and convective heat transfer coefficient (h) for single-tube HX is given by Equations 5.21 and 5.22. Wen et al. [78] investigated the thermal performance of ZnO-based water nanofluid through single-tube HX for several Re numbers. The Nusselt number Nu is increased up to 10.5% for 1.5% nanoparticle concentration compared with the base fluid. Amrollahi et al. [79] reported the thermal performance of MWCNTs-water nanofluid in a tube of 11.42 mm diameter for Re range 1592–4778. The heat transfer performance is improved by up to 25% compared with water. Pak et al. reported the turbulent flow of γ-Al_2O_3 and TiO_2 nanoparticles suspended in water through a single pipe. The Nusselt number for the fluid increases with volume concentration and Reynolds number. A new correlation between the Nusselt number, u , Reynolds number, Re (104–105), and Prandtl number, Pr (6.54–12.33) is proposed as given in Equation 5.22.

Hussein et al. [80] studied the thermal performance of MWCNTs-graphene/deionized water hybrid nanofluid, and the heat transfer coefficient increased by 43.4% with an increase in pressure drop by 11% for single-tube HX. Ding et al. also reported an augmented thermal performance of CNT nanofluid for laminar flow through a 4.5 mm diameter tube. Quantitatively, the Nusselt

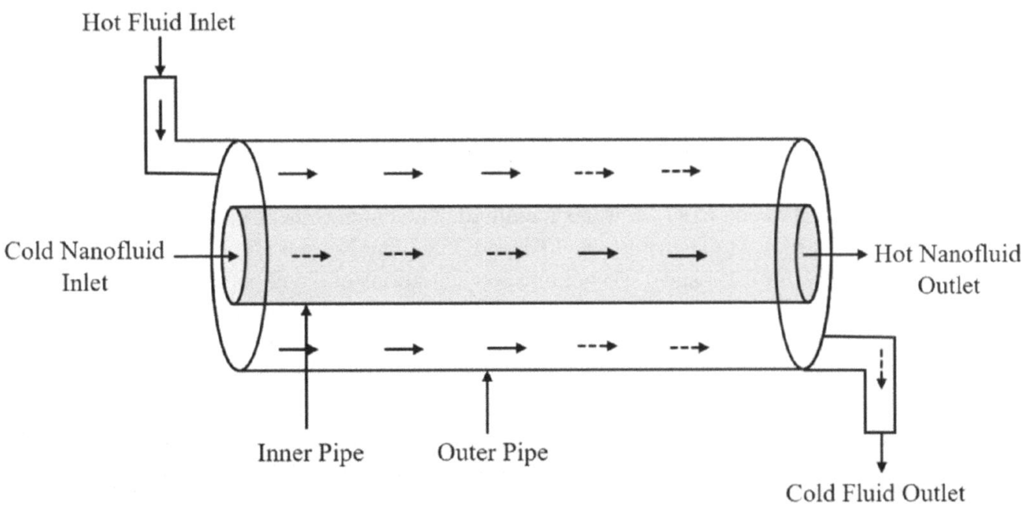

FIGURE 5.16 Schematic of the double-pipe heat exchanger.

number for single-tube HX increases by 5–15%, 15–30%, and 30–60% for 1%, 1–5%, and >5% volume concentration of nanoparticles. The corresponding improvement in heat transfer coefficient by using nanofluids is 10–50% for single-tube HX.

5.3.2.2 Double-Pipe Heat Exchangers

A double-pipe HX consists of two pipes placed concentric to each other. One fluid flows through the inner pipe while the other flows through the outer pipe as shown in Figure 5.16. It can operate in parallel or counterflow configurations. It is simple to design. However, the concentric arrangement can lead to a larger pressure drop.

The heat transfer coefficient (h) and Nusselt number (Nu) expression for heat transfer in a double tube is

$$h = \frac{k}{D} \times 1.7Re^{0.4} \tag{5.25}$$

HX using nanofluids is reported and given by expressions in Equations 4.25 and 4.26.

Where h is the convective heat transfer coefficient, k is the thermal conductivity, D is the diameter of the pipe, and Re is the Reynolds number.

$$Nu = 0.074Re^{0.707}Pr^{0.385}\Phi^{0.074} \tag{5.26}$$

Where Φ is the nanoparticle volume fraction and Pr is the Prandtl number.

Chun et al. [81] experimentally reported the convective heat transfer of nanofluids with three different alumina nanoparticles. The fluid flow is laminar inside the double-pipe HX system. An increase in heat transfer coefficient is observed. The increase in heat transfer coefficient may be due to high nanoparticle concentration. Duangthongsuk et al. [82] investigated the turbulent flow of 0.2% TiO_2 dispersed in H_2O through a double-tube counter-flow HX. The tube is 1.5 m long, with the inner and outer diameters of tubes one and two (8.13 mm, 9.53 mm) and (27.8 mm, 33.9 mm), respectively. The heat transfer coefficient with nanofluid flow increases by 6–11% compared with the base fluid.

Duangthongsuk et al. [83] also reported the turbulent flow of TiO_2 dispersed in H_2O through a double-tube counter-flow HX for 0.2–2% volume concentration. The correlation given by Pak et al. predicts well for 0.2% volume concentration but fails to predict 0.6–1% volume concentration. So, a new correlation for Nusselt number of TiO_2-water nanofluid with Reynolds number in the range 3000–18,000 and 0–1.0% volume concentration is developed as given in Equation 5.24. Asirvatham et al. [84] studied the laminar, transition, and turbulent flow of Ag-water nanofluid for 0.3–0.9% volume concentration of nanoparticles. The heat transfer coefficient increases by 69.3% for 0.9% silver nanoparticle concentration.

Quantitatively, the Nusselt number and heat transfer coefficient for double-tube HX increase by 25% and 20% for 0.5% of Al_2O_3 in water nanofluid. The enhancement in the value ofNu and$h$ for 2.0% concentration of Al_2O_3 in water is 40% and 35%. The Nusselt number increases by 15% and 35%, respectively, for 0.1% and 1.0% volume concentrations of Cu nanoparticles in water. The heat transfer coefficient is enhanced by 12% and 30% for nanofluids containing 0.1% and 1.0% Cu nanoparticles in water.

5.3.2.3 Plate Heat Exchangers

A plate HX utilizes a series of thin plates corrugated with gaskets fixed between them. The fluid flows through the channels formed by the plates to transfer heat from hot to cold fluid as shown in Figure 5.17. The feature design of a plate HX causes a good turbulent flow of fluids even at low flow rates. It can operate in parallel, counter, and crossflow configurations. The plate-type HX possesses

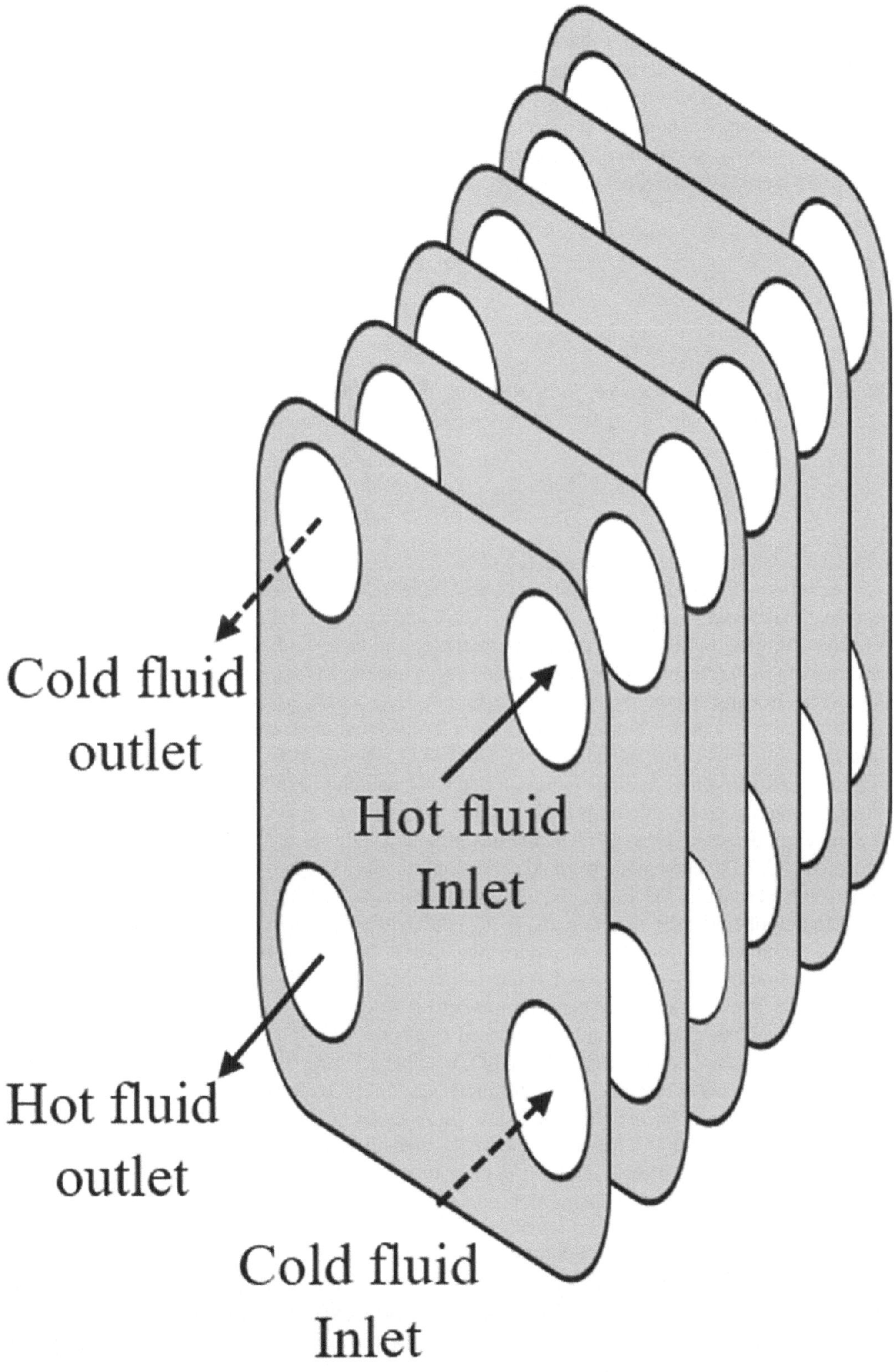

FIGURE 5.17 Plate-type heat exchanger.

a larger surface area, so the heat transfer coefficient is high. The compact and flexible design and ease of maintenance make them suitable for a wide range of applications.

The Nusselt number (Nu) and the heat transfer coefficient (h) are critical parameters that describe the heat transfer characteristics of the nanofluids through the plate HX. The Nusselt number for plate heat exchangers depends on the flow arrangement (counterflow or parallel flow) and the nature of the flow (laminar or turbulent). Turbulent flow is more common in a plate heat exchanger due to higher efficiency. The Gnielinski correlation can be used to calculate Nu for the turbulent flow of the nanofluid in a plate HX. The correlation is given by:

$$Nu = \frac{(f/8)(Re-1000)Pr}{1+12.7\left(\frac{f}{8}\right)^{\frac{1}{2}}\left(Pr^{\frac{2}{3}}-1\right)} \tag{5.27}$$

Where f is Darcy friction factor, Re is Reynolds number, and Pr is Prandtl number. The Darcy friction factor is estimated using the Colebrook equation for turbulent flow, which is given by:

$$\frac{1}{\sqrt{f}} = -2log_{10}\left(\frac{\varepsilon/D}{3.7}+\frac{2.51}{Re\sqrt{f}}\right) \tag{5.28}$$

Where ε/D is the relative toughness of the plate.

The expression for the heat transfer coefficient of nanofluid in a plate HX is the same as given in Equation 5.21. The enhancement in the values of Nu and h using nanofluids depends on the nanoparticle concentration and the material of nanoparticles and base fluid. Teng et al. reported a 9.19% improvement in the thermal behavior of carbon nanofluids when used in plate heat exchangers [85]. Huang et al. investigated the thermal performance of Al_2O_3 and MWCNT hybrid nanofluids in plate HXs and observed a higher thermal performance [86]. Kumar et al. examined nanofluids produced from various nanoparticles, including CNT, CuO/Al_2O_3 hybrid, Al_2O_3, TiO_2, ZnO, and CeO, using a PHX gasket type. Their findings indicated that CNT nanofluid exhibited the best thermal performance among the tested nanofluids [87].

Maré et al. discovered that CNT nanofluids provided better heat transfer enhancement in PHX compared with nanofluids made from Al_2O_3 nanoparticles [88]. Zheng et al. conducted a study on thermal enhancement in PHX using Fe_3O_4 nanofluids and found a 30.8% improvement at a particle concentration of 1.0% [89]. Additionally, Wang et al. performed an experimental study on enhancing the thermal performance of a miniature plate heat exchanger (PHE) with graphene nanofluids, composed of an EG/DW mixture and graphene nanoparticles, and reported a 4% improvement in thermal performance at a particle concentration of 1.0 wt% [90].

Quantitatively, the Nusselt number increased by approximately 10%, 25%, and 40% for 0.1%, 1.0%, and 3.0% volume concentrations of Al_2O_3 nanoparticle in water. The 0.5% and 2.0% volume concentration of CuO nanoparticle in water increases the Nu by 20% and 35%. The enhancement in the heat transfer coefficient using Al_2O_3 and CuO nanoparticles for plate HXs is the same (equal to percentage increase in the Nusselt number). The enhancement in the Nusselt number and heat transfer coefficient when using nanofluids in plate-type heat exchangers typically ranges from 10% to over 50%, depending on the nanoparticle concentration and the specific nanofluid used.

5.3.2.4 Shell and Tube Heat Exchangers

The shell and tube type HX consists of a shell and a tube. A series of tubes called tube bundles is fixed inside the shell. One fluid flows inside the tube bundles, while the other flows outside the tube bundle and inside the shell as shown in Figure 5.18. Shell and tube HXs operate in all three configurations. It offers a wide range of temperatures and pressures.

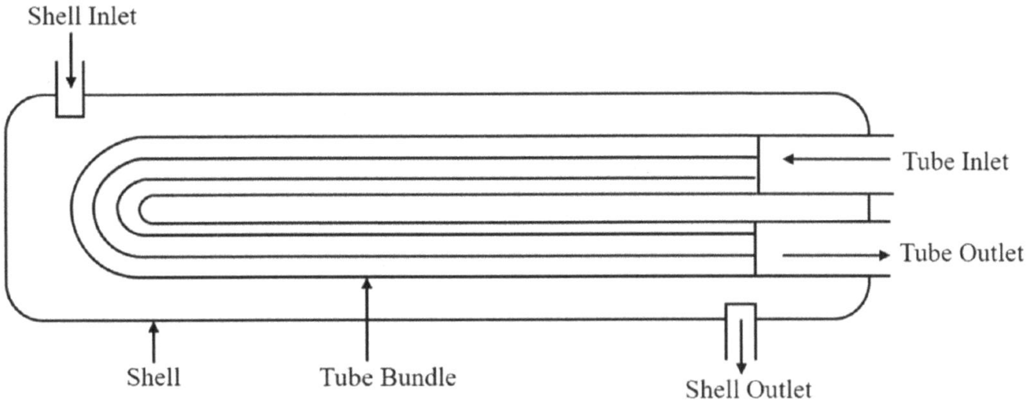

FIGURE 5.18 Shell and tube heat exchanger.

The heat transfer coefficient expression for nanofluid flow in shell and tube HXs is the same as given in Equation 4.21. The Nusselt number for shell and tube HXs is influenced by the flow properties, nanofluid properties, and geometry of the HX. For flow inside the tubes, the Dittus–Boelter correlations are commonly used, which are given by:

$$Nu = 0.023Re^{0.8}Pr^{0.4} \tag{5.29}$$

The flow on the shell side can be complex, involving crossflow, bypass flow, and possibly leakage flow. The Kern method is often used for an approximate solution, which is given by:

$$Nu = 0.36\left(\frac{d_e}{L}\right)^{1/3} Re^{0.55}Pr^{0.33} \tag{5.30}$$

Where d_e is the equivalent diameter of the shell side area, and L is the length of tube bundle.

Farajollahi et al. conducted an experimental investigation on horizontal shell and tube heat exchangers using two types of nanofluids (Al_2O_3/water and TiO_2/water) under turbulent flow conditions. The results indicated a significant enhancement in the heat transfer coefficient (HTC), with up to 20% improvement for Al_2O_3 nanofluid and up to 24% for TiO_2 nanofluid compared with the base fluid at the same Peclet number. Barzegarian et al. also studied Al_2O_3 nanofluid with various particle concentrations in shell and tube HXs, featuring a tube with a 5 mm inner diameter under laminar flow conditions [91]. They observed a considerable enhancement in heat transfer, which increased with particle concentration, reaching up to 29.8% for a 0.3 vol% concentration.

Shahrul et al. experimentally investigated the thermal performance of several nanofluids with different types of nanoparticles (ZnO, CuO, Fe_3O_4, TiO_2, and Al_2O_3) in shell and tube HXs. Their results showed enhancement in HTC and energy effectiveness for all studied nanofluid samples. The largest HTC enhancement was observed with Al_2O_3 nanofluid, while the smallest was with CuO nanofluid. The greatest energy effectiveness enhancement, 43%, was noted with ZnO nanofluid. Albadr et al. also conducted an experimental investigation on the thermal performance of Al_2O_3 nanofluids under turbulent flow conditions [67]. The use of nanofluids in shell and tube heat exchangers can lead to significant improvements in both the Nusselt number and the convective heat transfer coefficient, with typical enhancements ranging from 10% to 50%.

5.3.2.5 Multi-Channel/Micro-Channel Heat Exchangers

A multi-channel HX utilizes multiple parallel channels to facilitate heat transfer between fluids as shown in Figure 5.19. These types of HX are compact in design and useful for high heat transfer rate

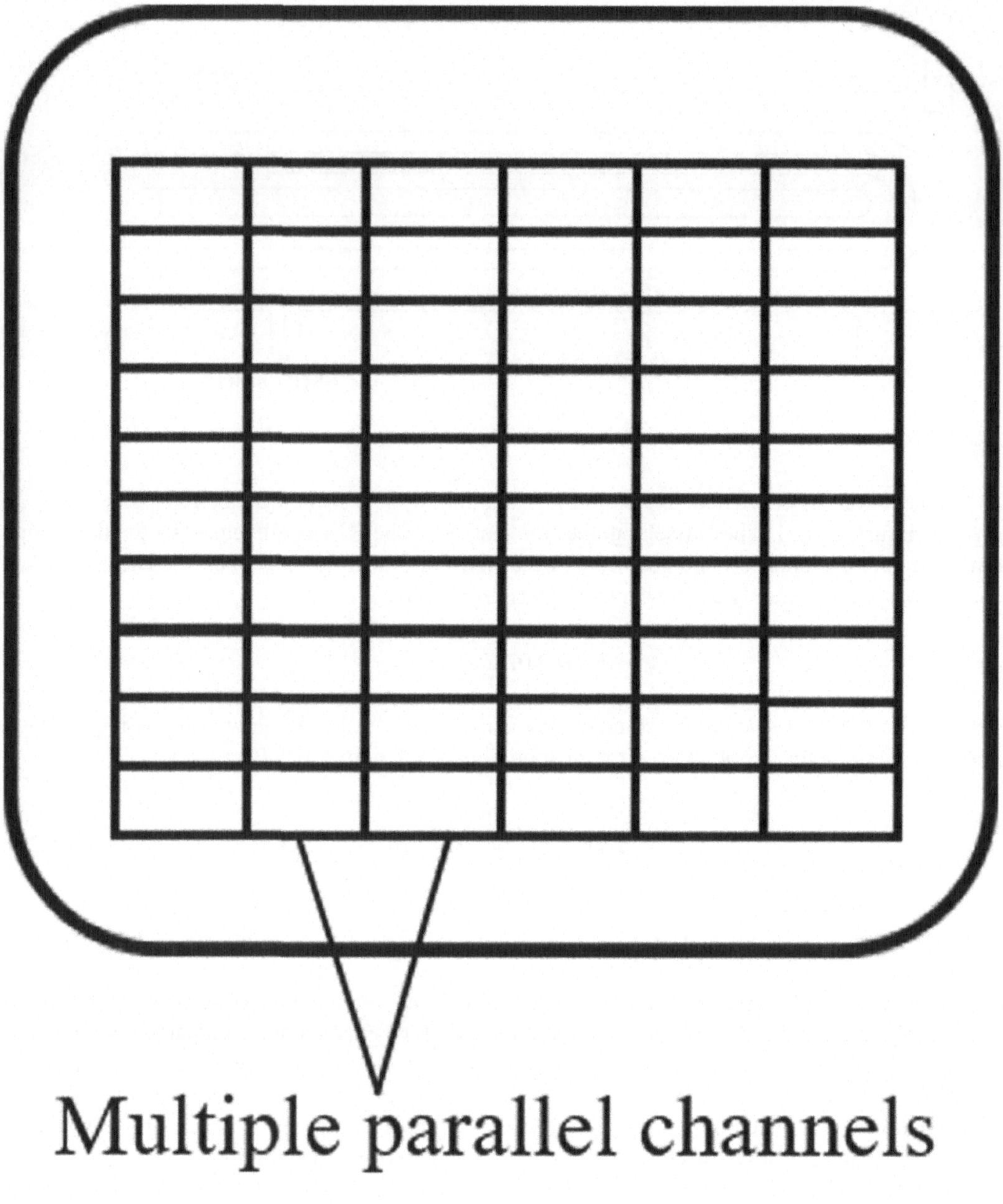

FIGURE 5.19 Multi-channel heat exchanger.

applications. A multi-channel HX is termed as a micro-channel HX when the size of the channel is in micrometers. The choice between micro-channel and multi-channel heat exchangers depends on the specific requirements of the application, including the need for precise temperature control, allowable pressure drop, fluid cleanliness, and cost considerations. Micro-channel HXs excel in high-performance, compact applications but at the cost of higher pressure drops and potential clogging issues. Multi-channel heat exchangers offer a more balanced approach, providing good heat transfer efficiency with less complexity and lower operational risks.

The Nusselt number correlation for multi-channel/micro-channel HXs is given by the Dittus–Boelter correlations (Equation 5.29).

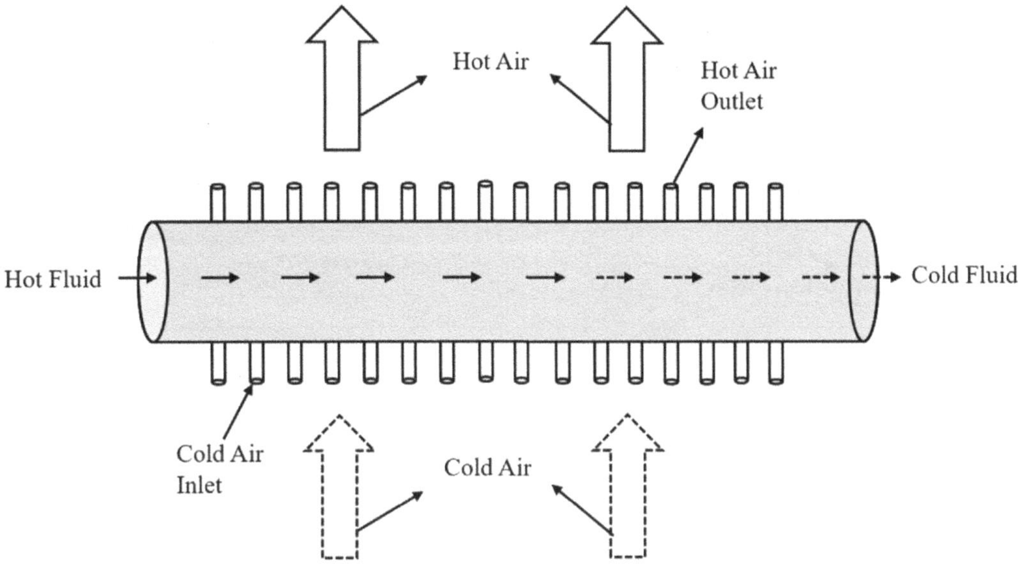

FIGURE 5.20 Air-cooled heat exchanger.

Sarafraz et al. studied the heat transfer behavior of MWCNTs nanofluids based on water flow in a heat sink heat exchanger with rectangular copper micro-channels using particle concentrations between 0.05% and 0.1% and Reynolds numbers below 1400 [92]. They observed an improvement in thermal performance of up to 29%, with higher HTC values particularly noticeable in the entrance area of the channels due to the hydraulic development of the flow.

Mohammadi et al. investigated the fluid flows of SiC nanofluids through a mini heat sink heat exchanger used as coolants for an electronic device. Their results showed a heat transfer improvement of 55.5% for 0.5 vol% SiC and 58.1% for 1.0 vol% SiC at a flow rate of 0.2 l/min, compared with the base fluid [93]. The heat transfer characteristics (Nu and h) improved up to 60% with the utilization of nanofluids in multi-channel/micro-channel HXs.

5.3.2.6 Air-Cooled Heat Exchangers

An air-cooled HX is a type of heat exchanger that uses air as the cooling medium to transfer heat from a process fluid to the surrounding environment as shown in Figure 5.20. This makes them particularly useful in situations where water is scarce or where the environmental impact of using water needs to be minimized. The nanofluids in air-cooled HXs can significantly enhance heat transfer performance due to their higher thermal conductivity and specific heat capacity.

The Nusselt number and heat transfer coefficient expression for nanofluids flowing in air-cooled HX are the same as given in Equations 5.27 and 5.19. Vermahmoudi et al. reported the enhancement of heat transfer characteristics of Fe_2O_3-water nanofluid in compact air-cooled HXs. Various concentrations of Fe_2O_3-water nanofluid (0.15–0.65 vol.%) were prepared and stabilized using 0.8 wt.% polyethylene glycol with the pH maintained at 11.1. The maximum enhancements in the overall heat transfer coefficient (13%) and Nusselt number (11.5%) are achieved for 0.65% volume concentration of Fe_2O_3 [94].

Teng et al. adopted MWCNT-water nanofluid with a surfactant of SDBS in an air-cooled HX. The heat transfer coefficient and Nusselt number increased by 7.77% and 7.53% for 0.25% volume concentration of MWCNT [95]. Leong et al. investigated ethylene glycol-based copper nanofluids in an automotive cooling system. An increase in heat transfer coefficient by 3.8% with a 2.0% volume concentration of Cu at a Reynolds number of 6000 for air and 5000 for coolants is observed [96].

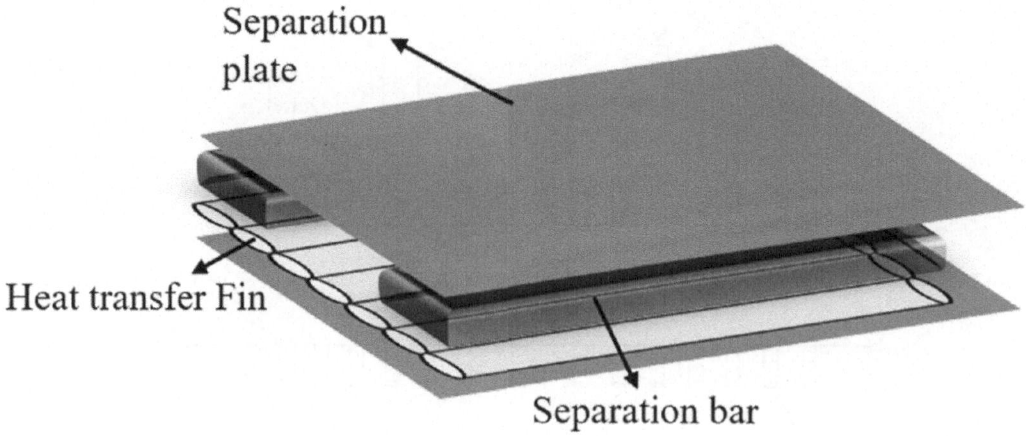

FIGURE 5.21 Plate and fin heat exchanger.

The overall enhancement in heat transfer characteristics (Nu and h) of air-cooled HX for Al_2O_3, CuO, SiO_2, and MWCNT nanoparticle-based nanofluid is approximately 10–30%, 15–25%, 5–15%, and 20–50%, respectively.

5.3.2.7 Plate and Fin Heat Exchangers

A plate and fin HX is a type of heat exchanger design that uses plates and finned chambers as shown in Figure 5.21 to transfer heat between fluids, most commonly gases. It is often categorized as a compact heat exchanger to emphasize its relatively high heat transfer surface area-to-volume ratio. These are widely used for aerospace and automotive applications due to their compact size and lightweight properties. These types of HX can handle a wide range of temperatures and pressures and therefore are suitable for gases, liquids, and even two-phase mixtures. These are prone to fouling due to the small size of channels and so heat transfer efficiency can be reduced. So, regular maintenance is essential to ensure the efficiency and longevity of plate and fin heat exchangers.

For laminar flow conditions, the Nusselt number correlation for nanofluids can be simplified as:

$$Nu = 1.86\left(Re.Pr\right)^{1/3}\left(\frac{k_{nf}}{k_b}\right)^p \tag{5.31}$$

Where k_{nf} is the thermal conductivity of nanofluid and k_b is the thermal conductivity of the base fluid. The exponent p is determined experimentally and depends on the type and concentration of nanoparticles.

For turbulent flow, a common form of the correlation is:

$$Nu = 0.23Re^{0.8}Pr^{0.4}\left(\frac{k_{nf}}{k_b}\right)^p \tag{5.32}$$

Suresh et al. experimentally studied Cu-Al_2O_3/water hybrid nanofluids. A fully laminar convective heat transfer flow is developed with the Reynolds number varying from 700 to 2400. After the critical experimental investigation, the average Nusselt number is increased by 10.94% for hybrid nanofluids when compared with pure water [97]. Aliabadi et al. investigated the effect of Cu-water nanofluids in a plate and fin heat exchanger with up to 0.4% nanoparticle volume concentration. The heat transfer enhancement increased up to 24.14% [98].

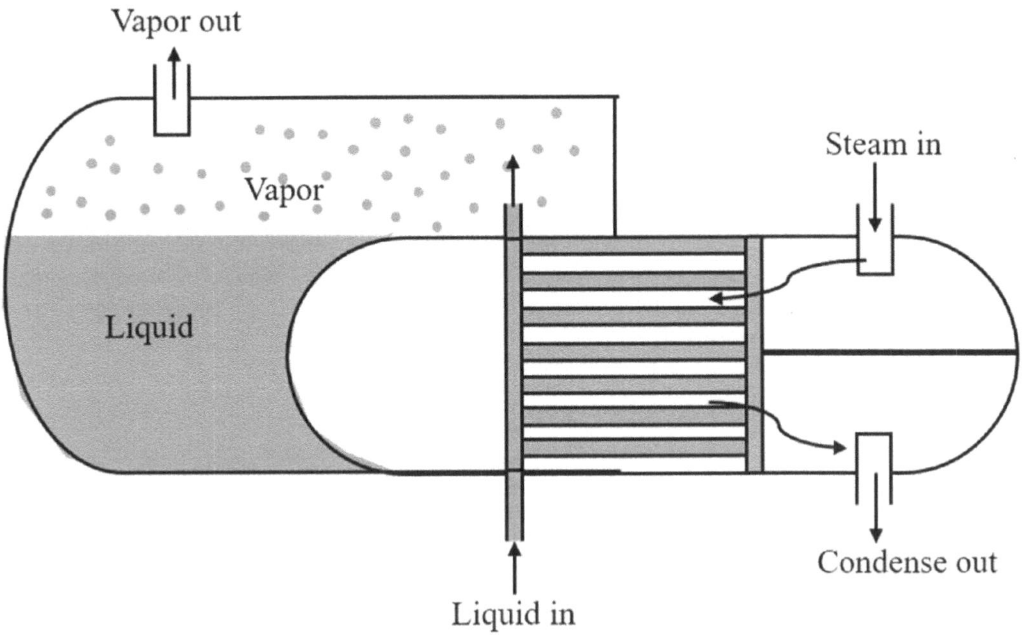

FIGURE 5.22 Phase-change heat exchanger.

Zamzamian et al. reported the effect of nanofluids on the forced convective heat transfer rate with the turbulent flow regime. The nanoparticles Al_2O_3 and CuO with a mean size of 20 nm were separately mixed with ethylene glycol with distinct volume concentrations varying from 0.1 to 1.0%. The heat transfer enhancement obtained from experiments was from 2% to 50% [99]. Huang et al. also reported the improved heat transfer characteristics for Al_2O_3-MWCNT hybrid nanofluid [100]. Several other studies also reported the improvement in heat transfer characteristics (Nu and h) of plate and fin HX by up to 5–40% with the use of nanofluids [101, 102].

5.3.2.8 Phase-Change Heat Exchangers

The phase-change HX utilizes the latent heat associated with a phase change to transfer heat between fluids (as shown in Figure 5.22). The most common phase changes involved are between liquid and vapor phases, such as evaporation and condensation. These heat exchangers are highly efficient in transferring large amounts of energy due to the significant energy involved in phase transitions. One of the fluids undergoes a phase change, absorbing or releasing latent heat, while the other fluid transfers this heat.

The expression for Nusselt number of nanofluid in phase-change HXs is given by [103]:

$$Nu = Nu_b \left(\frac{k_{nf}}{k_b} \right)^{0.33} \tag{5.33}$$

The enhancement of the Nusselt number (Nu) and heat transfer coefficient (h) in phase-change HXs when using nanofluids is due to several factors, such as increased thermal conductivity, modified surface wettability, and changes in fluid dynamics brought about by the presence of nanoparticles [104, 105]. This enhancement can be particularly significant in applications where phase changes are the primary mechanisms of heat transfer. Different nanoparticles (e.g., metal oxides like Al_2O_3, CuO, and TiO_2) and their concentrations have varied effects on thermal conductivity and boiling/

condensation characteristics. Nanoparticles can modify surface wettability, affecting nucleation sites in boiling and droplet dynamics in condensation [105].

Studies have reported Nusselt number enhancements ranging from 10% to 100% during boiling heat transfer when using nanofluids compared with base fluids. The heat transfer coefficient can increase significantly, sometimes up to two to three times the baseline values, depending on nanofluid concentration and the boiling regime (nucleate or film boiling) [106]. The enhancement in Nu during film condensation with nanofluids can be around 10% to 50%. This enhancement is attributed to better heat conduction through the liquid film and changes in droplet formation and removal on the condensing surface. The heat transfer coefficient can increase by a similar range, typically 20% to 100%, due to improved thermal conductivity and more efficient heat removal processes facilitated by nanoparticles [107, 108].

5.4 USES OF AI/ML IN ADVANCED HEAT TRANSFER ENHANCEMENT TECHNIQUES

Artificial intelligence (AI) and machine learning (ML) are revolutionizing the field of heat transfer, particularly in enhancing the performance of nanofluids. One of the significant applications of AI/ML is in the optimization of nanofluid formulations. AI and ML algorithms can optimize the composition and properties of nanofluids for maximum thermal conductivity and heat transfer efficiency. Studies have demonstrated the use of genetic algorithms and neural networks to identify optimal nanoparticle types, sizes, and concentrations, significantly improving thermal performance [109].

5.4.1 Optimization of Heat Exchanger Design and Performance

In the design and optimization of heat exchangers, ML algorithms evaluate various design parameters and operating conditions. Research has shown that techniques such as genetic algorithms (GAs) [110] and artificial neural networks (ANN) can identify the best configurations for enhanced heat transfer, considering the specific properties of nanofluids. GAs can search through a vast space of design parameters to find optimal solutions, while ANNs can model complex nonlinear relationships between variables, providing insights into the most efficient designs. For example, a GA might optimize the heat exchanger configuration by adjusting parameters such as fin spacing, tube diameter, and flow rates to achieve maximum heat transfer efficiency.

5.4.2 Data-Driven Modeling for Nanofluid HX Applications

Data-driven modeling using ML techniques is increasingly applied in HX applications with nanofluids. For instance, ANNs can be trained on experimental and simulation data to predict the thermal performance of heat exchangers with nanofluids. This approach helps in understanding how different nanofluid properties, such as nanoparticle concentration and size, influence heat transfer efficiency. ANNs can be trained on extensive datasets to predict the thermal performance of heat exchangers using nanofluids. For instance, an ANN model developed by Kalteh et al. [111] was able to predict the convective heat transfer coefficient of nanofluids with high accuracy. Their model considered variables such as nanoparticle concentration, size, and Reynolds number, showing an improvement in prediction accuracy compared with traditional methods. The general form of an ANN model for predicting heat transfer performance was represented as:

$$Q = f\left(T_{in}, T_{out}, m, C_p, \mu, k\right) \tag{5.34}$$

Where T_{in} and T_{out} are the inlet and outlet temperatures, m is the mass flow rate, Cp is the specific heat capacity, μ is the dynamic viscosity, and k is the thermal conductivity.

Support vector machines (SVMs) are effective in handling nonlinear relationships between input variables and the target output. For example, Ibrahim et al. [112] applied SVMs to predict the thermal conductivity of nanofluids, achieving a high correlation with experimental data. The SVM model demonstrated robustness in handling various operating conditions and nanoparticle concentrations. Random forest (RF) models, which consist of multiple decision trees, provide high accuracy and can handle complex interactions between variables. Si et al. [113] used RF to predict the heat transfer performance of nanofluid-based heat exchangers, considering factors such as nanoparticle type, concentration, and flow rate. Their model outperformed traditional regression techniques in terms of prediction accuracy and reliability.

5.4.3 Role of AI/ML in Predicting Heat Transfer Coefficients

AI/ML techniques play a crucial role in predicting the heat transfer coefficients of nanofluids in heat exchangers, both with and without phase change (boiling/evaporative) applications. Models like ANNs, GA, and SVMs are employed to analyze the complex interactions between nanofluid properties and heat transfer performance.

For instance, Nguyen et al. [114] developed an ANN model to predict the heat transfer coefficient of Al_2O_3-water nanofluids in a shell and tube heat exchanger. Their model achieved a prediction accuracy of 98%, demonstrating the capability of ANNs to capture the nonlinear relationships between the input parameters and the heat transfer coefficient. Another study by Mmohammadiun et al. [93] applied an ANN to predict the heat transfer coefficient of ZnO-water nanofluids in a double-pipe heat exchanger, achieving a correlation coefficient (R^2) of 0.95, showcasing the robustness of ANNs in various heat exchanger configurations.

In a study by Pourpasha et al. [115], GAs were used to optimize the heat transfer performance of a nanofluid-based plate heat exchanger. The optimized model showed a 15% improvement in heat transfer efficiency compared with conventional methods. Moreover, Wang et al. [90] used GAs to optimize the design parameters of a nanofluid-based micro-channel heat sink, achieving a 20% enhancement in heat transfer coefficient and a significant reduction in pressure drop.

Mohapatra et al. [116] applied SVMs to predict the heat transfer coefficient of CuO-water nanofluids in a micro-channel heat sink. The SVM model achieved high accuracy, with a correlation coefficient (R^2) of 0.97 compared with experimental data. Additionally, Behrozifard et al. [27] employed SVMs to predict the heat transfer coefficient of graphene-water nanofluids in a plate heat exchanger, demonstrating an R^2 of 0.96, thereby confirming the effectiveness of SVMs in diverse heat transfer scenarios.

In phase-change applications, such as boiling and evaporation, the role of AI/ML techniques becomes even more critical due to the complexity of the heat transfer mechanisms involved. For example, several developed ANN models were reviewed by Najmi and Hassan [117] to predict the boiling heat transfer coefficient of SiO_2-water nanofluids. These models successfully predicted the boiling heat transfer coefficient with an accuracy of 96%, highlighting the effectiveness of ANNs in phase-change applications. Similarly, Alawi et al. [118] utilized a deep learning approach to predict the evaporative heat transfer coefficient of Al_2O_3-water nanofluids, achieving a prediction accuracy of 94%, thus demonstrating the potential of AI techniques in enhancing phase-change heat transfer processes.

These models can accurately predict how changes in nanofluid composition and operating conditions affect the heat transfer coefficient, aiding in the design of more efficient heat exchangers.

5.4.4 Determining Thermophysical Properties of Nanofluids

These techniques are invaluable in determining the thermophysical properties of nanofluids, such as thermal conductivity, viscosity, and specific heat capacity. Extensive studies have shown that ML models trained on comprehensive experimental data can predict these properties with high accuracy. For instance, a regression model can be used to predict the thermal conductivity k of nanofluids as a function of temperature T and nanoparticle concentration ϕ:

$$k = a + bT + c\phi + dT\phi \tag{5.35}$$

where a, b, c, and d are the coefficients determined through training on experimental data. For example, Jana et al. [119] used an ANN to predict the thermal conductivity of alumina-water nanofluids and found that their model could accurately capture the experimental data within a margin of error of 2%. Similarly, Murshed et al. [120] developed an SVM model to predict the thermal conductivity of various nanofluids, demonstrating high accuracy across different temperature ranges and nanoparticle concentrations. Changdar [121] used a random forest model to predict the viscosity of titanium dioxide-water nanofluids and reported a prediction accuracy of 95% compared with experimental results. Their model successfully accounted for the nonlinear relationship between the parameters. In a study by Wang et al. [29, 36], an ANN was used to predict the specific heat capacity of copper oxide-water nanofluids. The model demonstrated a high correlation coefficient ($R^2 = 0.98$) with the experimental data, underscoring its reliability in practical applications. This predictive capability is essential for the development of new nanofluid formulations and for optimizing their performance in heat exchanger applications.

5.4.5 Real-Time Monitoring and Control

AI systems also enable real-time monitoring and control of heat transfer processes. By continuously analyzing data from sensors, AI can adjust operational parameters dynamically to maintain optimal performance, ensuring efficient heat transfer under varying conditions [122]. This real-time capability is particularly useful in managing the complex and dynamic nature of heat transfer systems utilizing nanofluids. For instance, a control system powered by AI might adjust the flow rate or temperature settings in response to real-time sensor data to optimize heat transfer efficiency.

5.4.6 Energy Efficiency Optimization

Energy efficiency optimization is another area where AI algorithms excel. By analyzing the relationship between operating conditions, system performance, and energy consumption, AI can recommend adjustments that minimize energy use while maintaining high heat transfer rates. Moreover, ML techniques diagnose issues in thermal systems by recognizing patterns in temperature, pressure, and flow rate data [123, 124]. This diagnostic capability helps in preventing faults and reducing maintenance costs. For example, an ML model could be trained to recognize early signs of fouling in heat exchangers based on historical sensor data, enabling proactive maintenance.

Ahmed et al. [125] developed an AI-based control system for a heat exchanger that adjusts operational parameters based on real-time data. Their system demonstrated a significant improvement in heat transfer efficiency and operational stability compared with traditional control methods. In another study, Godasiaei et al. [126] applied a deep learning model to predict and control the heat transfer performance of a nanofluid-based heat exchanger. The model was able to adapt to changing conditions in real-time, resulting in overall efficiency with R^2 0.91.

5.4.7 Simulation and Predictive Modeling

AI can create dynamic models that simulate the performance of heat transfer systems using nanofluids. These models can predict system behavior under various conditions, allowing for better planning and optimization of thermal management strategies. [127–129]. For example, computational fluid dynamics (CFD) simulations enhanced by ML algorithms can provide detailed insights into the flow and heat transfer characteristics of nanofluids in heat exchangers.

Thus, by integrating AI and ML with nanofluid technology, significant advancements in heat transfer efficiency and system reliability can be achieved

5.5 HEAT TRANSFER MODELING OF NANOFLUIDS

Modeling nanofluids involves creating mathematical and physical models that describe how nanoparticles interact with the base fluid and how these interactions affect the overall thermal properties of the suspension. This includes predicting key parameters such as thermal conductivity, viscosity, and convective heat transfer coefficients.

5.5.1 Classical Models

5.5.1.1 Effective Medium Theory

The effective medium theory (EMT) provides a fundamental approach to modeling the thermal conductivity of nanofluids. EMT simplifies the complex system of nanoparticles suspended in a base fluid as a homogeneous mixture. The theory assumes that the effective thermal conductivity (k_{eff}) of the nanofluid can be derived from the properties of the individual components—nanoparticles and base fluid. The general expression for the effective thermal conductivity is given by:

$$k_{eff} = k_f\left(1+\phi\frac{k_p-k_f}{k_p+2k_f}\right) \tag{5.36}$$

where k_f is the thermal conductivity of the base fluid, k_p is the thermal conductivity of the nanoparticles, and ϕ is the volume fraction of the nanoparticles [130]. This model provides a useful approximation but assumes uniform particle dispersion and spherical particles, which may not always reflect real conditions.

5.5.1.2 Maxwell–Garnett Model

The Maxwell–Garnett model, developed by James Clerk Maxwell and extended by Garnett, is another classical model used to describe the effective thermal conductivity of nanofluids. This model treats nanoparticles as spherical inclusions in a continuous base fluid matrix. The effective thermal conductivity k_{eff} is expressed as:

$$k_{eff} = k_f\left(\frac{k_p+2k_f+2\phi\left(k_p-k_f\right)}{k_p+2k_f-\phi\left(k_p-k_f\right)}\right) \tag{5.37}$$

This model is particularly useful for predicting the thermal conductivity of dilute nanofluids where particle interactions are minimal.

5.5.2 Modified Models

5.5.2.1 Brownian Motion Model

The Brownian motion model accounts for the effects of nanoparticles' thermal motion on heat transfer properties. It suggests that nanoparticles in suspension experience significant agitation due to thermal energy, which enhances the heat transfer characteristics of the fluid. The model indicates that smaller nanoparticles have a higher impact on thermal conductivity due to their increased Brownian motion. The effective thermal conductivity in this model is given by:

$$k_{eff} = k_f\left(1+\frac{3\phi\left(k_p-k_f\right)}{k_p+2k_f}\right) \tag{5.38}$$

where the term $\frac{3\phi\left(k_p-k_f\right)}{k_p+2k_f}$ incorporates the effects of Brownian motion [131] This model provides a more realistic prediction by considering the dynamic behavior of nanoparticles.

5.5.2.2 Non-Newtonian Fluid Model

Nanofluids often exhibit non-Newtonian behavior, particularly at high concentrations of nanoparticles. The non-Newtonian fluid model adjusts for this by modifying the effective viscosity and thermal conductivity calculations. For instance, the modified thermal conductivity expression accounts for the changes in flow dynamics and particle interactions:

$$k_{eff} = k_f\left(1+\frac{\phi\left(k_p-k_f\right)}{k_p+2k_f}\right) \tag{5.39}$$

In this model, additional parameters are introduced to account for non-Newtonian effects, such as the shear-thinning behavior observed in high-concentration nanofluids [132]. The model helps in understanding how alterations in viscosity and flow behavior affect overall thermal performance.

5.5.3 Experimental Validation

Experimental validation is crucial to confirm the accuracy of theoretical models and to understand real-world performance. Various studies have tested the predictions of these models against experimental data. Xie et al. [133] conducted experiments with alumina (Al_2O_3) and copper oxide (CuO) nanofluids, comparing experimental results with theoretical predictions. They found that while theoretical models generally predicted an enhancement in thermal conductivity, actual measurements often exceeded these predictions. This discrepancy was attributed to factors such as nanoparticle agglomeration and surface effects, which are not always accounted for in classical models. Lee et al. [134] explored the heat transfer performance of nanofluids in cooling systems and found that the enhanced heat transfer capabilities predicted by theoretical models were confirmed experimentally. However, their research highlighted the significant role of nanoparticle stability and dispersion in achieving the predicted performance. They noted that real-world applications often faced challenges such as particle settling and clumping, which impacted the effectiveness of nanofluids.

5.6 ENERGY EFFICIENCY AND PERFORMANCE ANALYSIS

Nanofluids, within conventional heat transfer fluids, have been recognized for their potential to enhance heat transfer capabilities. This improvement significantly impacts the energy efficiency of thermal systems in various industrial applications, such as automotive cooling, electronics thermal management, and renewable energy systems. This section focuses on analyzing the energy efficiency and performance of nanofluids compared with traditional fluids.

5.6.1 Energy Efficiency of Nanofluids

While nanofluids offer superior thermal performance, they often exhibit higher viscosity than base fluids, leading to increased pumping power requirements [135]. This higher viscosity can cause greater frictional resistance within the fluid, resulting in higher pressure drops and necessitating more energy for pumping. The balance between the thermal performance gains from using nanofluids and the additional energy required for pumping must be carefully analyzed to ensure overall efficiency. If the increased pumping power is not optimized correctly, it can offset the thermal performance benefits provided by nanofluids. For instance, the calculation of pumping power shows that increased viscosity leads to higher pressure drops, which directly translates into more energy consumption. This added energy demand can diminish the overall gains in thermal efficiency if not managed properly.

Given the potential drawbacks in pumping power, it becomes crucial to explore the performance of nanofluids in heat exchangers, where they can enhance the overall heat transfer rate. This enhancement can allow for smaller and more efficient designs, as the increased heat transfer efficiency leads to lower operating temperatures and significant energy savings. By using the effectiveness-NTU (number of transfer units) method to evaluate heat exchanger effectiveness, we can quantify performance improvements when using nanofluids. This method helps in understanding how efficiently the heat exchanger operates with nanofluids compared with conventional fluids, providing a clear metric for performance enhancement and highlighting the practical benefits that offset the pumping power concerns.

Economic considerations also play a critical role in the adoption of nanofluids. The cost of nanoparticles and the preparation of nanofluids can be substantial, necessitating a comprehensive cost-benefit analysis to determine their economic viability. While the initial investment in nanoparticles and preparation methods may be high, the long-term energy savings achieved through improved thermal efficiency can justify this expenditure. Particularly in applications with high energy consumption and significant heat transfer requirements, the initial costs can be offset by long-term savings, making the use of nanofluids a financially viable option. Therefore, evaluating the initial investment against long-term savings is essential for making informed decisions about the adoption of nanofluids in various thermal systems, ensuring that the benefits outweigh the costs.

5.6.2 Thermal Performance of Nanofluids

Nanofluids exhibit significantly enhanced thermal conductivity due to the high thermal conductivity of the nanoparticles dispersed within the base fluid. This enhancement is influenced by several factors, including the material, size, shape, and concentration of the nanoparticles. Metals such as copper and silver, metal oxides like alumina and titania, and carbon-based materials like graphene and carbon nanotubes all contribute to varying degrees of thermal conductivity improvements. The small size of nanoparticles increases the surface area-to-volume ratio, facilitating more efficient heat transfer. The shape of nanoparticles also plays a crucial role, with elongated shapes like rods or tubes often providing better thermal pathways compared with spherical particles. The concentration of nanoparticles must be optimized to balance the thermal conductivity gains against potential increases in viscosity, which could adversely affect flow dynamics [127, 136, 137].

The convective heat transfer coefficient, a critical parameter for evaluating thermal performance, is generally higher for nanofluids [138]. This is because the improved thermal conductivity enhances the fluid's ability to transfer heat, and the presence of nanoparticles modifies the flow dynamics, often resulting in more turbulent flow, which further enhances convective heat transfer.

The increased interaction between nanoparticles and the base fluid also contributes to this improvement by creating more effective thermal bridges within the fluid.

$$h_{nf} = h_f \left[1 + 2.628\varphi^{0.333} \left(\frac{k_{nf}}{k_f} \right)^{0.714} \left(\frac{\mu_{nf}}{\mu_f} \right)^{0.143} \left(\frac{c_{nf}}{c_f} \right)^{0.429} \right] \tag{5.40}$$

Where h_{nf} is the convective heat transfer coefficient for the nanofluid, h_f is the convective heat transfer coefficient for the base fluid, φ is the volume fraction of nanoparticles in the nanofluid, k_{nf} and k_f are the thermal conductivities of the nanofluid and base fluid, respectively. μ_{nf} and μ_f are the dynamic viscosities of the nanofluid and base fluid, respectively. c_{nf} and c_f are the specific heats of the nanofluid and base fluid, respectively.

Additionally, the Nusselt number, which represents the ratio of convective to conductive heat transfer, typically increases with the use of nanofluids. A higher Nusselt number indicates more efficient heat transfer due to improved convective mechanisms [139]. The interaction between nanoparticles and fluid molecules facilitates more effective energy transfer from the fluid to the heated surface, leading to enhanced convective heat transfer.

5.7 CHALLENGES AND LIMITATIONS

Despite the promising performance, several challenges and limitations need to be addressed before nanofluids can be widely adopted in industrial and commercial applications. These challenges range from technical issues related to the stability and consistency of nanofluids to economic and environmental concerns that impact their feasibility and sustainability. Understanding and overcoming these hurdles is crucial for the practical implementation and long-term success of nanofluids in diverse thermal management systems. A few important challenges are discussed in this section.

5.7.1 Stability and Durability

Ensuring the long-term stability of nanofluids is crucial for maintaining their enhanced properties. Agglomeration, where nanoparticles clump together, and sedimentation, where particles settle out of suspension, can significantly degrade performance over time. These phenomena can lead to uneven heat distribution and reduced thermal conductivity, negating the benefits of using nanofluids [140, 141]. The challenge lies in developing effective stabilization techniques, such as surfactants and surface modifications, to maintain uniform dispersion of nanoparticles [142]. Research on the use of surfactants has shown promise in improving the stability and thermal conductivity of nanofluids [143].

5.7.2 Pumping Power and Viscosity

Nanofluids often exhibit higher viscosity compared with base fluids, which can lead to increased pumping power requirements. This higher viscosity causes greater frictional resistance within the fluid, resulting in higher pressure drops and necessitating more energy for pumping. This additional energy consumption can offset the thermal performance benefits of nanofluids if not properly managed [144, 145]. Achieving a balance between improved heat transfer properties and the energy cost of increased viscosity is a significant challenge [146, 147].

5.7.3 Environmental and Health Considerations

The environmental impact and potential health risks associated with nanoparticles are critical concerns. Nanoparticles, due to their small size and high surface area, can be highly reactive and potentially toxic. They may pose risks to human health through inhalation, ingestion, or skin contact and can also lead to environmental contamination if not properly managed [148, 149]. Ensuring safe handling, disposal, and compliance with regulatory standards is essential to mitigate these risks. Developing comprehensive guidelines and regulations for the lifecycle management of nanofluids is necessary to ensure their safe use [150]. A comprehensive lifecycle analysis of nanofluids, from production to disposal, is also necessary to understand their overall energy efficiency and environmental impact. It includes evaluating the energy and resources required for nanoparticle production, the operational benefits and challenges during use, and the implications of disposal or recycling [151, 152]. Without this, the true cost and environmental impact of nanofluids may be underestimated.

5.7.4 Economic Viability

The production and preparation of nanofluids can be expensive, involving high costs for nanoparticles and specialized equipment for their dispersion. The economic challenge is to make nanofluids cost-effective for large-scale applications. While nanofluids offer enhanced thermal performance, the initial investment may be prohibitive for some industries [153]. A thorough cost-benefit analysis is required to assess whether the long-term energy savings and performance improvements justify the initial expenditures [154, 155].

5.7.5 Compatibility with Existing Systems

Integrating nanofluids into existing thermal systems can pose compatibility issues. Conventional systems may not be designed to handle the unique properties of nanofluids, such as increased viscosity or potential chemical reactivity [156]. Retrofitting existing systems to accommodate nanofluids can be complex and costly. Additionally, the long-term effects of nanofluids on system components, such as corrosion or material degradation, need to be thoroughly studied to ensure reliability and safety [157].

5.7.6 Lack of Standardization

The lack of standardized procedures for the synthesis, characterization, and testing of nanofluids poses a challenge. Inconsistent methodologies across different studies make it difficult to compare results and draw definitive conclusions about the performance and applicability of nanofluids [158]. Developing standardized protocols and guidelines is essential to advance the field and ensure the reliability and reproducibility of the results [159].

Current methods for producing nanofluids can be labor-intensive and costly, limiting their commercial viability. Innovations in manufacturing processes that can produce high-quality nanofluids at scale are needed to make them more accessible for industrial applications [160].

5.7.7 Thermal Conductivity Models

Existing models for predicting the thermal conductivity of nanofluids are often based on empirical data and may not accurately capture all the influencing factors. The complexity of interactions between nanoparticles and base fluids makes it challenging to develop universally applicable models [161]. Improved theoretical models that can more accurately predict the thermal conductivity of nanofluids under various conditions are needed to optimize their use [162].

5.7.8 Long-Term Performance and Degradation

The long-term performance and potential degradation of nanofluids under operational conditions remain areas of concern. Factors such as temperature fluctuations, chemical reactions, and mechanical stresses can impact the stability and performance of nanofluids over time. Long-term studies are necessary to understand these effects and develop strategies to mitigate performance degradation [163].

5.7.9 Heat Transfer Coefficient Enhancement

The relationship between nanoparticle concentration, fluid flow characteristics, and heat transfer performance is complex and not fully understood. Further research is needed to optimize nanofluid formulations and operating conditions to maximize heat transfer coefficient enhancement [164]. Understanding how different concentrations and flow conditions affect heat transfer can lead to better performance and efficiency in practical applications [165, 166].

5.8 FUTURE PERSPECTIVES AND RESEARCH DIRECTIONS

Despite the potential use of nanofluids, several challenges and limitations hinder their widespread adoption. Addressing these challenges requires ongoing research and innovation. The future of nanofluids depends on overcoming technical barriers, ensuring safety and environmental compliance, and optimizing their performance for specific applications. Here are the key areas for future research and development to realize the full potential of nanofluids.

Ensuring the long-term stability of nanofluids is crucial for maintaining their enhanced properties. Future research should focus on developing advanced stabilization techniques to prevent agglomeration and sedimentation. This includes exploring novel surfactants, surface modification methods, and hybrid stabilization approaches that combine multiple techniques. These innovations aim to maintain uniform dispersion of nanoparticles, preserving improved thermal properties and ensuring reliable performance over extended periods. Research can explore the development of multifunctional nanofluids that offer additional benefits beyond enhanced thermal conductivity. For instance, nanofluids with magnetic properties could be manipulated using magnetic fields to improve heat transfer or direct fluid flow. Similarly, nanofluids with antimicrobial properties could be used in medical and food processing applications to prevent microbial growth while enhancing thermal management. Such multifunctionalities could expand the applications of nanofluids, making them even more valuable.

Developing environmentally friendly and biodegradable nanoparticles is a key area for future research. Natural and biocompatible materials, such as cellulose nanocrystals or plant-based nanoparticles, could be explored as alternatives to traditional metal or oxide nanoparticles. These efforts aim to ensure that the use of nanofluids is sustainable and safe for both humans and the environment. Conducting comprehensive lifecycle assessments of nanofluids is essential to understand their full environmental and economic impact. These assessments will provide a holistic view of the sustainability of nanofluids and help in making informed decisions about their adoption.

Improving theoretical models to accurately predict the thermal and rheological properties of nanofluids is crucial. Future research can focus on developing comprehensive models that account for the complex interactions between nanoparticles and base fluids. These models can be validated against experimental data and cover a wide range of operating conditions. Developing scalable and cost-effective manufacturing processes for nanofluids is essential for their commercial viability. Research should focus on optimizing synthesis and dispersion techniques to reduce production costs without compromising quality. Innovations in nanomanufacturing, such as continuous flow synthesis and in-situ dispersion methods, could facilitate large-scale production of high-quality nanofluids, making them more accessible for industrial applications.

Developing clear health and safety regulations for the use of nanofluids is needed for their widespread adoption. Future researchers can work toward establishing guidelines for the safe handling, usage, and disposal of nanofluids. This includes conducting comprehensive toxicity studies, developing exposure limits, and creating standardized safety protocols to protect workers and the environment. Future research should prioritize application-specific optimization of nanofluids. This involves tailoring nanofluid properties to meet the specific requirements of different industries, such as automotive, electronics, renewable energy, and HVAC systems. By understanding the unique challenges and needs of each application, researchers can develop customized nanofluids that offer maximum performance benefits, thus enhancing their practical utility and effectiveness. Addressing these challenges through focused research and development efforts will pave the way for the widespread adoption of nanofluids, leading to enhanced thermal management solutions and more efficient energy systems. By continuing to innovate and explore new applications, nanofluids can play a critical role in advancing heat transfer technologies and contributing to a more sustainable future.

REFERENCES

1. K. V. Wong and O. De Leon, "Applications of nanofluids: Current and future," 2010. doi: 10.1155/2010/519659.
2. R. Saidur, K. Y. Leong, and H. A. Mohammed, "A review on applications and challenges of nanofluids," Apr. 01, 2011, *Elsevier Ltd*. doi: 10.1016/j.rser.2010.11.035.
3. S. A. Angayarkanni and J. Philip, "Review on thermal properties of nanofluids: Recent developments," Nov. 01, 2015, *Elsevier*. doi: 10.1016/j.cis.2015.08.014.
4. J. Philip and P. D. Shima, "Thermal properties of nanofluids," Nov. 15, 2012, *Elsevier B.V.* doi: 10.1016/j.cis.2012.08.001.
5. C. Dhand *et al.*, "Methods and strategies for the synthesis of diverse nanoparticles and their applications: A comprehensive overview," Nov. 26, 2015, *Royal Society of Chemistry*. doi: 10.1039/c5ra19388e.
6. S. Iravani, H. Korbekandi, S. V Mirmohammadi, and B. Zolfaghari, "Synthesis of silver nanoparticles: chemical, physical and biological methods," 2014. [Online]. Available: http://journals.lww.com/rips.
7. Z. Liu, R. Fu, and Y. Yuying, "Preparation and evaluation of stable nanofluids for heat transfer application," in *Advances in Nanofluid Heat Transfer*, Elsevier, 2022, pp. 25–57. doi: 10.1016/B978-0-323-88656-7.00013-1.
8. A. K. Tiwari, A. Kumar, and Z. Said, "Synthesis, characterization, and measurement techniques for the thermophysical properties of nanofluids," in *Advances in Nanofluid Heat Transfer*, Elsevier, 2022, pp. 59–93. doi: 10.1016/B978-0-323-88656-7.00012-X.
9. L. S. Sundar, E. V. Ramana, H. M. Ali, and A. C. M. Sousa, "Experimental correlations for Nusselt number and friction factor of nanofluids," in *Advances in Nanofluid Heat Transfer*, Elsevier, 2022, pp. 1–23. doi: 10.1016/B978-0-323-88656-7.00002-7.
10. H. Mercan, A. Celen, and T. Taner, "Thermophysical and rheological properties of unitary and hybrid nanofluids," in *Advances in Nanofluid Heat Transfer*, Elsevier, 2022, pp. 95–129. doi: 10.1016/B978-0-323-88656-7.00005-2.
11. Y. K. Leong and B. C. Ong, "Critical zeta potential and the Hamaker constant of oxides in water," *Powder Technol*, vol. 134, no. 3, pp. 249–254, Sep. 2003, doi: 10.1016/j.powtec.2003.08.054.
12. X. Li, D. Zhu, and X. Wang, "Evaluation on dispersion behavior of the aqueous copper nano-suspensions," *J Colloid Interface Sci*, vol. 310, no. 2, pp. 456–463, Jun. 2007, doi: 10.1016/j.jcis.2007.02.067.
13. S. S. Khaleduzzaman, M. R. Sohel, R. Saidur, and J. Selvaraj, "Stability of Al2O3-water nanofluid for electronics cooling system," in *Procedia Engineering*, Elsevier Ltd, 2015, pp. 406–411. doi: 10.1016/j.proeng.2015.05.026.
14. X. Yang and Z. Hua Liu, "A kind of nanofluid consisting of surface-functionalized nanoparticles," *Nanoscale Res Lett*, vol. 5, no. 8, pp. 1324–1328, 2010, doi: 10.1007/s11671-010-9646-6.
15. N. R. Karthikeyan, J. Philip, and B. Raj, "Effect of clustering on the thermal conductivity of nanofluids," *Mater Chem Phys*, vol. 109, no. 1, pp. 50–55, May 2008, doi: 10.1016/j.matchemphys.2007.10.029.
16. D. W. Oh, A. Jain, J. K. Eaton, K. E. Goodson, and J. S. Lee, "Thermal conductivity measurement and sedimentation detection of aluminum oxide nanofluids by using the 3ω method," *Int J Heat Fluid Flow*, vol. 29, no. 5, pp. 1456–1461, Oct. 2008, doi: 10.1016/j.ijheatfluidflow.2008.04.007.

17. N. Zhao, C. Qi, T. Chen, J. Tang, and X. Cui, "Experimental study on influences of cylindrical grooves on thermal efficiency, exergy efficiency and entropy generation of CPU cooled by nanofluids," *Int J Heat Mass Transf*, vol. 135, pp. 16–32, Jun. 2019, doi: 10.1016/j.ijheatmasstransfer.2019.01.106.
18. Z. Liu, Y. Yan, R. Fu, and M. Alsaady, "Enhancement of solar energy collection with magnetic nanofluids," *Thermal Science and Engineering Progress*, vol. 8, pp. 130–135, Dec. 2018, doi: 10.1016/j.tsep.2018.08.015.
19. A. K. Singh and V. S. Raykar, "Microwave synthesis of silver nanofluids with polyvinylpyrrolidone (PVP) and their transport properties," *Colloid Polym Sci*, vol. 286, no. 14–15, pp. 1667–1673, 2008, doi: 10.1007/s00396-008-1932-9.
20. E. Tang, G. Cheng, and X. Ma, "Preparation of nano-ZnO/PMMA composite particles via grafting of the copolymer onto the surface of zinc oxide nanoparticles," *Powder Technol*, vol. 161, no. 3, pp. 209–214, Feb. 2006, doi: 10.1016/j.powtec.2005.10.007.
21. D. Li, B. Hong, W. Fang, Y. Guo, and R. Lin, "Preparation of well-dispersed silver nanoparticles for oil-based nanofluids," *Ind Eng Chem Res*, vol. 49, no. 4, pp. 1697–1702, Feb. 2010, doi: 10.1021/ie901173h.
22. T. Matsoukas and S. Lotfizadeh, "Colloidal Thermal Fluids," 2014.
23. T. T. Loong and H. Salleh, "A review on measurement techniques of apparent thermal conductivity of nanofluids," in *IOP Conference Series: Materials Science and Engineering*, Institute of Physics Publishing, Aug. 2017. doi: 10.1088/1757-899X/226/1/012146.
24. M. Gupta, V. Singh, R. Kumar, and Z. Said, "A review on thermophysical properties of nanofluids and heat transfer applications," 2017, *Elsevier Ltd*. doi: 10.1016/j.rser.2017.02.073.
25. W. Yu, H. Xie, X. Wang, and X. Wang, "Significant thermal conductivity enhancement for nanofluids containing graphene nanosheets," *Phys Lett A: Gen At Solid State Phys*, vol. 375, no. 10, pp. 1323–1328, Mar. 2011, doi: 10.1016/j.physleta.2011.01.040.
26. H. A. Mintsa, G. Roy, C. T. Nguyen, and D. Doucet, "New temperature dependent thermal conductivity data for water-based nanofluids," *Int J Therm Sci*, vol. 48, no. 2, pp. 363–371, Feb. 2009, doi: 10.1016/j.ijthermalsci.2008.03.009.
27. M. P. Beck, Y. Yuan, P. Warrier, and A. S. Teja, "The effect of particle size on the thermal conductivity of alumina nanofluids," *J Nanoparticle Res*, vol. 11, no. 5, pp. 1129–1136, Jul. 2009, doi: 10.1007/s11051-008-9500-2.
28. M. Kole and T. K. Dey, "Role of interfacial layer and clustering on the effective thermal conductivity of CuO-gear oil nanofluids," *Exp Therm Fluid Sci*, vol. 35, no. 7, pp. 1490–1495, Oct. 2011, doi: 10.1016/j.expthermflusci.2011.06.010.
29. X. Wang, X. Xu, and S. U. S Choi, "Thermal Conductivity of Nanoparticle-Fluid Mixture," *J Thermophys Heat Transf*, vol. 13, no. 4, pp. 474–480, 1999, doi: 10.2514/2.6486.
30. G. M. Moldoveanu, A. A. Minea, G. Huminic, and A. Huminic, "Al2O3/TiO2 hybrid nanofluids thermal conductivity: An experimental approach," *J Therm Anal Calorim*, vol. 137, no. 2, pp. 583–592, Jul. 2019, doi: 10.1007/s10973-018-7974-4.
31. R. Vidhya, T. Balakrishnan, and B. S. Kumar, "Investigation on thermophysical properties and heat transfer performance of heat pipe charged with binary mixture based ZnO-MgO hybrid nanofluids," in *Materials Today: Proceedings*, Elsevier Ltd, 2020, pp. 3423–3433. doi: 10.1016/j.matpr.2020.09.284.
32. S. U. Ilyas, M. Narahari, J. T. Y. Theng, and R. Pendyala, "Experimental evaluation of dispersion behavior, rheology and thermal analysis of functionalized zinc oxide-paraffin oil nanofluids," *J Mol Liq*, vol. 294, Nov. 2019, doi: 10.1016/j.molliq.2019.111613.
33. G. Paul, T. Pal, and I. Manna, "Thermo-physical property measurement of nano-gold dispersed water based nanofluids prepared by chemical precipitation technique," *J Colloid Interface Sci*, vol. 349, no. 1, pp. 434–437, Sep. 2010, doi: 10.1016/j.jcis.2010.05.086.
34. T. P. Teng, Y. H. Hung, T. C. Teng, H. E. Mo, and H. G. Hsu, "The effect of alumina/water nanofluid particle size on thermal conductivity," *Appl Therm Eng*, vol. 30, no. 14–15, pp. 2213–2218, Oct. 2010, doi: 10.1016/j.applthermaleng.2010.05.036.
35. J. R. Satti, D. K. Das, and D. Ray, "Investigation of the thermal conductivity of propylene glycol nanofluids and comparison with correlations," *Int J Heat Mass Transf*, vol. 107, pp. 871–881, Apr. 2017, doi: 10.1016/j.ijheatmasstransfer.2016.10.121.
36. X. Q. Wang and A. S. Mujumdar, "Heat transfer characteristics of nanofluids: a review," Jan. 2007. doi: 10.1016/j.ijthermalsci.2006.06.010.
37. P. Keblinski, S. R. Phillpot, S. U. S. Choi, and J. A. Eastman, "Mechanisms of heat ¯ow in suspensions of nano-sized particles (nano¯uids)." [Online]. Available: www.elsevier.com/locate/ijhmt

38. J. A. Eastman, S. U. S. Choi, S. Li, W. Yu, and L. J. Thompson, "Anomalously increased effective thermal conductivities of ethylene glycol-based nanofluids containing copper nanoparticles," *Appl Phys Lett*, vol. 78, no. 6, pp. 718–720, Feb. 2001, doi: 10.1063/1.1341218.
39. S. U. S. Choi, Z. G. Zhang, W. Yu, F. E. Lockwood, and E. A. Grulke, "Anomalous thermal conductivity enhancement in nanotube suspensions," *Appl Phys Lett*, vol. 79, no. 14, pp. 2252–2254, Oct. 2001, doi: 10.1063/1.1408272.
40. L. Qiang, and Y. Xuan, "Convective heat transfer and flow characteristics of Cu-water nanofluid," *Sci China Technol Sci,* vol. 45, pp. 408–416, 2002, doi: 10.1360/02ye9047.
41. H. Xie, J. Wang, T. Xi, and Y. Liu, "Thermal conductivity of suspensions containing nanosized SiC particles," *Int J Thermophys,* vol. 23, pp. 571–580, 2002, doi: 10.1023/A:1015121805842.
42. S. M. S. Murshed, K. C. Leong, and C. Yang, "Investigations of thermal conductivity and viscosity of nanofluids," *International Journal of Thermal Sciences*, vol. 47, no. 5, pp. 560–568, May 2008, doi: 10.1016/j.ijthermalsci.2007.05.004.
43. H. Xie, H. Lee, W. Youn, and M. Choi, "Nanofluids containing multiwalled carbon nanotubes and their enhanced thermal conductivities," *J Appl Phys*, vol. 94, no. 8, pp. 4967–4971, Oct. 2003, doi: 10.1063/1.1613374.
44. S. K. Das, N. Putra, P. Thiesen, and W. Roetzel, "Temperature dependence of thermal conductivity enhancement for nanofluids," *J Heat Transfer*, vol. 125, no. 4, pp. 567–574, Aug. 2003, doi: 10.1115/1.1571080.
45. J. A. Eastman, S. U. S. Choi, S. Li, W. Yu, and L. J. Thompson, "Anomalously increased effective thermal conductivities of ethylene glycol-based nanofluids containing copper nanoparticles," *Appl Phys Lett*, vol. 78, no. 6, pp. 718–720, Feb. 2001, doi: 10.1063/1.1341218.
46. X. F. Li, D. S. Zhu, X. J. Wang, N. Wang, J. W. Gao, and H. Li, "Thermal conductivity enhancement dependent pH and chemical surfactant for Cu-H2O nanofluids," *Thermochim Acta*, vol. 469, no. 1–2, pp. 98–103, Mar. 2008, doi: 10.1016/j.tca.2008.01.008.
47. I. M. Shahrul, I. M. Mahbubul, S. S. Khaleduzzaman, R. Saidur, and M. F. M. Sabri, "A comparative review on the specific heat of nanofluids for energy perspective," 2014, *Elsevier Ltd.* doi: 10.1016/j.rser.2014.05.081.
48. B. C. Pak and Y. I. Cho, "Hydrodynamic and heat transfer study of dispersed fluids with submicron metallic oxide particles," *Experimental Heat Transfer*, vol. 11, no. 2, pp. 151–170, 1998, doi: 10.1080/08916159808946559.
49. Y. Xuan and W. Roetzel, "Conceptions for heat transfer correlation of nano¯uids," [Online]. Available: www.elsevier.com/locate/ijhmt.
50. L. S. Sundar, M. K. Singh, and A. C. M. Sousa, "Turbulent heat transfer and friction factor of nanodiamond-nickel hybrid nanofluids flow in a tube: An experimental study," *Int J Heat Mass Transf*, vol. 117, pp. 223–234, 2018, doi: 10.1016/j.ijheatmasstransfer.2017.09.109.
51. H. Tiznobaik and D. Shin, "Enhanced specific heat capacity of high-temperature molten salt-based nanofluids," *Int J Heat Mass Transf*, vol. 57, no. 2, pp. 542–548, 2013, doi: 10.1016/j.ijheatmasstransfer.2012.10.062.
52. I. Wole-Osho, E. C. Okonkwo, D. Kavaz, and S. Abbasoglu, "An experimental investigation into the effect of particle mixture ratio on specific heat capacity and dynamic viscosity of Al2O3-ZnO hybrid nanofluids," *Powder Technol*, vol. 363, pp. 699–716, Mar. 2020, doi: 10.1016/j.powtec.2020.01.015.
53. G. M. Moldoveanu and A. A. Minea, "Specific heat experimental tests of simple and hybrid oxide-water nanofluids: Proposing new correlation," *J Mol Liq*, vol. 279, pp. 299–305, Apr. 2019, doi: 10.1016/j.molliq.2019.01.137.
54. A. B. Çolak, O. Yıldız, M. Bayrak, A. Celen, A. S. Dalkılıç, and S. Wongwises, "Experimental study on the specific heat capacity measurement of water- based Al2O3-Cu hybrid nanofluid by using differential thermal analysis method," *Curr Nanosci*, vol. 16, no. 6, pp. 912–928, Nov. 2019, doi: 10.2174/15734137 15666191118105331.
55. J. R. Satti, D. K. Das, and D. Ray, "Specific heat measurements of five different propylene glycol based nanofluids and development of a new correlation," *Int J Heat Mass Transf*, vol. 94, pp. 343–353, Mar. 2016, doi: 10.1016/j.ijheatmasstransfer.2015.11.065.
56. Y. R. Sekhar and K. V. Sharma, "Study of viscosity and specific heat capacity characteristics of water-based Al2O3nanofluids at low particle concentrations," *J Exp Nanosci*, vol. 10, no. 2, pp. 86–102, Jan. 2015, doi: 10.1080/17458080.2013.796595.
57. C. V. Popa, C. T. Nguyen, and I. Gherasim, "New specific heat data for Al2O3 and CuO nanoparticles in suspension in water and Ethylene Glycol," *International Journal of Thermal Sciences*, vol. 111, pp. 108–115, Jan. 2017, doi: 10.1016/j.ijthermalsci.2016.08.016.

58. A. M. Hussein, R. A. Bakar, K. Kadirgama, and K. V. Sharma, "Experimental measurement of nanofluids thermal properties," *International Journal of Automotive and Mechanical Engineering*, vol. 7, no. 1, pp. 850–863, 2013, doi: 10.15282/ijame.7.2012.5.0070.
59. C. T. Nguyen *et al.*, "Viscosity data for Al2O3-water nanofluid-hysteresis: is heat transfer enhancement using nanofluids reliable?," *International Journal of Thermal Sciences*, vol. 47, no. 2, pp. 103–111, Feb. 2008, doi: 10.1016/j.ijthermalsci.2007.01.033.
60. J. B. Mena, A. A. Ubices De Moraes, Y. R. Benito, G. Ribatski, and J. A. R. Parise, "Extrapolation of Al2O3-water nanofluid viscosity for temperatures and volume concentrations beyond the range of validity of existing correlations," *Appl Therm Eng*, vol. 51, no. 1–2, pp. 1092–1097, 2013, doi: 10.1016/j.applthermaleng.2012.11.002.
61. K. S. Suganthi and K. S. Rajan, "Temperature induced changes in ZnO-water nanofluid: Zeta potential, size distribution and viscosity profiles," *Int J Heat Mass Transf*, vol. 55, no. 25–26, pp. 7969–7980, Dec. 2012, doi: 10.1016/j.ijheatmasstransfer.2012.08.032.
62. M. Sivasubramanian, A. Shylaja, and K. S. Rajan, "Thermo-physical properties of engineered dispersions of nano-sand in propylene glycol," *Colloids Surf A Physicochem Eng Asp*, vol. 449, no. 1, pp. 8–18, May 2014, doi: 10.1016/j.colsurfa.2014.02.040.
63. Y. F. Niu, W. L. Zhao, and Y. Y. Gong, "Preparation and thermophysical properties of SiO2 nanofluids," in *Advanced Materials Research*, 2013, pp. 125–128. doi: 10.4028/www.scientific.net/AMR.771.125.
64. J. Chevalier, O. Tillement, and F. Ayela, "Rheological properties of nanofluids flowing through microchannels," *Appl Phys Lett*, vol. 91, no. 23, 2007, doi: 10.1063/1.2821117.
65. P. K. Namburu, D. P. Kulkarni, D. Misra, and D. K. Das, "Viscosity of copper oxide nanoparticles dispersed in ethylene glycol and water mixture," *Exp Therm Fluid Sci*, vol. 32, no. 2, pp. 397–402, Nov. 2007, doi: 10.1016/j.expthermflusci.2007.05.001.
66. M. Saeedinia, M. A. Akhavan-Behabadi, and P. Razi, "Thermal and rheological characteristics of CuO-Base oil nanofluid flow inside a circular tube," *International Communications in Heat and Mass Transfer*, vol. 39, no. 1, pp. 152–159, Jan. 2012, doi: 10.1016/j.icheatmasstransfer.2011.08.001.
67. J. Albadr, S. Tayal, and M. Alasadi, "Heat transfer through heat exchanger using Al2O3 nanofluid at different concentrations," *Case Studies in Thermal Engineering*, vol. 1, no. 1, pp. 38–44, Oct. 2013, doi: 10.1016/j.csite.2013.08.004.
68. B. C. Pak and Y. I. Cho, "Hydrodynamic and heat transfer study of dispersed fluids with submicron metallic oxide particles," *Experimental Heat Transfer*, vol. 11, no. 2, pp. 151–170, 1998, doi: 10.1080/08916159808946559.
69. R. Vidhya, T. Balakrishnan, and B. S. Kumar, "Investigation on thermophysical properties and heat transfer performance of heat pipe charged with binary mixture based ZnO-MgO hybrid nanofluids," in *Materials Today: Proceedings*, Elsevier Ltd, 2020, pp. 3423–3433. doi: 10.1016/j.matpr.2020.09.284.
70. H. Yarmand *et al.*, "Graphene nanoplatelets-silver hybrid nanofluids for enhanced heat transfer," *Energy Convers Manag*, vol. 100, pp. 419–428, Aug. 2015, doi: 10.1016/j.enconman.2015.05.023.
71. V. V. Wanatasanappan, M. Z. Abdullah, and P. Gunnasegaran, "Thermophysical properties of Al2O3-CuO hybrid nanofluid at different nanoparticle mixture ratio: An experimental approach," *J Mol Liq*, vol. 313, Sep. 2020, doi: 10.1016/j.molliq.2020.113458.
72. O. Hozien, W. M. El-Maghlany, M. M. Sorour, and Y. S. Mohamed, "Experimental study on thermophysical properties of TiO2, ZnO and Ag water base nanofluids," *J Mol Liq*, vol. 334, Jul. 2021, doi: 10.1016/j.molliq.2021.116128.
73. N. Asokan, P. Gunnasegaran, and V. Vicki Wanatasanappan, "Experimental investigation on the thermal performance of compact heat exchanger and the rheological properties of low concentration mono and hybrid nanofluids containing Al2O3 and CuO nanoparticles," *Thermal Science and Engineering Progress*, vol. 20, Dec. 2020, doi: 10.1016/j.tsep.2020.100727.
74. P. Sivashanmugam, "Application of Nanofluids in Heat Transfer," in *An Overview of Heat Transfer Phenomena*, InTech, 2012. doi: 10.5772/52496.
75. A. K. Tiwar, V. Kumar, Z. Said, and H. K. Paliwal, "A review on the application of hybrid nanofluids for parabolic trough collector: Recent progress and outlook," Apr. 10, 2021, *Elsevier Ltd*. doi: 10.1016/j.jclepro.2021.126031.
76. K. S. Arjun and K. Rakesh, "Heat transfer enhancement using alumina nanofluid in circular micro channel," *J Eng Sci Technol*, vol. 12, no. 1, pp. 265–279, 2017.
77. A. Esmaeilnejad, H. Aminfar, and M. S. Neistanak, "Numerical investigation of forced convection heat transfer through microchannels with non-Newtonian nanofluids," *International Journal of Thermal Sciences*, vol. 75, pp. 76–86, 2014, doi: 10.1016/j.ijthermalsci.2013.07.020.

78. T. Wen, L. Lu, H. Zhong, and B. Shen, "Thermal properties measurement and performance evaluation of water/ZnO nanofluid in a mini channel with offset fins," *Int J Heat Mass Transf*, vol. 162, Dec. 2020, doi: 10.1016/j.ijheatmasstransfer.2020.120361.
79. A. Amrollahi, A. M. Rashidi, R. Lotfi, M. Emami Meibodi, and K. Kashefi, "Convection heat transfer of functionalized MWNT in aqueous fluids in laminar and turbulent flow at the entrance region," *International Communications in Heat and Mass Transfer*, vol. 37, no. 6, pp. 717–723, Jul. 2010, doi: 10.1016/j.icheatmasstransfer.2010.03.003.
80. A. A. Hussien, M. Z. Abdullah, N. M. Yusop, M. A. Al-Nimr, M. A. Atieh, and M. Mehrali, "Experiment on forced convective heat transfer enhancement using MWCNTs/GNPs hybrid nanofluid and mini-tube," *Int J Heat Mass Transf*, vol. 115, pp. 1121–1131, 2017, doi: 10.1016/j.ijheatmasstransfer.2017.08.120.
81. B.-H. Chun, H. U. Kang, and S. H. Kim, "Effect of alumina nanoparticles in the fluid on heat transfer in double-pipe heat exchanger system," *Korean J Chem Eng*, vol. 25, pp. 966–971, 2008, doi: 10.1007/s11814-008-0156-5.
82. W. Duangthongsuk and S. Wongwises, "Heat transfer enhancement and pressure drop characteristics of TiO2-water nanofluid in a double-tube counter flow heat exchanger," *Int J Heat Mass Transf*, vol. 52, no. 7–8, pp. 2059–2067, Mar. 2009, doi: 10.1016/j.ijheatmasstransfer.2008.10.023.
83. W. Duangthongsuk and S. Wongwises, "An experimental study on the heat transfer performance and pressure drop of TiO2-water nanofluids flowing under a turbulent flow regime," *Int J Heat Mass Transf*, vol. 53, no. 1–3, pp. 334–344, Jan. 2010, doi: 10.1016/j.ijheatmasstransfer.2009.09.024.
84. L. G. Asirvatham, B. Raja, D. Mohan Lal, and S. Wongwises, "Convective heat transfer of nanofluids with correlations," *Particuology*, vol. 9, no. 6, pp. 626–631, Dec. 2011, doi: 10.1016/j.partic.2011.03.014.
85. T. P. Teng, T. C. Hsiao, and C. C. Chung, "Characteristics of carbon-based nanofluids and their application in a brazed plate heat exchanger under laminar flow," *Appl Therm Eng*, vol. 146, pp. 160–168, Jan. 2019, doi: 10.1016/j.applthermaleng.2018.09.125.
86. D. Huang, Z. Wu, and B. Sundén, "Effects of hybrid nanofluid mixture in plate heat exchangers," *Exp Therm Fluid Sci*, vol. 72, pp. 190–196, Apr. 2016, doi: 10.1016/j.expthermflusci.2015.11.009.
87. V. Kumar, A. K. Tiwari, and S. K. Ghosh, "Effect of variable spacing on performance of plate heat exchanger using nanofluids," *Energy*, vol. 114, pp. 1107–1119, Nov. 2016, doi: 10.1016/j.energy.2016.08.091.
88. T. Maré, S. Halelfadl, O. Sow, P. Estellé, S. Duret, and F. Bazantay, "Comparison of the thermal performances of two nanofluids at low temperature in a plate heat exchanger," *Exp Therm Fluid Sci*, vol. 35, no. 8, pp. 1535–1543, 2011, doi: 10.1016/j.expthermflusci.2011.07.004.
89. D. Zheng, J. Wang, Z. Chen, J. Baleta, and B. Sundén, "Performance analysis of a plate heat exchanger using various nanofluids," *Int J Heat Mass Transf*, vol. 158, Sep. 2020, doi: 10.1016/j.ijheatmasstransfer.2020.119993.
90. Z. Wang, Z. Wu, F. Han, L. Wadsö, and B. Sundén, "Experimental comparative evaluation of a graphene nanofluid coolant in miniature plate heat exchanger," *International Journal of Thermal Sciences*, vol. 130, pp. 148–156, Aug. 2018, doi: 10.1016/j.ijthermalsci.2018.04.021.
91. R. Barzegarian, A. Aloueyan, and T. Yousefi, "Thermal performance augmentation using water based Al2O3-gamma nanofluid in a horizontal shell and tube heat exchanger under forced circulation," *International Communications in Heat and Mass Transfer*, vol. 86, pp. 52–59, Aug. 2017, doi: 10.1016/j.icheatmasstransfer.2017.05.021.
92. M. M. Sarafraz, V. Nikkhah, M. Nakhjavani, and A. Arya, "Fouling formation and thermal performance of aqueous carbon nanotube nanofluid in a heat sink with rectangular parallel microchannel," *Appl Therm Eng*, vol. 123, pp. 29–39, 2017, doi: 10.1016/j.applthermaleng.2017.05.056.
93. M. Mohammadi, A. Taheri, M. Passandideh-Fard, and M. Sardarabadi, "Electronic chipset thermal management using a nanofluid-based mini-channel heat sink: An experimental study," *International Communications in Heat and Mass Transfer*, vol. 118, Nov. 2020, doi: 10.1016/j.icheatmasstransfer.2020.104836.
94. Y. Vermahmoudi, S. M. Peyghambarzadeh, S. H. Hashemabadi, and M. Naraki, "Experimental investigation on heat transfer performance of Fe 2 O 3/water nanofluid in an air-finned heat exchanger," *European Journal of Mechanics, B/Fluids*, vol. 44, pp. 32–41, 2014, doi: 10.1016/j.euromechflu.2013.10.002.
95. T. P. Teng, Y. C. Hsu, W. P. Wang, and Y. B. Fang, "Performance assessment of an air-cooled heat exchanger for multiwalled carbon nanotubes-water nanofluids," *Appl Therm Eng*, vol. 89, pp. 346–355, Jul. 2015, doi: 10.1016/j.applthermaleng.2015.06.042.
96. K. Y. Leong, R. Saidur, S. N. Kazi, and A. H. Mamun, "Performance investigation of an automotive car radiator operated with nanofluid-based coolants (nanofluid as a coolant in a radiator)," *Appl Therm Eng*, vol. 30, no. 17–18, pp. 2685–2692, Dec. 2010, doi: 10.1016/j.applthermaleng.2010.07.019.

97. S. Suresh, K. P. Venkitaraj, P. Selvakumar, and M. Chandrasekar, "Effect of Al 2O 3-Cu/water hybrid nanofluid in heat transfer," *Exp Therm Fluid Sci*, vol. 38, pp. 54–60, Apr. 2012, doi: 10.1016/j.expthermflusci.2011.11.007.
98. M. Khoshvaght-Aliabadi, F. Hormozi, and A. Zamzamian, "Experimental analysis of thermal-hydraulic performance of copper-water nanofluid flow in different plate-fin channels," *Exp Therm Fluid Sci*, vol. 52, pp. 248–258, Jan. 2014, doi: 10.1016/j.expthermflusci.2013.09.018.
99. A. Zamzamian, S. N. Oskouie, A. Doosthoseini, A. Joneidi, and M. Pazouki, "Experimental investigation of forced convective heat transfer coefficient in nanofluids of Al2O3/EG and CuO/EG in a double pipe and plate heat exchangers under turbulent flow," *Exp Therm Fluid Sci*, vol. 35, no. 3, pp. 495–502, Apr. 2011, doi: 10.1016/j.expthermflusci.2010.11.013.
100. D. Huang, Z. Wu, and B. Sundén, "Effects of hybrid nanofluid mixture in plate heat exchangers," *Exp Therm Fluid Sci*, vol. 72, pp. 190–196, Apr. 2016, doi: 10.1016/j.expthermflusci.2015.11.009.
101. M. N. Pantzali, A. A. Mouza, and S. V. Paras, "Investigating the efficacy of nanofluids as coolants in plate heat exchangers (PHE)," *Chem Eng Sci*, vol. 64, no. 14, pp. 3290–3300, Jul. 2009, doi: 10.1016/j.ces.2009.04.004.
102. N. S. Pandya, H. Shah, M. Molana, and A. K. Tiwari, "Heat transfer enhancement with nanofluids in plate heat exchangers: A comprehensive review," May 01, 2020, *Elsevier Ltd.* doi: 10.1016/j.euromechflu.2020.02.004.
103. Y. Wang, K. H. Deng, B. Liu, J. M. Wu, and G. H. Su, "A correlation of nanofluid flow boiling heat transfer based on the experimental results of AlN/H2O and Al2O3/H2O nanofluid," *Exp Therm Fluid Sci*, vol. 80, pp. 376–383, Jan. 2017, doi: 10.1016/j.expthermflusci.2016.08.020.
104. S. M. You, J. H. Kim, and K. H. Kim, "Effect of nanoparticles on critical heat flux of water in pool boiling heat transfer," *Appl Phys Lett*, vol. 83, no. 16, pp. 3374–3376, Oct. 2003, doi: 10.1063/1.1619206.
105. S. J. Kim, I. C. Bang, J. Buongiorno, and L. W. Hu, "Surface wettability change during pool boiling of nanofluids and its effect on critical heat flux," *Int J Heat Mass Transf*, vol. 50, no. 19–20, pp. 4105–4116, Sep. 2007, doi: 10.1016/j.ijheatmasstransfer.2007.02.002.
106. C. Y. Park, Y. Jang, B. Kim, and Y. Kim, "Flow boiling heat transfer coefficients and pressure drop of FC-72 in microchannels," *Int J Multiphase Flow*, vol. 39, pp. 45–54, Mar. 2012, doi: 10.1016/j.ijmultiphaseflow.2011.11.002.
107. X. Fang, Y. Chen, H. Zhang, W. Chen, A. Dong, and R. Wang, "Heat transfer and critical heat flux of nanofluid boiling: A comprehensive review," 2016, *Elsevier Ltd.* doi: 10.1016/j.rser.2016.05.047.
108. M. S. Kamel, F. Lezsovits, and A. K. Hussein, "Experimental studies of flow boiling heat transfer by using nanofluids: A critical recent review," Dec. 01, 2019, *Springer Netherlands.* doi: 10.1007/s10973-019-08333-2.
109. K. Maqsood, A. Ali, S. Ilyas, S. Garg, M. D.- Chemosphere, and undefined 2022, "Multi-objective optimization of thermophysical properties of multiwalled carbon nanotubes based nanofluids," *Elsevier*, Accessed: Jun. 22, 2024. [Online]. Available: https://www.sciencedirect.com/science/article/pii/S0045653521021627.
110. S. Bagherzadeh, M. Sulgani, … V. N.-P. A. S., and undefined 2019, "By the new proposed multi-objective optimization/statistical model composed of 'ANN+ Genetic Algorithm' based on empirical data of CuO/paraffin nanofluid in a pipe," *Elsevier*, Accessed: Jun. 22, 2024. [Online]. Available: https://www.sciencedirect.com/science/article/pii/S0378437119306454
111. H. Safikhani, A. Abbassi, A. Khalkhali, and M. Kalteh, "Multi-objective optimization of nanofluid flow in flat tubes using CFD, Artificial Neural Networks and genetic algorithms," *Advanced Powder Technology*, vol. 25, no. 5, pp. 1608–1617, Sep. 2014, doi: 10.1016/j.apt.2014.05.014.
112. I. O. Alade, T. A. Oyehan, I. K. Popoola, S. O. Olatunji, and A. Bagudu, "Modeling thermal conductivity enhancement of metal and metallic oxide nanofluids using support vector regression," *Advanced Powder Technology*, vol. 29, no. 1, pp. 157–167, Jan. 2018, doi: 10.1016/j.apt.2017.10.023.
113. Y. Si *et al.*, "Prediction and evaluation of energy and exergy efficiencies of a nanofluid-based photovoltaic-thermal system with a needle finned serpentine channel using random forest machine learning approach," *Eng Anal Bound Elem*, vol. 151, pp. 328–343, Jun. 2023, doi: 10.1016/j.enganabound.2023.03.009.
114. C. T. Nguyen, G. Roy, C. Gauthier, and N. Galanis, "Heat transfer enhancement using Al2O3-water nanofluid for an electronic liquid cooling system," *Appl Therm Eng*, vol. 27, no. 8–9, pp. 1501–1506, Jun. 2007, doi: 10.1016/j.applthermaleng.2006.09.028.
115. H. Pourpasha, P. Farshad, and S. Zeinali Heris, "Modeling and optimization the effective parameters of nanofluid heat transfer performance using artificial neural network and genetic algorithm method," *Energy Reports*, vol. 7, pp. 8447–8464, Nov. 2021, doi: 10.1016/j.egyr.2021.10.121.

116. R. Mohapatra, S. Panda, and S. R. Mishra, "Exploring heat transfer enhancement: Machine learning predictions using artificial neural network for water-based Cu and CuO micropolar nanofluid transportation over a radiating surface," *Bionanoscience*, Jun. 2024, doi: 10.1007/s12668-024-01392-2.
117. M. Najmi and A. Hassan, "A review of pool boiling heat transfer properties by nanofluid," *J Teknol*, vol. 85, no. 3, pp. 1–13, May 2023, doi: 10.11113/jurnalteknologi.v85.18324.
118. O. A. Alawi *et al.*, "Development of optimized machine learning models for predicting flat plate solar collectors thermal efficiency associated with Al2O3-water nanofluids," *Eng Appl Artif Intell*, vol. 133, Jul. 2024, doi: 10.1016/j.engappai.2024.108158.
119. S. Jana, A. Salehi-Khojin, W. Z.-T. acta, and undefined 2007, "Enhancement of fluid thermal conductivity by the addition of single and hybrid nano-additives," *Elsevier*, Accessed: Jul. 19, 2024. [Online]. Available: https://www.sciencedirect.com/science/article/pii/S0040603107002675
120. S. Murshed, K. Leong, C. Y.-I. J. of thermal, and undefined 2005, "Enhanced thermal conductivity of TiO2—water based nanofluids," *ElsevierSMS Murshed, KC Leong, C YangInternational Journal of thermal sciences, 2005•Elsevier*, Accessed: Jul. 19, 2024. [Online]. Available: https://www.sciencedirect.com/science/article/pii/S129007290500013X
121. X. Wang, D. Z.-C. P. Letters, and undefined 2009, "Investigation of pH and SDBS on enhancement of thermal conductivity in nanofluids," *Elsevier*, Accessed: Jul. 19, 2024. [Online]. Available: https://www.sciencedirect.com/science/article/pii/S0009261409000475
122. S. Thanikodi, K. Singaravelu, C. Devarajan, V. Venkatraman, and V. Rathinavelu, "Teaching learning optimization and neural network for the effective prediction of heat transfer rates in tube heat exchangers," *doiserbia.nb.rs*, doi: 10.2298/TSCI190714438T.
123. H. Upreti, Z. Uddin, … A. P.-N. H. T., and undefined 2023, "Particle swarm optimization based numerical study for pressure, flow, and heat transfer over a rotating disk with temperature dependent nanofluid properties," *Taylor & Francis*, vol. 83, no. 8, pp. 815–844, 2023, doi: 10.1080/10407782.2022.2156412.
124. A. Starke, A. da S.-H. Transfer, P. B. Fundamentals, and undefined 2023, "A review on the applicability of machine learning techniques to the metamodeling of energy systems," *Taylor & Francis*, 2023, doi: 10.1080/10407790.2023.2280208.
125. N. Ahmed, M. Assadi, A. A. Ahmed, and R. Banihabib, "Optimal design, operational controls, and data-driven machine learning in sustainable borehole heat exchanger coupled heat pumps: Key implementation challenges and advancement opportunities," Jun. 01, 2023, *Elsevier B.V.* doi: 10.1016/j.esd.2023.04.004.
126. S. H. Godasiaei and A. J. Chamkha, "Advancing heat transfer modeling through machine learning: A focus on forced convection with nanoparticles," *Numeri Heat Transf A Appl*, 2023, doi: 10.1080/10407782.2023.2299734.
127. A. Jokar, … S. O. T., and undefined 2013, "Heat transfer and fluid flow analysis of nanofluids in corrugated plate heat exchangers using computational fluid dynamics simulation," *asmedigitalcollection.asme.org*, Accessed: Jun. 22, 2024. [Online]. Available: https://asmedigitalcollection.asme.org/thermalscienceapplication/article-abstract/5/1/011002/378931.
128. M. Uddin, M. Alam, M. R.-A. J. for S. and, and undefined 2017, "Natural convective heat transfer flow of nanofluids inside a quarter-circular enclosure using nonhomogeneous dynamic model," *Springer*, vol. 42, no. 5, pp. 1883–1901, May 2017, doi: 10.1007/s13369-016-2330-0.
129. P. Gunnasegaran *et al.*, "Simulation of heat transfer enhancement in nanofluids using dissipative particle dynamics," *Elsevier*, vol. 2012, p. 11, Apr. 2012, doi: 10.5402/2012/585496.
130. J. Buongiorno, "Convective transport in nanofluids," 2006, doi: 10.1115/1.2150834.
131. J. Koo, C. K.-J. of N. research, and undefined 2004, "A new thermal conductivity model for nanofluids," *SpringerJ Koo, C KleinstreuerJournal of Nanoparticle research, 2004•Springer*, vol. 6, no. 6, pp. 577–588, Dec. 2004, doi: 10.1007/s11051-004-3170-5.
132. S. Choi and J. Eastman, "Enhancing thermal conductivity of fluids with nanoparticles," 1995, Accessed: Jul. 27, 2024. [Online]. Available: https://www.osti.gov/biblio/196525.
133. I. Gonçalves *et al.*, "Thermal conductivity of nanofluids: A review on prediction models, controversies and challenges," *mdpi.comI Gonçalves, R Souza, G Coutinho, J Miranda, A Moita, JE Pereira, A Moreira, R LimaApplied Sciences, 2021•mdpi.com*, 2021, doi: 10.3390/app11062525.
134. J. Lee, I. M.-I. J. of H. and M. Transfer, and undefined 2007, "Assessment of the effectiveness of nanofluids for single-phase and two-phase heat transfer in micro-channels," *Elsevier*, Accessed: Jul. 27, 2024. [Online]. Available: https://www.sciencedirect.com/science/article/pii/S0017931006004649.

135. R. Vajjha, D. D.-I. journal of heat and mass transfer, and undefined 2012, "A review and analysis on influence of temperature and concentration of nanofluids on thermophysical properties, heat transfer and pumping power," *Elsevier*, Accessed: Jun. 22, 2024. [Online]. Available: https://www.sciencedirect.com/science/article/pii/S0017931012001974.
136. D. Mohite, A. Goyal, A. Singh, … M. A.-M. T., and undefined 2024, "Improvement of thermal performance through nanofluids in industrial applications: A review on technical aspects," *Elsevier*, Accessed: Jun. 22, 2024. [Online]. Available: https://www.sciencedirect.com/science/article/pii/S2214785324002712.
137. K. Cacua, R. Buitrago-Sierra, B. Herrera, E. Pabón, and S. M. S. Murshed, "Nanofluids' stability effects on the thermal performance of heat pipes: A critical review," *J Therm Anal Calorim*, vol. 136, no. 4, pp. 1597–1614, May 2019, doi: 10.1007/S10973-018-7787-5.
138. Y. Xuan, Q. L.-J. H. transfer, and undefined 2003, "Investigation on convective heat transfer and flow features of nanofluids," *asmedigitalcollection.asme.orgY Xuan, Q LiJ. Heat transfer, 2003•asmedigital collection.asme.org*, 2003, doi: 10.1115/1.1532008.
139. Z. Haddad, H. Oztop, E. Abu-Nada, A. M.-R. and Sustainable, and undefined 2012, "A review on natural convective heat transfer of nanofluids," *Elsevier*, Accessed: Jun. 22, 2024. [Online]. Available: https://www.sciencedirect.com/science/article/pii/S1364032112002602.
140. X. Q. Wang and A. S. Mujumdar, "Heat transfer characteristics of nanofluids: a review," Jan. 2007. doi: 10.1016/j.ijthermalsci.2006.06.010.
141. A. Ghadimi, R. Saidur, and H. S. C. Metselaar, "A review of nanofluid stability properties and characterization in stationary conditions," *Int J Heat Mass Transf*, vol. 54, no. 17–18, pp. 4051–4068, Aug. 2011, doi: 10.1016/j.ijheatmasstransfer.2011.04.014.
142. S. Das, S. Choi, W. Yu, and T. Pradeep, *Nanofluids: science and technology*. 2007. Accessed: Jul. 24, 2024. [Online]. Available: https://books.google.com/books?hl=en&lr=&id=QpbOpMSpXpYC&oi=fnd&pg=PR7&dq=Das,+S.+K.,+Choi,+S.+U.+S.,+Yu,+W.,+%26+Pradeep,+T.+(2007).+Nanofluids:+Science+and+Technology.&ots=fn8ZlIZfTO&sig=TYIQO4X-zZgVmZ8Q5Oq-2p_E3hA.
143. E. Assadpour and S. Mahdi Jafari, "A systematic review on nanoencapsulation of food bioactive ingredients and nutraceuticals by various nanocarriers," Oct. 28, 2019, *Taylor and Francis Inc.* doi: 10.1080/10408398.2018.1484687.
144. C. Kim and K. Kwak, "Viscosity of thermal conductivity of copper oxide nanofluid dispersed in ethylene glycol viscosity and thermal conductivity of copper oxide nanofluid dispersed in ethylene glycol," 2005. [Online]. Available: https://www.researchgate.net/publication/279887639.
145. S. Z. Heris, S. G. Etemad, and M. N. Esfahany, "Experimental investigation of oxide nanofluids laminar flow convective heat transfer," *International Communications in Heat and Mass Transfer*, vol. 33, no. 4, pp. 529–535, Apr. 2006, doi: 10.1016/j.icheatmasstransfer.2006.01.005.
146. H. Chen, Y. Ding, Y. He, and C. Tan, "Rheological behaviour of ethylene glycol based titania nanofluids," *Chem Phys Lett*, vol. 444, no. 4–6, pp. 333–337, Aug. 2007, doi: 10.1016/j.cplett.2007.07.046.
147. M. Kole and T. K. Dey, "Viscosity of alumina nanoparticles dispersed in car engine coolant," *Exp Therm Fluid Sci*, vol. 34, no. 6, pp. 677–683, Sep. 2010, doi: 10.1016/j.expthermflusci.2009.12.009.
148. G. Oberdörster, E. Oberdörster, and J. Oberdörster, "Nanotoxicology: An emerging discipline evolving from studies of ultrafine particles," Jul. 2005. doi: 10.1289/ehp.7339.
149. A. Nel, T. Xia, L. Mädler, and N. Li, "Toxic potential of materials at the nanolevel," [Online]. Available: https://www.science.org
150. M. J. Lourenço, J. Alexandre, C. Huisman, X. Paredes, and C. N. de Castro, "The balance between energy, environmental security, and technical performance: The regulatory challenge of nanofluidsc," *Nanomaterials*, vol. 11, no. 8, 2021, 1871, doi: 10.3390/nano11081871.
151. M. A. Kiser, P. Westerhoff, T. Benn, Y. Wang, J. Pérez-Rivera, and K. Hristovski, "Titanium nanomaterial removal and release from wastewater treatment plants," *Environ Sci Technol*, vol. 43, no. 17, pp. 6757–6763, Sep. 2009, doi: 10.1021/es901102n.
152. R. Hischier and T. Walser, "Life cycle assessment of engineered nanomaterials: State of the art and strategies to overcome existing gaps," *Science of the Total Environment*, vol. 425, pp. 271–282, May 2012, doi: 10.1016/j.scitotenv.2012.03.001.
153. M. Chandrasekar, S. Suresh, and T. Senthilkumar, "Mechanisms proposed through experimental investigations on thermophysical properties and forced convective heat transfer characteristics of various nanofluids - A review," Aug. 2012. doi: 10.1016/j.rser.2012.03.013.

154. J. A. Eastman, S. U. S. Choi, S. Li, W. Yu, and L. J. Thompson, "Anomalously increased effective thermal conductivities of ethylene glycol-based nanofluids containing copper nanoparticles," *Appl Phys Lett*, vol. 78, no. 6, pp. 718–720, Feb. 2001, doi: 10.1063/1.1341218.
155. M. Jama *et al.*, "Critical review on nanofluids: Preparation, characterization, and applications," 2016, *Hindawi Limited*. doi: 10.1155/2016/6717624.
156. L. S. Sundar, K. V. Sharma, M. T. Naik, and M. K. Singh, "Empirical and theoretical correlations on viscosity of nanofluids: A review," 2013, *Elsevier Ltd*. doi: 10.1016/j.rser.2013.04.003.
157. K. Wu *et al.*, "Corrosion behavior of brass in SCW-SDBS-TiO2 nanofluid," *J Alloys Compd*, vol. 855, Feb. 2021, doi: 10.1016/j.jallcom.2020.157483.
158. X. Wang, X. Xu, and S. U. S. Choi, "Thermal conductivity of nanoparticle-fluid mixture," *J Thermophys Heat Trans*, vol. 13, no. 4, pp. 474–480, 1999, doi: 10.2514/2.6486.
159. W. Yu, H. Xie, and W. Chen, "Experimental investigation on thermal conductivity of nanofluids containing graphene oxide nanosheets," *J Appl Phys*, vol. 107, no. 9, May 2010, doi: 10.1063/1.3372733.
160. W. T. Urmi, M. M. Rahman, K. Kadirgama, D. Ramasamy, and M. A. Maleque, "An overview on synthesis, stability, opportunities and challenges of nanofluids," in *Materials Today: Proceedings*, Elsevier Ltd, 2020, pp. 30–37. doi: 10.1016/j.matpr.2020.10.998.
161. J. Koo, C. K.-J. of N. research, and undefined 2004, "A new thermal conductivity model for nanofluids," *SpringerJ Koo, C KleinstreuerJournal of Nanoparticle research, 2004•Springer*, vol. 6, no. 6, pp. 577–588, Dec. 2004, doi: 10.1007/s11051-004-3170-5.
162. B. Pak, Y. C.-E. H. T. an International, and undefined 1998, "Hydrodynamic and heat transfer study of dispersed fluids with submicron metallic oxide particles," *Taylor & FrancisBC Pak, YI ChoExperimental Heat Transfer an International Journal, 1998•Taylor & Francis*, vol. 11, no. 2, pp. 151–170, 2007, doi: 10.1080/08916159808946559.
163. W. Yu, H. X.-J. of nanomaterials, and undefined 2012, "A review on nanofluids: preparation, stability mechanisms, and applications," *Wiley Online LibraryW Yu, H XieJournal of nanomaterials, 2012•Wiley Online Library*, vol. 2012, p. 17, 2012, doi: 10.1155/2012/435873.
164. D. Wen and Y. Ding, "Experimental investigation into convective heat transfer of nanofluids at the entrance region under laminar flow conditions," *Int J Heat Mass Transf*, vol. 47, no. 24, pp. 5181–5188, Nov. 2004, doi: 10.1016/j.ijheatmasstransfer.2004.07.012.
165. J. A. Eastman, S. U. S. Choi, S. Li, W. Yu, and L. J. Thompson, "Anomalously increased effective thermal conductivities of ethylene glycol-based nanofluids containing copper nanoparticles," *Appl Phys Lett*, vol. 78, no. 6, pp. 718–720, Feb. 2001, doi: 10.1063/1.1341218.
166. S. Z. Heris, S. G. Etemad, and M. N. Esfahany, "Convective heat transfer of a cu/water nanofluid flowing through a circular tube," *Experimental Heat Transfer*, vol. 22, no. 4, pp. 217–227, Dec. 2009, doi: 10.1080/08916150902950145.

6 Phase Change Materials in Heat Exchangers

*Mohd Naqueeb Shaad Jagirdar,
Kaustav Moni Goswami, and Hakeem Niyas*

6.1 INTRODUCTION

Phase change materials (PCMs) have recently gained popularity due to their high storage capacity in isothermal conditions. Energy can be stored as either sensible or latent heat. Water is a cost-effective and efficient way to store heat. However, low energy density leads to a large storage mechanism. PCMs have been used for their high latent heat of transition, energy density, and low cost [1]. PCMs have the capability to store significant amounts of heat or cold, even with minimal temperature fluctuations. PCM has a heat storage capacity that is approximately three to four times higher than the amount of heat accumulated as sensible heat in solids or liquids within a 20°C temperature range, which can provide a substantial benefit in certain applications, such as in the context of residential space conditioning [2]. In addition to conventional PCMs, researchers are also investigating shape-stabilized PCMs that can address the limitations of a typical PCM. Hence, PCMs possess numerous prospective uses, particularly in the realm of buildings that are energy efficient [3].

The world encounters a lot of solar energy and using it could substantially cut down on fossil fuel use. Connecting the times when energy is available and when it is needed is hard because it is not always accessible. So, it makes sense to use thermal energy storage (TES) to get more use out of this tremendous energy source [4]. Based on how they store energy, these materials are divided into three groups: chemical, sensible, and latent heat storage (LHS). A particularly difficult approach to energy storage is the chemical type, which includes various chemical reactions. Another problem is that these processes cannot be reversed completely in practical cases, and the chemicals are not stable [5]. Sensible heat storage (SHS) is the most prevalent kind of storage system due to its lower expenses and simplicity. However, due to its inadequate storage capacity, large systems are required. LHS is the most effective method for achieving a greater heat storage capacity without chemical reactions. High energy storage density is the salient feature of this system. However, the primary drawback of LHS is its low thermal conductivity and, thus, its lower heat transfer rate. Various research studies are being conducted in order to improve the transfer of heat [6].

There are three main types of PCMs: organic, inorganic, and eutectic, as classified in Figure 6.1. The major differences and pros and cons are briefly discussed in this section. Organic PCMs have been the subject of many successful tests in various fields, including space heating, electronic device cooling, solar heating, and automotive cabin cooling [7, 8]. The primary advantage of organic PCMs is that they do not separate into phases, and their latent heat of fusion stays the same. Organic PCMs have a low melting point, which allows them to be used in low-temperature situations. It takes a higher surface area for organic PCMs to move heat at the rate that is needed because they do not conduct heat well. Inorganic PCM is classified as salt hydrate and metallic [9]. These materials possess a considerable volumetric density for latent heat, better cyclic stability, and excellent thermal conductivity. Yet, salt hydrates are prone to phase separation. Because inorganic salts are extremely corrosive, the container must be made of highly corrosion-resistant materials,

 DOI: 10.1201/9781003509837-6

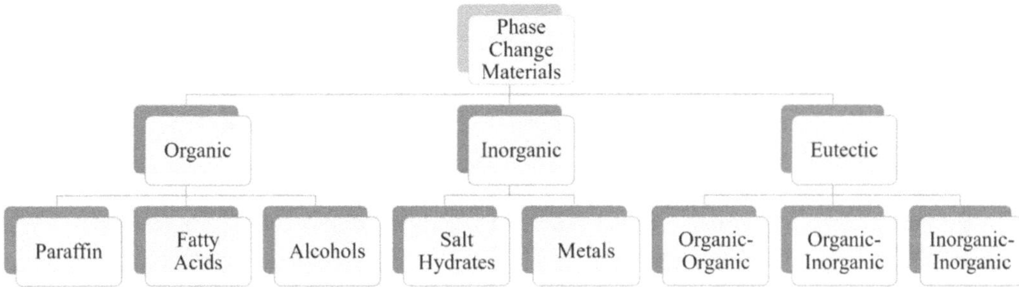

FIGURE 6.1 Classification of PCM [12].

which raises the total expense. Metallic PCMs can be used in very high temperature applications such as nuclear power plants [10].

The third type of PCM is eutectic. Eutectic PCM is a mixture of multiple substances, with all elementary substances changing phase consistently to form a crystallized mixture with a defined melting and solidification temperature. During liquefying and freezing, this type does not segregate into phases. They freeze into a mixture of crystals and melt instantaneously without separation [11]. This chapter reviews the impact of employing PCM on thermal energy systems' performance.

6.1.1 Principles of PCM

This section includes an extensive introduction to the functionality of LHS materials. Thermal energy can be preserved in both forms, SHS and LHS; nevertheless, a key distinction is that a temperature increase follows SHS, whereas LHS occurs at a steady temperature (Figure 6.2) [13]. In this instance, heat is utilized to transform the phase of the substance from solid to fluid at a fixed temperature, and it is conserved in the substance as latent heat. This flow is depicted in Figure 6.3 [14, 15]. As observed, during the phase transition process, the temperature stays the same (there's no shift in sensible heat) as its internal energy rises.

As illustrated in Figure 6.3, LHS is categorized into three distinct categories: solid-solid, solid-liquid, and liquid-gas [14]. The most appealing kind is solid-liquid, which can deliver a great energy storage density with a little volumetric change. The latent heat of fusion determines energy storage density and indicates how much energy is stored in the substance. This is the primary benefit of solid-liquid (D-E) PCMs, which can store a specific quantity of energy in a significantly low mass and size of the material. Solid-solid (B-C) transitions usually have a low latent heat. At the same time, liquid-gas (F-G) shifts are undesirable in closed systems owing to the substantial volumetric changes. The design parameter-based studies are important in incorporating PCM in HE. General considerations are (a) the ratio of shell diameter to tube diameter and (b) configuration.

(a) The ratio of shell diameter to tube diameter. This ratio has the greatest impact on the overall melting time, solidification time, and stored energy. As the ratio increases, the quantity of PCM also increases, increasing stored energy [16].
(b) Configurations can be categorized as horizontal, vertical, or inclined. The horizontal layout delivers the best results regarding the charging process. Conduction is the principal mode of heat transmission that occurs during the discharge process. This means that the configuration does not substantially impact the amount of time it takes for the solidification to occur [16].

Researchers have also considered nanoparticle-incorporated PCM to improve thermal performance. Ho and Gao [17] experimented to explore the effect of introducing alumina (Al_2O_3) nanoparticles

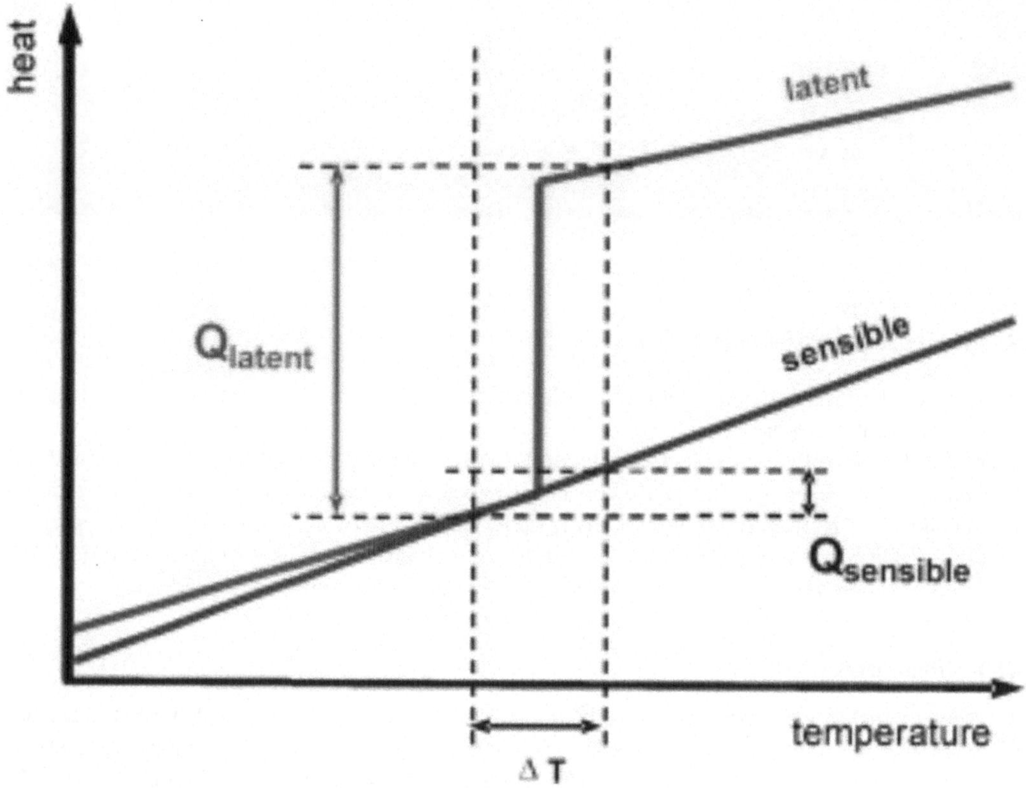

FIGURE 6.2 Operational mechanism of a storage system [13].

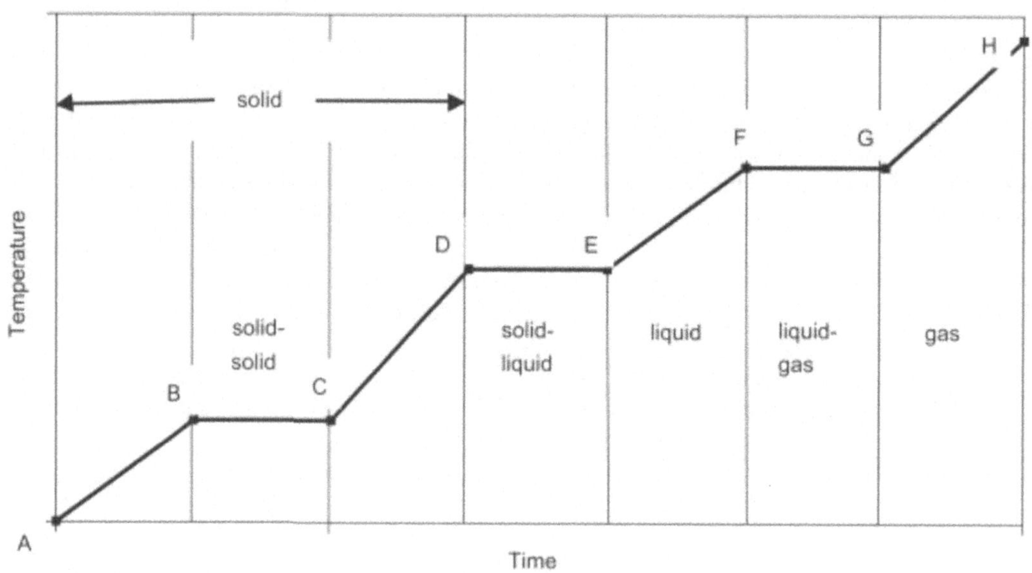

FIGURE 6.3 Thermal profile illustrating the phase changes based on temperature during the heating process of a material [14].

TABLE 6.1
Thermophysical Properties of PCM [18]

Property	Remark
Latent heat of fusion	Higher is preferable
Melting point	It should be within the desired operating temperature range
Specific heat	Higher is preferable
Density	Higher density results in a compact module

into n-octadecane on the paraffin's thermophysical characteristics. The qualities included latent heat of fusion, density, dynamic viscosity, and thermal conductivity. When they compared the same nanoparticle mass fraction, they discovered that the relative increases in dynamic viscosity were nearly 20% and more than 28% for the paraffin that contained 5% and 10% of alumina particles, respectively, at a temperature of 30°C. These relative increases appear to be much greater than the relative enhancements in thermal conductivity, which were around ten and four times higher.

6.1.2 Thermal Properties

The thermal properties of PCMs make them ideal for various energy storage and management applications. These properties include Latent heat, specific heat, thermal conductivity, etc. Some required thermophysical properties are given in Table 6.1 [18]. Most industrial PCMs used to store thermal energy are solids like metal, inorganic salt, and paraffin before they change phases. PCMs that store and release heat energy can do so in both directions because their molecular structures change from ordered to disordered. It is usually possible to divide solid materials' heat conduction into three groups: electronic heat conduction, phonon heat conduction, and photon heat conduction [19]. The three processes listed work with different types of medium: free electrons, lattice vibration waves, and electromagnetic radiation [20]. Photon heat conduction works mostly through the radiation conduction of photons, which is a part of the process that is not very strong for most substances. Because of this, most studies have been done on electronic and phonon thermal conduction mechanisms [21].

6.1.3 Heat Transfer

The efficient rate of heat transfer is the most crucial aspect of any heat-related phenomenon [22]. Various studies have been undertaken to improve this rate inside the thermal system and minimize its thermal losses to enhance its efficiency. One of the most crucial things about materials used in thermal devices is how well they transfer heat. One known limitation of organic PCMs is that they cannot transfer heat well because they have poor thermal conductivity. This means that melting and freezing take a bit longer, which drops both the heat flow and the amount of heat that can be stored. Altering the composition, including conductive phases, enclosing the material (encapsulation), increasing the surface area, and utilizing mixes of PCMs are some ways the issue of low heat conductivity has been addressed [23]. PCMs improve the heat transference of solar thermal systems by letting more solar energy be used, cutting down on heat loss, improving heat distribution, and storing more heat. According to most studies, the primary benefits of using PCM are better heat transfer and decreased heat loss [24]. It is suggested that these areas be worked on to improve heat transfer.

6.1.4 Thermal Conductivity

Thermal conductivity is a critical metric to consider, particularly if the application necessitates a lot of power. Yet overall, PCM has lesser thermal conductivities of around 0.2 W/mK (certain PCMs, like metallics, have substantially higher values). To attain adequate charging and discharging power, PCM's thermal conductivity must be improved or enhanced. This issue has been handled in different ways. Initially, focusing on the PCM directly, usually via the inclusion of tiny particles with high conductivity (nanoparticles), results in nano-enhanced PCM [25]. Despite the continual publication of several publications in this field of study, recent research has revealed that this strategy has not yielded excellent results. The next goal is to enhance the rates at which heat is transferred by employing geometric methods. This involves optimizing the design parameters and operational features of the system, with particular attention to the HE between the PCM and the external heat source or heat sink [26]. To achieve the intended behavior, the heat storage and transmission mechanism must be properly designed.

6.2 DESIGN CONSIDERATION

6.2.1 Heat Exchanger Systems

LHS systems can be seen as HEs, with different materials used as PCM and heat transfer fluids (HTFs). At times, the PCM stays in the HE, while at other times, it is used as the HTF as a slurry. There will be three main types of storage systems: those with HEs, those with direct contact, and those that are modular via encapsulation of PCM. Table 6.2 illustrates the different ways of utilizing the PCM in HEs.

Depending on the application, HTFs, and further criteria, different varieties of HEs can be identified. For example, PCM HEs are suitable for high-temperature industrial applications with high flue gas contamination. The shape of this HE allows for easy removal and cleaning of heat transfer surfaces. When temperatures exceed 650°C, tubular HEs with helical, U-tubes, and straight tubes are commonly utilized in other applications [28, 29]. High-temperature application entails using equipment in a process with high-temperature streams. Different HE based on their temperature limit are shown in Figure 6.4.

Contemporary HE designs incorporate basic geometric forms, such as plates, sheets, and circular tubes, as the primary components. The geometry constraint was mostly due to restrictions in the manufacturing process. Recently, innovative production methods, such as additive manufacturing (AM), have emerged. This technique involves the precise addition of material using laser and/or electron beams to create complex near-net shape parts without needing material removal. This presents an exciting opportunity to fabricate new HEs [30]. Using materials with low heat capacitance, modest thermal expansion coefficients, and high heat transmission coefficients is crucial. To prevent rapid fractures, materials must possess exceptional tensile and creep capabilities, favorable

TABLE 6.2
List of LHS System Ideas Dependent on the Type of HE [27]

When PCM Is Not Flowing/Circulating	Using PCM Flow (PCM Slurries/ Emulsion)
Compact flat plate HE	Tanks equipped with an internal HE
Shell-and-tube configurations with or without fins	Tanks connected to an external HE
Tanks containing macro-encapsulated PCM in various shapes, such as spheres, flat plates, and tubes	Indirect contact PCM-HTF
Direct contact PCM-HTF	

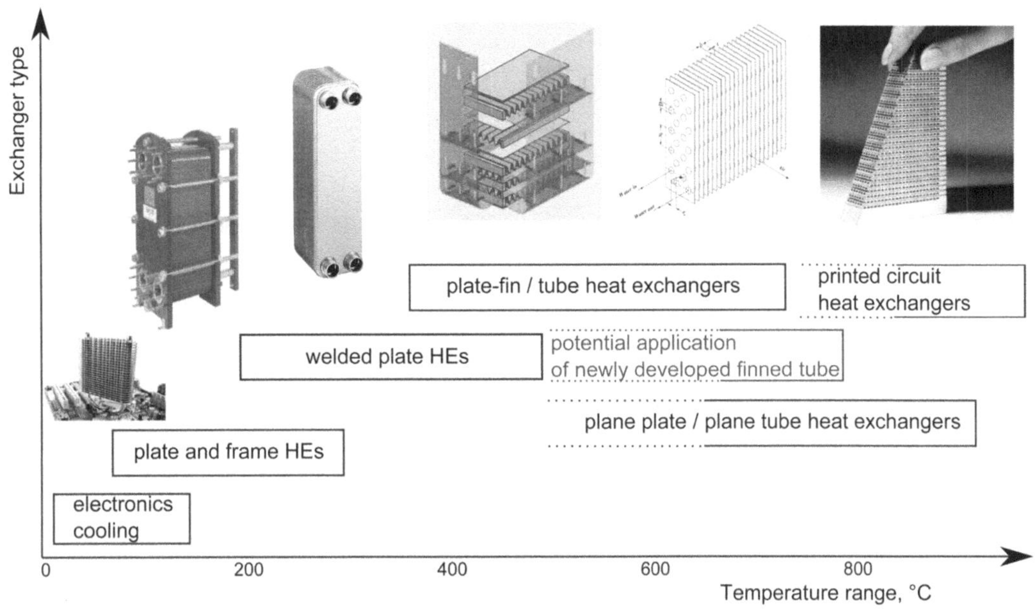

FIGURE 6.4 Common compact HEs and their temperature limits [29].

fatigue and fatigue-corrosion qualities, high fracture toughness, and impact strength. By employing compact HE with increased surface area-to-volume ratios, which can be accomplished by sophisticated manufacturing methods such as AM of metals, it is possible to overcome these limitations. Different materials are carbon steel, stainless steel, Al alloys, Cu alloys, Ni alloys, etc. [28]. Sabau [30] developed an innovative HE structure that included a compact arrangement of counterflow channels enclosed by plenums. The designs being evaluated utilize hexagonal packing of fluid flow pathways in the crossflow zone to optimize heat exchange efficiency while minimizing the amount of metal required. Angled end plates are used to modify fluid flow through the tubes and facilitate the AM process. The thermal performance of the HEs configurations was found to be excellent in terms of heat load. Different types of HEs and their settlements are listed in Table 6.3 [28].

As mentioned earlier, HTF is one of the criteria through which the applicability of HE can be determined. There are various properties required for the selection of HTF in HE. Some of these properties are listed in Table 6.4.

TABLE 6.3
Different Types of HEs and Their Characteristics [28]

Heat Exchanger	Heat Transfer	Pressure and Temperature Range	Modes
Plate and fin HE	Gas-gas	Withstand 90–200 bar pressure and temperature of 780°C	Cross, counter, cross-counter, and co-current flow
Helical coil	Gas-liquid	High temperature and pressure applications 690°C	The rotational stream further mixes the fluid
Finned coil	Fouling formation reduces heat transfer (disadvantage)	Very low to very high temperature applications: 50–750°C	Fins and stubbed surfaces were added for effective HT

TABLE 6.4
Required Properties of HTF in HE

Property	Remarks
Molten salts	Low viscosity
Liquid metal	Low viscosity
Gases	Low viscosity and low thermal conductivity

6.2.2 Storage with Heat Transfer via Internal Surface

The general heat transfer rate in these kinds of systems depends on how the HEs are built and how they operate (the temperature and flow rate of the HTF fluid) [31]. Increasing the heat transfer area leads to more power [32]. This method is used for this kind of storage because it makes the contact area between the fluid and PCM broader while also making the PCM thinner, which lowers its thermal resistance. Also, fins or longer surfaces are often added to help heat transfer even more [33]. Intensification of heat transfer attempts to reduce equipment size and improve performance. The result of intensification techniques involves one or more of the below-mentioned items [29]:

1. Lowered size of HE for a given load
2. Capacity increase for a conventional HE
3. Decreased approach temperature range
4. Limited pumping power

6.2.3 Types of Heat Exchangers

Figure 6.5 depicts the various HEs based on shape and type. Generally, plate-and-frame is the most prevalent form of storing energy. It functions like a typical exchanger, except one of the working fluids is a PCM. Multiple design configurations of HEs have been introduced and examined. Plate-and-frame HEs, also known as compact flat plate HEs, are widely utilized in various industries (Figure 6.5a). A thermal storage system with a high level of power can be built by substituting one of the fluids (50% of the system) with PCM, which will remain stationary between the plates. The input and exit of this section of the HE will be closed [34]. The high-power output of a conventional compact flat plate HE is due to the small distance between the plates and the presence of multiple plates, resulting in a large heat transfer surface area. This allows for efficient transfer of energy in terms of heat in the entering and leaving points of the PCM. Generally, the other kinds of HE systems have PCM filled in a tank and an HTF flowing via the inner HE. These are like regular shell-and-tube HE systems, but the PCM is on the section of the shell [35].

6.2.3.1 Type of Direct Contact

Storage with this direct contact allows HTF to flow directly across PCM, removing the HE wall's thermal barrier. The PCM and HTF must be insoluble, with differing densities, for the HTF to immerse or be suspended within the PCM in liquid form and establish interaction. Figure 6.6 depicts the working of direct contact systems. Figure 6.7 depicts the complicated arrangement of channels formed during the device's charging-discharging cycles. This demonstrates how nature determines the best flow pattern to minimize the drop in pressure function while increasing heat transfer rates.

6.2.3.2 Module Type

The module-type storage has a construction system similar to that of HE-type storage. The system adds the system of heat exchange to the PCM, which is usually in a tank; this storage system of

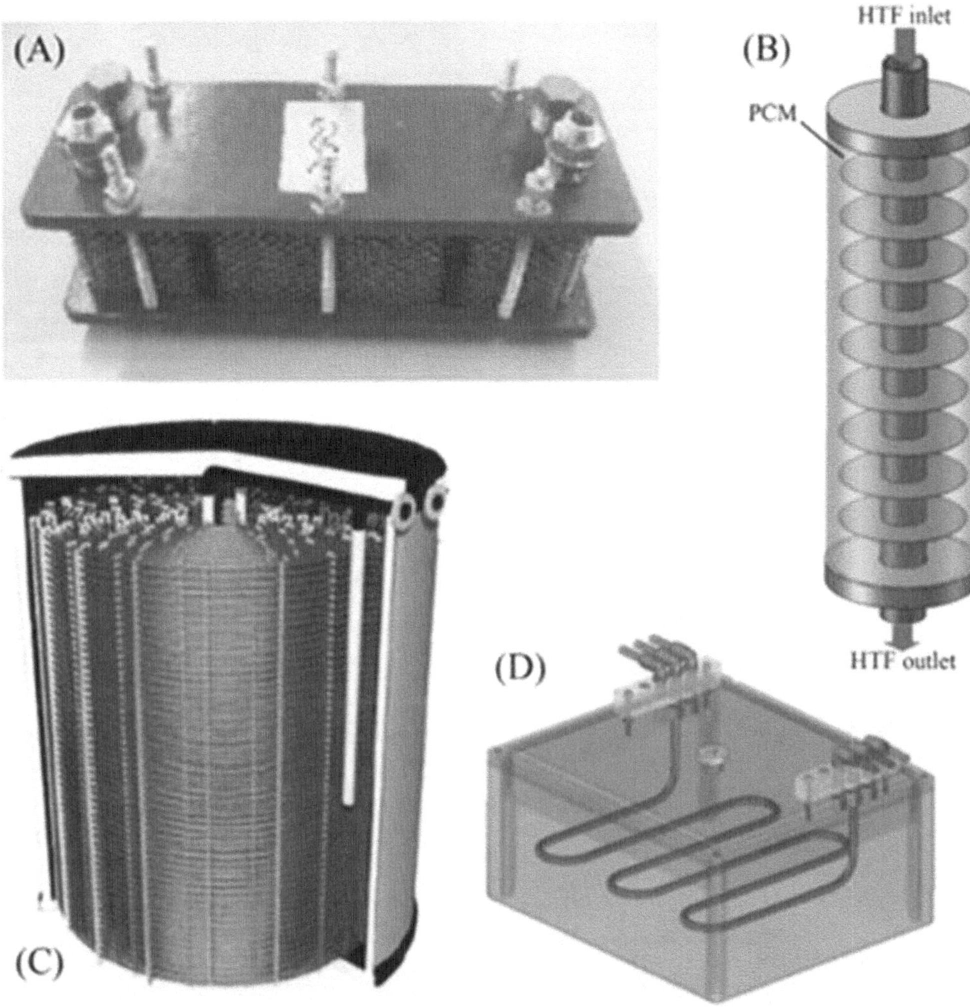

FIGURE 6.5 Illustrations for categories of HE. (a) Compact flat plate [34]. (b) Annular shell-and-tubes having fins [36]. (c) Coil in tank [2]. (d) Coil in shell [32].

module-type employs PCM that has been macro encapsulated in plates, spherical structures, tubular holders, and so on [39]. The HTF circulates through this encapsulation, all sealed within a container. The key benefits of employing macro-encapsulated PCM modules include production and marketing, where the manufacturer can tackle several difficulties of volumetric alteration, transfer of heat, and compatibility of materials in the PCM module [40]. This type of storage is also flexible in terms of design since modules can be made in advance in a range of sizes and forms. The pieces are also simple to arrange and ship. A few illustrations of this type of system are shown in Figure 6.8. The amount of surface area to volume plays a big role in how heat circulates through these systems. To raise this ratio, heat transfer will be much better because there will be more surface area for heat to move over and less overall energy to move. Because of this, the configuration of the system modules needs to be carefully executed and chosen based on needs and purpose.

PCM modules are often macro-encapsulated in bags, dimple sheets, flat plates, and spherical capsules. Module storage typically employs liquids as heat transfer fluids, such as water, water+salt, and water+glycol mixes, and have also been tested with air.

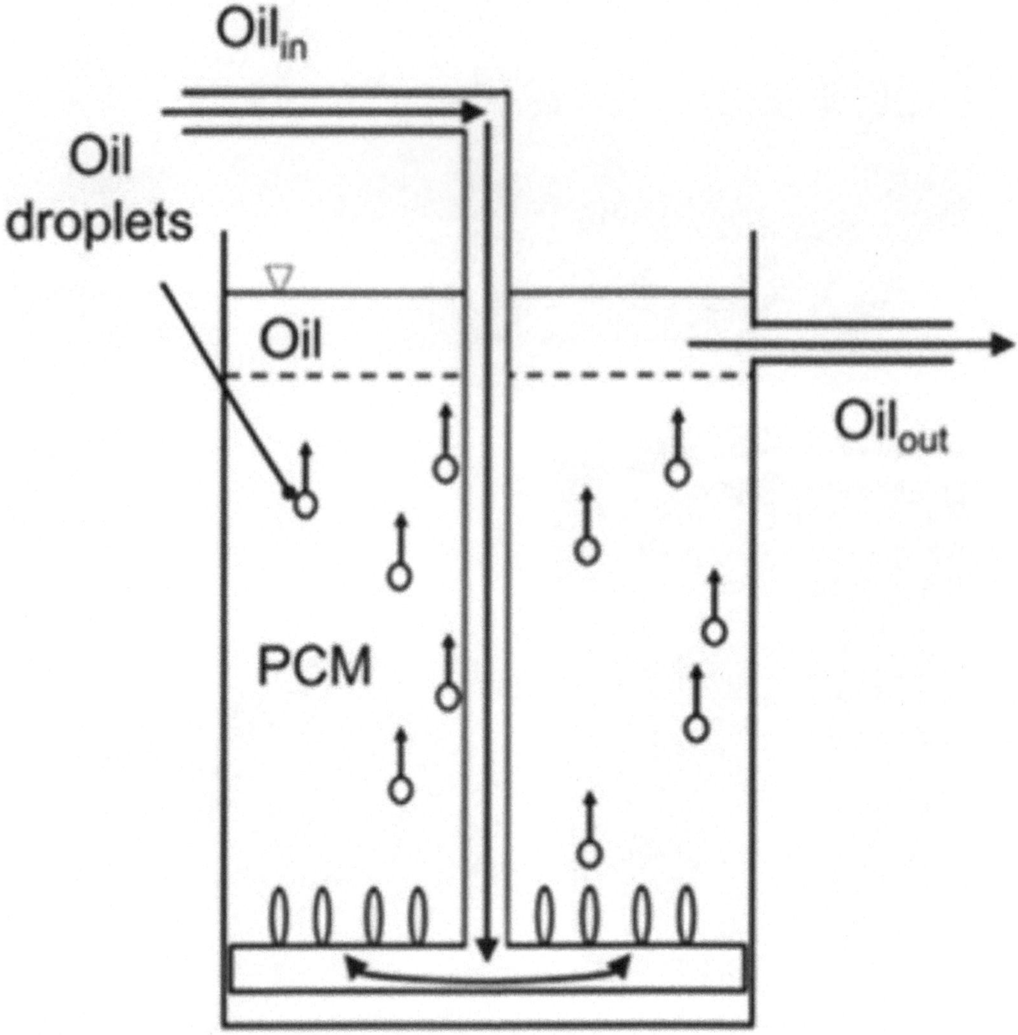

FIGURE 6.6 Typical direct contact storage system [37].

Flat plates, bags, or dimple sheets are commonly used for heat transfer with air because of their higher surface-to-volume ratio. Spherical modules on their own arrange themselves when filled in a tank. For spheres of equal size, the maximum achievable package density (PCM proportion of volume) is 74%. This is why module-type storages are sometimes known as hybrid storages [2].

6.2.4 Design Methodologies

Design methods are being created, studied, and used to find the ideal arrangement and general configuration of a storage system with PCM. However, it is essential to highlight that not all methods are suitable for every application due to the nonlinear nature of the function that relates enthalpy and temperature in PCM, which imposes significant constraints on conventional approaches. This section presents the design techniques that are currently being created and examined, as well as the ones that are most widely employed. The findings mostly rely on experimentation and experimental characterization, with some incorporation of computational modeling. Nevertheless, a fundamental aspect of advancing design techniques is the capacity to assess and contrast various LHS systems

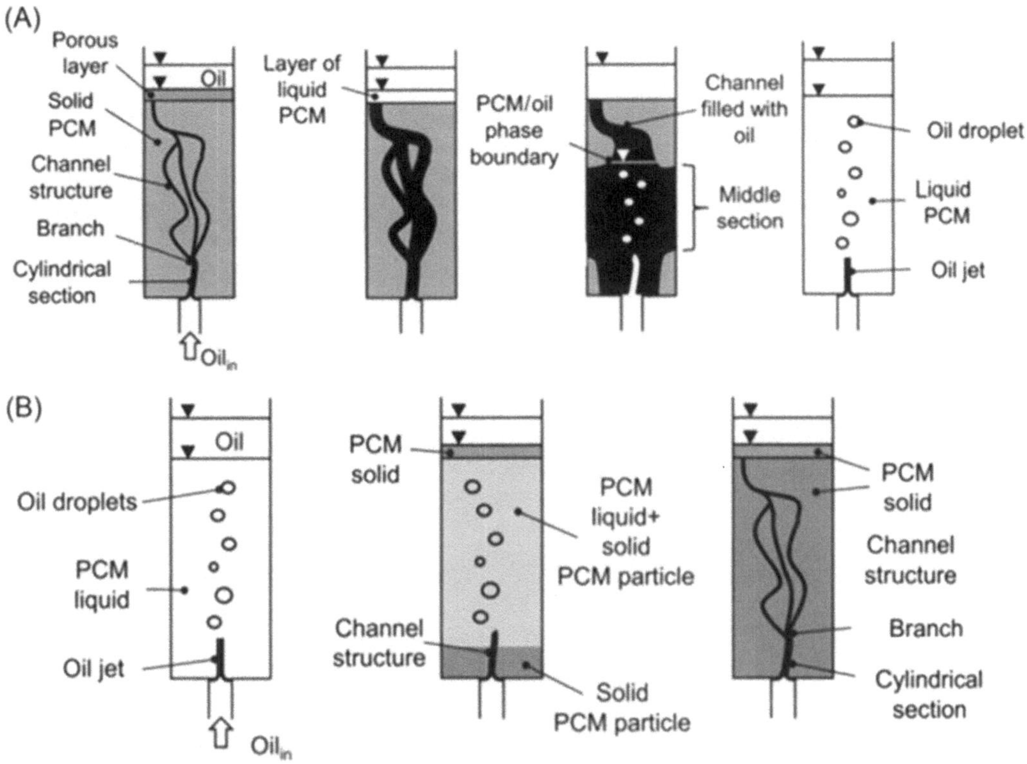

FIGURE 6.7 The advancement of the PCM-HTF combination in a direct contact heat storage system: (a) charging; (b) discharging [38].

(varying in size, internal configuration, PCMs, and operational parameters) based on power output, duration of operation, and other relevant factors. To ensure that the optimal design of the LHS system can be tailored to a particular application. The present focus of research is the development of comparison metrics, which is considered a highly significant topic of study.

6.2.5 Heat Transfer Storage Involves Replacing the Heat Storage Medium

If the PCM modules are sufficiently compact to move out of the storage with the HTF, then most of the stored heat can be expelled from the storage without necessitating any procedure for heat exchange. Here, the media for storing and transferring heat becomes the same fluid, which is a liquid and has a part that gathers heat by changing phases. This kind of part is called a phase change slurry or phase change liquid. Usually, they are made up of a base fluid, and PCM is added through emulsion or microencapsulation/shape stabilization, as shown in Figure 6.9 [44]. This makes the amount of surface area to volume the best it can be, and it also makes heat transfer much better. It is also important to note that some storage systems are made up of nothing but a PCM portion whose surfaces share the heat with the environment around them. This sort of storage generally cannot run off much power because there is a lot of space for heat to move. Because of this, they are mostly only useful for temperature control, where an object's temperature must stay the same [45]. Low heat transmission is acceptable in this scenario because the essential parameter is the amount and enthalpy of the PCM, which defines the system's heat storage capacity and the duration the good/body remains at a steady temperature.

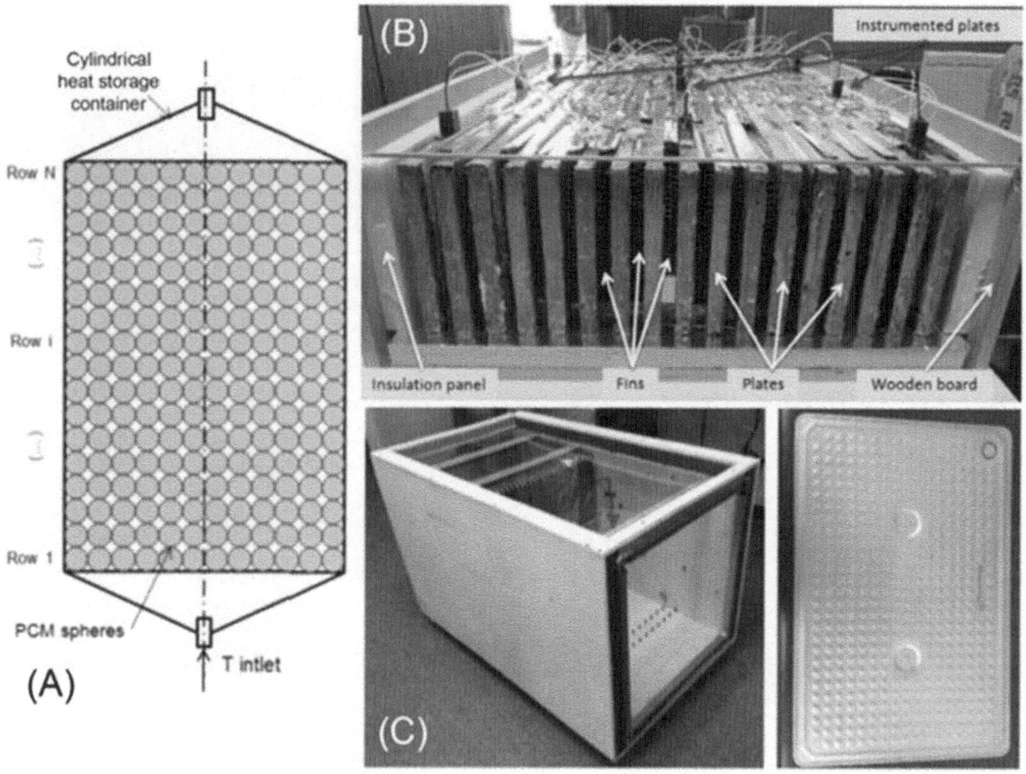

FIGURE 6.8 Examples of thermal energy storage systems using phase change materials (PCM) in different geometrical configurations: (a) packed bed system utilizing spherical PCM modules [41]; (b) air-PCM heat exchange system employing aluminum rectangular containers [42]; (c) air-PCM heat exchange system with PCM encapsulated in commercial flat plates.

6.3 ENVIRONMENTAL IMPACT

This section describes the environmental benefits of using phase transition materials, such as decreased levels of gases that contribute to the greenhouse effect, global temperature, and biocompatibility. The commercial and residential sectors account for 20% of the overall consumption of energy. From 2012 to 2040, both sectors are predicted to grow by 1.4% and 1.6% per year. Consequently, greenhouse gas GHG emissions will rise. To compensate for the increase in energy usage, heating and cooling technologies must be made more efficient. Llorach-Massana et al. [46] empirically analyzed the environmental impact of utilizing Rubitherm (RT) PCM on a root zone heating system. During the day, the PCM absorbs energy, reducing the requirements for energy in the case of a typical system. The decrease in greenhouse gas emissions was prioritized as a major sustainable contribution. This demonstrates that PCM can lower the emissions of CO_2 from traditional root zone systems utilized for heating. Falco et al. [47] suggested a novel PCM-based device for frigid storage applications at 5.5°C. The RT6HC eutectic metal was used as the PCM in the cold storage tank, which was set up in a standard chiller fan with a coil system. A preliminary test showed that a small office that needs 40 kW of cooling power could save 13–16% of its energy. Implementing a larger cold storage unit that works with a PCM will have a big impact on the world in terms of lowering the risk of global warming and using fossil fuels. The compatibility of the material with the surroundings is another environmental factor. Biodegradable materials can undergo decomposition into naturally occurring compounds. Hence, completely biodegradable substances do not exert any adverse ecological consequences. Nevertheless, even materials that can be broken down by natural processes may have a significant and enduring effect on the environment.

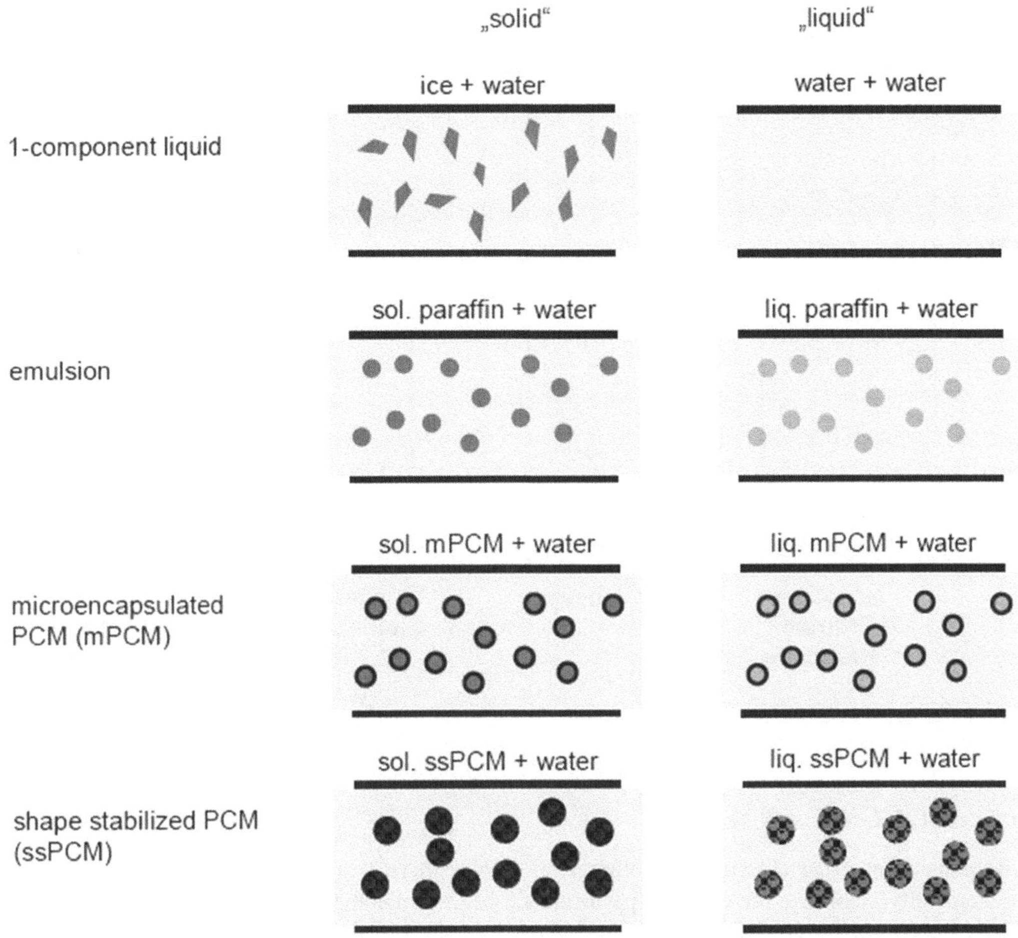

FIGURE 6.9 Different types of PCM slurries [2].

Certain biodegradable compounds may require several years or even decades to undergo complete decomposition. Various established testing methods, including ASTM 6400 and OECD 301, have been employed to assess the biodegradability of PCMs. Paraffin products exhibit a biodegradability range of 21–55%. Paraffin products require a significant amount of time, ranging from years to decades, to break down completely, and they will continue to have a detrimental effect on the environment over an extended period. Conversely, PCMs derived from vegetables are not carcinogenic and generally do not harm animals or plants, nor do they have any lasting adverse effects on the environment. An investigation of the biodegradability of biobased PCM products revealed that approximately 80% of these materials decompose after 28 days. This characteristic makes them highly desirable and environmentally benign. It was found that PCMs made a greenhouse heating system work better, saving energy and lowering the greenhouse's carbon output [48, 49].

The Environmental Cost Indicator (ECI) is a quantitative measure used to assess the environmental consequences of products or processes in terms of their monetary value, usually denominated in euros. The ECI for PCMs is shown in Figure 6.10. Aridi et al. [50] reviewed ECI for various PCMs, which can be utilized for applications in buildings. They compared ECI for octadecane, manganese nitrate hexahydrate, and coconut oil. They found high ECI for octadecane PCM. The inefficiency of coconut oil PCM is attributed to its low latent heat of fusion, which hinders its energy storage and discharge capabilities.

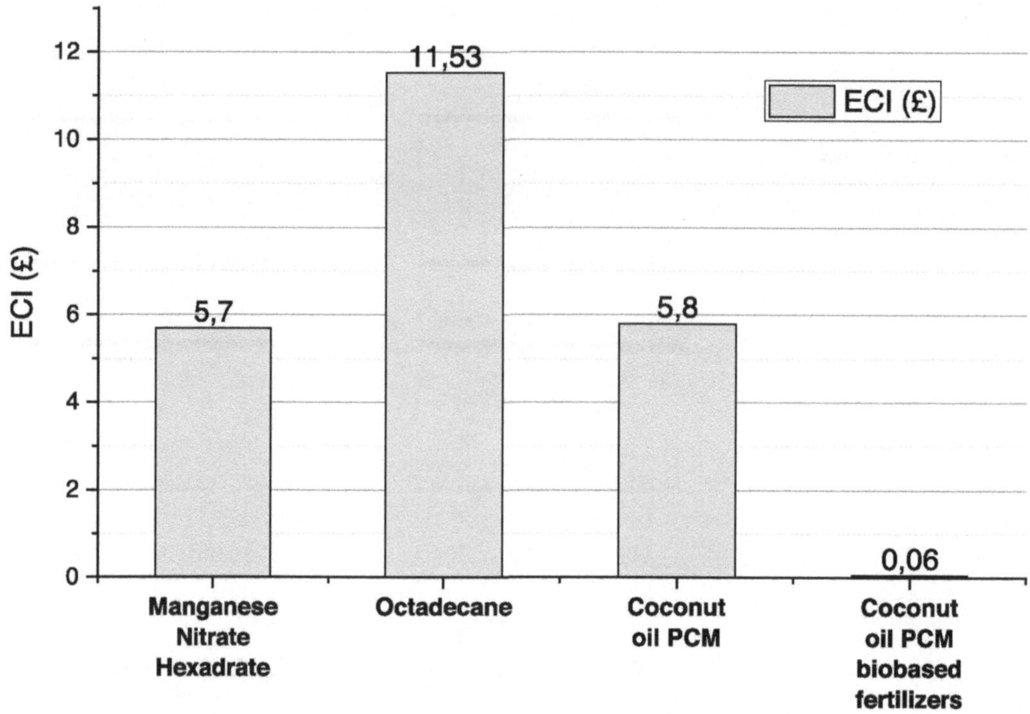

FIGURE 6.10 ECI for various PCMs [50].

6.4 ECONOMIC ANALYSIS

The main things that determine whether or not solar thermal systems are technologically and economically viable are the cost and performance of the collectors. Enhanced energy conversion, decreased consumption of energy, reduced energy loss, increased efficiency, climate variables, and the cost of parts are all things that affect the financial justification. The payback time is affected by all of the above. Several studies have looked at the costs of conventional systems [51, 52] and solar absorption systems that are combined with PCM. Godarzi et al. [53] employed an exergy-economic analysis and a genetic algorithm to come up with a PCM store for a 45.4 kW LiBr-H_2O system. The finding predicts that if PCM is used, the payback time goes from 0.61 to 1.13 years. This is because PCM is very expensive. Misra et al. [54] used exergetic cost and thermo-economic optimization to find the best way to amplify the LiBr-H_2O system. The cost of production went up by 3.5%, and spending went up by 12.5%. Even though the system costs more, the money saved on energy costs over time will cover the project's initial cost. An assessment [55] has been conducted to determine the efficiency of a combined air-to-water heat pipe and thermal storage fitted with PCM. The findings indicated a significant decrease of 50% in the volume of the heat storage tank while having a minimal effect on the provision of heat. The overall results demonstrate that a decrease of 520.8 PJ, which is comparable to 36% of the energy consumption by end-users, may be achieved if all 17,000 considered households successfully undergo a retrofit program [56]. A separate study [57] was conducted to assess the techno-economic factors of PCM in order to enhance the efficiency of a concentrated solar power plant operating at high temperatures. It was discovered that the ability to store more energy in a given space could compensate for the increased expense of the device. While studies demonstrate advancements in environmental protection, the utilization of PCM may not always be financially feasible. The primary factor is the prevailing market price of PCM. The price may decrease if governments implement subsidies or if there is a rise in demand, resulting in increased production volume and reduced product costs. Since not many PCMs are being made, the price will stay high as long as the market is not introduced to it on a large scale. This makes it less likely that it can make a profit right now.

6.5 BENEFITS OF PHASE CHANGE MATERIALS IN HEAT EXCHANGERS

Incorporating PCMs into HEs provides substantial advantages, augmenting their efficiency and efficacy in diverse applications. Integrating PCMs can also increase the efficiency of HEs by reducing heating and cooling requirements and lowering energy consumption and operating expenses. PCMs have higher energy storage densities than conventional thermal storage materials, enabling smaller HEs and energy storage systems [58]. When PCMs are incorporated into HEs, they can make thermal control systems much more efficient and effective. This section discusses the advantages and uses of PCMs in HEs.

6.5.1 Advantages of PCMs

6.5.1.1 Enhanced Energy Storage and Discharge Capabilities

PCMs can preserve and discharge significant amounts of energy when they undergo phase transitions, such as changing from a solid to a state of liquid and vice versa. Due to this phenomenon, enhanced control over temperature shifts can be achieved, which will result in the enhanced effectiveness of the TES system. Studies suggest that PCMs can increase the thermal storage capacity of heat exchangers by as much as half when compared with traditional materials [59, 60].

6.5.1.2 Improved Thermal Management

The main aim of PCMs to be used in heat exchangers is to ensure temperature stability by absorbing the surplus heat during the rise in temperature and releasing heat during the decrease in temperature. As a result, the system experiences enhanced temperature regulation and stability. Research has demonstrated that PCMs can effectively decrease fluctuations in temperature inside thermal management systems by upwards of 75% [61].

6.5.1.3 Enhanced Efficiency of the System

Through the integration of PCMs, HE systems can attain increased efficiency, reducing the need for extra energy inputs to maintain desired temperature levels. Consequently, there is a decrease in operational costs as well as energy consumption. The study conducted by Agyenim et al. [35] showed that heat exchangers with integrated PCM have the potential to deliver energy savings of up to 30% in thermal regulation systems.

6.5.1.4 Compact Design and Less Area

PCMs have a higher energy density, which gives an advantage of developing more condensed heat exchanger systems. This can be especially needful in situations where there is limited space available for system installation. The study conducted by Zhang et al. [62] has shown that using PCM can decrease the overall dimensions of thermal storage units by as much as 40% while maintaining performance.

6.5.2 Applications of PCMs

PCMs are utilized in heating, ventilation, and air conditioning systems to enhance the energy efficiency and thermal comfort of buildings. They assist in mitigating the maximum demand and overall energy usage of heating and cooling systems [63]. PCMs are employed in solar thermal energy systems to store surplus heat during daylight hours and subsequently release it throughout the night. This utilization enhances the efficiency and dependability of solar power plants. PCMs are used in electronic devices to regulate the dissipation of heat and maintain ideal operating temperatures, hence enhancing the performance and longevity of electronic parts [64]. Thermal management systems for batteries in hybrid and e-vehicles use PCMs to keep batteries at the ideal temperature, which improves their performance and extends their service life [65]. To ensure the safety and quality of pharmaceuticals and food, PCMs are used in cold chain logistics to maintain the temperature of perishable products during transportation [66].

6.6 USE OF AI/ML IN THERMAL BEHAVIOR CONTROL OF PCM

The most extensively used AI approaches, such as Artificial Neural Network (ANN), Support Vector Machine (SVM), Random Forest, Particle Swarm Optimization (PSO), and Genetic Algorithm (GA), can work without significant physical expertise. With the help of these machine-learning algorithms, the performance of the HE system can be predicted and precisely optimized [67]. The system operation can be effectively managed to achieve goals like user demand satisfaction, energy savings, increased renewable energy utilization ratio, and reduced environmental impacts [68]. PCMs present an exciting new frontier for thermal management system optimization with the potential for artificial intelligence (AI) and machine learning (ML) integration into thermal behavior control. These advancements can make PCM-based systems more efficient, flexible, and predictive. A variety of data-driven ML models can be used to evaluate the design and performance of thermal energy storage systems. Jawaid et al. [69] forecast daily mean sun irradiance using linear regression and other regression models. To improve system performance, they compared these models to an artificial neural network. Therefore, investigating the application of ML in PCM or solar thermal systems has the potential to create novel possibilities for improving energy storage and overall efficiency. In this section, areas such as optimization, predictive modeling, and fault detection are discussed.

6.6.1 Optimization and Predictive Modeling

Predictive models for the thermal behavior of PCMs under different conditions can be developed using AI and ML techniques. By optimizing PCM charging and discharging cycles, these models can boost system performance and energy efficiency. To better regulate heat, for example, deep learning methods can foretell phase transitions with a high degree of accuracy. Jagirdar et al. [70] used an adaptive neuro-fuzzy inference system to predict the charging time of a shell-and-tube type LHS system. They also found the effects of various properties of the PCM on charging times.

6.6.2 System of Adaptive Controls

Adaptive control systems powered by artificial intelligence can make real-time adjustments to the functioning of HEs based on PCM. Optimal temperature control is achieved by these systems, which may react to both external stimuli and internal system demands. Enhancing the efficiency of PCM systems in buildings can be achieved by adaptive control employing reinforcement learning, according to research [71].

6.6.3 Fault Detection and Diagnosis

In order to ensure reliability and decrease maintenance expenses, PCM systems can be equipped with machine-learning algorithms that can identify and fix faults. Artificial intelligence can anticipate breakdowns by examining trends in temperature data and spotting outliers. The operational reliability of the TES system can be improved using ML-based fault detection, according to studies [72].

6.6.4 Integration of Hybrid Systems

The integration of other green power harvesting systems, such as photovoltaic-based technology and power generation via wind, is the future of PCM technology. The interaction between these systems can be optimized by AI and ML, which can balance energy storage and utilization. This integration has the potential to result in energy solutions that are more flexible and environmentally friendly.

6.7 CONCLUSIONS AND FUTURE RECOMMENDATIONS

Several studies have reported on different types of HEs. The major conclusions of the chapter are listed below:

- Different HE types and their performance with PCMs must be analyzed, and thus, the effectiveness of HE systems will be enhanced.
- An outstanding characteristic of the compact flat plate HE is its high surface area-to-volume ratio, which greatly improves heat transfer efficiency. In applications demanding high power output, this design is very efficient as it can support several plates with narrow gaps, therefore creating a substantial heat transfer surface. The incorporation of PCM in this arrangement enables significant energy storage in a confined area, rendering it very efficient for applications with limited space, such as in electronic cooling or other compact renewable energy systems.
- Helical coil HEs facilitate efficient heat transfer by creating a rotational stream that mixes the fluid. This design is particularly effective in high-temperature and high-pressure applications. However, integrating PCMs into this design can be challenging, as the complexity of the coil may lead to an uneven phase change across the coil, potentially affecting thermal performance.
- Plate and fin HEs increase the surface area for heat transfer, are effective in gas-to-gas applications, and can withstand high pressures and temperatures. However, the addition of PCMs may be less effective in achieving uniform phase change, and the potential for fouling could reduce overall efficiency over time.
- Shell-and-tube HEs with fins, where the HTF is placed in the tube with PCM filling the shell, are well-suited for large-scale applications requiring high energy storage capacity. The fins increase the heat transfer area, ensuring effective phase change and energy storage. The ratio of shell diameter to tube diameter in shell-and-tube HEs plays a critical role in determining the overall melting and solidification time of the PCM.
- Direct contact HEs, where HTF flows directly through the PCM, offer excellent thermal performance due to the elimination of the thermal barrier typically presented by HE walls. However, their efficiency heavily depends on the compatibility of the PCM and HTF, particularly regarding density and phase separation.
- Module-type HEs, consisting of PCM encapsulated in various shapes with HTF circulating through, offer design flexibility and ease of manufacturing. This design ensures consistent phase change and thermal performance, making it suitable for customizable energy storage solutions.
- AI and ML algorithms can be leveraged to predict the thermal behavior of PCMs under various operational conditions. By analyzing vast datasets, these algorithms can identify the optimal material combinations, geometric configurations, and operating parameters that yield the highest efficiency.

One of the key areas for future development is the enhancement of thermal properties through material innovation. Nanoparticle-incorporated PCMs, where adding nanoparticles, such as alumina, to PCMs like n-octadecane can significantly improve thermal conductivity. This enhancement could address one of the primary limitations of traditional PCMs—low thermal conductivity—by enabling faster and more uniform phase transitions. As research progresses, we can expect to see the development of more sophisticated PCM formulations that combine high latent heat capacity with superior thermal conductivity, pushing the boundaries of what is possible in thermal energy storage. Incorporating AI into HE control systems offers the potential for real-time adaptive management, as highlighted in this chapter. These smart control systems can dynamically adjust operational parameters—such as flow rates and temperature settings—based on real-time data, ensuring that the HE operates at peak efficiency. For instance, in PCM-integrated HE, AI could optimize the charging and discharging cycles of PCM, reducing energy consumption and enhancing overall system reliability. AI-driven reinforcement learning could be employed to develop systems that learn and adapt over time, continuously improving their performance in response to changing environmental conditions or energy demands.

REFERENCES

1. M. N. A. Hawlader, M. S. Uddin, and M. M. Khin, "Microencapsulated PCM thermal-energy storage system," *Applied Energy*, vol. 74, no. 1, pp. 195–202, Jan. 2003, doi: 10.1016/S0306-2619(02)00146-0.
2. H. Mehling and L. F. Cabeza, *Heat and cold storage with PCM: An up to date introduction into basics and applications.* in Heat and Mass Transfer. Berlin, Heidelberg: Springer, 2008. doi: 10.1007/978-3-540-68557-9.
3. Y. P. Zhang, K. P. Lin, R. Yang, H. F. Di, and Y. Jiang, "Preparation, thermal performance and application of shape-stabilized PCM in energy efficient buildings," *Energy and Buildings*, vol. 38, no. 10, pp. 1262–1269, Oct. 2006, doi: 10.1016/j.enbuild.2006.02.009.
4. V. Palomba, V. Brancato, and A. Frazzica, "Experimental investigation of a latent heat storage for solar cooling applications," *Applied Energy*, vol. 199, pp. 347–358, Aug. 2017, doi: 10.1016/j.apenergy.2017.05.037.
5. M. S. Naghavi, K. S. Ong, M. Mehrali, I. A. Badruddin, and H. S. C. Metselaar, "A state-of-the-art review on hybrid heat pipe latent heat storage systems," *Energy Conversion and Management*, vol. 105, pp. 1178–1204, Nov. 2015, doi: 10.1016/j.enconman.2015.08.044.
6. F. S. Javadi, H. S. C. Metselaar, and P. Ganesan, "Performance improvement of solar thermal systems integrated with phase change materials (PCM), a review," *Solar Energy*, vol. 206, pp. 330–352, Aug. 2020, doi: 10.1016/j.solener.2020.05.106.
7. A. Sharma, V. V. Tyagi, C. R. Chen, and D. Buddhi, "Review on thermal energy storage with phase change materials and applications," *Renewable and Sustainable Energy Reviews*, vol. 13, no. 2, pp. 318–345, Feb. 2009, doi: 10.1016/j.rser.2007.10.005.
8. S. Kumar, A. Thyagarajan, and D. Banerjee, "An investigation on chevron plate heat exchanger filled with organic phase change material," presented at the ASME 2020 Heat Transfer Summer Conference collocated with the ASME 2020 Fluids Engineering Division Summer Meeting and the ASME 2020 18th International Conference on Nanochannels, Microchannels, and Minichannels, American Society of Mechanical Engineers Digital Collection, Sep. 2020. doi: 10.1115/HT2020-9122.
9. H. Niyas, C. R. C. Rao, and P. Muthukumar, "Performance investigation of a lab-scale latent heat storage prototype – Experimental results," *Solar Energy*, vol. 155, pp. 971–984, Oct. 2017, doi: 10.1016/j.solener.2017.07.044.
10. S. Kumar, R. B. Grover, P. K. Vijayan, and U. Kannan, "Numerical investigation on the effect of shrouds around an immersed isolation condenser on the thermal stratification in large pools," *Annals of Nuclear Energy*, vol. 110, pp. 109–125, Dec. 2017, doi: 10.1016/j.anucene.2017.06.022.
11. H. Niyas, S. Prasad, and P. Muthukumar, "Performance investigation of a lab–scale latent heat storage prototype – Numerical results," *Energy Conversion and Management*, vol. 135, pp. 188–199, 2017, doi: 10.1016/j.enconman.2016.12.075.
12. J. Lizana, R. Chacartegui, A. Barrios-Padura, and J. M. Valverde, "Advances in thermal energy storage materials and their applications towards zero energy buildings: A critical review," *Applied Energy*, vol. 203, pp. 219–239, Oct. 2017, doi: 10.1016/j.apenergy.2017.06.008.
13. M. Helm, C. Keil, S. Hiebler, H. Mehling, and C. Schweigler, "Solar heating and cooling system with absorption chiller and low temperature latent heat storage: Energetic performance and operational experience," *International Journal of Refrigeration*, vol. 32, no. 4, pp. 596–606, Jun. 2009, doi: 10.1016/j.ijrefrig.2009.02.010.
14. A. F. Regin, S. C. Solanki, and J. S. Saini, "Heat transfer characteristics of thermal energy storage system using PCM capsules: A review," *Renewable and Sustainable Energy Reviews*, vol. 12, no. 9, pp. 2438–2458, Dec. 2008, doi: 10.1016/j.rser.2007.06.009.

15. S. Kumar, A. Thyagarajan, and D. Banerjee, "Experimental investigation of thermal energy storage (TES) platform leveraging phase change materials in a chevron plate heat exchanger," presented at the ASME 2022 International Mechanical Engineering Congress and Exposition, American Society of Mechanical Engineers Digital Collection, Feb. 2023. doi: 10.1115/IMECE2022-96226.
16. L. Kalapala and J. K. Devanuri, "Influence of operational and design parameters on the performance of a PCM based heat exchanger for thermal energy storage – A review," *Journal of Energy Storage*, vol. 20, pp. 497–519, Dec. 2018, doi: 10.1016/j.est.2018.10.024.
17. C. J. Ho and J. Y. Gao, "Preparation and thermophysical properties of nanoparticle-in-paraffin emulsion as phase change material," *International Communications in Heat and Mass Transfer*, vol. 36, no. 5, pp. 467–470, May 2009, doi: 10.1016/j.icheatmasstransfer.2009.01.015.
18. H. Akeiber *et al.*, "A review on phase change material (PCM) for sustainable passive cooling in building envelopes," *Renewable and Sustainable Energy Reviews*, vol. 60, pp. 1470–1497, Jul. 2016, doi: 10.1016/j.rser.2016.03.036.
19. A. L. Loeb, "Thermal conductivity: VIII, a theory of thermal conductivity of porous materials," *Journal of the American Ceramic Society*, vol. 37, no. 2, pp. 96–99, 1954, doi: 10.1111/j.1551-2916.1954.tb20107.x.
20. M. Meschke, W. Guichard, and J. P. Pekola, "Single-mode heat conduction by photons," *Nature*, vol. 444, no. 7116, pp. 187–190, Nov. 2006, doi: 10.1038/nature05276.
21. J.-K. Yu, S. Mitrovic, D. Tham, J. Varghese, and J. R. Heath, "Reduction of thermal conductivity in phononic nanomesh structures," *Nature Nanotech*, vol. 5, no. 10, pp. 718–721, Oct. 2010, doi: 10.1038/nnano.2010.149.
22. S. Kumar, R. B. Grover, H. Yadav, P. K. Vijayan, U. Kannan, and A. Agrawal, "Experimental and numerical investigation on suppression of thermal stratification in a water-pool: PIV measurements and CFD simulations," *Applied Thermal Engineering*, vol. 138, pp. 686–704, Jun. 2018, doi: 10.1016/j.applthermaleng.2018.04.070.
23. F. S. Javadi, H. S. C. Metselaar, and P. Ganesan, "Performance improvement of solar thermal systems integrated with phase change materials (PCM), a review," *Solar Energy*, vol. 206, pp. 330–352, Aug. 2020, doi: 10.1016/j.solener.2020.05.106.
24. S. Kumar and D. Banerjee, "A review on phase change materials for sustainability applications by leveraging machine learning," in *Energy Consumption, Conversion, Storage, and Efficiency*, IntechOpen, 2024. doi: 10.5772/intechopen.114380.
25. G. Raam Dheep and A. Sreekumar, "Influence of nanomaterials on properties of latent heat solar thermal energy storage materials – A review," *Energy Conversion and Management*, vol. 83, pp. 133–148, Jul. 2014, doi: 10.1016/j.enconman.2014.03.058.
26. F. Agyenim, N. Hewitt, P. Eames, and M. Smyth, "A review of materials, heat transfer and phase change problem formulation for latent heat thermal energy storage systems (LHTESS)," *Renewable and Sustainable Energy Reviews*, vol. 14, no. 2, pp. 615–628, Feb. 2010, doi: 10.1016/j.rser.2009.10.015.
27. D. Groulx, A. Castell, and C. Solé, "Design of latent heat energy storage systems using phase change materials," in *Advances in Thermal Energy Storage Systems*, Elsevier, 2021, pp. 331–357. doi: 10.1016/B978-0-12-819885-8.00011-5.
28. S. Mahmoudinezhad, M. Sadi, H. Ghiasirad, and A. Arabkoohsar, "A comprehensive review on the current technologies and recent developments in high-temperature heat exchangers," *Renewable and Sustainable Energy Reviews*, vol. 183, p. 113467, Sep. 2023, doi: 10.1016/j.rser.2023.113467.
29. P. Stehlík, Z. Jegla, and B. Kilkovský, "Possibilities of intensifying heat transfer through finned surfaces in heat exchangers for high temperature applications," *Applied Thermal Engineering*, vol. 70, no. 2, pp. 1283–1287, Sep. 2014, doi: 10.1016/j.applthermaleng.2014.05.052.
30. A. S. Sabau *et al.*, "Design, additive manufacturing, and performance of heat exchanger with a novel flow-path architecture," *Applied Thermal Engineering*, vol. 180, p. 115775, Nov. 2020, doi: 10.1016/j.applthermaleng.2020.115775.
31. L. Kalapala and J. K. Devanuri, "Influence of operational and design parameters on the performance of a PCM based heat exchanger for thermal energy storage – a review," *Journal of Energy Storage*, vol. 20, pp. 497–519, Dec. 2018, doi: 10.1016/j.est.2018.10.024.
32. A. Patil and D. Groulx, "Experimental study of coil and shell phase change material heat exchanger: Early results," presented at the The Second Pacific Rim Thermal Engineering Conference, Hawaii, USA, 2019.
33. W. Ogoh and D. Groulx, "Effects of the number and distribution of fins on the storage characteristics of a cylindrical latent heat energy storage system: A numerical study," *Heat Mass Transfer*, vol. 48, no. 10, pp. 1825–1835, Oct. 2012, doi: 10.1007/s00231-012-1029-3.

34. M. Medrano, M. O. Yilmaz, M. Nogués, I. Martorell, J. Roca, and L. F. Cabeza, "Experimental evaluation of commercial heat exchangers for use as PCM thermal storage systems," *Applied Energy*, vol. 86, no. 10, pp. 2047–2055, Oct. 2009, doi: 10.1016/j.apenergy.2009.01.014.
35. F. Agyenim, N. Hewitt, P. Eames, and M. Smyth, "A review of materials, heat transfer and phase change problem formulation for latent heat thermal energy storage systems (LHTESS)," vol. 14, no. 2, Feb. 2010, doi: 10.1016/j.rser.2009.10.015.
36. X. Yang, Z. Lu, Q. Bai, Q. Zhang, L. Jin, and J. Yan, "Thermal performance of a shell-and-tube latent heat thermal energy storage unit: Role of annular fins," *Applied Energy*, vol. 202, pp. 558–570, Sep. 2017, doi: 10.1016/j.apenergy.2017.05.007.
37. S. Kunkel *et al.*, "Direktkontakt-Latentwärmespeicher zur Effizienzsteigerung von Wärmepumpen," *Chem Ing Tech*, vol. 90, pp. 234–240, 2018.
38. S. Kunkel *et al.*, "Channel formation and visualization of melting and crystallization behaviors in direct-contact latent heat storage systems," *International Journal of Energy Research*, vol. 44, no. 6, pp. 5017–5025, 2020, doi: 10.1002/er.5202.
39. M. F. Rahi, M. Arıcı, A. M. Abdulateef, and H. Niyas, "Numerical investigation and optimization of macro-encapsulated phase change material capsules in building roof slab," *Journal of Energy Storage*, vol. 68, p. 107806, Sep. 2023, doi: 10.1016/j.est.2023.107806.
40. E. Oró, L. Miró, C. Barreneche, I. Martorell, M. M. Farid, and L. F. Cabeza, "Corrosion of metal and polymer containers for use in PCM cold storage," *Applied Energy*, vol. 109, pp. 449–453, Sep. 2013, doi: 10.1016/j.apenergy.2012.10.049.
41. J. P. da Cunha and P. Eames, "Phase change materials to meet domestic hot water demand in the UK-A numerical study," *Journal of Fluid Flow, Heat and Mass Transfer (JFFHMT)*, vol. 4, no. 1, pp. 11–18, 2017.
42. M. E. Mankibi, N. Stathopoulos, N. Rezaï, and A. Zoubir, "Optimization of an Air-PCM heat exchanger and elaboration of peak power reduction strategies," *Energy and Buildings*, vol. 106, pp. 74–86, Nov. 2015, doi: 10.1016/j.enbuild.2015.05.023.
43. S. Wert, C. A. Cruickshank, and D. Groulx, "Characterization of an air-pcm energy storage design for air handling unit applications," in *Heat Transfer Summer Conference*, American Society of Mechanical Engineers, 2017, p. V001T09A005.
44. M. Delgado, A. Lázaro, J. Mazo, and B. Zalba, "Review on phase change material emulsions and microencapsulated phase change material slurries: Materials, heat transfer studies and applications," *Renewable and Sustainable Energy Reviews*, vol. 16, no. 1, pp. 253–273, Jan. 2012, doi: 10.1016/j.rser.2011.07.152.
45. P. Espeau, D. Mondieig, Y. Haget, and M. A. Cuevas-Diarte, "'Active' package for thermal protection of food products," *Packaging Technology and Science*, vol. 10, no. 5, pp. 253–260, 1997, doi: 10.1002/(SICI)1099-1522(199709/10)10:5<253::AID-PTS403>3.0.CO;2-L.
46. P. Llorach-Massana, J. Peña, J. Rieradevall, and J. I. Montero, "Analysis of the technical, environmental and economic potential of phase change materials (PCM) for root zone heating in Mediterranean greenhouses," *Renewable Energy*, vol. 103, pp. 570–581, Apr. 2017, doi: 10.1016/j.renene.2016.11.040.
47. M. De Falco, M. Capocelli, G. Losito, and V. Piemonte, "LCA perspective to assess the environmental impact of a novel PCM-based cold storage unit for the civil air conditioning," *Journal of Cleaner Production*, vol. 165, pp. 697–704, Nov. 2017, doi: 10.1016/j.jclepro.2017.07.153.
48. H. Benli and A. Durmuş, "Evaluation of ground-source heat pump combined latent heat storage system performance in greenhouse heating," *Energy and Buildings*, vol. 41, no. 2, pp. 220–228, Feb. 2009, doi: 10.1016/j.enbuild.2008.09.004.
49. F. Berroug, E. K. Lakhal, M. El Omari, M. Faraji, and H. El Qarnia, "Thermal performance of a greenhouse with a phase change material north wall," *Energy and Buildings*, vol. 43, no. 11, pp. 3027–3035, Nov. 2011, doi: 10.1016/j.enbuild.2011.07.020.
50. R. Aridi and A. Yehya, "Review on the sustainability of phase-change materials used in buildings," *Energy Conversion and Management: X*, vol. 15, p. 100237, Aug. 2022, doi: 10.1016/j.ecmx.2022.100237.
51. F. Calise, "Thermoeconomic analysis and optimization of high efficiency solar heating and cooling systems for different Italian school buildings and climates," *Energy and Buildings*, vol. 42, no. 7, pp. 992–1003, Jul. 2010, doi: 10.1016/j.enbuild.2010.01.011.
52. U. Eicker and D. Pietruschka, "Design and performance of solar powered absorption cooling systems in office buildings," *Energy and Buildings*, vol. 41, no. 1, pp. 81–91, Jan. 2009, doi: 10.1016/j.enbuild.2008.07.015.
53. A. A. Godarzi, M. Jalilian, J. Samimi, A. Jokar, and M. A. Vesaghi, "Design of a PCM storage system for a solar absorption chiller based on exergoeconomic analysis and genetic algorithm," *International Journal of Refrigeration*, vol. 36, no. 1, pp. 88–101, Jan. 2013, doi: 10.1016/j.ijrefrig.2012.08.028.

54. R. D. Misra, P. K. Sahoo, and A. Gupta, "Application of the exergetic cost theory to the LiBr/H2O vapour absorption system," *Energy*, vol. 27, no. 11, pp. 1009–1025, Nov. 2002, doi: 10.1016/S0360-5442(02)00065-8.
55. N. J. Kelly, P. G. Tuohy, and A. D. Hawkes, "Performance assessment of tariff-based air source heat pump load shifting in a UK detached dwelling featuring phase change-enhanced buffering," *Applied Thermal Engineering*, vol. 71, no. 2, pp. 809–820, Oct. 2014, doi: 10.1016/j.applthermaleng.2013.12.019.
56. S. R. Asaee, V. I. Ugursal, and I. Beausoleil-Morrison, "Techno-economic feasibility evaluation of air to water heat pump retrofit in the Canadian housing stock," *Applied Thermal Engineering*, vol. 111, pp. 936–949, Jan. 2017, doi: 10.1016/j.applthermaleng.2016.09.117.
57. "Techno-economic analysis of solar-assisted air-conditioning systems for commercial buildings in Saudi Arabia," *Renewable and Sustainable Energy Reviews*, vol. 54, pp. 1301–1310, Feb. 2016, doi: 10.1016/j.rser.2015.10.047.
58. A. Alamnia, *Integration of Phase Change Materials in Advancing Heat Exchangers for Enhanced Utilization of Variable Renewable Energy*. IntechOpen, 2024. doi: 10.5772/intechopen.1005216.
59. M. M. Farid, A. M. Khudhair, S. A. K. Razack, and S. Al-Hallaj, "A review on phase change energy storage: Materials and applications," *Energy Conversion and Management*, vol. 45, no. 9, pp. 1597–1615, Jun. 2004, doi: 10.1016/j.enconman.2003.09.015.
60. A. Sharma, V. V. Tyagi, C. R. Chen, and D. Buddhi, "Review on thermal energy storage with phase change materials and applications," *Renewable and Sustainable Energy Reviews*, vol. 13, no. 2, pp. 318–345, Feb. 2009, doi: 10.1016/j.rser.2007.10.005.
61. L. F. Cabeza, A. Castell, C. Barreneche, A. de Gracia, and A. I. Fernández, "Materials used as PCM in thermal energy storage in buildings: A review," *Renewable and Sustainable Energy Reviews*, vol. 15, no. 3, pp. 1675–1695, Apr. 2011, doi: 10.1016/j.rser.2010.11.018.
62. Y. Zhang, G. Zhou, K. Lin, Q. Zhang, and H. Di, "Application of latent heat thermal energy storage in buildings: State-of-the-art and outlook," *Building and Environment*, vol. 42, no. 6, pp. 2197–2209, Jun. 2007, doi: 10.1016/j.buildenv.2006.07.023.
63. "Review on thermal energy storage with phase change materials (PCMs) in building applications," *Applied Energy*, vol. 92, pp. 593–605, Apr. 2012, doi: 10.1016/j.apenergy.2011.08.025.
64. N. R. Jankowski and F. P. McCluskey, "A review of phase change materials for vehicle component thermal buffering," *Applied Energy*, vol. 113, pp. 1525–1561, Jan. 2014, doi: 10.1016/j.apenergy.2013.08.026.
65. S. A. Khateeb, S. Amiruddin, M. Farid, J. R. Selman, and S. Al-Hallaj, "Thermal management of Li-ion battery with phase change material for electric scooters: Experimental validation," *Journal of Power Sources*, vol. 142, no. 1–2, pp. 345–353, Mar. 2005, doi: 10.1016/j.jpowsour.2004.09.033.
66. A. Shukla, A. Sharma, and K. Kant, "Solar greenhouse with thermal energy storage: A review," *Curr Sustainable Renewable Energy Rep*, vol. 3, no. 3, pp. 58–66, Dec. 2016, doi: 10.1007/s40518-016-0056-y.
67. P. Anand, P. K. S. Tejes, B. K. Naik, and H. Niyas, "Design analysis and performance prediction of packed bed latent heat storage system employing machine learning models," *Journal of Energy Storage*, vol. 72, p. 108690, Nov. 2023, doi: 10.1016/j.est.2023.108690.
68. N. Ahmed, M. Assadi, A. A. Ahmed, and R. Banihabib, "Optimal design, operational controls, and data-driven machine learning in sustainable borehole heat exchanger coupled heat pumps: Key implementation challenges and advancement opportunities," *Energy for Sustainable Development*, vol. 74, pp. 231–257, Jun. 2023, doi: 10.1016/j.esd.2023.04.004.
69. F. Jawaid and K. NazirJunejo, "Predicting daily mean solar power using machine learning regression techniques," in *2016 Sixth International Conference on Innovative Computing Technology (INTECH)*, Aug. 2016, pp. 355–360. doi: 10.1109/INTECH.2016.7845051.
70. M. N. S. Jagirdar, G. Priyadarshi, B. K. Naik, and H. Niyas, "Implementation of adaptive neuro-fuzzy inference system in design and process optimization of latent heat storage system," *Journal of Energy Storage*, vol. 89, p. 111810, Jun. 2024, doi: 10.1016/j.est.2024.111810.
71. C. Zeng *et al.*, "Application of PCM-based thermal energy storage system in buildings: A state of the art review on the mathematical modeling approaches and experimental investigations," *J. Therm. Sci.*, vol. 31, no. 6, pp. 1821–1852, Nov. 2022, doi: 10.1007/s11630-022-1650-5.
72. Z. Wang, T. Hong, and M. A. Piette, "Building thermal load prediction through shallow machine learning and deep learning," *Applied Energy*, vol. 263, p. 114683, Apr. 2020, doi: 10.1016/j.apenergy.2020.114683.

7 Additive Manufacturing of Novel Heat Exchanger Configurations

Amit Choudhari, Ashish Kumar Gupta, Abhishek Kumar, Prathamesh Sapale, Aditya Rane, and Abhishek Tiwari

7.1 INTRODUCTION

Heat exchangers are integral components in various industrial applications. They transfer heat between two or more fluids without mixing them [1]. These devices operate based on thermodynamics and fluid mechanics principles, facilitating the efficient exchange of thermal energy to optimize system performance. The primary objective of a heat exchanger is to either absorb or dissipate heat, depending on the operational requirements, thereby enhancing the energy efficiency and thermal management of systems [2]. Fundamental types of heat exchangers include shell-and-tube, plate, and finned-tube configurations, each tailored to specific applications and performance criteria. Shell-and-tube heat exchangers, for instance, are widely utilized in industries due to their robustness and adaptability to high-pressure environments [3]. Plate heat exchangers, characterized by their compact design and high thermal efficiency, are commonly used in applications where space constraints are critical. Finned-tube heat exchangers, on the other hand, are optimized for air-cooled systems, providing extended surface areas to enhance heat transfer rates [4]. In the automotive industry, heat exchangers are critical for managing engine temperatures and ensuring optimal performance and longevity of the vehicle. Radiators and intercoolers are prime examples of automotive heat exchangers designed to dissipate heat from the engine and turbochargers, respectively [5]. The heating, ventilation, and air conditioning (HVAC) industry heavily relies on heat exchangers to maintain desired indoor climates [6, 7]. Heat exchangers in HVAC systems facilitate heat transfer between indoor and outdoor environments, contributing to energy-efficient heating and cooling solutions. Power generation is another critical area where heat exchangers are indispensable, particularly in thermal power plants used in condensers and feedwater heaters, to enhance the efficiency of the power cycle [8].

Conventional manufacturing methods for heat exchangers, such as machining, welding, and casting, impose significant limitations on design flexibility [9]. These traditional techniques often necessitate simple geometries to ensure manufacturability, resulting in suboptimal thermal performance and limited ability to tailor designs for specific applications [10]. Additionally, material constraints inherent in conventional processes restrict the use of advanced materials that could enhance heat transfer efficiency and durability. Manufacturing inefficiencies, including high material waste and lengthy production times, further hinder heat exchanger production's economic viability and environmental sustainability [11]. These challenges underscore the need for innovative manufacturing approaches to overcome these limitations and unlock new possibilities in heat exchanger design and performance.

Additive manufacturing (AM), commonly known as 3D printing [12], encompasses a suite of technologies fabricating components layer by layer directly from digital models [13]. Key AM technologies applicable to heat exchanger production include selective laser melting (SLM), electron beam melting (EBM), binder jetting, fused deposition modeling (FDM), and direct energy deposition

DOI: 10.1201/9781003509837-7

(DED) [14]. Each technique offers distinct advantages regarding material compatibility, geometric precision, and production speed [15]. AM enables the creation of intricate and optimized designs that are impossible or impractical with traditional manufacturing methods. By leveraging AM, manufacturers can significantly improve heat exchanger performance through enhanced surface area, complex internal geometries, and integrated multi-functional features [16, 17]. Additionally, AM reduces material waste and allows rapid prototyping and customization, offering a transformative potential for the heat exchanger industry.

The primary objective of this chapter is to provide a comprehensive assessment of the state-of-the-art additive manufacturing of heat exchangers, highlighting the advancements in design, materials, and performance enabled by AM technologies. This review aims to elucidate the transformative potential of AM in overcoming the limitations of conventional manufacturing methods and driving innovation in heat exchanger configurations. The scope of the chapter encompasses a detailed examination of various AM technologies, a comparative analysis of their capabilities and limitations, and a discussion of novel heat exchanger designs facilitated by AM. Furthermore, this chapter will address the material considerations specific to AM heat exchangers, evaluate their performance through experimental and computational studies, and identify the challenges and future directions in this burgeoning field. The structure of the chapter is organized to systematically cover these topics, providing a coherent and insightful narrative that advances the understanding of AM's role in the development of novel heat exchanger configurations.

7.2 ADDITIVE MANUFACTURING TECHNOLOGIES FOR HEAT EXCHANGER FABRICATION

Heat exchanger design has undergone a significant paradigm shift with the advent of AM technologies [18]. AM offers unprecedented freedom, unlike conventional manufacturing techniques that often limit design complexity and material utilization. Figure 7.1 illustrates the primary classification

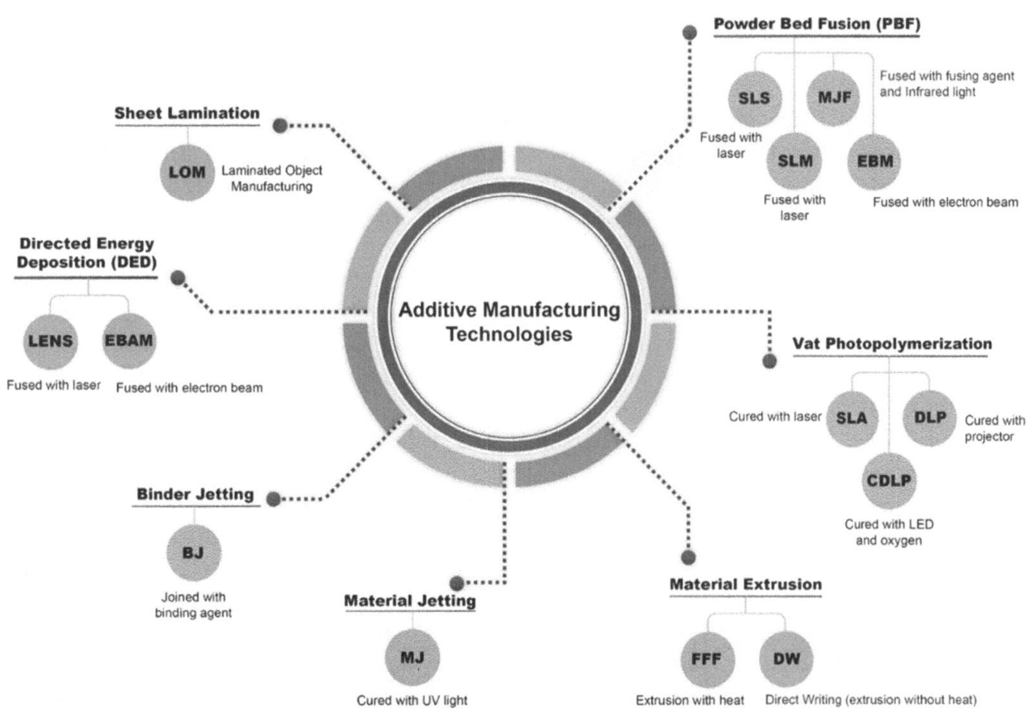

FIGURE 7.1 Classification of additive manufacturing technologies [19].

of AM technologies according to ASTM standards, which categorize AM into seven groups [19]. Vat photopolymerization techniques involve the selective solidification of liquid photopolymer resin layer by layer, while material jetting deposits droplets of photopolymer or other fluid materials onto a build platform [20]. Binder jetting (BJ) utilizes a liquid binding agent to bind powdered material particles, and powder bed fusion (PBF) techniques selectively apply thermal energy to fuse powdered material particles [21, 22]. DED involves the deposition of a material using a focused heat source, while sheet lamination bonds together layers of sheet material [23]. Finally, material extrusion, or fused deposition modeling, melts and deposits thermoplastic filament layer by layer [24]. Each category offers unique capabilities and applications, ranging from high-resolution prototyping to large-scale metal part production, providing a comprehensive overview of the diverse landscape of AM technologies. PBF, EBM, BJ, FDM, and DED equip researchers with the knowledge to select the optimal tool for their specific heat exchanger application. Table 7.1 provides a summary of the various additive manufacturing techniques that can be used for manufacturing heat exchangers with advantages and limitations.

SLM is a powerful tool for researchers due to its exceptional ability to fabricate intricate internal structures with high precision [25, 26]. This technology leverages a focused laser beam to meticulously melt and fuse metal powder layer by layer, akin to a three-dimensional sculptor. This meticulous approach is particularly advantageous for constructing complex flow channels and lattice structures within heat exchangers, ultimately leading to enhanced thermal performance through increased surface area optimization [27, 28]. Furthermore, SLM boasts compatibility with a diverse range of metallic materials, including aluminum, titanium, and nickel alloys, providing researchers with the necessary mechanical properties and thermal resistance for their heat exchanger designs [29, 30]. However, it is crucial to acknowledge the inherent limitations associated with this high-tech approach. SLM systems necessitate significant energy consumption during operation, and the build process can be time-consuming. Additionally, post-processing steps might be required to achieve the desired surface finish on the fabricated components. EBM can be considered the robust cousin of SLM [31, 32]. This technique employs an electron beam in a vacuum environment [33] to melt metal powder, rendering it exceptionally adept at handling high-temperature and reactive materials, particularly titanium alloys, which are susceptible to oxidation in conventional settings [34]. The EBM process yields parts that exhibit exceptional strength and can withstand extreme thermal conditions. The elevated processing temperatures employed during EBM also minimize residual stresses within the fabricated components [35]. Despite these advantages, EBM is limited to materials with good electrical conductivity, and the intricate vacuum chamber setup adds complexity and cost considerations to the process [36, 37]. Similar to SLM, achieving the optimal surface finish might necessitate post-processing steps.

Binder jetting is a compelling option for researchers seeking a production method that prioritizes speed and affordability [38]. This technique operates by selectively depositing a binding agent onto a powder bed, enabling the layer-by-layer creation of intricate structures from diverse materials, including metals [39], ceramics, and even polymers. The primary strengths of binder jetting lie in its high production speed and suitability for large-scale builds, making it an attractive choice for researchers focused on high-throughput manufacturing of heat exchangers [40]. Additionally, the minimal thermal stresses inherent to the binder jetting process often minimize the need for extensive post-processing steps [41]. However, achieving the desired density and mechanical strength level in the final parts might necessitate subsequent sintering and infiltration steps, potentially impacting the final material properties. The range of materials compatible with binder jetting suitable for high-performance heat exchanger applications might also be limited [42, 43]. FDM stands out as the workhorse of the AM industry, renowned for its user-friendliness and widespread adoption. This technique constructs parts through a straightforward process of extruding molten plastic filament layer by layer [44]. These attributes make FDM a favored approach for researchers during the initial stages of heat exchanger design, particularly for prototyping complex geometries at a reasonable cost. Additionally, FDM boasts a vast array of readily available plastic filaments,

TABLE 7.1
Summary of the Various Additive Manufacturing Techniques Used in Manufacturing Heat Exchangers [50–53]

AM Technique	Description	Advantages of Heat Exchangers	Limitations
Selective laser melting (SLM)	• A high-power laser is used to melt metal powder layer by layer to create complex geometries	• High precision and resolution for intricate features • Enables the creation of channels with high aspect ratios • Compatible with various high-performance metals	• Limited build volume • Relatively high cost per unit • Surface roughness can be a concern
Electron beam melting (EBM)	• Employs an electron beam to melt metal powder in a vacuum environment, enabling the processing of reactive materials	• Excellent for processing reactive metals like titanium and tantalum • High melting temperatures achievable for high-performance materials • Good thermal stability minimizes residual stress	• Requires a vacuum environment, increasing complexity and cost • Limited build volume compared with some other AM techniques
Binder jetting	• A process that selectively deposits a binding agent onto a powder bed, allowing for the creation of complex shapes from various materials, including metals, ceramics, and polymers	• Potential for low-cost, high-volume production of heat exchangers • Wide range of compatible materials • Can be suitable for larger build volumes	• Lower strength compared with metal melting techniques like SLM or EBM • Material properties can be anisotropic • Post-processing steps required for some materials
Fused deposition modeling (FDM)	• A filament-based method that extrudes molten material layer by layer to fabricate parts; primarily used with polymers	• Low cost and ease of use • Suitable for prototyping and creating complex geometries	• Limited to polymeric materials, which may not be suitable for high-temperature heat exchanger applications • Mechanical properties are generally lower than metals
Direct energy deposition (DED)	• A technique that utilizes a concentrated energy source to melt and deposit material onto a substrate; suitable for repairs and near-net-shape manufacturing	• Can be used with a broader range of materials than SLM or EBM, including metals, ceramics, and composites • High deposition rates possible • Potential for repairing or adding features to existing heat exchangers	• Surface roughness can be a challenge • Requires robust support structures for complex geometries

offering researchers material selection flexibility [45]. Nevertheless, it is essential to acknowledge the inherent limitations of FDM-produced parts. Compared with other AM techniques, FDM generally yields components with lower resolution and surface finish. Moreover, the mechanical properties of FDM parts are typically inferior, potentially restricting their use in high-temperature heat exchanger applications [46].

DED presents a unique proposition within the AM landscape. This technique leverages a concentrated energy source, such as a laser or electron beam, to melt and deposit material precisely onto a substrate [47]. This versatility empowers DED to function in two distinct capacities: creating entirely new parts layer by layer or repairing or adding features to existing components [48]. DED boasts impressive deposition rates, making it a strong contender for researchers tackling large-scale heat exchanger projects [49]. Furthermore, DED offers compatibility with a broader spectrum of materials compared with SLM or EBM, including metals, ceramics, and composites.

Each additive manufacturing technology offers unique advantages and limitations, and the selection should be tailored to the specific requirements and constraints of the heat exchanger application [54]. Therefore, the optimal additive manufacturing technique selection for heat exchanger fabrication depends on application requirements, including material properties, geometric complexity, and production volume [55]. SLM is frequently regarded as the most suitable technique for heat exchanger fabrication due to its ability to produce highly complex geometries with delicate features and superior mechanical properties [56]. SLM's precision facilitates the creation of intricate internal channels and lattice structures essential for enhancing heat transfer performance. BJ is promising for producing more significant components or when cost and production speed are critical factors [57]. Although the mechanical properties of parts produced via binder jetting may be inferior to those produced by SLM, its high production speed and scalability make it an attractive choice for specific applications [58]. EBM can be advantageous for high-performance applications requiring materials with high thermal and mechanical resilience. However, its use is limited by material conductivity and higher operational costs [59].

The choice of AM technology also plays a pivotal role in shaping heat exchangers' thermal performance and durability. This is primarily achieved through the unique ability of AM technologies, such as SLM and EBM, to fabricate complex internal structures. These structures, including lattice-type and biomimetic designs, significantly increase the surface area available for heat transfer, thereby enhancing the thermal performance of heat exchangers. For instance, researchers have demonstrated that heat exchangers manufactured using SLM can achieve significantly higher heat transfer coefficients due to the increased surface area and improved turbulence generated by intricate channel designs [60, 61]. AM technology plays a crucial role in optimizing the flow of fluids in heat exchangers, thereby reducing dead zones and enhancing convective heat transfer efficiency. This is achieved through the intricate channel designs enabled by AM processes, which ensure a smooth and efficient flow of fluids, thereby maximizing heat transfer [62, 63]. The durability of AM-produced heat exchangers is mainly influenced by the material properties and the specific AM process used. Materials like titanium and nickel alloys are popular in high-performance applications due to their high strength-to-weight ratios and excellent corrosion resistance [60, 64]. AM processes can influence material properties through parameters like layer thickness, energy density, and post-processing treatments. For example, the microstructure of materials produced by EBM can be more uniform than that produced by SLM, which impacts the fatigue resistance and lifespan of the heat exchanger under cyclic thermal loads [65].

Achieving optimal surface finishes on heat exchanger components manufactured via AM technologies is crucial, as surface roughness can significantly impact heat transfer efficiency and fluid flow dynamics. Different AM technologies address this challenge through various approaches and strategies, often involving specific process adjustments, material choices, and post-processing techniques. Post-processing is critical in achieving desired surface finishes regardless of the AM technology used. Techniques such as mechanical polishing, chemical etching, and shot peening reduce surface roughness and improve mechanical properties. For heat exchangers, where minimal surface roughness is paramount for adequate fluid flow and heat transfer, these post-processing steps are often necessary to meet stringent industry standards [66, 67]. SLM typically produces a relatively rough surface finish because of the small melt pool created by the laser, leading to the formation of balling effects and partially melted particles. To mitigate these issues, process parameters such as laser power, scanning speed, and hatch distance are meticulously optimized. For example, increasing the laser power or reducing the scan speed can reduce surface roughness by promoting better melting and flow of the metal powder [68–70]. Also, there are post-processing techniques like grinding and buffing to make the surface smoother. In the case of EBM, the high energy density of the electron beam allows for a smoother surface compared with SLM under similar conditions due to the preheating of the powder bed, which reduces residual stresses and minimizes warping. Nevertheless, EBM surfaces can still exhibit roughness due to the powder's nature and the beam's interaction, which can be addressed by adjusting the focus and intensity of the beam [71, 72]. DED is generally used to repair parts. The surface finish in DED processes can be refined by manipulating the deposition parameters, such as the feed rate and the travel speed of the deposition head. Additionally, DED allows for applying different materials to the surface for enhanced properties or improved finishes [73].

7.3 INNOVATIVE HEAT EXCHANGER DESIGNS ENABLED BY AM

7.3.1 Introduction to the Innovative Features of Heat Exchangers

AM enables innovative heat exchanger designs that significantly improve efficiency through enhanced geometric complexity and optimized material usage [56]. Traditional manufacturing techniques often limit the complexity of heat exchanger designs, resulting in suboptimal thermal performance. AM allows for creating intricate geometries, such as complex internal channels, lattice structures, and optimized surface textures, which maximize the surface area for heat transfer and improve fluid flow dynamics [50]. These advanced designs enhance heat exchangers' efficiency by increasing the heat transfer rate while minimizing pressure drop and material usage [74]. Furthermore, AM facilitates the integration of multi-functional features within a single component, reducing the need for assembly and enhancing thermal performance. The ability to precisely control the material distribution within the heat exchanger allows the use of advanced materials better suited for specific thermal and mechanical requirements [75]. This results in lighter, more efficient heat exchangers operating under more demanding conditions, improving overall energy efficiency and performance in aerospace, automotive, and power generation applications [76]. Residual stresses and thermal distortions are significant challenges in AM processes such as SLM and EBM, particularly affecting heat exchangers' structural integrity and performance. These phenomena result from rapid localized heating and cooling during the material deposition process, which creates steep temperature gradients within the component. The uneven expansion and contraction of the material can lead to warping, cracking, and residual stresses that compromise mechanical properties and thermal efficiency. Anwajler et al. [75] outline several mitigation strategies to address these issues centered around process optimization and post-processing techniques. Adjusting AM process parameters, such as laser power, scanning speed, and hatch spacing, is crucial for managing heat input and minimizing temperature gradients during fabrication. This approach aims to reduce the likelihood of thermal distortion and residual stresses by controlling the material cooling and solidification rate. Furthermore, integrating support structures during printing can provide mechanical stability and facilitate more uniform cooling, thereby reducing deformation. Post-processing methods, such as heat treatment, are also employed to relieve internal stresses and refine the material's microstructure, enhancing its mechanical and thermal properties [75].

Table 7.2 highlights innovative features like complex geometries, microchannel networks, embedded sensors, and hybrid materials that can significantly improve the effectiveness of heat exchangers by enhancing heat transfer efficiency, providing precise control, and leveraging advanced material properties [77, 78]. The innovative features described in Table 7.2 directly translate to enhanced heat exchanger effectiveness in real-world applications by leveraging advanced structural designs and integrating intelligent functionalities. Features like lattice structures and microchannel networks increase the surface area for heat transfer, allowing for more efficient thermal management [79]. Lattice structures minimize the pressure drop across the heat exchanger, reducing the energy required for fluid movement and enhancing overall system efficiency. Microchannel networks optimize this by providing precise control over fluid flow, which is crucial for applications requiring high heat transfer efficiency with minimal space, such as electronic cooling systems. Integrating embedded sensors and flow controls turns traditional heat exchangers into dynamic systems capable

TABLE 7.2
Summary of Basic Types of Features in Heat Exchangers to Improve Effectiveness

Feature	Description	Benefits
Complex Geometries and Enhanced Surface Areas		
Lattice structures	Intricate, open-cell structures are incorporated within the heat exchanger core.	• Maximized surface area for fluid interaction • Minimized pressure drop • Superior heat transfer efficiency
Microchannel networks	Dense networks of miniature channels within the heat exchanger.	• Significantly increased surface area for heat transfer • Precise control over channel size and orientation for optimized fluid flow • Enhanced heat exchanger performance
Integration of Multi-Functional Features		
Embedded sensors and flow controls	Miniature sensors are integrated directly into the heat exchanger structure, coupled with microfluidic components like valves.	• Real-time monitoring of temperature, pressure, and flow rates • Feedback control systems for optimized heat transfer and issue identification • Dynamic control of fluid flow paths for enhanced efficiency
Hybrid materials for enhanced thermal performance	Heat exchangers are fabricated with combinations of different materials within a single component.	• Exploitation of unique material properties (e.g., high thermal conductivity metal core, corrosion-resistant polymer shell) • Improved overall performance and functionality

of real-time monitoring and response [79–81]. These sensors track critical parameters such as temperature, pressure, and flow rates, enabling automated adjustments that optimize performance and prevent potential failures [82]. For example, real-time adjustments to coolant flows in power plants can significantly enhance efficiency and reduce the risk of overheating. Additionally, using hybrid materials allows heat exchangers to combine metals' thermal conductivity with polymers' corrosion resistance, providing tailored solutions that withstand harsh conditions while maintaining high performance. This is particularly beneficial in industries like chemical processing or offshore operations, where environmental conditions can degrade standard materials [82]. These enhancements improve the thermal efficiency of heat exchangers and contribute to longer service life and reduced maintenance needs, aligning with the demands of modern industrial applications for reliability and sustainability [83].

Lattice structures are intricate, open-cell configurations incorporated within the heat exchanger core [84]. These structures maximize the surface area for fluid interaction while minimizing pressure drop [85]. This design leads to superior heat transfer efficiency, as the increased surface area allows for more effective heat exchange between fluids. Microchannel networks consist of dense arrays of miniature channels within the heat exchanger [82]. These networks significantly increase the surface area for heat transfer, allowing precise control over channel size and orientation to optimize fluid flow. This results in enhanced heat exchanger performance, as the tiny channels provide more excellent surface contact between the fluid and the heat exchanger material. Miniature sensors and microfluidic components like valves are integrated directly into the heat exchanger structure. These embedded sensors enable real-time temperature, pressure, and flow rate monitoring, facilitating feedback control systems for optimized heat transfer and early issue identification [86]. Additionally, dynamic control of fluid flow paths enhances overall efficiency by adapting to changing operational conditions. Heat exchangers are fabricated using combinations of different materials within a single component to exploit unique properties. For instance, a high thermal conductivity metal core might be combined with a corrosion-resistant polymer shell. This hybrid

TABLE 7.3
Summary of Functionality Enhancement by Embedding Sensors and Flow Controls in Heat Exchangers [90, 91, 97]

Enhancement Feature	Technology Utilized	Functionality	Scientific Impact
Temperature and pressure monitoring	Embedded sensors	Continuously monitor thermal and pressure conditions within the heat exchanger.	Enables real-time optimization and control, safeguarding system integrity and operational efficiency.
Flow regulation	Automated control valves	Adjust fluid flow rates based on sensor feedback to maintain optimal conditions.	Optimizes thermal transfer rates and fluid dynamics, enhancing energy efficiency and reducing operational costs.
Predictive maintenance	Vibration and acoustic sensors	Detect abnormal patterns indicating potential mechanical failures.	Facilitates preemptive maintenance interventions, minimizing downtime and extending equipment lifespan.
Energy efficiency	Smart controls	Modulate heat exchanger operations based on varying ambient conditions and system demands.	Increases operational adaptability, ensuring energy is utilized only when necessary, thus conserving resources.
Enhanced heat transfer	Integrated sensors and controls	Measure and adjust parameters affecting heat transfer efficiency.	Maintains peak thermal efficiency through adaptive management of heat transfer surfaces and conditions.

approach improves overall performance and functionality, leveraging each material's best attributes to enhance thermal performance and durability [87–89].

7.3.2 Enhancement of the Features of Heat Exchangers

Table 7.3 outlines the significant enhancements in functionality achieved through integrating sensors and flow controls in heat exchangers, detailing the technology utilized, its functionality, and the resulting scientific impact. Embedded sensors for temperature [90] and pressure monitoring allow for continuous monitoring within the heat exchanger, enabling real-time optimization and control that safeguard system integrity and boost operational efficiency [91]. Automated control valves optimize thermal transfer rates and enhance energy efficiency by adjusting fluid flow rates according to sensor feedback, which also helps reduce operational costs [92, 93]. Additionally, integrating vibration and acoustic sensors facilitates predictive maintenance by identifying abnormal patterns that may signal mechanical failures, allowing for preemptive actions to extend the equipment's lifespan [94]. Smart controls enhance energy efficiency by modulating heat exchanger operations in response to changing ambient conditions and system demands, ensuring that energy is utilized judiciously and resources are conserved [95]. Integrated sensors and controls that enhance heat transfer maintain peak thermal efficiency by dynamically managing heat transfer surfaces and conditions, adapting to fluctuating operational needs to ensure optimal performance. These enhancements collectively elevate the operational capabilities of heat exchangers, making them more efficient, reliable, and adaptable to varied industrial demands [96].

Figure 7.2 illustrates the innovative design and operational principles of an integrated cross-media heat exchanger (iCMHX) [98]. In Figure 7.2(a), the double-pass iCMHX design demonstrates the cross-media approach, where metal wires act as fins for airside and liquid-side convective heat transfer. This configuration ensures effective heat transfer between the two media despite the polymer's low thermal conductivity. Figure 7.2(b) presents planar views with a staggered configuration

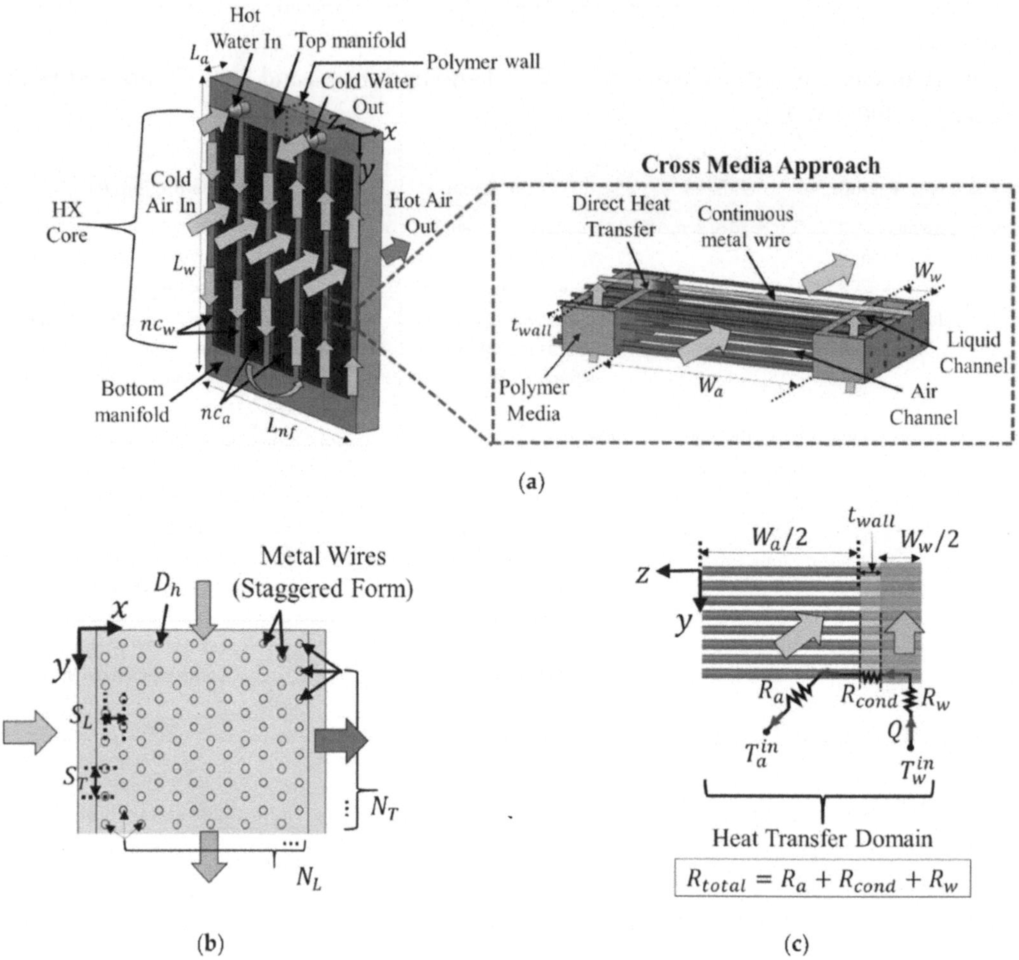

FIGURE 7.2 Cross sectional views of Heat exchangers: (a) A double-pass iCMHX design illustrating the cross-media approach; (b) planar views displaying a staggered configuration; (c) planar views depicting a simplified heat transfer domain [98].

of the metal wires, enhancing turbulence and improving heat transfer efficiency [98]. Figure 7.2(c) shows a simplified heat transfer domain, where heat transfer occurs in three stages: convection from hot water to the coated wires (Rw), conduction through the wires (Rcond), and convection from the cables to the cold air (Ra). The design also features integrated liquid manifolds and the option for thick polymer walls, which contribute to improved structural integrity and overall performance [99]. The double-pass iCMHX design effectively counters the low thermal conductivity of polymers by incorporating strategically placed metal wires within the polymer matrix, facilitating efficient heat transfer despite the inherent limitations of the polymer component. In an iCMHX configuration, metal wires serve as conduits for heat, bridging the gap between the air side and liquid side of the heat exchanger. These wires are laid in a staggered or inline configuration, enhancing the surface area available for heat exchange and improving the overall heat transfer efficiency of the system [98]. This setup allows for direct heat transfer from the hot liquid to the air through the metal wires, effectively bypassing the low thermal conductivity of the surrounding polymer. The design incorporates small diameter wires, which increase the heat transfer area relative to the mass of the wires, further optimizing the system's performance. Moreover, the double-pass flow arrangement within the iCMHX maximizes heat exchange by allowing the liquid to pass through the system twice, effectively doubling the opportunity for heat transfer [98]. This is particularly advantageous in applications requiring efficient cooling while maintaining compact system dimensions. The integration of metal and polymer materials, combined with the innovative design of the flow paths and the use of additive manufacturing techniques, results in a heat exchanger that is both high-performing and economically viable, tailored to meet the specific demands of liquid cooling applications in electronic devices [98, 100].

7.3.3 Material Considerations in AM Heat Exchangers

The selection of materials for AM heat exchangers is a critical aspect that significantly influences their performance, durability, and efficiency. A detailed summary is provided in Table 7.4. One of the primary considerations is thermal conductivity. High thermal conductivity materials, such as certain metals and metal alloys, are essential for effective heat transfer. These materials enable efficient heat dissipation from the heat source to the cooling medium, particularly important in applications requiring rapid temperature regulation, such as electronics cooling or aerospace systems. The ability to manage and transfer heat efficiently ensures that these systems operate within safe temperature ranges, thereby enhancing their performance and longevity [67].

Mechanical strength and durability are also paramount in material selection for AM heat exchangers. The chosen material must withstand the operational stresses and environmental conditions it will encounter. This includes resistance to fatigue, deformation, and potential impacts, ensuring the heat exchanger remains reliable and functional over time [101]. Durable materials are especially crucial in demanding applications like automotive or industrial processes, where consistent performance under varying conditions is necessary. The material's ability to maintain structural integrity under stress contributes to the overall reliability and safety of the system. Corrosion resistance is another vital consideration, particularly for heat exchangers exposed to harsh environments or corrosive fluids. Materials that resist corrosion, such as certain stainless steels and advanced polymers, help maintain the integrity and performance of the heat exchanger over time [102]. This resistance prevents degradation and potential failure, thereby extending the service life of the equipment. In environments where chemical exposure is a concern, selecting materials with high corrosion resistance is essential to ensure long-term functionality and safety. Manufacturability is equally important, as the chosen material must be compatible with the specific AM process being used, whether it is SLM, EBM, or another technique [103]. Considerations include material availability in powder or filament form, ease of processing, and the ability to achieve the desired geometric complexity and surface finish. The material must not only meet performance criteria but also be practical for

TABLE 7.4
Primary Criteria for the Selection of Material for Manufacturing Heat Exchangers

Selection Criteria	Need	AM Considerations
Heat transfer	High thermal conductivity	Metals (aluminum and copper) are considered, but printability and weight are also considered
Strength	Withstand pressure and stress	Aluminum alloys and titanium for high strength-to-weight ratio
Corrosion resistance	Fight off reactive fluids	Stainless steel and titanium for harsh environments
Printing ease	Compatible with your AM process	Powders or filaments are readily available for AM
Cost-effective	Balance performance and affordability	Consider trade-offs between material cost and post-processing needs
Lightweight	Minimize weight (e.g., aerospace)	Titanium or aluminum alloys are good options
Thermal expansion	Match other components	Minimize stress at material junctions

production within the constraints of the chosen AM technology [67]. Cost-effectiveness is always a significant factor in material selection. The material must balance performance and cost, ensuring the heat exchanger can be produced economically without compromising quality. This includes the raw material cost and the associated costs of processing, post-processing, and any additional treatments required. Balancing cost and performance ensures that the final product is both high-quality and affordable [29]. Weight is critical in applications where minimizing mass is essential, such as in aerospace or portable cooling systems. Lightweight materials like aluminum or advanced composites provide the necessary thermal and mechanical properties while reducing the overall weight of the heat exchanger. Reducing weight can lead to better fuel efficiency and ease of handling in various applications. Finally, the coefficient of thermal expansion (CTE) is crucial to ensure that the material's expansion and contraction rates are compatible with other materials used in the heat exchanger system [104]. Matching CTE values helps prevent thermal stresses and deformation during operation, maintaining structural integrity and performance stability across varying temperatures. This consideration is essential to ensure the longevity and reliability of the heat exchanger in different operational environments [102]. The material properties outlined in Table 7.3 are meticulously aligned with the rigorous demands of high-performance heat exchangers, particularly for critical aerospace and power generation applications [105]. Each property addresses specific operational challenges to optimize functionality and durability under extreme conditions. High thermal conductivity is essential for effective heat transfer and crucial in power generation and aerospace, where materials like aluminum and copper are favored due to their superior thermal properties [102]. However, aluminum's lower weight and adequate thermal performance make it preferable for aerospace applications, enhancing fuel efficiency and payload capacity. Strength is critical as heat exchangers must endure high pressures and stresses. Aluminum alloys and titanium are chosen for their excellent strength-to-weight ratios, providing necessary robustness while minimizing weight. This is particularly advantageous in aerospace, as it reduces structural load and increases efficiency. Corrosion resistance is vital in environments exposed to reactive fluids or harsh conditions. Materials must also be compatible with AM processes [106], requiring the selection of powders or filaments that are amenable to the specific AM technology, ensuring high-quality manufacturing outcomes. Cost-effectiveness involves balancing performance with material and processing costs, which is crucial in commercial aerospace and power generation where budget constraints are significant. Lastly, matching thermal expansion properties with other system components minimizes stress at material junctions [106]. This is crucial for maintaining structural integrity under varying thermal loads typical in aerospace and power generation applications. These material selections ensure that heat exchangers meet high performance, reliability, and durability demands in demanding sectors, optimizing both functionality and operational lifespan [107].

Table 7.5 provides an overview of popular metals used in the additive manufacturing of heat exchangers. Stainless steel is prominently featured due to its excellent mechanical properties and corrosion resistance. For example, SS 316L, frequently used and referenced in numerous studies, is known for its high corrosion resistance and good weldability, making it ideal for heat exchangers [108–110]. Other grades like SS 17-4 [111] and SS 347 [38] are valued for their high strength and high-temperature stability, respectively, while SS 304 is noted for its versatility and excellent forming and welding characteristics. Titanium alloys, particularly Ti-6Al-4V (Ti64) [112], are also highlighted in the table. This alloy is favored for its high strength, low weight, and excellent corrosion resistance, making it suitable for applications in aerospace and biomedical fields. Nickel alloys such as Inconel 718 and EOS NickelAlloy HX are recognized for their high strength and excellent corrosion resistance at high temperatures. Inconel 718 [113], widely referenced in research, is ideal for extreme environments such as jet engines and gas turbines. At the same time, EOS NickelAlloy HX [114] is specialized for heat exchangers due to its superior thermal properties and resistance to oxidation and corrosion. Nitinol (Ti-50at% Ni) is mentioned for its unique shape memory and super elastic properties, making it useful in medical devices and actuators. Cobalt chrome [115], valued for its high wear resistance and strength, is commonly used in dental and orthopedic implants. Aluminum

TABLE 7.5
Summary of the Popular Materials Used for Manufacturing Heat Exchangers using AM Technologies [50]

Material	Powder Bed Fusion	Cold Spray
Stainless steel (SS)	SS 316L [108–110] SS 17-4 [111] SS 347 [38]	SS 304 [133]
Titanium alloys	Ti-6Al-4V (Ti64) [112]	–
Cobalt chrome	C-Cr-Mo [115]	–
Nickel alloys	Inconel 718 [113] EOS NickelAlloy HX [114]	–
Aluminum alloys	AlSi10Mg [116–118] Al 6061 [119] Pure Al [134, 135] ALSi12 [120]	Pure Al [136, 137]
Copper	[121]	Copper [122]

alloys, including AlSi10Mg [116–118], Al 6061 [119], and AlSi12 [120], are noted for their good mechanical properties, being lightweight, and excellent castability. These alloys are extensively used in additive manufacturing for various applications. Pure aluminum is highlighted for its high thermal and electrical conductivity. Copper [121, 122], known for its excellent thermal and electrical conductivity, is ideal for heat exchangers and electrical components. Various Cu-based nanocomposite alloys have potential use for their high wear resistance and strength, superior thermal properties, and resistance to oxidation and corrosion [123–126]. It would be promising to see prototypes of similar copper-based nanocomposites produced by SLM and cold spray techniques shortly. The diverse array of metals listed in Table 7.5 for AM heat exchangers brings specific advantages and challenges influenced by their intrinsic properties and compatibility with AM processes. Stainless steel, widely used in AM due to its corrosion resistance and mechanical strength, is particularly beneficial for applications where longevity and resistance to harsh environments are crucial [67]. However, its thermal conductivity, while sufficient for many applications, is significantly lower than copper's, potentially limiting its effectiveness in applications requiring rapid heat dissipation [127]. Titanium alloys, like Ti-6Al-4V, offer exceptional strength-to-weight ratios and corrosion resistance, making them ideal for aerospace and biomedical applications where weight reduction and durability are critical. However, the high cost of titanium powder and processing difficulties due to its reactivity at high temperatures can make it less viable for broader industrial applications [128]. Nickel alloys, such as Inconel 718 [129], provide excellent high-temperature strength and corrosion resistance, suitable for extreme environments like aerospace engines. However, the same properties that make them advantageous also make them challenging to process in AM setups due to their toughness and high melting points, leading to higher production costs [102]. Aluminum alloys are favored for their lightweight and good thermal conductivity, which is advantageous in the automotive and aerospace industries. Nonetheless, their lower melting points and thermal conductivity can pose challenges during the laser melting process in AM, potentially leading to residual stresses and deformations [130]. Copper stands out for its superior thermal and electrical conductivity, making it excellent for heat exchangers and electrical applications. However, its high reflectivity and thermal conductivity can complicate laser-based AM processes, affecting energy absorption and heat distribution during manufacturing. While these materials offer significant benefits tailored to specific needs, their drawbacks regarding AM process compatibility and cost must be carefully considered.

The choice of material often involves a trade-off between desired properties and the practicalities of manufacturing, impacting the final product's performance and economic viability [131, 132].

Table 7.6 provides a summary of hybrid materials that significantly enhance heat exchangers' thermal performance by synergistically combining different materials' properties to optimize functionality and efficiency [138]. For instance, by integrating metals with high thermal conductivity, such as copper, with lightweight metals like aluminum, the heat exchangers benefit from rapid heat transfer capabilities while maintaining structural lightness [127]. This synergy optimizes heat transfer rates and reduces the system's overall weight, making it more efficient. Similarly, applying corrosion-resistant coatings like nickel on copper bases prolongs the life of heat exchangers by protecting them against environmental degradation and maintaining their high thermal performance. Furthermore, composite materials, such as metal matrix composites reinforced with ceramics, enable heat exchangers to operate effectively at higher temperatures by enhancing their temperature stability and mechanical strength. These strategic combinations of materials in hybrid heat exchangers lead to improved performance, cost-effectiveness, and durability, addressing specific operational demands of various industrial applications [139, 140].

Hybrid materials significantly enhance the thermal performance of heat exchangers by combining the advantages of different materials to overcome the limitations of any single one. For instance, integrating metals with high thermal conductivity, such as copper or aluminum, with materials

TABLE 7.6
Summary of the Synergic Effect of Combining Materials on the Performance of Heat Exchangers [127, 138–140]

Hybrid Material Feature	Materials Combined	Functionality	Effect
Enhanced thermal conductivity	Metals with high thermal conductivity (e.g., copper) combined with lightweight metals (e.g., aluminum).	Facilitates rapid heat transfer while maintaining structural lightness.	Increases overall thermal efficiency by leveraging one material's high conductivity with another's lightweight property, optimizing heat transfer rates and reducing system weight.
Corrosion resistance	Corrosion-resistant coatings or layers (e.g., nickel) on a thermally efficient base (e.g., copper).	Protects against chemical or environmental degradation while maintaining high thermal performance.	Extends the lifespan and reliability of heat exchangers operating in corrosive environments without compromising thermal efficiency.
High-temperature stability	Composites of ceramics and metals (e.g., metal matrix composites with ceramic reinforcements).	Endures higher operating temperatures without material degradation.	Enhances the operation of heat exchangers in high-temperature applications, providing stability and maintaining integrity under thermal stress.
Improved mechanical strength	Fiber-reinforced composites (e.g., carbon fiber-reinforced polymers) combined with metals.	Supports structural integrity under physical loads while facilitating heat dissipation.	Synergistically, fibers' high strength is combined with metals' thermal conductivity, resulting in a durable and efficient heat exchange system.
Cost-effectiveness	Economical materials (e.g., polymers) combined with selective high-performance materials (e.g., metal inserts).	Balances cost and performance for specific application requirements.	Allows for cost-effective manufacturing of heat exchangers by strategically using expensive materials only where high performance is critical.

like graphene, carbon nanotubes (CNTs), or ceramics, can result in superior heat transfer capabilities, especially at the micro or nanoscale. These combinations often lead to improved mechanical strength, durability, and corrosion resistance. By incorporating polymers or ceramics with reinforcing fillers like CNTs or metal particles, hybrid composites offer not only thermal efficiency but also lightweight and robust structures. Furthermore, hybrid materials allow for tailored thermal properties; for example, multi-layered structures can be engineered to provide directional heat flow, localized cooling, or insulation where needed. Surface-engineered hybrids, such as ceramic-metal or polymer-metal coatings, reduce fouling and corrosion, which helps maintain thermal performance over time.

The synergy between different materials plays a crucial role. A metal-ceramic hybrid combines the excellent thermal conductivity of metals with the chemical stability and high-temperature resistance of ceramics, making it ideal for extreme environments. A metal-polymer hybrid leverages the heat transfer capabilities of metals and the flexibility and corrosion resistance of polymers, resulting in lightweight systems suitable for portable or compact devices. Meanwhile, polymer matrices filled with nanomaterials like CNTs or graphene gain the processability of polymers and the extraordinary thermal and mechanical properties of nanofillers, enabling applications in miniaturized or flexible heat exchangers. Additionally, phase change materials (PCMs) enhanced with conductive additives, such as aluminum or carbon fibers, provide smart thermal regulation by efficiently storing and releasing heat. Altogether, hybrid materials offer a multifunctional solution for advanced heat exchanger designs, allowing for greater efficiency, adaptability, and longevity in a wide range of industrial and technological applications.

Table 7.7 provides a comprehensive overview of the critical factors to consider when selecting an AM technique for heat exchanger applications, focusing on the capabilities. In different AM processes such as SLM, DMLS, and EBM, material compatibility is paramount, as the choice of material influences properties like thermal conductivity, corrosion resistance, and mechanical strength,

TABLE 7.7
Summary of the Factors to be Considered When Selecting an AM Technique for a Specific Heat Exchanger Application [108, 141, 142]

Factor	Description	Impact on Selection
Material compatibility	Different AM techniques handle various materials. Material choice affects thermal conductivity, corrosion resistance, and mechanical properties.	Select an AM technique to process the desired material, such as SLM for metals like titanium and stainless steel.
Geometric complexity	AM excels in producing complex designs, which can enhance heat transfer efficiency.	Option for techniques like SLM or DMLS that manage complex geometries effectively.
Surface finish and resolution	Required surface quality impacts fluid dynamics and heat transfer.	Choose techniques with higher resolution and smoother finishes, such as SLM or DMLS.
Production volume and cost	Consider the scale of production and budget constraints.	For low volumes, AM is cost-effective; assess the scalability and cost for higher volumes.
Thermal and mechanical properties	The operating conditions dictate necessary properties like thermal stability and mechanical strength.	Evaluate which AM process delivers parts that meet the specific thermal and mechanical demands, like EBM, for high thermal resistance.
Post-processing requirements	Some AM parts require significant post-processing to achieve the desired qualities.	Factor in the extent and cost of post-processing when selecting the AM technique.

guiding the selection of an appropriate AM process for specific materials [108, 141, 142]. SLM is preferred for metals like titanium and stainless steel. The ability of AM to produce complex geometries is another crucial consideration, enhancing heat transfer efficiency; therefore, techniques such as SLM or DMLS that effectively manage intricate designs are favored. Surface finish and resolution are also significant, affecting fluid dynamics and heat transfer, making high-resolution techniques essential for achieving optimal performance. Production volume and cost considerations highlight AM's cost-effectiveness for low volumes, though scalability for higher volumes must be assessed. Thermal and mechanical property requirements, driven by specific operating conditions, necessitate choosing processes like EBM for parts requiring high thermal resistance. Lastly, post-processing requirements must be considered, as some AM parts need extensive post-processing to meet desired specifications, influencing the overall selection of the AM technique based on the balance between initial print quality and the extent of subsequent finishing needed [108, 141, 142].

SLM and EBM are powder bed fusion technologies that provide high precision and material properties but come with significant costs related to raw materials (metal powders), energy consumption, and machine depreciation [143]. The high-resolution capabilities of these methods allow for the creation of complex internal geometries that are not possible with conventional manufacturing techniques, leading to potentially better performance in heat exchanger applications [50]. However, the slow build rates and extensive post-processing requirements can limit their scalability and economic efficiency for mass production unless the added value of customization and performance optimization outweighs these costs [50]. On the other hand, binder jetting generally offers faster build times and lower material costs since it does not require high-power lasers or electron beams [144]. This technology can be more scalable and cost-effective for larger production volumes [144]. It involves binding powder material in layers and then sintering them in a furnace. This can lead to reduced mechanical properties but is offset by significantly lower production costs and faster throughput [144]. AM's initial setup and material costs can be higher than traditional manufacturing methods, such as brazing and machining. However, producing complex designs without expensive tooling or assembly can reduce overall costs in specific applications. Therefore, the cost-effectiveness of AM technologies depends on the specific requirements of the heat exchanger design, such as performance enhancements and being lightweight, which can justify the higher initial expenses through improved efficiency or reduced lifecycle costs [145, 146].

7.3.4 Case Study

In the study conducted by Sabau et al. [109], they present an innovative architecture that transitions fluid pathways between tube and shell flow within the heat exchanger. This novel configuration aims to enhance thermal performance and compactness using a counterflow sandwich plenum (CSP) design constructed using advanced additive manufacturing techniques. The proposed CSP HX design offers flexibility in fluid flow management and aims to overcome limitations associated with traditional fabrication methods. The key novelty of the design lies in its ability to switch fluid pathways from tube flow to shell flow or vice versa within a single heat exchanger. This transition is facilitated by a three-zone architecture consisting of two plenums filled with tube bundles and a solid core of counterflow channels sandwiched between them. The design employs non-circular cross-section tubes (triangular in this case) to achieve high packing density and enhanced heat transfer performance. This approach allows for a more compact and efficient HX than traditional shell-and-tube designs [147]. The utilization of additive manufacturing, specifically laser powder bed fusion (LPBF), enables the creation of complex geometries that were previously unachievable with conventional methods, thus opening new avenues for heat exchanger optimization [109, 148].

In the initial builds of additively manufactured heat exchangers using LPBF, Sabau et al. encountered several significant challenges that impacted the manufacturing process and the performance of the final products [109]. One of the primary issues was achieving adequate surface quality and structural integrity, which are critical for the effective heat transfer and mechanical strength required in heat exchangers. The intricate designs typical of heat exchangers often include complex geometries and overhangs that are difficult to produce with LPBF, necessitating support structures [109]. While essential during printing, these supports complicate post-processing steps such as removal, which can damage the component or leave residues that adversely affect heat transfer efficiency. Material selection also presented a challenge, as the materials ideal for heat exchanger applications based on their thermal and mechanical properties might not align well with the requirements of LPBF technology. Issues such as poor layer adhesion and increased porosity were common, impacting the durability and efficiency of the heat exchangers. Furthermore, the rapid cooling rates associated with LPBF can induce thermal stresses that lead to distortion and warping of the parts, requiring careful management of the printing parameters and post-processing treatments to mitigate these effects [109].

Figure 7.3(i) showcases the additively manufactured heat exchangers. These images underscore the iterative process of optimizing the manufacturing techniques and addressing initial build challenges such as leakage and structural integrity [109]. Figure 7.3(ii) provides detailed cross-sectional views normal to the tube axis, highlighting critical geometric parameters and structural features. This demonstrates the staggered arrangement that facilitates the transition between tube and shell flow, enhancing the overall heat transfer performance. Figure 7.3(iii-a) depicts the refrigerant and brine flow rates and absolute pressures. The refrigerant flow rate is approximately 0.0154 L/s, while the brine flow rate is around 0.23 L/s. The absolute pressure measurements indicate that the refrigerant maintains a pressure of approximately 1.08 MPa, and the brine sustains around 1.53 MPa. These values are crucial for assessing the operational stability and efficiency of the heat exchanger under various conditions [109]. Figure 7.3(iii-b) provides insights into the heat exchanger's temperature variations and differential pressures. The inlet temperature for the refrigerant is consistently at 20°C, while its outlet temperature shows a significant decrease of about 35°C, demonstrating substantial heat transfer efficiency. For the brine, the inlet temperature is stable at 160°C, and the outlet temperature decreases by approximately 6°C, indicating effective heat absorption. The differential pressures for the refrigerant and brine are also illustrated, with the refrigerant showing a variation within a 10% range of the mean value and the brine exhibiting significant differential pressure, reflecting the efficient fluid dynamics and heat transfer capabilities of the heat exchanger [109].

Figure 7.3(iv) compares the thermal performance of the baseline heat exchanger (case 0) with the CSP heat exchangers by examining the inlet and outlet temperatures for both refrigerant and brine [109]. Figure 7.3(iv-a) focuses on the refrigerant temperatures. For the baseline heat exchanger, the inlet temperature is set at 20°C, and the outlet temperature shows less variation compared with the CSP designs, indicating lower heat transfer efficiency. In contrast, the CSP heat exchangers maintain a consistent inlet temperature of 20°C. Still, their outlet temperatures exhibit a more significant reduction, often by around 35°C, signifying a higher thermal performance and more efficient heat transfer. Figure 7.3(iv-b) details the brine temperatures [109]. The inlet temperature for both the baseline and CSP heat exchangers is stable at 160°C. However, the outlet temperature for the baseline heat exchanger remains relatively high, reflecting less effective heat absorption [109]. On the other hand, the CSP heat exchangers demonstrate a significant temperature drop of about 6°C at the outlet, underscoring their superior ability to absorb and transfer heat compared with the baseline model.

7.4 CHALLENGES AND FUTURE RESEARCH IN AM HEAT EXCHANGER PRODUCTION

AM presents several challenges and limitations in the production of heat exchangers, spanning technical, economic, and regulatory domains. Achieving the necessary geometric precision and surface quality is a major technical hurdle [50]. Techniques like SLM and EBM often produce parts requiring extensive post-processing to meet surface smoothness and dimensional accuracy standards.

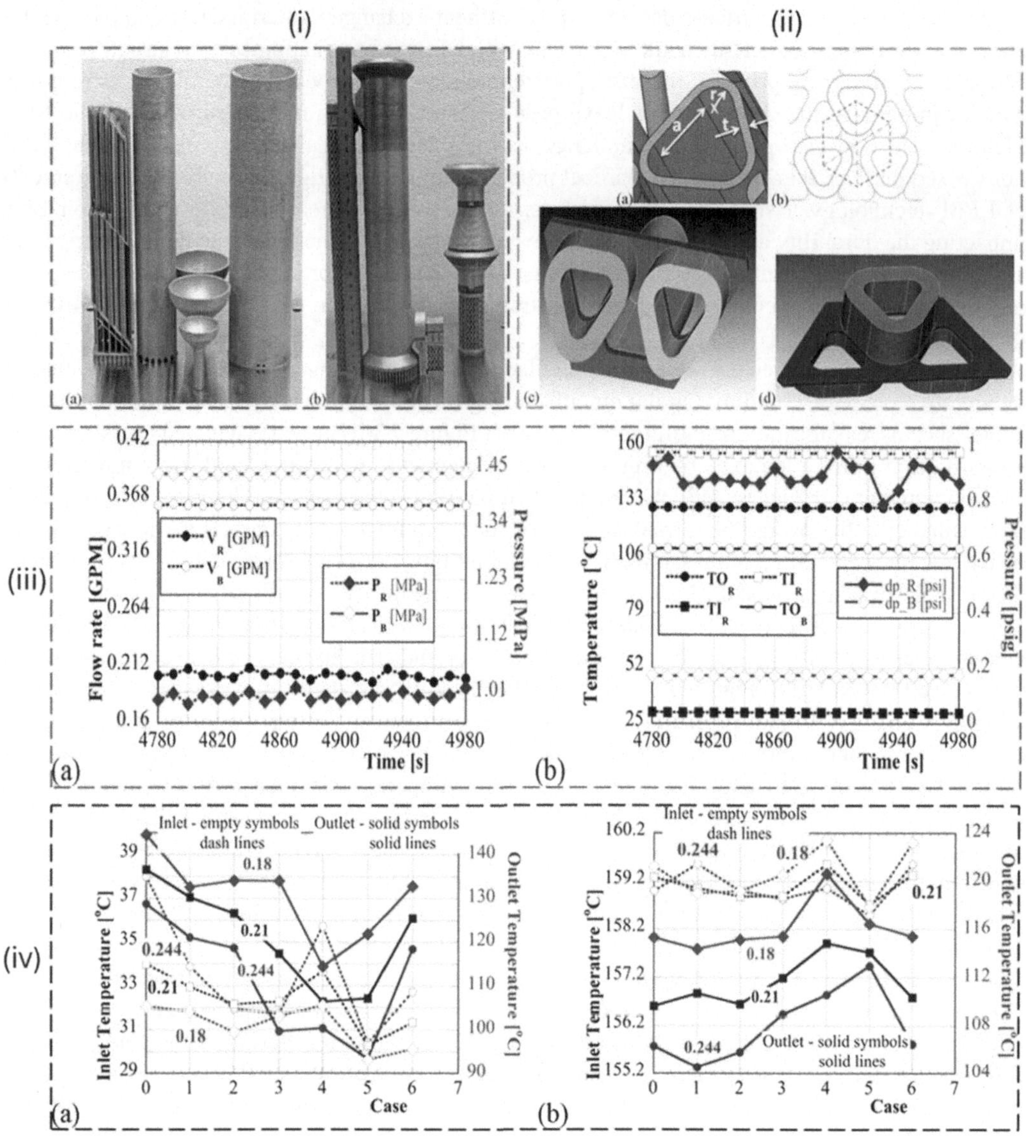

FIGURE 7.3 (i) Images of additively manufactured heat exchangers: (a) initial 3D-printed versions of HX01 and HX02 (tube thickness of 0.38 mm), displaying a partially printed heat exchanger to highlight the key design features, and (b) a subsequent build of HX1 (tube thickness of 0.61 mm) including two end-caps; (ii) cross-sections perpendicular to the tube axis highlighting: (a) three geometric parameters: tube wall thickness (t), corner radius of curvature (r), and side length (a); (b) unit cell (dashed line) used to assess the shell side hydraulic diameter; (c) and (d) close-up views of tubes on either side of a thin core; (iii) steady-state measurement data for HX1: (a) flow rates and absolute pressures, and (b) inlet and outlet temperatures along with differential pressures; (iv) the variation in temperature at inlet and outlet for the baseline heat exchanger and sandwiched between plenums (CSP) heat exchangers for (a) refrigerant and (b) brine [109].

Additionally, managing residual stresses and thermal distortions during the AM process is critical, as the rapid heating and cooling cycles can lead to warping, cracking, and other structural defects [67]. This issue is exacerbated by the potential for anisotropy in material properties due to the layer-by-layer construction. The economic feasibility of AM for heat exchanger production is constrained by high costs associated with AM equipment, materials, and energy consumption [81, 149,

150]. Techniques like SLM and EBM involve significant capital expenditure and operational costs. Furthermore, the slow production rates limit scalability, requiring substantial investment for mass production. Though more cost-effective, binder jetting often necessitates additional steps like sintering and infiltration to achieve desired mechanical properties, adding complexity and cost [151]. The lack of established standards and certification processes for AM parts creates uncertainty regarding their performance and reliability in critical applications [152]. Regulatory bodies are still developing guidelines to address AM-specific issues such as build orientation and post-processing effects [153, 154]. Ensuring consistent quality and repeatability is crucial, especially for heat exchangers in safety-critical industries like aerospace and power generation. The absence of standardized testing protocols complicates the certification process, necessitating extensive testing and qualification efforts by manufacturers. Addressing these challenges through continued research and collaboration is essential to fully realizing AM's potential in heat exchanger production. Table 7.8 provides future research directions that can be followed to improve the efficiency of heat exchangers.

AM holds significant potential to transform heat exchanger production's economic viability and environmental sustainability [141]. By facilitating complex, tailored designs, AM reduces material waste compared with traditional subtractive manufacturing techniques, which inherently involve cutting away large portions of raw material [156, 157]. This precision in material usage decreases the cost of raw materials and minimizes the environmental burden associated with extracting and

TABLE 7.8
Summary of the Research Opportunities Available for Producing Heat Exchangers Using Additive Manufacturing Technology [103, 155]

Future Direction/ Research Opportunity	Description	Potential Benefits
Development of novel materials designed explicitly for AM heat exchangers	• Explore materials with superior thermal conductivity, enhanced corrosion resistance, and tailored mechanical properties for AM processing	• Significantly improved heat transfer efficiency • Increased durability and lifespan of heat exchangers • Optimization of material properties for specific AM techniques
Advanced multi-material AM for heat exchangers	• Investigate techniques for seamlessly integrating dissimilar materials within a single build functionally, graded materials with varying properties throughout the structure	• Creation of heat exchangers with unique performance characteristics • Optimization of material properties for different sections of the heat exchanger (e.g., high thermal conductivity core, corrosion-resistant shell)
In-situ process monitoring and control for AM heat exchangers	• Develop real-time monitoring systems to track material properties during the AM process; identify and address potential defects	• Improved quality control and consistency of AM-produced heat exchangers • Reduced risk of failure and improved overall performance
Integration of sensors and microfluidic components directly into AM heat exchangers	• Refine techniques for embedding temperature, pressure, and flow rate sensors in microfluidic channels for active flow control	• Creation of intelligent heat exchangers that can self-regulate and optimize performance • Enhanced efficiency and adaptability of heat exchangers to varying operating conditions
Standardization of AM processes and qualification methods for heat exchangers	• Establish standardized protocols for AM heat exchanger production and reliable qualification methods to ensure performance and safety	• Streamlined adoption of AM technology in heat exchanger manufacturing • Increased confidence and broader acceptance of AM-produced heat exchangers in critical applications

processing these materials. Another pivotal advantage AM offers is efficiency in manufacturing. Producing components directly from digital models can significantly shorten production cycles and reduce energy consumption [157]. Moreover, AM's localized production capabilities could lessen the need for extended supply chains, thereby reducing the carbon footprint associated with transportation. Regarding recycling and reuse, AM technologies such as powder bed fusion create opportunities for using recycled material powders, enhancing the circular economy within industrial manufacturing processes. The potential to easily integrate recycled materials into the production cycle without significantly degrading material properties is a transformative aspect of AM that aligns with sustainable manufacturing principles. AM also favorably impacts heat exchanger lifecycle costs. The enhanced performance of AM-produced heat exchangers, owing to optimized designs for heat transfer efficiency, can lead to lower operational costs over time. Furthermore, the durability improvements afforded by AM can extend the service life of these components, reducing the frequency and costs associated with replacements [158].

Current regulatory frameworks are increasingly addressing the unique challenges of AM technologies in producing heat exchangers, particularly those used in safety-critical applications such as aerospace, automotive, and energy sectors [18]. These frameworks ensure that AM parts meet stringent safety, quality, and performance standards comparable to those manufactured through traditional methods. One of the primary challenges in regulatory oversight of AM technologies involves the variability in material properties and the reproducibility of parts [159]. AM processes like SLM or EBM can lead to anisotropy in mechanical properties due to layer-wise construction, different cooling rates, and possible defects such as porosity and residual stresses [159]. This variability makes it challenging to standardize testing protocols that assume material homogeneity and consistent manufacturing conditions as in traditional processes. Additionally, the lack of comprehensive databases on material properties for AM processes complicates the certification of AM parts. Most existing material standards and testing protocols are based on bulk materials and do not account for the microstructural differences inherent in materials processed by AM. Regulatory bodies and industry consortia are developing AM-specific standards and guidelines to address these issues. Organizations such as ASTM International and ISO have established committees to formulate standards that cover everything from powder properties to finished part testing [160]. Another obstacle is integrating in-situ monitoring and quality assurance during the AM build process. Standardizing such protocols is essential for certifying AM parts for safety-critical applications, ensuring consistent quality throughout production. However, developing universally accepted in-situ monitoring techniques that can be applied across different AM machines and materials remains a complex task [161]. Regulatory frameworks are evolving to address AM's challenges, but significant work remains in standardizing testing protocols and certification processes. This includes the development of detailed material property databases, the establishment of reliable in-situ monitoring techniques, and the integration of these elements into a cohesive regulatory structure that ensures the safety and reliability of AM-produced heat exchangers in critical applications [162].

7.5 CONCLUSION

Additive Manufacturing has demonstrated significant potential in revolutionizing the design and production of heat exchangers. This review highlighted various AM technologies, such as SLM, EBM, and binder jetting, which create complex geometries and customized material properties unattainable through traditional manufacturing methods. The integration of innovative features, such as lattice structures and multi-material constructs, has led to notable enhancements in heat exchangers' thermal efficiency and mechanical performance. Despite these advancements, several challenges remain, including achieving high geometric precision, managing residual stresses, and ensuring economic feasibility for large-scale production. Future research should focus on developing novel materials specifically tailored for AM processes, advancing multi-material AM techniques, and integrating real-time process monitoring to ensure quality control. Additionally, embedding sensors and microfluidic components directly into AM-produced heat exchangers could lead to intelligent systems capable of self-regulation and optimized performance. Standardization of AM processes and qualification methods is also crucial to increase the adoption of AM technology in critical applications. Therefore, AM has opened new frontiers in heat exchanger design and performance; continued interdisciplinary research and collaboration are essential to overcome existing limitations and fully harness the benefits of this transformative manufacturing technology.

REFERENCES

1. Shah, R.K. and D.P. Sekulic, *Fundamentals of heat exchanger design*. 2003: John Wiley & Sons.
2. Bhutta, M.M.A., et al., CFD applications in various heat exchangers design: a review. *Applied Thermal Engineering*, 2012. **32**: pp. 1–12.
3. Lee, H., *Thermal design: heat sinks, thermoelectrics, heat pipes, compact heat exchangers, and solar cells*. 2022: John Wiley & Sons.
4. Stewart, S.W., S.V. Shelton, and K.A. Aspelund, Finned-tube heat exchanger optimization methodology. *Heat Transfer Engineering*, 2005. **26**(7): pp. 22–28.
5. Lee, D.H., et al., Improvement of heat transfer with perforated circular holes in finned tubes of air-cooled heat exchanger. *International Communications in Heat and Mass Transfer*, 2012. **39**(2): pp. 161–166.
6. Talkar, S., A. Choudhari, and P. Rayar, Building envelope optimization and cost-effective approach in HVAC to support smart manufacturing. In: Vasudevan, H., Kottur, V., and Raina, A. (eds), *Proceedings of international conference on intelligent manufacturing and automation. Lecture notes in mechanical engineering*. 2020: Springer, Singapore. https://doi.org/10.1007/978-981-15-4485-9_31.
7. Choudhari, A., et al., Designing and prototyping for conservation and effective utilization of waste heat from air conditioner. In *IOP conference series: materials science and engineering* (Vol. 1104, No. 1, p. 012007). 2021: IOP Publishing. https://doi.org/10.1088/1757-899X/1104/1/012007.
8. Zohuri, B., *Application of compact heat exchangers for combined cycle driven efficiency in next generation nuclear power plants: a novel approach*. 2015: Springer.
9. Anozie, A.N. and O.J. Odejobi, The search for optimum condenser cooling water flow rate in a thermal power plant. *Applied Thermal Engineering*, 2011. **31**(17–18): pp. 4083–4090.
10. Chauhan, S.S. and S. Khanam, Enhancement of efficiency for steam cycle of thermal power plants using process integration. *Energy*, 2019. **173**: pp. 364–373.
11. Vance, D., et al., Estimation of and barriers to waste heat recovery from harsh environments in industrial processes. *Journal of Cleaner Production*, 2019. **222**: pp. 539–549.
12. Gupta, A.K., et al., Custom implants and beyond: the biomedical potential of additive manufacturing. *Archives of Case Reports*, 2024. **8**(2). https://doi.org/10.29328/journal.acr.1001092024.
13. Ian Gibson, I.G., *Additive manufacturing technologies 3D printing, rapid prototyping, and direct digital manufacturing*. 2015: Springer.
14. Razavykia, A., et al., An overview of additive manufacturing technologies—a review to technical synthesis in numerical study of selective laser melting. *Materials*, 2020. **13**(17): p. 3895.
15. Sun, C., et al., Additive manufacturing for energy: a review. *Applied Energy*, 2021. **282**: p. 116041.
16. Zhao, H. and G. Zong, Metal additive manufacturing. *Materials in Advanced Manufacturing*, 2022: pp. 269–300.
17. Wang, H., et al., Laser-processed functional surface structures for multi-functional applications-a review. *Journal of Manufacturing Processes*, 2024. **116**: pp. 247–283.
18. Tofail, S.A.M., et al., Additive manufacturing: scientific and technological challenges, market uptake and opportunities. *Materials Today*, 2018. **21**(1): pp. 22–37.
19. Rafiee, M., R.D. Farahani, and D. Therriault, Multi-material 3D and 4D printing: a survey. *Advanced Science*, 2020. **7**(12): p. 1902307.
20. Huang, Y., et al., Additive manufacturing: current state, future potential, gaps and needs, and recommendations. *Journal of Manufacturing Science and Engineering*, 2015. **137**(1): p. 014001.
21. Mirzababaei, S. and S. Pasebani, A review on binder jet additive manufacturing of 316L stainless steel. *Journal of Manufacturing and Materials Processing*, 2019. **3**(3): p. 82.
22. Popov, V.V., et al., Powder bed fusion additive manufacturing using critical raw materials: a review. *Materials*, 2021. **14**(4): p. 909.

23. Singh, A., S. Kapil, and M. Das, A comprehensive review of the methods and mechanisms for powder feedstock handling in directed energy deposition. *Additive Manufacturing*, 2020. **35**: p. 101388.
24. Bakır, A.A., R. Atik, and S. Özerinç, Mechanical properties of thermoplastic parts produced by fused deposition modeling: a review. *Rapid Prototyping Journal*, 2021. **27**(3): pp. 537–561.
25. Gao, B., et al., A review of research progress in selective laser melting (SLM). *Micromachines*, 2022. **14**(1): p. 57.
26. Pathak, A., et al., Application of laser technology in the mechanical and machine manufacturing industry. In *Laser-based technologies for sustainable manufacturing* (pp. 107–155). 2023: CRC Press.
27. Chen, F., et al., Heat transfer efficiency enhancement of gyroid heat exchanger based on multidimensional gradient structure design. *International Communications in Heat and Mass Transfer*, 2023. **149**: p. 107127.
28. Kumar, A., A. Kumar, and A. Kumar, *Laser-based technologies for sustainable manufacturing*. 2023: CRC Press.
29. Yap, C.Y., et al., Review of selective laser melting: Materials and applications. *Applied Physics Reviews*, 2015. **2**(4): p. 041101.
30. Kumar, A., et al., *Physics of laser–matter interaction in laser-based manufacturing, in laser-based technologies for sustainable manufacturing* (pp. 45–54). 2023: CRC Press.
31. Salmon, F., et al., Optimizing performance for cooling electronic components using innovative heterogeneous materials. *Applied Energy*, 2024. **362**: p. 122983.
32. Kumar, A., A. Kumar, and A. Kumar, *Introduction to optics and laser-based manufacturing technologies, in laser-based technologies for sustainable manufacturing* (pp. 1–43). 2023: CRC Press.
33. Gupta, A., et al., Design and analysis of vertical vacuum fryer. In: Vasudevan, H., Kottur, V., and Raina, A. (eds), *Proceedings of international conference on intelligent manufacturing and automation. Lecture notes in mechanical engineering*. 2020: Springer. https://doi.org/10.1007/978-981-13-2490-1_13.
34. Son, K.N., et al., Design of multifunctional lattice-frame materials for compact heat exchangers. *International Journal of Heat and Mass Transfer*, 2017. **115**: pp. 619–629.
35. Bartlett, J.L. and X. Li, An overview of residual stresses in metal powder bed fusion. *Additive Manufacturing*, 2019. **27**: pp. 131–149.
36. Mahale, T.R. *Electron beam melting of advanced materials and structures: Mass customization, mass personalization*. PhD dissertation, Industrial Engineering, North Carolina State University. Available at: https://repository.lib.ncsu.edu/items/480a96b7-e0de-422f-b467-8ff07b8e35ed
37. Choudhari, A., et al. Design and development of vacuum frying machine for the production of high-quality fried products. In Proceedings of International conference on intelligent manufacturing and automation: ICIMA. 2022: Springer. https://doi.org/10.1007/978-981-19-7971-2_50.
38. Zhang, S. and W. Seiya, *3D printing as an alternative manufacturing method for the micro gas turbine heat exchanger*. Royal Institute of Technology (KTH) and Duke University, 2015.
39. Gupta, A.K., et al., Advances in Nickel-Containing High-Entropy Alloys: From Fundamentals to Additive Manufacturing. Materials, 2024. **17**(15): p. 3826.
40. Li, M., et al., Metal binder jetting additive manufacturing: a literature review. *Journal of Manufacturing Science and Engineering*, 2020. **142**(9): p. 090801.
41. Chen, H. and Y.F. Zhao, Process parameters optimization for improving surface quality and manufacturing accuracy of binder jetting additive manufacturing process. *Rapid Prototyping Journal*, 2016. **22**(3): pp. 527–538.
42. Konda Gokuldoss, P., S. Kolla, and J. Eckert, Additive manufacturing processes: selective laser melting, electron beam melting and binder jetting—Selection guidelines. *Materials*, 2017. **10**(6): p. 672.
43. Lores, A., et al., A review on recent developments in binder jetting metal additive manufacturing: materials and process characteristics. *Powder Metallurgy*, 2019. **62**(5): pp. 267–296.
44. Dudek, P., FDM 3D printing technology in manufacturing composite elements. *Archives of Metallurgy and Materials*, 2013. **58**(4): pp. 1415–1418.
45. Dey, A. and N. Yodo, A systematic survey of FDM process parameter optimization and their influence on part characteristics. *Journal of Manufacturing and Materials Processing*, 2019. **3**(3): p. 64.
46. Wickramasinghe, S., T. Do, and P. Tran, FDM-based 3D printing of polymer and associated composite: A review on mechanical properties, defects and treatments. *Polymers*, 2020. **12**(7): p. 1529.
47. Ahn, D.-G., Directed energy deposition (DED) process: state of the art. *International Journal of Precision Engineering and Manufacturing-Green Technology*, 2021. **8**: pp. 703–742.
48. Svetlizky, D., et al., Laser-based directed energy deposition (DED-LB) of advanced materials. *Materials Science and Engineering: A*, 2022. **840**: p. 142967.

49. Spirrett, F., *Investigations into glass additive manufacturing by selective laser melting and directed energy deposition*. University of Nottingham, 2021.
50. Niknam, S.A., M. Mortazavi, and D. Li, Additively manufactured heat exchangers: a review on opportunities and challenges. *The International Journal of Advanced Manufacturing Technology*, 2021. **112**(3): pp. 601–618.
51. Klein, E., et al., *A review of recent advances in additively manufactured heat exchangers*. 17th International Refrigeration and Air Conditioning Conference at Purdue, 2018.
52. Jiménez, M., et al., Additive manufacturing technologies: an overview about 3D printing methods and future prospects. *Complexity*, 2019. **2019**(1): p. 9656938.
53. Fawaz, A., et al., Topology optimization of heat exchangers: a review. *Energy*, 2022. **252**: p. 124053.
54. Gao, W., et al., The status, challenges, and future of additive manufacturing in engineering. *Computer-Aided Design*, 2015. **69**: pp. 65–89.
55. Gardan, J., Additive manufacturing technologies: state of the art and trends. *Additive Manufacturing Handbook*, 2017: pp. 149–168.
56. Jafari, D. and W.W. Wits, The utilization of selective laser melting technology on heat transfer devices for thermal energy conversion applications: A review. *Renewable and Sustainable Energy Reviews*, 2018. **91**: pp. 420–442.
57. Kirrolia, A., N.R. Bishnoi, and R. Singh, Microalgae as a boon for sustainable energy production and its future research & development aspects. *Renewable and Sustainable Energy Reviews*, 2013. **20**: pp. 642–656.
58. Chaudhary, S., et al., Recent advances in additive manufacturing, applications and challenges for dentistry: a review. *ACS Biomaterials Science & Engineering*, 2023. **9**(7): pp. 3987–4019.
59. Bidare, P., et al., Porosity, cracks, and mechanical properties of additively manufactured tooling alloys: a review. *Advances in Manufacturing*, 2022. **10**(2): pp. 175–204.
60. Ho, J.Y., K.C. Leong, and T.N. Wong, Additively-manufactured metallic porous lattice heat exchangers for air-side heat transfer enhancement. *International Journal of Heat and Mass Transfer*, 2020. **150**: p. 119262.
61. Abeykoon, C., Compact heat exchangers–Design and optimization with CFD. *International Journal of Heat and Mass Transfer*, 2020. **146**: p. 118766.
62. Menni, Y., A. Azzi, and A. Chamkha, Enhancement of convective heat transfer in smooth air channels with wall-mounted obstacles in the flow path: A review. *Journal of Thermal Analysis and Calorimetry*, 2019. **135**: pp. 1951–1976.
63. Zheng, N., et al., A review on single-phase convective heat transfer enhancement based on multi-longitudinal vortices in heat exchanger tubes. *Applied Thermal Engineering*, 2020. **164**: p. 114475.
64. Muley, A., et al. *Recent advances in thin-wall additively manufactured heat exchangers*. American Society of Mechanical Engineers.
65. Byiringiro, J., et al., Heat transfer improvement using additive manufacturing technologies: a review. *Archives of Materials Science and Engineering*, 2023. **123**(1): pp. 30–41.
66. Peng, X., et al., A review of post-processing technologies in additive manufacturing. *Journal of Manufacturing and Materials Processing*, 2021. **5**(2): p. 38.
67. Kaur, I. and P. Singh, State-of-the-art in heat exchanger additive manufacturing. *International Journal of Heat and Mass Transfer*, 2021. **178**: p. 121600.
68. Charles, A., et al., Effect of process parameters on the generated surface roughness of down-facing surfaces in selective laser melting. *Applied Sciences*, 2019. **9**(6): p. 1256.
69. Wang, Z., et al., Optimization of processing parameters and establishment of a relationship between microstructure and mechanical properties of SLM titanium alloy. *Optics & Laser Technology*, 2019. **112**: pp. 159–167.
70. Gupta, A.K., et al., Laser irradiation-induced nanoscale surface transformations in strontium titanate. *Crystals*, 2022. **12**(5): p. 624.
71. Galati, M., P. Minetola, and G. Rizza, Surface roughness characterisation and analysis of the electron beam melting (EBM) process. *Materials*, 2019. **12**(13): p. 2211.
72. Gupta, A.K., S. Gupta, and R. Sachan, Laser irradiation induced atomic structure modifications in strontium titanate. *Jom*, 2022. **74**: pp. 143–150.
73. Fayazfar, H., et al., An overview of surface roughness enhancement of additively manufactured metal parts: a path towards removing the post-print bottleneck for complex geometries. *The International Journal of Advanced Manufacturing Technology*, 2023. **125**(3): pp. 1061–1113.
74. Downing, D., et al., Heat transfer in lattice structures during metal additive manufacturing: numerical exploration of temperature field evolution. *Rapid Prototyping Journal*, 2020. **26**(5): pp. 911–928.

75. Anwajler, B., Potential of 3D printing for heat exchanger heat transfer optimization—sustainability perspective. *Inventions*, 2024. **9**(3): p. 60.
76. Tang, Y. and Y.F. Zhao, A survey of the design methods for additive manufacturing to improve functional performance. *Rapid Prototyping Journal*, 2016. **22**(3): pp. 569–590.
77. Gu, D., et al., Material-structure-performance integrated laser-metal additive manufacturing. *Science*, 2021. **372**(6545): p. eabg1487.
78. Morciano, M., et al., 3D printed lattice metal structures for enhanced heat transfer in latent heat storage systems. *Journal of Energy Storage*, 2023. **65**: p. 107350.
79. Padrão, D., et al., New structure-performance relationships for surface-based lattice heat sinks. *Applied Thermal Engineering*, 2024. **236**: p. 121572.
80. Qian, B., et al., Microchannel liquid-cooled heat exchanger based on a nonuniform lattice: study on structure calculation, formation process, and boiling heat transfer performance. *Materials*, 2021. **14**(23): p. 7248.
81. Kumar, S., A. Thyagarajan, and D. Banerjee, *Experimental investigation of thermal energy storage (TES) platform leveraging phase change materials in a Chevron plate heat exchanger.* 2022: American Society of Mechanical Engineers.
82. Sajjad, U., et al., Manufacturing and potential applications of lattice structures in thermal systems: a comprehensive review of recent advances. *International Journal of Heat and Mass Transfer*, 2022. **198**: p. 123352.
83. Ozguc, S., L. Pan, and J.A. Weibel, Topology optimization of microchannel heat sinks using a homogenization approach. *International Journal of Heat and Mass Transfer*, 2021. **169**: p. 120896.
84. Maloney, K.J., et al., Multifunctional heat exchangers derived from three-dimensional micro-lattice structures. *International Journal of Heat and Mass Transfer*, 2012. **55**(9–10): pp. 2486–2493.
85. Kaur, I. and P. Singh, Conjugate heat transfer in lattice frame materials based on novel unit cell topologies. *Numerical Heat Transfer, Part A: Applications*, 2022. **82**(12): pp. 788–801.
86. Mohammed, H.A., et al., Heat transfer and fluid flow characteristics in microchannels heat exchanger using nanofluids: a review. *Renewable and Sustainable Energy Reviews*, 2011. **15**(3): pp. 1502–1512.
87. Jiang, P.-X., et al., Thermal–hydraulic performance of small scale micro-channel and porous-media heat-exchangers. *International Journal of Heat and Mass Transfer*, 2001. **44**(5): pp. 1039–1051.
88. Dixit, T. and I. Ghosh, Review of micro-and mini-channel heat sinks and heat exchangers for single phase fluids. *Renewable and Sustainable Energy Reviews*, 2015. **41**: pp. 1298–1311.
89. Gupta, A.K., et al., Preparation effect of mould systems on microstructure and mechanical properties of spheroidised graphite iron. *International Journal of Engineering Research and Applications (IJERA)*, 2016. **6**(4): pp. 68–73.
90. Norris, A., M. Saafi, and P. Romine, Temperature and moisture monitoring in concrete structures using embedded nanotechnology/microelectromechanical systems (MEMS) sensors. *Construction and Building Materials*, 2008. **22**(2): pp. 111–120.
91. Kisner, R., et al., *Embedded sensors and controls to improve component performance and reliability conceptual design report.* 2012: Oak Ridge National Lab.(ORNL), Oak Ridge, TN (United States).
92. Bhaskarwar, T.V., S.S. Giri, and R.G. Jamakar. *Automation of shell and tube type heat exchanger with PLC and LabVIEW.* 2015: IEEE.
93. Svensson, E., et al., *Design of heat exchanger networks with good controllability.* CIT Industriell Energi AB, 2018.
94. Shi, L., et al., Thermal and acoustic performance of silencing heat exchanger for engine waste heat recovery. *Applied Thermal Engineering*, 2022. **201**: p. 117711.
95. Utne, I.B., T. Brurok, and S. Larsen, Monitoring the mechanical integrity of heat exchangers. *Process Safety Progress*, 2011. **30**(4): pp. 328–333.
96. Alam, T. and M.-H. Kim, A comprehensive review on single phase heat transfer enhancement techniques in heat exchanger applications. *Renewable and Sustainable Energy Reviews*, 2018. **81**: pp. 813–839.
97. Sung, W.-T. and Y.-C. Hsu, Designing an industrial real-time measurement and monitoring system based on embedded system and ZigBee. *Expert Systems with Applications*, 2011. **38**(4): pp. 4522–4529.
98. Kailkhura, G., et al., Numerical and experimental study of a novel additively manufactured metal-polymer composite heat-exchanger for liquid cooling electronics. *Energies*, 2022. **15**(2): p. 598.
99. Khan, M.G. and A. Fartaj, A review on microchannel heat exchangers and potential applications. *International Journal of Energy Research*, 2011. **35**(7): pp. 553–582.
100. Gupta, A.K., B. Mallik, and D. Roy, Structure property correlation of in situ reinforced Al–based metal matrix composite via stir casting. *Materials Performance and Characterization*, 2020. **9**(1): pp. 25–35.

101. Saltzman, D., et al., Design and evaluation of an additively manufactured aircraft heat exchanger. *Applied Thermal Engineering*, 2018. **138**: pp. 254–263.
102. Zhang, X., et al., *Recent developments in high temperature heat exchangers: a review. Frontiers in Heat and Mass Transfer,* 2018. **11**: 18.
103. McDonough, J.R., A perspective on the current and future roles of additive manufacturing in process engineering, with an emphasis on heat transfer. *Thermal Science and Engineering Progress*, 2020. **19**: p. 100594.
104. Chen, H., et al., Low thermal expansion metal composite-based heat spreader for high temperature thermal management. *Materials & Design*, 2021. **208**: p. 109897.
105. Zhang, C. and V. Gümmer, High temperature heat exchangers for recuperated rotorcraft powerplants. *Applied Thermal Engineering*, 2019. **154**: pp. 548–561.
106. Careri, F., et al., Additive manufacturing of heat exchangers in aerospace applications: a review. *Applied Thermal Engineering*, 2023: p. 121387.
107. Carozza, A., Heat exchangers in the aviation engineering. *Heat Exchangers—Advanced Features and Applications*, 2017: pp. 149–166.
108. Romei, F., A.N. Grubišić, and D. Gibbon, Manufacturing of a high-temperature resistojet heat exchanger by selective laser melting. *Acta Astronautica*, 2017. **138**: pp. 356–368.
109. Sabau, A.S., et al., Design, additive manufacturing, and performance of heat exchanger with a novel flow-path architecture. *Applied Thermal Engineering*, 2020. **180**: p. 115775.
110. Jafari, D., W.W. Wits, and B.J. Geurts. *An investigation of porous structure characteristics of heat pipes made by additive manufacturing.* 2017: IEEE.
111. Alsulami, M., et al., Design complexity and performance analysis in additively manufactured heat exchangers. *The International Journal of Advanced Manufacturing Technology*, 2020. **110**(3): pp. 865–873.
112. Thompson, S.M., et al., Additive manufacturing of heat exchangers: a case study on a multi-layered Ti–6Al–4V oscillating heat pipe. *Additive Manufacturing*, 2015. **8**: pp. 163–174.
113. Arie, M.A., et al., Experimental characterization of heat transfer in an additively manufactured polymer heat exchanger. *Applied Thermal Engineering*, 2017. **113**: pp. 575–584.
114. Stimpson, C.K., et al., Effectiveness measurements of additively manufactured film cooling holes. *Journal of Turbomachinery*, 2018. **140**(1): p. 011009.
115. Gerstler, W.D. and D. Erno. Introduction of an additively manufactured multi-furcating heat exchanger. 2017: IEEE.
116. Stimpson, C.K., et al., Roughness effects on flow and heat transfer for additively manufactured channels. *Journal of Turbomachinery*, 2016. **138**(5): p. 051008.
117. Ventola, L., et al., Rough surfaces with enhanced heat transfer for electronics cooling by direct metal laser sintering. *International Journal of Heat and Mass Transfer*, 2014. **75**: pp. 58–74.
118. Fasano, M., et al., Passive heat transfer enhancement by 3D printed Pitot tube based heat sink. *International Communications in Heat and Mass Transfer*, 2016. **74**: pp. 36–39.
119. Ho, J.Y., K.K. Wong, and K.C. Leong, Saturated pool boiling of FC-72 from enhanced surfaces produced by selective laser melting. *International Journal of Heat and Mass Transfer*, 2016. **99**: pp. 107–121.
120. Lohan, D.J., E.M. Dede, and J.T. Allison, Topology optimization for heat conduction using generative design algorithms. *Structural and Multidisciplinary Optimization*, 2017. **55**: pp. 1063–1077.
121. Singer, F., et al., Additively manufactured copper components and composite structures for thermal management applications. 2017: IEEE.
122. Perry, J., et al., Pin fin array heat sinks by cold spray additive manufacturing: economics of powder recycling. *Journal of Thermal Spray Technology*, 2019. **28**: pp. 144–160.
123. Roy, D., et al., Synergistic effect of Nb and Zr additions on the structure-property relationships of nano-crystalline Cu processed by mechanical alloying and hot pressing. *Journal of Alloys and Compounds*, 2021. **854**: p. 157174.
124. Roy, D., et al., Synergistic effect of Nb and Zr addition in thermal stabilization of nano-crystalline Cu synthesized by ball milling. *Materials Letters*, 2020. **271**: p. 127780.
125. Roy, D., et al., Stable nanocrystalline structure attainment and strength enhancement of Cu base alloy using bi-modal distributed tungsten dispersoids. *Philosophical Magazine*, 2022. **102**(3): pp. 189–209.
126. Chakraborty, S., et al., Studies on nano-metal dispersed Cu-Cr matrix composite. *Materials Letters*, 2019. **257**: p. 126739.
127. Han, X.-H., et al., A review of metal foam and metal matrix composites for heat exchangers and heat sinks. *Heat Transfer Engineering*, 2012. **33**(12): pp. 991–1009.

128. Kishawy, H.A. and A. Hosseini, Titanium and titanium alloys. *Machining Difficult-To-Cut Materials: Basic Principles and Challenges*, 2019: pp. 55–96.
129. Battaglia, F., et al., Experimental characterization of an additively manufactured inconel 718 heat exchanger for high-temperature applications. *Energies*, 2023. **16**(10): p. 4156.
130. Khare, S., et al., Selection of materials for high temperature latent heat energy storage. *Solar Energy Materials and Solar Cells*, 2012. **107**: pp. 20–27.
131. Zahri, N.A.M., *Mechanical and microstructure analysis of copper to copper foam brazing using Copper-Tin Nickel-Phosphorus amorphous filler alloys*. 2021.
132. Benallou, A., Heat exchangers for electronic equipment cooling. In: Castro Gómez, L., Velázquez Flores, V.M., & Navarrete Procopio, M. (Eds.), *Heat Exchangers*. 2021: IntechOpen.
133. Assaad, J., et al., Novel stacked wire mesh compact heat exchangers produced using cold spray. *Journal of Thermal Spray Technology*, 2011. **20**: pp. 1192–1200.
134. Esarte, J., et al., Optimizing the design of a two-phase cooling system loop heat pipe: wick manufacturing with the 3D selective laser melting printing technique and prototype testing. *Applied Thermal Engineering*, 2017. **111**: pp. 407–419.
135. Ameli, M., et al., A novel method for manufacturing sintered aluminium heat pipes (SAHP). *Applied Thermal Engineering*, 2013. **52**(2): pp. 498–504.
136. Dupuis, P., et al., Heat transfer and flow structure characterization for pin fins produced by cold spray additive manufacturing. *International Journal of Heat and Mass Transfer*, 2016. **98**: pp. 650–661.
137. Cormier, Y., et al., Additive manufacturing of pyramidal pin fins: height and fin density effects under forced convection. *International Journal of Heat and Mass Transfer*, 2014. **75**: pp. 235–244.
138. Hussein, A.M., Thermal performance and thermal properties of hybrid nanofluid laminar flow in a double pipe heat exchanger. *Experimental Thermal and Fluid Science*, 2017. **88**: pp. 37–45.
139. Hussain, A.R.J., et al., Review of polymers for heat exchanger applications: factors concerning thermal conductivity. *Applied Thermal Engineering*, 2017. **113**: pp. 1118–1127.
140. Wang, Q., et al., A review on application of carbonaceous materials and carbon matrix composites for heat exchangers and heat sinks. *International Journal of Refrigeration*, 2012. **35**(1): pp. 7–26.
141. Thomas, D.S. and S.W. Gilbert, Costs and cost effectiveness of additive manufacturing. *NIST Special Publication*, 2014. **1176**: p. 12.
142. Sekulic, D.P. and R.K. Shah, *Fundamentals of heat exchanger design*. 2023: John Wiley & Sons.
143. Ladani, L. and M. Sadeghilaridjani, Review of powder bed fusion additive manufacturing for metals. *Metals*, 2021. **11**(9): p. 1391.
144. Kelly, J.P., et al., Binder jet additive manufacturing of ceramic heat exchangers for concentrating solar power applications with thermal energy storage in molten chlorides. *Additive Manufacturing*, 2022. **56**: p. 102937.
145. Costabile, G., et al., Cost models of additive manufacturing: a literature review. *International Journal of Industrial Engineering Computations*, 2017. **8**(2): pp. 263–283.
146. Kadir, A.Z.A., Y. Yusof, and M.S. Wahab, Additive manufacturing cost estimation models—a classification review. *The International Journal of Advanced Manufacturing Technology*, 2020. **107**: pp. 4033–4053.
147. Mohammadzadeh, A.M., B. Jafari, and K. Hosseinzadeh, Comprehensive numerical investigation of the effect of various baffle design and baffle spacing on a shell and tube heat exchanger. *Applied Thermal Engineering*, 2024. **249**: p. 123305.
148. Kumar, S., A. Thyagarajan, and D. Banerjee. *An investigation on chevron plate heat exchanger filled with organic phase change material*. 2020: American Society of Mechanical Engineers.
149. Patel, A.K., Advancements in heat exchanger design for waste heat recovery in industrial processes. *World Journal of Advanced Research and Reviews*, 2023. **19**(3): pp. 137–152.
150. Kumar, S. and K. Rathore, Renewable energy for sustainable development goal of clean and affordable energy. *International Journal of Materials Manufacturing and Sustainable Technologies*, 2023. **2**(1): pp. 1–15.
151. Ziaee, M. and N.B. Crane, Binder jetting: a review of process, materials, and methods. *Additive Manufacturing*, 2019. **28**: pp. 781–801.
152. Bae, C.-J., A.B. Diggs, and A. Ramachandran, Quantification and certification of additive manufacturing materials and processes. *Additive Manufacturing*, 2018, Elsevier. pp. 181–213.
153. Chen, Z., et al., A review on qualification and certification for metal additive manufacturing. *Virtual and Physical Prototyping*, 2022. **17**(2): pp. 382–405.
154. Kumar, S., et al., *Experimental investigations on thermal stratification in a large pool of water with immersed isolation condenser*. 2020: American Society of Mechanical Engineers.

155. Elsheikh, A.H., et al., Applications of heat exchanger in solar desalination: current issues and future challenges. *Water*, 2022. **14**(6): p. 852.
156. Ingarao, G. and P.C. Priarone, A comparative assessment of energy demand and life cycle costs for additive-and subtractive-based manufacturing approaches. *Journal of Manufacturing Processes*, 2020. **56**: pp. 1219–1229.
157. Iqbal, A., et al., Readiness of subtractive and additive manufacturing and their sustainable amalgamation from the perspective of Industry 4.0: a comprehensive review. *The International Journal of Advanced Manufacturing Technology*, 2020. **111**: pp. 2475–2498.
158. Gopal, M., H.G. Lemu, and E.M. Gutema, Sustainable additive manufacturing and environmental implications: literature review. *Sustainability*, 2022. **15**(1): p. 504.
159. Berdine, G.G., M. DiPaola, and M. Weinberg, Economic and regulatory perspectives on additive manufacturing. *3D Printing in Orthopaedic Surgery*, 2019, Elsevier, pp. 41–48.
160. Monzón, M.D., et al., Standardization in additive manufacturing: activities carried out by international organizations and projects. *The International Journal of Advanced Manufacturing Technology*, 2015. **76**: pp. 1111–1121.
161. Everton, S.K., et al., Review of in-situ process monitoring and in-situ metrology for metal additive manufacturing. *Materials & Design*, 2016. **95**: pp. 431–445.
162. Tähtinen, H., *Regulations concerning additive manufacturing*. Master's thesis, JAMK University of Applied Sciences, 2023.

8 Microchannel Heat Exchangers

Nancy Maurya, Naman Jain, Tapasvi Bhatt, Sunil Kumar, and Suvanjan Bhattacharyya

8.1 INTRODUCTION

8.1.1 Overview of Microchannel Heat Exchangers

Heat exchangers play a crucial role in our day-to-day lives and are essential for everyday comfort and convenience. These are the devices that are used to transfer heat between two or more fluids on the principle of forced convection. From heating our homes and water to cooling our refrigerators and air conditioning systems, heat exchangers contribute to energy efficiency in industrial processes, transportation, and power generation [1–4]. With them, maintaining optimal temperatures and efficient energy usage is easy. Within a heat exchanger, the movement of fluids through the channels and the exchange of heat between these fluids and the walls of the channels are two critical phenomena. With the ever-increasing need for energy and the world transitioning toward making device sizes smaller and more efficient, the demand for compact and lightweight heat exchangers capable of delivering efficient heat transfer rates has increased [5–7]. Microchannels are devices where their diameter is typically less than 1 mm, which enables the heat transfer surface densities to reach levels of 10,000 m^2 or more [5, 8]. This miniaturization race has led to advancements in intensified heat transfer in thermofluid applications, wherein the objective is to achieve greater heat transfer efficiency from smaller heat exchangers [5, 9].

8.1.2 Historical Development and Applications

Significant advancements in microfabrication technology have driven advancements in microchannel heat exchangers. Early work involved building microchannels in silicon wafers using anisotropic wet chemical etching and sawing, but these methods imposed significant restrictions on channel geometry [10]. Later, various advanced micromachining techniques were developed, including anisotropic dry etching and other high aspect ratio techniques, which removed many process-induced constraints on microchannel design [10]. The thermohydraulic performance of microchannels has been extensively studied, and they are now increasingly employed in various applications due to their high thermal performance and efficiency. The market has witnessed a notable shift toward enhancing the energy efficiency of cooling products, with microchannel heat exchangers emerging as a significant contributor to this trend. Evolving from conventional fin and tube heat exchangers, microchannel variants offer a compelling blend of efficiency, cost-effectiveness, and superior corrosion resistance, particularly within the air-cooled chiller sector. Research is ongoing to understand the liquid and vapor phase interactions during two-phase flow in microchannels, and future directions include the widespread usage of microchannels in micro-electromechanical devices, microscale sensors and actuators, advanced high heat flux removal systems, and biomedical applications.

8.1.3 Advantages and Challenges

There are several advantages of microchannel heat exchangers. Some of them are stated as follows:

 DOI: 10.1201/9781003509837-8

- They can achieve 20–40% greater overall heat transfer compared with traditional designs due to the high surface area to volume ratio of the small channels [11].
- They are 10–30% smaller in face area and up to 60% lighter than conventional heat exchangers of similar capacity [11]. The compact size is enabled by the thin profile of the microchannel tubes.
- They have a higher heat transfer coefficient due to their brazed assembly, which enhances heat transfer performance. Additionally, the reduced scale effects in microchannel heat exchangers can enhance fluid compressibility and viscous forces, further improving heat transfer ratios [5].

However, these devices pose certain challenges:

- High-pressure drops associated with these devices are one of the primary concerns. Due to their smaller flow passages, microchannel heat exchangers experience higher fluid friction, which necessitates higher pumping power to maintain desired flow rates [12]. This challenge becomes pronounced in applications where energy efficiency is vital, as elevated pressure drops can compromise overall system efficiency.
- The intricate design and small channel networks require expensive manufacturing processes, advanced machinery, and skilled labor. This complexity poses a challenge, limiting their integration in applications where cost-effectiveness is a crucial factor [12]. Additionally, the elevated maintenance and repair costs due to the complex design add to the overall expenses.
- Because of their very small sizes and intricacies, they are limited in their scalability. This restricts their widespread adoption in various applications, particularly in industries with slim profit margins or stringent budget considerations.

8.2 FUNDAMENTAL CONCEPTS

8.2.1 Heat Transfer Mechanisms in Microchannels

Heat transfer in the microchannels can happen using two methods:

1. Single-phase flow
2. Two-phase flow

These are elaborated on in the following sections.

8.2.1.1 Single-Phase Flow

Surface forces are what are assumed to be the main differentiating criteria between macro- and microchannels. Considering microchannels, the effect of surface forces is massive owing to their large surface area to volume ratio. Van der Waals forces, electrostatic forces, and steric forces are the three primary forces mainly observed in a micro surface that have an effect on the surface tension of the liquids. Heat transfer in a microchannel is a conjugate phenomenon with conduction in solid to convection in liquid. Hence, it becomes difficult to develop one governing equation to justify the flow. A classical fin approach is mostly taken into consideration with certain assumptions as 1D heat transfer, constant convective heat transfer coefficient, and uniform fluid temperature [13].

8.2.1.2 Two-Phase Flow

Two-phase flow in microchannels is mainly governed by surface tension, viscosity, and inertia force. Many different authors have proposed several equations or correlations to define the two-phase flow, however, it is not as easy to be able to identify it. Surface roughness and wettability are two characteristics that play a major role in understanding two-phase flow. An experiment was

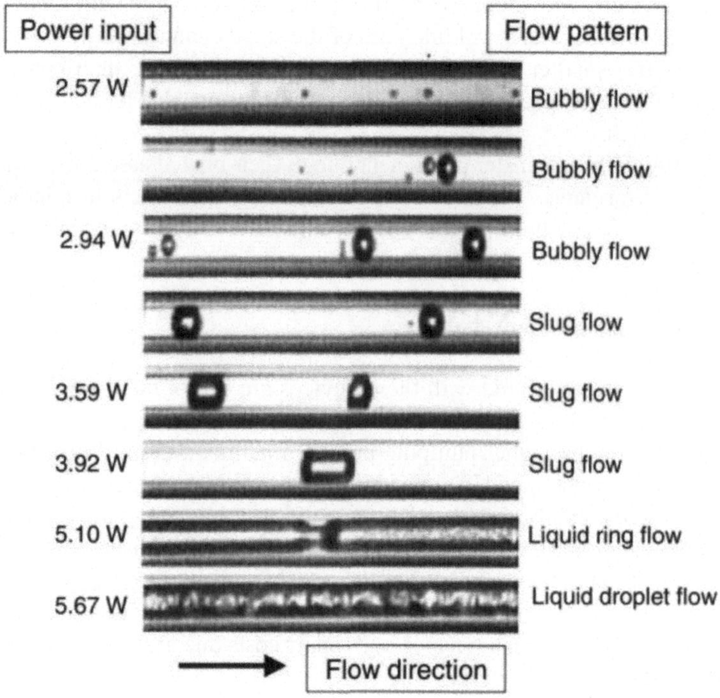

FIGURE 8.1 Steam-water two-phase flow using a quartz tube section [14].

conducted by Serizawa et al. [14], using steam-water and air-water with the inner diameter of the test section to be 20, 25, and 100 µm. Injection syringes were installed on the outlet of the test section and it was reportedly visualized using a quartz tube section (Figures 8.1 and 8.2). The transition map is given in Figure 8.3.

For the test section of the steam-water experiment, a 50 µm i.d., 10 mm long silica capillary tube that was heated by direct current through a stainless-steel tube was employed. The pure water was circulated into the test portion by applying pressure to the water reservoir. The experiments were carried out at a low liquid velocity J_L of 2.47 cm/s by adjusting the heat input to the preheater.

Thus, it was evident that stable annular flow was achieved because the tube wall was considerably more easily wetted in such circumstances. Furthermore, a foamy or wispy annular flow was also noticed on the tube wall where tiny bubbles were caught in the thin liquid film flow. This is a result of the water film expanding and its increased wettability, which allows it to engulf tiny bubbles without shattering. However, in contaminated tubes, the creation of a dry space between the gas slug and the tube wall at very low flow rates was noted. Thus, it was established that the wettability of the fluids and the tube has a significant impact on two-phase flow structures in ultrasmall channels. As a result, surface contamination is a critical factor that heavily influences the transitions of two-phase flow patterns.

The following equation is implemented for heat transfer between the microchannel and the working fluid:

$$Q = m C_p \left(T_{out} - T_{in}\right) \tag{8.1}$$

Where Q is the heat transfer amount,

m is the mass flow rate of working fluid, T_{in} is the inlet temperature of working fluid,
T_{out} is the outlet temperature of working fluid, and
C_p is the specific heat of working fluid.

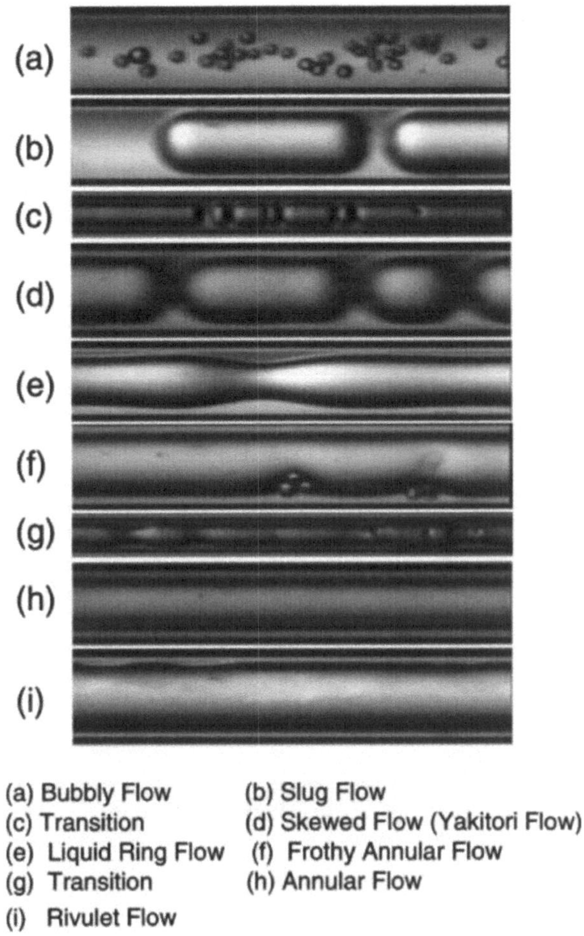

FIGURE 8.2 Air-water two-phase flow using a quartz tube of 100 μm.

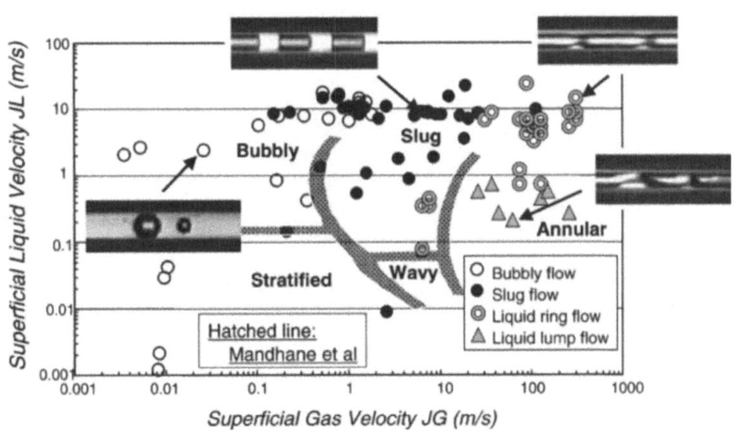

FIGURE 8.3 Flow transition map.

The Joule effect is utilized for the heat applied to the microchannel block

$$Q = V.I \tag{8.2}$$

Where V and I are the voltage applied to the heater and the current passing through the heating element's circuit. The convective heat transfer between fluid and microchannel is given by

$$Q_{conv} = \frac{Q}{A} = \frac{m^{\cdot}C_p\left(T_{out} - T_{in}\right)}{N.I\left(2\eta H + W\right)} \tag{8.3}$$

Where N, H, and l are the number of channels, the height of the microchannels, and the length of each channel,

while W is the width of each channel. The fin's efficiency is given by:

$$\eta = \frac{\tanh\left(mH\right)}{mH} \tag{8.4}$$

And,

$$m = \sqrt{\frac{2h_{av}}{W.\,k}} \tag{8.5}$$

In Equation 8.5), k is the copper's thermal conductivity and *hav* is the average heat transfer coefficient, and is calculated using the equation:

$$h_{av} = \frac{1}{l'}\int_0^{l'} h_z dz \tag{8.6}$$

Where l' is the distance from the entrance region of the microchannel to a specific location for which the HTC is calculated:

$$h_{az} = \frac{Q_{conv}}{T_{w,z} - T_{b,z}} \tag{8.7}$$

Where $T_{w,z}$ is the surface temperature of the microchannel at each location and $T_{b,z}$ is the mean temperature of the fluid. It is calculated by the given equation:

$$T_{w,z} = T_{th} - Q''_{conv}\frac{s}{k} \tag{8.8}$$

Where s is the small conductive gap between the surface of each channel and the exact location of each thermocouple. The average temperature of fluid is calculated with the following equation:

$$T_{b,z} = \frac{z}{l}T_{out} + \left(1 - \frac{z}{l}\right)T_{in} \tag{8.9}$$

Where z and l are the axial positions of the thermocouples and the length of the microchannel, respectively.

Equation 8.10 will be used to calculate the thermal-hydraulic performance of the microchannel,

$$Thermo-hydraulic\ performance = \frac{Nu}{Nu_w}\left(\frac{f_w}{f}\right)^{0.3} \tag{8.10}$$

Where Nu_w and f_w represent wall Nusselt number and wall friction factor.

Here, the friction factor will be calculated with the following equation:

$$f = \frac{\Delta P_{nf}}{\left(\dfrac{l}{D_{hyd.}}\right)\left(\dfrac{2u^2}{\rho_{nf}}\right)} \tag{8.11}$$

Where, l is the length of microchannel,

u is the average bulk velocity (m/s),

and $D_{hyd.}$ is the hydraulic diameter of each microchannel.

8.2.2 Fluid Flow Characteristics

The fluid flow is assumed to be steady, incompressible, and irrotational. It is governed by the continuity, momentum, and energy equations that are listed as follows [13]:

- Continuity equation

$$\nabla.(\rho V) = 0 \tag{8.12}$$

- Momentum equation

$$\nabla.(\rho VV) = -\nabla P + (\nabla \tau_{ij}) + F_k \tag{8.13}$$

where F_k represents the total magnetic volume force obtained from the equation

$$F_k = \frac{1}{2}\mu_o \chi_m (1 + \chi_m) \nabla\left(\overline{H\vec{H}}\right) + \mu_o \chi_m \vec{H}\left[\left(\vec{H}.\nabla\right)\chi_{m'}\right] \tag{8.14}$$

Where χ_m is the magnetic susceptibility as a function of temperature.

- Energy equation

$$\nabla.\left(\rho V C_p \Delta Temp\right) = \nabla.\left(k \nabla Temp\right) \tag{8.15}$$

8.2.3 Thermophysical Properties of Working Fluids

The following equations are used to calculate properties such as density, specific heat, thermal conductivity, and viscosity of the working fluids used in various papers [13].

$$\rho_{nf} = (1 - \varphi)\rho_f + \varphi\rho_p \tag{8.16}$$

$$C_{p,nf} = \frac{(1 - \varphi)\rho_f C_{p.f} + \varphi\rho_p C_{p,p}}{\rho_{nf}} \tag{8.17}$$

$$\mu_{nf} = \frac{1}{(1 - \varphi)^{2.5}} \times \mu_f \tag{8.18}$$

$$K_{nf} = \frac{K_P + (SH-1)K_f - (SH-1)\varphi(K_f - K_P)}{K_P + (SH-1)K_f + \varphi(K_f - K_P)} \tag{8.19}$$

The physical properties of the base fluid, water, in the present analysis, are assumed to be a function of temperature. They are provided as follows [13]:

$$\begin{aligned}\rho_f &= 2.446 * 10^3 - 20.674Temp + 0.11576Temp^2 \\ &-3.12895\times10^{-4}Temp^3 + 4.0505\times10^{-7}Temp^4 \\ &-2.0546\times10^{-10}Temp^5\end{aligned} \tag{8.20}$$

$$\mu_f = 2.414\times10^{-5}\times10\left[\frac{247}{Temp-140}\right] \tag{8.21}$$

$$K_f = -1.13 + 9.71\times10^{-3}Temp - 1.31\times10^{-5}Temp^2 \tag{8.22}$$

The thermophysical properties of nanoparticles are listed in Table 8.1, and thermophysical properties of the base fluids are listed in Table 8.2.

8.3 EXPERIMENTAL METHODS FOR MICROCHANNEL HEAT EXCHANGERS

8.3.1 Experimental Setup and Instrumentation

Experimental investigations of microchannel heat exchangers require a carefully designed test rig with precise control and measurement capabilities. The key components of a typical experimental setup are as follows:

- *Fluid supply system*. The fluid supply system for the microchannel test section ensures a controlled flow of the working fluid (gas or liquid). It includes a high-pressure tank or cylinder for storing the fluid, a high-efficiency particle filter to prevent contaminants, and

TABLE 8.1
Properties of Nanoparticles

S. No	Property	Aluminum	Copper	Silver
1	Thermal conductivity (W/mK)	237	400	429
2	Density (ρ_p) kg/m^3	2710	8933	10,500
3	Specific heat (C_P) J/kg K	900	385	234

TABLE 8.2
Properties of Base Fluids

S. No	Property	Water	Liquid Sodium	Ethylene Glycol
1	Thermal conductivity (KL) W/Mk	0.605	76	0.252
2	Density (ρ_f) kg/m^3	997.1	880	1111
3	Specific heat (C_f) J/kg K	4179	1300	2415
4	Dynamic viscosity (μ_0) Kg/m3	0.001003	0.34	0.0157

a mass flow rate controller, such as a thermal mass flow meter or critical flow venturi, for precise regulation. A preheater or precooler adjusts the fluid's temperature before it enters the test section.

- *Microchannel test section.* The microchannel test section is the core of our setup for studying heat transfer and fluid flow. It includes the microchannel heat exchanger, whose design and fabrication are crucial. Heating or cooling systems, such as electrical heaters or hot/cold fluid loops, are used based on our objectives. Proper insulation is essential to minimize external heat interference and ensure precise measurements.
- *Instrumentation and data acquisition.* Accurate measurement is crucial for evaluating microchannel heat exchangers. Instrumentation includes differential pressure transducers for measuring pressure drops, absolute pressure sensors for inlet and outlet pressures, and temperature sensors like thermocouples or resistance temperature detectors (RTDs) placed strategically.

8.3.2 Data Analysis and Interpretation

Careful analysis and interpretation of experimental data are essential:

- *Data processing.* In signal conditioning, the raw signals from sensors, such as thermocouples, pressure transducers, and flow meters, are processed.. Particle image velocimetry (PIV) or laser Doppler velocimetry (LDV) visualize and quantify flow patterns. Infrared cameras are used for mapping temperature distributions on the surface. A high-precision data acquisition system records and processes sensor signals for real-time monitoring and analysis.
- *Calibration and validation.* Proper calibration and validation are essential for reliable data. Sensor calibration ensures precise measurements against traceable standards. System validation involves tests with known conditions or benchmark cases to verify performance. A comprehensive uncertainty analysis quantifies the propagation of uncertainties from individual measurements to the calculated metrics, ensuring data reliability.

8.3.3 Measurement Techniques

Accurate measurement techniques are essential for evaluating microchannel heat exchanger performance:

- *Flow rate measurement.* We measure flow rates using mass or volumetric flow meters, providing crucial data for assessing efficiency. Mass flow meters, such as thermal mass flow meters or critical flow venturis, directly measure the mass flow rate of the working fluid. Alternatively, when the density of the working fluid is known or can be accurately determined, volumetric flow meters like turbine meters or positive displacement meters measure the volumetric flow rate, which can then be converted to mass flow rate using the known density. Both methods provide essential data for evaluating the performance and efficiency of microchannel heat exchangers in various applications.
- *Pressure measurement.* Pressure measurement is vital for determining pressure drops and pumping power requirements. Differential pressure transducers, known for their high precision and low measurement uncertainty, measure the pressure drop across the microchannel heat exchanger. These transducers provide accurate readings, particularly suitable for low-pressure applications. Additionally, absolute pressure sensors measure the inlet and outlet pressures of the microchannel heat exchanger. These measurements are crucial for calculating the overall pressure drop and ensuring the working fluid remains within the

desired phase (liquid or gas). Together, these techniques enable precise assessment of fluid dynamics and energy requirements in microchannel heat exchanger systems.

- *Temperature measurement.* Accurate temperature measurements are crucial for determining the heat transfer performance of the microchannel heat exchanger. Thermocouples are widely used due to their robustness, wide temperature range, and cost-effectiveness. They are typically placed at the inlet, outlet, and within the channels or on the surface. RTDs offer higher accuracy and stability, though they are more expensive and slower to respond. Infrared thermography provides a map of temperature distributions on the surface of the microchannel heat exchanger, offering valuable insights into its heat transfer characteristics.
- *Flow visualization.* Flow visualization techniques study the flow patterns within microchannels, providing insights into heat transfer and fluid flow phenomena. PIV [15, 16], a non-intrusive optical method, uses laser illumination and high-speed imaging to track tracer particles in the working fluid, enabling the quantitative measurement of velocity fields. LDV, also non-intrusive, measures the velocity of tracer particles by analyzing the Doppler shift of scattered laser light, providing accurate point-wise velocity measurements. LDV is more complex and expensive compared with PIV. Additionally, dye injection techniques visualize flow patterns by injecting colored dye into the working fluid and observing flow behavior using high-speed imaging or other visualization methods. Together, these techniques facilitate detailed analysis of flow dynamics within microchannels, crucial for optimizing heat transfer efficiency and understanding fluid behavior and noise, filtering unwanted frequencies, or converting signals to appropriate units. This ensures data accuracy for further analysis. Calibration corrections are applied using calibration curves or equations obtained during the calibration process, accounting for any non-linearities or offsets, enhancing data accuracy. Additionally, uncertainty propagation quantifies the overall uncertainty in performance metrics such as heat transfer coefficients and friction factors. These steps are critical for ensuring reliable experimental data, enabling precise evaluation of heat transfer and fluid flow characteristics in microchannel heat exchangers and other setups.
- *Performance metric calculations.* Heat transfer coefficients, quantifying the rate of heat transfer between the working fluid and microchannel walls, are calculated using energy balance equations and measured temperature differences. These coefficients provide insights into the efficiency of heat transfer within the microchannel heat exchanger. Friction factors represent the pressure drop due to viscous effects and are calculated from the measured pressure drop and known flow conditions. This helps assess energy losses and pumping power requirements. Additional metrics, such as overall heat transfer rate, effectiveness, and pumping power requirements, are derived from the measured data and calculated coefficients and friction factors. These metrics collectively provide a comprehensive assessment of heat transfer efficiency and fluid flow characteristics, essential for optimizing design and operational parameters.
- *Comparison and validation.* Experimental results are compared with theoretical predictions based on established correlations or models for heat transfer and fluid flow in microchannels, validating the data and identifying any deviations from theoretical expectations. Additionally, computational fluid dynamics (CFD) simulations or other numerical models simulate heat transfer and fluid flow phenomena in different areas of research [16,17]. Experimental results are compared with these numerical predictions to validate model accuracy and identify discrepancies. Sensitivity analysis investigates the impact of parameters such as channel geometry, flow conditions, and material properties on performance. This aids in optimizing design and operational parameters for enhanced performance. Together, these methods provide a comprehensive understanding of the thermal-hydraulic behavior of microchannel heat exchangers.

- *Data visualization and reporting.* Experimental data and calculated performance metrics of microchannel heat exchangers are presented in graphical form, such as graphs, charts, and contour plots, facilitating effective visualization and interpretation. These visual representations provide insights into heat transfer and fluid flow characteristics within the microchannels. Additionally, empirical correlations may be developed based on experimental data to create simplified models predicting performance under various operating conditions. These correlations aid in understanding system behavior and design optimization.

The various experimental methods for microchannel heat exchangers are summarized in Table 8.3.

8.4 NUMERICAL METHODS

After performing a thorough literature review, it was observed that CFD analysis of microchannel heat exchangers can be used to determine properties and performance characteristics [21]. Many studies have been published showing that results obtained from CFD analysis are satisfactory—validating the results obtained from experimental results.

Performing CFD analysis is inexpensive and is a good way to understand expected results from experimental setups.

8.4.1 Computational Fluid Dynamics Approaches

Once a CAD model has been made, followed by its discretization (meshing), the most important part is left—setting up the CFD simulation that matches our needs [17, 22]. CFD is based on fundamental governing equations, namely, the continuity equation and the Navier–Stokes equation [23, 24]. There are several techniques to simulate microchannel heat exchangers:

- *Direct numerical simulation* is used to simulate the microscale fluid flow and heat transfer. This approach involves solving the Navier–Stokes and energy equations directly, without any simplifying assumptions or models, to obtain detailed information about flow behavior and heat transfer within the microchannel. It provides the grounds for high-fidelity simulations, allowing accurate predictions and improved understanding of fluid flow characteristics. However, one must be aware of the significant computational power required for large-scale simulation as it solves complex equations and generates large amounts of data, which requires advanced data techniques to extract the meaningful insights [1].
- *Multiphase flow modeling* involves simulating the simultaneous flow of two or more immiscible fluids within the system. This might include liquid-liquid, liquid-gas, or any three-phase flow [25–28]. Within this flow modeling, one may use it to model flows such as Eulerian–Eulerian, which models each phase as a continuous fluid [27, 28], Eulerian–Lagrangian, which models one phase as continuous fluid and other phase as discrete particles [27, 28], or volume of fluid (VOF), which uses a volume fraction to track the interface between the phases [27]. In each of the models mentioned, the Navier–Stokes equation is solved at least once.
- *Conjugate heat transfer simulation* is a type of thermal analysis that involves the exchange of thermal energy between solid and fluid domains. This is commonly used in electronics cooling, heat exchangers, and so on. It solves the coupled heat transfer equations in both the solid and fluid domains simultaneously. By using this, we can understand the heat transfer between the fluid flowing through the microchannels and the surrounding solid walls [21, 29]. This can help to optimize the design of the microchannel heat exchanger for improved thermal performance and efficiency. However, it is more computationally intensive than fluid-only simulations.

TABLE 8.3
Summary of Experimental Setups for Microchannel Heat Exchangers

Authors	Materials Used
Denkenberger et al. [18], Song and Bullard [19] 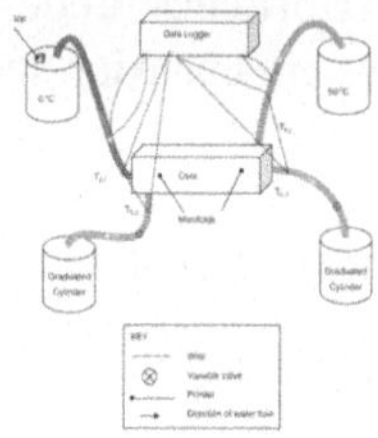	Consists of fluid supply system (high-pressure tank, filters, flow controllers), microchannel test section (heat exchanger, heating/cooling system), instrumentation (pressure/temperature sensors, flow visualization), and data acquisition system. Proper calibration and validation are crucial.
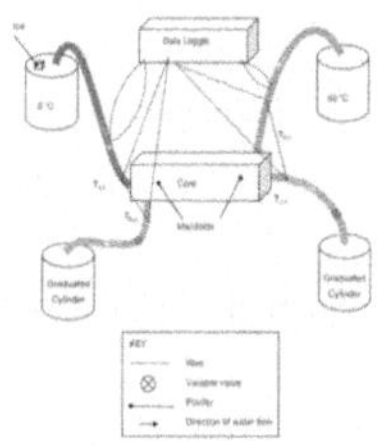	Mass flow meters (thermal or critical flow venturi) or volumetric flow meters are used to measure the flow rate of the working fluid.
Denkenberger et al. [18], Song and Bullard [19] 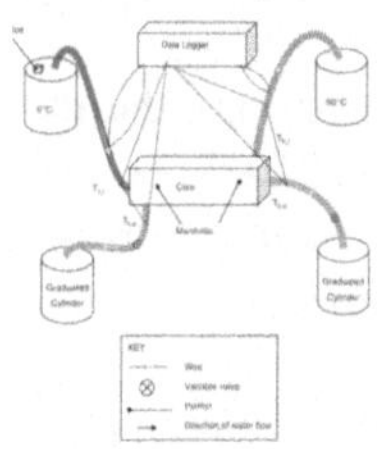	Differential pressure transducers for pressure drop across microchannels. Absolute pressure sensors for inlet/outlet pressures.
Denkenberger et al. [18], Song and Bullard [19]	
Denkenberger et al. [18], Song and Bullard [19], Huang et al. [20]	Thermocouples or resistance temperature detectors at inlet, outlet, and within microchannels. Infrared thermography for surface temperature mapping.
Denkenberger et al. [18], Song and Bullard [19], Huang et al. [20]	Particle image velocimetry, laser Doppler velocimetry, or dye injection techniques are used to visualize flow patterns within microchannels.
Denkenberger et al. [18], Song and Bullard [19], Huang et al. [20]	Signal conditioning, calibration corrections, uncertainty propagation, calculation of heat transfer coefficients, friction factors, and other performance metrics. Comparison with theoretical predictions, numerical simulations, and sensitivity analysis.
Denkenberger et al. [18], Song and Bullard [19], 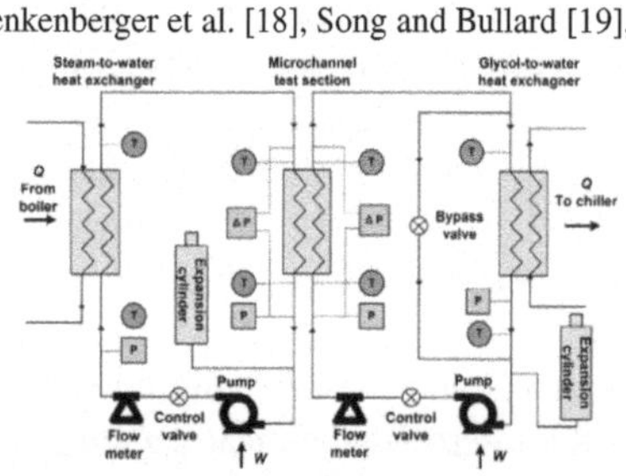 Kromer et al. [21]	Graphical representation, empirical correlations, technical reports, and publications.

- *Microscale heat transfer* models can incorporate slip flow conditions and rarefaction effects, which are important when considering microchannel heat exchangers. Within this model, we have several approaches, such as modifying continuum models for slip and jump conditions [29], employing the Boltzmann transport equation for the transport of thermal energy [29, 30], or utilizing molecular dynamics or Monte Carlo simulations to note the behavior of each atom or molecule [29]. However, we must consider the complexity and scalability issues while using these models as they are computationally intensive.

Once the simulation technique is finalized, we must think about the solver type. They are categorized into two types: steady-state solver and transient solver. While steady-state solvers solve for conditions that do not change over time, transient solvers solve for time-dependent conditions. Steady-state solvers are often used when we do not require the interim results and are only concerned about the result achieved at steady state, while transient solvers are used to get results after each specified time duration. One can also consider the use of a pseudo-transient solver that is used to accelerate the convergence of steady-state simulations by introducing the transient terms in the governing equations, thereby dampening the time variations and improving the stability of the solution.

8.4.2 Modeling Heat Transfer and Fluid Flow

Once the mesh has been created and is sufficiently refined, and the CFD approach has been chosen, we need to set up the physical model. There are various software that we can use, including ANSYS Fluent, ANSYS CFX, COMSOL, and OpenFoam. Now we need to choose the flow regime for the fluid flow, which can be laminar, k-ε turbulence or transition model, or k-ω turbulence or transition model. Table 8.4 shows the models and characteristics of each.

Next, we must choose the mode of heat transfer, which includes convection (natural or forced), conduction, and very rarely radiation. We need to specify the boundary conditions such as the velocity, pressure and temperature, and the thermal and fluid flow conditions such as slip factor, heat, thermal insulation, and so on. Now what is left is for us to provide the initial guesses for the fluid flow and temperature to start the simulation and for the solver to solve the governing equations, which are continuity, momentum, and energy equations.

8.4.3 Simulation Results and Validation

While the simulation results can be validated with different research papers, it seemed best to validate against an experimental setup. The earliest simulation results were validated with the experimental setup and then continued. A comparison with experimental results is thus essential to confirm accuracy while performing the validation study.

There are various correlations that can be used to validate the results, one of which is proposed by Wu and Little [31] for microchannels as presented in Table 8.5.

TABLE 8.4
Characteristics of Different Models

Model	Characteristics
Laminar	Smooth fluid flow where viscous forces dominate, no turbulence.
k-ε turbulence	Solves two transport equations for turbulent kinetic energy (k) and dissipation rate (ε).
k-ε transition	Incorporates transition modeling to improve laminar-turbulent transition.
k-ω turbulence	Solves for k and specific dissipation rate (ω), which enhances accuracy near walls.
k-ω transition	Improves the k-ω model with additional equations for the development of turbulence in boundary layers.

TABLE 8.5
Correlations Used in Different Papers for the Nusselt Number to Establish Heat Transfer

Author	Dimension of the Test Section	Formula or Correlation Used	Remark
Peiyi and Little [31]	6 cm × 1.4 cm × 0.2 cm	• $Nu = 0.00222 Pr^{0.4} Re^{1.09}$ • $f = 0.165(3.48 - log Re)^{0.24} + (0.081 \pm 0.007)$	Re > 3000 • 900 < Re < 3000 The transition from laminar to turbulent depends on the surface roughness. The higher the roughness, the quicker the turbulent phase is attained
Kays and Crawford [32]	Rectangular	$Nu_{fd} = 8.235\left(1 - \frac{1.833}{\alpha} + \frac{3.767}{\alpha^2} - \frac{5.814}{\alpha^3} + \frac{5.361}{\alpha^4} - 2/\alpha^5\right)$	• Fully developed • Re < 2200
Incropera and Dewitt [32]	Circular	$Nu = 1.86\left(\frac{RePrD}{L}\right)^{\frac{1}{3}}\left(\frac{\mu_f}{\mu_w}\right)^{0.14}$	• Re < 2200 • Simultaneously developing
Shah and London [32]	Circular	$Nu = \begin{cases} 1.953\left(RePr\frac{D}{L}\right)^{1/3} & \left(RePr\frac{D}{L}\right) \geq 33.33 \\ 4.364 + 0.0722 RePr\frac{D}{L} & \left(RePr\frac{D}{L}\right) < 33.33 \end{cases}$	• Thermally developing laminar (constant wall heat flux)

8.5 APPLICATIONS AND CASE STUDIES

With the advent of more compact electronics, automotives, and photovoltaic cells with fast processing, the resulting heat must be easily dissipated and cooled to allow for longevity. In addition, biomedical devices need safe operating conditions for humans. Cooling for mechanical devices and the high precision required for human operations have made these small and compact devices necessary in micro and mini configurations. With advancements in technologies, these micro- and minichannels have gained popularity. It is important to understand the impact that these high-end devices have on their surroundings, and because of this, the shift in focus of the world toward clean energy for engineering and human mortality for medicine is on the rise. These devices primarily have applications in electronics, the automotive industry, renewable energy, and biomedical areas [33]. The following sections explain each of the above-mentioned topics in detail.

8.5.1 Electronics Cooling

The idea behind using microchannels in compact and high-end devices was first brought into the picture by Tuckerman and Pease [34], who used water as the working fluid and dissipated a high heat flux of 790 W/cm^2 using these microchannels with fins. These fins are credited as being one of the most prolific ways of heat dissipation, owing to their support in increasing the heat transfer surface area for these microchannels. This is one way to lower the temperature of electronic devices. One other method is to identify the microchannels as a porous medium and use Darcy's law to explain the flow [35] and improve heat dissipation. For a microchannel, the heat is mainly released using conduction as its main method; hence, to obtain a better heat transfer, the simplest method involves raising the solid conductivity of the substance. This can be done by the use of micro heat pipes [36], which can be manufactured by using the vapor deposition technique on either side of the microchannel [36]. The primary applications are mainly found in micro-electro-mechanical systems and microfluidic device settings. The commercial applications of these devices have yet to hit the market due to a lack of awareness and not understanding the concept of heat exchangers and fluid flow behavior in microchannels.

8.5.2 Automotive Applications

Keeping in mind the environmental effects of exhaust gases of vehicles due to petrol and diesel, the world is shifting its focus to much cleaner fuel, primarily toward electric vehicles and hybrid electric vehicles. These vehicles are now fitted with batteries, which have heat exchangers to help their cooling [37, 38]. Comparing these heat exchangers with the microchannel added heat sink, it is evident from Table 8.6 that given the compact nature of battery spacing in the vehicles, the microchannel has been identified as the most efficient in cooling.

The heat transfer area of the microchannel varies linearly with the channel's hydraulic diameter, whereas the heat transfer rate varies parabolically with the hydraulic diameter. Consequently, the heat transfer rate increases as the unit's compactness increases, making it ideal for the microchannels to cool electric vehicles' batteries.

8.5.3 Renewable Energy Systems

When talking about renewable energy systems and microchannel applications, there is no better example than photovoltaic (PV) cells, as they are the primary example of clean renewable energy [39, 40]. Through studies, it has been observed that absorbed solar radiation is directly proportional to the output power of PV cells, and due to this absorbed solar radiation, the temperature of the PV cells rises, causing a decline in the efficiency of the cell [41–43]. This temperature rise can be maintained using an efficient cooling system employing microchannels and a coolant. This dissipated heat

TABLE 8.6
Comparison of Copper Tube Aluminum Heat Exchanger with Microchannel Heat Exchanger

Heat Exchanger	Size/mm	Net Mass (kg)	Refrigerant Charge (kg)	Air Volume (m^3/h)	Fan Impeller Diameter (mm)	Frontal Area (m^2)
Copper tube aluminum heat exchanger	2050 × 775 × 21.65	14.48	3.29	5605	560	1.558
Microchannel heat exchanger	1350 × 720 × 18	9.16	1.59	3652	445	0.972

can be used as a source of thermal energy to power up other appliances. The main reason for using microchannels in the cooling of these cells is that these cells are already manufactured in a micro size, so the use of the microchannels adheres to the compactness necessary for the design considerations of the cells. An experimental study was conducted by Rahimi et al. [44] using water as a coolant and microchannels at a low Re number of 70. It was observed that a 30% increase in power output was noted as compared with the non-cooled situations. A study was conducted on the effect of a microchannel heat sink coupled with a concentrated PV system on the overall performance of a concentrated PV thermal system by Radwan et al. [45]. It was observed that there was a significant reduction in temperature of the cell with the thermal efficiency reaching a maximum of 62%.

8.5.4 Biomedical Devices

The use of microchannels in biomedical applications is mostly in microfluidics. Molecular adsorption, optical properties, machinability, surface charge, electroosmotic flow mobility, and electrical and thermal conductivity are a few of the properties that play a crucial role in the field of microfluidics [46]. Making use of one of its properties, a study of the motion of the blood cells has been conducted using microchannels with a dimension of 100 μm, and it is a prime example of the application of microchannels in biomedical fields. Microchannels have also been used to control the liquid flow by making the substrate have surface free energy patterns under spontaneous flow conditions [33].

Authors	Application	Dimension of the Test Section	Working Fluid/ Material Used	Remark
Radwan et al. [45]	Photovoltaic cells	2.5 cm × 100 μm (l × h)	Tedlar	Concentrated photovoltaic systems show the best effect by using the microchannel for cooling, where the reduction in solar cell temperature is significant
Ramos-Alvarado et al. [47]	Electronic cooling	• Serpentine A and B = 0.9 mm × 1.2 mm (width × depth) • Parallel I-A and B = 0.95 mm × 1.2 mm • Parallel II-A and B = 1.2 mm × 1.2 mm • Distributors A and B = 2.84 mm × 1.3 mm	De-ionized water	Distribution A and B configuration is preferred for a standardized heat sink to be used

Authors	Application	Dimension of the Test Section	Working Fluid/ Material Used	Remark
Deng et al. [48]	Automotive cooling	• Engine coolant = 96.6 mm × 2 mm × 0.5 mm • R134a = 84.3 mm × 2 mm × 0.5 mm • Natural gas = 13 mm × 2 mm × 0.3 mm	Water, engine coolant, and R134a	Validation of applied extrapolation, symmetric methods resulting in the reduction of computational cost
Li et [49]	Automotive air conditioning	235mm × 2.88mm	CO_2 gaseous	Prediction errors on the gas cooler capacity and the refrigerant side pressure drop were ±5 and ±8%, respectively, indicating good performance at high frontal air velocity
Harris et al. [50]	Automotive cooling	500 μm × 1.2 mm (width × depth)	Water-glycol	Cross-flow heat exchangers in radiators can achieve greater heat transfer, hence are better suited to be used for cooling vehicles
Yao et al. [46]	Microfluidic devices	Alveolar-capillary layer coated with extracellular matrix	Separated PDMS microchannel	Stacking of microchannels to allow for the easy interaction of liquid with layers

8.6 PERFORMANCE ENHANCEMENT TECHNIQUES

Microchannel heat exchangers are vital components in various industrial and technological applications due to their high heat transfer efficiency and compact size. However, to meet the ever-increasing demands for improved performance and efficiency, several enhancement techniques have been developed. The following sections explore some of these techniques, focusing on surface modifications, nanofluids and hybrid fluids, and enhanced flow regimes.

8.6.1 SURFACE MODIFICATIONS

Surface modifications play a crucial role in enhancing the heat transfer performance of microchannel heat exchangers. One of the most effective methods is roughening the channel walls. Rough surfaces increase the surface area available for heat transfer and promote turbulence in the fluid flow, which enhances the overall heat transfer coefficient. Various techniques, such as mechanical roughening, chemical etching, and coating with high thermal conductivity materials, have been explored to achieve the desired surface texture.

In addition to roughening, coating the channel walls with thin layers of materials that have high thermal conductivity can significantly improve heat transfer performance. For instance, applying a layer of graphene or carbon nanotubes can reduce thermal resistance and increase the heat transfer rate [51]. Coatings can also provide a protective barrier against corrosion and fouling, thereby extending the lifespan of the heat exchanger.

Research has shown that specific surface modifications, such as adding micro-fins or dimples, can further enhance heat transfer. These structures disrupt the boundary layer and promote better mixing of the fluid, leading to higher heat transfer rates [52]. Overall, surface modifications are a versatile and effective means of improving the performance of microchannel heat exchangers.

8.6.2 Nanofluids and Hybrid Fluids

Nanofluids, which are suspensions of nanoparticles in a base fluid, have emerged as a promising solution for enhancing the thermal conductivity and heat transfer performance of microchannel heat exchangers. The inclusion of nanoparticles, such as metal oxides, carbon nanotubes, or graphene, increases the thermal conductivity of the base fluid, resulting in improved heat transfer characteristics.

One of the key advantages of nanofluids is their ability to enhance heat transfer without requiring significant changes to the heat exchanger design. Studies have demonstrated that nanofluids can increase the heat transfer coefficient by up to 30% compared with conventional fluids. This improvement is attributed to the increased thermal conductivity of the fluid and the enhanced thermal dispersion effects caused by the Brownian motion of the nanoparticles.

Hybrid fluids, which combine different types of nanoparticles, offer further improvements in heat transfer performance. For example, a hybrid fluid containing both metal and metal oxide nanoparticles can leverage the high thermal conductivity of metal nanoparticles and the stability of metal oxide nanoparticles. This synergistic effect can result in even higher heat transfer rates and more efficient thermal management.

However, the use of nanofluids and hybrid fluids also presents challenges, such as potential clogging of microchannels, increased pressure drop, and long-term stability issues. Therefore, ongoing research is focused on optimizing the formulation and stability of these advanced fluids to ensure their practical applicability in microchannel heat exchangers [53].

8.6.3 Enhanced Flow Regimes

Optimizing the channel geometry and flow conditions can lead to enhanced flow regimes that significantly improve the heat transfer performance of microchannel heat exchangers. One approach is to introduce secondary flows, which disrupt the primary flow and promote better mixing of the fluid. This can be achieved by designing channels with specific geometries, such as zigzag patterns or spiral structures, that induce secondary flow effects.

Vortex generators are another effective method for enhancing flow regimes. These are small structures placed within the channels that create vortices in the fluid flow. The generated vortices increase fluid mixing and reduce thermal boundary layer thickness, leading to higher heat transfer coefficients. Vortex generators can be integrated into the channel walls or placed as inserts, depending on the specific application requirements. Additionally, altering the flow conditions, such as flow rate and fluid properties, can also enhance heat transfer performance. For example, operating at higher Reynolds numbers can transition the flow from laminar to turbulent, which significantly enhances heat transfer due to increased mixing and reduced thermal resistance. However, this approach must be balanced with the potential increase in pressure drop and energy consumption. Another method for enhancing the heat transfer is the use of phase change materials, which are gaining popularity in several fields [7, 43].

Advanced computational techniques, such as CFD, are increasingly used to optimize channel geometries and flow conditions. These tools allow for detailed analysis and simulation of different designs, enabling engineers to identify the most effective configurations for enhancing heat transfer performance [40].

8.7 CONCLUSION

8.7.1 Summary of Key Points

Microchannel heat exchangers have become important because of their high efficiency and compactness.

- Experimental setups, precise instrumentation, and CFD simulations provide valuable insights into performance, validating results and enhancing designs.
- There are various applications, such as electronics cooling, automotive systems, renewable energy, and biomedical devices, that showcase versatility and efficiency in diverse fields.
- Performance is enhanced through surface modifications, nanofluids, hybrid fluids, and optimized flow regimes, improving overall heat transfer rates and effectiveness.

8.7.2 Future Research Directions

- Stronger microchannel heat exchangers can be built to reduce material constraints, such as exploring materials with superior thermal properties—nanostructured and composite coatings as a possible field.
- Investigating novel geometries and flow patterns using advanced CFD simulations, artificial intelligence, and machine learning for better efficiency and reduced pressure drops.
- As the world is transitioning toward greener and cleaner technology, microchannel heat exchangers can be used for enhancing the efficiency of renewable energy systems.

8.7.3 Final Thoughts

Microchannel heat exchangers offer efficient thermal management solutions across industries and integrating experimental and numerical methods enhances performance understanding. They have huge potential to advance compact, sustainable thermal systems through innovation and interdisciplinary collaboration.

8.8 NOMENCLATURE

Q	Amount of heat (J)
M	Mass flow rate of working fluid (m^3/s)
T_{in}	Inlet temperature of working fluid (K)
T_{out}	Outlet temperature of working fluid (K)
C_p	Specific heat capacity (J/kg/K)
V	Voltage (V)
I	Current (A)
N	Number of channels
H	Height of the channel (m)
W	Width of the channel (m)
l	Length of minichannel (m)
h	Fin efficiency
k	Thermal conductivity of copper (W/mK)
l'	Distance from the entrance region of the microchannel to a specific location (m)
h_{av}	Average heat transfer coefficient (W/m^2K)
$T_{w,z}$	Surface temperature of the microchannel at each location (K)
$T_{b,z}$	The mean temperature of the fluid (K)
Q_{conv}	Convective heat transfer (J)
s	Conductive gap between the surface of each channel and the exact location of each thermocouple (m)
z	Axial positions of thermocouples (m)
Nu_w	Wall Nusselt number
f_w	Wall friction factor

u	The average bulk velocity (m/s)
D_{hyd}	The hydraulic diameter of each microchannel (m)
ΔP_{nf}	Pressure difference of nanofluid
ρ	Density (kgm^{-3})
χ_m	Magnetic Susceptibility
V	Velocity (m/s)
F_k	Total Force (N)
μ_0	Dynamic Viscosity net of magnetic field (Pas)
$\overrightarrow{H}$	Magnetic Field Intensity (T)
τ_{ij}	Shear stress (N/m^2)
φ	Nanoparticle Concentration
Subscripts	
b	Bulk
f	Fluid
nf	Nanofluid
np	Nanoparticle

REFERENCES

1. Gao J, Hu Z, Yang Q, Liang X, Wu H. Fluid flow and heat transfer in microchannel heat sinks: Modelling review and recent progress. *Therm Sci Eng Prog [Internet].* 2022;29:101203. Available from: https://doi.org/10.1016/j.tsep.2022.101203.
2. Li M, Gao X, Li H, Sang J, Nie P, Fang W, et al. Investigation of flow and heat transfer characteristics in microchannels with fins. *Machines.* 2023;11:154.
3. Kumar S, Vijayan PK, Kannan U, Sharma M, Pilkhwal DS. Experimental and computational simulation of thermal stratification in large pools with immersed condenser. *Appl Therm Eng [Internet].* 2017;113:345–61. Available from: https://doi.org/10.1016/j.applthermaleng.2016.10.175.
4. Kumar S, Grover RB, Vijayan PK, Kannan U. Annals of nuclear energy numerical investigation on the effect of shrouds around an immersed isolation condenser on the thermal stratification in large pools. *Ann Nucl Energy [Internet].* 2017;110:109–25. Available from: https://doi.org/10.1016/j.anucene.2017.06.022.
5. Li H, Li Y, Huang B, Xu T. Numerical investigation on the optimum thermal design of the shape and geometric parameters of microchannel heat exchangers with cavities. *Micromachines.* 2020;11:721.
6. Kumar S, Banerjee D. A review on phase change materials for sustainability applications by leveraging machine learning. *Energy Consum Conversion, Storage, Effic.* 2024. Available from: http://dx.doi.org/10.5772/intechopen.114380.
7. Kumar S, Thyagarajan A, Banerjee D. Experimental investigation of thermal energy storage (Tes) platform leveraging phase change materials in a chevron plate heat exchanger. *ASME Int Mech Eng Congr Expo Proc.* 2022;8:1–7.
8. Mokrane M, Bourouis M. Heat transfer and fluid flow characteristics in a micro heat exchanger employing warm nanofluids for cooling of electronic components. *Energies.* 2024;17:2383.
9. Sarkar J, Bhattacharyya S. Application of graphene and graphene-based materials in clean energy-related devices Minghui. *Arch Thermodyn.* 2012;33:23–40.
10. Kandlikar SG, Grande WJ. Evolution of microchannel flow passages-thermohydraulic performance and fabrication technology. *Heat Transf Eng.* 2003;24:3–17.
11. Mudawar I. Two-phase microchannel heat sinks: Theory, applications, and limitations. *J Electron Packag.* 2011;133. Available from: https://doi.org/10.1115/AJTEC2011-44005
12. Rostami S, Nadooshan AA, Raisi A, Bayareh M. Effect of using a heatsink with nanofluid flow and phase change material on thermal management of plate lithium-ion battery. *J Energy Storage [Internet].* 2022;52:104686. Available from: https://doi.org/10.1016/j.est.2022.104686.
13. Bhattacharyya S, Vishwakarma DK, Dey K, Yasmin H, Vaziri R, Sharifpur M. Numerical study on thermal–hydraulic characteristics in a inclined mini-channel for solar PV panel cooling with external magnetic field and magnetic nanofluid. *J Magn Magn Mater [Internet].* 2024;590:171626. Available from: https://doi.org/10.1016/j.jmmm.2023.171626.

14. Caney N, Louahlia-Gualous H, Bontemps A. Two-phase flow in microchannels. *Microscale Nanoscale Heat Transf Anal Des Appl.* 2016;26:445–78.
15. Barth T, Scholz P, Wierach P. Flow control by dynamic vane vortex generators based on piezoceramic actuators. *AIAA J.* 2011;49:921–31.
16. Kumar S, Grover RB, Yadav H, Vijayan PK, Kannan U, Agrawal A. Experimental investigations on thermal stratification in a large pool of water with immersed isolation condenser. *Int Conf Nucl Eng Proceedings, ICONE.* 2020;3:1–7.
17. Kumar S, Klassen M, Klassen D, Hardin R, King MD. Dispersion of sneeze droplets in a meat facility indoor environment – Without partitions. *Environ Res [Internet].* 2023;236:116603. Available from: https://doi.org/10.1016/j.envres.2023.116603
18. Denkenberger DC, Brandemuehl MJ, Pearce JM, Zhai J. Expanded microchannel heat exchanger: Design, fabrication, and preliminary experimental test. *Proc Inst Mech Eng Part A J Power Energy.* 2012;226:532–44.
19. Song S, Bullard CW. Experimental and simulation analysis of microchannel evaporators. *Univ Illinois [Internet].* 2002;Tech. Rep. Available from: http://www.ideals.illinois.edu/handle/2142/13406
20. Huang B, Li H, Xia S, Xu T. Experimental investigation of the flow and heat transfer performance in micro-channel heat exchangers with cavities. *Int J Heat Mass Transf.* 2020;159:120075.
21. Kromer DA, Huning AJ, Garimella S. I2S-LWR microchannel heat exchanger design and experimental validation. *Ann Nucl Energy [Internet].* 2020;145:105710. Available from: https://doi.org/10.1016/j.anucene.2017.08.043
22. Kumar S, Grover RB, Yadav H, Vijayan PK, Kannan U, Agrawal A. Experimental and numerical investigation on suppression of thermal stratification in a water-pool : PIV measurements and CFD simulations by. *Appl Therm Eng [Internet].* 2018; Available from: https://doi.org/10.1016/j.applthermaleng.2018.04.070.
23. Dilbaz F, Selimefendigil F, Öztop HF. Lithium-ion battery module performance improvements by using nanodiamond-FE3O4 water/ethylene glycol hybrid nanofluid and fins. *J Therm Anal Calorim [Internet].* 2022;147:10625–35. Available from: https://doi.org/10.1007/s10973-022-11269-9.
24. Qu W, Mudawar I. Experimental and numerical study of pressure drop and heat transfer in a single-phase micro-channel heat sink. *Int J Heat Mass Transf.* 2002;45:2549–65.
25. Anjaneya G, Sunil S, Kakkeri S, Math MM, Vaibhav MN, Solaimuthu C, et al. Numerical simulation of microchannel heat exchanger using CFD. *Int J Interact Des Manuf [Internet].* 2023;18:5847–63. Available from: https://doi.org/10.1007/s12008-023-01376-8.
26. Batzdorf S, Gambaryan-Roisman T, Stephan P. Direct numerical simulation of the microscale fluid flow and heat transfer in the three-phase contact line region during evaporation. *J Heat Transfer.* 2018;140. Available from: https://doi.org/10.1115/1.4038191.
27. Panda K, Hirokawa T, Huang L. Design study of microchannel heat exchanger headers using experimentally validated multiphase flow CFD simulation. *Appl Therm Eng [Internet].* 2020;178:115585. Available from: https://doi.org/10.1016/j.applthermaleng.2020.115585.
28. Kumar V, Vikash, Nigam KDP. Multiphase fluid flow and heat transfer characteristics in microchannels. *Chem Eng Sci [Internet].* 2017;169:34–66. Available from: http://dx.doi.org/10.1016/j.ces.2017.01.018.
29. Croce G, Coppola MA. ICNMM2015-48128. 2017;1–8.
30. Asiva Noor Rachmayani. No 主観的健康感を中心とした在宅高齢者における 健康関連指標に関する共分散構造分析Title. 2015;6.
31. Peiyi W, Little WA. Measurement of friction factors for the flow of gases in very fine channels used for microminiature Joule-Thomson refrigerators. *Cryogenics (Guildf).* 1983;23:273–7.
32. Incropera FP, Dewitt DP. *Fundamentals of Heat and Mass Transfer.* 1998; John Wiley & Sons, Inc.
33. Prakash S, Kumar S. Fabrication of microchannels: A review. *Proc Inst Mech Eng Part B J Eng Manuf.* 2015;229:1273–88.
34. Tuckerman DB, Pease RFW. High-performance heat sinking for VLSI. *IEEE Electron Device Lett.* 1981;EDL-2:126–9.
35. Rashidi S, Bovand M, Esfahani JA. Heat transfer enhancement and pressure drop penalty in porous solar heat exchangers: A sensitivity analysis. *Energy Convers Manag [Internet].* 2015;103:726–38. Available from: http://dx.doi.org/10.1016/j.enconman.2015.07.019.
36. Mallik AK, Peterson GP, Weichold MH. Fabrication of vapor-deposited micro heat pipe arrays as an integral part of semiconductor devices. *J Microelectromechanical Syst.* 1995;4:119–31.
37. Ohadi M, Choo K, Dessiatoun S, Cetegen E. Next generation microchannel heat exchangers. *SpringerBriefs Appl. Sci. Technol.* 2013. Available from: https://doi.org/10.1007/978-1-4614-0779-9.

38. Jhariya M, Dewangan AK, Moinuddin SQ, Kumar S, Ahmad A, Yadav AK. Research progress on efficient battery thermal management system (BTMs) for electric vehicles using composite phase change materials with liquid cooling and nanoadditives. *J Therm Anal Calorim [Internet].* 2024; Available from: https://doi.org/10.1007/s10973-024-13752-x.
39. Mohammed Adham A, Mohd-Ghazali N, Ahmad R. Thermal and hydrodynamic analysis of microchannel heat sinks: A review. *Renew Sustain Energy Rev [Internet].* 2013;21:614–22. Available from: http://dx.doi.org/10.1016/j.rser.2013.01.022.
40. Kumar S, Rathore K. Renewable Energy for sustainable development goal of clean and affordable energy. *International Journal of Materials, Manufacturing and Sustainable Technologies.* 2023;2:1–15.
40. Rathore K, Kumar S. Net zero emissions for our future generations through renewables: A brief review. *Int J Energy Resour Appl [Internet].* 2023;2:39–56. Available from: https://docs.google.com/uc?export=download&id=1sYggkUhoD5SlVLoCJ5wzFA57dWESwD8A.
41. Souayeh B, Bhattacharyya S, Hdhiri N, Alam MW. Selection of best suitable eco-friendly refrigerants for HVAC sector and renewable energy devices. *Sustainability.* 2022;14:11663.
42. Pathak SK, Sharma PO, Goel V, Bhattacharyya S, Aybar H, Meyer JP. A detailed review on the performance of photovoltaic/thermal system using various cooling methods. *Sustain Energy Technol Assessments [Internet].* 2022;51:101844. Available from: https://doi.org/10.1016/j.seta.2021.101844.
43. Dawahdeh A, Sharadga H, Kumar S. Novel MPPT controller augmented with neural network for use with photovoltaic systems experiencing rapid solar radiation changes. *Sustainability.* 2024;16:1021.
44. Rahimi M, Karimi E, Asadi M, Valeh-e-Sheyda P. Heat transfer augmentation in a hybrid microchannel solar cell. *Int Commun Heat Mass Transf [Internet].* 2013;43:131–7. Available from: http://dx.doi.org/10.1016/j.icheatmasstransfer.2013.02.002.
45. Radwan A, Ookawara S, Ahmed M. Analysis and simulation of concentrating photovoltaic systems with a microchannel heat sink. *Sol Energy [Internet].* 2016;136:35–48. Available from: http://dx.doi.org/10.1016/j.solener.2016.06.070.
46. Yao P, Schneider GJ, Prather DW. Three-dimensional lithographical fabrication of microchannels. *J Microelectromechanical Syst.* 2005;14:799–805.
47. Ramos-Alvarado B, Li P, Liu H, Hernandez-Guerrero A. CFD study of liquid-cooled heat sinks with microchannel flow field configurations for electronics, fuel cells, and concentrated solar cells. *Appl Therm Eng [Internet].* 2011;31:2494–507. Available from: http://dx.doi.org/10.1016/j.applthermaleng.2011.04.015.
48. Deng Y, Menon S, Lavrich Z, Wang H, Hagen CL. Design, simulation, and testing of a novel microchannel heat exchanger for natural gas cooling in automotive applications. *Appl Therm Eng [Internet].* 2017;110:327–34. Available from: http://dx.doi.org/10.1016/j.applthermaleng.2016.08.193.
49. Li W, Peng X, Xiao M, Garg A, Gao L. Multi-objective design optimization for mini-channel cooling battery thermal management system in an electric vehicle. *Int J Energy Res.* 2019;43:3668–80.
50. Harris C, Despa M, Kelly K. Design and fabrication of a cross flow micro heat exchanger. *J Microelectromechanical Syst.* 2000;9:502–8.
51. Bahiraei M, Heshmatian S. Electronics cooling with nanofluids: A critical review. *Energy Convers Manag [Internet].* 2018;172:438–56. Available from: https://doi.org/10.1016/j.enconman.2018.07.047.
52. Wen D, Ding Y. Experimental investigation into convective heat transfer of nanofluids at the entrance region under laminar flow conditions. *Int J Heat Mass Transf.* 2004;47:5181–8.
53. Jacobi AM, Shah RK. Heat transfer surface enhancement through the use of longitudinal vortices: A review of recent progress. *Exp Therm Fluid Sci.* 1995;11:295–309.

9 Computational Modelling of Heat Exchangers

Chaitanya R. Mali

9.1 INTRODUCTION

The role of computational fluid dynamics (CFD) is crucial in the field of heat exchanger (HEX) modelling. Conventionally used analytical techniques for HEX design, such as log mean temperature difference (LMTD) and number of transfer units (NTU), are associated with several assumptions, such as constant heat transfer coefficient (HTC), uniform flow distribution, constant fluid properties, and single-phase flow. These assumptions cannot apply to complex systems/geometries, transient operating conditions, multicomponent systems with varying fluid properties, multiphase flow scenarios, and so on [1–3]. Sensitivity of the HEX design to assumptions of conventional design methods leads to prototype developments for performance testing for different design iterations [4, 5]. In the past few decades, with advancements in numerical techniques and the evolution of computer hardware, CFD emerged, and is an essential and efficient tool for the thermal-hydraulic performance evaluation and design optimization of HEXs. Depending on the objective of the thermal analysis, such as in large heat transport loops, numerical tools like RELAP are used to study overall system behaviour and model heat exchanger performance [6, 7]. In contrast, for detailed design and performance simulations of heat exchangers, CFD tools are more critical. Theoretically, CFD discretizes the system/domain under consideration into grids/mesh elements and applies governing equations, capturing the physics of the process using different numerical methods (for example, the finite volume method or the finite element method) to discrete mesh elements to obtain numerical solutions in terms of different flow variables such as velocity, pressure, temperature, and other flow parameters [8].

CFD provides the flexibility of testing/modelling several complex prototype designs of HEXs and various operating and boundary conditions at the pre-fabrication stage, minimizing cost and time constraints. In addition, CFD offers several advantages in HEX designs, such as detailed, comprehensive insights into thermal hydraulics of the system, capturing flow non-uniformity and secondary flows, and the precise evaluation of thermal performance by capturing hydrodynamic and thermal boundary layer effects [20–22].

Detailed CFD analysis of HEX is useful for addressing different issues like flow and temperature maldistribution, fouling, and pressure drop, which can severely affect thermal-hydraulic performance of HEXs [23, 24]. Figure 9.1 shows an overview of applications of CFD modelling for different types of HEXs. Figure 9.1 shows that HEX modelling is not limited to single-phase studies; multiphase modelling of HEX with and without phase change is found relevant in cases of direct contact heat exchangers, steam generators, evaporators, and so on. One can refer to Hewitt et al. [25] to understand the broad spectrum of HEX classification.

Besides several advantages and extensive applications, CFD modelling of HEXs is associated with some limitations. Modelling a complex, time-dependent chaotic phenomenon called turbulence requires high-fidelity numerical methods and a sufficiently resolved mesh to capture precise physics, which is computationally expensive in terms of time, space, and cost [26, 27]. The design and optimization of HEXs using CFD is a vast activity that involves a simulation matrix

DOI: 10.1201/9781003509837-9

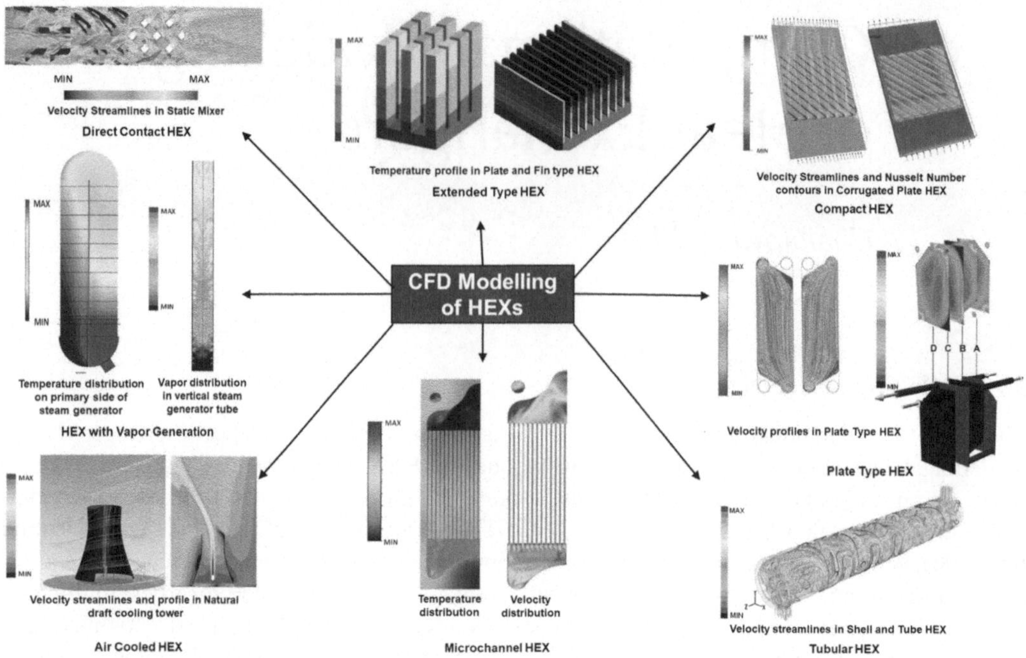

FIGURE 9.1 Overview of applications of CFD modelling for different types of HEXs (all figures are reprinted with copyright permissions from references [9–19]).

of a wide array of operating conditions and geometrical designs. Performing a simulation for each set of arrays requires mesh regeneration-regulation, resetting of boundary conditions, and extensive post-processing. So, design optimization activity is tedious and encumbering. Preprocessing (geometry definition, mesh discretization, and regulation), problem setup (settling out convergence issues, code execution, intermediate solution assessment, parameter setting), and post-processing activity (results, analysis, and interpretation) in the CFD requires significant human interception and knowledge, which is susceptible to errors and irregularities in the design optimization activity of HEXs [28–30]. Artificial intelligence (AI)/machine learning (ML)-based CFD has evolved in the past decade to overcome limitations of CFD modelling of HEXs. AI/ML refers to revolutionary integration between computer science and data analysis. AI/ML-based CFD offers several advantages over the conventional CFD approach. AI/ML techniques can enhance turbulence modelling by analyzing large-scale flow datasets to extract patterns and correlations associated with turbulence, providing insights into turbulent structures, a complex interaction between turbulent eddies and surrounding fluid, transition phenomena, and flow instabilities, resulting in enhanced precision in physics capture [31–34]. ML algorithms can optimize CFD simulations by reducing the number of required grid points and iterations, reducing computational cost [35, 36]. AI/ML techniques enable automation in preprocessing and post-processing CFD tasks such as mesh generation and regulation, parameter tuning, and results analysis, which results in faster turnaround time and accuracy [36–38]. Applications of AI/ML-based CFD are also associated with challenges in terms of data mining, data consistency, the generalizability of the model, interpretability, ease of use, and extended system development time and efforts [28, 35]. However, steering progress is happening in AI/ML-based CFD approaches in different types of HEXs, such as compact HEXs, inclined tube HEXs, shell and tube HEXs, plate and frame HEXs, and many more [36, 39].

This chapter mainly focuses on applications of CFD for HEX modelling. In Section 9.2, CFD modelling approaches, along with governing physics, are mentioned. Section 9.2 also explains different AI/ML approaches/techniques for HEX design and optimization. Section 9.3 discusses CFD modelling requirements, such as the type of mesh and mesh criteria and boundary conditions for HEX modelling applications. Section 9.4 discusses the different case studies and practical applications of CFD for modelling HEXs. Section 9.4 also highlights the case studies that show applications of AI/ML-based CFD for HEX design and optimization. Section 9.5 concludes with an overview and outlook of CFD modelling of HEXs.

9.2 MODELLING APPROACHES FOR HEXS

Heat exchangers play a key role in engineering systems and power production, including solar, geothermal, and biomass energy [40, 41]. Thus, a solid understanding of modelling approaches is essential. The modelling of the particular type of HEX system is associated with theoretical modelling (involving a concept of energy, force, etc.) and experimental analysis (involving physical observations). Theoretical modelling led to the construction of conservation laws of mass, energy, and momentum, and equations of the state of the fluid under consideration. Similarly, experimental analysis led to the development of empirical/semi-empirical correlations, for example, curve fitting activity – a combination of theoretical and experimental analysis results in the development of a mathematical model that approximates the actual system. Most mathematical models representing the given HEX system are non-linear and dynamic. Hence, the exact analytical solution of the mathematical model is very difficult or impossible. Thus, mathematical models are approximated to numerical/computational models where different numerical methods are used to obtain numerical solutions. In this section, basic governing or Navier–Stokes equations of particular HEX systems are discussed with continuum fluid approximation. The approximation of continuum fluid refers to the mean free path of the constituent molecules of the fluid being smaller than the physical length scale of the geometry under consideration.

Modelling of HEXs can be categorized into two streams, single-phase modelling and multiphase flow modelling based on heat transfer mechanisms. The multiphase flow modelling approach is essential in condensers, boilers, steam generators, evaporators, and radiators used in air conditioning. The single-phase modelling approach is useful for a wide range of HEX applications and is an important step towards understanding the complex physics related to HEXs. Along with single-phase and multiphase numerical modelling of HEXs, this section discusses AI/ML-based CFD approaches such as data-driven modelling, physical modelling, and hybrid modelling, which are associated with different algorithms.

9.2.1 The Single-Phase Modelling Approach

Basic governing equations for single-phase flow can be derived using the Lagrangian or Eulerian frames of reference. In the Lagrangian frame of reference, an arbitrary control/material volume can be considered impenetrable, and the velocity of the control volume surface is equal to the local fluid velocity. The control volume is stationary and penetrable in the Eulerian frame of reference. Basic governing conservation equations are independent of choices of frame of reference, particularly in the case of single-phase modelling. However, these choices are sub-categorized into different streams in multiphase flow. Considering the Eulerian frame of reference for fluid of density $\rho(x,t)$ and velocity $u(x,t)$ at position x and time t, conservation equations are given below.

9.2.1.1 Conservation of Mass

$$\frac{\partial \rho}{\partial t} + \nabla \cdot (\rho V) = S_m \tag{9.1}$$

In Equation 9.1, the first term on the left-hand side (LHS) represents the time rate of mass change per unit volume, i.e., accumulation; the second term represents mass transfer per unit volume due to velocity gradients, i.e., convection. The first term on the right-hand side (RHS) represents the volumetric source term. In the case of single-phase flow, there is no net generation or destruction of mass. The second term can be expanded explicitly to its gradient and divergence operators as

$$\nabla\cdot(\rho V)=\left(\frac{\partial(\rho u)}{\partial x}+\frac{\partial(\rho v)}{\partial y}+\frac{\partial(\rho w)}{\partial z}\right) \tag{9.2}$$

For incompressible flow (constant density ρ), Equation 9.1 is deduced to Equation 9.3

$$\nabla\cdot(V)=0 \tag{9.3}$$

9.2.1.2 Conservation of Momentum

The momentum conservation equation relies on Newton's second law of motion, i.e., $F = ma$ where 'F' is the force, 'm' is mass, and 'a' is acceleration. The fluid motion results from different forces acting on the fluid, which can be categorized into body forces and surface forces. Body forces that act on the entire mass of the fluid include gravitational force and electromagnetic force. Similarly, surface forces acting on the fluid surfaces include pressure and viscous forces. Viscous forces are further categorized into shear and normal stresses resulting from frictional losses. The governing momentum equation can be written as

$$\frac{\partial}{\partial t}(\rho V)+\nabla\cdot(\rho VV)=-\nabla\sigma+\rho g+F \tag{9.4}$$

In Equation 9.4, the first term in the LHS represents the time rate of change of momentum per unit volume, i.e., momentum accumulation term, and the second term represents the rate of change of momentum due to convection per unit volume, i.e., velocity gradients are responsible for fluid momentum.

The first term in the RHS of Equation 9.4 represents the total stress term that arises out of small-scale molecular interactions [42, 43] and it is decomposed as

$$\nabla\sigma=-\nabla p+\nabla\cdot\tau \tag{9.5}$$

In Equation 9.5, ∇p represents the rate of change of momentum per unit volume due to pressure gradients; this term results from heat sources, pumps, and so on. $\nabla\tau$ is the rate of change of momentum per unit volume due to the viscous shear stress term. $\nabla\tau$ is a tensor term that consists of nine components. The viscous stress term is responsible for the dampening of fluid motion. Considering Newtonian fluids, shear stress is directly proportional to velocity gradients, i.e., fluid deformation rate as per Newton's law of viscosity. The simplified model, which gives the relationship between stress tensor and velocity components, is given below

$$\tau_{ij}=\begin{vmatrix}\tau_{xx} & \tau_{xy} & \tau_{xz}\\ \tau_{yx} & \tau_{yy} & \tau_{yz}\\ \tau_{zx} & \tau_{zy} & \tau_{zz}\end{vmatrix}=\mu\left(\nabla V+\nabla V^{T}\right)+\left(\kappa-\frac{2}{3}\mu\right)\delta_{ij}(\nabla\cdot V) \tag{9.6}$$

In Equation 9.6, $\mu\left(\nabla V+\nabla V^{T}\right)$ represents the strain rate tensor term where μ is the coefficient of viscosity, and superscript 'T' is the transpose of the tensor term. In the case of non-Newtonian fluids, for example, Bingham plastic fluid and viscoelastic fluids, the molecular viscosity term is replaced by the effective viscosity term where viscosity is expressed in terms of different models such as the power law model, the Herschel–Bulkley model, or the temperature-dependent viscosity model. κ is the bulk viscosity coefficient, which is zero for ideal gases. The second term on the LHS of Equation 9.6 is zero for incompressible fluids. The term $-\frac{2}{3}\mu$ arises out of the Stokes hypothesis. δ_{ij} is the Kronecker delta function, i.e., $\delta_{ij}=1$ when $i=j$ and $\delta_{ij}=0$ if $i\neq j$.

The second term on the RHS of Equation 9.4 represents the effect of gravitational force on momentum change per unit volume, which is essential for buoyancy-driven flows. The third term on the RHS of Equation 9.4 is any external force if it exists. The external force term is mostly unavoidable in the case of multiphase flow modelling, which will be explained in the next sub-section.

9.2.1.3 Conservation of Energy

Total energy is the summation of internal or thermal energy and kinetic/mechanical energy. The equation of energy conservation relies on the well-known first law of thermodynamics, i.e., the rate of change of internal energy inside the fluid element is equal to the summation of the net flux of heat added into the fluid and the rate of work done on the fluid element due to body and surface forces. Net flux of heat arises from the conduction and volumetric heating of the fluid element due to absorption, emission, or radiation.

$$\frac{\partial}{\partial t}\left[\rho\left(e+\frac{V^2}{2}\right)\right]+\nabla\cdot\left[\rho\left(e+\frac{V^2}{2}\right)V\right]=-\nabla\left[p\cdot V\right]+\nabla\cdot\left[k\nabla T+\left(\tau\cdot V\right)-\left(\sum_j h_k j_k\right)\right]+S_h \tag{9.7}$$

In Equation 9.7, the first term on the LHS represents the rate of change of total energy, i.e., accumulation. The second term represents a change in the total energy due to fluid convection. The first term on the RHS of Equation 9.7 represents work done on the fluid by pressure forces. The second, third, and fourth terms on the RHS represent the rate of change of total energy by conduction, viscous dissipation (viscous forces), and diffusive mass fluxes (species diffusion), respectively. The last term on the RHS is a volumetric heat source term, like the user-defined heat source term. Internal energy '*e*' is defined in terms of reduced pressure as

$$e=h-\frac{p}{\rho} \tag{9.8}$$

Here, *h* is an enthalpy, and it is given as

$$h=\sum_j m_j h_j \tag{9.9}$$

Further, considering incompressible flow

$$h_j=\int_{T_{ref}}^{T} Cp_j dT \tag{9.10}$$

Where T_{ref} is a reference temperature and Cp_j is specific heat of the species j at constant pressure. Hence, total energy Equation 9.7 can be re-written in terms of enthalpy as

$$\frac{\partial}{\partial t}[\rho h]+\nabla\cdot[\rho hV]=-\nabla[p\cdot V]+\nabla\cdot\left[k\nabla T+(\tau\cdot V)-\left(\sum_j h_k j_k\right)\right]+S_h \tag{9.11}$$

9.2.1.4 Conservation of Species

The species conservation equation can be useful for HEX systems where properties of fluid vary with temperature and composition; it is given below:

$$\frac{\partial}{\partial t}(\rho Y_m)+\nabla\cdot(\rho V Y_m)=-\nabla\cdot(j_m)+S_m \tag{9.12}$$

In Equation 9.12, Y_m is the mass fraction, the first term on the LHS represents the rate of change of species per unit volume, i.e., species accumulation terms, the second term on the LHS represents the rate of change in species mass fraction due to convection per unit volume. The first term on the RHS of Equation 9.12 represents the rate of change of species mass fraction due to diffusive fluxes. Diffusive mass fluxes include diffusion due to concentration gradients and thermal gradients. The expression of diffusive mass flux is given below:

$$j_m=-\rho D_m \nabla m_k - D_{T,m}\frac{\nabla T}{T} \tag{9.13}$$

Here, D_m is the mass diffusion coefficient for species m and $D_{T,m}$ is the thermal mass diffusion coefficient

9.2.1.5 Introduction to Turbulence Modelling

Most industrial HEX systems are characterized by turbulent flow due to efficient mixing and heat transfer capabilities [3]. Turbulence is an irregular condition where velocity and other fluid properties show random fluctuations with time and space coordinates. However, the precise definition of turbulence is difficult to establish. Random fluctuations are three-dimensional special characters, unsteady, strongly diffusive, non-linear, and contain a mass of eddies of various sizes [44]. Richardson [45] introduced the concept of energy cascading in turbulence, illustrated in Figure 9.2a. Different types of eddies exist in flow and can be categorized based on spatial and temporal scales. Referring to Figure 9.2b, large-scale eddies can be termed integral eddies, carrying almost

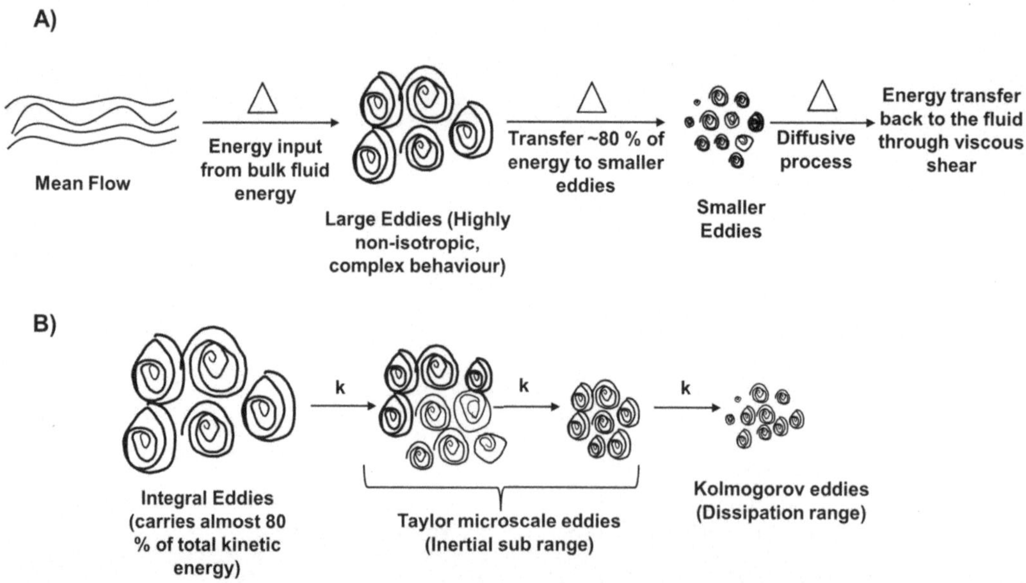

FIGURE 9.2 Concept of energy cascading and variation in size of eddies during cascading in turbulence modelling. (a) Concept of energy cascading in turbulence; (b) types of eddies in energy cascading.

80% of total kinetic energy (k) and responsible for the highest degree of mixing within the fluid, resulting in a rapid diffusive process. Vorticity interactions are responsible for kinetic energy dissipation. Dissipation of kinetic energy from integral eddies results in Taylor microscale eddies, which fall in the category of inertial subrange where inertial effects are dominant and viscous effects are negligible. Further diffusion of Taylor eddies resulted in Kolmogorov eddies. Kolmogorov length scale is the smallest eddy length scale where eddy motion is independent of mean flow and large eddies, i.e., smaller eddies are statistically isotropic and only depend on viscous effects, provided that turbulence is in equilibrium. Kolmogorov length scale can be specified in terms of the relationship between kinematic viscosity (ν) and rate of dissipation of turbulent kinetic energy (ε).

$$\eta = \left(\frac{\nu^3}{\varepsilon} \right)^{1/4} \tag{9.14}$$

Direct solution of governing equations, i.e., direct numerical simulation (DNS), is required to model turbulent flow to resolve all turbulent scales (including the Kolmogorov length scale). DNS modelling can give detailed information about turbulent interactions and all closures. However, to resolve eddies up to the smallest length scale, extremely fine mesh and the smallest time step size are required, resulting in increased computational expenses in terms of time and cost. In addition, the DNS modelling approach in HEX simulations is not feasible. Here, the concept of time averaging comes into the picture. Statistically, in turbulent flow, instantaneous local flow velocities are the addition of mean flow velocity, i.e., time-averaged flow velocity, and fluctuating velocity, which is instantaneous fluctuating velocity around mean flow velocity. Instantaneous pressure component decomposition can be done similarly.

Figure 9.3 illustrates turbulent velocity components. Applying Reynolds averaging rules to velocity and pressure and scalar components and decomposing each flow variable in governing equations results in a Reynolds averaged Navier–Stokes (RANS) equation. The RANS turbulence modelling

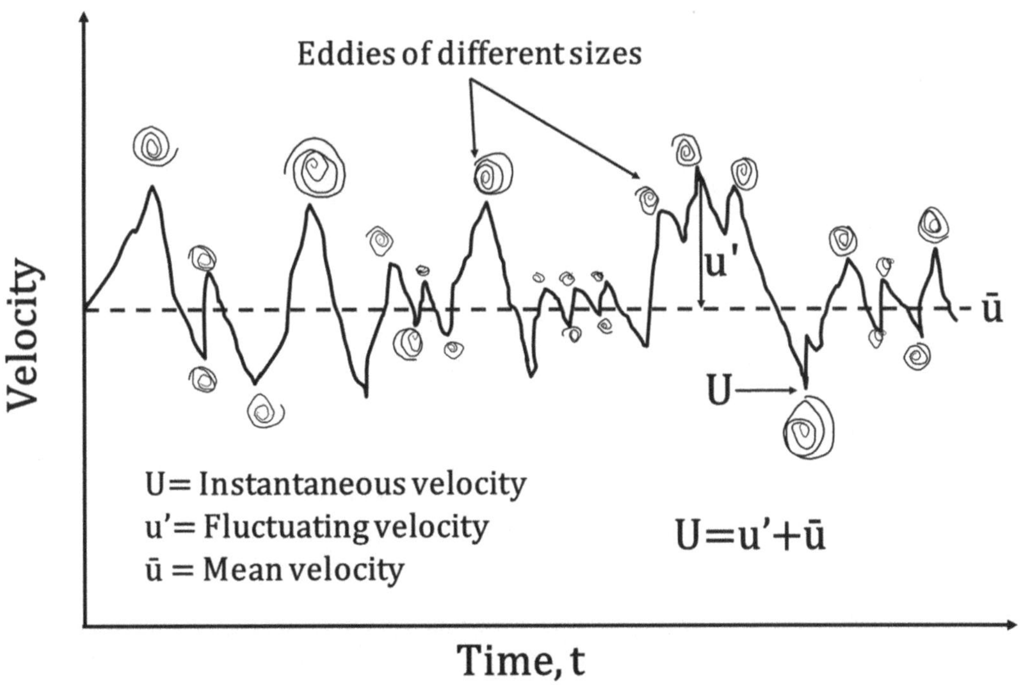

FIGURE 9.3 Turbulent velocity behaviour and components.

approach can model the entire turbulent flow without resolving it; hence, the RANS approach is computationally less expensive and more useful in the case of HEX simulations. The intermediate approach between the DNS and RANS approach is the large eddy simulation (LES) approach. In the case of the LES approach, integral eddies and Taylor microscale eddies are resolved up to a user-defined eddy size, Δ_{LES}. However, microscale eddies smaller than Δ_{LES}. are modelled. The following equation can briefly summarize the performance of RANS, LES, and DNS.

$$\begin{gathered} \text{Faster: RANS} > \text{LES} > \text{DNS} \\ \text{Detailed: DNS} > \text{LES} > \text{RANS} \end{gathered} \tag{9.15}$$

9.2.2 Multiphase Modelling Approach

Multiphase flow, especially two-phase flow, is common in most industrial HEXs, especially steam generators in nuclear or thermal power plants, and evaporators, reboilers/waste heat reboilers, and chillers in chemical industries. Multiphase flow also exists in direct contact HEX, like static mixers and stirred tank chemical reactors. Multiphase HEX systems are mainly preferred because of enhanced heat transfer rates, flexibility in handling diverse fluid phases, and improved process efficiency. A detailed classification of multiphase HEX with applications has been mentioned by Hewitt et al. [25].

Understanding the flow regime in multiphase flow modelling of HEX systems is crucial for the prediction of thermal-hydraulic performance and pressure drop. Flow regimes/patterns in multiphase HEX systems can be preliminarily classified based on thermodynamic phases such as gas-gas, liquid-liquid, gas-liquid, solid-liquid-gas, and so on. Gas-liquid system is the most common configuration in different multiphase industrial HEX systems [25]. In line with gas-liquid HEX systems, separate flow regime maps are available in the literature for vertical tubes, horizontal tubes, inclined tubes, cross-flow shell and tube HEX, and so on [25, 46, 47]. Figure 9.4 shows typical flow regime maps in vertical and horizontal tubes for the reader's understanding. CFD modelling of multiphase HEX systems is complicated due to inadequate information and understanding challenges. Knowledge of flow regime/pattern (usually expressed empirically in terms of phase velocities and volume fraction), physics of the interaction between different phases (drag-non-drag, surface tension forces), and modelling of phase-change processes (evaporation-condensation, boiling models, interfacial heat transfer correlations) involves substantial empiricism. In addition, changes in material properties and their effect on each physical parameter need to be established. Besides all these challenges, multiphase modelling (with and without phase change) is positively evolving and implemented in different industrial HEX systems [48–53].

Multiphase flow can be topologically categorized into dispersed, mixed, and separated flow, which creates HEX systems with a matrix of flow regimes and flow topology. Hence, judicious selection of models and development of CFD methodology is necessary to address complicated physics associated with multiphase HEX systems. Multiphase HEX system can be modelled using the following approaches:

1. Eulerian-Eulerian (E-E) approach. Based on interpenetrating continua of two phases, where governing equations are solved for each phase separately.
2. Eulerian-Lagrangian (E-L) approach. Tracks the particles of each dispersed phase separately. The motion of the dispersed phase is governed by the Lagrangian framework, and the continuous phase motion is governed by the Eulerian framework.
3. Volume of fluid (VOF) approach. Tracks volume fraction of each phase throughout the domain and solves a single set of governing equations with mixture properties.

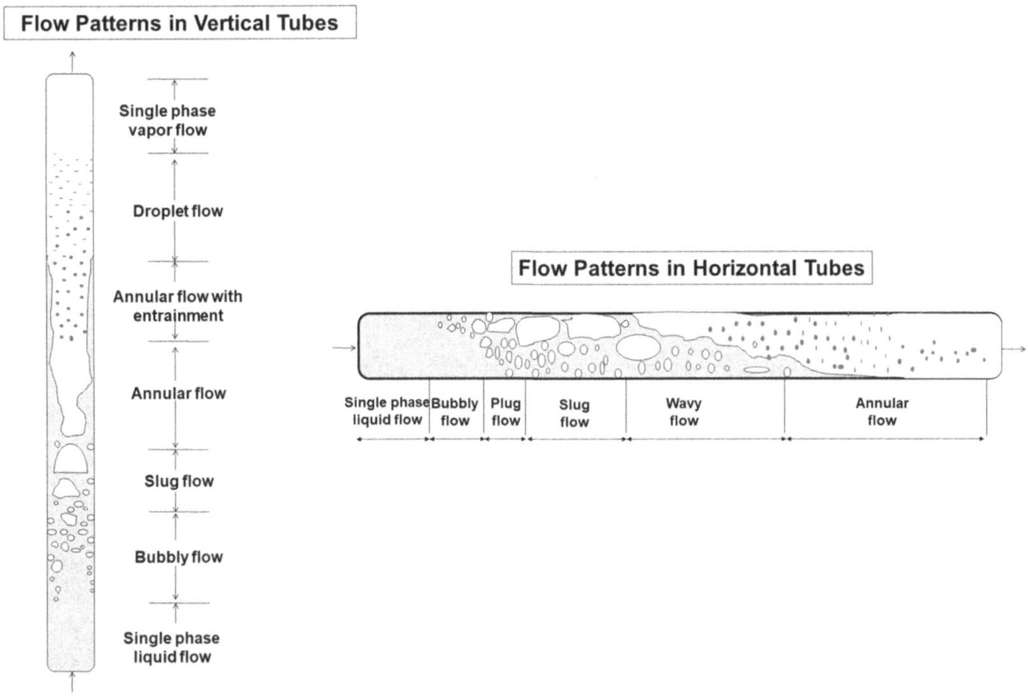

FIGURE 9.4 Typical flow regime maps in vertical and horizontal tubes [46].

Each modelling approach is discussed along with governing equations and subsets of models to gain insights and apply them in multiphase HEX systems.

9.2.2.1 The Eulerian-Eulerian Modelling Approach

The E-E approach is based on the assumption of an interpenetrating continuum in which each phase is treated as a continuous medium, i.e., both phases exist everywhere in the domain (so-called control volume) and are interpenetrating. This approach utilizes the concept of secondary phase volume fraction and governing equations, i.e., continuity, momentum, and energy conservation equations are solved separately for each phase. The concept of ensemble averaging can be used at a conceptual level to instantaneous governing equations and to address the heterogenicity and dynamic nature of the two-phase flow. This approach allows the capture of physical interaction between two phases with the help of different closure models of interphase transport. The E-E approach is more relevant for multiphase HEX systems where the dispersed phase volume fraction is significant (greater than 10 %) [54]. The concept of phasic volume fraction in the E-E approach – the volume fraction equation denotes the summation of the volume fraction of 'n' number of phases that exists in a control volume equals unity.

$$\sum_{k=1}^{n} \alpha_k = 1 \tag{9.16}$$

The governing equations for the E-E approach are as follows.

Continuity equation:

$$\frac{\partial}{\partial t}(\alpha_k \rho_k) + \nabla.(\alpha_k \rho_k V_k) = (m_{kj} - m_{jk}) \tag{9.17}$$

In Equation 9.17, subscript '*k*' stands for phase. The first term on the LHS represents the accumulation of mass for phase *k*. The second term represents mass transfer per unit volume due to convection for phase *k*. The RHS of Equation 9.17 represents interphase mass transfer, i.e., transfer of mass from phase '*k*' to phase '*j*' and vice-versa.

Momentum equation:

$$\frac{\partial}{\partial t}(\alpha_k \rho_k V_k) + \nabla.(\alpha_k \rho_k V_k V_k) = \\ -\alpha_k \nabla P + \alpha_k \rho_k g + \nabla.\left(\alpha_k \mu_k \left(\nabla V_k + (\nabla V_k)^T\right)\right) + \left(m_{kj} V_k - m_{jk} V_k\right) + F_k \tag{9.18}$$

In Equation 9.18, the first term on the LHS is the accumulation of momentum for phase '*k*, the second term represents the rate of change of momentum due to convection for phase '*k*'. The first term on the RHS of Equation 9.18 represents the pressure force term, where pressure is shared by all phases, the second term represents the gravity force, and the third term is a combination of viscous stresses and turbulent stresses. The fourth term on the RHS of Equation 9.18 represents momentum transfer between phases due to mass transfer; it is more relevant in the case of phase-change operations. The last term represents the interphase momentum exchange force due to the interaction between different phases. Knowledge of physics related to the interfacial momentum exchanger is essential for the precise modelling of HEX systems. Interfacial force is composed of drag ($F_{D,k}$) and non-drag forces such as lift force ($F_{L,k}$), wall lubrication force ($F_{WL,k}$), turbulent dispersion force ($F_{TD,k}$), and virtual mass force ($F_{VM,k}$).

$$F_k = F_{D,k} + F_{L,k} + F_{WL,k} + F_{TD,k} + F_{VM,k} \tag{9.19}$$

Table 9.1 shows a mathematical representation of different interfacial forces with a brief definition. Drag force is one of the critical forces in multiphase HEX simulations, and it directly influences performance by influencing velocity profile and pressure drop [1–3]. There are ample empirical, semi-empirical, and mechanistic correlations of drag force coefficients for a single solid particle, a single fluid particle (droplet/bubble), multi-solid particles, and multi-fluid particles that are available in the literature and are well summarized and listed in Joshi et al. [44]. Similarly, non-drag forces like lift, wall lubrication, and turbulent dispersion forces are necessary to predict precise local flow behaviour.

Energy equation:

$$\frac{\partial}{\partial t}(\rho_k \alpha_k h_k) + \nabla.(\rho_k \alpha_k V_k h_k) = \\ -\nabla \cdot q_k + (\tau_k : \nabla V_k) + \alpha_k \frac{dP_k}{dt} + \left(m_{kj} h_{kj} - m_{jk} h_{jk}\right) + Q_{kj} + S_h \tag{9.20}$$

The energy equation is written in the form of enthalpy balance in Equation 9.20. In Equation 9.20, the first term on the RHS represents a change in enthalpy due to the conduction for phase '*k*', the second term represents change in enthalpy due to viscous dissipations, the third term represents a change in enthalpy due to pressure forces, and the fourth term is change in enthalpy due to interfacial mass transfer between two phases, i.e., enthalpy change due to evaporation and condensation. The fifth term represents interfacial heat transfer between two phases. The sixth term is the source term; it can be heat-added or -removed from the system. For example, the source term can be external wall heat flux in the case of flow boiling simulations in a once-through steam generator or exothermic heat generated by a chemical reaction.

TABLE 9.1
Interfacial Forces [24, 55]

Interaction Force	Model Equation	Remarks
Drag force	$F_D = \frac{1}{2} C_D \rho_k \lvert V_j - V_k \rvert \times (V_j - V_k) A_P$ A_P is the projected area of the dispersed phase C_D is the drag force coefficient	Offers resistance to the motion of bodies in flow due to shear and viscous forces on particles
Lift force	$F_L = F_{inviscid} + F_{vortices} = -C_L \alpha_j \rho_k (V_j - V_k) \times (\nabla V_k)$ C_L is the lift force coefficient	Offers lateral motion to particles in shear flow due to the velocity gradient and negligible viscosity in a turbulent liquid flow field or shear flow
Wall lubrication force	$F_{WL} = C_{WL} \alpha_j \rho_k \lvert V_r \rvert^2 n$ C_{WL} is the wall lubrication force coefficient $\lvert V_r \rvert$ is relative velocity between two phases tangential to the wall surface n is the normal vector pointing away from the wall	Exists in gas-liquid flow. Asymmetric drainage of liquid near the wall pushes the bubble away from the wall
Turbulent dispersion force	$F_{TD} = -C_{TD} K_{jk} \frac{\mu_k^t}{\rho_k Sc} \left(\frac{\nabla \alpha_j}{\alpha_j} - \frac{\nabla \alpha_k}{\alpha_k} \right)$ C_{TD} is the turbulent dispersion force coefficient K_{jk} is the ensemble average of the fluctuating component of the drag force between two phases Sc is the turbulent Schmidt number	Accounts for turbulent fluctuations in the liquid phase and their effect on dispersed phase motion
Virtual mass force	$F_{VM} = C_{VM} \rho_k \alpha_j \alpha_k \frac{D(V_j - V_k)}{Dt}$ C_{VM} is the virtual mass force coefficient $\frac{D(\ )}{Dt}$ denotes the material derivative	Accounts for the added mass effect due to the acceleration of the continuous phase resulting from the relative acceleration between two phases

Phase-change operations like boiling, evaporation-condensation, and cavitation need to be modelled in case of multiphase HEXs. Figure 9.5 shows a phase-change modelling matrix in the E-E approach. Detailed explanation of phase-change modelling can be found elsewhere [24, 55–57].

The E-E approach can be used in a simplified form in which slip velocities between two phases are neglected, i.e., two phases travelling with the same velocities to form a classic homogeneous multiphase model. A single set of governing equations is solved for the mixture (Equations 9.17, 9.18, and 9.20 are solved for the mixture), resulting in a significant reduction of computational time and cost without compromising overall simulation accuracy. Algebraic expressions can be used to capture the slip between two phases. E-E mixture modelling can be used in HEX cases where complex phase interaction is involved, there is inadequate information on interfacial forces, and limited computational resources are available.

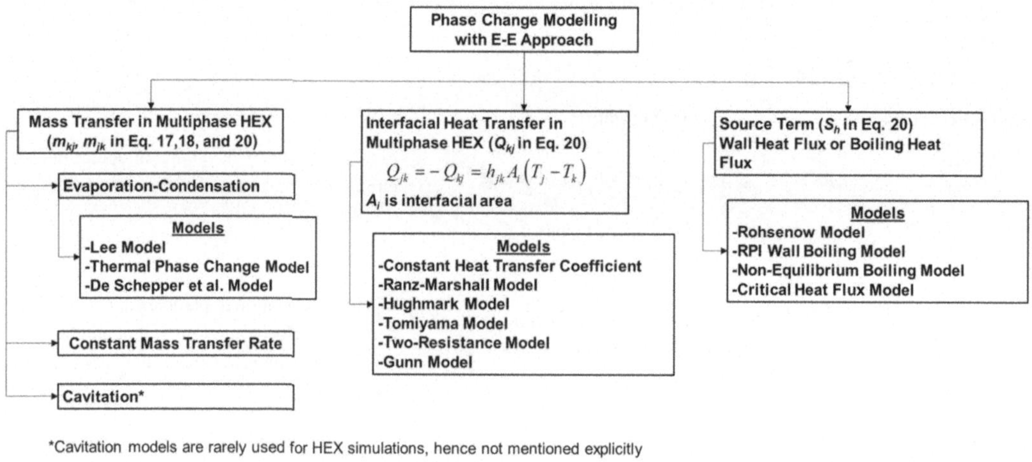

FIGURE 9.5 A phase-change modelling matrix in the E-E approach [56].

9.2.2.2 The Eulerian-Lagrangian Modelling Approach

The E-L approach is preferable in multiphase HEX simulations where the dispersed phase volume fraction is less than 10%. The E-L approach is preferable for solid particles dispersed in gas or liquid in HEX applications like spray cooling, applications, droplet flow in steam generators, condensers, boilers, fluidized bed heat exchangers, and combustion heat exchangers [58, 59]. In the E-L approach, the continuous phase is modelled by the Eulerian framework and the dispersed phase is modelled by the Lagrangian framework, where tracking of a large number of particle trajectories (with force balance) is possible. This approach can be used with and without considering particle interactions, depending on objectives. However, capturing phase interaction for a large number of particles is computationally expensive. The equation of motion of the dispersed phase can be solved for particle trajectories.

$$m_j \frac{dV_j}{dt} = F_{P,J} + F_{D,J} + F_{L,j} + F_{G,j} + F_{VM,j} \tag{9.21}$$

Force balance over a single particle is used to obtain the velocity field, which can be used for particle trajectories. Governing equations of the continuous phase can be written similarly to Equations 9.17, 9.18, and 9.20. In the E-L approach, particle flow in the flow field can be modelled based on one-way or two-way coupling. One-way coupling assumes that the effect of dispersed phase motion on the continuous phase flow field is negligible. However, in a two-way coupled approach, coupling between continuous and dispersed phases occurs through interphase momentum, heat exchange terms, and volume fraction. Mass exchange from the dispersed phase on the continuous phase can be expressed in terms of different heat and mass transfer mechanisms such as droplet vaporization, droplet boiling, and devolatilization [56]. Particle erosion arising out of particle-wall interaction is defined using the erosion model under the E-L framework. It is one of the essential applications in HEX modelling [60–62].

9.2.2.3 The Volume of Fluid Modelling Approach

The VOF approach tracks the motion of all phases in each computational cell throughout the domain, indirectly tracking the moving and deforming interface between phases. The phase volume fraction, also referred to as the advection marker function, can be used to track the motion of all phases in the domain [63]. This approach assumes that the summation of the volume fraction of phases in each computational cell in the domain is unity (refer to Equation 9.16), i.e., the computational cell either fully occupied with phase '*k*' or unoccupied with phase '*k*'; the computational cell partially occupied with '*k*' phase and other phases with interface in between. VOF solves a single set of governing equations with mixture properties (to account for fluid property change at the interface) and appropriate jump boundary conditions at the interface. The governing equation can be written similar to Equations 9.1, 9.4, and 9.11 with mixer properties. The phase volume fraction equation to track the interface between '*n*' number of phases in a particular HEX system under consideration is given as

$$\frac{\partial \alpha_k}{\partial t} + \nabla.(\alpha_k V_k) = \sum_{j=1}^{n} (m_{jk} - m_{kj}) \tag{9.22}$$

Mixture properties for a two-phase HEX system can be defined as

$$\rho_{mix} = \sum \alpha_k \rho_k \qquad \mu_{mix} = \sum \alpha_k \mu_k \tag{9.23}$$

Mass averaged quantities, especially specific heat, temperature, are defined as

$$Cp_{mix} = \frac{\sum Cp_k \alpha_k \rho_k}{\sum \alpha_k \rho_k} \qquad T_{mix} = \frac{\sum T_k \alpha_k \mu_k}{\sum \alpha_k \mu_k} \tag{9.24}$$

The effect of surface tension along the interface between phases, generally referred to as interfacial surface tension, is necessary to account for in case of compact HEXs or micro HEXs simulations with a high surface-to-volume ratio and small flow passages [64–66]. Surface tension force can be modelled with the continuous surface force model by Brackbill et al. [67], and is given as

$$F = 2\sigma \alpha_j \kappa n \tag{9.25}$$

Here, α_j is the volume fraction of dispersed phase, n is surface normal vector, and κ is surface curvature. Here,

$$n = \frac{\nabla \alpha_j}{|\nabla \alpha_j|} \quad and \quad \kappa = -\nabla n \tag{9.26}$$

The VOF modelling approach can be used in boiling and condensation applications in steam generator tubes, passive decay heat removal systems of nuclear reactors, micro-channels, and flow distribution optimization in micro-channel/compact HEXs [65, 66, 68–72]. Resolving interface is crucial in VOF simulations, which require sufficiently refined mesh and lower time step size, which is computationally expensive. However, the ability of VOF to resolve interface can be used to extract different interfacial closures required for E-E and E-L approaches. Recent advancements in multiphase CFD solvers led to the development of the multi-fluid VOF model, which combines the advantages of the E-E and VOF models and can be helpful in many HEX simulations [73–75].

9.2.3 AI/ML-Based CFD Modelling Approaches

AI-based CFD combines human intelligence techniques (such as learning, decision-making, problem-solving, etc.) with CFD modelling of the HEX system under consideration. ML is the branch of AI where systems learn from the data, map the data, and make predictions/decisions. The need for and importance of incorporating AI/ML in CFD have already been discussed. AI algorithms are categorized into supervised, unsupervised, and semi-supervised learning. These algorithms come under the single umbrella of ML and are further branched into a wide range of categories. A supervised learning (SL) algorithm aims to develop the functional relationship or mapping between input data and equivalent output labelled data based on learning from a dataset of samples with an unknown relationship between input and output labelled data (generally referred to as training data). The developed functional relationship can be used to make predictions/decisions about new data. Training or optimization of data is done by minimizing the difference between calculated and actual variables' values, represented as a loss function. SL is further categorized into classification and regression algorithms. Classification algorithms are used to categorize discrete-unstructured data into different classes based on the mapping of labelled training datasets. The filtered data in different classes can be used to develop a continuous functional relationship using regression algorithms [76–79]. Unsupervised learning (UL) does not require labelled data and a training process. Typical tasks of UL include clustering and dimensionality reduction. The clustering technique groups similar data points together based on their characteristics/feature similarity, which can be used to find unknown correlations and patterns in the dataset. Dimensionality reduction reduces the number of characteristic dimensions in the dataset without loss of information; in simple words, it can be used to simplify the data and remove noise/irrelevant information in the data, like the commonly used fast Fourier transform (FFT) technique. Semi-supervised learning (SSL) combines both labelled and unlabelled data. SSL can be used when there is limited availability of labelled data and enough availability of unlabelled data [80, 81]. Reinforcement learning (RL), a branch of SSL, can combine SL and UL with interactive learning from the environment and past experiences [76].

The background knowledge of the AI/ML framework, particularly in HEX modelling as mentioned earlier, is necessary to understand the usage of AI/ML techniques for CFD-based modelling of HEX. Commonly used AI/ML-based CFD approaches, particularly in HEX modelling, are discussed in the following sub-sections.

9.2.3.1 Integration of the Data-Driven Model with CFD

The traditional approach of designing and optimizing particular HEX systems using CFD deals with the limitation of computational expenses, complex system analysis (non-linearities and turbulence), and so on, as mentioned previously. Data-driven model (DDM) is a computational or mathematical model representing the physics of the process using an AI/ML framework based on data. The integration of DDM with CFD is a new architecture called simulation digital twin (SDT) that addresses CFD's limitations and accelerates CFD simulations. Considering the example of the design and optimization of HEX, parametric CFD studies need to be performed on HEX to generate data for the construction of DDM. The scope of the parametric studies can be defined by analyzing the parameters that influence the desired output. Accordingly, a bounded input parameter set is created, and a number of CFD simulations can be performed to generate the data. Researchers have performed parametric studies with different mathematical and statistical techniques like response surface methodology (RSM), fractional factorial, optimal Latin hypercube design (Opt LHD), Monte-Carlo sampling, Taguchi experimental design, and so on [82–85] Data generated from the CFD simulations can be used as input parameters for the AI/ML framework. Different AI/ML frameworks such as artificial neural networks (ANNs), deep neural networks (DNNs), convolutional neural networks (CNNs), random forest (RF), and polynomial regression can be used at this stage to construct DDMs [30, 82, 86–88]. The ANN is the most commonly and popularly used SL technique in HEX CFD applications [89–91].

ANNs are computational models inspired by the human brain's structure and functions. It consists of interconnected nodes of organized layers, as shown in Figure 9.6. In Figure 9.6, the input layer consists of input data from parametric studies. Input data from parametric HEX CFD simulations can be temperatures, Reynolds numbers, mass flow rates, and Prandtl numbers on the hot and cold sides of HEXs. The ANN can be applied for various engineering problems such as development of materials, solar energy systems, heat exchanger performance, and improvements [92, 93]. The input data of each input node is associated with a certain weight to study the effect of each input parameter on HEX performance. Hidden layers allow ANNs to establish linear/non-linear

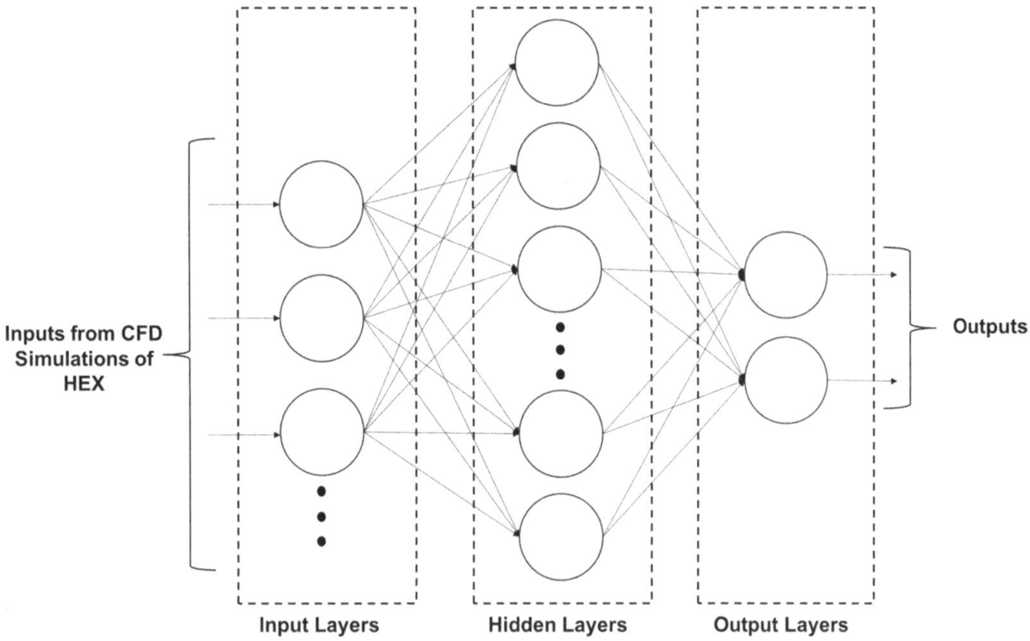

FIGURE 9.6 A typical artificial neural network structure.

relationships/patterns in the input data and extract relevant features for making accurate predictions. Construction of the ANN model requires training input and output data by optimizing weights assigned to each input node by a trial-and-error method until the required accuracy is achieved. Different training techniques like Newton's method, the quasi-Newton method, backpropagation, gradient descent, Levenberg–Marquardt, conjugate direction method, and regularization can be used to train the ANN, i.e., minimizing activation functions [94–98]. The number of hidden layers can be optimized based on the required accuracy. Output layers have required output parameters, such as heat transfer coefficients, thermal efficiency, and so on, to predict HEX performance. The established SDT can be used to evaluate new operating conditions and develop a Pareto front in case of geometrical design optimization. A typical flow chart of the ANN framework can be found in Aradag et al. and Mohanraj et al. [15, 91]. Integration of the DDM with CFD can be found relevant to a wide range of HEX applications such as (1) modelling boiling heat transfer in the pool and flow boiling scenarios [88, 99]; (2) development of a thermal simulation model in an air-to-earth HEX [100]; (3) Nusselt number correlation development in the case of a fin-tube bundle HEX [101]; (4) prediction of the thermal-hydraulic behaviour of a cooling fin [30]; (5) prediction of thermal-hydraulic performance of a gasketed plate HEX; and (6) performance evaluation of a micro-channel HEX [16]; and many more.

9.2.3.2 Genetic Algorithm-Based CFD Approach

Apart from the ANN approach, a genetic algorithm (GA) can be relevant in many AI/ML-based CFD modelling of HEXs. GA is an SL/SSL optimization technique for cases involving a broad spectrum of complex data and multiple objectives. The typical workflow of GA is shown in Figure 9.7. Considering the geometrical design optimization of plate and fin type HEXs, an initial population of possible solutions includes different sets of geometrical parameters (x) such as fin height, fin pitch, fin thickness, and so on. Design of experiments (DOE) studies can be performed to obtain the most crucial/relevant parameters and establish a relationship between design parameters and objective functions for optimization study, which reduces the number of CFD simulations and computational efforts. In general, DOE studies reduce the search space by putting constraints. However,

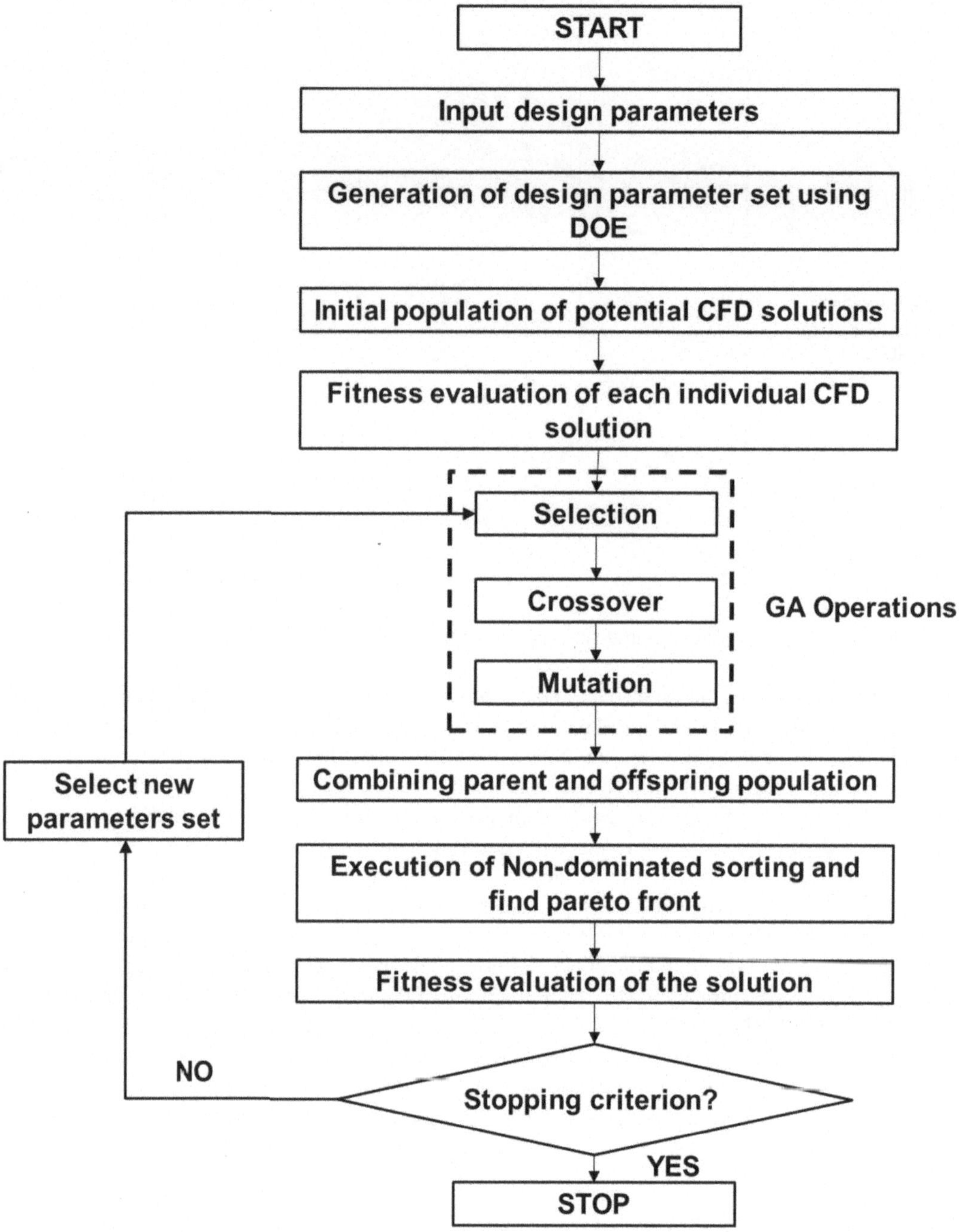

FIGURE 9.7 Workflow of a genetic algorithm.

more constraints result in a loss of feasibility and freedom of movement in the parametric space [102]. Mathematical and statistical techniques used to perform DOE are mentioned in earlier sections. A fitness evaluation of each geometrical parameter set (xi) can be done by evaluating objective functions (for example, Nusselt number '*Nu*', Reynolds number 'Re', and friction factor '*f*' evaluation for different sets of geometries) through CFD simulations.

GA operations consist of selecting an elite set of geometric parameters, which can also be referred to as a parent set based on fitness value (the maximum value of '*Nu*' and minimum value of '*f*' is desired in HEX optimization). Crossover refers to the exchange of information between selected sets to create offspring, and mutation refers to random changes introduced into offspring genetic information. Crossover and mutation over the parent set create a new parametric set called the offspring population. Single-point, multipoint, and simulated binary crossover are the most commonly used crossover techniques in the case of GA-based optimization of HEXs [102–104]. Objective functions are again evaluated for offspring members, i.e., new parameter set using CFD simulations. Then, the offspring population is combined with the new population. In HEX applications, GA is advantageous for multi-objective geometrical design and mesh optimization studies of HEXs [84, 105, 106]. Multi-objective optimization refers to the simultaneous improvement/optimization of all objective functions. Simultaneous optimization involves combining multiple objectives into a single objective by assigning weights. Second-generation non-dominated sorting genetic algorithm (NSGA-II) is a popular GA used for the multi-objective optimization of HEXs. The name defines sorting out and eliminating non-dominated parents and offspring population. The NSGA-II method is based on Pareto optimality, which involves generating a Pareto front, i.e., a set of Pareto optimal solutions (80% outcomes from 20% inputs) through genetic modifications such as crossover and mutation operations. The final outcome of NSGA-II is to find the optimal Pareto front for the desired objective function representing an optimum set of geometric parameters for a particular type of HEX. Finally, check for the condition of the desired criterion of optimization; if the criterion is not satisfied, select a new parent set of parameters and perform the GA operations. The GA-based CFD approach is used for geometrical optimization of plate and fin HEXs [84, 107], optimization of helical baffles of shell and tube HEX [108], optimization of Nusselt number and pumping power in micro-channel heat sinks [109], and so on.

9.2.3.3 Hybrid ANN-GA-Based CFD Approach

GA-based CFD optimization requires multiple CFD simulations to evaluate fitness function, which is computationally expensive in terms of time and resources, especially in the case of 3D-transient simulations of HEXs [105, 110]. In addition, GA techniques used for multi-objective optimization are stochastic in nature, which may lead to uncertainty in finding possible trade-offs or global optima effectively [111, 112]. To address the limitations of GA-based CFD optimization and to obtain an efficient-robust global optimal solution, a comprehensive hybrid approach integrating CFD, ANN, and GA is introduced for multi-objective optimization of HEXs and reactor applications [85, 113–115]. Combining ANN with GA also leverages the strength of both methods. The ANN-GA-based CFD approach can be explained by considering the example of the design and optimization of a shell and tube HEX with staggered baffles by Saijal and Danish [116]. Parametric CFD simulations are performed on five design parameters of HEX to study their effect on objective outputs, i.e., minimize the pressure drop ΔP and maximize the heat transfer Q. Taguchi experimental design DOE technique is used to reduce the number of CFD simulations without compromising the optimization activity. Two ANNs are constructed for two desired output parameters ΔP and Q. The input layer of ANN consists of five design parameters, and the output layer consists of one output parameter; 70% of the data obtained from parametric CFD simulations is used for training, 15% is used for validation of ANN, and the remaining 15% is used for the testing of the ANN. Trained ANNs of ΔP and Q are used as an objective function for multi-objective optimization using NSGA-II. ANN-based fitness functions, i.e., $\Delta P = \min\{ANN_{\Delta P}(design\ parameters)\}$ and $Q = -\min\{ANN_Q(design\ parameters)\}$, are created and runs the design input parameters through both the objective functions; accordingly, the parent population is selected for GA operations. Crossover and mutations are performed by the tuning of mutation rate, crossover rate, and number of populations. GA operations are performed until the optimal solution of the Pareto front for ΔP and Q corresponding to optimal design parameters is generated (refer to Figure 9.7 for the working of the GA).

In a recent development, a physics-informed neural network (PINN) combined with deep and transfer learning (TL) was used to solve partial differential equations (PDEs) and ordinary differential equations (ODEs) of governing equations of HEX systems [37]. This hybrid science-guided ML approach accelerates the HEX design by removing the need for mesh generation and unwanted training time (through transfer learning), resulting in stochastic optimization of HEX. Though the PINN-TL approach provides reliability and predictability, it is associated with algorithmic complexity, and successful implementation requires expertise both in ML and the physical domain. Though meshless simulations are a current trend of interest, mesh-based CFD simulations are still inevitable in HEX simulations. The information related to mesh and boundary conditions for CFD is discussed in the upcoming section.

9.3 A PREVIEW OF CFD MESH AND BOUNDARY CONDITIONS FOR HEX SIMULATIONS

9.3.1 Computational Mesh

A CFD mesh is a computational domain, i.e., finite volume comprising discrete computational points or nodes. A computational mesh to solve numerical simulation is analogous to sensors that analyse the actual physical systems. A good quality mesh is needed to reflect actual physics and precisely represent the system behaviour, especially in the case of HEX simulations, where capturing flow and heat transfer behaviour is crucial. There are several parameters that can be used to assess mesh quality. However, aspect ratio, skewness, and orthogonal quality are commonly used to assess mesh quality in practice. The aspect ratio is the ratio of the maximum edge length to the minimum edge length [22, 117]. Aspect ratio ensures a smooth transition of flow variables between nodes and jumps, or large gradients should be avoided. Orthogonal quality measures the angle between cell faces and cell edges, and the orthogonal quality of one is ideal [56]. Skewness measures the deviation of the shape of the given element from the perfectly hexahedral element of the same volume. Skewness is one of the crucial mesh quality parameters responsible for numerical instabilities and errors; hence, a mesh with skewness values less than 0.9 is always preferred for CFD simulations [22, 117]. Mesh can be divided into structured and unstructured mesh. Figure 9.8 shows different types of meshes used for HEX simulations. Hex-core and poly-hex-core meshes maintain hexahedral cells at the core and tetrahedral or polyhedral cells near the wall.

Nowadays, polyhedral mesh is commonly preferred as it combines the advantages of hexahedral and tetrahedral cells. Polyhedral mesh offers less numerical diffusion, resulting in increased precision, reduction in cell count, numerical stability, and well-approximated gradients, particularly in the case of HEX simulations [118]. RANS turbulence modelling approach is the most preferred in the case of HEX simulations. In the case of heat transfer simulations, capturing the near-wall effect is crucial to account for the thicknesses of the hydrodynamic boundary layer and thermal boundary layer in the case of the RANS modelling approach for a precise solution. Hence, the criterion of $y+$ ($y^+ = \left[(\rho\sqrt{\tau_w/\rho}y)/\mu\right]$, here τ_w is wall shear stress) is referred to for grid allocation near the wall. Wilcox and others [119] recommended $y+=5$ for the first computational node near the wall in the case of the k-ω turbulence model to capture viscosity-affected regions (viscous sublayer and buffer layer) near the wall (and capture the wall effects). Different RANS turbulence models have different $y+$ criteria. The first grid from the wall should be at $y+>30$ for the standard and RNG k-ε models [120, 121].

Similarly, $y+<1.5$ is preferred for the realizable k-ε model. HEX simulations are more likely to deal with complex and large flow domains, resulting in a large mesh count. Hence, optimization of grid size is necessary. Figure 9.9 shows biasing in mesh and inflation layers near the wall. Mesh biasing can be used near the wall to maintain the desired $y+$ and reduce computational nodes, as shown in Figure 9.9a. Hinge and Patwardhan [21] used inflation layers near the wall of the wire coil insert to capture the wall effects, and coarse mesh in the rest of the domain in the case of heat transfer augmentation studies shown in Figure 9.9b.

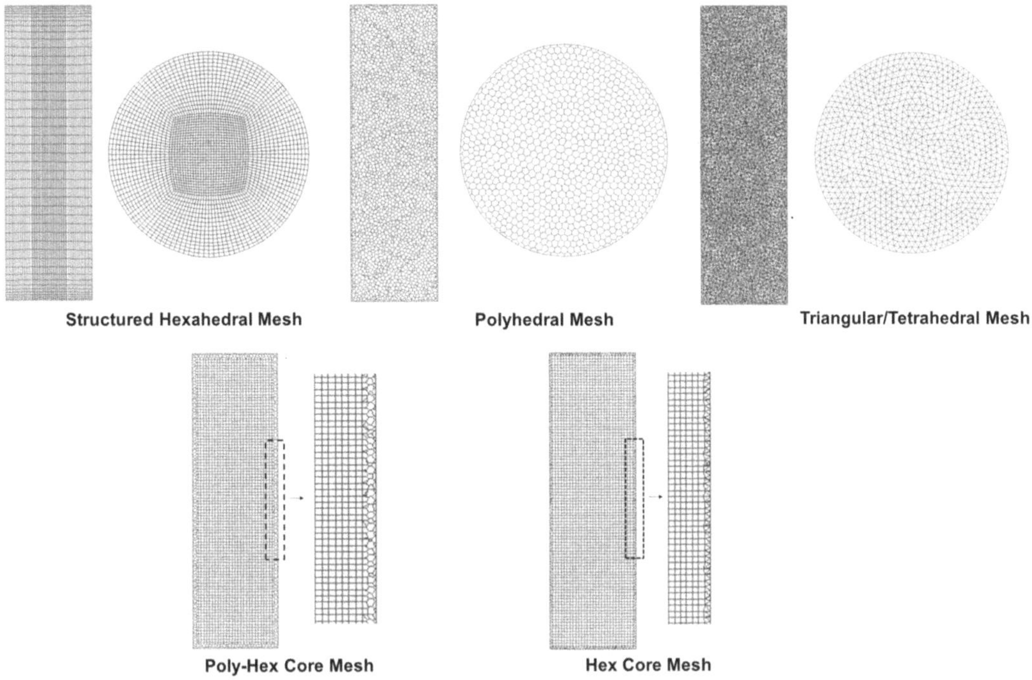

FIGURE 9.8 Different types of meshes used for HEX simulations.

9.3.2 Boundary Conditions and Setup

Understanding the underlying physics targeting the particular objectives of HEX design and optimization is crucial for effective CFD modelling. Hence, the appropriate selection of boundary conditions and initial conditions highlighting the underlying physics plays a vital role. Incorrect boundary and initial conditions immediately result in code crashes or red flags similar to possibilities of divergence due to poor mesh quality. In the case of HEX simulations, initial flow and heat transfer conditions to satisfy desired objectives of design and optimization can be achieved with different types of boundary conditions. Commonly used boundary conditions are given below.

Inlet boundary conditions. Information about velocity or mass flow rates, temperatures, and other variables, such as volume fraction and species fraction, can be given at the inlets (hot and cold side inlets) of the HEX system. Velocity and mass flows should be specified normal to the inlet surface. The inlet piping effect from bends and pipe fittings can be minimized using knowledge of fully developed flow. Oil and gas industry simulations prefer considering 10D upstream piping length to achieve fully developed flow at the inlet [123, 124]. The pressure inlet boundary condition can be used when information about the inlet velocity component is unknown.

Outlet boundary conditions. Outlet boundary conditions are used to model the flow exit. The pressure outlet is a commonly used outlet boundary condition in most HEX simulations [21, 125] where static pressure is defined at the outlet boundary. Other variables, such as volume fraction and temperature, can also be specified at the outlet in case of backflow. Outflow boundary conditions can also be used at the outlet when details of pressure and velocity are unknown. The outflow boundary condition assumes zero gradient (normal to outlet boundary) for all variables except pressure.

Wall boundary conditions. The no-slip boundary condition is normally employed at the wall, which is nothing but fluid wholly attached to the wall, and the fluid velocity equals the velocity of the wall (i.e., zero shear). For HEX simulations, thermal conditions can be specified on the wall in the form of a fixed temperature/temperature profile at the wall surface, specific heat flux, known value of heat transfer coefficient and free stream temperature, external temperature, and

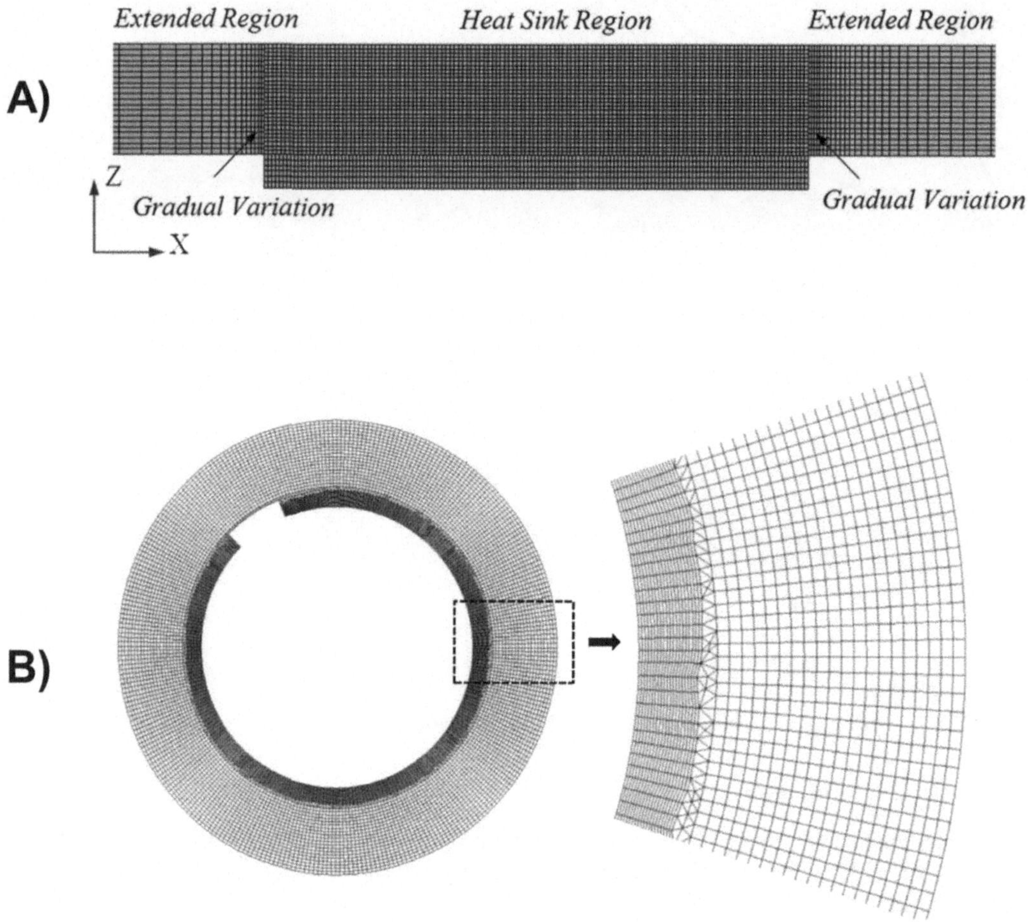

FIGURE 9.9 (a) Mesh biasing in a plate and frame heat sink [122]; (b) inflation layers to capture wall effects [21].

so on. External emissivity and radiation temperatures can be defined at the wall when radiation is involved. In the case of E-L simulations, different boundary condition types such as trap, reflect, and escape need to be specified at the wall. In addition, roughness parameters and erosion modelling parameters need to be specified at the wall in case of erosion prediction in the HEX.

Symmetry/periodic boundary conditions. HEX simulations deal with complex and large domain sizes, resulting in large mesh counts and increased computational costs, as mentioned earlier. Domain size can be reduced/optimized by identifying geometric symmetry or geometric periodicity/repetitiveness in the flow field, and symmetry and periodic boundary conditions can be used in such scenarios without compromising underlying physics. The symmetry boundary condition considers zero normal velocity gradient at the symmetry plane/axis and zero gradients of all variables except velocity at the symmetry plane. Hence, there are no convective and diffusive fluxes across the symmetry plane. Mali et al. [24] used axis-symmetric boundary conditions for modelling flow boiling in steam generator tubes. Two-dimensional axis-symmetric simulations can be used at the initial stage for model development and testing and to generate proof of concept studies. Similarly, a periodic boundary condition is a cyclic boundary condition where rotational or translational periodicity in the geometry can be addressed. Hinge and Patwardhan [21] used periodic boundary condition for a heat transfer augmentation study with wire coil inserts to reduce the domain size and expedite the simulation.

Apart from choosing appropriate boundary conditions, using appropriate numerical methods and relaxation parameters is essential for a converged stabilized numerical solution. Detailed information on numerical methods for flow and heat transfer can be found in Patankar [126]. An appropriate time step size setting is far more critical for a converged stabilized solution. Courant flow number (CFL) is used to guide the time step size to avoid truncation or numerical errors arising out of a large time step size. An ANN-based CFD expert system developed by Rubio et al. [127] is used to optimize overall computational time by making precise predictions about the number of iterations and time required for the convergence of aircraft systems where the input layer contains CFL, mesh size, number of processors, and Mach number.

9.4 APPLICATIONS OF CFD IN HEXS

This section highlights the applications of CFD in different HEX designs and optimization along with CFD-based AI/ML applications of HEXs. Thermal-hydraulic performance analysis using CFD involves predicting crucial parameters of HEXs, mainly HTC, friction factor, fouling factor, pressure drop, temperature and flow distribution, and so on. In multiphase HEXs, prediction of void fraction, flow regimes, boiling and condensation characteristics, and critical heat flux (CHF) is crucial. Hence, applications of CFD modelling in HEX are discussed according to the prediction of crucial HEX parameters with the help of selective literature studies.

9.4.1 Prediction of the Heat Transfer Coefficient

The literature shows that in experiments and modelling, both approaches are applied to study heat transfer enhancement and predict the heat transfer coefficient [128, 129]. Alperen and Sertec [109] developed the Nusselt number '*Nu*' correlation for micro-channel heat sinks using CFD-based multi-objective optimization using GA (NSGA-II). The authors observed that the Reynolds number (Re=500) and aspect ratio (micro-channel height to width ratio, 2.13) are crucial parameters for estimating *Nu* from NSGA-II optimization, where CFD plays a significant role in executing the objective function *Nu*. As heat flux is applied at the bottom of the heat sink, a lower aspect ratio results in a higher value of *Nu* at all Re. Similarly, Daneshparvar and Beigzadeh [108] optimized helical baffles in shell and tube HEX with parametric (helix angle, baffle pitch, and mass flow rate) CFD simulations. They established the optimized correlations of HTC correlation and pressure drop, where the constants of the correlations are obtained from GA-based multi-objective optimization. The authors observed that increased baffle pitch results in increased HTC due to increased fluid collision with shell walls and baffles. Liu et al. [84] also used NSGA-II to optimize objective functions, such as the Colburn factor (depending on *Nu*, Re, and *Pr*) and friction factor, using design parameters of a plate and fin HEX. DOE-backed CFD simulations were performed to execute objective functions. Analysis of Pareto optimal design parameters showed a 12.83% increase in Colburn factor and a 26.91% decrease in friction factor. Kanaris et al. [17] demonstrated the effectiveness of CFD simulations to develop *Nu* and friction factor (with ±10% deviation) for a coregulated plate HEX. The authors studied the effect of different corrugation shapes on predicting friction factors and heat transfer augmentation with successful experimental validation. Their Figure 1 shows the velocity streamlines and Nusselt number contours results in a corrugated plate HEX obtained from CFD simulations. Corrugations on plates generate swirling flow and are responsible for turbulence shear stress and generation, resulting in increased *Nu*. Shi et al. [130] accurately predicted HTCs for nucleate boiling and post-dry-out conditions inside the vertical pipe of a once-through steam generator. HTC predicted from numerical simulations showed a maximum 17.9% deviation with results obtained from empirical correlations. The authors observed increased bubble density near the wall in the case of nucleate boiling conditions that were responsible for the increase in HTC; a similar observation was highlighted by Mali et al. [24] in the case of CFD modelling of full-range flow boiling phenomenon. Shabanian et al. [131] performed an experimental and CFD evaluation of an air-cooled HEX equipped with classic, butterfly, and twisted tape inserts to predict HTC and friction factors. This study is a classic example of the combination of CFD studies of direct contact type HEX and air-cooled HEX, but it is more focused on the different inserts' heat transfer augmentation effect. CFD results showed that butterfly type inserts provide enhanced heat transfer augmentation compared with other types of inserts due to increased turbulent intensity resulting from the swirling effect. Studies mentioned in this sub-section highlight the capability of CFD modelling in predicting HTC for different types of HEXs.

9.4.2 Prediction of Friction Factors and Pressure Drop

Pal et al. [132] evaluated the shell side flow dynamics of shell and tube HEXs with and without baffles using CFD; the authors studied the effect of shell diameters on velocity and temperature distribution. The authors observed that the predicted shell side pressure drop agrees with that predicted by the conventional, popularly used Bell–Delaware method with ±10% deviation. It can be suggested that baffle spacing, or the number of baffles, be decreased to improve flow distribution over the tube bundle, which will increase pressure drop. El Maakoul et al. [133] developed a friction factor correlation for a helical baffle annulus type of double-pipe HEX using CFD. The authors analyzed the pressure drop for different baffle spacings. They observed that a significant contribution to pressure drop arises from maximum velocity for low baffle spacing, resulting in an average friction factor on the annulus side, i.e., the dominant dynamic component of pressure drop. Abeykoon [5] highlighted the importance of CFD in the design and optimization of a number of tubes and baffles of a single-pass shell and tube HEX to develop a compact HEX. The author considered six design and optimization cases for CFD based on the preliminary theoretical design of a HEX. The author found 1.81% and −8.03% deviation in pressure drop predicted by CFD with theoretical correlations. Salimi et al. [134] stated higher accuracy of ANNs compared with GA in predicting friction factor (0.139% mean relative error for the ANN and 4.104% for GA) in twisted tape inserts in serpentine channels. Geometrical parameters, such as pitch length hydraulic diameter, are considered as input to ANNs and GA to predict the objective/target function of friction factor and *Nu*. CFD played a crucial role in the execution and validation of the objective function friction factor. Gullapalli and Sundén [135] effectively used CFD for pressure drop prediction in a compact brazed plate HEX, where the effect of different turbulence models on pressure drop was studied, and observed pressure drop predictions were independent of the turbulence model used. Altaf et al. [18] extensively used CFD to optimize the thermal performance of plate-fin heat sinks, and different heat sink geometries were considered for evaluation. The authors observed that an increase in slots results in approximately a 9% reduction in pressure drop across the heat sink. Mali et al. [24] developed a subcooled nucleate flow boiling model with an appropriate section of boiling parameter models and phase interaction models to predict pressure drop across a 13.4 m vertical tube at 15 MPa pressure. Predicted two-phase pressure drop results are successfully validated with experimental data of Liu et al. [136] with a 2–10% deviation. The authors exclusively mentioned that pressure drop prediction in subcooled nucleate boiling significantly depends on the prediction of the vapour volume fraction.

9.4.3 Prediction of Fouling

Fouling in HEXs reduces heat transfer efficiency, increases pressure drops, and increases pumping power, increasing operational costs. Hence, analysis and prediction of fouling is essential for maintaining HEX efficiency and longevity. HEXs in food processing industries are more likely to be subjected to fouling issues due to the deposition of food material on the heating surface. Choi et al. [137] carried out 3D CFD simulations to predict fouling inside a plate and frame HEX during the milk pasteurization process (heating and cooling of milk for a certain period to eliminate microorganisms present). The user-defined function (UDF) of the reaction kinetics model representing the fouling reaction scheme is incorporated into the solver to predict fouling. Results showed that the predicted fouling rate of CFD matches the experimental results with approximately 7% deviation. In addition, mass fraction contours of deposited protein on the plate and frame HEX surface showed an analogy with temperature distribution as a characteristic of the Arrhenius equation. Tang et al. [138] numerically investigated fly ash deposition on industrial waste heat recovery HEX tube bundles. The authors performed complex two-phase E-L simulations (with DPM simulations with the discrete random walk model) and a fly ash fouling model to simulate fly ash behaviours such as collision, rebound, deposition, and removal. In addition, the authors used a dynamic mesh approach to predict fouling morphology and study the effect of fouling on the flow field and particle trajectories. The authors mentioned that the fouling rate reduced with particle diameter and higher flue gas velocity due to smaller deposition and a more significant removal rate. The numerically predicted fouling mass was successfully validated with experimental observations. Xu et al. [139] used the E-E two-fluid and particle deposition models to simulate particle fouling characteristics on the furnace tube. Analysis of the deposition height for different sets of operating conditions showed a maximum deviation of 15.58% from experimental results. The authors explicitly studied the effect of particle deposition mechanisms. They stated the dominance of Brownian, eddy deposition, and thermophoretic force on smaller particles due to the irregular thermal motion of smaller particles. However, larger particles were greatly influenced by gravity forces. Liu et al. [140] performed subcooled flow boiling CFD simulations (using the E-E model in combination with the wall heat flux partitioning model) to study the effect of different fouling layer thicknesses on crucial parameters such as HTC, vapour fraction, and pressure drop of two-phase flow boiling in a vertical tube. The authors observed a 50 times reduction in *Nu* due to the deposited fouling layer in the case of subcooled flow boiling. Paz et al. [141] developed a CFD methodology to predict the thickness of fouling layer deposition in an exhaust gas recirculation system (used to reduce NO_x emissions) with integration of an external algorithm in the CFD solver. Sequential utilization of semi-empirical fouling and dynamic mesh models was used to recreate the evolution of the depositing fouling layer by capturing the precise movement of fouling – the fluid interface. The authors observed the highest fouling thickness in the regions of low wall shear stress and high thermal gradients. Fouling prediction based on different AI techniques is also reported by different researchers where experimental data is used for training and validation in most HEX studies [142, 143]. It can be seen from different studies that researchers used different ways and techniques to predict fouling in HEXs, and methods are still evolving.

9.4.4 Prediction of Multiphase Parameters

Apart from the prediction of HTC, fouling, pressure drop for HEX design and optimizations, prediction of volume fraction, and CHF, flow regimes are crucial in the case of multiphase HEXs, especially in steam generators, evaporators, and boilers. Precise prediction of these parameters is essential to ensure the operational safety of the HEX. Selective studies showing the capabilities of CFD and AI/ML-based CFD in the prediction of multiphase parameters in HEX are mentioned in this section. Liu et al. [144] vigorously explored the capability of the E-E two-fluid model and wall boiling models for the prediction of subcooled nucleate boiling (prediction of vapour distribution and temperature), CHF, dry out, and post-dry out (prediction of wall temperature/wall heat flux) in vertical tubes. Zhang et al. [145] used a similar approach for predicting CHF in vertical tubes, where the authors observed less than 15% deviation with experimental results. Mali et al. [24, 55] made substantial efforts to predict thermal hydraulics and pressure drop characteristics in vertical steam generator tubes. The authors developed a new methodology to model the full-range flow boiling phenomenon, which covers all flow regimes in a 23 m long vertical steam generator tube at high pressure (172 bar). The authors highlighted the accuracy of the CFD model over conventional 1D models in predicting thermal-hydraulic characteristics. Catsoulis et al. [99] used a CFD-based DDM approach, where different ML algorithms were used for predictive modelling based on data generated from CFD parametric studies. The authors developed a workflow to produce an SDT and simulated different multiphase flow cases. In particular, the authors precisely developed an SDT to predict the vapour volume fraction and gas velocity profile for the experimental results of DEBORA flow boiling. They highlighted the advantage of DDM in CFD by evaluating the required computational speed, where they observed a speed up in the efficiency by 2.1e+5.

9.5 SUMMARY

This chapter highlighted the need and advantages of CFD for HEX design and optimizations. In addition, an association of AI/ML techniques with CFD in HEX modelling to address the limitations of CFD in terms of computational expenses, precision, and operation was pointed out. Different approaches for modelling HEXs and the information on governing equations describing underlying physics were discussed. A single-phase modelling approach with turbulent flow is the most common and widely used in HEX applications. Modelling turbulence is essential for quantitative predictions in HEXs and captures precise flow physics. The basics of the turbulence phenomenon, along with the essence of different turbulence modelling approaches, were discussed in this chapter. Multiphase HEX systems like boilers, evaporators, and waste heat recovery units are preferred because of enhanced heat transfer rates and improved process efficiency. E-E, E-L, and VOF approaches for modelling multiphase flow were explained with closures of phase interaction and interfacial mass transfer. AI/ML-based CFD modelling requires an understanding of fundamental algorithms (supervised, unsupervised, and semi-supervised learning) and knowledge of the AI/ML framework; based on this background information, DDM-based, GA-based, and hybrid ANN-GA-based CFD approaches were explained in detail with applications in HEX modelling. However, AI/ML-based CFD is also limited in data mining, training time, data consistency, and ease of use. The pragmatic CFD approach in HEX involves appropriateness in mesh and boundary conditions. Information about different types of meshes and boundary conditions was explicitly mentioned as a crucial part of CFD execution. Finally, CFD and AI/ML-based CFD applications in HEX design and optimizations were discussed to highlight CFD capability.

REFERENCES

1. D.P. Sekulic, R.K. Shah, *Fundamentals of heat exchanger design*, John Wiley & Sons, 2023.
2. S. Kakaç, H. Liu, A. Pramuanjaroenkij, *Heat exchangers*, CRC Press, 2002. https://doi.org/10.1201/9781420053746.
3. T.L. Bergman, A.S. Lavine, F.P. Incropera, D.P. DeWitt, *Introduction to heat transfer*, John Wiley & Sons, 2011.
4. M.M. Aslam Bhutta, N. Hayat, M.H. Bashir, A.R. Khan, K.N. Ahmad, S. Khan, CFD applications in various heat exchangers design: A review, *Appl Therm Eng* 32 (2012): 1–12. https://doi.org/10.1016/j.applthermaleng.2011.09.001.
5. C. Abeykoon, Compact heat exchangers – Design and optimization with CFD, *Int J Heat Mass Transf* 146 (2020): 118766. https://doi.org/10.1016/j.ijheatmasstransfer.2019.118766.
6. S. Kumar, R.B. Grover, P.K. Vijayan, U. Kannan, Numerical investigation on the effect of shrouds around an immersed isolation condenser on the thermal stratification in large pools, *Ann Nucl Energy* 110 (2017): 109–125. https://doi.org/10.1016/j.anucene.2017.06.022.
7. S. Kumar, P.K. Vijayan, U. Kannan, M. Sharma, D.S. Pilkhwal, Experimental and computational simulation of thermal stratification in large pools with immersed condenser, *Appl Therm Eng* 113 (2017): 345–361. https://doi.org/10.1016/j.applthermaleng.2016.10.175.
8. H.K. Versteeg, *An introduction to computational fluid dynamics the finite volume method, 2/E*, Pearson Education India, 2007.
9. F. Azizi, W. Abou-Hweij, N. Lebaz, N. Sheibat-Othman, A numerical evaluation of flows through an SMX-Plus mixer, *Chem Eng Res Des* 178 (2022): 382–394. https://doi.org/10.1016/j.cherd.2021.12.030.
10. B. Hasanpour, M.S. Irandoost, M. Hassani, R. Kouhikamali, Numerical investigation of saturated upward flow boiling of water in a vertical tube using VOF model: Effect of different boundary conditions, *Heat and Mass Transfer* 54 (2018): 1925–1936. https://doi.org/10.1007/s00231-018-2289-3.

11. X. Zhao, M. Wang, G. Wu, J. Zhang, W. Tian, S. Qiu, G.H. Su, The development of high fidelity Steam Generator three dimensional thermal hydraulic coupling code: STAF-CT, *Nucl Eng Technol* 53 (2021): 763–775. https://doi.org/10.1016/j.net.2020.07.043.
12. A. Klimanek, M. Cedzich, R. Białecki, 3D CFD modeling of natural draft wet-cooling tower with flue gas injection, *Appl Therm Eng* 91 (2015): 824–833. https://doi.org/10.1016/j.applthermaleng.2015.08.095.
13. A.A. Hussain, B. Freegah, B.S. Khalaf, H. Towsyfyan, Numerical investigation of heat transfer enhancement in plate-fin heat sinks: Effect of flow direction and fillet profile, *Case Stud Therm Eng* 13 (2019): 100388. https://doi.org/10.1016/j.csite.2018.100388.
14. F.C.C. Galeazzo, R.Y. Miura, J.A.W. Gut, C.C. Tadini, Experimental and numerical heat transfer in a plate heat exchanger, *Chem Eng Sci* 61 (2006): 7133–7138. https://doi.org/10.1016/j.ces.2006.07.029.
15. S. Aradag, Y. Genc, C. Turk, Comparative gasketed plate heat exchanger performance prediction with computations, experiments, correlations and artificial neural network estimations, *Eng Appl Comput Fluid Mech* 11 (2017): 467–482. https://doi.org/10.1080/19942060.2017.1314870.
16. C. Kamsuwan, X. Wang, L.P. Seng, C.K. Xian, R. Piemjaiswang, P. Piumsomboon, K. Manatura, M. Kaewbumrung, Y. Pratumwal, S. Otarawanna, B. Chalermsinsuwan, Enhancing performance of polymer-based microchannel heat exchanger with nanofluid: A computational fluid dynamics-artificial neural network approach, *S Afr J Chem Eng* 46 (2023): 361–375. https://doi.org/10.1016/j.sajce.2023.09.001.
17. A.G. Kanaris, A.A. Mouza, S. V. Paras, Flow and heat transfer prediction in a corrugated plate heat exchanger using a CFD code, *Chem Eng Technol* 29 (2006): 923–930. https://doi.org/10.1002/ceat.200600093.
18. K. Altaf, A. Tariq, S.W. Ahmad, G. Hussain, T.A.H. Ratlamwala, H.M. Ali, Thermal and hydraulic analysis of slotted plate fins heat sinks using numerical and experimental techniques, *Case Stud Therm Eng* 35 (2022): 102109. https://doi.org/10.1016/j.csite.2022.102109.
19. E. Ozden, I. Tari, Shell side CFD analysis of a small shell-and-tube heat exchanger, *Energy Convers Manag* 51 (2010): 1004–1014. https://doi.org/10.1016/j.enconman.2009.12.003.
20. K. Derakhshanpour, R. Kamali, M. Eslami, Effect of rib shape and fillet radius on thermal-hydrodynamic performance of microchannel heat sinks: A CFD study, *Int Commun Heat Mass Transfer* 119 (2020): 104928. https://doi.org/10.1016/j.icheatmasstransfer.2020.104928.
21. S.P. Hinge, A.W. Patwardhan, Thermal-hydraulic performance of an annular pipe with square wire coil inserts using computational fluid dynamics, *Ind Eng Chem Res* 59 (2020): 3887–3903. https://doi.org/10.1021/acs.iecr.9b04192.
22. B. Andersson, R. Andersson, L. Håkansson, M. Mortensen, R. Sudiyo, B. Van Wachem, *Computational fluid dynamics for engineers*, Cambridge University Press, 2011.
23. P.G. Rousseau, E.Z. Gwebu, Modelling of a superheater heat exchanger with complex flow arrangement including flow and temperature maldistribution, *Heat Transfer Eng* 40 (2019): 862–878. https://doi.org/10.1080/01457632.2018.1446816.
24. C.R. Mali, V. Vinod, A.W. Patwardhan, New methodology for modeling pressure drop and thermal hydraulic characteristics in long vertical boiler tubes at high pressure, *Prog Nucl Energy* 113 (2019). https://doi.org/10.1016/j.pnucene.2019.01.026.
25. G.F. Hewitt, G.L. Shires, T. Bott, *Process heat transfer*, Begell House, 1994.
26. Z. Wu, B. Zhang, H. Yu, J. Ren, M. Pan, C. He, Q. Chen, Accelerating heat exchanger design by combining physics-informed deep learning and transfer learning, *Chem Eng Sci* 282 (2023): 119285.
27. R. Vinuesa, S.L. Brunton, The potential of machine learning to enhance computational fluid dynamics, ArXiv Preprint ArXiv:2110.02085 (2021): 1–13.
28. V. Panwar, S.K. Vandrangi, S. Emani, Artificial intelligence-based computational fluid dynamics approaches, in: *Hybrid computational intelligence*, Elsevier, 2020, 173–190. https://doi.org/10.1016/B978-0-12-818699-2.00009-3.
29. G.M. Antonova, Pattern recognition methods as applied to optimize systems represented by simulation models, *Pattern Recognit Image Anal* 22 (2012): 69–81. https://doi.org/10.1134/S105466181201004X.
30. A. Larrañaga, S.L. Brunton, J. Martinez, S. Chapela, J. Porteiro, Data-driven prediction of the performance of enhanced surfaces from an extensive CFD-generated parametric search space, *Mach Learn Sci Technol* 4 (2023): 25012.
31. S.L. Brunton, J.L. Proctor, J.N. Kutz, Discovering governing equations from data by sparse identification of nonlinear dynamical systems, *Proc Natl Acad Sci* 113 (2016): 3932–3937. https://doi.org/10.1073/pnas.1517384113.
32. S. Beetham, J. Capecelatro, Formulating turbulence closures using sparse regression with embedded form invariance, *Phys Rev Fluids* 5 (2020): 084611. https://doi.org/10.1103/PhysRevFluids.5.084611.

33. J. Ling, A. Kurzawski, J. Templeton, Reynolds averaged turbulence modelling using deep neural networks with embedded invariance, *J Fluid Mech* 807 (2016): 155–166. https://doi.org/10.1017/jfm.2016.615.
34. K. Duraisamy, G. Iaccarino, H. Xiao, Turbulence modeling in the age of data, *Annu Rev Fluid Mech* 51 (2019): 357–377. https://doi.org/10.1146/annurev-fluid-010518-040547.
35. A. Beck, M. Kurz, A perspective on machine learning methods in turbulence modeling, *GAMM-Mitteilungen* 44 (2021). https://doi.org/10.1002/gamm.202100002.
36. L.-T. Zhu, X.-Z. Chen, B. Ouyang, W.-C. Yan, H. Lei, Z. Chen, Z.-H. Luo, Review of machine learning for hydrodynamics, transport, and reactions in multiphase flows and reactors, *Ind Eng Chem Res* 61 (2022): 9901–9949. https://doi.org/10.1021/acs.iecr.2c01036.
37. Z. Wu, B. Zhang, H. Yu, J. Ren, M. Pan, C. He, Q. Chen, Accelerating heat exchanger design by combining physics-informed deep learning and transfer learning, *Chem Eng Sci* 282 (2023): 119285.
38. B. Wang, J. Wang, Application of artificial intelligence in computational fluid dynamics, *Ind Eng Chem Res* 60 (2021): 2772–2790. https://doi.org/10.1021/acs.iecr.0c05045.
39. J. Zou, T. Hirokawa, J. An, L. Huang, J. Camm, Recent advances in the applications of machine learning methods for heat exchanger modeling—a review, *Front Energy Res* 11 (2023). https://doi.org/10.3389/fenrg.2023.1294531.
40. K. Rathore, S. Kumar, Net zero emissions for our future generations through renewables: A brief review, *Int J Energy Res Apps* 2 (2023): 39–56. https://doi.org/10.56896/IJERA.2023.2.1.004.
41. S. Kumar, K. Rathore, Renewable energy for sustainable development goal of clean and affordable energy, *Int J Mater Manuf Sustain Technol* 2 (2023): 1–15. https://doi.org/10.56896/IJMMST.2023.2.1.001.
42. S. Bale, T.P. Liyana-Arachchi, F.R. Hung, Molecular dynamics simulation of single-walled carbon nanotubes inside liquid crystals, *Mol Simul* 42 (2016): 1242–1248. https://doi.org/10.1080/08927022.2016.1174859.
43. J.O. Hirschfelder, C.F. Curtiss, R.B. Bird, *The molecular theory of gases and liquids*, John Wiley & Sons, 1964.
44. J.B. Joshi, K. Nandakumar, A.W. Patwardhan, A.K. Nayak, V. Pareek, M. Gumulya, C. Wu, N. Minocha, E. Pal, M. Kumar, V. Bhusare, S. Tiwari, D. Lote, C. Mali, A. Kulkarni, S. Tamhankar, *Computational fluid dynamics*, 2019. https://doi.org/10.1016/B978-0-08-102337-2.00002-X.
45. L.F. Richardson, *Weather prediction by numerical process*, University Press, 1922.
46. J.G. Collier, J.R. Thome, *Convective boiling and condensation*, Clarendon Press, 1994.
47. J.M. Mandhane, G.A. Gregory, K. Aziz, A flow pattern map for gas—liquid flow in horizontal pipes, *Int J Multiphase Flow* 1 (1974): 537–553. https://doi.org/10.1016/0301-9322(74)90006-8.
48. M. Dawson, A pragmatic approach for simulating the fluid flow and heat transfer within industrial evaporator vessels, *Int J Heat Mass Transf* 135 (2019): 907–924. https://doi.org/10.1016/j.ijheatmasstransfer.2019.01.130.
49. F. Tahir, A. Mabrouk, M. Koc, Review on CFD analysis of horizontal falling film evaporators in multi effect desalination plants, *Desalin Water Treat* 166 (2019): 296–320. https://doi.org/10.5004/dwt.2019.24487.
50. F. Zidouni, E. Krepper, R. Rzehak, S. Rabha, M. Schubert, U. Hampel, Simulation of gas–liquid flow in a helical static mixer, *Chem Eng Sci* 137 (2015): 476–486. https://doi.org/10.1016/j.ces.2015.06.052.
51. K.J. Farrell, CFD approach for predicting entrainment in kettle reboilers, in: *Proceeding of 4th Thermal and Fluids Engineering Conference*, Begellhouse, Connecticut, 2019, 287–296. https://doi.org/10.1615/TFEC2019.cmd.027542.
52. J. Shi, L. Yang, B. Sun, Y. Li, X. Yu, P. Zhang, Numerical calculation of the thermal-hydraulic characteristics in a once-through steam generator based on a coupled boundary, *Appl Therm Eng* 148 (2019): 734–746. https://doi.org/10.1016/j.applthermaleng.2018.11.096.
53. M.K. Awasthi, N. Dutt, A. Kumar, S. Kumar, Electrohydrodynamic capillary instability of Rivlin–Ericksen viscoelastic fluid film with mass and heat transfer, *Heat Transfer* 53 (2024): 115–133. https://doi.org/10.1002/htj.22944.
54. V. V Ranade, *Computational flow modeling for chemical reactor engineering*, Academic Press, 2002.
55. C.R. Mali, V. Vinod, A.W. Patwardhan, Comparison of phase interaction models for high pressure subcooled boiling flow in long vertical tubes, *Nucl Eng Design* 324 (2017). https://doi.org/10.1016/j.nucengdes.2017.09.010.
56. A. Fluent, *Theory guide 19.2*, Ansys Inc, 2018.

57. S.-J. Li, L.-T. Zhu, X.-B. Zhang, Z.-H. Luo, Recent advances in CFD simulations of multiphase flow processes with phase change, *Ind Eng Chem Res* 62 (2023): 10729–10786. https://doi.org/10.1021/acs.iecr.3c00706.
58. V. Anisiuba, H. Ma, A. Silaen, C. Zhou, Computational studies of air-mist spray cooling in continuous casting, *Energies (Basel)* 14 (2021): 7339. https://doi.org/10.3390/en14217339.
59. D.I. Kolaitis, M.A. Founti, A comparative study of numerical models for Eulerian–Lagrangian simulations of turbulent evaporating sprays, *Int J Heat Fluid Flow* 27 (2006): 424–435. https://doi.org/10.1016/j.ijheatfluidflow.2006.01.002.
60. G.A. Novelletto Ricardo, M. Sommerfeld, Comprehensive Euler/Lagrange modelling including particle erosion for confined gas-solid flows, *Particuology* 84 (2024): 209–235. https://doi.org/10.1016/j.partic.2023.06.010.
61. H.M. Badr, M.A. Habib, R. Ben-Mansour, S.A.M. Said, S.S. Al-Anizi, Erosion in the tube entrance region of an air-cooled heat exchanger, *Int J Impact Eng* 32 (2006): 1440–1463. https://doi.org/10.1016/j.ijimpeng.2005.01.008.
62. W. Gao, Y. Li, L. Kong, Numerical investigation of erosion of tube sheet and tubes of a shell and tube heat exchanger, *Comput Chem Eng* 96 (2017): 115–127. https://doi.org/10.1016/j.compchemeng.2016.10.005.
63. S. Kumar, M. King, R.G. Hardin, Modeling of flows through porous media, in: *Computational Fluid Flow and Heat Transfer*, New York: CRC Press, 2024, 133–158. https://doi.org/10.1201/9781003465171-8.
64. S.G. Kandlikar, Heat transfer mechanisms during flow boiling in microchannels, *J Heat Transfer* 126 (2004): 8–16. https://doi.org/10.1115/1.1643090.
65. S. Ben Saad, C. Gentric, J.-F. Fourmigué, P. Clément, J.-P. Leclerc, CFD and experimental investigation of the gas–liquid flow in the distributor of a compact heat exchanger, *Chem Eng Res Design* 92 (2014): 2361–2370. https://doi.org/10.1016/j.cherd.2014.02.002.
66. K. Panda, T. Hirokawa, L. Huang, Design study of microchannel heat exchanger headers using experimentally validated multiphase flow CFD simulation, *Appl Therm Eng* 178 (2020): 115585. https://doi.org/10.1016/j.applthermaleng.2020.115585.
67. J.U. Brackbill, D.B. Kothe, C. Zemach, A continuum method for modeling surface tension, *J Comput Phys* 100 (1992): 335–354. https://doi.org/10.1016/0021-9991(92)90240-Y.
68. Z. Liu, B. Sunden, J. Yuan, VOF modeling and analysis of filmwise condensation between vertical parallel plates, *Heat Transf Res* 43 (2012): 47–68. https://doi.org/10.1615/HeatTransRes.2012004376.
69. Z.-Y. Wu, H. Wang, W.-H. Cai, Y.-Q. Jiang, Numerical investigation of boiling heat transfer on the shell-side of spiral wound heat exchanger, *Heat and Mass Transfer* 52 (2016): 1973–1982. https://doi.org/10.1007/s00231-016-1867-5.
70. S. Al-rifai, Y. Cao, Multiphase modeling of heat and mass transfer inside transport membrane condenser (TMC) tube bundle, *Int J Heat Mass Transf* 214 (2023): 124429. https://doi.org/10.1016/j.ijheatmasstransfer.2023.124429.
71. S. Kumar, P.K. Vijayan, R.B. Grover, U. Kannan, H. Yadav, A. Agrawal, Experimental investigations on thermal stratification in a large pool of water with immersed isolation condenser, in: *Volume 3: Student Paper Competition; Thermal-Hydraulics; Verification and Validation*, American Society of Mechanical Engineers, 2020. https://doi.org/10.1115/ICONE2020-16647.
72. S. Kumar, R.B. Grover, H. Yadav, P.K. Vijayan, U. Kannan, A. Agrawal, Experimental and numerical investigation on suppression of thermal stratification in a water-pool: PIV measurements and CFD simulations, *Appl Therm Eng* 138 (2018): 686–704. https://doi.org/10.1016/j.applthermaleng.2018.04.070.
73. G. Cˇerne, S. Petelin, I. Tiselj, Coupling of the interface tracking and the two-fluid models for the simulation of incompressible two-phase flow, *J Comput Phys* 171 (2001): 776–804. https://doi.org/10.1006/jcph.2001.6810.
74. O.Y. Shonibare, K.E. Wardle, Numerical investigation of vertical plunging jet using a hybrid multifluid–VOF multiphase CFD solver, *Int J Chem Eng* 2015 (2015): 1–14. https://doi.org/10.1155/2015/925639.
75. M. Akhlaghi, V. Mohammadi, N.M. Nouri, M. Taherkhani, M. Karimi, Multi-fluid VoF model assessment to simulate the horizontal air–water intermittent flow, *Chem Eng Res Design* 152 (2019): 48–59. https://doi.org/10.1016/j.cherd.2019.09.031.
76. A. Beck, M. Kurz, A perspective on machine learning methods in turbulence modeling, *GAMM-Mitteilungen* 44 (2021). https://doi.org/10.1002/gamm.202100002.
77. B. Wang, J. Wang, Application of artificial intelligence in computational fluid dynamics, *Ind Eng Chem Res* 60 (2021): 2772–2790. https://doi.org/10.1021/acs.iecr.0c05045.

78. Z.-H. Zhou, A brief introduction to weakly supervised learning, *Natl Sci Rev* 5 (2018): 44–53. https://doi.org/10.1093/nsr/nwx106.
79. T. Miyato, S.-I. Maeda, M. Koyama, S. Ishii, Virtual adversarial training: A regularization method for supervised and semi-supervised learning, *IEEE Trans Pattern Anal Mach Intell* 41 (2019): 1979–1993. https://doi.org/10.1109/TPAMI.2018.2858821.
80. H. Zhao, J. Zheng, W. Deng, Y. Song, Semi-supervised broad learning system based on manifold regularization and broad network, *IEEE Trans Circ Syst I: Regular Papers* 67 (2020): 983–994. https://doi.org/10.1109/TCSI.2019.2959886.
81. Gao Huang, Shiji Song, J.N.D. Gupta, Cheng Wu, Semi-supervised and unsupervised extreme learning machines, *IEEE Trans Cybern* 44 (2014): 2405–2417. https://doi.org/10.1109/TCYB.2014.2307349.
82. M. Hatami, M. Jafaryar, D.D. Ganji, M. Gorji-Bandpy, Optimization of finned-tube heat exchangers for diesel exhaust waste heat recovery using CFD and CCD techniques, *Int Commun Heat Mass Transf* 57 (2014): 254–263. https://doi.org/10.1016/j.icheatmasstransfer.2014.08.015.
83. T.-C. Hung, H.-C. Chen, D.-S. Lee, H.-H. Fu, Y.-T. Chen, G.-P. Yu, Optimal design of a concentric heat exchanger for high-temperature systems using CFD simulations, *Appl Therm Eng* 75 (2015): 700–708. https://doi.org/10.1016/j.applthermaleng.2014.09.079.
84. C. Liu, W. Bu, D. Xu, Multi-objective shape optimization of a plate-fin heat exchanger using CFD and multi-objective genetic algorithm, *Int J Heat Mass Transf* 111 (2017): 65–82. https://doi.org/10.1016/j.ijheatmasstransfer.2017.03.066.
85. K.K. Saijal, T. Danish, Design optimization of a shell and tube heat exchanger with staggered baffles using neural network and genetic algorithm, *Proc Inst Mech Eng C J Mech Eng Sci* 235 (2021): 5931–5946. https://doi.org/10.1177/09544062211005797.
86. N. Flynn, X. Qian, Heat sink design optimization via GAN-CNN combined deep-learning, in: *Volume 3A: 49th Design Automation Conference (DAC)*, American Society of Mechanical Engineers, 2023. https://doi.org/10.1115/DETC2023-116429.
87. H. Keramati, F. Hamdullahpur, M. Barzegari, Deep reinforcement learning for heat exchanger shape optimization, *Int J Heat Mass Transf* 194 (2022): 123112. https://doi.org/10.1016/j.ijheatmasstransfer.2022.123112.
88. Y. Liu, N. Dinh, Y. Sato, B. Niceno, Data-driven modeling for boiling heat transfer: Using deep neural networks and high-fidelity simulation results, *Appl Therm Eng* 144 (2018): 305–320. https://doi.org/10.1016/j.applthermaleng.2018.08.041.
89. M. Ghalandari, M. Irandoost Shahrestani, A. Maleki, M. Safdari Shadloo, M. El Haj Assad, Applications of intelligent methods in various types of heat exchangers: a review, *J Therm Anal Calorim* 145 (2021): 1837–1848. https://doi.org/10.1007/s10973-020-10425-3.
90. J. Zou, T. Hirokawa, J. An, L. Huang, J. Camm, Recent advances in the applications of machine learning methods for heat exchanger modeling—a review, *Front Energy Res* 11 (2023). https://doi.org/10.3389/fenrg.2023.1294531.
91. M. Mohanraj, S. Jayaraj, C. Muraleedharan, Applications of artificial neural networks for thermal analysis of heat exchangers – a review, *Int J Ther Sci* 90 (2015): 150–172. https://doi.org/10.1016/j.ijthermalsci.2014.11.030.
92. S. Kumar, D. Banerjee, A review on phase change materials for sustainability applications by leveraging machine learning, in: *Energy Consumption, Conversion, Storage, and Efficiency*, IntechOpen, 2024. https://doi.org/10.5772/intechopen.114380.
93. A. Dawahdeh, H. Sharadga, S. Kumar, Novel MPPT controller augmented with neural network for use with photovoltaic systems experiencing rapid solar radiation changes, *Sustainability* 16 (2024): 1021. https://doi.org/10.3390/su16031021.
94. Y. Lecun, L. Bottou, Y. Bengio, P. Haffner, Gradient-based learning applied to document recognition, *Proc IEEE* 86 (1998): 2278–2324. https://doi.org/10.1109/5.726791.
95. D.E. Rumelhart, G.E. Hinton, R.J. Williams, Learning representations by back-propagating errors, *Nature* 323 (1986): 533–536. https://doi.org/10.1038/323533a0.
96. T. Hastie, J. Friedman, R. Tibshirani, *The elements of statistical learning*, New York: Springer, 2001. https://doi.org/10.1007/978-0-387-21606-5.
97. X. Li, J. Shi, X. Dong, J. Yu, A new conjugate gradient method based on Quasi-Newton equation for unconstrained optimization, *J Comput Appl Math* 350 (2019): 372–379. https://doi.org/10.1016/j.cam.2018.10.035.
98. J.K. Liu, Y.M. Feng, L.M. Zou, A spectral conjugate gradient method for solving large-scale unconstrained optimization, *Comput Math Appl* 77 (2019): 731–739. https://doi.org/10.1016/j.camwa.2018.10.002.

99. S. Catsoulis, J.-S. Singh, C. Narayanan, D. Lakehal, Integrating supervised learning and applied computational multi-fluid dynamics, *Int J Multiphase Flow* 157 (2022): 104221.
100. J. Zhang, F. Haghighat, Development of artificial neural network-based heat convection algorithm for thermal simulation of large rectangular cross-sectional area Earth-to-Air Heat Exchangers, *Energy Build* 42 (2010): 435–440. https://doi.org/10.1016/j.enbuild.2009.10.011.
101. K. Lindqvist, Z. Wilson, E. Næss, N. Sahinidis, A machine learning approach to correlation development applied to fin-tube bundle heat exchangers, *Energies (Basel)* 11 (2018): 3450. https://doi.org/10.3390/en11123450.
102. L. Daróczy, G. Janiga, D. Thévenin, Systematic analysis of the heat exchanger arrangement problem using multi-objective genetic optimization, *Energy* 65 (2014): 364–373. https://doi.org/10.1016/j.energy.2013.11.035.
103. Z.J. Zhai, Y. Xue, Q. Chen, Inverse design methods for indoor ventilation systems using CFD-based multi-objective genetic algorithm, *Build Simul* 7 (2014): 661–669. https://doi.org/10.1007/s12273-014-0179-2.
104. Y. Ge, Q. He, Y. Lin, W. Yuan, J. Chen, S.-M. Huang, Multi-objective optimization of a mini-channel heat sink with non-uniform fins using genetic algorithm in coupling with CFD models, *Appl Therm Eng* 207 (2022): 118127. https://doi.org/10.1016/j.applthermaleng.2022.118127.
105. R. Hilbert, G. Janiga, R. Baron, D. Thévenin, Multi-objective shape optimization of a heat exchanger using parallel genetic algorithms, *Int J Heat Mass Transf* 49 (2006): 2567–2577. https://doi.org/10.1016/j.ijheatmasstransfer.2005.12.015.
106. M. Darvish Damavandi, M. Forouzanmehr, H. Safikhani, Modeling and Pareto based multi-objective optimization of wavy fin-and-elliptical tube heat exchangers using CFD and NSGA-II algorithm, *Appl Therm Eng* 111 (2017): 325–339. https://doi.org/10.1016/j.applthermaleng.2016.09.120.
107. R. Song, M. Cui, Single- and multi-objective optimization of a plate-fin heat exchanger with offset strip fins adopting the genetic algorithm, *Appl Therm Eng* 159 (2019): 113881. https://doi.org/10.1016/j.applthermaleng.2019.113881.
108. M.R. Daneshparvar, R. Beigzadeh, Multi-objective optimization of helical baffles in the shell-and-tube heat exchanger by computational fluid dynamics and genetic algorithm, *Energy Rep* 8 (2022): 11064–11077. https://doi.org/10.1016/j.egyr.2022.08.249.
109. Y. Alperen, C. Sertac, Multi objective optimization of a micro-channel heat sink through genetic algorithm, *Int J Heat Mass Transf* 146 (2020): 118847. https://doi.org/10.1016/j.ijheatmasstransfer.2019.118847.
110. S. Baodong, W. Lifeng, L. Jianyun, C. Heming, Multi-objective optimization design of a micro-channel heat sink using adaptive genetic algorithm, *Int J Numer Meth Heat Fluid Flow* 21 (2011): 353–364. https://doi.org/10.1108/09615531111108512.
111. K. Deb, A. Pratap, S. Agarwal, T. Meyarivan, A fast and elitist multiobjective genetic algorithm: NSGA-II, *IEEE Trans Evol Comput* 6 (2002): 182–197. https://doi.org/10.1109/4235.996017.
112. T. Back, *Evolutionary algorithms in theory and practice: evolution strategies, evolutionary programming, genetic algorithms*, Oxford University Press, 1996.
113. S.A. Abdollahi, A. Basem, A. Alizadeh, D.J. Jasim, M. Ahmed, A.J. Sultan, S.F. Ranjbar, H. Maleki, Combining artificial intelligence and computational fluid dynamics for optimal design of laterally perforated finned heat sinks, *Results Eng* 21 (2024): 102002. https://doi.org/10.1016/j.rineng.2024.102002.
114. K. Sharifi, M. Sabeti, M. Rafiei, A.H. Mohammadi, A. Ghaffari, M.H. Asl, H. Yousefi, A good contribution of computational fluid dynamics (CFD) and GA-ANN methods to find the best type of helical wire inserted tube in heat exchangers, *Int J Therm Sci* 154 (2020): 106398. https://doi.org/10.1016/j.ijthermalsci.2020.106398.
115. J.O.B. Lira, H.G. Riella, N. Padoin, C. Soares, Computational fluid dynamics (CFD), artificial neural network (ANN) and genetic algorithm (GA) as a hybrid method for the analysis and optimization of micro-photocatalytic reactors: NOx abatement as a case study, *Chem Eng J* 431 (2022): 133771. https://doi.org/10.1016/j.cej.2021.133771.
116. K.K. Saijal, T. Danish, Design optimization of a shell and tube heat exchanger with staggered baffles using neural network and genetic algorithm, *Proc Inst Mech Eng C J Mech Eng Sci* 235 (2021): 5931–5946. https://doi.org/10.1177/09544062211005797.
117. A.B. Cubit, *CUBIT 15.3 user documentation*, Sandia National Laboratories: Albuquerque, 2017.
118. M. Sosnowski, J. Krzywanski, K. Grabowska, R. Gnatowska, Polyhedral meshing in numerical analysis of conjugate heat transfer, *EPJ Web Conf* 180 (2018): 02096. https://doi.org/10.1051/epjconf/201818002096.
119. D.C. Wilcox, others, *Turbulence modeling for CFD*, La Canada, CA: DCW Industries, 1998.

120. B.E. Launder, *DB Spalding mathematical models of turbulence*, London: Academic Presse, 1972.
121. V. Yakhot, S.A. Orszag, S. Thangam, T.B. Gatski, C.G. Speziale, Development of turbulence models for shear flows by a double expansion technique, *Phys Fluids A* 4 (1992): 1510–1520. https://doi.org/10.1063/1.858424.
122. F. Zhou, I. Catton, Numerical evaluation of flow and heat transfer in plate-pin fin heat sinks with various pin cross-sections, *Numeri Heat Transf A Appl* 60 (2011): 107–128. https://doi.org/10.1080/10407782.2011.588574.
123. A. Singh, S. Sarkar, J.M. Spritzer, A.O. Okeremi, CFD based erosion analysis for design of subsea hardware in a high flow rate gas field, in: *All Days*, OTC, 2014. https://doi.org/10.4043/24710-MS.
124. F.N. Mouketou, A. Kolesnikov, Modelling and simulation of multiphase flow applicable to processes in oil and gas industry, *Chem Prod Process Model* 14 (2019). https://doi.org/10.1515/cppm-2017-0066.
125. M. Darbandi, M.-S. Abdollahpour, M. Hasanpour-Matkolaei, A new developed semi-full-scale approach to facilitate the CFD simulation of shell and tube heat exchangers, *Chem Eng Sci* 245 (2021): 116836. https://doi.org/10.1016/j.ces.2021.116836.
126. S. V. Patankar, *Numerical heat transfer and fluid flow*, CRC Press, 2018. https://doi.org/10.1201/9781482234213.
127. G. Rubio, E. Valero, S. Lanzan, Computational fluid dynamics expert system using artificial neural networks, *Int J Comput Infor Eng* 6 (2012): 413–417.
128. S. Kumar, A. Thyagarajan, D. Banerjee, An investigation on Chevron plate heat exchanger filled with organic phase change material, in: *ASME 2020 Heat Transfer Summer Conference*, American Society of Mechanical Engineers, 2020. https://doi.org/10.1115/HT2020-9122.
129. S. Kumar, A. Thyagarajan, D. Banerjee, Experimental investigation of thermal energy storage (TES) platform leveraging phase change materials in a chevron plate heat exchanger, in: *Volume 8: Fluids Engineering; Heat Transfer and Thermal Engineering*, American Society of Mechanical Engineers, 2022. https://doi.org/10.1115/IMECE2022-96226.
130. J. Shi, B. Sun, W. Han, G. Zhang, Y. Li, L. Yang, Numerical simulation of dryout and post-dryout heat transfer in a straight-pipe once-through steam generator, *Appl Therm Eng* 105 (2016): 132–141. https://doi.org/10.1016/j.applthermaleng.2016.05.145.
131. S.R. Shabanian, M. Rahimi, M. Shahhosseini, A.A. Alsairafi, CFD and experimental studies on heat transfer enhancement in an air cooler equipped with different tube inserts, *Int Commun Heat Mass Transf* 38 (2011): 383–390. https://doi.org/10.1016/j.icheatmasstransfer.2010.12.015.
132. E. Pal, I. Kumar, J.B. Joshi, N.K. Maheshwari, CFD simulations of shell-side flow in a shell-and-tube type heat exchanger with and without baffles, *Chem Eng Sci* 143 (2016): 314–340. https://doi.org/10.1016/j.ces.2016.01.011.
133. A. El Maakoul, A. Laknizi, S. Saadeddine, A. Ben Abdellah, M. Meziane, M. El Metoui, Numerical design and investigation of heat transfer enhancement and performance for an annulus with continuous helical baffles in a double-pipe heat exchanger, *Energy Convers Manag* 133 (2017): 76–86. https://doi.org/10.1016/j.enconman.2016.12.002.
134. S. Salimi, M.J. Mozoun, R. Beigzadeh, Heat transfer enhancement of serpentine channels with twisted tape insert by computational fluid dynamics and artificial intelligence, *Can J Chem Eng* 102 (2024): 1322–1340. https://doi.org/10.1002/cjce.25108.
135. V.S. Gullapalli, B. Sundén, CFD Simulation of heat transfer and pressure drop in compact brazed plate heat exchangers, *Heat Transf Eng* 35 (2014): 358–366. https://doi.org/10.1080/01457632.2013.828557.
136. W. Liu, T. Hidesada, K. Takase, H. Hayafune, S. Futagami, N. Kisohara, Thermal-Hydraulic Experiments under high pressure condition, in: *18th International Conference on Nuclear Engineering: Volume 4, Parts A and B*, ASMEDC, 2010, 1101–1109. https://doi.org/10.1115/ICONE18-30167.
137. W. Choi, S. Jun, L.T. Nguyen, N. Rungraeng, H. Yi, S. Balasubramanian, V.M. Puri, J. Lee, 3-D Milk fouling modeling of plate heat exchangers with different surface finishes using computational fluid dynamics codes, *J Food Process Eng* 36 (2013): 439–449. https://doi.org/10.1111/j.1745-4530.2012.00684.x.
138. S.-Z. Tang, F.-L. Wang, Q. Ren, Y.-L. He, Fouling characteristics analysis and morphology prediction of heat exchangers with a particulate fouling model considering deposition and removal mechanisms, *Fuel* 203 (2017): 725–738. https://doi.org/10.1016/j.fuel.2017.03.049.
139. Z. Xu, J. Li, Z. Han, Numerical study of particle fouling deposition on heat transfer surface, *ESS* 1 (2022): 44–52. https://doi.org/10.1016/j.enss.2021.11.001.
140. X. Liu, X. Zhang, T. Lu, K. Mahkamov, H. Wu, M. Mirzaeian, Numerical simulation of sub-cooled boiling flow with fouling deposited inside channels, *Appl Therm Eng* 103 (2016): 434–442. https://doi.org/10.1016/j.applthermaleng.2016.04.041.

141. C. Paz, E. Suárez, M. Conde, J. Vence, Development of a computational fluid dynamics model for predicting fouling process using dynamic mesh model, *Heat Transf Eng* 41 (2020): 199–207. https://doi.org/10.1080/01457632.2018.1522108.
142. V.R. Radhakrishnan, M. Ramasamy, H. Zabiri, V. Do Thanh, N.M. Tahir, H. Mukhtar, M.R. Hamdi, N. Ramli, Heat exchanger fouling model and preventive maintenance scheduling tool, *Appl Therm Eng* 27 (2007): 2791–2802. https://doi.org/10.1016/j.applthermaleng.2007.02.009.
143. E. Davoudi, B. Vaferi, Applying artificial neural networks for systematic estimation of degree of fouling in heat exchangers, *Chem Eng Res Design* 130 (2018): 138–153. https://doi.org/10.1016/j.cherd.2017.12.017.
144. H. Li, S.A. Vasquez, H. Punekar, R. Muralikrishnan, Prediction of boiling and critical heat flux using an eulerian multiphase boiling model, in: *Volume 6: Fluids and Thermal Systems; Advances for Process Industries, Parts A and B*, ASMEDC, 2011, 463–476. https://doi.org/10.1115/IMECE2011-65539.
145. R. Zhang, T. Cong, W. Tian, S. Qiu, G. Su, Prediction of CHF in vertical heated tubes based on CFD methodology, *Prog Nucl Energy* 78 (2015): 196–200. https://doi.org/10.1016/j.pnucene.2014.10.001.

10 Fouling and Corrosion Mechanisms and Monitoring Techniques

Sikta Panda and Gaurab Kumar Ghosh

10.1 INTRODUCTION

Heat exchangers (HEs) are commonly employed to efficiently transport heat across different mediums, which include steam, gas, vapors, and different liquids. Heat exchangers have found extensive applications in various industries, including automobiles, aerospace, power plants, condensers, industrial cooling systems, heat pipes, desulfurization units, ethanol vaporizers, compressors, plants producing fertilizers, ammonia, styrene, and petrochemicals, sulfur recovery units, and hydrocracker units. Heat exchangers are available in tubular and plate configurations. A broad view of the types of industrial heat exchanges on the basis of their geometry, mechanism, arrangement, and state of fluid is shown in Figure 10.1.

The process of selecting materials for heat exchangers involves careful evaluation of key factors such as resistance to corrosion and fouling, thermal conductivity, strength of the material, and cost [1]. The primary obstacle lies in attaining a high level of refractory properties, as heat exchangers are subjected to severe conditions and significant fluctuations in operational factors. Potential failures in HEs may arise as a result of faults that are introduced into pipes and tubing at several phases, including production, utilization, investigation, transport, and warehousing. Additionally, failures can occur during the beginning, regular operations, and termination of the heat exchanger. Typical causes of failures comprise fatigue, creep, corrosion, fouling, scaling, salt deposition, vibration, and weld defects. These failures can be caused by improper choice of materials or tube design, failure to follow endorsed operational guidelines, and/or negligence by humans [2]. Heat exchangers fail at several points, potentially causing tube failures. Some of the more prevalent ones are tube corrosion, which is the most common source of oxidation in shell and tube heat exchangers with carbon steel tubes [3]. Second, contact with steam, vapor, or water hammer may cause the rupture

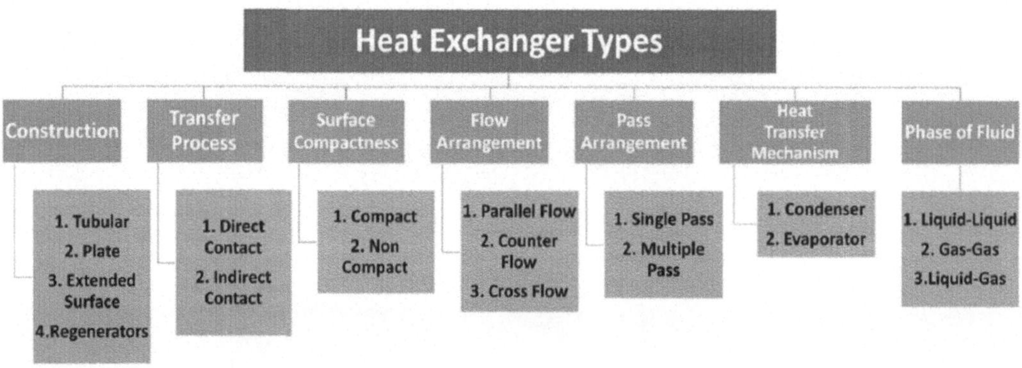

FIGURE 10.1 Classification of heat exchangers.

DOI: 10.1201/9781003509837-10

or collapse of the heat exchanger's shell or tubes [4]. The third important factor is thermal fatigue; based on accumulated stresses from continuous thermal cycles and a high temperature differential, tubing could be subject to tearing and cracking, which is known as thermal fatigue, causing the tubing to burst [5]. Vibrations and/or resonances may also cause harm to the heat exchanger's tube integrity [6].

Fouling and corrosion are prevalent issues encountered in heat exchangers, posing substantial impacts on their efficiency and durability. In order to comprehend the processes of fouling and corrosion in heat exchangers, it is crucial to acknowledge that they are separate yet interrelated phenomena [7]. Fouling refers to the build-up of undesirable deposits on the surface involved in heat transfer, which can include scale, sediments, or bacterial growth [8, 9]. Accumulations of these deposits can diminish the efficiency of heat transfer and raise the pressure drop, resulting in a decline in the operation of the heat exchanger. Corrosion is the process by which metal surfaces deteriorate as a result of electrochemical processes occurring under the influence of water [7]. Monitoring methods for fouling and corrosion in heat exchangers are essential for identifying and reducing these problems. Monitoring approaches encompass routine inspections and maintenance schedules, alongside the utilization of sophisticated technologies including sensors, probes, and non-destructive testing methods. In addition, employing corrosion-resistant materials and coatings can effectively mitigate or reduce the effects of corrosion in heat exchangers [10].

To summarize, fouling and corrosion pose substantial obstacles in heat exchangers, which can have a detrimental impact on their efficiency and durability. Hence these problems demand careful monitoring, prediction, and remediation for uninterrupted working. Therefore, in the upcoming sections of this chapter, the focus will be concentrated on the monitoring and prediction of failures in heat exchangers; moreover, a special emphasis is put on the prevention and remediation mechanisms for these failures. The visualization of heat exchanger components, the flow dynamics of the fluids, and the arrangement of tubes inside heat exchangers provide better insights into their failure modes and mechanisms. Figure 10.2 represents a schematic diagram of an industrial heat exchanger.

10.2 FOULING AND CORROSION MECHANISMS IN HEAT EXCHANGERS

For the smooth operation and elongated service period of heat exchangers, failure mechanisms must be analyzed, and measures should be undertaken to mitigate them. But before jumping into

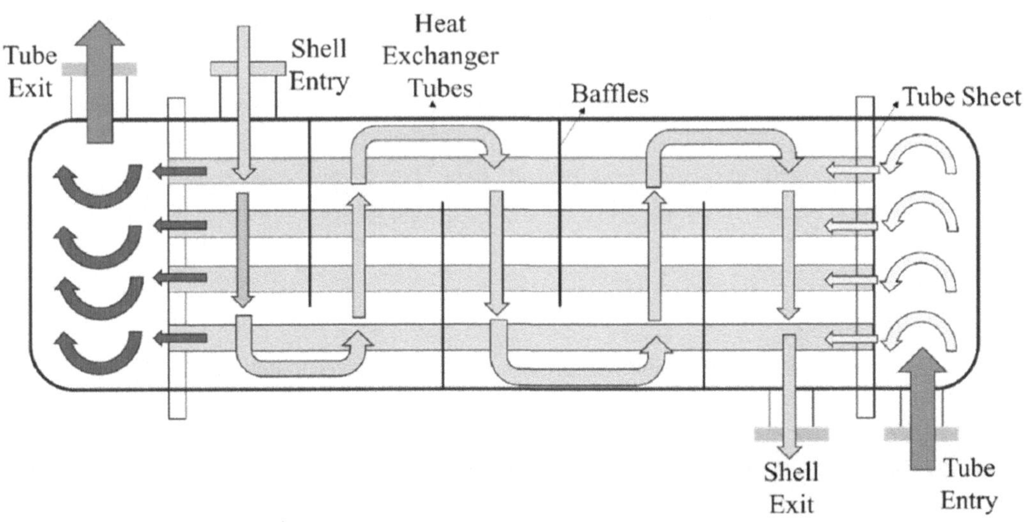

FIGURE 10.2 Graphical representation of an industrial heat exchanger.

the monitoring methods for fouling and corrosion in HEs, it is important to learn about the sources that cause this to occur.

10.2.1 Mechanisms of Fouling in HEs

The following are a few important mechanisms for fouling occurrences.

10.2.1.1 Sedimentation

Particulate or sedimentation fouling is the process of suspended particulates being deposited onto a surface. Various substances such as tube scales, corrosion-induced substances, sand, dirt, and so on can accumulate on HE tube surfaces. Sediment particulates frequently become trapped or confined within the fouling layers. They can induce additional fouling pathways and serve as catalysts. Maintaining elevated flow velocities may help reduce debris deposits and fouling; the lowest velocity of 1 m/s is recommended for the stream in the tubes and along the length of the shell. There are various techniques available for determining the minimal velocity required to suspend particles in a fluid, given the composition of the fluid and particles. The fouling mechanism is clearly known, and no other factors contribute to it. Employing elevated speeds requires an assessment of the possibility of erosion and vibration caused by fluid movement [11, 12].

10.2.1.2 Chemical Reactions

Chemical reaction fouling processes are characterized by their rates following an exponential (Arrhenius) relation with temperature. Asphaltene fouling is the result of the attachment of asphaltenes, found in crude oil, to surfaces operating at elevated temperatures, especially in cases of sulfide outer layers. Asphaltenes undergo thermal decomposition to become coke when exposed to temperatures exceeding 340°C. Asphaltene fouling due to crude oil can occur when the temperature of the wall rises above 300°C and the shear stress at the wall falls below 10 Pa [11]. Coking, another form of chemical reaction-induced fouling, is the formation and adhesion of an insoluble hydrocarbon from the flowing stream. Insoluble gum production occurs when oxygen undergoes chemical interactions with hydrocarbon fluids through autoxidation processes. Metal ions accelerate autoxidation processes. Oxygen saturation of petroleum can result in a tenfold increase in fouling rate. To prevent the production of insoluble gum and subsequent fouling, it is advisable to reduce air exposure, store hydrocarbons in the presence of an inert gas, or utilize antioxidants. The process of forming large molecules or macromolecules by combining smaller components called monomers is known as polymerization. This is usually achieved through condensation or free radical processes.

10.2.1.3 Crystallization

Crystallization fouling, often known as scaling, is a prevalent type of fouling. This commonly appears in cooling water, however, it may additionally be found in other flows. Crystallization fouling is the result of a chemical species, usually a salt, being excessively concentrated over the surface [13]. Two categories of salt solubility prevail, i.e., normal and inverted. Salt saturation with normal solubility rises consistently as temperature rises, while in salts with inverse solubility, the relation is reversed.

10.2.1.4 Biofouling

Biofouling occurs due to the adhesion and growth of various bacterial colonies and other organic matter. They generally tend to occur in the presence of water as a medium in HEs [14]. Biological growth generally ceases to exist beyond 25°C. However, some biological growth may be persistent and reduce the efficiency of the HE [15]. To sum up the prevailing fouling mechanisms, a schematic diagram is shown in Figure 10.3.

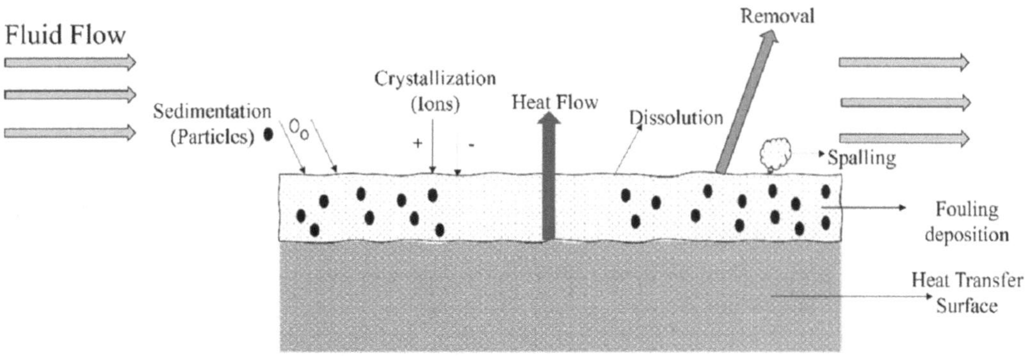

FIGURE 10.3 Fouling mechanism on the heat transfer surface.

10.2.2 Mechanisms of Corrosion in HEs

After learning about the causes of fouling in HE, it is important to know about the sources and mechanisms of corrosion in HE.

10.2.2.1 Galvanic Corrosion

Galvanic corrosion is the primary type of chemically induced corrosion in breakdowns in heat exchangers. Galvanic corrosion can occur in various locations, including gaskets, rivets, and welded joints. Galvanic corrosion frequently arises due to the presence of different metals welded together and operating inside a sensitive electrolyte [16]. This leads to the selective reaction of the most reactive metal, whereas the reaction of the other component (which is more resistant to corrosion) decelerates or ceases entirely. It can be concluded that galvanic corrosion caused by the voltage differential in different metals fused together in a highly conductive environment is a source of HE failure. Figure 10.4 shows a schematic diagram of a galvanic corrosion mechanism on a substrate.

10.2.2.2 Pitting Corrosion

Pitting corrosion typically initiates and spreads rapidly within an unusually brief timeframe, resulting in heat exchanger tubes failing prematurely and leading to substantial harm in the HE. This type of corrosion occurs exclusively when aggressive anions (chlorides) are present. Pitting corrosion

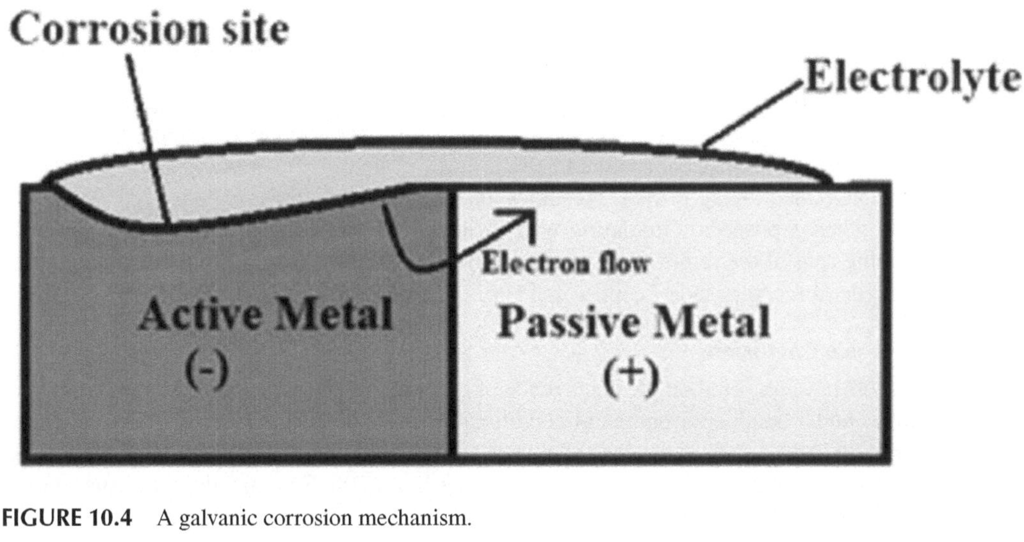

FIGURE 10.4 A galvanic corrosion mechanism.

arises in a specific area and enables the formation of microscopic pits. This type of corrosion is usually intensified by slower streams or static conditions. The extent of pitting is influenced by various elements such as temperature, pH level, composition of materials, microstructure, and the presence of reactive ions [17]. There is extensive documentation on pitting corrosion that occurs in HEs that are constructed of steel, nickel alloys, copper alloys, and titanium [18]. These accidents are caused by pitting assaults in an environment that contains chloride ions.

10.2.2.3 Fretting Corrosion

Fretting corrosion occurs when wear and corrosion combine to cause damage. Vibration contact arises when there is interaction between tightly fitting surfaces that are under high load, such as baffles and tubes in shell and tube heat exchangers [1]. If two metal interfaces move against each other, the friction between them weakens the protective coating, exposing the metal surfaces to a corrosive environment. Tubes composed of Cu alloys, Fe alloys, Ni alloys, and Ti alloys are known to undergo failures due to fretting corrosion [19].

10.2.2.4 Erosion Corrosion

Erosion corrosion mostly happens on the surface of heat exchanger tubes as a result of the passage of fluid through them. This causes aggravation when the fluid is mixed with other particulates that may cause abrasion [20]. The wear of the metal surface is caused by the motion of solid particles, especially in circumstances in which the fluid is compelled to change its direction or when there is a strong shear stress [21]. The native oxide layer of the material substrate, which protects its surface, is eliminated either through the force exerted by particles or, in certain situations, by the automatic flow of the fluid (known as impingement attack) [22]. Erosion can potentially happen near the extremities of the tubes or at points where the tubes are bent. Erosion corrosion may be characterized by the formation of small craters over the surface that are devoid of any corrosion products and have a shiny appearance. Pits typically occur following a localized attack that has disrupted the protective oxide layer [23]. Erosion corrosion is characterized by a distinct shape that follows a certain direction and often leads to the formation of grooves or pits in a horseshoe pattern. The tip of the groove or pit points in the direction of the current flow [24]. The presence of fouling/corrosion products causes partial blockage in the tubes, resulting in a reduced inner diameter and increased velocity. Localized turbulent flow typically removes deposits and beneath deposits, resulting in significant erosion and weakening of tubes [24]. The velocity of the fluid is of utmost importance in determining the performance of heat exchanger tube materials.

10.2.2.5 Dealloying or Selective Leaching

The least corrosion-resistant component is predominantly corroded in the case of dealloying or selective leaching compared with other elements. Dealloying is a phenomenon that happens in certain alloys when they are exposed to hostile environmental changes. Dealloying is influenced by various parameters, such as metallurgical and environmental variables (beta phase attack and temperature). Dezincification, which refers to the process of dealloying copper-zinc alloys, is the prevalent example of dealloying in HEs. HE tubes made from yellow brass experienced significant dezincification when exposed to circulating water with a high salt concentration. The process of selective leaching caused the removal of zinc from the copper alloy, leading to the formation of a porous and fragile film composed of copper and copper oxide [24].

10.2.2.6 Crevice Corrosion

Many researchers have reported on the occurrence of crevice corrosion causing leakages from heat exchanger tubes under harsh environmental conditions. Crevice corrosion typically occurs in concealed and confined areas, such as the space between a tube and a tube sheet, at sealed joints, underneath accumulated fouling and deposits, and within bolt holes [18]. Specifically, areas susceptible to this corrosion include stationary zones and intricate geometric structures having sharp contours.

An example of crevice corrosion can be the presence of sediments in the spaces between tubes and baffles, which creates narrow gaps that may trap tiny amounts of fluid, leading to increased corrosion of passivated material [18].

10.2.2.7 Stress Corrosion Cracking

Repeated thermal cycling can cause fatigue in shell and tube heat exchangers. Fatigue can initiate the formation of cracks, beginning with subtle striations that are difficult to detect but gradually become more noticeable and eventually lead to the full separation of the material. This is known as stress corrosion cracking (SCC). This happens at the junction where the tubes connect to the tube sheet [25]. Additionally, there have been documented instances of SCC caused by high temperature stress along the axis of the tube. These cracks occur owing to repetitive temperature changes, which create strains in the tube wall. Fatigue typically arises when the thermal pressure in the tube at the connection with the tube sheet exceeds the fatigue limit, leading to the development of a crack around the circumference of the tube [25]. Tube-to-tube sheet joints are primarily formed through expansion, welding, or a combination of both techniques [26]. The expansion method is commonly used for tube-to-tube sheet connections in situations where pressure gradients are not sufficiently large to result in leakage. The tube and shell must have a secure and airtight connection, as well as sufficient strength to endure a 50-unit axial load.

Due to the extreme temperature and pressure involved in the working of a heat exchanger, expanded joints are not suitable. Therefore, tube-to-tube sheet joints are often welded. Nevertheless, these welded connections represent the most vulnerable region and are a significant factor contributing to the failure of heat exchangers [27]. The failure scenario is determined by the local stress levels, which result from the applied loads as well as thermal and residual stress. The corrosion mechanisms mentioned in the above sections (10.2.2.2–10.2.2.6) generally occur as a combination of two or more mechanisms. Figure 10.5 shows the combined effect of pitting, fretting, erosion, and crevice corrosion.

10.3 MONITORING TECHNIQUES

In order to prevent technical issues related to the obstruction of heat exchangers (such as blockage and congestion of tubes, deterioration of resources, decrease in energy efficiency, and production hindrance due to the need for safeguarding and cleaning), multiple techniques are employed for investigating and administering fouling growth [29]. Predictive maintenance, troubleshooting fouling progress, and measurements using probes are some commonly used monitoring methods.

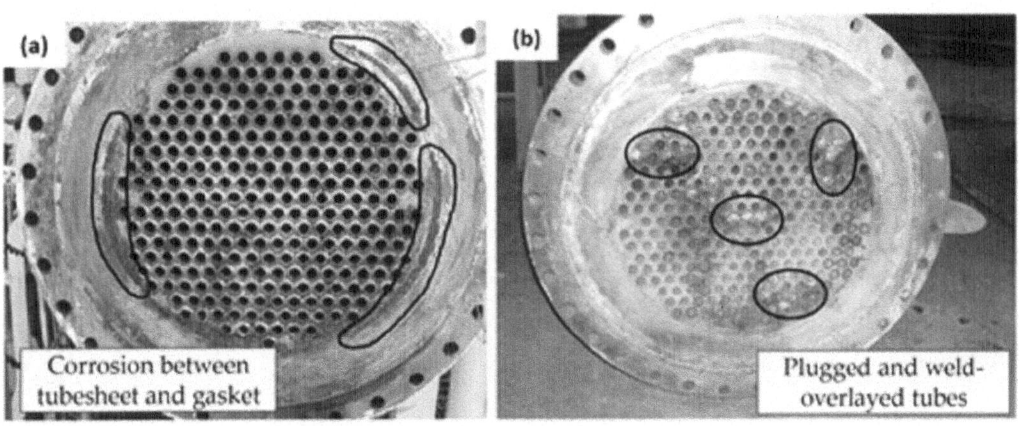

FIGURE 10.5 Failure in a heat exchanger due to corrosion phenomenon (courtesy: open source [28]).

10.3.1 Predictive Maintenance

There are multiple methods for creating a predictive maintenance strategy, which can be done using either thermography or visual inspection.

Thermography enables the measurement of infrared emissions, which are essentially heat, to determine the operational status of installations. Predictive maintenance is performed with the aid of devices operating on infrared light or waves. Infrared thermometers are specifically intended to accurately detect the surface temperature at a specific, limited location within a system or installation. Due to the confinement of the measurement to a singular spot, this method presents constrained alternatives. Second, linear scanners measure temperature along a certain line, resulting in one-dimensional data. While this approach provides a broader perspective, its use is constrained. Lastly, imagers are devices that study the entirety of the emissions originating from the source. Typically, these cameras are infrared cameras. While these approaches are successful and innovative, they only allow for the identification of fouling issues and are limited to qualitative analysis due to the various inaccuracies in temperature measurement using radiometry.

Conversely, visual inspection of industrial facilities also assures predictive maintenance. Here, the user can personally identify the extent of fouling on the exchangers in their installation based on their own experience. Therefore, this regular procedure offers subjective yet essential information for establishing a maintenance program. The underlying premise shared by these strategies is the systematic monitoring of the installation's condition. They offer varying degrees of information regarding the fouling condition of the exchangers but provide minimal details about heat transfer.

10.3.2 Troubleshooting

This includes conducting measurements directly at the terminals of the exchanger. In industrial applications, various techniques are employed to identify and track the development of fouling [29]. Typically, three strategies are employed.

10.3.2.1 Computation of the Heat Transfer Coefficient

Fouling is a phenomenon that leads to a reduction in the rate at which heat is transferred, resulting in a decrease in the output temperature of the product. To achieve a desired fixed output temperature for a product, the temperature of heating source (THS) is raised with its former temperature lowered. This method can be utilized to quantify the presence and magnitude of fouling: as the THS increases, the fouling becomes more pronounced. The mechanism of heat transfer is governed by several factors such as overall heat transfer coefficient, heat flux, and fouling resistance. Fouling resistance depends on the thermal conductivity of the product, the heating source, and the fouling layer. The quantity of heat that is transmitted from the heating source to the product is denoted by Q. This is computed by evaluating the total heat transfer coefficient U.

$$U = \frac{Q}{A\Delta T_{lm}} = \frac{\dot{m}C_P\Delta T}{A\Delta T_{lm}} \tag{10.1}$$

Where $\dot{m}$, C_P, ΔT, A, and ΔT_{lm}represent mass flow rate, specific heat capacity of the product, temperature change of the product, cross-sectional area, and the logarithmic mean temperature difference between the product and heating medium, respectively. The presence of fouling incorporates an additional resistance (R_f), which depends on the overall heat transfer coefficient, fouling thickness (t_f), and fouling thermal conductivity (K_f). Hence, this can be computed from Equation 10.1 by comparing a clean (U_o) and a fouled heat exchanger (U_f).

$$R_f = \frac{1}{U_f} - \frac{1}{U_o} = \frac{t_f}{K_f} \tag{10.2}$$

Thermal resistance of a material is its ability to obstruct the passage of heat through it. R_{th} is a parameter that quantifies the amount of heat that is dissipated during the process of heating. It varies inversely with the thermal conductivity (K_f) and can be represented as

$$R_{th} = \frac{1}{K_f} \tag{10.3}$$

10.3.2.2 Observing Pressure Difference

Pressure difference between the intake and output of heat exchangers is constantly monitored. Fouling in a flow channel causes a reduction in the mean square area, resulting in a pressure drop while maintaining a steady flow rate. Riverol and Napolitano [30] employed this as an input variable in estimating fouling through computational methods like artificial neural networks (ANNs). Typically, pressure drop is observed in conjunction with other methods such as temperature measurements [31]. Martin established a connection between heat and mass transfer in HEs that have intermittent patterns, such as cross-corrugated plate and tube bundles. He also considered the pressure drop (Δp) using the generalized Lévêque equation. The Hagen number, denoted as H_g, was developed as a means to forecast heat and mass transfer across various applications and heat exchanger configurations in terms of density (ρ), flow direction (Δz), outer tube diameter (d), and kinematic viscosity (ϑ).

$$H_g = \left(\frac{1}{\rho}\right)\left(\frac{\Delta p}{\Delta z}\right)\frac{d^3}{\vartheta^2} \tag{10.4}$$

10.3.2.3 Measurement of Electrical Parameters

Electrical characteristics are monitored in situations involving electric heating or electrodes. For example, when precipitation occurs, it causes a change in salt concentration, which alters the electric conductivity. A technique was devised to assess fouling by quantifying the electrical parameters [32]. Chen et al. [32] calculated the electrical resistance (δ R) with the help of stainless-steel electrodes. The thermal resistance (R_{th}) was then determined using the heat flux, the temperature of the fouled wall's backside surface, and the bulk temperature of the fluid. The discrepancy between the resistance of a non-fouled reference section and a fouled section was calculated and linked to R_{th} using the constants a and b.

$$R_{th} = a\delta R + b\left(\delta R\right)^2 \tag{10.5}$$

Guérin et al. [33] observed the development of fouling in a plate heat exchanger by incorporating electrodes to examine the impact of Reynolds number, concentration of calcium, and temperature build-up by whey protein concentrate (WPC). The deposit thickness was measured by a pneumatic lifting apparatus in which the weight of the plate was calculated before and after the test. Kirchhoff's rule separates the resistance of the fluid and the resistance of the deposit under fixed operating circumstances.

10.3.3 Measurement with Probes

- In this technique, the accumulation of unwanted substances is observed by means of a probe. The devices are categorized into four primary classifications based on their manner of operation:
- *Optical equipment.* Probes equipped with optical means, such as lasers, are sufficient to accurately measure the thickness of the fouling layer. Nevertheless, the aforementioned

strategies are limited to clinical experiments conducted under highly regulated circumstances, making it challenging to apply them to real-world industrial settings.

- *Mass measurement devices.* These devices enable the precise measurement of mass increase, allowing for quantitative assessment of the deposit. This approach necessitates the regulation of numerous factors, including the temperature of the wall and gas.
- *Devices equipped with flowmeters.* They observe the heat flow reduction over time due to blockage. Therefore, they enable the tracking of changes in fouling resistance over time. However, these probes are incapable of providing measurements for both the thickness and profile of the deposit. Until now, this particular sort of device has primarily been examined in situations where the predominant method of heat transfer is by radiation.

The monitoring of corrosion phenomenon is mainly conducted by the following online techniques.

10.3.3.1 Linear Polarization Resistance for Corrosion Rate Measurement

Linear polarization resistance (LPR) is an electrochemical approach and is the sole corrosion monitoring technology that enables uninterrupted and on-site measurement of corrosion rates. However, this technique is only applicable to liquids that are conductive as electrolytes; its response and data acquisition characteristics supersede other kinds of corrosion monitoring methods. Polarization resistance is an effective technique for quickly detecting corrosion issues and taking corrective measures, which helps extend the lifespan of the plant and reduce unexpected periods of inactivity. The approach is most efficient when used as a persistent monitoring system. When a metal or alloy electrode is immersed in a conductive liquid possessing sufficient oxidizing power, it will undergo corrosion via an electrochemical process. This method entails two concurrent and mutually reinforcing reactions. At locations where oxidation occurs, metal ions will move from the metal surface into the surrounding solution, resulting in an excess of free electrons on the surface. The surplus electrons will migrate toward adjacent cathodic sites, where they will be used by oxidizing species present in the corrosive liquid. An illustration of the electrochemical process taking place on a solid metal surface is depicted in Figure 10.6.

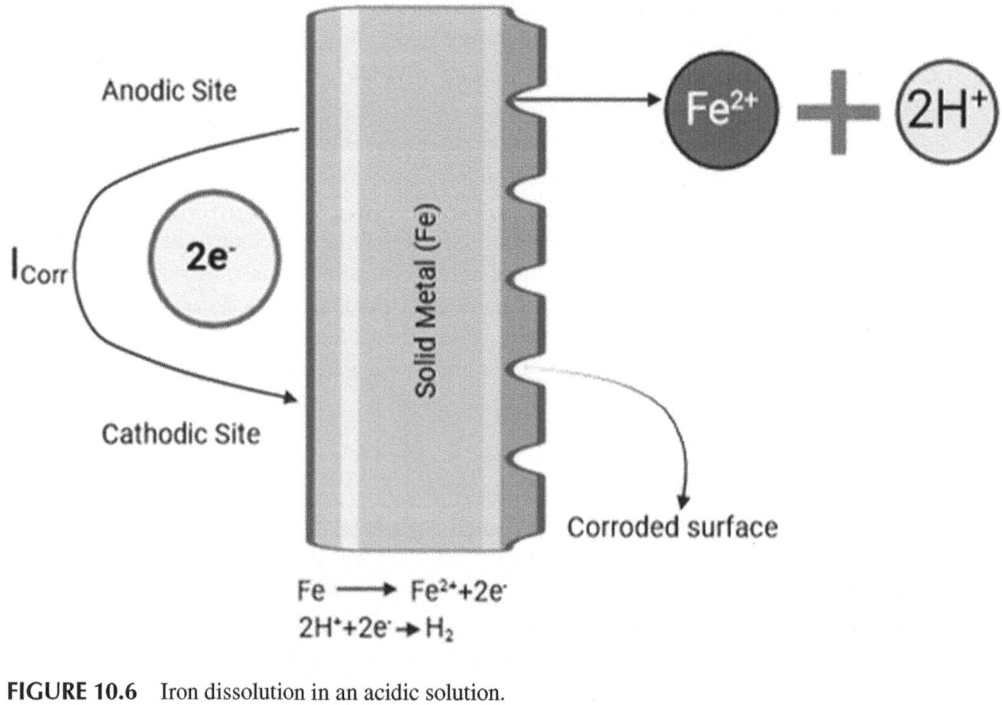

FIGURE 10.6 Iron dissolution in an acidic solution.

The open circuit potential (OCP) refers to the state of equilibrium displayed by a material in the absence of any current flow. It serves as an indicator of the stability of the electrochemical system. This phenomenon can also be referred to as corrosion potential (E_{corr}). At this specific point, the rate of dissolution from the anode is precisely equivalent to the rate of the cathodic reaction, resulting in no net current flowing through the electrode. The corrosion current (I_{corr}) is the value of either the cathodic or anodic current at OCP. With the help of I_{corr}, the corrosion rate can be determined using a modified form of Faraday's law:

$$Corrosion\,rate = \frac{I_{corr} \times W}{A \times D} \times 128.67 \tag{10.6}$$

Where *W*, *A*, and *D* represent the equivalent weight (g), area (cm^2), and density (g/cm^3) of the corroding metal (g)

10.3.3.2 Measuring Biofilm Growth

The growth of the biofilm is assessed using the particular electrochemical probe BIOX [34], which is a unique tubular electrochemical "battery." The current density of this probe exhibits an augmentation when the biofilm builds up on a stainless-steel cathode. The biofilm on the surface of the BIOX probe alters the electrochemical properties at the interface between the metal and the biofilm, resulting in an increased rate of oxygen reduction reactions at the cathode. The electrochemical sensors exhibit exceptional sensitivity to the emergence of the biofilm in its first stages. The electrochemical sensor detects the presence of the biofilm by analyzing a small amount of biofilm that has formed on a small percentage of the electrode surface. At a bacterial density of around 10^7 bacteria/cm^2, the electrical signal becomes saturated, resulting in a signal rise of almost 0.5 V compared with the initial value. Since 1990, industrial BIOX systems consisting of a dedicated computer, flow meter, and temperature meter have been used in numerous thermal power stations to enhance antifouling treatments [35].

10.3.3.3 Chlorination Treatment

The corrosion phenomenon can also be monitored through chlorination treatment followed by the BIOX probe. A sensitive cathode is deployed to capture the activity of powerful oxidants and biofilm growth at the cathodic site. The oxidant species (i.e., chlorine and bromine) act as supplementary cathodic processes, augmenting the flow of galvanic current through the electrochemical probe, similar to the biofilm. The detection of oxidants is quick; however, the response of biofilms is delayed as the bacterial colonization on the electrode surface requires adequate time. Under normal operating settings, the BIOX software allows for a straightforward differentiation of the contributions of both phenomena by observing the probe's output responses over time [36].

10.4 METHODS FOR THE PREDICTION OF FAILURE

If a single apparatus malfunctions inside an industrial unit, it can halt the entire manufacturing process. It can further result in increased labor hours, project delays, and process deferrals, and together lead to significant financial losses. It is important to regularly assess the condition of the instruments, particularly the heat exchanger, in order to ensure their longevity. Any atypical disruption or malfunction of the device can lead to its failure. Even a slight disregard for the irregularity can result in instrument failure. In recent times, various machine learning (ML) algorithms such as linear/multi linear regression, time-series forecasting, logistic regression, cluster analysis, auto associative kernel regression (AAKR), support vector machines (SVM), artificial neural networks (ANNs), deep neural networks (DNNs), random decision forests, and so on have demonstrated significant potential in accurately predicting instrument failure beforehand [37–40]. A group of researchers presented a very reliable deep neural network model capable of forecasting fouling in

HE [41]. They utilized historical operating data to train a neural network with two hidden layers. The network was then used to forecast the capacity of restricting fouling for water and gas flows, which were transformed into the logarithmic domain. The researchers presented a complex neural network structure to predict overall effect as well as individual effect on resistance to fouling. An integrated system using hybrid AI and ML techniques, together with an acoustic emission system, was utilized to detect leaks at an early stage by analyzing fluid flow and temperature profiles [42].

The corrosion current density, a direct measure of corrosion rate from a given surface, was forecasted employing an ANN accompanied by the imperialist competitive algorithm (ICA) [43]. Through examining six distinct ML methods, i.e., random choice forests, gradient boosting, highly randomized trees, adaptive boosting, k-nearest neighbors, and an external neural network, Aláiz-Moretón et al. [44] discovered faults in a geothermal heat exchanger. Additionally, a support vector machine was employed to predict the performance of the heat exchangers [45].

In the subsequent section, a case study is discussed where a predictive model was suggested to forecast the failure of an industrial heat exchanger (PE-2-E-400). The Weibull model approach was adopted to fit the data of tube thickness over a period of 14 years.

10.4.1 A Case Study

An industrial heat exchanger of a renowned plant was studied by Mohd Fikram [46]. The tubes were procured from M/S Benteler, Germany. The specifications of the tube conformed to ASTM standards. The thickness of the tube was recorded in those regions where the main reason for failure was corrosion or erosion. The data was obtained from the plant from 1991 to 2006. The cumulative distribution function was plotted to predict the data and other parameters (scale and shape factor) were also calculated to infer on the reliability of the system. An initial inspection of data over the period 1991–2005 revealed that the tube thickness was reduced by one-third of the original thickness. *Weibull WinSmith* software was employed to obtain the reliability plots. To run the simulation, the reliability function ($F(x)$) was computed from the following equation:

$$F(x) = 1 - e^{-\left(\frac{x}{\alpha}\right)^{\beta}} \tag{10.7}$$

Where α and β represent scale and shape factor, respectively

The pair (β, α) is sufficient to define the reliability function. The value of β is computed from the slope of a Weibull plot. The values of 41.15 and 40.78 correspond to the data points for 2005 and 1991, respectively. The authors have used these data to plot a scatter matrix plot (Figure 10.7) to infer the percentage of errors between the actual and predicted scale factors (characteristic tube thickness), which increase over years of operation. Its prediction can be effectively done by a regression model.

Recently, with the growing popularity of machine learning approaches such as ANNs, the accuracy of the prediction can further be increased with improved output characteristics. Heat transfer parameters such as Nusselt number, friction factor, pressure drop, and boiling and condensation coefficients have been precisely predicted by various machine learning techniques and are well cited for enriching the existing literature [42–45].

10.5 PREVENTION AND REMEDIATION MECHANISMS

10.5.1 Fouling Remediation

It is important to frequently cleanse the HEs in order to uphold their effectiveness. The cleaning process of the HE can be broadly categorized into two sets: online and offline cleaning methods [47]. It is possible to perform the cleaning online while the heat exchanger is in operation to ensure

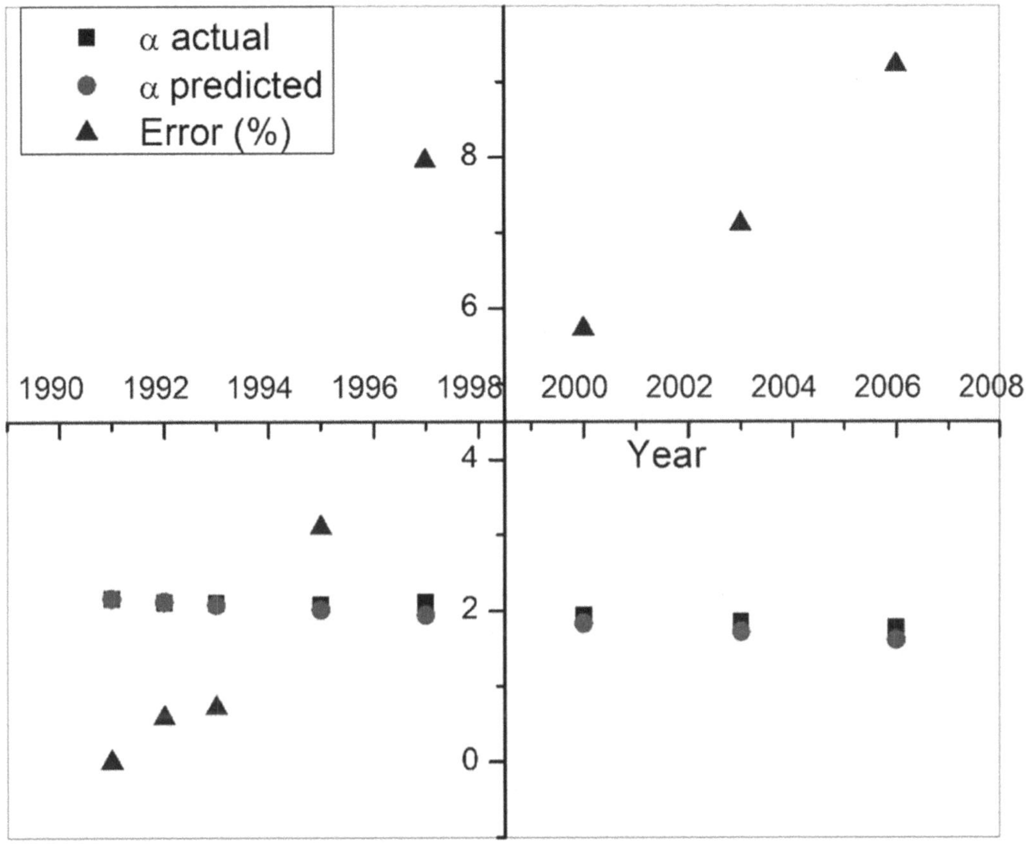

FIGURE 10.7 Computation of errors in a prediction from two sets of data.

optimal equipment performance without any interruptions. This allows for longer periods between maintenance shutdowns. In several other instances, it is important to carry out a thorough cleaning of the equipment via an offline cleaning process, which is performed at the time of the installation's stop phases.

10.5.1.1 Online Cleaning

Typically, online cleaning involves the utilization of mechanical and/or chemical techniques. The cleaning techniques mentioned are specifically tailored for the tubes, and dismantling is uncalled for [29]. One of the primary benefits of online cleaning is uninterrupted operation of the HE, minimizing the need for maintenance downtimes. Suddenly halting the process incurs additional expenses for installing a new heat exchanger or undertaking costly modifications, without any guarantee that the tubes will be free of fouling. The online mechanical cleaning methods include:

1. The inner surfaces of tubes are cleaned by circulating sponge rubber balls, brushes, or a stretched spring [29]. These procedures have the capability to inhibit the build-up of particulate matter, create biofilm, accumulate scales, and deposit corrosion products. They are exclusively suitable for flowing within the interior of tubes.
2. *Acoustic blow-molding.* This method utilizes sound waves (vibrations) at both high and low frequencies to remediate fouling issues on HEs. This is a prophylactic method rather than a therapeutic one. One advantage of this method is that it is not harmful to the surfaces it is applied to. However, it is not very good at removing stubborn and sticky deposits, which are often linked with slagging.

3. *Flow reversal.* This technique enables the removal of stuck particles in the entry collectors.
4. *Overspeed.* This technique enables the identification of operating circumstances that are similar to clean conditions. The overspeed method eliminates the outer deposit, but it also traps bacteria on the surface, promoting the creation of more deposits and speeding up the deposit growth process.
5. *Thermal shock.* This method involves rapidly altering the temperature, which leads to the shattering of the foulant layer and the potential for it to peel off.
6. *Radiation.* UV and gamma rays have long been explored for the purpose of sterilizing water contaminated with microorganisms.

Online chemical cleaning methods include inoculation of remediating chemicals inside the flow, scale inhibitors, and adopting chlorination processes to combat biofouling [29].

10.5.1.2 Offline Cleaning

Offline cleaning involves the cessation of performance and the cleansing of the HE, which may be accomplished using chemical or mechanical methods. This procedure can be applied without the need to disassemble the heat exchangers. However, it typically requires admittance to the internal interfaces. It is advisable to contemplate the setting up of a "standby" HE that allows for the cleaning of the clogged HE to still allow output. Mechanical and chemical methods include:

1. *Pressurized fluid.* This technique has been empirically validated and is widely employed across several sectors. Removing deposits within the tubes of the heat exchanger is an efficient method. Nevertheless, there is always a potential risk involved for surface wear, particularly for the degradation of the protective coating intended for corrosion resistance. The process of high-pressure water flushing involves removing loosened particles and repeating the process until clean surfaces are achieved.
2. *Use of tools.* This method is employed to eliminate even highly resistant deposits within the tubes of the HE. The removal equipment consists of motors (pneumatic or electric), hole-punches, and hydraulic guns. These devices can be used on rotating shafts, such as drills, cutting and polishing tools, and brushes. The materials used to build these devices can vary, including steel, nylon, and brass, depending on the material of the HE tubes and the type of deposit.
3. Chemical cleansing involves the use of acid solutions, chlorinated solvents, aromatic solvents, and oxidant solutions, which act as scale inhibitors [29].

10.5.2 Corrosion Remediation

The corrosion issue in industrial heat exchangers is mainly countered by offline design considerations and corrective action.

10.5.2.1 Offline Design Modification

To prevent excessive fouling, one can manage the fluid velocity, choose appropriate materials, regulate operating constraints, select the right HE types, and consider the shape. The reduction of fouling can be achieved through the implementation of a distinctive baffle and tube configuration. During the design stage, it is necessary to assign a predetermined value for fouling resistance, which is dependent on the kind of fluid. This is done even though fouling is a hypothesis that varies with time. The maintenance schedule and operating variables of the HE are contingent on the fluid type and the fouling factor of its design. To avoid unexpected shutdowns for cleaning purposes, it is necessary to create a maintenance schedule [48, 49]. TEMA is a well-known and commonly used source of fouling characteristics for designing heat exchangers [50].

The measures to mitigate SCC problems involve: (i) using greater caution at the end of tube rolling to minimize additional wall thickness decrease and extrusion; (ii) ensuring enough venting of the condenser to eliminate the stagnant area below the top tube sheet, ensuring no empty areas for vapor entrapment and maintaining constant wetting of the tubes to prevent alternating wet and dry conditions; and (iii) a heat treatment or shot peening process can be applied after rolling the tube. This process helps to decrease the residual stress to a level where SCC is more likely to occur [51].

In order to prevent operational (start and shutdown) issues, it is necessary to flush all newly installed or previously used heat exchangers before starting them up. This process will effectively remove any debris that may have become stuck in the pipe system. Prior to initiating the system, it is imperative to introduce cold fluid into the heat exchanger following the cleaning process. It is crucial to commence the system gradually and cautiously to avoid any harm to the heat transfer fluid.

The suggested measures for protection against crevice corrosion comprise the following: (i) avoiding any gaps or openings in the structural design; (ii) ensuring that the welding joints have complete dissemination of the weld metal; (iii) preventing the accumulation of sediments by evading protruded geometries where sediments can collect during the design process; (iv) modifying the design to eliminate any gaps or openings; (v) selecting the appropriate construction material; (vi) using nonabsorbent gaskets whenever possible; and (vii) regularly cleaning the heat exchangers to remove deposits and debris [52].

10.5.2.2 Corrective Action

Galvanic and erosion corrosion can be effectively mitigated through the careful selection of materials, meticulous design, proper coating, and diligent maintenance [53–57]. To significantly reduce the likelihood of pitting corrosion, it is crucial to control the chloride content and/or choose suitable materials. Dealloying in heat exchangers is primarily mitigated through careful material selection. The addition of small amounts of tin, arsenic, antimony, and phosphorus significantly increases the resistance of materials like brass. This enhanced resistance is particularly observed in alloys such as brass. These materials, in addition to Cu-Ni, are regarded as more resistant to dealloying compared with brass.

10.6 FUTURE SCOPE AND DIRECTION

Emerging technologies have been successful in predicting failures in heat exchangers and may thereby reduce the replacement cost with minimal maintenance and overhauling. However, there are some limitations with these technologies such as data overfitting and limited application in real-time situations. The existing literature centers around the prediction of failure modes in advance to take corrective actions. However, correlating the different dimensionless numbers provides more insights into actual flow behavior inside a heat exchanger. The flow pattern inside a heat exchanger follows a mixed flow regime from laminar to turbulent, and a single heat transfer characteristic is insufficient to depict its failure model. Hence, the failure of an industrial heat exchanger can also be reduced by efficient design optimization, which may reduce periodic predictive maintenance. A paradigm shift from a predictive to a preventive model is likely to come up in the future.

10.7 CONCLUSIONS

Fouling and corrosion negatively impact the thermohydraulic efficiency of HEs. In general, the heat exchanger's heat transmission proportion is greatly reduced. This occurs as a result of the products having low heat conductivity. In addition, the large deposition of these products might obstruct fluid movement and increase the pressure difference. Therefore, gaining a thorough comprehension of the factors that lead to fouling and corrosion, and the methods to prevent them, has several prospects for corrosion experts and engineers. This chapter provided an overview of fouling and corrosion in HEs, including their mechanisms and factors that influence the rates of fouling and corrosion. The effect of fouling and corrosion on the working of a system was deliberated. Methods for monitoring fouling and corrosion and strategies for mitigating fouling and corrosion were summarized. An exploration of emerging technologies in fouling/corrosion management was also outlined.

REFERENCES

1. Ali, M., Ul-Hamid, A., Alhems, L. M., & Saeed, A. (2020). Review of common failures in heat exchangers–Part I: Mechanical and elevated temperature failures. *Engineering Failure Analysis*, *109*, 104396.
2. Schwartz, M. P. (1982). *Four types of heat exchanger failures*. ITT Bell & Gosset.
3. Panahi, H., Eslami, A., Golozar, M. A., & Laleh, A. A. (2020). An investigation on corrosion failure of a shell-and-tube heat exchanger in a natural gas treating plant. *Engineering Failure Analysis*, *118*, 104918.
4. Milivojevic, S., Stevanovic, V. D., & Maslovaric, B. (2014). Condensation induced water hammer: Numerical prediction. *Journal of Fluids and Structures*, *50*, 416–436.
5. Laurent, M., Estevez, R., Fabrègue, D., & Ayax, E. (2016). Thermomechanical fatigue life prediction of 316L compact heat exchanger. *Engineering Failure Analysis*, *68*, 138–149.
6. Goyder, H. G. D. (2002). Flow-induced vibration in heat exchangers. *Chemical Engineering Research and Design*, *80*(3), 226–232.
7. Berradja, A. (2019). *Electrochemical techniques for corrosion and tribocorrosion monitoring: fundamentals of electrolytic corrosion*. Corrosion Inhibitors.
8. Ghosh, D. P., Hassan, M., Dennis, S. J., Sharma, D., Elhashimi, M. A., & Abbasi, B. (2024). Anti-fouling rotating polymer-based heat exchanger for zero liquid discharge humidification-dehumidification desalination. *Water Research*, *258*, 121749.
9. Ilyunin, O., Bezsonov, O., Rudenko, S., Serdiuk, N., Udovenko, S., Kapustenko & Arsenyeva, O. (2024). The neural network approach for estimation of heat transfer coefficient in heat exchangers considering the fouling formation dynamic. *Thermal Science and Engineering Progress*, *51*, 102615.
10. Ahmed, H. (2024). A review on corrosion of all aluminium microchannel heat exchangers. doi:10.5772/intechopen.1005180.
11. Jradi, R., Marvillet, C., & Jeday, M. R. (2022). Fouling in industrial heat exchangers: Formation, detection and mitigation. In *Heat Transfer-Fundamentals, Enhancement and Applications*. IntechOpen.
12. Shell & Tube Heat Exchanger Cleaning. Apex engineering products. Available online: https://www.apexengineeringproducts.com/shell-and-tube-heat-exchanger cleaning/ (accessed on January 8, 2023).
13. Berce, J., Zupančič, M., Može, M., & Golobič, I. (2021). A review of crystallization fouling in heat exchangers. *Processes*, *9*(8), 1356.
14. Frota, M. N., Ticona, E. M., Neves, A. V., Marques, R. P., Braga, S. L., & Valente, G. (2014). On-line cleaning technique for mitigation of biofouling in heat exchangers: A case study of a hydroelectric power plant in Brazil. *Experimental Thermal and Fluid Science*, *53*, 197–206.
15. Pohl, S., Madzgalla, M., Manz, W., & Bart, H. J. (2015). Biofouling on polymeric heat exchanger surfaces with E. coli and native biofilms. *Biofouling*, *31*(9–10), 699–707.
16. Mousavian, R. T., Hajjari, E., Ghasemi, D., et al. (2011). Failure analysis of a shell and tube oil cooler. *Engineering Failure Analysis*, *18*, 202–211.
17. Ezuber, H., Alshater, A., Nisar, S. O., et al. (2018). Effect of surface finish on the pitting corrosion behavior of sensitized AISI 304 austenitic stainless-steel alloys in 3.5% NaCl solutions. *Surface Engineering and Applied Electrochemistry*, *54*, 73–80.
18. Yang, Z. G., Gong, Y., & Yuan, J. Z. (2012). Failure analysis of leakage on titanium tubes within heat exchangers in a nuclear power plant Part I: Electrochemical corrosion. *Materials and Corrosion*, *63*, 7–17.
19. Chandra, K., Kain, V., Sinha, S. K., & Gujar, H. G. (2020). Metallurgical investigation of a heat-exchanger tube of 70/30 cupronickel failed by fretting corrosion. *Engineering Failure Analysis*, *116*, 104756.
20. Syrett, B. C., & Coit, R. L. (1983). Causes and prevention of power plant condenser tube failures. *Materials Performance*, *22*(2), 44–50.
21. Thulukkanam, K. (2000). Heat exchanger design handbook. *Heat Exchanger Design Handbook*. https://doi.org/10.1201/9781420026 870/.
22. Davis, J. R. (2001). *Copper and copper alloys*. ASM International.
23. Kuźnicka, B. (2009). Erosion–corrosion of heat exchanger tubes. *Engineering Failure Analysis*, *16*, 2382–2387.

24. Klein, U., Zunkel, A., & Eberle, A. (2014). Breakdown of heat exchangers due to erosion corrosion and fretting caused by inappropriate operating conditions. *Engineering Failure Analysis, 43*, 271–280.
25. Liu, L., Ding, N., Shi, J., et al. (2016). Failure analysis of tube-to-tubesheet welded joints in a shell-tube heat exchanger. *Case Studies in Engineering Failure Analysis, 7*, 32–40.
26. Wei, X. L., & Ling, X. (2015). Investigation of welded structures on mechanical properties of 304L welded tube-to-tubesheet joints. *Engineering Failure Analysis, 52*, 90–96.
27. Yokell, S. (1982). Heat exchanger tube to tubesheet connections. *Chemical Engineering, 75*, 78–94.
28. Ardy, H., Albatros, T., & Sumboja, A. (2023). Failure analysis of duplex stainless steel for heat exchanger tubes with seawater cooling medium. *Metals, 13*(7), 1182.
29. Demasles, H., Mercier, P., Tochon, P., & Thonon, B. (2007). *Guide de l'encrassement deséchangeurs de chaleur.* France: Editions GRETh.
30. Riverol, C., & Napolitano, V. (2005). Estimation of fouling in a plate heat exchanger through the application of neural networks. *Journal of Chemical Technology &Biotechnology, 80*(5), 594–600.
31. Martin, H. (2002). The generalized Lévêque equation and its practical use for the prediction of heat and mass transfer rates from pressure drop. *Chemical Engineering Science, 57*(16), 3217–3223.
32. Chen, X. D., Li, D. X. Y., Lin, S. X. Q., & Özkan, N. (2004). On-line fouling/cleaning detection by measuring electric resistance-equipment development and application to milk fouling detection and chemical cleaning monitoring. *Journal of Food Engineering, 61*(2), 181–189.
33. Guérin, R., Ronse, G., Bouvier, L., Debreyne, P., & Delaplace, G. (2007). Structure and rate of growth of whey protein deposit from in situ electrical conductivity during fouling in a plate heat exchanger. *Chemical Engineering Science, 62*(7), 1948–1957.
34. Mollica, A., & Cristiani, P. (2002). On-line biofilm monitoring by electrochemical probe 'BIOX'. *IWA Journal Water Science and Technology, 47*(5), 45–49.
35. Cristiani, P (2005). Solutions to fouling in power station condensers. *Applied Thermal Engineering, 25*(16), 2630–2640.
36. Carvalho, L. M., & Cristiani, P. (2011). Experiences of on-line monitoring of microbial corrosion and antifouling on copper alloys condenser tubes. In *Proceedings of International Conference on Heat Exchangers Fouling and Celaning, Crete Island, Greece.*
37. Taqvi, S. A. A., Zabiri, H., Tufa, L. D., et al. (2021). A review on datadriven learning approaches for fault detection and diagnosis in chemical processes. *ChemBioEng Reviews, 8*, 239–259. https://doi.org/10.1002/CBEN.20200 0027.
38. Sohaib, M., Kim, C.-H., & Kim, J.-M. (2017). A hybrid feature model and deep-learning-based bearing fault diagnosis. *Sensors, 7*(12), 876. https://doi.org/10.3390/s17122876.
39. Bode, G., Thul, S., Baranski, M., & Müller, D. (2020). Real-world application of machine-learning-based fault detection trained with experimental data. *Energy*, 198, 117323. https://doi.org/10.1016/j.energy.2020.117323.
40. Feyli, B., Soltani, H., Hajimohammadi, R., et al. (2022). A reliable approach for heat exchanger networks synthesis with stream splitting by coupling genetic algorithm with modified quasilinear programming method. *Chemical Engineering Science, 248*, 117140. https://doi.org/10.1016/j.ces.2021.117140.
41. Sundar, S., Rajagopal, M. C., Zhao, H., et al. (2020). Fouling modelling and prediction approach for heat exchangers using deep learning. *International Journal of Heat and Mass Transfer, 159*, 120112. https://doi.org/10.1016/j.ijheatmasstransfer.2020.120112.
42. Ahn, B., Kim, J., & Choi, B. (2019). Artificial intelligence-based machine learning considering flow and temperature of the pipeline for leak early detection using acoustic emission. *Engineering Fracture Mechanics, 210*, 381–392. https://doi.org/10.1016/j.engfr acmech.2018.03.010.
43. Sadowski, L., & Nikoo, M. (2014). Corrosion current density prediction in reinforced concrete by imperialist competitive algorithm. *Neural Computing and Applications, 257*(25), 1627–1638. https://doi.org/10.1007/S00521-014-1645-6.
44. Aláiz-Moretón, H., Castejón-Limas, M., Casteleiro-Roca, J. L., et al. (2019). A fault detection system for a geothermal heat exchanger sensor based on intelligent techniques. *Sensors, 19*(12), 2740. https:// doi. org/ 10. 3390/ S1912 2740.
45. Sivaprakash, M., Haribabu, K., Sathish, T., et al. (2020). Support vector machine for modelling and simulation of heat exchangers. *Thermal Science, 24*, 499–503. https://doi.org/10.2298/TSCI190419398M.
46. Bin Mat Yazid, M. F. (2008). *Development of a failure prediction model for heat exchanger tube.* Doctoral dissertation, Universiti Teknologi Petronas.
47. Muller-Steinhagen, H., Malayeri, M. R., & Watkinson, P. (2001). Heat exchanger fouling: Mitigation and cleaning strategies. *HeatTransfer Engineering, 32*, 189–219.

48. Chenoweth, J. M. (1988). General design of heat exchangers for fouling conditions. *Fouling Science and Technology*, 477–494. https://doi.org/10.1007/978-94-009-2813-8_32.
49. Garrett-Price, B. A., Smith, S. A., Watts, R. L., & Knudsen, J. G. (1985). *Fouling of heat exchangers: Characteristics, costs, prevention, control, and removal.* Saddle River: Noyes Pubns.
50. Standards of the Tubular Exchanger Manufacturers Association (2007). TEMA, 9th edn. New York.
51. Ahmed, T. M., Alfantazi, A., Budac, J., & Freeman, G. (2009). Failure analysis of 316L stainless steel tubing of the high-pressure still condenser. *Engineering Failure Analysis, 16*, 1432–1441. https://doi.org/10.1016/J.Engfa Ilanal.2008.09.015.
52. Syrett, B. C., & Coit, R. L. (1983). Causes and prevention of power plant condenser tube failures. *Materials Performance*, *22*(2), 44–50.
53. Panda, S., Kazama, M., Kawai, T., Biswas, C. K., & Paul, S. (2022). Controlled surface modification of Ti6Al4V using biomimetic mineralization via thermo-chemical route improves bioactivity. *Ceramics International*, *48*(8), 11286–11297. https://doi.org/10.1016/j.ceramint.2021.12.351.
54. Panda, S., Kazama, M., Kawai, T., Biswas, C. K., & Paul, S. (2022). Rare transition metal doped hydroxyapatite coating prepared via microwave irradiation improved corrosion resistance, biocompatibility and anti-biofilm property of titanium alloy. *Journal of Alloys and Compounds*, *918*, 165662, https://doi.org/10.1016/j.jallcom.2022.165662.
55. Panda, S., Kazama, M., Kawai, T., Biswas, C. K., & Paul, S. Influence of strontium and niobium on the physical and biological performance of hydroxyapatite as a bioactive coating on implant materials. *Ceramics International*, *48*(22), 33256–33266. https://doi.org/10.1016/j.ceramint.2022.07.268.
56. Panda, S., Biswas, C. K., & Paul, S. (2022). *Anodization of implantable metal and alloy surfaces: Purpose and scope, surface engineering: methods and application.* CRC Press-Taylor & Francis.
57. Panda, S., Biswas, C. K., & Paul, S. (2023). *Techniques of biopolymer and bio-ceramic coatings on prosthetic implants, mechanical engineering in biomedical application: Bio-materials, implant design, Bio-3D printing, computational, tissue & biofluid mechanics.* Scrivener Publishing, Wiley.

11 Heat Exchanger Cleaning Methods

Prathamesh P. Khandekar and Manishkumar D. Yadav

11.1 INTRODUCTION

The entire chemical industry depends on clean heat exchangers to maintain the efficient working of a plant. Heat exchangers are used for cooling and heating the feed using a cooling and heating medium. During operation, the equipment receives depositions of materials, which affect the heat transfer of the equipment. The accumulation of undesired substances on the surface is defined as fouling. Fouling is seen in both crude preheaters and overhead condensers. Heat exchangers may deteriorate due to fouling; hence, they must be withdrawn from service for cleaning or replaced by another heat exchanger. The overall design of heat exchangers may significantly affect fouling, but, in most cases, fouling occurs despite good design, effective operation, and maintenance. As a result, heat exchangers and associated equipment must be cleaned periodically. Once the equipment is cleaned, the surface will be almost bare metal, providing the most effective heat transfer. The cleaning techniques used to remove the foulants from heat exchangers, both on the tube side and the shell side, are classified into two categories: mechanical and chemical cleaning. These methods are further classified as online or offline. The cleaning process employed while the plant is still operating is referred to as online; when the plant is shut down to clean the heat exchangers, it is referred to as offline. Factors such as cost, effectiveness, thoroughness, speed, and accessibility are taken into consideration when selecting any particular method. Before selecting any method, the composition of fouling deposits must be analyzed. This method is helpful in identifying the source of major foulants; on that basis, we can select chemical or mechanical methods. X-ray diffraction, X-ray spectrometry, and optical emission spectroscopy are some of the techniques used (Figure 11.1).

The fouling layers on the inside and outside surfaces increase with time. The fouling layer has lower thermal conductivity than the metal layer, as a result, the overall thermal resistance increases. There is an increase in surface roughness, increasing the pressure drop across heat exchangers. In pharmaceutical products, the process fluid or product can become contaminated. To avoid fouling, the heat exchangers are oversized, which increases the capital expenditure. The entire heat exchanger unit needs to be removed multiple times and there is a loss in production (Thulukkanam, 2024).

11.2 TYPES OF FOULING IN HEAT EXCHANGERS

11.2.1 Crystallization Fouling

The salts dissolved in the heat exchange medium crystallize on the heat transfer surface, resulting in scaling. This occurs because the solution becomes supersaturated, meaning the amount of soluble solid phase in the solution is greater than the equilibrium concentration. Supersaturation occurs due to the evaporation of solvents, heating a solution of inverse solubility salts, and cooling the solution below its solubility point. Also, supersaturation may occur in the fluid bulk, leading to crystal formation. The formed crystals then act as a fouling agent and can act as a seed for others. The salts that precipitate on the hot surfaces cause more fouling problems during heat transfer. Usually, the

DOI: 10.1201/9781003509837-11

FIGURE 11.1 Fouling in heat exchangers (Zeyninejad Movassag et al., 2013).

scales contain calcium carbonate ($CaCO_3$) or calcium sulfate ($CaSO_4$). Calcium carbonate nucleates in different forms, namely calcite, aragonite, and vaterite, and of these three, calcite is the most thermodynamically stable. All three polymorphs have different temperature solubility dependencies. At temperatures above 50°C, aragonite is formed; at temperatures below 40°C, vaterite is formed. Calcite is always present, and the other two are transformed into calcite over time. Calcium sulfate also appears in three forms: namely, gypsum or dihydrate ($CaSO_4.2H_2O$), calcium sulfate hemihydrate ($CaSO_4.0.5H_2O$), and anhydrite ($CaSO_4$). Gypsum is easily precipitated in the temperature range of 40–98°C, while others precipitate above 98°C. Once the supersaturation is met, the process of scaling occurs in a specific sequence of events; the initial heat transfer remains unchanged and is known as the delay phase. Nuclei for crystallization are formed in the areas of local supersaturation during the delay phase. Once the nuclei are formed, a concentration gradient is generated in the system. The concentration gradients initiate the diffusion of foulant from the bulk phase to the heat transfer surface. As a result, the overall heat transfer coefficient decreases due to the thermal resistance of the fouling layer.

11.2.2 Biofouling

River water or groundwater is frequently used for cooling in industries, the moderate temperature throughout all the different process sections enhances biofouling. There is an accumulation of biofilms and their inherent metabolic products (Frota et al., 2014). Biofilms are organized and highly hydrated communities of microorganisms attached to the heat transfer surface. Bacteria usually predominate in biofilm layers, and some of them produce extracellular polymers (extracellular polymeric substance) and easily attach to solid surfaces. The process of developing the biofilm is divided into two categories: promoting biomass production and opposing the development of such biolayers.

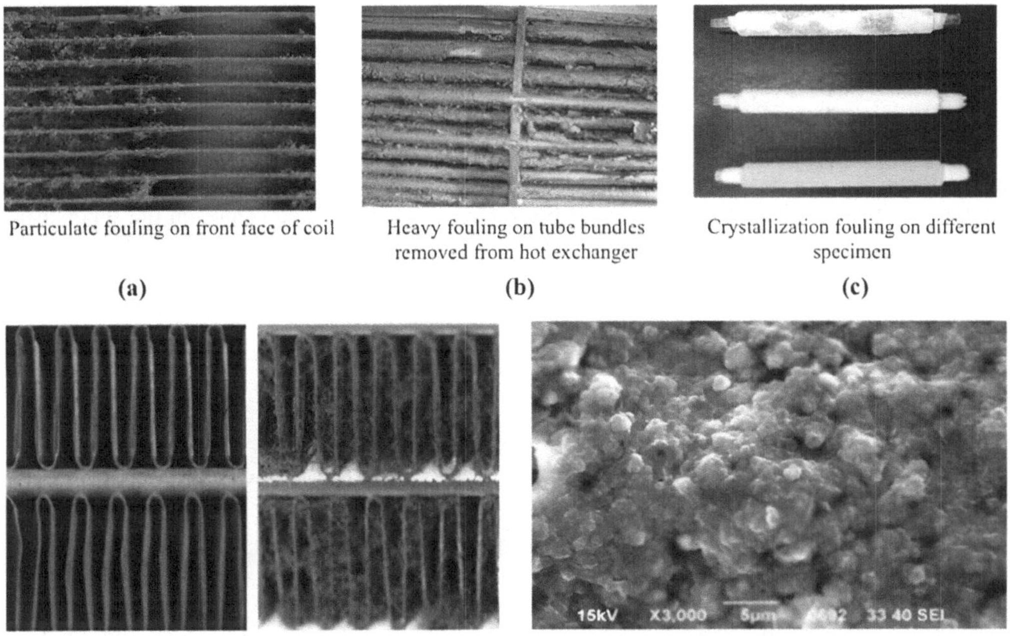

Particulate fouling on front face of coil (a)

Heavy fouling on tube bundles removed from hot exchanger (b)

Crystallization fouling on different specimen (c)

Heat exchanger fouled with dust (d)

Bio-fouling morphology (e)

FIGURE 11.2 Types of fouling in heat exchangers (Awais & Bhuiyan, 2019).

Initially, there is scaling on the surface, then the microorganisms are attached on the deposit surface and the attached microbes produce extracellular polymers, followed by mass transfer of nutrients to the biofilm-liquid interface, consumption of nutrients/oxygen, formation and release of metabolic products, and lastly, there is the removal of biomass from the biofilm (Melo & Flemming, 2010).

The removal of biofilm from the surface occurs by erosion and sloughing off. In erosion, there is continuous detachment of small amounts of biomass from the film, while in cases of sloughing off, there are random detachments of a macroscopic number of biofilms. The biofilm tends to reach up to a maximum thickness of some millimeters, hence the balance between the two sets of processes, i.e., biofilm production and biofilm removal will determine the overall rate of biofilm buildup (Melo & Flemming, 2010) (Figure 11.2).

11.2.3 Chemical Reaction Fouling

Chemical reaction fouling is caused because of three types of reaction: autoxidation, polymerization, and thermal decomposition. Asphaltene precipitation, oxidative polymerization, and coke formation occur due to chemical reactions in the heat transfer medium. Reaction with the metal can cause corrosion fouling. Autooxidation of hydrocarbons is considered a major source of unwanted deposits in heat exchangers, with the formation of unwanted gums in heat exchanger networks affecting heat transfer. Autoxidation occurs in the presence of excess oxygen with the formation of peroxy radicals. Oxygen radicals act as a precursor, and when the concentration of these radicals is low, the deposition is controlled by the transport of reactants to the hot metal surface. On the contrary, when the concentration of radicals is large, the deposition rate is controlled by the mass transfer rate of the radicals. In deoxygenated systems, fouling occurs due to thermal decomposition and polymerization (Watkinson & Wilson, 1997).

11.2.4 Gas-Side Fouling of Heat Exchangers

In steam boilers, waste heat is recovered from flue gases using heat transfer equipment. The flue gas contains fly ash, viscous substances, and a small amount of SO_3. Fly ash deposition by slagging or fouling is the main reason for boiler maintenance. The deposits formed in the radiant section are called slag deposits, while the deposits formed in the convective pass on the steam tube are called fouling deposits. The SO_3 in the fly ash reacts with water vapor, forming sulfuric acids, and when the temperature is below the dew point of the sulfuric acid, it will condense on the tube surface, causing corrosion. The fly ash deposits form an insulation layer affecting the rate of heat transfer (Li et al., 2017) (Figure 11.3).

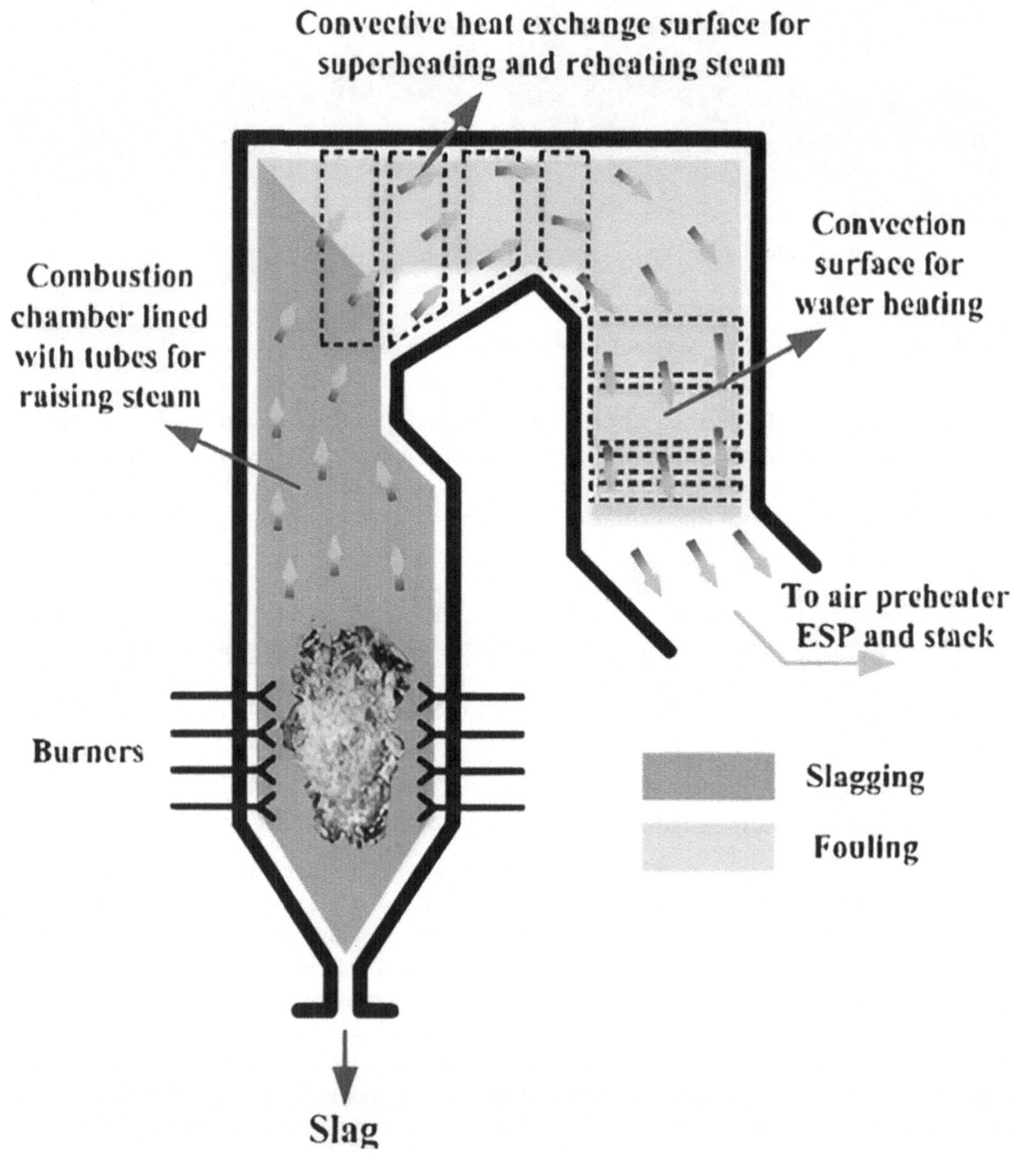

FIGURE 11.3 Gas-side fouling (Li et al., 2017).

11.2.5 Corrosion Fouling

The corrosion product formed on a heat transfer surface as a result of corrosion of the surface will affect the heat transfer and the fluid friction characteristics. The corrosion products are formed when exposed to an oxygenated atmosphere, resulting in the formation of a galvanic cell where flowing fluid acts as an electrolyte. The corrosion involves two simultaneous electrochemical reactions: one happening at the anode and the other at the cathode (Somerscales, 1997).

$$M \rightarrow M^{Z^+} + ze^- \tag{11.1}$$

$$\frac{z}{4}O_2 + \frac{z}{2}H_2O + ze^- \rightarrow z\left(OH^-\right) \tag{11.2}$$

Reaction 11.1 occurs at the anode and Reaction 11.2 occurs at the cathode. On the removal of the fluid, the metal hydroxide loses its constituent water, and the deposit consists of metal oxides. The rate of change in fouling thermal resistance depends on the rate of the transport of oxygen dissolved in water to the heat transfer surface.

So, the rate of growth of fouling is given by

$$\frac{dm_g}{d\theta} = K_1 \frac{dm_1}{d\theta} \tag{11.3}$$

Here the $\frac{dm_g}{d\theta}$ = rate of growth of fouling deposits,

K_1 is the yield coefficient, and

$\frac{dm_1}{d\theta}$ is the rate of oxygen transport (Figure 11.4).

Figure 11.5 depicts electrochemical cell formation. Corrosion can be of different types and are classified as uniform, localized, or pitting (Somerscales, 1997).

11.3 MECHANICAL CLEANING

Irrespective of the material of construction of heat exchanger tubes, the most effective way to ensure efficient heat transfer is to clean the tube consistently. This will increase the shelf life of the heat exchanger. After the deposits in the tubes are removed, the bare metal surface is exposed to the

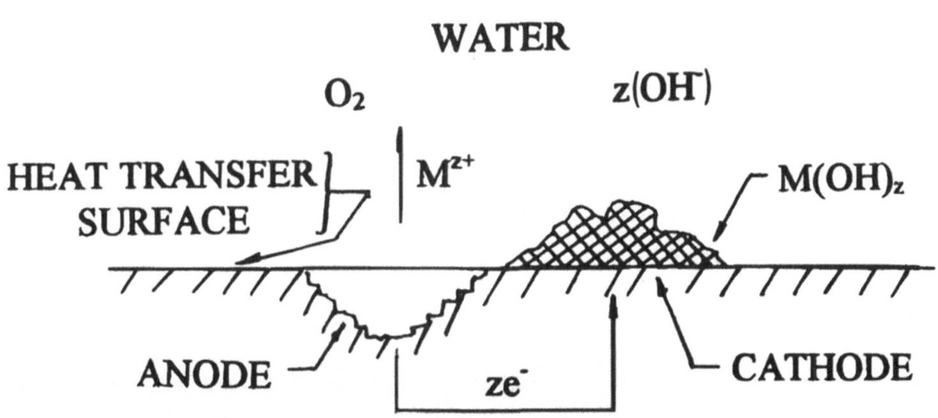

FIGURE 11.4 The corrosion fouling mechanism (Somerscales, 1997).

FIGURE 11.5 Corrosion fouling of heat exchangers (Zeyninejad Movassag et al., 2013).

heat transfer fluid, providing the highest level of heat transfer. Cleaning can be performed online or offline, however, the most effective cleaning method is offline mechanical cleaning (Kishore et al., 2018).

11.4 OFFLINE MECHANICAL CLEANING

11.4.1 Hydro-Blasting

High-pressure water in the pressure range of 10,000–40,000 psi is used to remove unwanted deposits. Water jets are most effective on the external surfaces of the tube where there is easy accessibility for passing water. Delivery of water can be through multijet sprays or high-pressure water lances. In cases of hard deposits, abrasive particles are added with the water stream. Hydro-blasting keeps the heat exchanger offline and is labor intensive (Davis, 2015).

11.4.2 Jet Cleaning

Jet cleaning is used to remove scales in heat exchangers. The jet contains pressurized water in a pressure range depending on the type of impurity; for the removal of various types of fouling, specific pressure ranges are utilized to achieve optimal cleaning efficiency. Airborne contaminants in air-cooled heat exchangers are effectively removed at a pressure range of 2–4 bar. Soft deposits, including mud, loose rust, and algae, require cleaning pressures within the range of 40–120 bar. The removal of heavy organic deposits and polymeric fouling necessitates higher pressures, typically between 300 and 400 bar. The descaling of tube-side surfaces demands the highest pressures, ranging from 300 to 700 bar. These pressure parameters are critical to ensuring efficient fouling mitigation while minimizing damage to the equipment (Köhler et al., 2013).

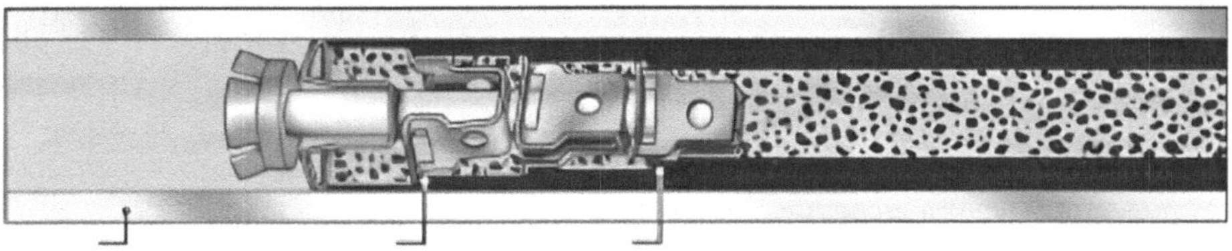

FIGURE 11.6 A scraper-type cleaner (Müller-Steinhagen et al., 2011).

11.4.3 Scraper-Type Tube Cleaners

Scraper-type cleaners, such as the CONCO tube cleaner, are all-purpose tube cleaners for condensers and condensers. They are mostly used for hard deposits like calcium carbonate, corrosion byproducts, and micro/macro fouling. Cleaners are designed in such a way that the blades are mounted on a spindle at one of the ends, while on the other end, there is a plastic disk that allows a jet of water to propel the cleaner through the tube. These cleaners are shot with a water gun at 300 psi through the tube, plowing off all the accumulated deposits. Water flushes the removed matter out of the cleaner, which can be collected for analysis later. These metal cleaners are designed with a controlled spring-loaded cutting edge; the diameter of the cutting surfaces should be close to the diameter of the tube (Saxon & Putman, 2003).

Figure 11.6 perfectly represents the working mechanism of a scraper-type tube cleaner.

11.4.4 Projectile Cleaning

The methods mentioned in the above sections require hazardous chemicals and large amounts of water. In projectile cleaning, a pneumatic launcher propels the projectile, a dart made up of nylon, into the tube. The projectile has a diameter larger than that of the tube, and as it travels through the tube, it removes deposits formed due to fouling (Kohli, 2018).

Figure 11.7 illustrates the projectile cleaning methodology. A pneumatic launcher with a nozzle is used to compress the projectile. Once compressed through the nozzle, the projectile expands

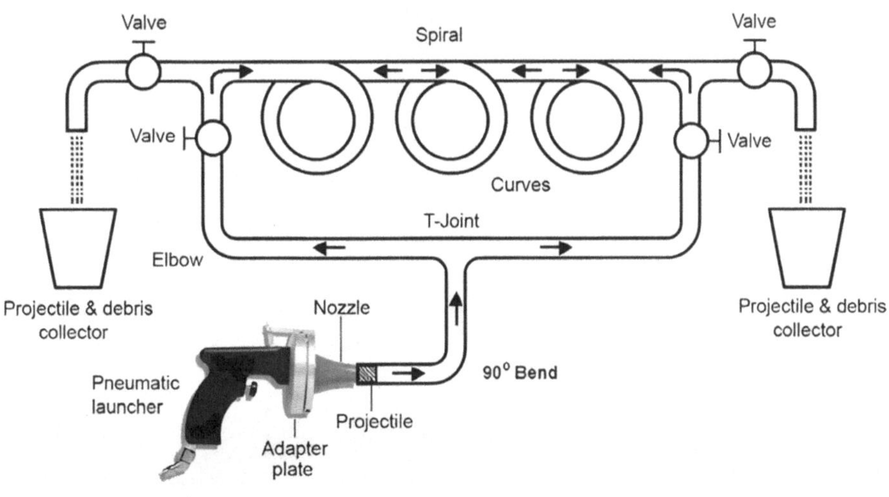

FIGURE 11.7 Projectile cleaning (Kohli, 2018).

against the inner walls of the tube (Jalalirad et al., 2013). The projectile is made of elastic material. The friction on the surface, combined with the propulsion force from the compressed gas, allows the projectile to move at an approximate speed of 15 m/s—removing all the deposits. The projectile used is made up of virgin foam or rebonded foam. The density of the projectile should be sufficiently high so that the compressed gas cannot penetrate inside the projectile, but not large enough so that the projectile becomes incompressible (Kohli, 2018).

Figure 11.8 shows the condition of the tube and the projectile before and after cleaning.

The nozzle has a smooth internal surface to avoid frictional losses. At both ends of the tube there is a projectile collector that is not only used to collect the projectiles but also to assess the internal conditions of the tube and determine whether any additional cleaning procedure is required. Depending on the working condition of the projectile, it may be cleaned or reused. In pharma industries, reusing of the projectile is avoided.

FIGURE 11.8 The condition of a tube and a projectile before and after cleaning (Kohli, 2018).

11.5 THERMAL CLEANING OF HEAT EXCHANGERS

The thermal cleaning of heat exchangers is not a new initiative; however, it is not commonly used in industries. The Thermo-Clean group has developed an innovative way of cleaning heat exchangers using thermal cleaning methods. Pyrolysis is a process of thermal decomposition of organic material in an oxygen-deficient atmosphere. At a temperature above 450°C, all the organic material is converted into homogeneous residue that contains pyrolysis gases, pyrolysis oil, and a carbon-rich residue. The released gases are reused to recover energy. Pyrolysis ovens are used for this process, with the average size ranging from 1 m^3 to 100 m^3. The heat exchangers are loaded onto a forklift and brought into the oven chamber. Once in the chamber, the oxygen level is gradually reduced by up to 8%, then the temperature is slowly increased to as high as 450°C, depending on the nature of fouling and the material of the heat exchanger. After achieving sufficient temperature, the equipment is subjected to such conditions for a residence time of only two seconds.

After pyrolysis, only 1–5% of deposits remain inside the heat exchanger. Controlling temperature and oxygen concentration are monitored using different thermocouples and sensors. The produced heat is circulated internally due to excellent internal circulation. In jet cleaning, it hardly reaches the bends, therefore, as a result, some remnants remain behind, causing heat exchanger fouling in a short time. Thermal cleaning oxidizes the deposits accumulated in such bends, providing much better cleaning. When the tube is fixed with the shell, there is no access around the tubes to pass jets, and thermal cleaning can be effective.

Static mixers are used to increase the mixing and heat transfer inside the tube. With fixed static mixers, there is no space to penetrate water for cleaning applications. To clean such substances, thermal treatment followed by jet cleaning is used.

11.6 ONLINE MECHANICAL METHODS

Offline methods require the shutdown of the unit operation in order to clean the heat exchanger. Online methods provide us the flexibility to remove the accumulated deposits during equipment operation.

11.6.1 Sponge Rubber Ball Cleaning Systems

The online condenser tube cleaning system continuously cleans the heat exchanger tubes during the cooling operation. The ball is fed into the fluid entering the tubes before entering the tube bundle heat exchanger. The ball has a diameter slightly greater than the tube diameter. Due to the oversizing of the balls, as the ball passes through the tube, it presses against the tube wall, wiping off the deposit (Nyemba et al., 2018) (Figure 11.9).

On average, such systems require a ball recirculation pump, ball strainer, ball collector, ball injector, and a control panel. The recirculation pump is a special type of centrifugal pump designed to transport cleaning balls. The strainer section separates the cleaning balls from the fluid. All this equipment is necessary to ensure there is no blockage caused by the ball. In order to clean the tubes' entire surface, a certain force is required so that the accumulated deposits can move. As a result, an oversized clean ball is preferred. The differential pressure, pressure loss by cooling water via the heat exchanger, affects the size of the ball. When the ball moves through the tube, it is not only pushed by the pressure difference but also perfused by the cooling water. The porous nature of the ball is the reason behind this behavior. Depending on the type of deposit, the nature of the ball changes (Kleinebrahm, 2017).

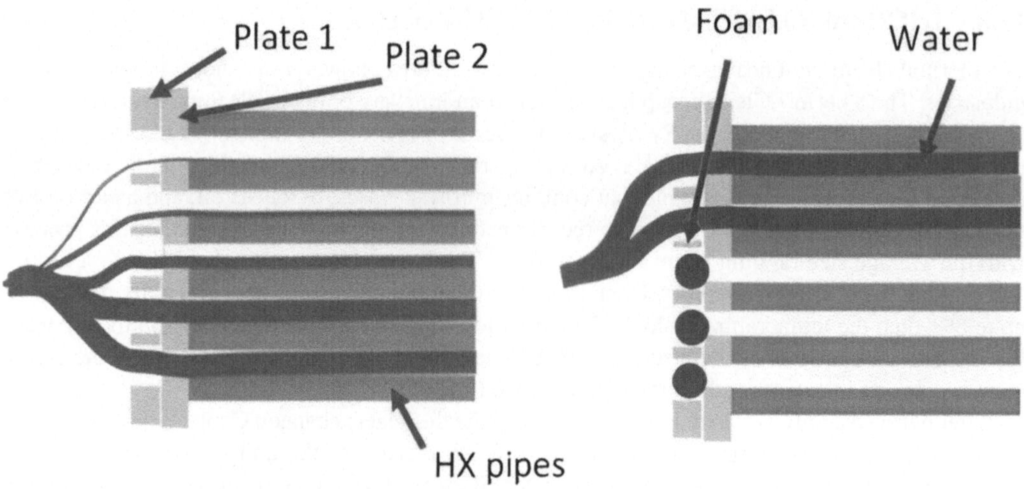

FIGURE 11.9 Self-cleaning heat exchangers (Brooks & Roy, 2022).

11.6.2 Insert Technology

Heat exchanger tube inserts are used in industry to improve heat transfer and mitigate fouling by means of a mechanical effect. The three types of inserts most commonly used are Spirelf, Turbotal, and Fixotal systems. The Spirelf is a fine wire with springs threaded through the tubes and held in place by straight wires. When fluid enters, the Spirelf vibrates axially and radially, not allowing any nucleation sites to form. This reduces the formation of deposits and breaks the boundary layer on the tube side, improving heat transfer (Figure 11.10).

Turbotal is a rigid helicoidal form in the tubes. The Turbotal is held at the inlet of the tube by a fixing device, which provides it freedom to rotate during operation. The rotation increases the heat transfer coefficient by increasing turbulence. Fixotal consists of a wire coil that is inserted in every tube. This coil has a swirl effect, which is the primary reason for improving turbulence. It increases the fluid flow velocity inside the tube. Turbotal and Spirelf are typically used in a crude preheat train, debottlenecking installations in crude distillation units, hydrotreaters, and feed-effluent exchangers. When a Spirelf device is inserted inside a reboiler in the slug flow zone, it reduces large bubble size formation.

11.6.3 Self-Cleaning Heat Exchangers

Self-cleaning heat exchanger technology works on the principle of a fluidized bed. The fouling liquid is fed from the top, with specially designed inlet and outlet channels. Solid beads are fed into the tubes, and due to the internal flow distribution system, beads are uniformly distributed along

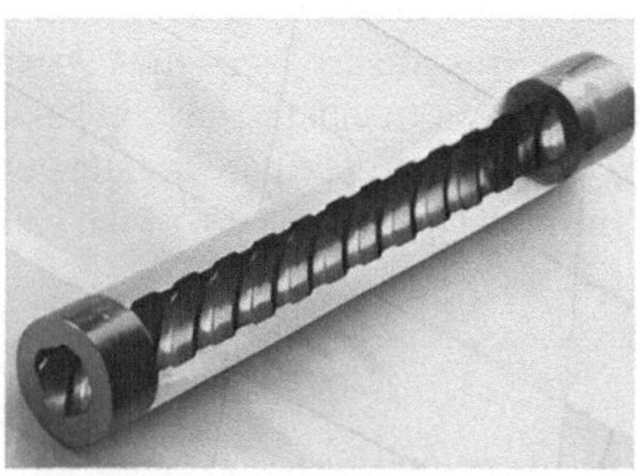

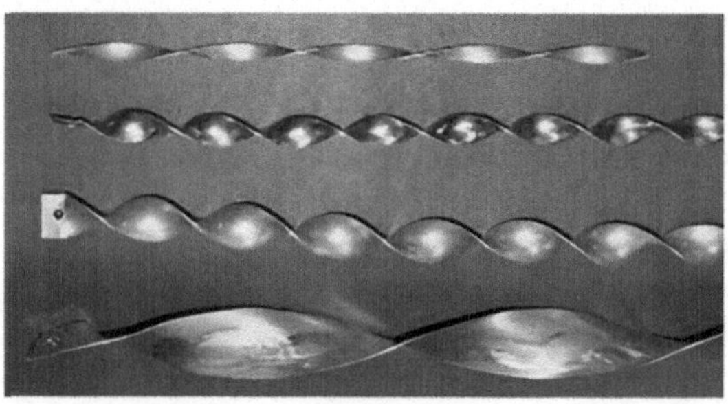

FIGURE 11.10 Spirelf (Popov et al., 2012).

the tube and are suspended throughout the internal surface of the bundle. The particles pass through the tube by upward flow of liquid, and as the particles move, they cause a scraping, sweeping effect on the heat exchange tube. Due to this action, the accumulated deposits are removed alongside the flow of particles. The beads are generally metal wires, ceramic balls, and galls beads, with a range of 1–4 mm in outer diameter. At the outlet, there is a solid-liquid separator that separates the beads and returns them to the inlet (Klaren & De Boer, n.d.)

EM baffle (expanded baffle) is a rigid piece of cold rolled metal that is slit and expanded. EM baffles provide advantages similar to nonsegmental baffles, such as a reduction of fouling and no vibrations, as well as lower costs than many nonsegmental baffles. Adding EM baffles to self-cleaning heat exchangers improves the heat transfer coefficient and reduces the total pumping power required on both sides (Klaren & De Boer, n.d.).

Figure 11.11 shows the (1) hot fluid outlet, (2) cooling water outlet, (3) solid beds, (4) springs, (5) cooling water inlet, and the (6) mesh, hot fluid inlet (Wu et al., 2022).

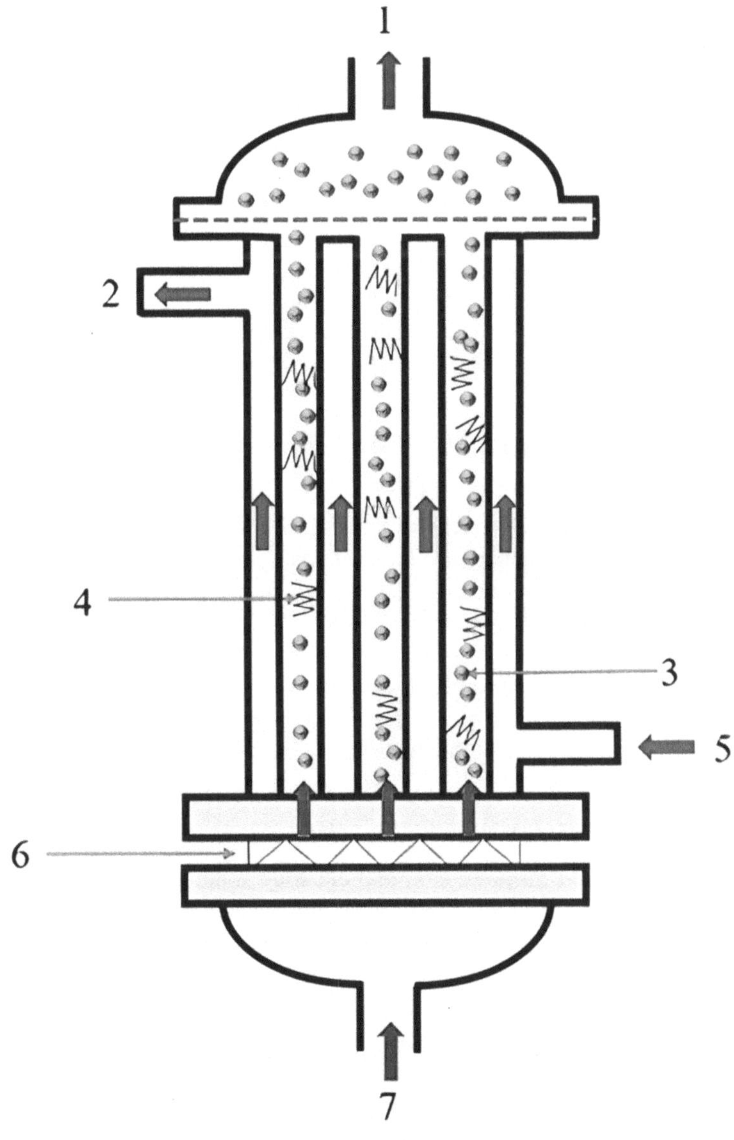

FIGURE 11.11 Fluidized bed cleaning (Wu et al., 2022).

11.7 CHEMICAL METHODS

Chemical cleaning of heat exchangers is only used when other methods are not satisfactory. In chemical cleaning, the role of the chemical is to dissolve or loosen the deposits. Chemical cleaning methods are mostly offline and are also known as clean-in-place. Detergents and sterilizing agents are commonly used as they are cost-effective and the unit does not need to be opened. In some cases, if the metal is not compatible with the cleaning solution, then inhibitors are added; hence, knowledge regarding the deposits must be available beforehand. There is always the possibility of equipment damage from corrosion, hence chemical cleaning must be assessed with a corrosion test beforehand. Chemical cleaning solutions include mineral acids, organic acids, alkaline bases, organic solvents, oxidizing agents, and reducing agents. Chemical cleaning follows three basic steps: chemicals are flushed to remove debris, followed by heating and water circulation, and finally, cleaning chemicals and inhibitors are injected along with circulating water. After sufficient time, the entire cleaning solution is discharged, and the system is flushed using water. The metal surface is again flushed with water to remove traces of chemicals (Thulukkanam, 2024).

Foulants such as iron oxide can be cleaned using inhibited hydrofluoric acid, hydrochloric acid, mono-ammoniated citric acid, sulfamic acid, or EDTA. Calcium and magnesium scales are typically removed with inhibited hydrochloric acid, citric acid, or EDTA. For oil or light greases, sodium hydroxide or trisodium phosphate are effective cleaning agents. Heavy organic deposits require chlorinated or aromatic solvents, followed by thorough rinsing to ensure complete removal. Coke deposits can be addressed using alkaline solutions of potassium permanganate. These cleaning methods are tailored to the specific nature of each fouling substance.

In many cases, chemical cleaning has certain merits such as uniform cleaning, and as there is no need to dismantle the entire unit, inaccessible areas can be accessed. However, it has certain disadvantages as well, as the chemicals used are hazardous, hence they need to be disposed of with proper safety concerns. Noxious gases can be emitted during cleaning due to chemical reactions between deposits and residue.

11.8 SOLVENT-BASED CHEMICAL CLEANING

Organic solvents are quite effective in removing deposits as they have a medium-to-high organic content. They are preferred when the concentration of the organic solvent is greater than 30%. They are cheap as they are available in abundance in refineries.

11.8.1 Chemical Solutions

Mineral acids like diluted HCl are circulated through tubes to remove scales caused due to cooling water. Chemical solutions are used when the organic content is less than 30%. A good cleaning agent is expected to remove 60–80% of the deposit.

11.8.2 Steam-Injection Cleaning

In this method, a concentrated mix of steam and cleaning solution is passed into a fast-moving stream. The stream atomizes the chemical, ensuring good contact with the metal surface.

The techniques discussed in the aforementioned sections are the traditional methods used in industrial practice. Due to the inclusion of big data analytics in almost every aspect of industry, machine learning (ML) and artificial intelligence (AI) have been extensively explored in the chemical industry for making decisions with respect to predicting the occurrence of fouling, as well as the appropriate methodology for mitigation. In subsequent sections, AI and ML are followed by their application to understand fouling in heat exchangers.

11.9 GENETIC ALGORITHMS

Genetic algorithms (GAs) are stochastic search methods that lead a population toward an optimum, using the principles of evolution and natural genetics. With proper encoding, we can use GAs to solve linear and nonlinear problems. The cleaning scheduling problem is a nonlinear discrete decision-making problem of a combinatorial nature where continuous and binary decision variables co-exist, leading to a mixed integer nonlinear programming problem (Al Ismaili et al., 2019).

Figure 11.12 depicts the different operations of a GA. Here our optimized fitness functions are heat transfer coefficient and total cost. Bidabadi et al. (2013) provide an example that optimizes a plate heat exchanger. Thakore & Bhatt (2007) provide design details of a plate heat exchanger. In a plate heat exchanger, there are four nonlinear constraints including pressure drops in both hot and cold streams.

The hot stream pressure drop constraint is

$$\Delta P_h - \frac{1.45 L V_h^2 \rho_h}{1705} = 0 \tag{11.4}$$

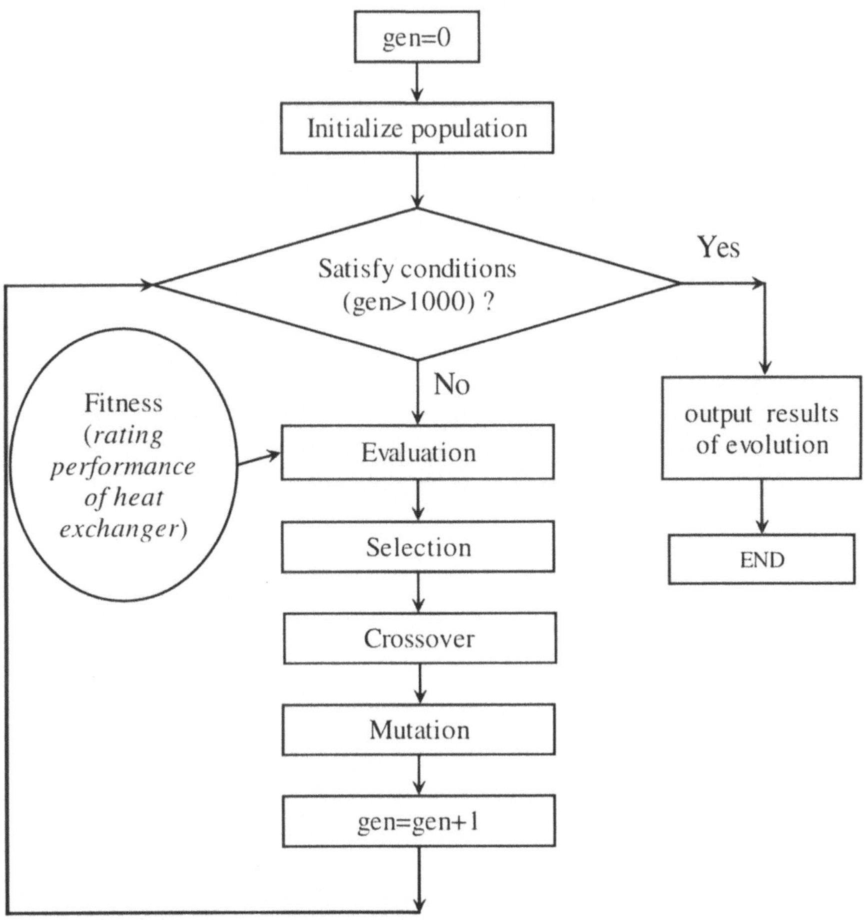

FIGURE 11.12 Working of the genetic algorithm (Xie et al., 2008).

The cold stream pressure drop constraint is

$$\Delta P_c - \frac{1.45 L V_c^2 \rho_c}{1705} = 0 \tag{11.5}$$

The outer diameter constraint is

$$D_s^2 - \left[C^2 + \left(15.36L\left(b_h + b_c + t\right)\right)\right] = 0 \tag{11.6}$$

The heat load constraint is

$$Q - \left(2U.A.\Delta LMTD\right) = 0 \tag{11.7}$$

The genetic algorithm runs over a population of 50–100 iterations. The program calculates values at 50 random points as genes or individuals enter the algorithm. The produced values are compared with the parent value, i.e., initial values, and children are produced with one parent or both parents' vectors (Bidabadi et al., 2013). In the upcoming generation, people more qualified or who have more fitness values are selected as parents. The algorithm will stop once the stop condition is reached. GAs are probabilistic in nature, therefore two runs of a GA with the exact same setting could lead to two different results.

11.10 ARTIFICIAL NEURAL NETWORKS

From the above section, we know that fouling depends on various variables, including the geometry of the heat transfer surface and the operating conditions of fluids. Due to the dependence of all such values, it is quite challenging to develop an empirical formula. Artificial neural networks (ANNs) can be valuable tools to model such complicated systems and they can provide accurate estimates of fouling. Himm (2000) supports the claim of using ANNs in solving chemical engineering problems. ANNs are designed to model the behavior of the human brain. ANNs are considered as nonlinear empirical models that are especially useful in making predictions in time and categorizing data (Himm, 2000).

Figure 11.13 shows a unit structure in an ANN which is referred to as a node, and this node is an analog to a single neuron in the human brain. It receives signals from other nodes or from the source, where the input is referred to as the weight. The weight signals are summed to the result,

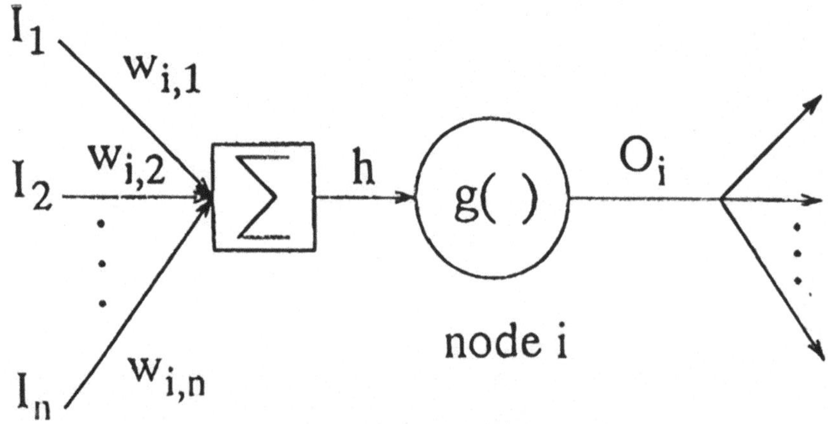

FIGURE 11.13 The structure of a single processing node (Himm, 2000).

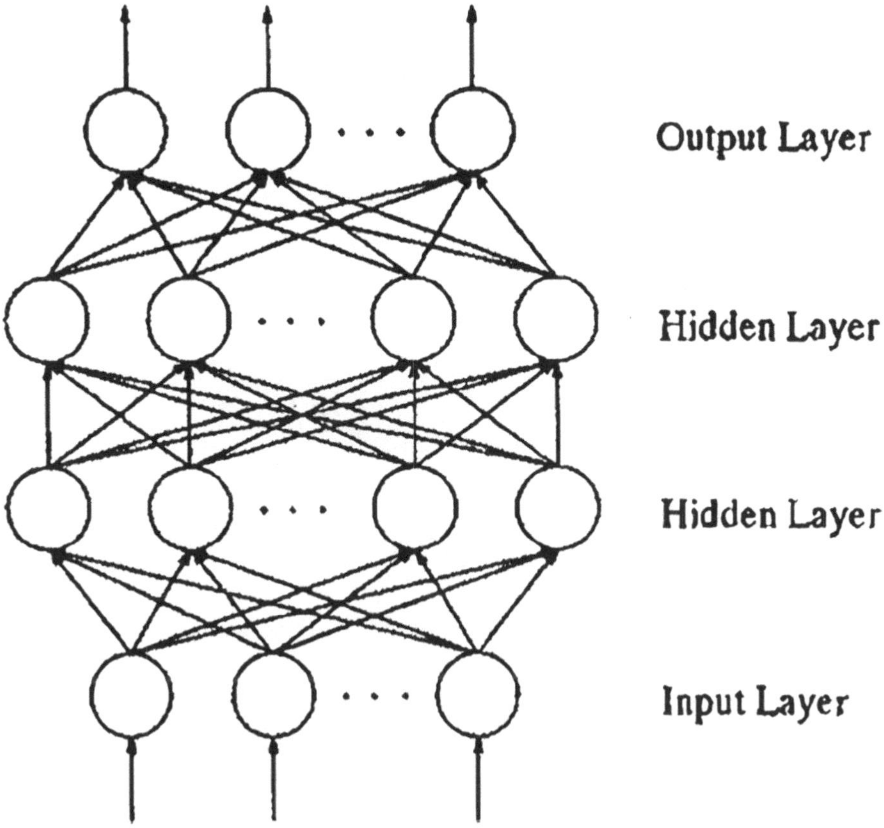

FIGURE 11.14 The structure of a layered neural network (Himm, 2000).

which is called activation h, and are transferred to a transfer function g, which can be a mathematical function or a simple bounded differential equation. The resulting outputs of the node are inputs for other nodes (Aguel et al., 2019) (Figure 11.14).

A collection of interconnected nodes forms an ANN. Input layers are nodes receiving signals from the source, while the output layer nodes return the signals to the external source. Hidden nodes or the hidden layer are the nodes that do not receive or send signals to external sources. The architecture of an ANN contains few layers and some dedicated neurons for the transfer function (Figure 11.15).

To get a better understanding, we consider a two-layer ANN network (Dheenamma et al., 2019).

$$z_i = x_1 wl + x_2 w_2 + x_3 w_3 \, i + bi \; for \, i = 1 - 2$$

$$z_i = represents\,the\,input\,value$$

$$x_1, x_2, x_3 = input\ vector$$

$$w_1, w_2 w_3 = weights\ connecting\ neuron$$

$$i^{th} = hidden\,neuron\ ,b_i = bias\,to\,the\,hidden\,neuron$$

$$Z_i = f(z)_i$$

$$Z_i = is\,the\,output\,of\,ith\,hidden\,neuron$$

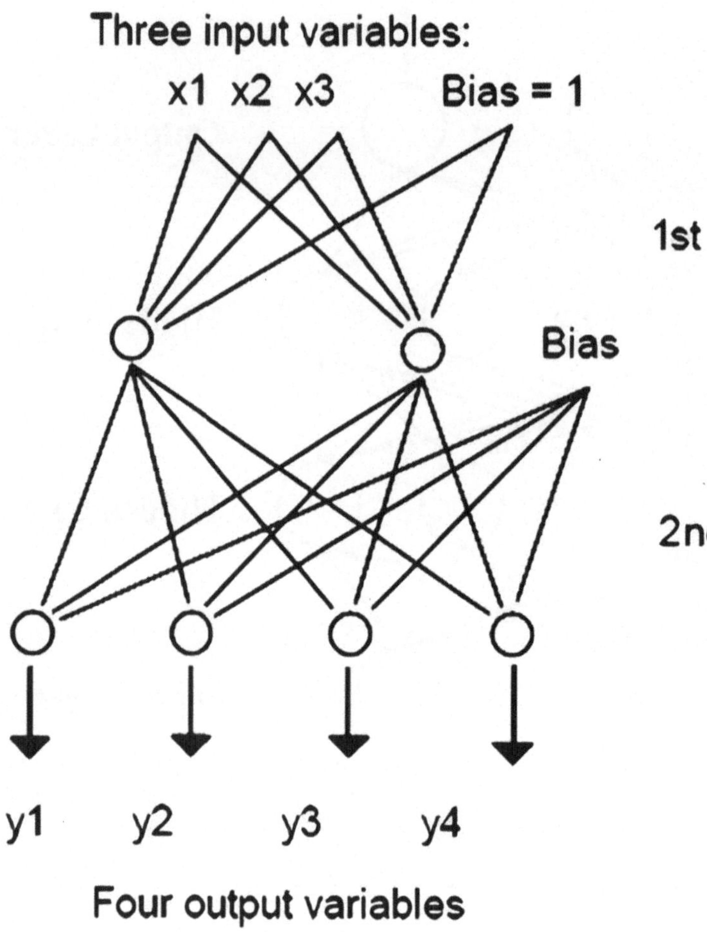

FIGURE 11.15 Two-layered ANN (Dheenamma et al., 2019).

$$y_j = f\left(\left(\sum_{i=1}^{2} Z_i * v_{ij}\right) + b_i\right) for\ j = 1-4$$

y_j = *represent* output of the value of jth neuron

f is the transfer function

These networks need to be trained on collected datasets using a learning algorithm in order to estimate the dependent output variables. During training of the neural network, the weights and biases are calculated to get the output value. The input to these systems is from training datasets. Based on the output values and the target values, the network parameters are repeatedly adjusted until the output value and target values closely match. The development of an ANN follows four steps (Dheenamma et al., 2019):

1. Data collection and division of the data
2. Setting the neural network architecture and training the neurons
3. Finding the optimum ANN architecture
4. Evaluating the ANN model

TABLE 11.1
Specifications of Corrugated PHE (Dheenamma et al., 2019)

Specifications	Values
Plate length (m)	0.4250
Plate width (m)	0.1250
Thickness of plate (m)	0.0008
Thickness of channel (m)	0.0040
Inlet port diameter (m)	0.0320
Heat transfer area (m^2)	0.2656

11.11 APPLICATIONS OF ANNS

Consider a plate heat exchanger (PHE) with seven corrugated stainless-steel patterns. The heat exchanger has equally distributed passes for both cold and hot side fluids. Non-Newtonian fluid carboxy methyl cellulose (CMC) and xanthan gum (XG) are used as water additives with varying 0.1–0.6% w/w (Table 11.1).

11.11.1 Step 1: Data Collection and Division of Data

The data is collected from experimentation or from the operating cycles in a plant. The collected raw data contains data in different ranges due to the conditions in which it is generated. Hence, it is necessary to treat the dataset prior to modeling. In order to remove outliers, median centering is used. In median centering, the median values are determined and subtracted from each column matrix. Scaling is used when we have data in different units. Principal component analysis is an analytical technique used as a stage to build a new compressed database containing independent latent variables. In the case of a large industrial operating dataset, projection to latent structures can be used. It analyzes and extracts meaningful information from correlated and noisy datasets (Aguel et al., 2019).

11.11.2 Step 2: Setting the Neural Network Architecture and Training Neurons

The model is developed with four input parameters: cold and hot fluid, Reynolds number (N_{Re}), and Prandtl number (N_{Pr}); these four are considered because geometry and flow effects use Reynolds number while fluid properties and temperature effects use Prandtl number. In addition, effectiveness factor (ε), frictional factor of XG (f_{xg}), CMC (f_{cmc}), water (f_w), and U overall heat transfer coefficient were also considered as output variables. Figures 11.16 and 11.17 show models for both XG and CMC. The transfer function applied for input-hidden and hidden-output layers are tan-sigmoid functions (Dheenamma et al., 2019).

11.11.3 Step 3: Evaluating the ANN Model

The evaluation is done by measuring the mean square error of the training data fed into the model regression values. Once the mean square error is obtained, we can stop the training. The model generated predicts four output variables for the XG and CMC solution.

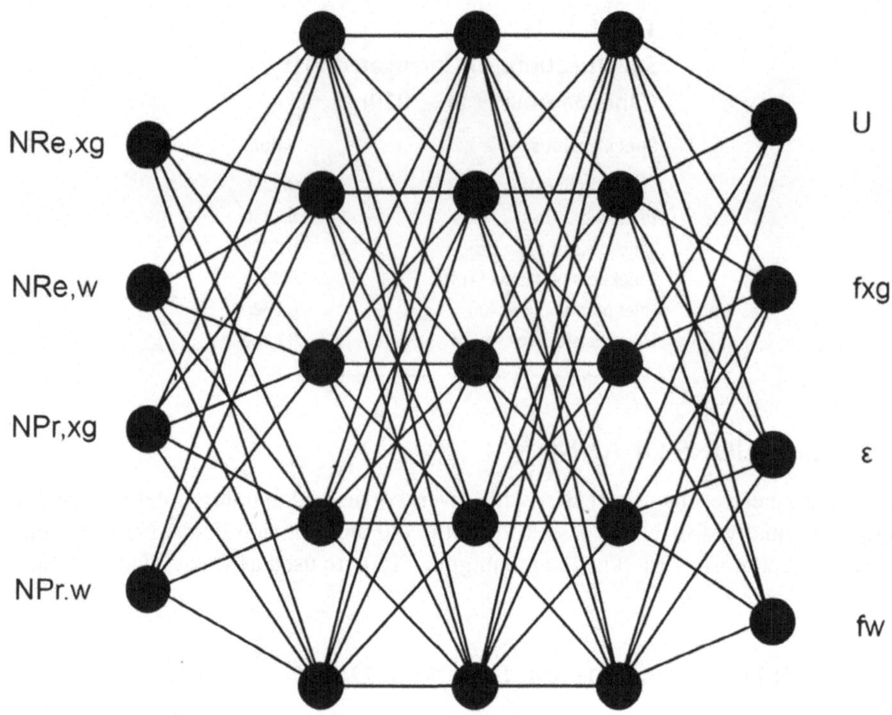

FIGURE 11.16 ANN model for XG (Dheenamma et al., 2019).

11.11.4 Step 4: Finding Optimal ANN Architecture

There is no theoretical method to obtain optimal ANN architecture. A trial-and-error approach is usually employed and we investigated performance factor (*R*), standard deviation (σ), and the percentage absolute error (*PARE*) to determine the optimum ANN model.

$$R = \left(\frac{1}{N}\right)\sum_{i-1}^{N}\left(\frac{A}{B}\right) \tag{11.8}$$

Here *R* is the performance value, *A* is the experimental value, *B* is the predicted value, and *N* is the number of datasets (Dheenamma et al., 2019).

$$\sigma = \sqrt{\left(\frac{1}{N}\right)\sum_{i=1}^{N}\left(\left(\frac{A}{B}\right) - R\right)^2} \tag{11.9}$$

$$PARE = \frac{A-B}{A} * 100 \tag{11.10}$$

11.12 CONCLUSIONS

Fouling remains a major challenge in heat exchanger operations, causing decreased efficiency, higher energy usage, and potential equipment damage. Although traditional techniques such as chemical cleaning and mechanical brushing have been used for years, the increasing complexity

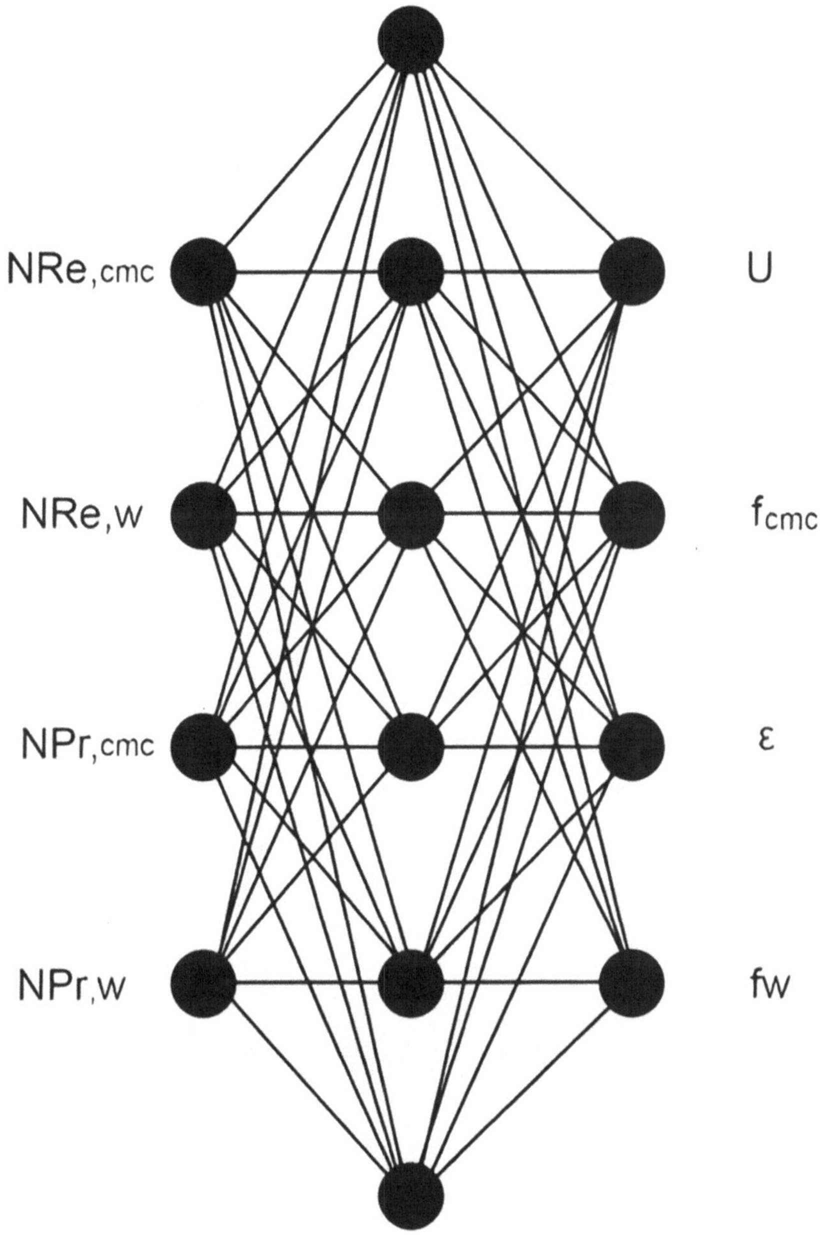

FIGURE 11.17 ANN model for CMC (Dheenamma et al., 2019).

of fouling mechanisms and the demand for more effective solutions have driven the exploration of innovative strategies. AI has emerged as a promising tool for tackling fouling issues. AI-powered systems can process vast amounts of data from sensors and control systems to detect early signs of fouling, forecast fouling rates, and optimize cleaning schedules. ML algorithms can also be employed to create predictive models for fouling prevention, allowing for proactive actions before fouling becomes critical.

Nevertheless, the successful application of AI in fouling management requires a solid data infrastructure, domain knowledge, and careful consideration of factors like sensor precision, data quality, and computational resources. As AI technology continues to evolve, its role in mitigating heat exchanger fouling is expected to significantly enhance system performance and lower operational costs. In summary, effective fouling management in heat exchangers demands a comprehensive approach that integrates traditional cleaning techniques with advanced technologies like AI. By harnessing the potential of AI, more efficient, predictive, and data-driven solutions can be achieved to address this ongoing challenge.

REFERENCES

Aguel, S., Meddeb, Z., & Jeday, M. R. (2019). Parametric study and modeling of cross-flow heat exchanger fouling in phosphoric acid concentration plant using artificial neural network. *Journal of Process Control, 84*, 133–145. https://doi.org/10.1016/j.jprocont.2019.10.001.

Al Ismaili, R., Lee, M. W., Wilson, D. I., & Vassiliadis, V. S. (2019). Optimisation of heat exchanger network cleaning schedules: incorporating uncertainty in fouling and cleaning model parameters. *Computers and Chemical Engineering, 121*, 409–421. https://doi.org/10.1016/j.compchemeng.2018.11.009.

Awais, M., & Bhuiyan, A. A. (2019). Recent advancements in impedance of fouling resistance and particulate depositions in heat exchangers. *International Journal of Heat and Mass Transfer, 141*, 580–603. https://doi.org/10.1016/j.ijheatmasstransfer.2019.07.011.

Bidabadi, M., Sadaghiani, A. K., & Azad, A. V. (2013). Spiral heat exchanger optimization using genetic algorithm. In *Scientia Iranica B* (Issue 5). www.scientiairanica.com.

Brooks, S., & Roy, R. (2022). Design and complexity evaluation of a self-cleaning heat exchanger. *International Journal of Heat and Mass Transfer, 191*, 122725. https://doi.org/10.1016/j.ijheatmasstransfer.2022.122725.

Davis, J. L. (2015). Heat exchanger cleaning methods. In *Energy Management and Efficiency for the Process Industries* (pp. 139–142). Wiley. https://doi.org/10.1002/9781119033226.ch11

Dheenamma, M., Soman, D. P., Muthamizhi, K., & Kalaichelvi, P. (2019). In pursuit of the best artificial neural network configuration for the prediction of output parameters of corrugated plate heat exchanger. *Fuel, 239*, 461–470. https://doi.org/10.1016/j.fuel.2018.11.034.

Frota, M. N., Ticona, E. M., Neves, A. V., Marques, R. P., Braga, S. L., & Valente, G. (2014). On-line cleaning technique for mitigation of biofouling in heat exchangers: a case study of a hydroelectric power plant in Brazil. *Experimental Thermal and Fluid Science, 53*, 197–206. https://doi.org/10.1016/j.expthermflusci.2013.12.006.

Himm, D. M. (2000). Applications of artificial neural networks in chemical engineering. *Korean Institute of Chemical Engineers, 17*, 4.

Jalalirad, M. R., Abd-Elhady, M. S., & Malayeri, M. R. (2013). Cleaning action of spherical projectiles in tubular heat exchangers. *International Journal of Heat and Mass Transfer, 57*(2), 491–499. https://doi.org/10.1016/j.ijheatmasstransfer.2012.10.071.

Kishore, P. S., Kumar, R., & Vamsi Venkata, N. (2018). Comparative study of mechanical and chemical methods for surface cleaning of a marine shell-and-tube heat exchanger. *Heat Transfer - Asian Research, 47*(3), 520–530. https://doi.org/10.1002/htj.21316.

Klaren, D. G., & De Boer, E. F. (n.d.). *Revamping existing severely fouling conventional heat exchangers into a self-cleaning (fluidised bed) configuration: new developments and examples of revamps*. www.heatexchanger-fouling.com.

Kleinebrahm, R. (2017). Mechanical online system for cleaning heat exchanger tubes by Sponge Rubber Balls (Taprogge System). *Kleinebrahm, R*, 240–247.

Köhler, H., Störk S. H., M. Müller., & M. Jacobi-Peukert, M. (2013). Optimization approach for efficient cleaning with impinging jets: Influence of nozzle diameter, pressure and nozzle distance. In *Heat Exchanger Fouling and Cleaning* (pp. 106–118). Budapest, Hungary.

Kohli, R. (2018). Applications of projectiles for nonaqueous cleaning of interior surfaces of tubes. In *Developments in surface contamination and cleaning: applications of cleaning techniques volume 11* (pp. 765–792). Elsevier. https://doi.org/10.1016/B978-0-12-815577-6.00020-7.

Li, M. J., Tang, S. Z., Wang, F. long, Zhao, Q. X., & Tao, W. Q. (2017). Gas-side fouling, erosion and corrosion of heat exchangers for middle/low temperature waste heat utilization: A review on simulation and experiment. *Applied Thermal Engineering, 126*, 737–761. https://doi.org/10.1016/j.applthermaleng.2017.07.095.

Melo, L., & Flemming, H.-C. (2010). Mechanistic aspects of heat exchanger and membrane biofouling and prevention. In *The science and technology of industrial water treatment* (pp. 365–380). CRC Press. https://doi.org/10.1201/9781420071450-c18.

Müller-Steinhagen, H., Malayeri, M. R., & Watkinson, A. P. (2011). Heat exchanger fouling: Mitigation and cleaning strategies. *Heat Transfer Engineering, 32*(3–4), 189–196. https://doi.org/10.1080/01457632.2010.503108.

Nyemba, W. R., Chinguwa, S., Zimba, I., & Mbohwa, C. (2018). Development of an effective self-cleaning system to minimize fouling in heat exchangers. Proceedings of the *International Conference on Industrial Engineering and Operations Management, (November)*, 403–414.

Popov, I. A., Gortyshov, Y. F., & Olimpiev, V. V. (2012). Industrial applications of heat transfer enhancement: the modern state of the problem (a Review). *Thermal Engineering, 59*(1), 1–12. https://doi.org/10.1134/S0040601512010119.

Saxon, G. E., & Putman, R. E. (n.d.). *The practical application and innovation of cleaning technology for heat exchangers.* http://dc.engconfintl.org/heatexchanger/40.

Somerscales, E. F. C. (1997). Fundamentals of corrosion fouling. *Experimental Thermal and Fluid Science, 14*(4), 335–355. https://doi.org/10.1016/S0894-1777(96)00136-7.

Thakore, S. B., & Bhatt, Bharat I. (2007). *Introduction to process engineering and design.* McGraw Hill Education.

Thulukkanam, K. (2024). *Heat exchangers: Operation, performance, and maintenance.* CRC Press.

Watkinson, A. P., & Wilson, D. I. (1997). Chemical reaction fouling: A review. *Experimental Thermal and Fluid Science, 14*(4), 361–374..

Wu, C., Yang, H., He, X., Hu, C., Yang, L., & Li, H. (2022). Principle, development, application design and prospect of fluidized bed heat exchange technology: comprehensive review. *Renewable and Sustainable Energy Reviews, 157*, 112023. https://doi.org/10.1016/j.rser.2021.112023.

Xie, G. N., Sunden, B., & Wang, Q. W. (2008). Optimization of compact heat exchangers by a genetic algorithm. *Applied Thermal Engineering, 28*(8–9), 895–906. https://doi.org/10.1016/j.applthermaleng.2007.07.008.

Zeyninejad Movassag, S., Nemati Taher, F., Razmi, K., & Tasouji Azar, R. (2013). Tube bundle replacement for segmental and helical shell and tube heat exchangers: performance comparison and fouling investigation on the shell side. *Applied Thermal Engineering, 51*(1–2), 1162–1169. https://doi.org/10.1016/j.applthermaleng.2012.10.025.

12 Fouling and Corrosion Prevention Strategies

Arya Shah, Karan Sharma, Vaibhav More, M.D. Yadav, and Sunil Kumar

12.1 INTRODUCTION

Natural corrosion and inevitable occurrence are the degradation of materials, particularly metals, caused by chemical reactions with environmental factors, including oxygen, water, and acids. They are an immense concern in all types of industries because they can cause apparatus failure, reduced efficiency, and higher costs for upkeep. The traditional methods to prevent corrosion consist of choosing the right material, protective coatings, corrosion inhibitors, cathodic protection, and proper maintenance practices. These measures mitigate corrosion impact either by limiting exposure to corrosive agents or protecting material surfaces through physical protection. Besides these general corrosion challenges, industrial systems face several fouling forms.

Fouling is described as unwanted deposit, biofilm accumulation on the surface, and less efficient heat transfer and increased risk of corrosion, which is also highly disruptive in heat exchangers where both corrosion and fouling simultaneously affect the performance. Thus, corrosion and fouling prevention is essential for achieving a long life with good operational efficiency of heat exchangers. Various methods for controlling corrosion and fouling in heat exchangers, including material, surface treatments, and adjustments in operation, in which the integrated approach is thus further underlined as critical for optimizing performance and achieving reliability, are discussed in this chapter.

12.1.1 DEFINITION OF CORROSION

Examples of material degradation, usually with metallic constituents, often show corrosion in heat exchangers, as produced through electrochemical interactions that originated from the working environment inside the heat exchanger system. Most industrial processes feature basic elements such as the transfer of heat from one fluid to another without letting them mix and require heat exchangers. Such systems are usually subjected to conditions that might include fluid flow, temperature fluctuations, and corrosive agent exposures (e.g., chlorides, acids, or oxygen), which can accelerate material degradation [1]. Corrosion in heat exchangers is determined by several factors; these include the material of construction, the chemical nature of fluids, the hydrodynamic nature of fluid flow, temperature, pressure, and even heat exchanger geometry and design [2].

Uniform corrosion is where the metal corrodes evenly across a surface. Localized corrosion includes pitting, crevice corrosion, and galvanic corrosion. Pitting corrosion is a type of localized corrosion where local anodic sites on the material surface undergo rapid dissolution leading to the formation of small pits that weaken the material [3]. Crevice corrosion takes place in confined areas, such as between the tubes of a heat exchanger and their tube sheet supports, where oxygen and ions may be starved, thus promoting corrosion. Galvanic corrosion results from two dissimilar metals being in electrical contact within an electrolyte such that one metal will corrode more readily than another [4].

DOI: 10.1201/9781003509837-12

Corrosion is peculiar in heat exchangers, since, apart from weakening the strength and further longevity of the material, thermal performance is also involved. Corrosion may roughen heat transfer surfaces and thus increase fouling, reduce the efficiency of heat transfer, and raise maintenance costs. For example, corrosion inside the inner sides of the tubes in a heat exchanger containing multiple tubes may cause thinning of the walls of the tubes that ultimately results in the failure of the tube or leakage, leading to unsafe and inefficient operation of a system [3, 4]. Various approaches based on the operating conditions for the system are used to prevent corrosion in heat exchangers. Some of the common methods employed are corrosion-resistant materials such as stainless steel, titanium, or alloys with high resistance to chloride-induced stress corrosion cracking. Protective coatings, epoxy or ceramic-based layers, between the metal surface and the corrosive fluid, create a barrier. The use of corrosion inhibitors—chemicals added to the heat exchange fluid to reduce the rate of oxidation—is another widely used technique in closed-loop cooling systems. Cathodic protection is a technique that employs a sacrificial anode for protecting a metal surface from corrosion; it is particularly effective in systems exposed to seawater or brine [5]. Understanding the particular corrosion mechanisms operating in heat exchangers is therefore crucial for the design of effective prevention strategies that will ensure system reliability, efficiency, and long life.

12.1.2 Definition of Fouling

Fouling in heat exchangers refers to the deposition of unwanted matter, biofilms, or particulates on the surfaces of the heat transfer that inhibit the efficiency of thermal energy exchange. Fouling occurs when fluid passing through the heat exchanger is filled with suspended solids, particulate matter, or dissolved substances that, over time, adhere to tube surfaces, plates, or contact areas. Such deposits include mineral scales, organic substances, biofilms, and corrosion byproducts [6]. Fouling increases thermal resistance, lowers heat-transfer efficiency, and often raises operational costs owing to the increased energy needed for the desired heat exchange rates [7].

Fouling may be classified as per the nature of the deposits. The basic type is scaling, which occurs due to the deposition of dissolved salts as their temperature or concentration increases. Biofouling is characterized by a trend of growth of microorganisms within the fluid, either bacteria, algae, or fungi, often stimulated by warmth or a nutrient-rich environment [8]. Particulate fouling occurs as solid particles from the fluid deposit on the surfaces of the heat exchanger [9]. Fouling is particularly severe in heat exchangers because it reduces both heat transfer performance and accelerates corrosion, leading to higher maintenance requirements [10]. The strategies for effective fouling management include cleaning, chemical inhibitors, and material selection, all of which are necessary for the maintenance of operational efficiency and the longevity of heat exchanger systems.

12.1.3 The Economic Impact of Corrosion and Fouling

Considerable money in terms of increased operation expenses, loss of efficiency, and premature failure of equipment is generated by corrosion and fouling in exchangers. Heat exchangers play a crucial role in many industries, and corrosion or fouling events have huge consequences. The economic losses primarily arise from weakening metal components [11]. These may cause leakage and system shutdowns, along with expensive repairs or replacements. For instance, corrosion of heat exchanger tubes and failures require costly replacement and result in unproductive lost time. Corrosion damage also impairs heat-transfer efficiency, and systems have to use extra energy to maintain the required temperatures [12]. This increase in energy use also increases the costs of operation. Increased risk of structural failure is another consequence of corrosion; hence, there is a need for more frequent inspection, which further compounds maintenance costs [13].

Fouling, however, adds to the problems, as the efficiency of heat transfer goes down. Thermal resistance increases as deposits keep forming on the surfaces of the heat exchanger; hence, the system will be less efficient. The system requires more energy to provide the same rates of heat transfer. Additionally, fouling increases the demand for maintenance such as cleaning, descaling, and chemical treatment. In the worst-case scenarios, if not handled, fouling can cause the system to fail, a cause for unplanned shutdowns and production losses [14]. Corrosion and fouling shorten a heat exchanger's lifespan and raise the cost of operations because they increase energy usage, maintenance, and repair of a plant, thus affecting the business bottom line in industrial systems.

12.2 THERMODYNAMICS AND KINETICS OF CORROSION AND FOULING

NACE International defines corrosion as the deterioration of a substance or its properties because of a reaction with its environment. All materials deteriorate in a conducive environment [15]. We usually speak about corrosion concerning the deterioration and degradation of metals and alloys in the surrounding environment, as metals and alloys are essential for constructing basic structures around us. Metals and alloys are required to build buildings, electronics, automobile components, reactors, piping systems, and many other systems where tensile strength, electrical conductivity, or thermal conductivity is required. Some favorable physicochemical properties of exploited metals and alloys are tensile strength, thermal conductivity, electrical conductivity, malleability, and ductility. These properties change as the material degrades due to interactions with the environment.

Out of all metals used for various purposes, iron is by far the most widely studied and used due to its abundance, strength, and versatility. Iron alloyed with many other elements can form sheets, pipes, rods, and other valuable geometries. Specifically, in the chemical and allied industries, iron and steel of various grades are used to make reactors, heat exchangers, columns, piping systems, fittings, civil structure supports, and much more. Suppose the environment around the equipment used to process some material causes deterioration. In that case, it reduces the efficiency of the process and may pose a safety hazard, which may lead to loss of life, damage to health, property, and environment, and economic loss. The chemical and allied industry deals with many hazardous and corrosive chemicals. This makes the environment around reactors, heat exchangers, separators, and so on—i.e., wherever metals and alloys are present—conducive to corrosion. Understanding corrosion science and how construction materials interact with their surroundings is essential.

Specifically, in heat exchangers, the construction material, due to the nature of the process carried out by the equipment, i.e., heating and cooling, experiences cyclic thermal stresses, which aid in the corrosion process. It is also required to heat chemicals that corrode metal. The selection of materials for construction and mitigation of corrosion becomes of utmost importance to ensure safe operating conditions. The most prevalent medium used for cooling is water, which aids corrosion by being the electrolyte in which ions can readily dissolve.

12.2.1 Electrochemical Theory of Corrosion

The mechanisms by which corrosion deteriorates metals can be mainly classified as follows:

1. Electrochemical corrosion
2. Flow-induced phase change-induced corrosion
3. Chemical corrosion
4. Environment-assisted fracture and stress corrosion [14]

The first two are essential because we are mainly dealing with flow-based systems. To understand electrochemical corrosion, we need to understand some important definitions:

12.2.1.1 Anode

The structure from which conventional current enters the electrolyte or the structure where corrosion occurs. The anode is the region where oxidation occurs [15].

12.2.1.2 Cathode

The structure from where conventional current leaves the electrolyte and enters the metal or from where electrons enter the electrolyte. The cathode is the region where reduction occurs [15].

12.2.1.3 Electrolyte

Any liquid that can transfer ions from one location to another can be termed as an electrolyte. Further, the solvent's acidic, alkaline, or neutral pH is essential [15].

12.2.1.4 External Path

The region in the metal that connects the anode and cathode electrically [15]. The electrochemical corrosion rate mainly depends on two factors: thermodynamic instability, which depends on the corrosion cell's electromotive force (emf), and the kinetic hindrance offered by the system under corrosion [16]. Thermodynamic instability for metals is illustrated by the electrochemical series, which ranks metals based on their tendency to oxidize, making them more likely to corrode [16]. We rank metals based on their standard electrode potential. The standard electrode potential refers to the inherent potential of the metal electrode to either oxidize or reduce when it is in equilibrium with its ions at 298 K, 1 atm, and 1M concentration. The emf, in contrast, refers to the entire electrochemical cell potential. If the standard potential of the metal is less than −0.414V, the metal has higher thermodynamic stability and can corrode without oxygen or any other oxidant. Such metals include K, Na, Li, Ce, Al, Fe, and Ti [16].

Suppose the standard potential of the metal lies between −0.414V and 0V. In that case, the metals have moderate thermodynamic instability and corrode in aqueous media in the presence of oxygen and without oxygen in an acidic medium. Such metals include Cd, Pb, Ni, Mo, and so on [16]. If the standard potential of the metal lies between 0V and 0.815V, the metals have intermittent thermodynamic instability and are stable in acidic or neutral media without oxygen. They corrode in these media in the presence of oxygen or an oxidant [16]. If the standard potential of the metal lies beyond 0.815V, the metals have high thermodynamic stability and do not corrode in neutral media in the presence of oxygen. They can corrode only in acidic media when oxidizers or complex formers are present. Such metals include Pt, Ir, Au, and so on [16]. Even though aluminum is placed higher than iron in the electrochemical series, it has a standard electrode potential that is less (more negative) than iron. Still, the rate of corrosion observed for aluminum is 0.096 mils per year in the E25 fuel blend, while in the same condition, the rate of corrosion observed for iron in the same condition was 0.297 mils per year [17]. Why does this happen? This happens as aluminum forms a protective non-porous oxide layer, which prevents the electrolyte and oxidant from reacting with the metal surface.

So, looking at the above definitions and observed phenomena, we can say that for corrosion to take place, we need a corrosion triangle to form—we need an oxidizer, electrolyte, and exposed metal surface. It is analogous to the fire triangle—we need heat, oxygen, and fuel for fire to exist. As we know, to stop a fire, we need to eliminate the components of the fire triangle; similarly, to eliminate corrosion, we need to eliminate any of the three elements of the corrosion triangle.

12.2.2 Fouling in Heat Exchangers

While corrosion mainly affects the structural integrity of heat-exchanging equipment, fouling mainly affects the efficiency of the heat exchanger. Fouling is primarily characterized by material that adheres to or accumulates at the heat-exchanging surface. Revising some basics of heat transfer, adding another layer to the existing system adds another heat transfer resistance and decreases the overall heat transfer coefficient. The thermal conductivity of foulants is considerably lower than that of construction materials like metals and alloys [18]. This can make the foulant layer the heat transfer controlling step in the process. Over time, this may cause significant economic and technical difficulties. Fouling in heat exchangers is a multifaceted issue that needs to be classified based on the type of foulant and fouling mechanism.

12.2.2.1 Particulate Fouling

When the process or utility has a high total suspended solids (TSS) content, the solids accumulate at the heat transfer surface. Particulate fouling can be characterized by the formation of deposits and a significant reduction in the heat transfer coefficients of the heat exchanger that were previously observed [19]. This reduction in heat transfer coefficient happens due to the low thermal conductivity of the particulate foulant.

12.2.2.2 Biological Fouling

Biofouling or biological fouling occurs when algae, bacteria, or microorganisms stick to the heat transfer surface. The layer formed by these microorganisms that have adhered to the surface is known as a biofilm. The biofilm has a very low thermal conductivity, which makes it an insulating layer on the heat transfer surface, decreasing the overall efficiency of the heat exchanger [20].

12.2.2.3 Scaling Fouling

Scale formations are mainly associated with systems where hard water, where concentrations of calcium and magnesium ions are high, is used as a utility. When calcium and magnesium form salts with anions like carbonate and so on, they have an inverse solubility relationship with temperature, i.e., their solubility in water decreases with an increase in temperature [21]. A very thin layer of scale can drastically increase the energy consumption of the heat exchanger [21].

12.2.2.4 Chemical Fouling

If the substance in the process fluid or the utility fluid reacts with the surface of the heat exchanger, it forms a residual layer and gives rise to chemical fouling. This kind of fouling is prevalent when the heat exchanger operates near the thermal decomposition temperature of the fluid. If not taken care of, over-heating of the heat exchanger may pose an operational hazard [22]

12.2.2.5 Corrosion Fouling

Corrosion alters the surface of the heat-exchanging surface. It makes the surface rougher, making it easier for foulants to adhere, further increasing the corrosion rate. Such phenomena, where fouling and corrosion occur together, can cause localized heat zones and compromise the structural integrity of the heat exchanger [23].

12.2.3 DLVO Theory of Fouling

The Derjaguin–Landau–Verwey–Overbeek (DLVO) theory explains the interactions between charged particles in fluid media. The DLVO theory was primarily used to predict the stability of colloidal systems.

For a given system, the DLVO theory says that the total interaction energy is expressed as a sum of electrostatic repulsion forces and van der Waals interaction forces of attraction [24]. This model can also account for hydrogen bonding, acid-base interactions, hydrophobic forces, and so on, which affect the stability of colloidal suspensions [25]. The DLVO theory can help us understand microbial adhesion and particulate aggregation and their stability in various environments. In the context of heat exchangers, the DLVO theory helps explain how the attractive and repulsive forces acting on the particle or microorganism affect the fouling rate, wherein the foulants form crystals, biofilms, and particulate aggregates.

The DLVO theory has been useful in predicting the fouling propensity of $CaSO_4$ in plate heat exchangers [26]. This modeling technique helps us develop fouling propensity indicators that can help us design and operate heat exchangers optimized to reduce fouling risks.

12.3 CORROSION MITIGATION TECHNIQUES

Effective corrosion control in heat exchangers is pivotal for extending operational lifespan, ensuring safety, and maintaining performance within industrial systems exposed to corrosive environments. The interaction between fluids and structural materials often accelerates degradation, necessitating the implementation of both traditional and innovative strategies. Conventional methods, such as utilizing corrosion-resistant alloys, applying protective coatings, and introducing corrosion inhibitors, tend to offer limited and reactive solutions. These strategies often fall short in addressing the multifaceted and dynamic nature of corrosion, which is influenced by variables such as temperature, pressure, flow velocity, pH levels, and fluid chemistry. Recent advancements incorporating artificial intelligence (AI) and machine learning (ML) are revolutionizing corrosion mitigation by enabling more predictive and data-driven approaches. AI/ML algorithms can process extensive datasets from real-time monitoring systems, uncovering complex patterns and allowing for more accurate predictions of corrosion behavior under varying operational conditions. These technologies enhance predictive maintenance by forecasting corrosion progression based on historical data, thus optimizing maintenance schedules and preventing premature system failures. Additionally, ML-based models can simulate corrosion rates for different materials, informing the design of heat exchangers with superior resistance. Furthermore, AI assists in optimizing corrosion inhibitor formulations, analyzing their effectiveness across a spectrum of operational and environmental parameters, and providing tailored solutions for specific use cases. These AI-driven advances not only reduce operational downtime and maintenance expenses but also ensure the long-term structural integrity of heat exchangers in demanding industrial environments, ultimately enhancing their sustainability and performance.

12.3.1 Process Modifications for Corrosion Control

Process modifications offer a proactive approach to mitigating corrosion by altering operational parameters that affect the interaction between materials and corrosive agents. In industries where equipment operates under extreme conditions, these modifications aim to minimize corrosion rates by carefully adjusting variables such as temperature, pressure, flow velocity, pH, and chemical composition of working fluids. For instance, lowering the operating temperature can decelerate chemical reactions responsible for corrosion, while regulating flow velocity can prevent erosion-corrosion—where high fluid velocities wear down protective surfaces. Other strategies may involve modifying the chemical composition of process streams, such as the addition of corrosion inhibitors or pH adjustment, to suppress corrosive reactions. Since these factors are often interdependent and exhibit non-linear behaviors, process modifications require careful consideration of the specific materials, fluid properties, and system dynamics involved, making corrosion mitigation a complex, multi-variable challenge.

The application of AI/ML technologies in process optimization offers a significant enhancement in mitigating corrosion by providing a more data-driven and predictive methodology. Conventional approaches to process optimization typically rely on empirical, trial-and-error methods, which may not fully capture the complex, interdependent variables influencing corrosion. AI/ML enables the integration of large-scale datasets from real-time sensors, allowing for more nuanced control and optimization of operating conditions.

12.3.1.1 Dynamic Real-Time Process Control

AI/ML systems can be integrated into the control frameworks of heat exchangers, enabling real-time adjustments to operational parameters based on changing conditions. Through continuous monitoring, AI-driven systems can detect deviations from optimal corrosion-minimizing conditions and autonomously adjust variables such as flow velocity, inhibitor dosing, or system temperature.

This dynamic control ensures that the system remains within a safe operational range, preventing sudden increases in corrosion rates due to unexpected process variations.

12.3.1.2 Anomaly Detection and Early Intervention

AI and ML techniques, particularly those based on advanced pattern recognition, are capable of identifying early signs of corrosion or system inefficiencies that traditional monitoring systems may miss. By analyzing extensive datasets from sensors (monitoring variables like pressure, temperature, pH, and flow rate), these algorithms can detect subtle shifts indicative of early stage corrosion or process inefficiencies. This early detection allows for timely corrective actions, preventing significant damage and extending the operational life of the equipment.

12.3.1.3 Multi-Variable Non-Linear Optimization

Corrosion mitigation is inherently a multi-variable optimization problem, where numerous factors such as temperature, pressure, and fluid chemistry interact in complex, non-linear ways. AI/ML models excel at solving these types of optimization challenges, continuously refining process parameters based on real-time data to maintain optimal operating conditions. Machine learning models can iteratively improve their predictions and recommendations, ensuring that the heat exchanger operates in conditions that minimize corrosion while maximizing overall efficiency.

12.3.1.4 AI-Enhanced Process Simulation and Design

AI/ML can also enhance the simulation of process modifications by integrating with computational fluid dynamics (CFD) models to predict how changes in process conditions will influence long-term corrosion rates. This capability allows for the identification of potential corrosion hotspots and the optimization of heat exchanger designs to reduce susceptibility to corrosion. By simulating various operational scenarios, engineers can design systems that are inherently more resistant to corrosion-related degradation.

12.3.1.5 Corrosion Inhibitor Optimization

AI/ML models are also instrumental in optimizing the application of corrosion inhibitors. By analyzing performance data from various inhibitors under diverse operational conditions, machine learning algorithms can identify the most effective inhibitors for specific environments. Moreover, these systems can dynamically adjust inhibitor dosing in response to real-time feedback from process sensors, ensuring efficient use of inhibitors and maximizing corrosion protection. This targeted approach reduces waste and improves the overall effectiveness of corrosion control strategies.

12.3.2 Use of Additives

The application of chemical additives, specifically corrosion inhibitors, is a critical method used in various industrial processes to mitigate corrosion, particularly in systems like heat exchangers, pipelines, and other fluid-handling equipment. Corrosion inhibitors are substances that, when added in small concentrations, significantly reduce the corrosion rate of metallic surfaces exposed to corrosive environments. They function by disrupting the electrochemical reactions that drive corrosion, either by forming protective films on the metal surface or by altering the environment's chemistry to reduce corrosivity. The application of additives is discussed in the following subsections.

12.3.2.1 Film-Forming Inhibitors

These inhibitors (e.g., phosphates, silicates) form a physical barrier that isolates the metal surface from corrosive species such as oxygen, water, and ions (e.g., chlorides). This film may consist of organic or inorganic compounds that adsorb onto the surface, creating a hydrophobic layer, reducing the interaction between the metal and the corrosive medium.

12.3.2.2 Anodic and Cathodic Inhibitors

Anodic inhibitors (e.g., chromates, molybdates) work by passivating the metal surface, creating a stable oxide layer that protects the underlying metal. In contrast, cathodic inhibitors (e.g., zinc salts, sulfites) slow down the reduction reactions, such as hydrogen evolution or oxygen reduction, thus minimizing the cathodic half of the corrosion reaction. These inhibitors alter the kinetics of the electrochemical reactions that contribute to metal dissolution.

12.3.2.3 pH Modifiers

The pH of the fluid can significantly affect corrosion rates. Acidic conditions tend to accelerate corrosion, especially for metals like steel. pH modifiers such as alkaline additives (e.g., sodium hydroxide) can neutralize acids in the fluid, thereby reducing the overall corrosivity and protecting the metal surfaces. Conversely, acidic inhibitors (e.g., sulfur-based compounds) can stabilize protective oxide layers in alkaline environments.

12.3.2.4 Organic and Inorganic Additives

Organic inhibitors, such as amines, thiols, or imidazolines, adsorb onto the metal surface through nitrogen, sulfur, or oxygen atoms, forming a dense, hydrophobic film that prevents water or oxygen from contacting the metal surface. Inorganic inhibitors, such as phosphates or zinc compounds, tend to precipitate onto the surface, forming protective layers that block corrosive ions.

12.3.2.5 Volatile Corrosion Inhibitors

Volatile corrosion inhibitors (VCIs) are used in environments where direct application of inhibitors is challenging, such as in closed systems or equipment storage. These inhibitors vaporize and condense onto metal surfaces, forming protective films in the presence of moisture. VCIs are particularly effective in preventing atmospheric corrosion in storage environments.

There are significant challenges in the application of the additives. The effectiveness of corrosion inhibitors depends on several factors, including their concentration, compatibility with the fluid medium, operating temperature, pressure, and fluid dynamics. Inappropriate dosing can lead to overconsumption (increased cost and potential negative side effects like scaling) or under-application (ineffective protection and higher corrosion rates). Furthermore, the real-time variability of operating conditions introduces complexity in maintaining the optimal concentration of additives over the lifecycle of the system.

Following modern approaches, the application of AI and ML offers innovative solutions for overcoming the challenges associated with the use of additives in corrosion mitigation. These technologies enable the optimization of corrosion inhibitors by leveraging large datasets generated from real-time monitoring, historical records, and operational conditions. AI/ML models can process these vast datasets to predict corrosion behavior, optimize additive usage, and dynamically adjust dosing strategies based on real-time feedback.

12.3.2.6 Predictive Modeling for Additive Selection and Dosing

AI/ML models can be trained on extensive historical datasets that include information about fluid properties (e.g., temperature, pH, flow rates), system material properties, and corrosion inhibitor performance data. By analyzing this data, ML algorithms can predict the optimal type of inhibitor and the precise dosage required for specific operating conditions. This approach moves beyond empirical dosing strategies by providing a data-driven selection process, ensuring that inhibitors are always applied in the most effective concentrations, minimizing waste and ensuring continuous protection.

12.3.2.7 Real-Time Adaptive Dosing

AI-powered systems integrated with real-time sensors (e.g., for temperature, pressure, flow rate, pH, and oxidation-reduction potential) can continuously monitor the operating environment. ML algorithms can process this data to dynamically adjust the dosage of additives based on current system conditions. For instance, if the system detects an increase in temperature or a drop in pH, AI models can autonomously increase the concentration of inhibitors to maintain corrosion protection. This capability ensures that the dosing remains optimal even as operating conditions fluctuate, reducing the risk of under-protection or overuse of chemicals.

12.3.2.8 Corrosion Rate Prediction and Performance Monitoring

AI/ML algorithms can be used to predict the future performance of additives by analyzing sensor data streams from corrosion monitoring systems. These systems (e.g., electrochemical sensors, resistance probes) provide real-time corrosion rate measurements, which can be fed into predictive models. ML models can learn from past data to forecast when the effectiveness of an inhibitor diminishes, prompting timely adjustments in dosing or additive selection. This allows for proactive intervention before significant corrosion occurs, enhancing the system's long-term performance.

12.3.2.9 Optimization of Inhibitor Formulation

The composition of corrosion inhibitors plays a crucial role in their effectiveness under specific environmental conditions. By leveraging AI/ML, chemical engineers can optimize inhibitor formulations by testing various combinations of chemicals in silico (computer-based simulations). ML models can be trained on experimental data to identify the most effective inhibitor formulations for different materials, environmental conditions, and process parameters. This reduces the need for costly and time-consuming laboratory testing while accelerating the development of new, more effective corrosion inhibitors.

12.3.2.10 Multi-Variable Optimization of Process Conditions

The effectiveness of corrosion inhibitors is often influenced by multiple interacting variables such as temperature, pressure, flow velocity, and fluid chemistry. AI/ML models excel in handling complex, non-linear optimization problems where multiple factors must be balanced. By simultaneously optimizing both the process conditions and the additive dosing, AI can ensure that the system operates under conditions that maximize corrosion protection while minimizing operational costs. For instance, AI/ML models can optimize flow rates and temperature to improve inhibitor performance without compromising the overall efficiency of the heat exchanger or process system.

12.3.2.11 Anomaly Detection and Preventive Maintenance

Machine learning models, particularly those focused on anomaly detection, can continuously analyze real-time sensor data to detect early signs of corrosion or deviations in inhibitor performance. AI-based systems can detect subtle changes in data that may indicate an upcoming failure or inefficiency, triggering preventive maintenance or adjustments in the corrosion protection strategy. This approach reduces the risk of unexpected equipment failure and extends the operational life of industrial systems by maintaining optimal corrosion protection.

The implementation of chemical additives for corrosion mitigation, when combined with AI/ML technologies, represents a sophisticated, data-driven approach to optimizing corrosion protection in industrial systems. AI/ML enables predictive modeling, real-time dosing adjustments, anomaly detection, and multi-variable optimization, ensuring that inhibitors are applied in the most effective and efficient manner. Furthermore, AI/ML algorithms provide enhanced predictive capabilities, allowing operators to forecast corrosion behavior and make proactive adjustments to both additive formulations and process conditions. As these technologies continue to evolve, their integration into corrosion mitigation strategies will lead to more reliable, cost-effective, and sustainable solutions for managing corrosion in industrial systems.

12.3.3 Coating Techniques

Coating techniques are extensively employed in various industries to mitigate corrosion by providing a physical and/or chemical barrier that isolates the underlying metal from corrosive agents. The effectiveness of coatings depends on multiple factors, including the composition, thickness, porosity, adhesion strength, and the environmental and operational conditions. Different types of coatings are used depending on the nature of the application, the type of metal, and the expected exposure to corrosive environments. The primary categories of coatings used for corrosion protection include organic, inorganic, metallic, and advanced nanotechnology-based coatings.

12.3.3.1 Organic Coatings

Organic coatings, such as epoxy, polyurethane, and acrylic-based systems, are commonly applied to metals due to their excellent adhesion and flexibility, which provides robust corrosion resistance. Epoxy coatings, in particular, are highly cross-linked polymers that form dense, impermeable films, significantly limiting the diffusion of corrosive agents such as water, oxygen, and ions (e.g., chloride ions). Polyurethane coatings provide additional benefits, including abrasion resistance and UV stability, which make them suitable for outdoor applications. Organic coatings can also include corrosion inhibitors as additives, which provide an additional chemical mechanism of protection.

12.3.3.2 Inorganic Coatings

Inorganic coatings are composed of silicates, phosphates, ceramics, or other non-organic compounds. Zinc-rich silicate coatings, for instance, offer a dual mechanism of protection—acting as a barrier and providing cathodic protection by sacrificing the zinc to protect the steel substrate. These coatings are ideal for high-temperature applications where organic coatings might degrade. Inorganic ceramic coatings, such as alumina or silica, offer exceptional resistance to heat, chemical attack, and mechanical wear. However, the application and curing processes for inorganic coatings require precise control to prevent defects such as porosity or cracking, which can compromise protection.

12.3.3.3 Metallic Coatings

Metallic coatings, such as galvanizing (zinc coating), thermal spraying, and aluminizing, provide a high level of protection, especially in corrosive environments like marine or industrial atmospheres. Zinc coatings offer sacrificial protection where the zinc corrodes preferentially to the steel, thus protecting the substrate. Techniques such as hot-dip galvanizing or electroplating allow the formation of metallic layers that can provide both barrier and sacrificial protection. Metallic coatings are often applied using processes such as thermal spraying, physical vapor deposition (PVD), or chemical vapor deposition (CVD), which deposit a uniform metal layer onto the substrate.

12.3.3.4 Conversion Coatings

Conversion coatings, such as phosphate and chromate coatings, chemically alter the surface of the metal to form a thin protective layer. These coatings serve as a primer for further painting or coating and enhance corrosion resistance by passivating the surface, reducing the metal's reactivity. Phosphate coatings are frequently used in conjunction with organic topcoats in automotive and industrial applications, while chromate coatings are employed to enhance corrosion resistance in aluminum and magnesium alloys.

12.3.3.5 Nanocoatings

Nanotechnology-based coatings are gaining prominence due to their ability to provide superior protection by filling micro-cracks and surface defects, thereby preventing localized corrosion phenomena such as pitting or crevice corrosion. Nanocoatings typically involve the application of nanoparticles (e.g., silica, zinc oxide, or titanium dioxide) that enhance the mechanical strength and corrosion resistance of traditional coatings. These coatings can be applied as thin films via sol-gel processes, chemical vapor deposition, or electrochemical methods. Their nano-scale structure offers enhanced properties such as self-healing, hydrophobicity, and superhydrophobicity, which further reduces the risk of moisture penetration.

While coatings provide effective corrosion protection, their performance can degrade over time due to environmental factors such as UV exposure, mechanical wear, thermal cycling, or chemical exposure. Issues such as cracking, delamination, and porosity can develop, leading to localized corrosion. Optimizing the coating's composition, thickness, and application method is critical to ensuring long-term corrosion resistance, which requires precise control over the coating process and in-depth monitoring of the coating's integrity throughout its service life.

Following modern approaches, the application of AI and ML offers innovative solutions to address the challenges of optimizing and monitoring coating techniques used for corrosion mitigation. These technologies enable the processing and analysis of large amounts of data from various sources, such as sensor networks, environmental conditions, and historical performance records, providing real-time insights into coating performance. AI/ML can improve coating selection, application, and maintenance by predicting coating degradation, optimizing process parameters, and dynamically adjusting coating formulations.

12.3.3.6 AI/ML-Driven Coating Material Selection

Selecting the most appropriate coating material for corrosion mitigation requires the consideration of numerous factors, including the type of metal substrate, the environment, and mechanical requirements. AI/ML algorithms can analyze historical data on coating performance under different environmental conditions (e.g., humidity, temperature, salinity, chemical exposure) to predict the most effective coating materials. By processing large datasets that include various coatings and operational conditions, ML models can identify optimal coatings with enhanced corrosion resistance, guiding engineers to select materials that have demonstrated long-term durability in similar conditions.

12.3.3.6.1 Predictive Maintenance and Coating Degradation Forecasting

Coating degradation is influenced by factors such as UV exposure, mechanical abrasion, chemical exposure, and thermal cycling. AI/ML algorithms can use real-time data from corrosion sensors, such as electrochemical impedance spectroscopy (EIS) and non-destructive evaluation (NDE) techniques, to detect early signs of coating degradation. By analyzing trends in sensor data, predictive models can forecast when coatings will likely fail, allowing for proactive maintenance or reapplication before corrosion begins to penetrate the substrate. This predictive capability reduces unexpected failures and extends the life of the coated asset.

12.3.3.6.2 Optimizing Coating Thickness and Application Parameters

The thickness of a coating plays a critical role in its effectiveness for corrosion protection. Thicker coatings may provide better barrier protection but may be prone to mechanical stress, cracking, or delamination. AI/ML algorithms can analyze data from past applications to predict the optimal thickness for a given coating material and environment. These models can also account for the application method (e.g., spraying, dipping, thermal spraying) and environmental conditions during application, such as humidity and temperature, which impact curing and adhesion. By optimizing these parameters, AI/ML ensures the coating adheres properly and provides maximum protection.

12.3.3.6.3 *Real-Time Monitoring and Adaptive Control of Coating Performance*

AI systems integrated with real-time monitoring sensors can continuously assess coating performance in-service. These sensors provide data on environmental factors (temperature, humidity, salinity) and mechanical stresses, as well as direct measurements of the coating's integrity. AI/ML models can detect anomalies, such as cracks, delamination, or increased porosity, that signal impending coating failure. By detecting these early warning signs, AI can trigger maintenance alerts or suggest changes in operational conditions (e.g., reducing mechanical loads or adjusting temperature) to extend the coating's lifespan.

12.3.3.6.4 *Optimization of Coating Formulations Using AI/ML*

Developing new, more effective coating formulations is a time-consuming and resource-intensive process. AI/ML models can accelerate this process by analyzing datasets of various coating formulations and their performance characteristics. For example, machine learning can predict how different combinations of polymers, metallic particles, and nanoparticles (such as graphene or carbon nanotubes) will perform under specific conditions. By simulating the behavior of these formulations, AI/ML can help develop coatings that offer superior mechanical, chemical, and thermal resistance. This optimization allows for rapid experimentation and the creation of custom-tailored coatings for specific industrial applications.

12.3.3.6.5 *Multi-Variable Optimization of Coating Process Parameters*

The application of coatings involves several variables that must be precisely controlled, including surface preparation, temperature, curing time, and humidity levels during application. AI/ML models can optimize these multi-variable processes by analyzing their complex interactions and effects on coating performance. For instance, AI can predict the optimal surface roughness required for maximum coating adhesion while ensuring that the curing temperature and time are adjusted to minimize internal stresses that could lead to cracking. This results in coatings that are more robust, durable, and better suited to withstand the operational stresses of their specific environments.

12.3.3.6.6 *Failure Analysis and Root Cause Detection Using AI/ML*

Coating failures are often complex, involving multiple contributing factors such as improper surface preparation, environmental exposure, or material weaknesses. AI/ML models can analyze failure data and identify patterns that are not readily apparent to human operators. For example, ML algorithms can detect correlations between coating failures and specific environmental conditions, curing processes, or substrate materials. These insights can guide modifications in the coating application process or material selection to prevent future failures. By identifying the root causes of failure, AI/ML improves coating reliability and performance.

Coating techniques for corrosion mitigation are essential for protecting metal surfaces in industrial systems, and the integration of AI/ML technologies offers significant improvements in the selection, application, and maintenance of these coatings. AI/ML can optimize material selection, predict coating degradation, and enable real-time monitoring and adaptive control of coating performance. Furthermore, AI/ML-driven simulations, combined with multi-variable optimization of process parameters, allow for the development of custom coating formulations that provide superior protection in specific environments. As AI/ML technologies continue to evolve, their integration into coating strategies will lead to more reliable, efficient, and sustainable corrosion protection solutions, ensuring the long-term integrity of critical industrial assets.

12.4 FOULING MITIGATION TECHNIQUES

Improving energy efficiency in industrial processes requires the use of advanced heat transfer equipment, such as compact heat exchangers (CHEs) with high heat transfer coefficients. Plate heat exchangers (PHEs) are widely adopted due to their compact size, lower cost, and efficient heat transfer capabilities. These exchangers feature intricately corrugated plates that create a high level of turbulence, significantly enhancing heat transfer while also reducing the tendency for fouling compared with traditional shell-and-tube units.

Nevertheless, fouling remains a critical issue that impairs heat exchanger efficiency by forming insulating layers on heat transfer surfaces. The growth of fouling begins when foulants such as minerals, biological organisms, or particles in the fluid adhere to heat transfer surfaces. Initially, these deposits form a thin layer that gradually thickens, creating thermal resistance and reducing the heat transfer rate. As the fouling layer grows, it restricts fluid flow, leading to higher pressure drops and greater energy consumption for pumping. The adhesion and growth of foulants depend on factors like fluid velocity, surface roughness, temperature, and the chemical nature of the fluids in contact. Fouling is responsible for up to 2.5% of global CO_2 emissions and affects up to 90% of heat exchangers. It can be categorized into several types: corrosion, biological, chemical reaction, freezing, particulate, and crystallization [27]. In PHEs, crystallization and particulate fouling are most prevalent, particularly when heating water solutions in applications like power generation and desalination [28]. The roughness and restricted cross-section caused by deposits in compact exchangers exacerbate these effects, leading to higher pressure drops.

Mitigation techniques are therefore essential and include both online and offline methods. Online techniques, such as mechanical cleaning with brushes or sponge balls, chemical treatments, and ultrasonic cleaning, minimize interruptions. Offline methods, like high-pressure water jets and manual scraping, are more thorough but require system shutdowns. Additionally, preventive strategies, such as using antifouling coatings and optimizing flow conditions, are critical. Advances in real-time monitoring and predictive maintenance further help sustain efficiency and minimize the environmental impact of heat exchanger operations.

12.4.1 Use of Additives

Fouling in heat exchangers is the unwanted deposition of materials on heat transfer surfaces, leading to reduced thermal efficiency, increased pressure drop, and higher operational costs. Common types of fouling include particulate, crystallization (scaling), biofouling, chemical reaction, and corrosion fouling. These deposits act as thermal insulators and reduce the overall heat transfer coefficient, necessitating frequent maintenance and higher energy input for effective operation.

To combat fouling, the use of chemical additives has become a widely adopted solution due to their effectiveness in minimizing deposit formation and extending the operational lifespan of heat exchangers. This section explores the types of additives used, their mechanisms of action, and best practices for their application.

12.4.1.1 Additive Types and Mechanisms

12.4.1.1.1 Scale Inhibitors

Scale inhibitors are designed to prevent the deposition of minerals like calcium carbonate ($CaCO_3$) and calcium sulfate ($CaSO_4$), which are common in systems using hard water. These additives, such as phosphonates and polyphosphates, work by interfering with the nucleation and growth of mineral crystals, thereby keeping them in solution. They achieve this by binding to crystal growth sites or by altering the crystal structure to form a softer, more easily removable scale. These inhibitors reduce the precipitation tendency of dissolved salts by stabilizing supersaturated solutions and preventing crystal lattice formation. At low concentrations, they act as threshold inhibitors, significantly reducing the formation of hard scale layers.

12.4.1.1.2 Dispersants

Dispersants help control particulate fouling by keeping suspended particles, like dust, rust, or organic matter, dispersed in the fluid. This prevents particles from settling on heat exchanger surfaces. Common dispersants include surfactants that lower surface tension, allowing particles to remain suspended and reducing their adherence to surfaces. By reducing particle agglomeration, dispersants facilitate the removal of foulants via normal fluid flow, which is especially beneficial in systems with high particulate loads or cooling water systems prone to sludge buildup.

12.4.1.1.3 Corrosion Inhibitors

Corrosion products can lead to fouling by depositing oxides or corroded metal particles on heat exchanger surfaces. Corrosion inhibitors, such as chromates, zinc, and phosphates, create a protective film on metal surfaces to prevent corrosion reactions. Organic inhibitors like benzotriazole are commonly used in copper heat exchangers to form a passivating layer, preventing copper oxidation. These inhibitors work by either passivating the metal surface or by creating a hydrophobic barrier, thus reducing the availability of reactive sites for corrosion. This not only prevents fouling but also extends the life of the equipment.

12.4.1.1.4 Nanofluids

Nanofluids are engineered fluids that contain nanoparticles. Nanoparticles have been proven to disrupt the aggregation of foulants [30]. Foulants must aggregate to form a layer covering the heat transfer area to affect the heat exchanger's performance. Investigations of copper oxide nanoparticles have been carried out on oil foulants, and it has been proven that they significantly alter fouling characteristics [30].

12.4.1.1.5 Biocides and Biodispersants

Biological fouling (biofouling) results from the accumulation of microorganisms, algae, and other biological matter. Biocides, such as chlorine and bromine, are used to kill microorganisms, while biodispersants help break down biofilms, making it easier for the fluid flow to wash away the biological deposits. Biocides disrupt cellular functions in microorganisms, while biodispersants break the extracellular polymeric substances that bind biofilms to surfaces. This dual approach is crucial for systems using untreated water, like seawater cooling systems.

12.4.2 Process Modifications

Integrating fouling as a design parameter enables us to detect fouling early and helps us develop processes that minimize fouling and the heat transfer area required [31]. More robust modeling of fouling in heat exchangers can also account for maintenance and operational costs and can be optimized for the same [32]. Process modeling can be done for flow rate optimization, real-time modeling, and cleaning schedule optimization. Incorporating DLVO theory in the modeling helps us understand and incorporate the actual behavior of foulants and their interaction with the heat transfer surface. Traditional design methodology assumes fouling factors to be constant [33]. Engineers can gain valuable insights into fouling characteristics using DLVO theory, which accounts for variability, making it a more accurate model [34].

12.4.2.1 Flow Rate Optimization

The shear exerted at the heat-exchanging surface is proportional to the flow rate of fluid passing through the heat exchanger. High shearing force can help remove the adhered foulant. However, high flow rates correspond to high pressure drops across the equipment. A dynamic optimization framework can be employed to minimize utility costs for an optimum flow rate [35]. Optimizing flow rates through heat exchangers and cleaning schedules can extend the operation time while keeping fouling rates low [32].

12.4.2.2 Real-Time Modeling

Heat exchanger systems are prone to fouling and require regular cleaning, and they may not be done at regular intervals. Measuring the fouling rates in real-time can help schedule clean-ups efficiently, reducing downtime and increasing efficiency. Developing a monitoring system using multi-sensor technology to analyze the heat exchanger performance is essential to real-time modeling [36]. This also helps in reducing unscheduled maintenance. AI/ML models can be incorporated to increase the efficiency of real-time modeling and set better predictive models based on existing systems.

12.4.2.3 Cleaning Optimization

Including operational variables in optimizing cleaning schedules could essentially minimize the operational impacts of fouling and optimize cleaning frequency [34]. This modeling strategy helps maintain the optimal performance of heat exchangers while helping to save operational costs [37]. Process optimization methods prioritize preventive care, treating fouling as a key design parameter and considering system dynamics to ensure efficient and safe operations. Cleaning techniques, like backwashing and flushing, reverse fluid flow for a short period of time for high flow rates. This dislodges the foulant layer that has accumulated over the heat transfer surface. Similar to ultrasound treatments, this method of fouling mitigation is mainly studied for membrane systems [38] but can also be utilized for heat-exchanging systems.

12.4.2.4 Chemical Treatment

When we see the causes of fouling, fouling mainly occurs when the foulant adheres to or crystallizes on the heat transfer surface. One of the main ways to eliminate fouling in heat exchangers is by eliminating the foulant properties that either adhere or agglomerate to the heat transfer surface or by itself. To reduce the tendency to adhere to the heat transfer surface, the surface tension of the fluid is reduced. To prevent agglomeration, we need to add chemicals that prevent agglomerate formation. Chemical additives to mitigate fouling are widely used in oil refining and food processing.

12.4.2.5 Ultrasound

Ultrasonic waves break the foulant layer on a heat-exchanging surface due to the high-frequency pressure waves that the material experiences. Ultrasonic treatments are effective in breaking down organic deposits and biofilms. Ultrasonic treatments are mainly studied for controlling membrane fouling [39] but can also be employed for heat exchangers.

12.4.3 Types of Coatings

Fouling is a major challenge in heat exchanger operations, reducing efficiency and increasing operational costs. Surface coatings have emerged as a robust solution for fouling mitigation, employing strategies to reduce adhesion, promote easy removal, or resist the accumulation of contaminants [29]. Coating enhances the surfaces of heat exchangers to resist or minimize fouling while maintaining thermal and mechanical stability. The choice of coating depends on the nature of foulants and the operational environment, and the available options are categorized into metal-based, ceramic-based, polymer-based, and composite coatings.

12.4.3.1 Metal-Based

Titanium (Ti) coatings are frequently used due to their excellent resistance to corrosion and fouling. Titanium naturally forms a passive oxide layer (TiO_2) that acts as a physical barrier against corrosive substances and foulants. Coating techniques like PVD create thin, uniform layers with high thermal stability. Studies have shown that titanium coatings can reduce the induction period for fouling by up to 40% and facilitate the removal of crystallization-based deposits like $CaCO_3$, especially in seawater environments [40]. Hydrophobic Ti coatings exhibit contact angles of around 57°, which helps repel water-based foulants. In a study involving magnetron-sputtered Ti coatings on stainless steel tubes, the heat transfer coefficient (HTC) increased by 25%, and fouling resistance decreased by up to 60% compared with uncoated surfaces. Another experiment using anodized Ti surfaces demonstrated a 93.3% reduction in thermal fouling resistance in oil-contaminated water systems due to enhanced water-surface interactions that mitigated oil adhesion [41].

Iron-aluminum coatings with micro/nanostructures are being researched as a possible solution for an antifouling coating. Research shows that by modifying the topography and using special surface treatments, the surface can significantly mitigate adhesion to foulants while not increasing the heat transfer resistance [42].

Ni-P coatings are applied using electroless deposition, which ensures uniform thickness across complex geometries without the need for an external electric current. The primary advantage of Ni-P coatings is their smooth surface, which minimizes sites for foulant nucleation and adhesion. However, they are prone to microbial corrosion, particularly in environments rich in sulfate-reducing bacteria. To address this, composite coatings incorporating elements like boron nitride (BN) have been developed, significantly enhancing antifouling properties. Ni-P coatings have been shown to reduce fouling thermal resistance by 83.3% on aluminum fins in air-cooled condensers [43]. Additionally, surfaces coated with Ni-P-BN composites exhibited a 70% reduction in fouling compared with binary Ni-P coatings [44, 45]. These improvements are attributed to the lower surface energy and reduced wall friction achieved through BN incorporation.

12.4.3.2 Ceramic-Based

TiO_2 coatings are extensively studied for their antifouling properties. TiO_2 exists in multiple polymorphs, including anatase and rutile, each exhibiting unique fouling resistance characteristics. The anatase phase, for example, is effective in delaying crystallization fouling, while the rutile phase is more stable at high temperatures [46]. The deposition of TiO_2 coatings via techniques like liquid phase deposition (LPD) or high-power impulse magnetron sputtering (HiPIMS) ensures smooth surfaces with low porosity. In studies involving single-phase convective heat transfer (SPCHT), amorphous TiO_2 coatings delayed $CaCO_3$ deposition by up to 50 times compared with uncoated surfaces [47]. The HTC increased by 30% during boiling heat transfer experiments on TiO_2 nanotube-coated surfaces, owing to enhanced bubble nucleation and reduced fouling [48].

SiO_2 coatings are valued for their chemical inertness and low surface energy, making them ideal for preventing both crystallization and organic fouling [49]. Sol-gel methods are commonly used to apply SiO_2 coatings, offering precise control over surface properties and achieving pinhole-free layers. Hydrophobic SiO_2 surfaces, with contact angles exceeding 85°, facilitate self-cleaning in condensation applications, where water droplets easily remove foulants [50].

12.4.3.3 Polymer-Based

PTFE, or Teflon, is a fluoropolymer with one of the lowest coefficients of friction (~0.16) and exceptional temperature resistance. It is widely used to mitigate scaling and organic fouling. To enhance its low conductivity, PTFE can be combined with conductive fillers like carbon nanotubes (CNTs) or graphene [51]. Composite PTFE coatings with multi-wall carbon nanotubes (MWCNTs) have demonstrated a 70.3% reduction in fouling mass during milk pasteurization processes, compared with uncoated stainless steel [52]. Another study revealed a 33.82% reduction in asymptotic thermal fouling resistance for PTFE-coated plate heat exchangers handling $CaCO_3$-rich water. PEG-based hydrophilic coatings create a hydration layer that acts as a physical barrier to foulants. These coatings are particularly effective in preventing biofouling and crystallization fouling [53]. The performance of PEG coatings depends on parameters like graft density and thickness. Higher graft densities provide better antifouling resistance by increasing steric hindrance and hydration layer stability. PEG coatings with optimized graft densities reduced $CaCO_3$ surface coverage by up to

85% compared with untreated surfaces in stagnant heating conditions [54]. Furthermore, these coatings retained their antifouling efficacy over extended operational periods, making them suitable for industrial cooling systems.

12.4.3.4 Composite-Based

Combining the flexibility of polymers with the strength and thermal stability of ceramics creates versatile antifouling surfaces. For instance, PTFE-TiO_2 composites not only improve mechanical stability but also enhance hydrophobicity [55, 56]. These coatings have been used in geothermal applications to combat oil and salt fouling, reducing thermal resistance by over 70%. Hydrophobic coatings are mainly used in systems where water is the heat transfer fluid. The effect of these coatings was studied for a gasketed heat exchanger, wherein it was found that hydrophobic coatings were effective against fouling. There was a 25.6% reduction in the heat transfer coefficient compared with an uncoated surface [57].

Composite coatings such as PTFE-TiO_2 have shown a 20-fold reduction in fouling rates in dairy processing heat exchangers [58]. Additionally, sol-gel-derived SiO_2-TiO_2 composites have demonstrated a fivefold reduction in fouling thermal resistance in tap water heat exchangers, maintaining constant heat transfer rates over 400 hours of operation [59]. DLC coatings are known for their exceptional hardness, smoothness, and low surface energy. These coatings are particularly effective in environments with high shear forces, such as combined heat and power (mCHP) units. Modified DLC, incorporating elements like silicon or fluorine, enhances antifouling properties and thermal stability. Silica-fluoropolymers are sol-gel composite materials that effectively mitigate mineral fouling [60]. Sol-gel forms smooth and hydrophobic coatings that are hydrophobic, which are effective in reducing fouling [60]. Sol-gel coatings pose only minimal heat transfer resistance, making them advantageous

12.4.3.5 Upcoming Coatings

The development of smart coatings that can adapt to changing environmental conditions or self-heal after damage is an exciting frontier. These coatings utilize microcapsules embedded with healing agents, which release upon mechanical damage, restoring the coating's integrity. However, challenges such as cost, scalability, and long-term durability remain under extreme conditions. While polymer and composite coatings offer significant fouling resistance, concerns about the environmental impact of fluoropolymers and the toxicity of some metal-based coatings limit their widespread use. Research is ongoing to develop eco-friendly, high-performance alternatives that provide efficient heat transfer without compromising sustainability.

12.5 CONCLUSION

To conclude, the prevention of fouling and corrosion in industrial systems, particularly in heat exchangers, is critical to sustaining long-term operational efficiency and reliability. The multifaceted nature of these challenges necessitates an integrated approach that combines advanced materials, surface treatments, chemical additives, and optimized operational strategies. Traditional methods, such as the use of corrosion-resistant materials, protective coatings, and chemical inhibitors, remain fundamental yet must be augmented with modern techniques. Innovations like AI/ML-driven predictive maintenance and real-time monitoring are redefining how industries combat these persistent issues, offering proactive and data-driven solutions.

The electrochemical principles underlying corrosion and the complex mechanisms of fouling highlight the importance of a deep understanding of material-environment interactions. Mitigation strategies need to be adaptable to varying conditions and customized based on the specific industrial environment. As industrial processes evolve, emerging technologies and novel materials—such as nanocoatings and eco-friendly inhibitors—hold promise for more sustainable and efficient solutions. Ensuring the longevity and efficiency of heat exchangers will continue to depend on the thoughtful integration of these diverse strategies, driven by ongoing research and technological advancements.

REFERENCES

1. B. El Ibrahimi, J. V. Nardeli, and L. Guo, "An Overview of Corrosion," 2021, pp. 1–19. doi: 10.1021/bk-2021-1403.ch001.
2. W. Faes *et al.*, "Corrosion and corrosion prevention in heat exchangers," *Corrosion Reviews*, vol. 37, no. 2, pp. 131–155, Mar. 2019, doi: 10.1515/corrrev-2018-0054.
3. B. Kuźnicka, "Erosion–corrosion of heat exchanger tubes," *Engineering Failure Analysis*, vol. 16, no. 7, pp. 2382–2387, Oct. 2009, doi: 10.1016/j.engfailanal.2009.03.026.
4. M. Ali *et al.*, "Corrosion-related failures in heat exchangers," *Corrosion Reviews*, vol. 39, no. 6, pp. 519–546, Dec. 2021, doi: 10.1515/corrrev-2020-0073.
5. M. Rezaei, Z. Mahidashti, S. Eftekhari, and E. Abdi, "A corrosion failure analysis of heat exchanger tubes operating in petrochemical refinery," *Engineering Failure Analysis*, vol. 119, p. 105011, Jan. 2021, doi: 10.1016/j.engfailanal.2020.105011.
6. H. Müller-Steinhagen, "Cooling-water fouling in heat exchangers," 1999, pp. 415–496. doi: 10.1016/S0065-2717(08)70307-1.
7. C. Sandu and D. Lund, "Minimizing fouling in heat exchanger design," *Biotechnology Progress*, vol. 1, no. 1, pp. 10–17, Mar. 1985, doi: 10.1002/btpr.5420010105.
8. L. F. Melo and M. M. Pinheiro, "Biofouling in heat exchangers," in *Biofilms — Science and Technology*, Dordrecht: Springer Netherlands, 1992, pp. 499–509. doi: 10.1007/978-94-011-1824-8_44.
9. M. Awais and A. A. Bhuiyan, "Recent advancements in impedance of fouling resistance and particulate depositions in heat exchangers," *International Journal of Heat and Mass Transfer*, vol. 141, pp. 580–603, Oct. 2019, doi: 10.1016/j.ijheatmasstransfer.2019.07.011.
10. H. Müller-Steinhagen, M. R. Malayeri, and A. P. Watkinson, "Fouling of heat exchangers-new approaches to solve an old problem," *Heat Transfer Engineering*, vol. 26, no. 1, pp. 1–4, Jan. 2005, doi: 10.1080/01457630590889906.
11. A. M. Pritchard, "The economics of fouling," in *Fouling Science and Technology*, Dordrecht: Springer Netherlands, 1988, pp. 31–45. doi: 10.1007/978-94-009-2813-8_3.
12. A. Sheikh, "A risk based heat exchanger analysis subject to fouling Part II: economics of heat exchangers cleaning," *Energy*, vol. 25, no. 5, pp. 445–461, May 2000, doi: 10.1016/S0360-5442(99)00081-X.
13. L. Gomes da Cruz, E. M. Ishiyama, C. Boxler, W. Augustin, S. Scholl, and D. I. Wilson, "Value pricing of surface coatings for mitigating heat exchanger fouling," *Food and Bioproducts Processing*, vol. 93, pp. 343–363, Jan. 2015, doi: 10.1016/j.fbp.2014.05.003.
14. S. K. Ogbonnaya and O. O. Ajayi, "Fouling phenomenon and its effect on heat exchanger: a review," *Frontiers in Heat and Mass Transfer*, vol. 9, Dec. 2017, doi: 10.5098/hmt.9.31.
15. R. Norsworthy, "Understanding corrosion in underground pipelines: basic principles," in *Underground Pipeline Corrosion*, Elsevier, 2014, pp. 3–34. doi: 10.1533/9780857099266.1.3.
16. J. Wang, L. Sun, H. Li, R. Ding, and N. Chen, "Prediction model of fouling thickness of heat exchanger based on TA-LSTM structure," *Processes*, vol. 11, no. 9, p. 2594, Aug. 2023, doi: 10.3390/pr11092594.
17. Z.-B. Zhang, C. Yu, Y. Liu, and L.-H. Cao, "A novel method for calculating effective thermal conductivity of particulate fouling," *Thermal Science*, vol. 25, no. 1 Part B, pp. 421–431, 2021, doi: 10.2298/TSCI190308308Z.
18. J. Wang, "Prediction model of fouling thickness of heat exchanger based on TA-LSTM structure," *Processes*, 2023, doi: 10.3390/pr11092594.
19. Z. Zhang, C. Yu, Y. Liu, and L. Cao, "A novel method for calculating effective thermal conductivity of particulate fouling," *Thermal Science*, 2021, doi: 10.2298/tsci190308308z.
20. S.-H. Seol, A. A. Serageldin, and O. K. Kwon, "Experimental research on a heat pump applying a ball-circulating type automatic fouling cleaning system for fish farms," *Energies (Basel)*, 2020, doi: 10.3390/en13225856.
21. L. D. Tijing, H. Y. Kim, D. H. Lee, C. S. Kim, and Y. I. Cho, "Physical water treatment using RF electric fields for the mitigation of CaCO3 fouling in cooling water," *International Journal of Heat and Mass Transfer*, 2010, doi: 10.1016/j.ijheatmasstransfer.2009.12.009.
22. O. O. Ajayi and S. Ogbonnaya, "Fouling phenomenon and its effect on heat exchanger: a review," *Frontiers in Heat and Mass Transfer*, 2017, doi: 10.5098/hmt.9.31.

23. T. Ardsomang, J. W. Hines, and B. R. Upadhyaya, "Heat exchanger fouling and estimation of remaining useful life," *Annual Conference of the PHM Society*, 2021, doi: 10.36001/phmconf.2013.v5i1.2773.
24. D. Zhao *et al.*, "Salt-responsive phenol formaldehyde resin: changes of interface energy on the aggregation process," *Langmuir*, 2022, doi: 10.1021/acs.langmuir.2c02610.
25. X. Liu *et al.*, "Contribution of extracellular polymeric substances (EPS) to the sludge aggregation," *Environmental Science & Technology*, 2010, doi: 10.1021/es9016766.
26. "CFD analysis of gypsum crystallization fouling in 2D plate heat exchangers," *International Journal of Advanced Technology and Engineering Exploration*, 2022, doi: 10.19101/ijatee.2021.874616.
27. M. Awais and A. A. Bhuiyan, "Recent advancements in impedance of fouling resistance and particulate depositions in heat exchangers," *International Journal of Heat and Mass Transfer*, vol. 141, pp. 580–603, Oct. 2019, doi: 10.1016/J.IJHEATMASSTRANSFER.2019.07.011.
28. P. Kapustenko, J. J. Klemeš, and O. Arsenyeva, "Plate heat exchangers fouling mitigation effects in heating of water solutions: a review," *Renewable and Sustainable Energy Reviews*, vol. 179, p. 113283, Jun. 2023, doi: 10.1016/J.RSER.2023.113283.
29. A. Goswami, S. C. Pillai, and G. McGranaghan, "Micro/Nanoscale surface modifications to combat heat exchanger fouling," *Chemical Engineering Journal Advances*, vol. 16, p. 100519, Nov. 2023, doi: 10.1016/J.CEJA.2023.100519.
30. H. S. Majdi, H. A. Alabdly, B. O. Hasan, and M. M. Hathal, "Oil fouling in double-pipe heat exchanger under liquid-liquid dispersion and the influence of copper oxide nanofluid," *Heat Transfer-Asian Research*, 2019, doi: 10.1002/htj.21473.
31. J. C. Lemos, A. L. H. Costa, and M. J. Bagajewicz, "Design of shell and tube heat exchangers considering the interaction of fouling and hydraulics," *Aiche Journal*, 2022, doi: 10.1002/aic.17586.
32. A. Nakao, A. Valdman, A. L. H. Costa, M. J. Bagajewicz, and E. M. Queiroz, "Incorporating fouling modeling into shell-and-tube heat exchanger design," *ndustrial & Engineering Chemistry Research*, 2017, doi: 10.1021/acs.iecr.6b03564.
33. D. Butterworth, "Design of shell-and-tube heat exchangers when the fouling depends on local temperature and velocity," *Applied Thermal Engineering*, 2002, doi: 10.1016/s1359-4311(02)00025-x.
34. E. M. Ishiyama, A. Heins, W. R. Paterson, L. Spinelli, and D. I. Wilson, "Scheduling cleaning in a crude oil preheat train subject to fouling: incorporating desalter control," *Applied Thermal Engineering*, 2010, doi: 10.1016/j.applthermaleng.2010.04.027.
35. B. C. G. Assis *et al.*, "Dynamic optimization of the flow rate distribution in heat exchanger networks for fouling mitigation," *Industrial & Engineering Chemistry Research*, 2015, doi: 10.1021/acs.iecr.5b0045 3.
36. H. Luo, J. Chen, Z. Qi, Y. Su, and N. Yang, "Heat exchanger fouling state assessment and distributed monitoring system development," 2023, doi: 10.1117/12.2680853.
37. A. C. Caputo, P. M. Pelagagge, and P. Salini, "Joint economic optimization of heat exchanger design and maintenance policy," *Applied Thermal Engineering*, 2011, doi: 10.1016/j.applthermaleng.2010.12.033.
38. C. D. Peters, D. Li, Z. Mo, N. P. Hankins, and Q. She, "Exploring the limitations of osmotically assisted reverse osmosis: membrane fouling and the limiting flux," *Environmental Science & Technology*, 2022, doi: 10.1021/acs.est.2c00839.
39. M. Qasim, N. N. Darwish, S. Mhiyo, N. A. Darwish, and N. Hilal, "The use of ultrasound to mitigate membrane fouling in desalination and water treatment," *Desalination*, 2018, doi: 10.1016/j.desal.2018.04.007.
40. C. S. Oon *et al.*, "Heat transfer and fouling deposition investigation on the titanium coated heat exchanger surface," *Powder Technology*, vol. 373, pp. 671–680, Aug. 2020, doi: 10.1016/J.POWTEC.2020.07.010.
41. L. C. T. Shoute *et al.*, "Threshold hydrophobicity for inhibition of salt scale formation on SAM-modified titania nanotube arrays," *Applications of Surface Science*, vol. 473, pp. 282–290, Apr. 2019, doi: 10.1016/J.APSUSC.2018.11.173.
42. Z. R. He, D. Wang, Z. Fan, Y. Chen, S. Li, and C. Mo, "Fabrication of Fe–Al coatings with micro/nanostructures for antifouling applications," *Coatings*, 2020, doi: 10.3390/coatings10090902.
43. B. Zhao, X. Wang, Y. Xu, B. Liu, S. Cao, and Q. Zhao, "Reduction of dust deposition in air-cooled condensers in thermal power plants by Ni–P-based coatings," *Clean Technologies and Environmental Policy*, vol. 23, no. 6, pp. 1727–1736, Aug. 2021, doi: 10.1007/S10098-021-02055-6/FIGURES/9.
44. A. Al-Janabi, M. R. Malayeri, and H. Müller-Steinhagen, "Experimental fouling investigation with electroless Ni–P coatings," *International Journal of Thermal Sciences*, vol. 49, no. 6, pp. 1063–1071, Jun. 2010, doi: 10.1016/J.IJTHERMALSCI.2009.05.009.
45. A. Al-Janabi and M. R. Malayeri, "Nano-coated surfaces to mitigate fouling in thermal water services," *Desalination Water Treat*, vol. 55, no. 11, pp. 2909–2917, Sep. 2015, doi: 10.1080/19443994.2014.940205.

46. J. Guillot *et al.*, "Combining a molecular modelling approach with direct current and high power impulse magnetron sputtering to develop new TiO2 thin films for antifouling applications," *Applications of Surface Science*, vol. 333, pp. 186–193, Apr. 2015, doi: 10.1016/J.APSUSC.2015.01.199.
47. W. Yan, W. Lin-lin, and L. Ming-yan, "Antifouling and enhancing pool boiling by TiO_2 coating surface in nanometer scale thickness," *AIChE Journal*, vol. 53, no. 12, pp. 3062–3076, Dec. 2007, doi: 10.1002/aic.11345.
48. Y. Lv, M. Y. Liu, L. F. Hui, A. N. Pavlenko, A. S. Surtaev, and V. S. Serdyukov, "Heat transfer and fouling rate at boiling on superhydrophobic surface with TiO2 nanotube-array structure," *Journal of Engineering Thermophysics*, vol. 28, no. 2, pp. 163–176, Apr. 2019, doi: 10.1134/S1810232819020012.
49. C. Ning, L. Mingyan, and Z. Weidong, "Fouling and corrosion properties of SiO_2 coatings on copper in geothermal water," *Industrial & Engineering Chemistry Research*, vol. 51, no. 17, pp. 6001–6017, May 2012, doi: 10.1021/ie202091b.
50. A. Goswami, S. C. Pillai, and G. McGranaghan, "Surface modifications to enhance dropwise condensation," *Surfaces and Interfaces*, vol. 25, p. 101143, Aug. 2021, doi: 10.1016/j.surfin.2021.101143.
51. M. D. Yadav, K. Dasgupta, A. W. Patwardhan, and J. B. Joshi, "High performance fibers from carbon nanotubes: synthesis, characterization, and applications in composites—a review," *Industrial & Engineering Chemistry Research*, vol. 56, no. 44, pp. 12407–12437, Nov. 2017, doi: 10.1021/acs.iecr.7b02269.
52. N. Rungraeng, Y.-C. Cho, S. H. Yoon, and S. Jun, "Carbon nanotube-polytetrafluoroethylene nanocomposite coating for milk fouling reduction in plate heat exchanger," *Journal of Food Engineering*, vol. 111, no. 2, pp. 218–224, Jul. 2012, doi: 10.1016/j.jfoodeng.2012.02.032.
53. W. Li, K. Zhou, R. M. Manglik, G.-Q. Li, and A. E. Bergles, "Investigation of CaCO3 fouling in plate heat exchangers," *Heat and Mass Transfer*, vol. 52, no. 11, pp. 2401–2414, Nov. 2016, doi: 10.1007/s00231-016-1752-2.
54. J. E. Friis *et al.*, "Evaluation of surface-initiated polymer brush as anti-scaling coating for plate heat exchangers," *Progress in Organic Coatings*, vol. 136, p. 105196, Nov. 2019, doi: 10.1016/j.porgcoat.2019.06.042.
55. Y. Lv, M. Liu, and Y. Xu, "Corrosion and fouling behaviors on modified stainless steel surfaces in simulated oilfield geothermal water," *Protection of Metals and Physical Chemistry of Surfaces*, vol. 54, no. 3, pp. 526–535, May 2018, doi: 10.1134/S2070205118030322.
56. C. Yanhai, R. Lu, M. Xianliang, Y. Jinyong, and T. Xianhua, "The effect of PTFE addition on mechanical and anti-corrosion properties of coating of heat exchangers," *Materials Research Express*, vol. 6, no. 8, p. 085207, May 2019, doi: 10.1088/2053-1591/ab2222.
57. S. Dağlı and M. Özbey, "experimental investigation of the effect of hydrophobic coating on heat transfer and gasketed plate heat exchanger performance," *Proceedings of the Institution of Mechanical Engineers Part E Journal of Process Mechanical Engineering*, 2023, doi: 10.1177/09544089231156309.
58. M. Bangavadi Munivenkatappa *et al.*, "Mitigation of fouling during milk processing in polytetrafluoroethylene-titanium dioxide coated plate heat exchanger," *Journal of Food Process Engineering*, vol. 45, no. 6, Jun. 2022, doi: 10.1111/jfpe.13836.
59. V. Oldani, G. Sergi, C. Pirola, and C. L. Bianchi, "Use of a sol-gel hybrid coating composed by a fluoropolymer and silica for the mitigation of mineral fouling in heat exchangers," *Applied Thermal Engineering*, vol. 106, pp. 427–431, Aug. 2016, doi: 10.1016/j.applthermaleng.2016.06.014.
60. V. Oldani, G. Sergi, C. Pirola, and C. L. Bianchi, "Use of a Sol-Gel hybrid coating composed by a fluoropolymer and silica for the mitigation of mineral fouling in heat exchangers," *Applied Thermal Engineering*, 2016, doi: 10.1016/j.applthermaleng.2016.06.014.

13 Experimental Study and ANN-Based Predictions

Thermo-Hydraulic-Exergetic Performance of Heat Exchangers with Three Fluids in Two Thermal Communications

Vikas Bargah, Taraprasad Mohapatra, Hillol Joardar Sudhansu S. Mishra, and Sudhansu S. Sahoo

Nomenclature	
A	Surface area, m^2
c	Specific heat, kJ/Kg-K
d	Diameter, m
d_c	Coil diameter, m
D_S	Diameter of shell, m
f	Friction factor
h	Convective heat transfer coefficient, W/ m^2.K
j	Colburn factor
k	Thermal conductivity, W/m-K
$\dot{m}$	Mass flow rate, kg/sec
Nu	Nusselt number
Pr	Prandlt number
Q	Rate of heat transfer, W
Re	Reynolds number
T_b,i	Uniform average temperature at inlet, °C
T	Temperature, ° C
V	Velocity, m/sec
Y	Dimensionless length
Greek symbols	
ρ	Density, kg/m^3
μ	viscosity, N-S/m^2
ε_E	Exergy efficiency, %

Subscripts	
1	Inlet
2	Outlet
avg	Average
c	Coil
cr	Critical
cu	Copper
e	Equivalent
i	Inner
im	Innermost tube

Abbreviations	
DTHE	Double tube heat exchanger
FEM	Finite element method
IMTD	Integral mean temperature difference
MVHR	Mechanical ventilation with heat recovery
NTU	Number of the transfer unit
RSWPS	Room-sealed wood pellet stove
TCTHE	Triple concentric tube heat exchanger
TF1, TF2, and TF3	Thermal fluids 1,2, and 3
UHT	Ultra heat treatment

DOI: 10.1201/9781003509837-13

13.1 INTRODUCTION

Heat exchangers (HX) are crucial components in numerous industrial processes and are extensively utilized in the HVAC industry, the food industry, chemical factories, petroleum refineries, the nuclear industry, renewable energy systems, heating industries, and thermal energy storage systems (Kumar et al., 2017; Rathore and Kumar, 2023; Kumar, Thayagrajan et al., 2022). Recently, researchers have been concentrating on developing heat exchangers that are more efficient, cost-effective, and require minimal maintenance. As a result, various heat transfer enhancement techniques (such as active and passive) are currently being developed for heat exchangers, taking into account factors such as product features, construction, flow patterns, and fluid contact methods. One of the passive techniques is to add a helical coil to a double-pipe heat exchanger forming a three-fluid heat exchanger. The current heat exchanger employs this idea for improved performance. Further, analyzing heat exchangers theoretically requires making several assumptions and dealing with complex equations, while experimental methods tend to be costlier due to the significant initial investment needed to set up the experiments (Shah and Sekulic, 2003). To address these challenges, artificial neural network (ANN) models have been created for simulating, optimizing, and predicting the performance of thermal systems that incorporate heat exchangers (Yang, 2008;; Thibault et al., 1991; Kumar and Banerjee, 2024). ANNs establish correlations based on training data, eliminating the need for specific analytical equations and system descriptions. They can effectively correlate the nonlinear parameters inherent in heat transfer processes within heat exchangers with minimal errors. Some of the already published works underlying the motivation during the analysis are included in the following section.

To calculate the axial temperature distribution of fluids in a triple tube heat exchanger, as well as the overall heat transfer coefficients, a more general method was created. For comparison, findings from theoretical twin-tube heat exchangers were also tabulated. The data collected have been used to both conceptually explain and illustrate the benefits of utilizing a TTHE versus a DTHE (Batmaz, 2003). Both the analytical and finite element methods have been used to study the impact of heat leakage to the cold fluid in a three-fluid heat exchanger for a cryogenic application requiring thermal interaction between all three fluids. The given results provide significant insights into the behavior of the hot fluid in this particular kind of heat exchanger (Krishna et al., 2012). For three fluids, temperature distribution data have been collected by both analytical and experimental methods. Additionally evaluated is the impact of volume flow rate variation on the three-fluid heat exchangers' performance. When the flow rate of various fluids in a three-fluid heat exchanger increased, the overall heat transfer coefficient was shown to rise in each case (Mohapatra et al., 2017a). For varying flow rates and inlet temperatures, the three-fluid heat exchangers' properties related to pressure drop and heat transfer were evaluated. The total heat transfer coefficient rises and the effectiveness of heat transfer from the helical tube side fluid to the outer annulus side fluid falls with an increase in the volumetric flow rates of both fluids in both flow configurations. Additionally, it is noted that in both flow configurations, the overall heat transfer coefficient and effectiveness for heat transfer from the helical tube side fluid to the outer annulus side fluid rise with an increase in the helical tube side fluid inlet temperature (Mohapatra et al., 2019c). An analytical model of a three-fluid heat exchanger was used to forecast how various design factors would affect the heat exchanger's performance. Taguchi-based optimization is used to determine the ideal values of these parameters to maximize heat transfer and minimize pressure drop. There is good conformance between the analytical approach's results and the literature when they are validated and compared (Mohapatra et al., 2019b). To find the optimal values of input variables for maximal heat transfer and minimal irreversibility loss in a Three Fluids Heat Exchanger (TFHE), multi-response optimization research employing the desirability function approach was conducted. It was found that the smallest entropy generation number and the greatest tube side Nusselt number are anticipated for the TFHE to function optimally (Mohapatra et al., 2019a). By feeding hot water or thermal fluid from a solar flat plate collector via the triple fluid heat exchanger's helical tube to warm incoming cold air and water in the innermost pipe and outer annulus, the heating cost can be significantly reduced. The triple fluid heat exchanger is studied through experimentation, and its validity is confirmed by contrasting the experimental findings with previous research. There is a fair degree of agreement between the

> experimental and published values (Mohapatra et al., 2017b). Another study suggested a novel multi-fluid heat exchanger for the simultaneous heating of water and air using solar energy. The brazed helix tube is made from a helical tube and precision brazed between coil turns. Experimental testing was done to evaluate the thermo-hydraulic performance of the current novel multi-fluid heat exchanger (NMFHE) for hot water and air heating in homes (Almasri et al., 2024a, 2024b). The new three-fluid heat exchanger model can design more efficient exchangers, optimize existing ones, and study their behavior to identify new design opportunities (Shrivastava and Ameel, 2004a). The design and selection of three-fluid heat exchangers require careful consideration of the relative thermal capacities of the three fluid streams and the desired engineering goals. A multi-objective optimization approach may be necessary to design a heat exchanger that meets all the desired objectives (Shrivastava and Ameel, 2004b). As deposits of milk fouling accumulate on the heating surface, the efficiency of a dairy heat exchanger decreases. As a result of the increased barrier to heat movement, the temperature of the milk output drops as fouling thickness increases. Several models have been proposed to predict the temperature of the milk output and fouling thickness in a heat exchanger (Nema and Datta, 2006). A double-pipe TFHE was investigated using both numerical and experimental methods. Modeling approaches, such as finite volume methods and finite element methods, are widely used to simulate engineering problems across various domains, including the development and performance improvement of heat transport systems (Kumar et al., 2018, 2023, 2024). The heat exchanger's heat transfer parameter was assessed using a computational fluid dynamics (CFD) tool for a range of flow rates, tube diameters, temperature-dependent viscosity, and residence times (Rennie and Raghavan, 2005, 2006, 2007, 2010). TFHEs with two or three thermal communications were examined in terms of their fundamentals. The study delves deeply into the creation of dimensionless groups and how to use them to create mathematical models (Sekulić et al., 1995). Two sections of theoretical research were done on a triple concentric tube heat exchanger. The mathematical modeling is covered in the first section, while performance and design calculations are covered in the second. In the end, the three tubes' relative sizes were determined for the best heat exchanger performance (Ünal, 1998, 2001). Two thermal communications and three-fluid co-current heat exchangers were studied to produce integral mean temperature difference (IMTD) equations. The outcome demonstrates that, for three-fluid heat exchangers, the devised formula has a higher convergence rate than the current Log Mean Temperature Difference (LMTD) model
>
> **(Zhao and Li, 2012)**

In addition to conventional energy generation methods, including nuclear energy, heat exchangers play a critical role in renewable energy (e.g., solar energy, geothermal energy, and biomass energy) generation as well (Kumar et al., 2020; Kumar and Rathore, 2024). Analyzing heat exchangers theoretically requires making several assumptions and dealing with complex equations, while experimental methods tend to be costlier due to the significant initial investment needed to set up the experiments (Shah et al., 2003). To address these challenges, ANN models have been created for simulating, optimizing, and predicting the performance of thermal systems that incorporate heat exchangers (Yang, 2008; Sen et al., 2000). ANNs establish correlations based on training data, eliminating the need for specific analytical equations and system descriptions. They can effectively correlate the nonlinear parameters inherent in heat transfer processes within heat exchangers with minimal errors. A numerical model to study a three-dimensional, steady-state, conjugate heat transfer on OpenFOAM was developed. Comparison between the performance of four TPMS-structured heat exchangers and conventional compact heat exchangers across four dimensions—heat transfer, flow resistance, volume, and weight—was made. Furthermore, the flow properties of TPMS-structured heat exchangers were examined (Yan et al., 2023). An experimental examination was conducted to evaluate the heat transfer performance of an NMFHE concerning variation in the flow rate of heat transfer fluids HF_1 (hot water), HF_2 (normal water), and inlet temperature of HF_1 (hot water) (Almasari et al., 2023). An experimental and optimization study was conducted to examine the thermo-hydraulic performance of metal oxide nanofluid flow through the helical coil tube of a three-fluid heat exchanger. The experiment considered parametric investigations with nanofluid volume fraction (1%, 5%, and 10%), flow rate (100 LPH, 150 LPH, and 200 LPH), and inlet temperature (80°C, 120°C, and 160°C). The results of the experiment show that using CuO-water nanofluid at an intake temperature of 160°C and 5% volume percentage resulted in a significant improvement in heat transfer efficacy, with a maximum value of 0.884 (Sarangi et al., 2024).

13.1.1 Research Gap

- Very little research has been conducted on three-fluid heat exchangers utilizing helical coil addition for simultaneous water and space heating.
- Limited research has been conducted on heat exchangers in various fluid flow patterns and series heat transfer configurations. Parallel heat transfer is rarely studied.
- To reduce the number of experiments, an ANN is employed for the current heat exchanger, which is rare.

13.2 EXPERIMENTAL SETUP

13.2.1 HETF Test Section

For the current study, a HETF test section with a length of 1.8 m was fabricated as shown in Figure 13.1. The test section comprised the following parts:

- An outermost straight PVC tube with an inner diameter of 0.134 m and a thickness of 0.0092 m.
- An innermost straight tube of 0.072 m inner diameter and 0.0015 m thickness.
- A helical copper tube, 22.83 m in length with 0.0095 m inner diameter, 0.03 m pitch, 0.008 m thickness, and 60 turns positioned between the two straight tubes.

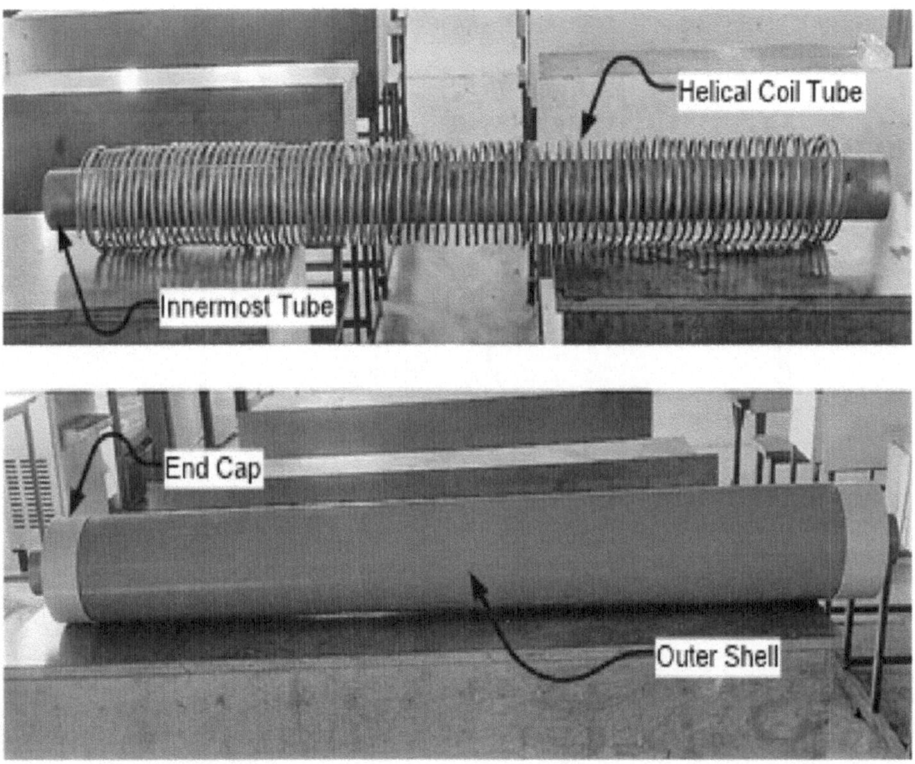

FIGURE 13.1 HETF test section.

13.2.1.1 Description of setup

The experiment setup used for the current study comprised five main parts: a central heat exchanger with three fluids (HETF) test section, TF1 flow circuit, TF2 flow circuit, temperature measurement unit, and an axial fan for TF3 flow, as shown in Figure 13.2 The detailed description of individual parts is mentioned in the subsequent sub-section.

The heating fluid TF1 flows through the outer annulus of the helical tube, which is the space between the two concentric tubes, deducting the volume of the helical tube. TF2 flows through the helical tube and TF3 flows through the innermost tube. A two thermal communication is established between the three fluids, where the heating fluid TF1 transfers its energy to the other two fluids.

13.2.1.2 TF1 Flow Circuit

The TF1 flow circuit contains a TF1 tank, incorporated with an immersion heater to heat TF1, and a 0.5 HP, Kirloskar centrifugal pump-rotameter assembly. The centrifugal pump is used to supply TF1 from the TF1 tank to the inlet of the outermost tube. The rotameter is used to measure the rate of flow of TF1. With the necessary piping arrangements, TF1 returns to the TF1 tank to complete the circuit.

13.2.1.3 TF2 Flow Circuit

The TF2 flow circuit contains a TF2 tank and a 0.5 HP Kirloskar centrifugal pump-rotameter assembly. The centrifugal pump is used to supply TF2 to the inlet of the helical tube. The rotameter is used to measure the rate of flow of TF2. With the necessary piping arrangements, the TF2 returns to the TF2 tank.

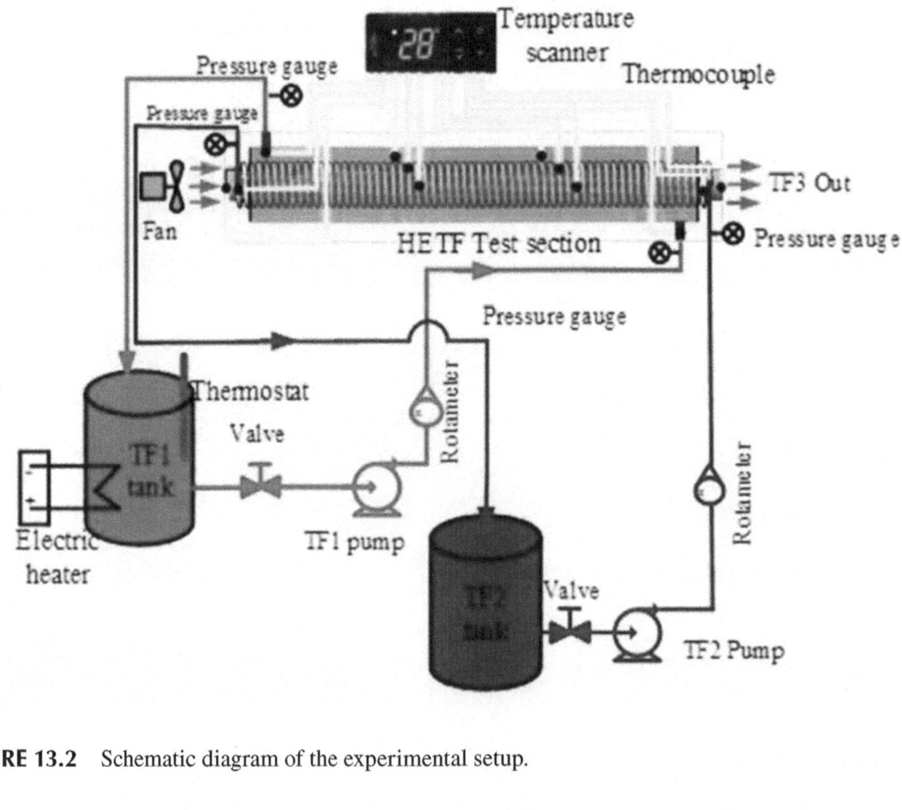

FIGURE 13.2 Schematic diagram of the experimental setup.

13.2.1.4 Axial Fan

A 4" axial fan is provided at the bore of the innermost tube for TF3 flow. The flow is regulated with the help of a regulator nob having three speeds calibrated with the help of an anemometer.

13.2.1.5 Temperature Measurement Unit

For measuring the temperature, a 4 K type thermocouple was inserted at different locations along the length of the HETF. A display panel having both automatic and manual displays was provided to enable the users to note down the temperatures.

13.2.2 Experimental Procedure

- First, the three fluids were circulated across the test segment of the HETF for 30–35 minutes to reach a steady-state condition for both flow arrangements to acquire an experimental temperature profile along the length of the HETF.
- Three distinct flow rates for TF1 (i.e., 100, 200, and 300 LPH), TF2 (i.e., 50, 100, and 150 LPH), and TF3 (i.e., 14,657, 29,314, and 43,971 LPH) were chosen to study the effect of volume flow rates of the three fluids on thermo-hydraulic and exergetic performance of the HETF.
- At four different locations along the length of the HETF, the temperature was measured with the help of thermocouples and was noted down.
- As a result, a graph between temperature and axial distance was plotted.
- To obtain the temperature distribution of the parallel flow design, the directions of TF1, regular water, and TF3 flow were all kept constant. Whereas, in the counter-flow arrangement, TF1 flowed opposite to the direction of TF2 and TF3.

13.3 DATA REDUCTION

Referring to Mohapatra et al. (2019c).

Colburn factor, *j*:

The Colburn factor for TF1 is calculated as

$$j_{TF1} = \frac{Nu_{TF1}}{\mathrm{Re}_{TF1}.\mathrm{Pr}_{TF1}} \tag{13.1}$$

The Colburn factor for TF2 is calculated as

$$j_{TF2} = \frac{Nu_{TF2}}{\mathrm{Re}_{TF2}.\mathrm{Pr}_{TF2}} \quad - (13.2)$$

The Colburn factor for TF3 is calculated as

$$j_{TF3} = \frac{Nu_{TF3}}{\mathrm{Re}_{TF3}.\mathrm{Pr}_{TF3}} \tag{13.3}$$

Friction factor, *f*:

The friction factor for TF1 is calculated as

$$f_{TF1} = \frac{d_{se}.2.\Delta p_{TF1}}{L_s \rho_{TF1}.u^2_{TF1}} \tag{13.4}$$

The friction factor for TF2 is calculated as

$$f_{TF2} = \frac{d_{ci}.2.\Delta p_{TF2}}{L_c \rho_{TF2}.u^2_{TF2}} \tag{13.5}$$

The friction factor for TF3 is calculated as

$$f_{TF3} = \frac{d_{im}.2.\Delta p_{TF3}}{L_s \rho_{TF3}.u^2_{TF3}} \tag{13.6}$$

JF FACTOR, *JF*:

The *JF* factor for TF1, TF2, and TF3 are calculated using the following formula:

$$JF = \frac{j}{f^{1/3}} \tag{13.7}$$

Where j is the Colbourn factor and f is the friction factor.

EXERGY EFFICIENCY:

$$\varepsilon_E = 1 - \frac{Exergy_{destroyed}}{Exergy_{supplied}} \tag{13.8}$$

13.4 ARTIFICIAL NEURAL NETWORKS

ANNs draw inspiration from the structure and function of biological neural networks in the human brain. However, ANNs are not mere imitations; they are sophisticated computational models utilized extensively in machine learning for tasks including classification, regression, clustering, and prediction. Similar to the brain, an ANN comprises numerous interconnected processing units, or neurons, collaborating to accomplish specific tasks. Each neuron receives inputs from other neurons or external source processes; these inputs use a transformation function and generate an output signal transmitted to other neurons. Various factors can affect the accuracy of an ANN model, including the choice of training algorithm, network architecture (configuration), activation functions employed, initial weights and biases, and the size and quality of the training dataset. The notation "3-n-4" denotes a configuration with three neurons in the input layer, an unknown number of neurons in the hidden layer denoted by "n," and four neurons in the output layer. Multi-layer perceptron ANNs offer several advantages, such as versatility and nonlinearity. Additionally, they are known for their interpretability, scalability, and ability to make fast predictions compared with other types of ANNs. The proposed two inlet temperatures (60°C and 80°C) condition for three input parameters (V_h, V_n, and V_a) with four experimental output responses (JF_h, JF_n, JF_a, and E_e), which have a nonlinear relation and are used for training, testing, and validation. The root mean square error (RMSE) and the corresponding correlation coefficients were employed as the judgment criteria for the ANN model. Various architectures were considered for training, where 70% of the data was utilized for training, 15% for testing, and another 15% for validation purposes. The Levenberg–Marquardt (LM) backpropagation supervised algorithm was chosen due to its ability to minimize the MSE between the predicted outputs of a neural network and the actual outputs. MATLAB R2016a was employed for creating, training, and testing the model to predict output response behavior. Performance during training was evaluated using MSE, while mean absolute percentage error (MAPE) was utilized for assessing performance during testing, as depicted in Equations 13.1 and 13.2.

$$MSE = \frac{1}{N}\sum_{n=1}^{N}(Actual - \Pr edicted)^2 \tag{13.9}$$

$$MAPE = \frac{1}{N}\sum_{n=1}^{N}\frac{|Actual - \Pr edicted|}{Actual}\times 100 \tag{13.10}$$

The ANN structure for both inlet temperatures, 60°C and 80°C, is shown in Figure 13.3. The input layer is the first layer of the network and receives the input data, which can be pre-processed before being fed into the network. Pre-processing steps, such as normalization or standardization, are typically applied to the input data to ensure that the data is in a suitable format for the neural network. The hidden layers are the layers between the input layer and the output layer. They are responsible for learning the underlying patterns in the data. The number of hidden layers and the number of neurons in each hidden layer are important hyperparameters of a neural network. In this case, two hidden layers are considered: "Hidden Layer 1" and "Hidden Layer 2." The chosen ANN architecture with three input nodes, two hidden layers (five nodes in the first hidden layer and nine nodes in the second hidden layer), and four output nodes is designed to balance complexity and performance. This structure allows the network to effectively learn and generalize from the data while avoiding overfitting and maintaining computational efficiency. The output layer has four nodes, corresponding to the four outputs required by the problem. Each node in the output layer produces one of the final predicted values. For instance, in a regression task, each node might output a continuous value, while in a classification task, each node might represent a different class probability.

Training an ANN involves adjusting the weights of the connections between neurons to minimize the error in the network's predictions. The process includes the following steps:

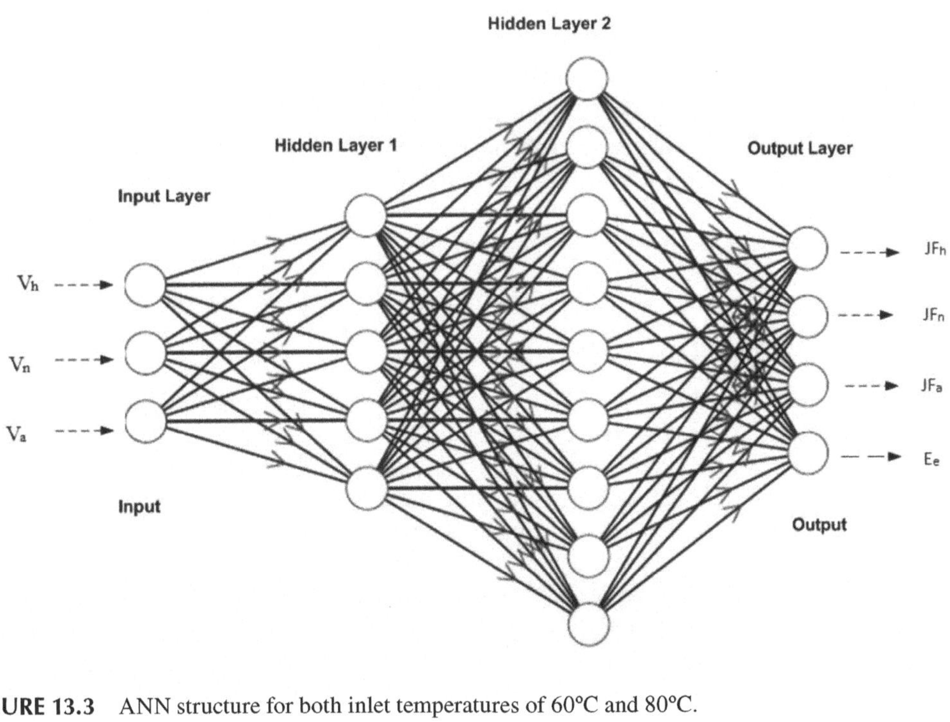

FIGURE 13.3 ANN structure for both inlet temperatures of 60°C and 80°C.

1. *Initialization.* The weights are initialized randomly or using specific initialization methods.
2. *Forward propagation.* The input data passes through the network, layer by layer, with each neuron applying an activation function to the weighted sum of its inputs.
3. *Loss calculation.* The network's output is compared to the true values using a loss function. Common loss functions include MSE for regression tasks and cross-entropy loss for classification tasks.
4. *Backpropagation.* The error is propagated back through the network to update the weights using the LM supervised algorithm. This step involves calculating the gradients of the loss function concerning each weight.
5. *Iteration.* Steps two to four are repeated for many iterations (epochs) until the network converges, i.e., the loss function reaches a minimum value or stops decreasing significantly.

Training ANNs involves addressing several challenges related to overfitting, gradient issues, hyperparameter tuning, computational resources, data quality, training time, and interpretability. By employing various techniques and strategies to tackle these challenges, it is possible to train effective and robust neural network models.

The choice between ANNs and other ML methods depends on the specific requirements of the task, including the nature of the data, the need for interpretability, computational resources, and the desired performance. ANNs excel in handling complex, high-dimensional data and capturing intricate patterns but require more computational resources and are less interpretable. Other methods like decision trees, Support Vector Machine (SVM), K-Nearest Neighbors (KNN), regression models, and Gradient Boosting Machine (GBM) offer varying trade-offs in terms of simplicity, interpretability, and performance.

13.5 RESULTS AND DISCUSSIONS

13.5.1 Effect of Flow Rates on Exergy Efficiency

Figure 13.4 depicts the effect of TF1 flow rate, TF2 flow rate, and velocity of TF3 on exergy efficiency at three distinct temperatures of 60°C, 70°C, and 80°C, respectively. Figure 13.4(a) shows the effect of TF1 flow rate on exergy efficiency. It is evident from the diagram that as the volume flow rate is raised by 200%, the exergy efficiency is reduced by 1.41%, 0.28%, and 2.06% at 60°C, 70°C, and 80°C, respectively. Figure 13.4(b) shows the effect of TF2 flow rate on exergy efficiency. It is evident from the diagram that as the volume flow rate is raised by 200%, the exergy efficiency is reduced by 5.8%, 6.9%, and 8.2% at 60°C, 70°C, and 80°C, respectively. Similarly, Figure 13.4(c) shows the effect of the velocity of TF3 on exergy efficiency. It is evident from the diagram that as the thermal fluid velocity is raised by 200%, the exergy efficiency is reduced by 1.2%, 0.2%, and 0.1% at 60°C, 70°C, and 80°C, respectively.

13.5.2 Effect of Flow Rates on *JF* Factor

Figure 13.5 shows the effects of TF1, TF2, and TF3 on the thermo-hydraulic performance of a HETF. In all potential flow configurations, the volume flow rates of the TF1, TF2, and TF3 have a significant impact on an HETF's thermo-hydraulic performance. The Colburn factor, j, and friction factor are selected for evaluating the thermodynamic and hydrodynamic parameters of an HETF to demonstrate how the volume flow rate influences the thermo-hydraulic performance of the HETF. To evaluate the performance of an HETF, we want a higher Colburn factor and a lower friction factor, which is very difficult to achieve at the same time due to the use of various heat transfer augmentation techniques, which result in a higher pressure drop, raising the friction factor as well.

The incorporation of a helical tube between two straight tubes results in a larger pressure drop. As a result, j and f are not always useful and have restrictions. This is accomplished using a

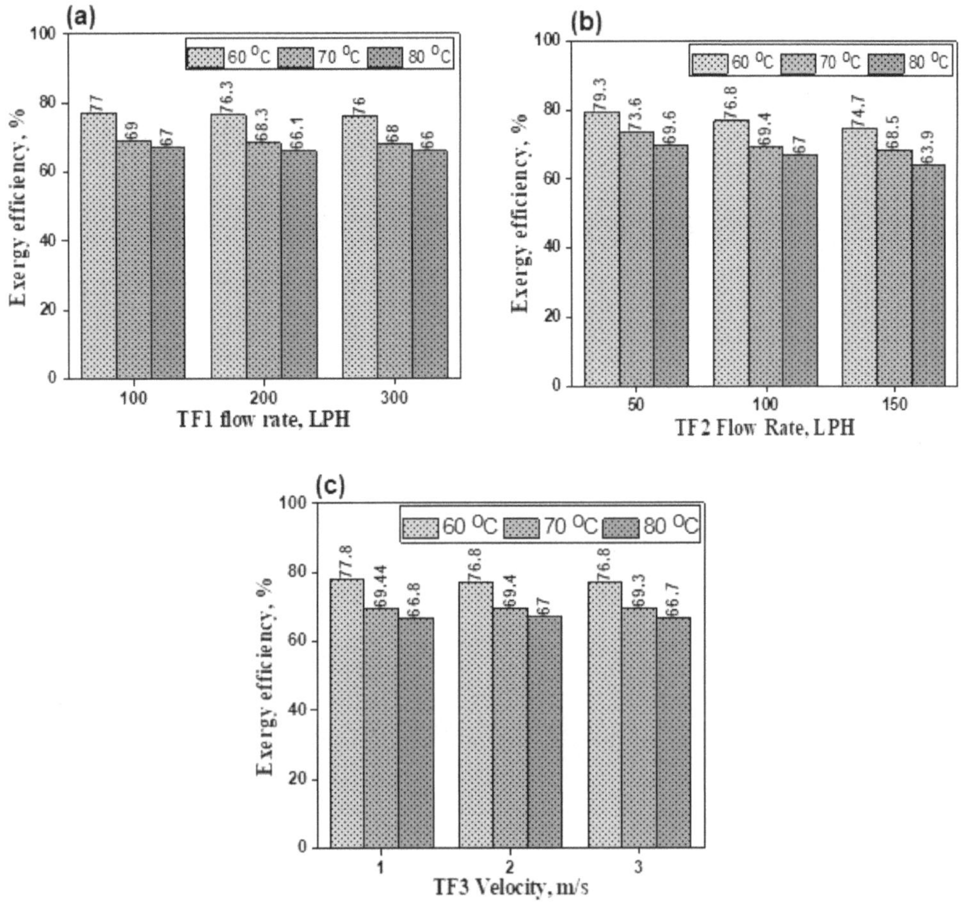

FIGURE 13.4 Effect of (a) TF1 flow rate, (b) TF2 flow rate, and (c) TF3 velocity on exergy efficiency.

non-dimensional quantity known as the *JF* factor, which is the ratio of the Colburn factor to the cube root of the friction factor. A higher *JF* value is required for optimal heat exchanger performance. Figure 13.5(b) shows that raising the TF2 flow rate by 200% reduces the *JF* factor by 65%, 66.6%, and 50% at 60°C, 70°C, and 80°C, respectively. Similarly, Figure 13.5(c) shows that raising the TF3 velocity by 200% reduces the *JF* factor by 5.88%, 3.08%, and 10.3% at 60°C, 70°C, and 80°C, respectively.

13.5.3 Prediction Through ANN

The label on the y-axis of Figure 13.6 mentions "Gradient." In machine learning, the gradient is used to calculate how much the model's output changes in response to a small change in the weights of its connections. It is essentially a measure of how well the model is learning. The y-axis values appear to be much smaller in magnitude (ranging from 10^{-7} to 10^{-6}) compared with Figure 13.6(b). This suggests that the gradients in Figure 13.6(a) are much smaller, possibly indicating the model has converged or reached a state where the gradients are close to zero. The x-axis represents "Epochs," which are iterations over the entire training dataset. The training state for the ANN model is displayed in Figure 13.6(a) and Figure 13.6(b), at epoch 10 for both inlet temperatures of 60°C and 80°C.

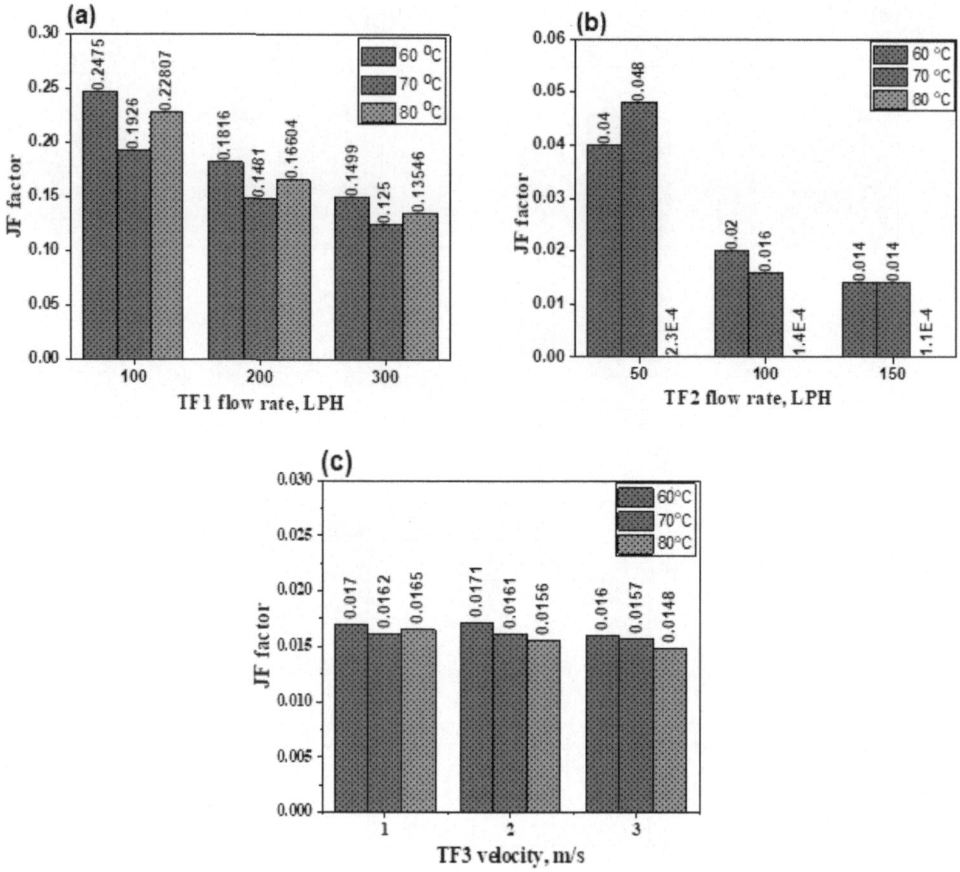

FIGURE 13.5 Effect of (a) TF1 flow rate, (b) TF2 flow rate, and (c) TF3 velocity on *JF* factor.

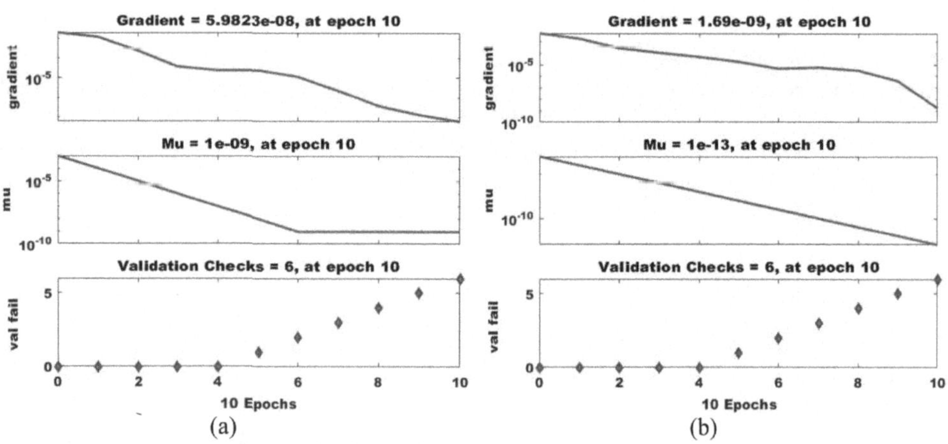

FIGURE 13.6 The training state for the ANN model for inlet temperatures (a) 60°C and (b) 80°C.

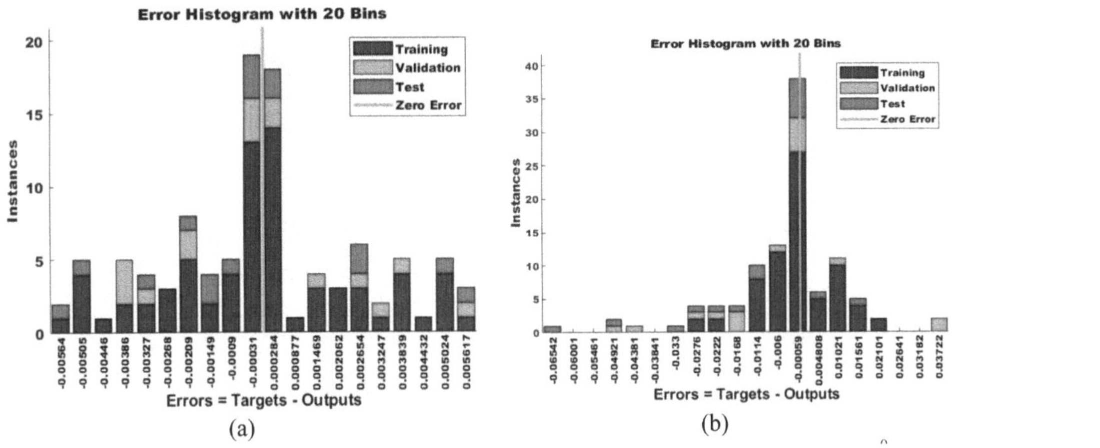

FIGURE 13.7 Histogram for the ANN model for inlet temperatures (a) 60°C and (b) 80°C.

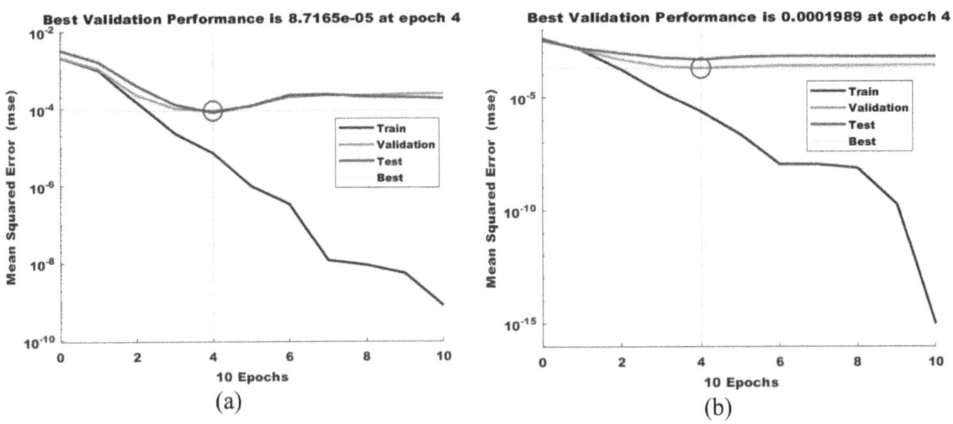

FIGURE 13.8 MSE plot for the ANN model for inlet temperatures (a) 60°C and (b) 80°C.

The range of errors on the x-axis is wider in Figure 13.7(b) compared with Figure 13.7(a). This indicates that the errors in Figure 13.7(b) have a larger magnitude overall. In other words, the model in Figure 13.7(b) might be making larger mistakes in the data compared with the model in Figure 13.7(a). The distribution of errors in Figure 13.7(a) appears to be more centered on zero error, with a smoother curve. This suggests that the model in Figure 13.3(a) is performing better and making fewer errors. The distribution in Figure 13.7(b) seems to be more spread out, with a possibly higher concentration of errors further away from zero. This indicates that the model in Figure 13.7(b) might be struggling to learn the patterns in the data.

Figure 13.8(a) shows a curve labeled "Train" that dips to a minimum MSE value around epoch 4, and then slightly increases. Also, a curve labeled "Validation" shows a similar trend but remains consistently higher than the training curve throughout. This suggests that the model might be generalizing well, as the training and validation errors are close. Figure 13.8(b) indicates a curve labeled "Train" that also dips to a minimum MSE value around epoch 4 but then increases more significantly than in Figure 13.8(a). The curve labeled "Validation" starts higher than the training curve and keeps increasing throughout. This larger gap between training and validation errors suggests that the model might be overfitting.

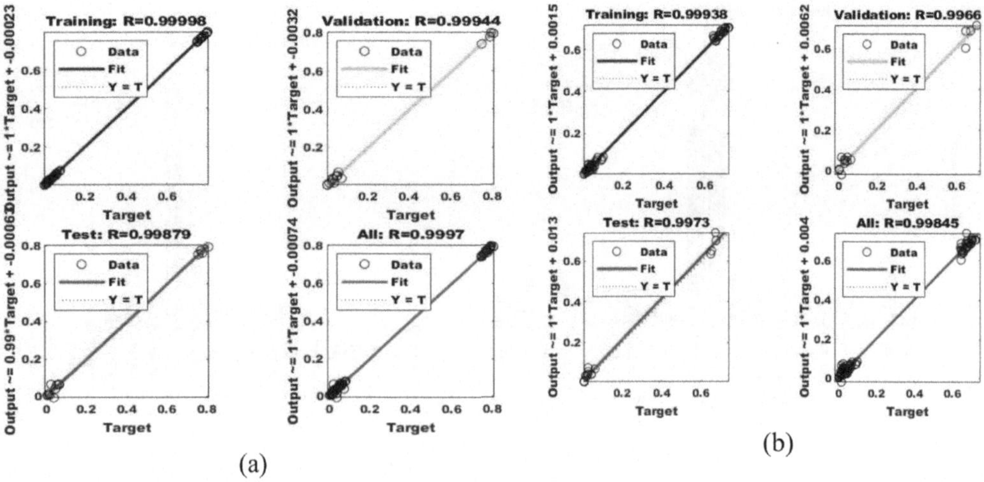

FIGURE 13.9 ANN for optimal fitness for inlet temperatures (a) 60°C and (b) 80°C.

LM backpropagation, a supervised algorithm, is used to achieve the very high R (0.9997 and 0.9984) values, as shown in Figure 13.9. This algorithm is utilized because this is the quickest and most reliable of the other algorithms, even though it requires more work and memory.

13.6 CONCLUSIONS AND SCOPE FOR FUTURE WORK

13.6.1 Conclusions

In this chapter, the heating efficiency of an HETF was investigated experimentally for changes in the volume flow rate of various fluids and flow arrangements. By contrasting the findings of the current analysis with those found in previously published work, the findings were confirmed. The present method and the methods that were mentioned in previously published work showed a very high degree of agreement. Along the axial length of the HETF test section, data to obtain experimental temperature profiles were calculated for the three fluids. The effect of changes in the volume flow rate of three different fluids on the performance of the HETF is discussed below.

1. By raising the flow rate of TF1 from 100 liter/hour to 300 liter/hour, the exergy efficiency was reduced by 1.2% for the three temperatures of 60°C, 70°C, and 80°C.
2. By raising the flow rate of TF2 from 50 liter/hour to 150 liter/hour, the exergy efficiency was reduced by 5.8%, 6.92%, and 8.2% for the temperatures of 60°C, 70°C, and 80°C, respectively.
3. By raising the velocity of TF3 from 1 m/s to 3 m/s, the exergy efficiency was reduced by 1.3%, 0.2%, and 0.15% for the temperatures of 60°C, 70°C, and 80°C, respectively.
4. By raising the flow rate of TF1 from 100 liter/hour to 300 liter/hour, the *JF* factor was reduced by 39.43%, 35.1%, and 40.6% for the temperatures of 60°C, 70°C, and 80°C, respectively.
5. By raising the flow rate of TF2 from 50 liter/hour to 150 liter/hour, the *JF* factor was reduced by 65%, 40.8%, and 52.17% for the temperatures of 60°C, 70°C, and 80°C, respectively.
6. By raising the velocity of TF3 from 1 m/s to 3 m/s, the *JF* factor was reduced by 5.9%, 7.4%, and 10.3% for the temperatures of 60°C, 70°C, and 80°C, respectively.

7. The ANN prediction analysis was successfully completed. A more favorable prediction is obtained for an inlet temperature of 60°C.

13.6.2 Future Scope

1. The current HETF is tested experimentally, which is time-consuming. Hence, an analytical and numerical model could be developed to save time.
2. The effect of the innermost tube diameter and the helical tube diameter on the performance of the HETF can be studied.
3. The current HETF suggests heating of space and water simultaneously. The same HETF can also be utilized in a refrigeration process.
4. The current HETF could be tested with nanofluids having better thermo-physical properties than water and providing a better heat transfer.
5. Multi-response optimization of different heat transfer characteristics can be done using optimization tools.
6. With the necessary arrangements, the current HETF can have three thermal communications instead of two.

The current HETF can be used in many industries, where simultaneous heating of two fluids is required, for example, the chemical and petrochemical industries.

ACKNOWLEDGMENT

The authors declare that they have no known competing financial interests or personal relationships.

DATA AVAILABILITY STATEMENT

The supporting data for the findings of this study are available on request from the corresponding author.

REFERENCES

Almasri, B., Mishra, S.S. and Mohapatra, T., Thermo-hydraulic performance augmentation in residential heating applications using a novel multi-fluid heat exchanger with helical coil tube insertion. *Proceedings of the Institution of Mechanical Engineers, Part A: Journal of Power and Energy*, 238(2), 2024a, 334–347.

Almasri, B., Mohapatra, T., Joardar, H., and Mishra, S. S. Experimental investigation and multi-objective optimization of a novel multi-fluid heat exchanger performances using response surface methodology and genetic algorithm. *Journal of the Brazilian Society of Mechanical Sciences and Engineering*, 46(7), 2024b, 400.

Almasri, B., Mohapatra, T., and Mishra, S. S. Experimental investigation and performance optimization of thermo-hydraulic and exergetic characteristics of a novel multi-fluid heat exchanger. *Journal of Thermal Analysis and Calorimetry*, 148(24), 2023, 14051–14068.

Batmaz, E. *Overall Heat Transfer Coefficients and Axial Temperature Distribution of Fluids in a Triple Tube Heat Exchanger*, November 6, 2003.

Krishna, V., Hegde, P. G., Subramanian, N., and Seetharamu, K. N., Effect of ambient heat-in-leak on the performance of a three fluid heat exchanger, for cryogenic applications, using finite element method. *International Journal of Heat and Mass Transfer*, 55(21–22), 2012, 5459–5470.

Kumar, S., and Banerjee, D. *PCMs for Sustainability Applications by Leveraging Machine Learning, Phase Change Materials (PCMs) – Recent Advances, New Perspectives, and Applications.* Intech Open Publication, 2024.

Kumar, S., and Rathore, K. Renewable energy for sustainable development goal of clean and affordable energy. *International Journal of Materials, Manufacturing and Sustainable Technologies*, 2(1), 2024, 1–15.

Kumar, S., Vijayan, P.K., Kannan, U., Sharma, M., and Pilkhwal, D.S., Experimental and computational simulation of thermal stratification in large pools with immersed condenser. *Applied Thermal Engineering*, 113, 2017, 345–361.

Kumar, S., Grover, R.B., Yadav, H., Vijayan, P.K., Kannan, U., and Agrawal, A. Experimental and numerical investigation on suppression of thermal strati fi cation in a water-pool: PIV measurements and CFD simulations. *Applied Thermal Engineering*, 138, 2018, 686–704.

Kumar, S., Grover, R.B., Yadav, H., Vijayan, P.K., Kannan, U., and Agrawal, A. Experimental investigations on thermal stratification in a large pool of water with immersed isolation condenser. *International Conference on Nuclear Engineering Proceedings, ICONE*, 3, 2020, 1–7.

Kumar, S., Thyagarajan, A., and Banerjee, D. Experimental investigation of thermal energy storage (TES) platform leveraging phase change materials in a chevron plate heat exchanger. *International Mechanical Engineering Congress and Exposition*, 2022, 1–7.

Kumar, S., Klassen, M., Kalassen, D., Hardin, R., and King, M. Dispersion of sneeze droplets in a meat facility indoor environment - without partitions. *Environmental Research*, 236, 2023, 116603.

Kumar, S., Hardin, R., and King, M. *Modeling of Flows Through Porous Media, Computational Fluid Flow and Heat Transfer: Advances, Design, Control and Applications*. Taylor & Francis, CRC Press Publication, 2024.

Mohapatra, T., Padhi, B. N., and Sahoo, S. S. Experimental investigation of convective heat transfer in an inserted coiled tube type three fluid heat exchanger. *Applied Thermal Engineering*, 117, 2017a, 297–307.

Mohapatra, T., Padhi, B. N., Sahoo, S. S., and Pramanik, R. N. Performance analysis of three fluid heat exchanger used in solar flat plate collector system. *Energy Procedia*, 109, 2017b, 322–330.

Mohapatra, T., Sahoo, S. S., and Padhi, B. N. Analysis, prediction and multi-response optimization of heat transfer characteristics of a three fluid heat exchanger using response surface methodology and desirability function approach, *Applied Thermal Engineering*, 151, 2019a, 536–555.

Mohapatra, T., Padhi, B. N., and Sahoo, S. S. Analytical investigation and performance optimization of a three fluid heat exchanger with helical coil insertion for simultaneous space heating and water heating. *Heat and Mass Transfer/Waerme- Und Stoffuebertragung*, 55(6), 2019b, 1723–1740.

Mohapatra, T., Ray, S., Sahoo, S. S., and Padhi, B. N. Numerical study on heat transfer and pressure drop characteristics of fluid flow in an inserted coiled tube type three fluid heat exchanger. *Heat Transfer—Asian Research*, 48(4), 2019c, 1440–1465.

Nema, P. K. and Datta, A. K. Improved milk fouling simulation in a helical triple tube heat exchanger. *International Journal of Heat and Mass Transfer*, 49(19–20), 2006, 3360–3370.

Rathore, K. and Kumar, S. Net zero emissions for our future generations through renewables: a brief review. *International Journal of Energy Resources Applications (IJERA)*, 2(1), 2023, 22–43.

Rennie, T. J. and Raghavan, V. G. S. Experimental studies of a double-pipe helical heat exchanger. *Experimental Thermal and Fluid Science*, 29(8), 2005, 919–924.

Rennie, T. J. and Raghavan, V. G. S. Numerical studies of a double-pipe helical heat exchanger. *Applied Thermal Engineering*, 26(11–12), 2006, 1266–1273.

Rennie, T. J. and Raghavan, V. G. S. Thermally dependent viscosity and non-Newtonian flow in a double-pipe helical heat exchanger. *Applied Thermal Engineering*, 27(5–6), 2007, 862–868.

Rennie, T. J. and Raghavan, V. G. S. Numerical analysis of the lethality and processing uniformity in a double-pipe helical heat exchanger. *Chemical Engineering and Processing: Process Intensification*, 49(7), 2010, 672–679.

Sarangi, A., Mohapatra, T., Mishra, S. S., Sahoo, S. S., and Mallik, R. K. Thermo-hydraulic performance of metal oxide nanofluid flow through the helical coil tube of a three-fluid heat exchanger: an experimental and optimization study. *Energy Sources, Part A: Recovery, Utilization, and Environmental Effects*, 46(1), 2024, 5974–5993.

Sekulić, D. P. and Shah, R. K. Thermal design theory of three-fluid heat exchangers. *Advances in Heat Transfer*, 26(C), 1995, 219–328.

Sen, M. and Yang, K. T. Applications of artificial neural networks and genetic algorithms in thermal engineering. *CRC Handbook of Thermal Engineering*, 2000, 620–661.

Shah, R. K. and Sekulic, D. P. *Fundamentals of heat exchanger design*. John Wiley & Sons; 2003.

Shrivastava, D. and Ameel, T. A. Three-fluid heat exchangers with three thermal communications. Part A: general mathematical model. *International Journal of Heat and Mass Transfer*, 47(17–18), 2004a, 3855–3865.

Shrivastava, D. and Ameel, T. A. Three-fluid heat exchangers with three thermal communications. Part B: effectiveness evaluation. *International Journal of Heat and Mass Transfer*, 47(17–18), 2004b, 3867–3875.

Thibault, J. and Grandjean, B. P. A neural network methodology for heat transfer data analysis. *International Journal of Heat and Mass Transfer*, 34(8), 1991, 2063–2070.

Ünal, A. Theoretical analysis of triple concentric-tube heat exchangers Part 1: mathematical modelling. *International Communications in Heat and Mass Transfer*, 25(7), 1998, 949–958.

Ünal, A. Theoretical analysis of triple concentric-tube heat exchangers Part 2: case studies, *International Communications in Heat and Mass Transfer*, 28(2), 2001, 243–256.

Yan, Kaixin, Junwei Wang, and Hongwu Deng. Numerical investigation into thermo-hydraulic characteristics and mixing performance of triply periodic minimal surface-structured heat exchangers. *Applied Thermal Engineering*, 230, 2023, 120748.

Yang, K. Artificial Neural Networks (ANNs): A new paradigm for thermal science and engineering. ASME. *Journal of Heat Transfer*, 130(9), 2008, 093001.

Zhao, M. and Li, Y. New integral-mean temperature difference model for thermal design and simulation of parallel three-fluid heat exchanger. *International Journal of Thermal Sciences*, 59, 2012, 203–213.

14 Enhancing Heat Exchanger Efficiency and Innovation Through AI and Machine Learning

Shashikant Gupta, Subrata Jana,
Shreesha Yogish Rao, and Dinkar Verma

14.1 INTRODUCTION

Heat exchangers play a vital role in numerous industries, serving as critical components in various thermal systems. Their importance stems from their ability to efficiently transfer thermal energy between fluids or between a fluid and a solid surface. The effective design and optimization of heat exchangers directly impact the overall performance, efficiency, and energy consumption of these systems. Heat exchangers are classified based on several criteria, including flow arrangement, construction, heat transfer mechanism, and process. Flow arrangement includes parallel flow, counter-flow, and cross-flow, each affecting heat transfer efficiency. Construction types include shell-and-tube, plate, and air-cooled heat exchangers, differing in design and application. Heat transfer mechanisms classify them into direct contact, where fluids mix, and indirect contact, where they are separated by a barrier. Finally, the process classification distinguishes between single-phase, where fluids remain in the same phase, and two-phase, where phase changes occur, such as in condensers and evaporators. These classifications guide the selection of heat exchangers for specific industrial uses, optimizing performance and efficiency.

14.1.1 The Importance of Heat Exchangers in Various Industries

Heat exchangers are ubiquitous in a wide range of industries, including power generation, chemical processing, refrigeration, aerospace, automotive, and heating, ventilation, and air conditioning (HVAC) systems (Shah and Mueller 1998, Webb 2000, Firouzfar et al. 2011, Hundy 2016, Kumar et al. 2020a, 2022, Thulukkanam 2024). In the power industry, heat exchangers are essential for steam generation, condensation, and cooling processes in thermal power plants, nuclear reactors, and renewable energy systems (Gholap and Khan 2007, Luo et al. 2016, Gomez-Garcia et al. 2017, Pandey et al. 2017, Tian et al. 2017, Kumar and Rathore 2023, Rathore et al. 2023). Chemical processing plants heavily rely on heat exchangers for heating, cooling, and energy recovery operations during various chemical reactions and separations (Gilmour 1960).

The refrigeration and HVAC industries utilize heat exchangers for air conditioning, refrigeration cycles, and heat pump applications. In the aerospace and automotive sectors, heat exchangers play a critical role in thermal management systems, ensuring efficient cooling of engines, electronics, and other components. Furthermore, heat exchangers are employed in a variety of other applications, such as waste heat recovery, cryogenic systems, and desalination processes (Timmerhaus and Schoenhals 1974, Noie-Baghban and Majideian 2000, Hatami et al. 2014, Popov et al. 2019, Elsheikh et al. 2022).

DOI: 10.1201/9781003509837-14

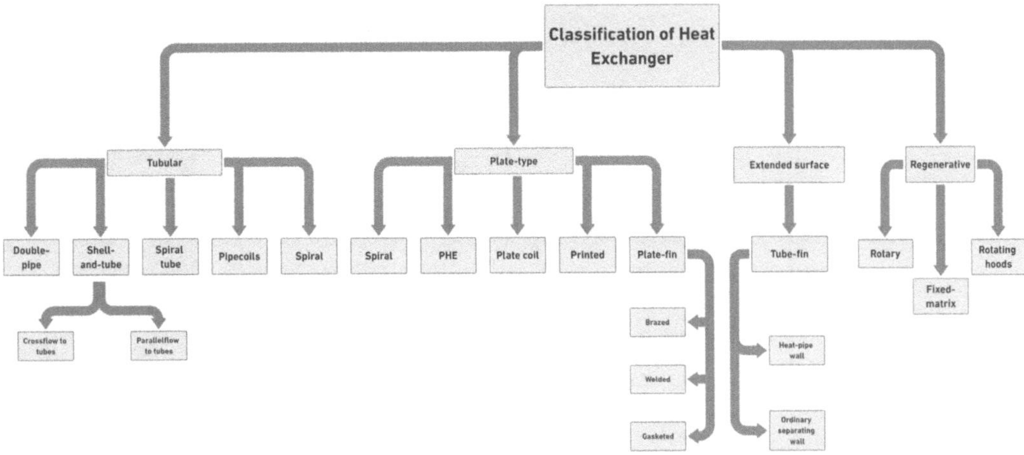

FIGURE 14.1 Classification of heat exchangers.

14.1.2 Challenges in Designing and Optimizing Heat Exchangers

Designing and optimizing heat exchangers presents several challenges that must be addressed to achieve desired performance levels and meet industry-specific requirements. Figure 14.1 illustrates a detailed hierarchical classification scheme for various types of heat exchangers, categorized based on structural and operational features. One significant challenge is the complexity of heat transfer mechanisms, which often involve intricate interactions between conduction, convection, and radiation heat transfer modes (Cavallini 2017, Zhang et al. 1997, Shah and Sekulic 1998). Additionally, the presence of phase change processes, such as condensation or boiling, can further complicate the design process, as found in passive safety systems (Kumar et al. 2017a, 2018, 2020b).

Another challenge lies in the trade-offs between various design objectives, such as maximizing heat transfer rates, minimizing pressure drops, and optimizing material usage and manufacturing costs (Hesselgreaves et al. 2016). Balancing these competing objectives requires careful consideration and optimization techniques. Furthermore, the integration of heat exchangers into larger systems introduces additional constraints and requirements, such as space limitations, weight restrictions, and compatibility with existing infrastructure (Shabgard et al. 2015, Kumar and Banerjee 2023). These factors necessitate the development of compact and efficient heat exchanger designs tailored to specific applications.

14.1.3 The Role of Numerical Simulations and Optimization in Heat Exchanger Design

Numerical simulations and optimization techniques have emerged as powerful tools in addressing the challenges associated with heat exchanger design and optimization (Selamat et al. 2016, Zheng et al. 2017, Kumar et al. 2015, 2017b). Computational fluid dynamics (CFD) simulations enable detailed analysis of fluid flow and heat transfer phenomena within heat exchangers, providing insights into temperature distributions, pressure drops, and overall thermal performance (Bhutta et al. 2012, Prashant and Pravin 2018, Abeykoon 2020, Sundén 2007). By coupling CFD simulations with optimization algorithms, designers can explore a vast design space and identify optimal configurations that maximize performance while satisfying various constraints (De Bellis and Catalano 2012, Liu et al. 2017). These optimization techniques can encompass geometrical parameters, flow conditions, and material properties, leading to enhanced heat transfer rates, reduced pressure losses, and improved overall efficiency.

Moreover, numerical simulations facilitate the investigation of complex heat transfer phenomena, such as conjugate heat transfer, phase change processes, and turbulent flow conditions, which

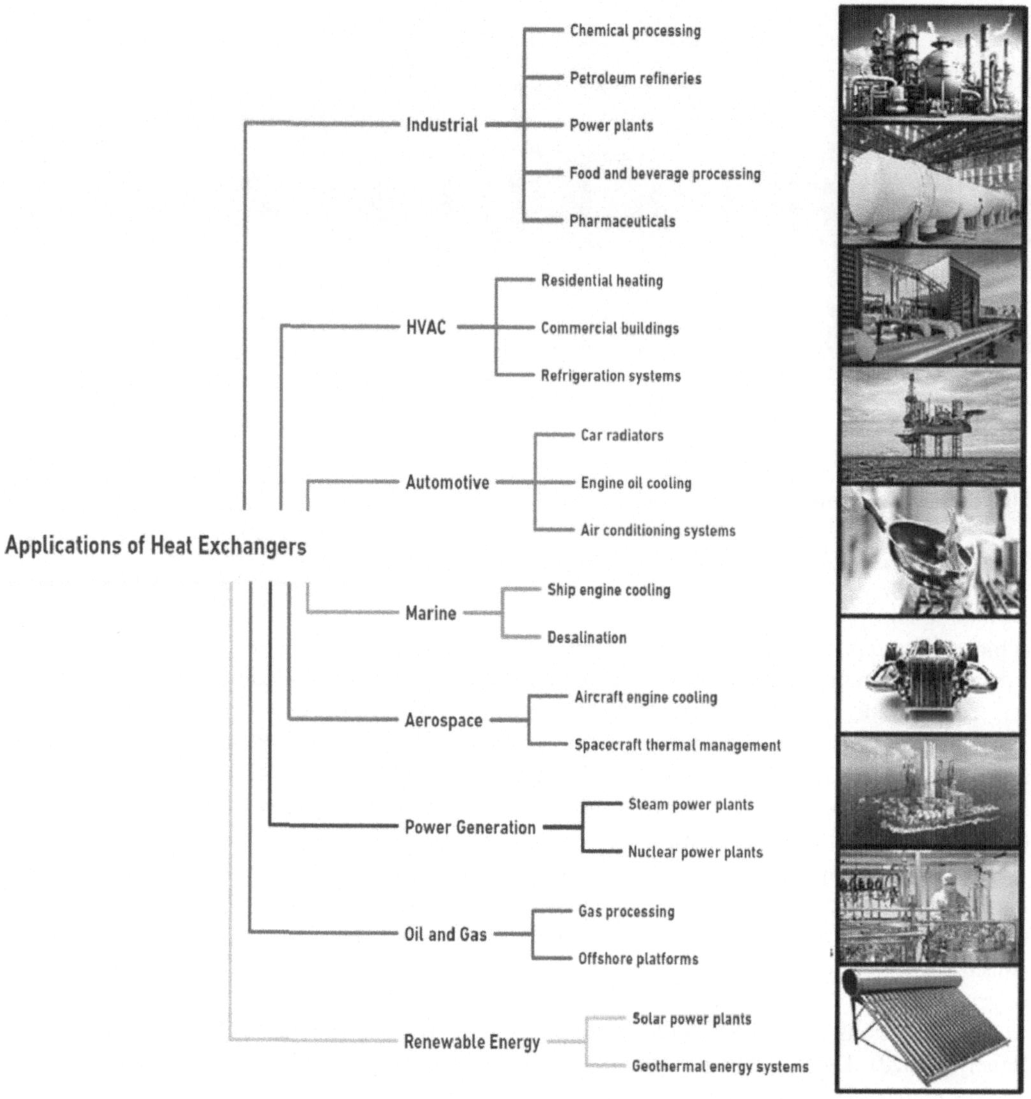

FIGURE 14.2 Various applications of heat exchangers.

are often challenging to study experimentally. These simulations allow for the exploration of innovative heat exchanger designs, including novel fin geometries, additive manufacturing techniques, and the incorporation of advanced materials or surface modifications. Figure 14.2 presents diverse practical applications of heat exchangers across industries including HVAC, automotive, marine, aerospace, and renewable energy sectors. Additionally, numerical simulations and optimization techniques contribute to the development of surrogate models and reduced-order models, enabling rapid design exploration and optimization processes (Yeşilyurt 1995, Longatte et al. 2013, Verma et al. 2018, Huang et al. 2022, Verma and Iyer, 2023). These models can be integrated with machine learning algorithms and artificial intelligence (AI) techniques, further enhancing the efficiency and accuracy of heat exchanger design and optimization workflows.

By leveraging numerical simulations and optimization techniques, engineers and researchers can gain valuable insights into the intricate dynamics of heat transfer processes, enabling the development of highly efficient and cost-effective heat exchanger solutions tailored to specific applications and industry requirements.

14.2 NUMERICAL SIMULATIONS FOR HEAT EXCHANGER PERFORMANCE EVALUATION

14.2.1 Overview of Numerical Methods for Heat Transfer Analysis

Numerical simulations have become indispensable tools for evaluating the performance of heat exchangers and analyzing complex heat transfer phenomena. These simulations rely on various numerical methods that discretize and solve the governing equations of fluid flow and heat transfer, enabling accurate predictions of temperature distributions, flow patterns, and thermal performance metrics. The governing equation of the numerical analysis is the Navier–Stokes equation, which is based on fluid flow combined with heat flow and is given by:

$$\rho\left(\frac{\partial v}{\partial t}+v.\left(v.\nabla\right)v\right)+\nabla p-\nabla.\left(\mu\nabla v\right)=\rho f,\nabla.v=0. \tag{14.1}$$

This equation relates the velocity vector (v), pressure field (p), density (ρ), coefficient of viscosity (μ), and volume acceleration (f).

The choice of numerical method to solve Equation 14.1 depends on factors such as the complexity of the problem, the desired accuracy, computational resources, and the ability to handle different types of geometries and boundary conditions. Several numerical methods have been widely employed in the field of heat transfer analysis, each with its own strengths and limitations.

14.2.2 Finite Difference Method

The finite difference method (FDM) is one of the oldest and most widely used numerical techniques for solving partial differential equations (PDEs) governing heat transfer and fluid flow problems. In FDM, the computational domain is discretized into a grid of nodes, and the governing PDEs are approximated by algebraic finite difference equations at each node.

The FDM approximates the derivatives in the heat transfer equations using difference equations. Consider the one-dimensional heat conduction equation:

$$\frac{\partial T}{\partial t}=\alpha\frac{\partial^2 T}{\partial x^2} \tag{14.2}$$

where T is the temperature, t is time, x is the spatial coordinate, and α is the thermal diffusivity.

In FDM, the spatial domain is discretized into a grid with points $x_i = i\Delta x$, and time is discretized into steps $t^n = n\Delta t$. The partial derivatives are approximated as:

$$\frac{\partial T}{\partial t}\approx\frac{T_i^{n+1}-T_i^n}{\Delta t}, \tag{14.3}$$

and

$$\frac{\partial^2 T}{\partial x^2}\approx\frac{T_{i+1}^n-2T_i^n+T_{i-1}^n}{\left(\Delta x\right)^2}. \tag{14.4}$$

Substituting these approximations into the heat conduction equation gives the finite difference equation:

$$\frac{T_i^{n+1}-T_i^n}{\Delta t}=\alpha\frac{T_{i+1}^n-2T_i^n+T_{i-1}^n}{\left(\Delta x\right)^2}. \tag{14.5}$$

Rearranging this, we obtain:

$$T_i^{n+1} = T_i^n + \frac{\alpha \Delta t}{(\Delta x)^2}\left(T_{i+1}^n - 2T_i^n + T_{i-1}^n\right). \tag{14.6}$$

This equation can be iteratively solved to find the temperature distribution over time. FDM is relatively simple to implement and can handle structured grids efficiently. However, it can encounter challenges when dealing with complex geometries or irregular boundaries, where the implementation of boundary conditions becomes more complicated. FDM is particularly suitable for simulating heat conduction problems and simple convection problems with regular geometries.

14.2.3 Finite Volume Method

The finite volume method (FVM) is a popular numerical technique for solving fluid flow and heat transfer problems, particularly in CFD applications. In FVM, the computational domain is divided into a finite number of control volumes or cells, and the governing equations are discretized and solved for each control volume. Consider the heat conduction equation:

$$\frac{\partial T}{\partial t} = \alpha \nabla^2 T. \tag{14.7}$$

For a control volume V_i with surface S_i, integrating the heat conduction equation over V_i and applying the divergence theorem yields:

$$\int_{V_i} \frac{\partial T}{\partial t} dV = \alpha \int_{S_i} \nabla T \cdot dA. \tag{14.8}$$

Discretizing in time, the volume integral becomes:

$$\frac{T_i^{n+1}V_i - T_i^n V_i}{\Delta t} = \alpha \sum_{faces} \left(\nabla T \cdot A\right)_{face}. \tag{14.9}$$

Approximating the fluxes at the faces, we get:

$$\frac{T_i^{n+1} - T_i^n}{\Delta t} = \frac{\alpha}{V_i} \sum_{faces} \left(\nabla T \cdot A\right)_{face}. \tag{14.10}$$

This can be solved iteratively to update the temperature T_i. One of the key advantages of FVM is its ability to conserve physical quantities, such as mass, momentum, and energy, ensuring that the numerical solutions satisfy the underlying conservation laws. FVM is well-suited for handling complex geometries and irregular meshes, making it a versatile choice for simulating heat exchangers with intricate geometries. Additionally, FVM can accommodate different types of meshes, including structured and unstructured grids, allowing for efficient mesh generation and local refinement in regions of interest. This flexibility makes FVM a powerful tool for simulating a wide range of heat exchanger configurations and flow conditions.

14.2.4 Finite Element Method

The finite element method (FEM) is a widely used numerical technique for solving partial differential equations in various engineering disciplines, including heat transfer and fluid mechanics. In FEM, the computational domain is discretized into a mesh of finite elements, and the governing equations are approximated using piecewise continuous functions defined over these elements.

Consider the weak form of the heat conduction equation:

$$\int_{\Omega} \frac{\partial T}{\partial t} w \, d\Omega + \alpha \int_{\Omega} \nabla T \cdot \nabla w \, d\Omega = 0, \tag{14.11}$$

where w is a test function. Discretizing T and w using basis functions ϕ_j:

$$T \approx \sum_j T_j \phi_j. \tag{14.12}$$

Substituting into the weak form and integrating by parts, we obtain the system of equations:

$$\sum_j \left(\int_{\Omega} \phi_i \phi_j \, d\Omega \right) \frac{dT_j}{dt} + \alpha \sum_j \left(\int_{\Omega} \nabla \phi_i \cdot \nabla \phi_j \, d\Omega \right) T_j = 0. \tag{14.13}$$

This system can be written as:

$$\boldsymbol{M} \frac{d\boldsymbol{T}}{dt} + \alpha \boldsymbol{K} \boldsymbol{T} = 0, \tag{14.14}$$

where $\boldsymbol{M}$ is the mass matrix, $\boldsymbol{K}$ is the stiffness matrix, and $\boldsymbol{T}$ is the temperature vector. This system of ordinary differential equations can be solved using time integration methods.

FEM excels in handling complex geometries and irregular domains, making it well-suited for simulating heat exchangers with intricate designs or intricate fin structures. Additionally, FEM can efficiently handle different types of boundary conditions, including those involving thermal radiation, phase change, and conjugate heat transfer. One of the advantages of FEM is its ability to adapt the mesh resolution to capture steep gradients or localized phenomena, ensuring accurate results while minimizing computational costs. However, FEM can be computationally more expensive than other methods, especially for large-scale problems or when dealing with complex, time-dependent simulations.

14.2.5 Lattice Boltzmann Method (LBM)

The lattice Boltzmann method (LBM) is a relatively newer numerical approach for simulating fluid flow and heat transfer problems. Unlike traditional methods that solve the Navier–Stokes equations directly, LBM is based on the kinetic theory of gases and utilizes particle distributions to model fluid behavior. LBM solves the discrete Boltzmann equation on a lattice grid. The Boltzmann equation for the distribution function $f_i(\boldsymbol{x},t)$ is:

$$f_i(\boldsymbol{x} + \boldsymbol{c}_i \Delta t, t + \Delta t) - f_i(\boldsymbol{x},t) = -\frac{\Delta t}{\tau} \left(f_i(\boldsymbol{x},t) - f_i^{eq}(\boldsymbol{x},t) \right), \tag{14.15}$$

where $\boldsymbol{c}_i$ are the lattice velocities, τ is the relaxation time, and f_i^{eq} is the equilibrium distribution function. The macroscopic variables (density, velocity, and temperature) are obtained from the moments of the distribution function:

$$\rho = \sum_i f_i, \tag{14.16}$$

$$\rho \boldsymbol{u} = \sum_i f_i \boldsymbol{c}_i, \tag{14.17}$$

$$\rho T = \sum_i f_i \left(\frac{|\boldsymbol{c}_i - \boldsymbol{u}|^2}{2} \right). \tag{14.18}$$

The LBM approach provides a direct simulation of fluid flow and heat transfer by evolving the distribution functions on the lattice grid, making it particularly suitable for complex boundary conditions and multiphase flows. LBM has gained popularity due to its inherent parallelism, making it well-suited for efficient implementation of modern parallel computing architectures. It can handle complex geometries and boundary conditions with relative ease, making it an attractive choice for simulating heat exchangers with intricate designs or porous media. Additionally, LBM naturally incorporates microscopic interactions and can readily handle multiphase and multicomponent flows, which are common in many heat exchanger applications. However, LBM may require fine grid resolutions to accurately capture boundary layers and small-scale phenomena, potentially increasing computational costs.

14.2.6 Comparison and Selection of Appropriate Numerical Methods

The selection of the appropriate numerical method for simulating heat exchanger performance depends on various factors, including the complexity of the problem, the desired accuracy, computational resources, and the ability to handle specific phenomena or geometries.

For simple heat conduction problems or convection problems with regular geometries, FDM can be a suitable choice due to its simplicity and efficiency. FVM is a versatile method that can handle a wide range of fluid flow and heat transfer problems (Kumar and King 2022, Kumar et al. 2023a, 2023b), making it a popular choice in CFD simulations for heat exchangers with complex geometries or intricate flow patterns.

FEM is particularly well-suited for simulating heat exchangers with intricate geometries, irregular domains, or complex boundary conditions involving thermal radiation, phase change, or conjugate heat transfer. Its ability to adapt mesh resolution and handle different types of boundary conditions makes it a powerful tool for accurate simulations.

LBM offers advantages for simulating complex geometries, porous media, and multiphase or multicomponent flows, and its inherent parallelism allows for efficient simulations on modern computing architectures. However, it may require finer grid resolutions, potentially increasing computational costs.

In practice, a combination of numerical methods may be employed to leverage their respective strengths and overcome their limitations. For instance, FVM can be coupled with FEM for simulating conjugate heat transfer problems, where FVM handles the fluid flow and FEM solves the heat conduction in solid domains. Ultimately, the selection of the appropriate numerical method should consider factors such as the problem complexity, desired accuracy, computational resources, and the ability to capture specific phenomena or geometries. Additionally, the availability of validated and efficient numerical codes, as well as the expertise and experience of the research team, play crucial roles in the successful implementation and interpretation of numerical simulations for heat exchanger performance evaluation.

14.3 SIMULATION TECHNIQUES AND TOOLS

14.3.1 Commercial Computational Fluid Dynamics Software

Commercial CFD software packages are widely used in the simulation and analysis of heat exchangers. These software tools offer robust and user-friendly interfaces, along with a wide range of advanced modeling capabilities and pre-processing and post-processing tools. Some of the widely used commercial CFD software in the field of heat exchanger simulations include ANSYS Fluent, ANSYS CFX, Siemens Star-CCM+, and CD-adapco STAR-CD (ANSYS Inc. 2023, Siemens PLM Software 2023). These commercial software packages typically employ FVM or FEM for discretizing and solving the governing equations of fluid flow and heat transfer (Sohrabi et al. 2024, Kishan et al. 2020). They provide comprehensive modeling tools for simulating various physical phenomena, such as turbulence, multiphase flows, conjugate heat transfer, and radiation heat transfer, which are crucial for accurate heat exchanger simulations (Debtera et al. 2018, Skočilas and Palaziuk 2015, Padoin et al. 2014, Papukchiev et al. 2020, Končar and Klobučar 2018).

One of the key advantages of commercial CFD software is the availability of user-friendly graphical user interfaces (GUIs) and advanced pre-processing and post-processing tools. These tools facilitate the creation of complex geometries, mesh generation, assignment of boundary conditions, and visualization of simulation results. Additionally, many commercial packages offer specialized modules or add-ons tailored for specific applications, such as heat exchanger design and optimization. However, commercial CFD software often comes with significant licensing costs, which may be a limiting factor for some research groups or organizations with limited budgets. Furthermore, the source code of these software packages is typically proprietary, limiting the ability to modify or extend the underlying numerical algorithms or models.

14.3.2 Open-Source CFD Codes

Open-source CFD codes offer an alternative to commercial software, providing access to the source code and enabling greater flexibility and customization. These codes are typically developed and maintained by academic or research communities, with contributions from users worldwide. Some popular open-source CFD codes used for heat exchanger simulations include OpenFOAM, SU2, and deal.II. Open-source CFD codes often employ advanced numerical methods, such as FVM, FEM, or LBM, and provide a wide range of modeling capabilities for simulating various physical phenomena relevant to heat exchanger applications. One of the key advantages of open-source codes is the ability to modify and extend the source code to incorporate new models, numerical schemes, or specialized applications. Additionally, open-source CFD codes can be freely distributed and used without licensing costs, making them attractive options for researchers and organizations with limited budgets. However, open-source codes may require more effort in terms of learning, setup, and user support, as they often lack user-friendly interfaces and the comprehensive documentation provided by commercial software.

14.3.3 In-House Code Development and Customization

In certain cases, researchers or engineers may develop their own in-house CFD codes tailored to specific heat exchanger applications or research objectives. This approach offers the highest level of flexibility and control, allowing for the implementation of specialized numerical methods, models, or optimization algorithms that may not be readily available in commercial or open-source software (Dixon et al. 2010, Baik et al. 2014, Seo et al. 2020). The development of in-house CFD codes typically involves implementing numerical methods, such as FDM, FVM, FEM, or LBM, and incorporating relevant physical models for simulating fluid flow, heat transfer, and other relevant phenomena. This process requires a deep understanding of numerical methods, programming skills, and knowledge of the underlying physics and mathematical models.

One of the key advantages of in-house code development is the ability to integrate customized models, optimization algorithms, or specialized applications directly into the simulation framework. This approach can be particularly useful for exploring novel heat exchanger designs, investigating complex physical phenomena, or developing advanced optimization strategies. However, in-house code development can be time-consuming and resource-intensive, requiring significant effort in code development, verification, and validation. Additionally, maintaining and updating in-house codes can be challenging, especially in rapidly evolving research fields or when dealing with complex applications.

TABLE 14.1
List of Different Simulation Methods with Advantages and Limitations

Method	Advantages	Limitations
Finite difference method (FDM)	• Simple implementation • Effective for structured grids • Efficient for simple geometries	• Less flexible for complex geometries • Can be less accurate for irregular boundaries • Grid-dependent accuracy
Finite volume method (FVM)	• Conserves mass, momentum, and energy • Well-suited for complex geometries • Flexible grid handling	• More complex implementation compared with FDM • Requires careful handling of fluxes at cell faces
Finite element method (FEM)	• Highly flexible for complex geometries • Suitable for a wide range of problems • High accuracy	• Computationally intensive • More complex meshing • Requires specialized knowledge for implementation
Lattice Boltzmann method (LBM)	• Suitable for complex boundary condition • Easy parallelization • Naturally handles multi-phase flows	• Requires fine lattices for accurate results • Limited to relatively simple physical models • High computational cost for fine grids
Direct numerical simulation (DNS)	• Provides highly accurate results • Captures all scales of turbulence • No modeling assumptions required	• Extremely computationally intensive • Limited to low Reynolds number flows • Impractical for large systems
Large eddy simulation (LES)	• Resolves large-scale turbulence • More accurate than RANS for unsteady flows • Suitable for high Reynolds numbers	• Still computationally intensive • Requires sub-grid-scale modeling • Complex implementation
Reynolds-averaged Navier–Stokes (RANS)	• Less computationally intensive • Suitable for industrial applications • Can handle high Reynolds numbers	• Relies on turbulence models • Less accurate for complex, unsteady flows • May require calibration of models
Boundary element method (BEM)	• Efficient for problems with infinite or semi-infinite domains • Reduces problem dimensionality	• Limited to linear problems • Complex implementation for nonlinear or time-dependent problems

14.3.4 High-Performance Computing and Parallel Computing

Heat exchanger simulations, particularly those involving complex geometries, turbulent flows, or multiphysics phenomena, can be computationally intensive and may require significant computational resources. High-performance computing (HPC) and parallel computing techniques play a crucial role in enabling efficient and scalable simulations for these demanding applications. HPC systems, such as supercomputers or high-performance clusters, provide massive computational power by combining multiple processors or nodes, allowing for the parallelization of computationally intensive simulations. Parallel computing techniques, such as domain decomposition methods or particle-based parallelization, enable the efficient distribution of computational tasks across multiple processors or cores, significantly reducing the overall simulation time (Rao et al. 2020, Shadloo 2021).

Many commercial and open-source CFD codes are designed to leverage parallel computing capabilities, allowing for efficient simulations on HPC systems (Afzal et al. 2017, Garcia-Gasulla et al. 2020). Additionally, in-house code development often incorporates parallel computing strategies to take advantage of modern computing architectures and accelerate simulations. Parallel computing techniques can be broadly classified into two categories: shared-memory parallelism and distributed-memory parallelism. Shared-memory parallelism is suitable for multi-core or multi-processor systems, where multiple threads or processes share a common memory space. Distributed-memory parallelism, on the other hand, is designed for cluster or grid computing environments, where multiple nodes communicate via message-passing interfaces (MPI) to coordinate and exchange data during simulations.

The choice of parallel computing strategy depends on various factors, including the characteristics of the problem, the numerical method employed, the available hardware resources, and the scalability requirements of the simulation. Efficient parallelization can significantly reduce computational times, enabling the exploration of larger design spaces, higher-resolution simulations, or the incorporation of more complex physical models. Thus, high-performance computing and parallel computing techniques are essential for enabling efficient and scalable simulations of heat exchangers, especially in cases involving complex geometries, turbulent flows, or multiphysics phenomena. By leveraging these techniques, researchers and engineers can address computationally demanding problems, explore novel designs, and gain deeper insights into the intricate heat transfer processes within heat exchangers.

14.4 MODELING COMPLEX HEAT TRANSFER PHENOMENA

Heat transfer phenomena in heat exchangers can involve various complexities, including the coupling of different heat transfer modes, phase changes, turbulent flow conditions, and radiation effects. Accurately capturing these complex phenomena is crucial for reliable performance predictions and optimizations. This section delves into the mathematical modeling and numerical techniques employed to simulate these intricate heat transfer processes.

14.4.1 Conjugate Heat Transfer

Conjugate heat transfer refers to the simultaneous occurrence of conduction heat transfer within solid domains and convective heat transfer in fluid domains, with a coupled interface condition at the solid-fluid boundary. This phenomenon is prevalent in heat exchangers, where the thermal energy is transferred from one fluid to another through a solid wall or extended surfaces (fins).

The governing equations for conjugate heat transfer involve the energy equation for the fluid domain and the heat conduction equation for the solid domain, coupled at the solid-fluid interface through appropriate boundary conditions. The fluid flow is typically modeled using the Navier–Stokes equations, while the solid domain is governed by the heat conduction equation.

In the fluid domain, the energy equation can be expressed as:

$$\rho_f\left(\partial T / \partial t+u\cdot\nabla T\right)=\nabla\cdot\left(k_f\nabla T\right)+S_T, \tag{14.19}$$

where ρ_f is the fluid density, T is the temperature, u is the fluid velocity vector, k_f is the fluid thermal conductivity, and S_T represents any additional heat sources or sinks.

In the solid domain, the heat conduction equation is given by:

$$\rho_s c_s\left(\partial T / \partial t\right)=\nabla\cdot\left(k_s\nabla T\right)+S_T, \tag{14.20}$$

where ρ_s is the solid density, c_s is the specific heat capacity of the solid, k_s is the solid thermal conductivity, and S_T represents any internal heat sources or sinks.

At the solid-fluid interface, the coupled boundary conditions typically involve the continuity of temperature and heat flux:

$$T_f = T_s, \tag{14.21}$$

$$k_f\left(\partial T / \partial n\right)_f = k_s\left(\partial T / \partial n\right)_s, \tag{14.22}$$

where the subscripts f and s denote the fluid and solid domains, respectively, and $\partial/\partial n$ represents the derivative in the direction normal to the interface.

Numerical techniques such as FVM or FEM can be employed to discretize and solve the coupled system of equations, accounting for the interface conditions. Specialized solution algorithms or domain coupling strategies may be required to handle the different physical domains and ensure accurate and stable solutions.

14.4.2 Phase Change Heat Transfer

Phase change heat transfer, involving phenomena such as condensation, evaporation, melting, or solidification, plays a crucial role in many heat exchanger applications. These processes involve the release or absorption of latent heat during phase transitions, which can significantly impact the overall heat transfer performance. The mathematical modeling of phase change heat transfer requires the consideration of additional conservation equations for the liquid and vapor/solid phases, as well as appropriate interfacial conditions and source terms to account for the latent heat effects.

One common approach is to employ the enthalpy-porosity formulation, which treats the mushy zone (the region where both liquid and solid phases coexist) as a pseudo-porous medium. The governing equations include the continuity, momentum, and energy equations, along with an additional equation for the liquid fraction or solid fraction. The energy equation in the enthalpy-porosity formulation can be expressed as:

$$\partial\left(\rho H\right) / \partial t + \nabla\cdot\left(\rho u H\right) = \nabla\cdot\left(k \nabla T\right) + S_L, \tag{14.23}$$

where ρ is the density, H is the enthalpy, u is the velocity vector, k is the thermal conductivity, and S_L represents the latent heat source term, which accounts for the phase change phenomena.

TABLE 14.2
List of Different Simulation Tools with Important Specifications

Simulation Tools	Specifications	Limitations
ANSYS Fluent	• Advanced turbulence models • Multi-phase flow capabilities • Heat transfer and radiation modeling	• High computational cost • Steep learning curve • Expensive licensing
ANSYS CFX	• Robust solver for turbo-machinery • Coupled heat transfer and fluid flow analysis • High accuracy for rotating machinery	• High cost • Less user-friendly than some competitors • Requires extensive training
COMSOL Multiphysics	• Multiphysics simulations • User-friendly interface • Extensive material library	• Limited CFD capabilities compared with specialized tools • High license cost • Moderate computational demand
STAR-CCM+	• Integrated CFD and CAD modeling • Extensive meshing capabilities • Support for multiphysics simulations	• High licensing fees • Complex user interface • Requires significant computational resources
OpenFOAM	• Open-source and free • Highly customizable • Large user community and support	• Requires significant expertise to use • Less polished user interface • Limited commercial support

The liquid fraction or solid fraction is typically modeled using an additional transport equation or algebraic relationships based on the phase equilibrium conditions. Numerical methods such as FVM or FEM can be employed to discretize and solve the coupled system of equations, considering the phase change effects and associated boundary conditions. Specialized techniques, such as the enthalpic method or the temperature-transforming model, may be used to handle the nonlinearities and discontinuities associated with phase change processes.

14.4.3 Turbulent Flow and Heat Transfer

Turbulent flow conditions are commonly encountered in heat exchangers, particularly in applications involving high flow rates or complex geometries (Aly 2014). Turbulence can significantly enhance heat transfer rates but also introduces additional complexities in the modeling and simulation processes. The Navier–Stokes equations, along with the continuity and energy equations, govern the fluid flow and heat transfer in turbulent regimes. However, directly solving these equations for turbulent flows (known as direct numerical simulation, DNS) is computationally prohibitive for most practical applications due to the wide range of spatial and temporal scales involved.

To overcome this challenge, turbulence modeling approaches are employed, which introduce additional transport equations or algebraic models to account for the effects of turbulence on the mean flow and heat transfer characteristics. One widely used approach is the Reynolds-averaged Navier–Stokes (RANS) equations, where the governing equations are time-averaged, and the turbulent fluctuations are modeled using additional transport equations for turbulence quantities, such as the turbulent kinetic energy and dissipation rate (k-ε model) or the specific dissipation rate (k-ω model) (Galvis et al. 2008, Patel et al. 2015, Oprițoiu 2016, Dogan et al. 2019, Menni et al. 2022). Alternatively, large eddy simulation (LES) techniques can be employed, where the large-scale turbulent motions are directly resolved, and the smaller, sub-grid-scale eddies are modeled using appropriate sub-grid-scale models.

The energy equation in turbulent flows includes additional terms to account for the turbulent heat flux and its interaction with the mean flow and temperature fields. Turbulence models, such as the Reynolds analogy or the algebraic turbulent heat flux models, are often employed to capture these effects. Numerical methods, such as FVM or FEM, combined with appropriate turbulence modeling approaches, are used to solve the coupled system of equations governing turbulent flow and heat transfer in heat exchangers.

14.4.4 Radiation Heat Transfer

Radiation heat transfer can play a significant role in certain heat exchanger applications, particularly those involving high temperatures or participating media (e.g., combustion systems, solar thermal systems, or industrial furnaces). Accurately modeling radiation heat transfer is crucial for reliable performance predictions and optimizations. The governing equation for radiation heat transfer is the radiative transfer equation (RTE), which describes the propagation of radiative intensity through a participating medium. The RTE is an integro-differential equation that accounts for the emission, absorption, and scattering of radiation by the medium.

The RTE can be expressed as:

$$(\Omega\cdot\nabla)I(r,\Omega)=\kappa_a\left[I_b(r)-I(r,\Omega)\right]+\kappa_s\int(4\pi)I(r,\Omega')\Phi(\Omega,\Omega')d\Omega', \tag{14.24}$$

where $I(r,\Omega)$ is the radiative intensity at position r and direction Ω, κ_a and κ_s are the absorption and scattering coefficients, respectively, I_b is the blackbody intensity, and Φ is the scattering phase function.

Numerical techniques for solving the RTE include the discrete ordinate method (DOM), the FVM, the discrete transfer method (DTM), and the Monte Carlo method (MCM). These methods discretize the spatial, angular, and spectral domains to obtain numerical solutions for the radiative intensity field.

The radiative heat source term in the energy equation accounts for the divergence of the radiative heat flux and can be expressed as:

$$S_r = -\nabla \cdot qr = \kappa_a \int (4\pi)\left[I_b(r) - I(r,\Omega)\right] d\Omega, \tag{14.25}$$

where qr is the radiative heat flux.

Coupling the RTE solution with the fluid flow and heat transfer equations allows for the accurate modeling of radiation heat transfer in heat exchangers, accounting for the interaction between radiation and other heat transfer modes (conduction and convection).

14.4.5 Multiscale Modeling Approaches

Many heat transfer phenomena in heat exchangers involve multiple spatial and temporal scales, ranging from microscopic interactions to macroscopic system-level behavior. Figure 14.3 depicts multiscale modeling approaches, categorizing methods into macroscopic, mesoscopic, and microscopic scales for heat transfer and system analysis. Multiscale modeling approaches aim to bridge these different scales, providing a comprehensive understanding of the underlying physical processes and enabling accurate performance predictions (Xie et al. 2021). One commonly employed multiscale modeling approach is the volume-averaging technique, which homogenizes the microscopic details within a representative elementary volume (REV) to obtain effective transport properties and governing equations at the macroscopic scale. This approach is particularly useful for modeling porous media or heterogeneous media. Here are the basic mathematical details.

For heat exchangers, the REV, (V_{REV}), is chosen such that it includes sufficient microscopic details (e.g., fluid passages, solid matrix) to capture the local heat transfer characteristics, yet is small enough to ensure homogeneity at the macroscopic scale. Consider the microscopic temperature field $(T(x))$. The volume-averaged temperature over the REV, $(\langle T \rangle)$, is:

$$\langle T \rangle = \frac{1}{V_{\text{REV}}} \int_{V_{\text{REV}}} T(x)\, dV. \tag{14.26}$$

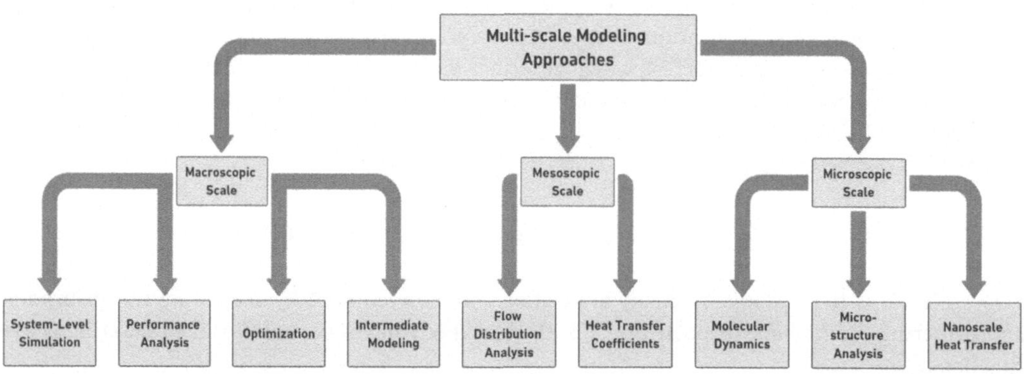

FIGURE 14.3 Multiscale modeling approaches.

For the microscopic heat flux vector $\left(q(x)\right)$, the volume-averaged heat flux is:

$$\langle q \rangle = \frac{1}{V_{\text{REV}}} \int_{V_{\text{REV}}} q(x)\,dV. \tag{14.27}$$

The microscopic heat conduction equation in a stationary medium is given by:

$$\nabla \cdot q + q_{\text{gen}} = \rho c_p \frac{\partial T}{\partial t}, \tag{14.28}$$

where $q = -k\nabla T$ (Fourier's law), q_{gen} is the heat generation term, A is the density, c_p is the specific heat capacity, and k is the thermal conductivity. To derive the macroscopic heat conduction equation, we volume-average the microscopic equation over the REV.

$$\left\langle \rho c_p \frac{\partial T}{\partial t} \right\rangle = \rho c_p \frac{\partial T}{\partial t}. \tag{14.29}$$

The average heat generation is:

$$\langle q_{\text{gen}} \rangle = \frac{1}{V_{\text{REV}}} \int_{V_{\text{REV}}} q_{\text{gen}}\,dV. \tag{14.30}$$

The average divergence of heat flux is given by:

$$\langle \nabla \cdot q \rangle = \nabla \cdot \langle q \rangle. \tag{14.31}$$

Using Fourier's law for the volume-averaged heat flux, we get:

$$\langle q \rangle = -k_{\text{eff}} \nabla \langle T \rangle, \tag{14.32}$$

where k_{eff} is the effective thermal conductivity, accounting for the heterogeneities within the REV.

The effective thermal conductivity k_{eff} is determined by homogenizing the thermal properties within the REV:

$$k_{\text{eff}} = \frac{1}{V_{\text{REV}}} \int_{V_{\text{REV}}} k(x)\,dV. \tag{14.33}$$

Combining all the averaged terms, the macroscopic heat conduction equation becomes:

$$\rho c_p \frac{\partial \langle T \rangle}{\partial t} = \nabla \cdot \left(k_{\text{eff}} \nabla \langle T \rangle\right) + \langle q_{\text{gen}} \rangle. \tag{14.34}$$

Another multiscale modeling approach is the heterogeneous multiscale method (HMM), which couples microscale and macroscale models through a two-way exchange of information. The macroscale model captures the overall system behavior, while the microscale model provides detailed information about the underlying physics at selected points or regions of interest. This coupling allows for the accurate representation of localized phenomena while maintaining computational efficiency at the system level.

Hybrid methods, combining continuum-based models with molecular dynamics (MD) or direct simulation Monte Carlo (DSMC) techniques, can also be employed for multiscale modeling. In these approaches, continuum models are used for the bulk regions, while MD or DSMC simulations are employed to capture detailed molecular-level interactions in regions of interest, such as near solid-fluid interfaces or in rarefied gas flows. For multiphase or multicomponent systems, population balance modeling (PBM) can be coupled with continuum models to account for the evolution of particle size distributions or species concentrations. PBM involves solving transport equations for the number density or mass density of particles or species, incorporating phenomena such as nucleation, growth, aggregation, and breakage.

Multiscale modeling approaches can also involve the integration of machine learning (ML) techniques or data-driven models as reported for the photovoltaic system (Sharadga et al. 2024). For example, ML methods can be used to develop surrogate models or reduced-order models that capture the essential physics at different scales, enabling efficient simulations and optimizations. The choice of multiscale modeling approach depends on the specific heat transfer problem, the scales of interest, and the computational resources available. Effective multiscale modeling requires a deep understanding of the underlying physical processes, careful model formulation, and efficient numerical techniques for coupling different scales and physical phenomena.

By employing multiscale modeling approaches, researchers and engineers can gain valuable insights into the intricate heat transfer mechanisms occurring at various spatial and temporal scales within heat exchangers. These insights can inform the design and optimization of heat exchangers, leading to improved performance, efficiency, and tailored solutions for specific applications.

14.5 OPTIMIZATION ALGORITHMS FOR IMPROVING THERMAL PERFORMANCE

Optimization algorithms play a crucial role in enhancing the thermal performance of heat exchangers by identifying optimal designs, operating conditions, or control strategies. This section explores various optimization techniques employed in the field of heat exchanger design and optimization, including mathematical formulations, gradient-based algorithms, derivative-free methods, evolutionary algorithms, and hybrid approaches.

14.5.1 Mathematical Formulation of Optimization Problems

The mathematical formulation of an optimization problem is a fundamental step in applying optimization algorithms to improve the thermal performance of heat exchangers. An optimization problem typically involves minimizing or maximizing an objective function subject to constraints imposed by physical laws, design requirements, or operational limits.

The general form of a constrained optimization problem can be expressed as:

Minimize $f(x)$ (Objective function)

Subject to:

$g_j(x) \leq 0, j = 1, 2, \ldots, m$ (Inequality constraints)

$h_k(x) = 0, k = 1, 2, \ldots, p$ (Equality constraints)

$x_l \leq x \leq x_u$ (Bound constraints)

Where:

$f(x)$ is the objective function to be minimized or maximized, typically representing the thermal performance metric (e.g., heat transfer rate, pressure drop, or energy efficiency).

x is the vector of design variables or decision variables (e.g., geometrical parameters, flow rates, or material properties).

$g_j(x)$ and $h_k(x)$ are the inequality and equality constraints, respectively, representing physical laws, design requirements, or operational limits.

x_l and x_u are the lower and upper bounds of the design variables, respectively.

The objective function and constraints may involve complex mathematical expressions or numerical simulations, such as CFD or finite element analysis (FEA), to accurately model the heat transfer phenomena and capture the relevant physics.

14.5.2 Gradient-Based Optimization Algorithms

Gradient-based optimization algorithms are widely used in heat exchanger design and optimization due to their efficiency and convergence properties. These algorithms rely on the gradient (first-order derivative) or higher-order derivatives of the objective function and constraints to guide the search direction toward the optimal solution. One of the most popular gradient-based algorithms is the sequential quadratic programming (SQP) method, which solves a sequence of quadratic programming sub-problems by approximating the objective function and constraints using quadratic models. The sub-problems are solved iteratively, and the solution is updated based on the gradient and Hessian information.

Other gradient-based algorithms include the method of moving asymptotes (MMA), the generalized pattern search (GPS), and the interior-point method (IPM). These algorithms employ different strategies for handling constraints, updating the search direction, and ensuring convergence to the optimal solution. The efficiency of gradient-based algorithms depends on the availability and accuracy of gradient information, which can be obtained through analytical differentiation, finite difference approximations, or adjoint sensitivity analysis techniques. However, these algorithms may struggle with non-convex or discontinuous objective functions and constraints, which can lead to convergence to local optima or failure to find feasible solutions.

14.5.3 Derivative-Free Optimization Algorithms

Derivative-free optimization algorithms, also known as direct search or pattern search methods, do not require explicit gradient information. These algorithms are particularly useful when the objective function or constraints are non-differentiable, discontinuous, or computationally expensive to evaluate, making gradient calculations challenging or impractical. One widely used derivative-free algorithm is the Nelder–Mead simplex method, which is based on the concept of a simplex (a geometric shape with n+1 vertices in an n-dimensional space). The algorithm iteratively modifies the simplex by reflecting, expanding, contracting, or shrinking its vertices, guided by the objective function values at these vertices.

Other derivative-free algorithms include mesh adaptive direct search (MADS), dividing rectangles (DIRECT), and the particle swarm optimization (PSO) method. These algorithms employ various strategies for sampling the design space, generating candidate solutions, and updating the search direction based on objective function evaluations. While derivative-free algorithms can handle complex and non-differentiable objective functions and constraints, they may require a larger number of function evaluations compared with gradient-based methods, which can be computationally expensive for complex heat exchanger simulations.

14.5.4 Evolutionary Algorithms

Evolutionary algorithms (EAs) are a class of optimization techniques inspired by natural evolution and biological processes, such as genetic inheritance, mutation, and selection. These algorithms operate on a population of candidate solutions, which are iteratively improved through mechanisms like crossover, mutation, and selection based on fitness values (objective function evaluations). One of the most widely used EAs is the genetic algorithm (GA), which mimics the principles of natural selection and genetic recombination. In GAs, candidate solutions (individuals) are represented as strings or vectors of design variables (chromosomes or genotypes). The algorithm iteratively creates new generations of solutions by applying genetic operators (crossover and mutation) to the fittest individuals from the previous generation.

Another popular EA is the differential evolution (DE) algorithm, which employs a vector-based mutation strategy and crossover operations to generate new candidate solutions. DE has been successfully applied to various heat exchanger optimization problems due to its simplicity, robustness, and ability to handle non-differentiable and multimodal objective functions. Other evolutionary algorithms include evolution strategies (ES), genetic programming (GP), and covariance matrix adaptation evolution strategy (CMA-ES). These algorithms differ in their representation of solutions, mutation strategies, and selection mechanisms but share the common principle of mimicking natural evolution to explore the design space and converge toward optimal solutions. Evolutionary algorithms are particularly well-suited for handling complex, non-convex, and multimodal optimization problems, as they can explore the design space more globally and are less likely to get trapped in local optima. However, they can be computationally expensive, especially for large-scale problems or when combined with high-fidelity simulations.

14.5.5 Hybrid Optimization Techniques

In many practical heat exchanger optimization problems, the objective functions and constraints may exhibit various complexities, such as non-convexity, multimodality, discontinuities, or high computational costs. To address these challenges, hybrid optimization techniques that combine the strengths of different algorithms have been developed. One common hybrid approach is the combination of gradient-based and derivative-free methods, where a gradient-based algorithm is used for local refinement and convergence, while a derivative-free method is employed for global exploration and handling non-differentiable or discontinuous regions of the design space.

Another hybrid technique is the coupling of evolutionary algorithms with local search methods or gradient-based algorithms. In this approach, the evolutionary algorithm performs a global search and generates promising candidate solutions, which are then refined using local search or gradient-based methods to improve convergence and accuracy. Hybrid algorithms can also involve the integration of surrogate models or response surface approximations with optimization algorithms. Surrogate models are computationally inexpensive approximations of the true objective function or constraints constructed using techniques such as kriging, radial basis functions (RBF) or artificial neural networks (ANNs). These surrogate models can be used to guide the optimization process, reducing the number of costly high-fidelity simulations required.

Hybrid optimization techniques can also incorporate ML methods, such as reinforcement learning or Bayesian optimization, to enhance the efficiency and effectiveness of the optimization process. ML algorithms can be used to learn the behavior of the objective function or constraints, adapt the search strategy, or guide the selection of promising candidate solutions. The choice of a specific hybrid optimization technique depends on the characteristics of the optimization problem, the available computational resources, and the trade-off between exploration and exploitation of the design space. Effective hybrid strategies can leverage the strengths of different algorithms to improve convergence, handle complexities, and achieve better overall performance in heat exchanger optimization applications.

14.6 VERIFICATION AND UNCERTAINTY QUANTIFICATION

Numerical simulations and optimization techniques play a crucial role in the design and performance evaluation of heat exchangers. However, it is essential to ensure the accuracy and reliability of these computational tools by employing rigorous verification and uncertainty quantification (UQ) methods. This section discusses various approaches for code verification, solution verification, uncertainty quantification techniques, sensitivity analysis, and model validation and calibration.

14.6.1 Code Verification and Solution Verification

Code verification is the process of ensuring that the computational model accurately represents the governing mathematical equations and physical laws. It involves a systematic examination of the numerical algorithms, discretization schemes, and implementation of the code to identify and eliminate coding errors, numerical deficiencies, and algorithmic inconsistencies. One commonly used approach for code verification is the method of manufactured solutions (MMS). In this method, an analytical solution is constructed by prescribing a known solution to the governing equations and substituting it back into the equations to obtain the corresponding source terms or forcing functions. The computational model is then executed with these manufactured source terms, and the numerical solution is compared against the prescribed analytical solution to assess the accuracy and convergence rates.

Solution verification, on the other hand, focuses on estimating the numerical errors and uncertainties associated with a specific simulation result. It involves quantifying the discretization errors introduced by the numerical approximations and assessing the effects of mesh resolution, iterative convergence criteria, and other numerical parameters on the solution accuracy. One widely used solution verification technique is the grid convergence index (GCI) method, based on the Richardson extrapolation approach. In this method, simulations are performed on multiple grid resolutions, and the numerical solution is extrapolated to the theoretical zero-grid spacing limit using a higher-order approximation. The GCI method provides an estimate of the numerical uncertainty in the solution due to discretization errors and can guide mesh refinement strategies. Other solution verification techniques include the use of adjoint-based error estimation, residual-based error estimation, and output-based adaptive mesh refinement strategies.

14.6.2 Uncertainty Quantification Techniques

Numerical simulations and optimization processes for heat exchangers often involve various sources of uncertainty, such as uncertain input parameters, model form uncertainties, and numerical approximations. UQ techniques aim to characterize the propagation of these uncertainties through the computational models and quantify their impact on the predicted outputs or objective functions. One widely used UQ approach is the MCM, which involves performing multiple simulations with randomly sampled input parameters from their respective probability distributions. The resulting output distributions provide statistical measures of uncertainty, such as mean values, standard deviations, and confidence intervals.

While the MCM is conceptually straightforward and applicable to a wide range of problems, it can be computationally expensive, especially for high-dimensional input spaces or when combined with computationally intensive simulations. To address this limitation, various variance reduction techniques, such as importance sampling, stratified sampling, and Latin hypercube sampling, can be employed to improve the efficiency of the MCM simulations. Another UQ approach is the stochastic collocation method, which involves constructing surrogate models or response surfaces based on a set of deterministic simulations at carefully chosen collocation points. These surrogate models can then be used to rapidly evaluate the output distributions and propagate uncertainties at a lower computational cost compared with direct MCM simulations.

Polynomial chaos expansions (PCEs) and generalized polynomial chaos (gPC) techniques are also widely used for UQ in heat transfer and fluid dynamics applications. These methods employ orthogonal polynomial bases to represent the stochastic input parameters and output quantities, allowing for efficient propagation of uncertainties through the computational models.

14.6.3 Sensitivity Analysis

Sensitivity analysis is an essential tool for identifying the most influential input parameters or design variables that significantly impact the performance metrics or objective functions of interest. This information can guide the prioritization of efforts for uncertainty reduction, model calibration, or design optimization. Local sensitivity analysis techniques, such as the finite difference method or the direct differentiation method, provide estimates of the local sensitivities by perturbing individual input parameters and evaluating the corresponding changes in the output quantities. These methods are computationally efficient but may not capture global nonlinear behavior or parameter interactions.

Global sensitivity analysis methods, such as Sobol' indices or the Fourier amplitude sensitivity test (FAST), provide a more comprehensive assessment of parameter sensitivities by exploring the entire input parameter space. These methods decompose the output variance into contributions from individual parameters and their interactions, allowing for the identification of the most significant factors and their relative importance. Sensitivity analysis can be coupled with UQ techniques to quantify the contribution of input uncertainties to the overall output uncertainty and prioritize efforts for uncertainty reduction or model calibration.

14.6.4 Model Validation and Calibration

Model validation and calibration are essential steps in ensuring the accuracy and predictive capabilities of computational models for heat exchanger simulations and optimization. Model validation involves assessing the agreement between model predictions and experimental data or real-world observations, while model calibration aims to improve the model's accuracy by adjusting uncertain parameters or correcting systematic biases as depicted in the Figure 14.4.

In the context of heat exchanger simulations, model validation typically involves comparing predicted temperature distributions, heat transfer rates, pressure drops, or other performance metrics against experimental measurements obtained from well-characterized test cases or prototype designs. Statistical measures, such as root-mean-square errors, coefficient of determination (R^2), or hypothesis testing, can be employed to quantify the agreement between model predictions and experimental data.

If discrepancies between model predictions and experimental data are identified, model calibration techniques can be employed to improve the model's accuracy. This may involve adjusting uncertain model parameters, such as boundary conditions, material properties, or empirical correlations, within their respective uncertainty ranges to minimize the discrepancies with experimental data. Bayesian calibration methods, which combine prior information about the model parameters with experimental data, are widely used for model calibration. These methods employ Bayes' theorem to update the probability distributions of the uncertain parameters based on the observed data, providing a systematic approach for parameter estimation and uncertainty quantification. Other calibration techniques include optimization-based approaches, where the model parameters are adjusted to minimize an objective function that quantifies the discrepancy between model predictions and experimental data, subject to constraints imposed by physical laws or parameter bounds. Iterative model validation and calibration cycles may be necessary to ensure the accuracy and robustness of the computational models across a range of operating conditions and design configurations relevant to heat exchanger applications. Effective verification, uncertainty quantification, sensitivity analysis, and model validation and calibration practices are essential for building confidence in numerical simulations and optimization results for heat exchanger design and performance evaluation. These methods provide a systematic framework for assessing and improving the accuracy, reliability, and predictive capabilities of computational models, enabling informed decision-making and design optimizations in the field of heat exchanger technologies.

14.7 CHALLENGES AND LIMITATIONS

While numerical simulations and optimization techniques have revolutionized the design and analysis of heat exchangers, several challenges and limitations remain. Addressing these challenges is crucial for achieving accurate and reliable results, as well as pushing the boundaries of heat exchanger performance. This section explores computational resource requirements, modeling

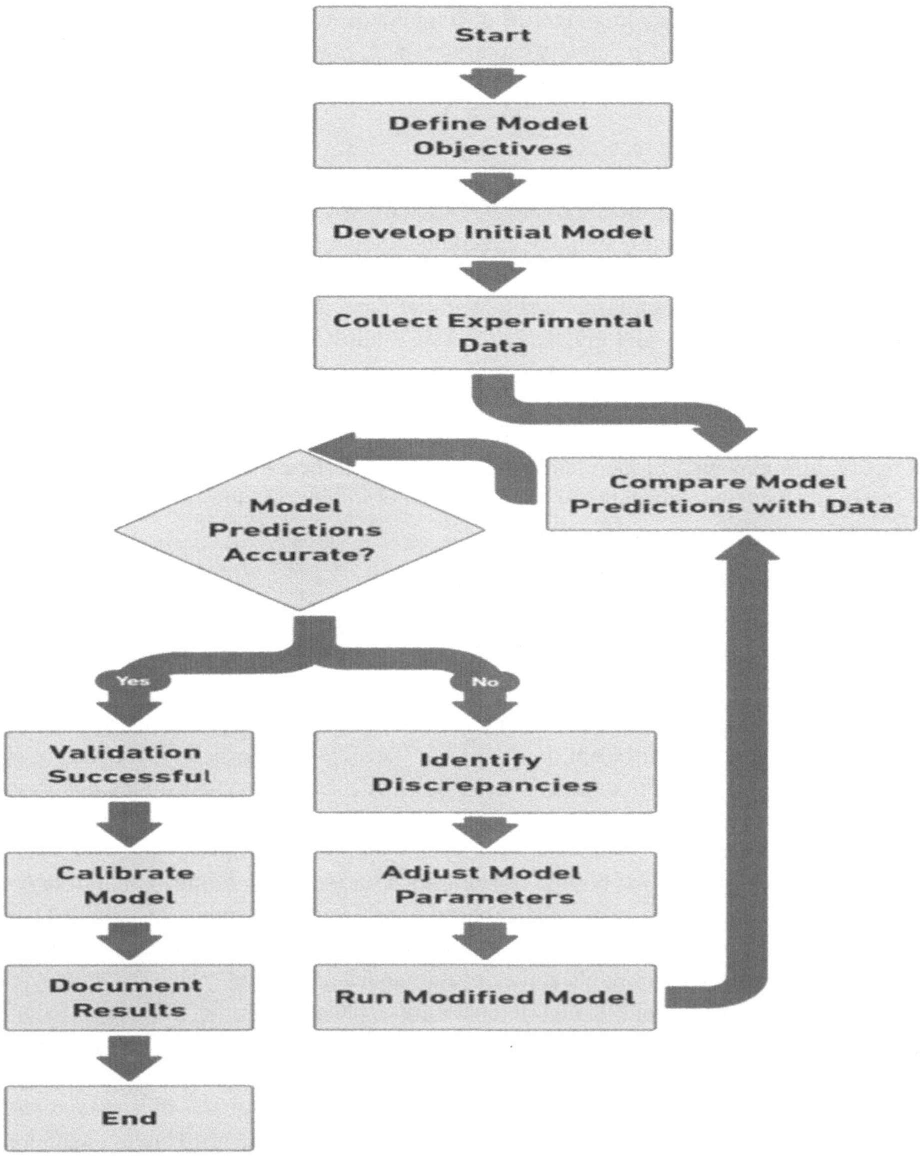

FIGURE 14.4 A flowchart for model validation with essential steps.

complexities and assumptions, numerical stability and convergence issues, and the availability of experimental data for validation. We describe a few more technical difficulties below.

14.7.1 Computational Resource Requirements

Numerical simulations and optimization processes for heat exchangers can be computationally intensive, particularly when dealing with complex geometries, multiphysics phenomena, or high-fidelity models. Computational resource requirements can pose significant challenges in terms of computational power, memory, and storage capacity. High-fidelity CFD simulations, which solve the Navier–Stokes equations along with additional transport equations for turbulence, energy, and species, can be extremely demanding in terms of computational resources. These simulations often require fine mesh resolutions to accurately capture boundary layers, turbulent eddies, and intricate flow structures, leading to large computational domains with millions or billions of grid cells or finite elements.

Multiphysics simulations, which couple multiple physical phenomena such as fluid flow, heat transfer, phase change, and radiation, further increase the computational complexity and resource requirements. Additionally, optimization processes that involve iterative simulations or population-based algorithms can significantly amplify the computational burden. To address these computational challenges, researchers and engineers often rely on HPC resources, such as supercomputers, clusters, or cloud-based computing platforms. Parallel computing techniques, including domain decomposition methods, MPI, and graphics processing unit (GPU) acceleration, are employed to distribute the computational workload across multiple processors or nodes, enabling efficient simulations and optimization processes.

However, access to HPC resources can be limited or costly, particularly for smaller research groups or organizations with limited budgets. Additionally, efficient parallelization and load balancing can be challenging, especially for complex geometries or non-uniform mesh distributions.

14.7.2 Modeling Complexities and Assumptions

Accurate modeling of heat transfer phenomena in heat exchangers often involves complex physical processes and interactions between multiple transport mechanisms. Capturing these complexities can be challenging and may require simplifying assumptions or approximations, potentially introducing errors or uncertainties in the simulation results. One of the key modeling challenges is the accurate representation of turbulent flow and heat transfer. Turbulence models, such as RANS or LES approaches, rely on approximations and empirical correlations, which may not accurately capture the intricate turbulent structures and their effects on heat transfer, particularly in complex geometries or flow regimes.

Phase change processes, such as boiling, condensation, or melting/solidification, introduce additional complexities in modeling the interfacial phenomena, phase transitions, and coupled heat and mass transfer mechanisms. Simplifying assumptions or empirical correlations are often employed, which may limit the accuracy or applicability of the models in certain scenarios. Radiation heat transfer, which can be significant in high-temperature applications or participating media, adds another layer of complexity. Accurately modeling the radiative transfer equation, accounting for emission, absorption, and scattering by the medium, can be computationally demanding and may require approximations or simplifications. Additionally, modeling the effects of surface roughness, fouling, or material degradation on heat transfer performance can be challenging, as these phenomena often involve complex microscale or multiscale interactions that are difficult to capture in macroscale simulations.

14.7.3 Numerical Stability and Convergence Issues

Numerical simulations of heat transfer phenomena in heat exchangers can encounter stability and convergence issues, particularly when dealing with complex flow patterns, steep gradients, or coupled nonlinear equations. These issues can lead to divergent solutions, oscillations, or non-physical results, undermining the reliability and accuracy of the simulations. One common source of numerical instabilities is the presence of high-aspect-ratio cells or highly skewed meshes, which can arise due to complex geometries or boundary layer resolution requirements. These mesh irregularities can introduce numerical diffusion or dispersion errors, leading to inaccurate solution fields or convergence difficulties.

Coupled nonlinear equations, such as those governing conjugate heat transfer, phase change processes, or radiation heat transfer, can also pose challenges for numerical solvers. The nonlinearities and coupling between different physical phenomena can lead to convergence issues, oscillations, or non-physical solutions, requiring careful treatment and specialized solution techniques (Verma et al. 2016, Verma et al. 2017a, 2017b). In optimization processes, numerical instabilities or convergence issues can be exacerbated by the iterative nature of the algorithms and the search for optimal solutions near constraint boundaries or design limits. Gradient-based optimization algorithms may encounter difficulties in the presence of non-differentiable or discontinuous objective functions or constraints, while population-based algorithms may struggle with premature convergence or stagnation in complex design spaces.

Addressing numerical stability and convergence issues often requires careful mesh generation, adaptive mesh refinement strategies, stabilization techniques (e.g., upwinding, artificial dissipation), or specialized solvers and solution algorithms. Additionally, thorough verification and validation processes, including code verification, solution verification, and uncertainty quantification, are essential to ensure the reliability and accuracy of numerical simulations.

14.7.4 Availability of Experimental Data for Validation

Experimental data plays a crucial role in validating numerical simulations and optimization results for heat exchangers. However, obtaining high-quality experimental data can be challenging due to various factors, including the complexity of experimental setups, measurement uncertainties, and the limitations of instrumentation and measurement techniques. Conducting well-controlled experiments for heat exchanger performance evaluation often requires sophisticated test facilities, precise flow control and metering systems, and accurate temperature, pressure, and flow rate measurements. Ensuring uniform inlet conditions, minimizing external disturbances, and accounting for heat losses or environmental factors can be challenging, introducing uncertainties in the experimental data.

Furthermore, measuring detailed flow fields, temperature distributions, or local heat transfer coefficients within heat exchangers can be extremely difficult, particularly in compact or intricate geometries. Intrusive measurement techniques, such as thermocouples or hot-wire anemometry, may alter the flow patterns or introduce disturbances, while non-intrusive techniques, like particle image velocimetry (PIV) or infrared thermography, have their own limitations in terms of spatial resolution, optical access, or measurement uncertainties. The availability of experimental data can also be limited by the specific operating conditions, fluid properties, or heat exchanger configurations of interest. Many experimental studies focus on simplified geometries or specific applications, which may not be directly applicable to the design or optimization problem at hand.

In the absence of comprehensive experimental data, researchers and engineers often rely on a combination of theoretical analysis, numerical simulations, and limited experimental validation. However, this approach can introduce uncertainties and potential discrepancies between numerical predictions and real-world performance. To address the limitations in experimental data availability, collaborative efforts between research institutions, industry partners, and multidisciplinary teams are essential. Sharing experimental data, establishing benchmarking cases, and developing standardized testing protocols can facilitate the validation and improvement of numerical models and optimization techniques for heat exchanger applications.

Despite the challenges and limitations, continuous advancements in computational resources, modeling techniques, numerical algorithms, and experimental methods are driving the field of heat exchanger simulations and optimization forward. Addressing these challenges through interdisciplinary collaborations, innovative approaches, and rigorous validation practices is crucial for achieving accurate and reliable results, enabling the design and optimization of highly efficient and tailored heat exchanger solutions for diverse applications.

TABLE 14.3
List of Important Challenges

Challenge	Description
Heat transfer coefficient prediction	Accurately predicting heat transfer coefficients under various operating conditions is difficult, impacting performance and energy efficiency.
Pressure drop calculation	Balancing low pressure drop with high heat transfer efficiency requires precise modeling to avoid excessive energy consumption.
Material selection	Choosing materials that can withstand extreme temperatures and corrosive environments while maintaining cost-effectiveness is complex.
Geometric design optimization	Designing geometries that maximize surface area for heat transfer while minimizing volume, weight, and manufacturing complexity involves trade-offs.
Thermal and mechanical integrity	Ensuring that heat exchangers maintain structural integrity under thermal stresses and mechanical loads is critical for reliability and safety.
Multi-objective optimization	Simultaneously optimizing multiple objectives, such as thermal performance, pressure drop, and cost, requires advanced optimization algorithms.
Simulation accuracy	Ensuring the accuracy of numerical simulations, particularly CFD, to reliably predict real-world performance is essential but challenging.
Cost efficiency	Achieving high performance without significantly increasing costs demands careful consideration of all design and material choices.
Environmental and sustainability concerns	Designing for minimal environmental impact and long-term sustainability adds another layer of complexity to the optimization process.

14.8 CASE STUDIES AND PRACTICAL APPLICATIONS

The numerical simulations and optimization techniques discussed in the previous sections have been successfully applied to various practical applications in the field of heat exchanger design and analysis. This section presents a selection of case studies and practical applications, highlighting the diverse range of problems addressed and the benefits achieved through the implementation of these advanced computational methods.

14.8.1 Optimization of Heat Exchanger Geometries

The optimization of heat exchanger geometries is a fundamental aspect of enhancing thermal performance while minimizing pressure drop, material usage, and manufacturing costs. Numerical simulations and optimization algorithms have been extensively employed to explore and optimize various geometrical parameters, including fin shapes, tube configurations, and flow channel designs. One notable case study involves the optimization of fin geometries in plate-fin heat exchangers. Researchers have employed CFD simulations coupled with optimization algorithms, such as genetic algorithms or particle swarm optimization, to investigate the effects of fin shapes, fin densities, and fin patterns on heat transfer and pressure drop characteristics. These studies have led to the development of novel fin geometries, such as wavy fins, offset strip fins, or vortex generator fins, which can significantly enhance heat transfer while minimizing pressure drop penalties. Another example is the optimization of tube configurations in shell-and-tube heat exchangers. CFD simulations and multi-objective optimization techniques have been used to explore various tube layouts, including inline, staggered, or rotated configurations, to achieve optimal heat transfer performance while minimizing fouling and flow-induced vibrations. These studies have resulted in improved designs that balance heat transfer enhancement, pressure drop reduction, and structural integrity considerations.

14.8.2 Enhancement of Heat Transfer Using Additives or Surface Modifications

In addition to geometrical optimizations, researchers have explored the use of additives or surface modifications to enhance heat transfer in heat exchangers. Numerical simulations and optimization techniques have played a crucial role in investigating the effects of these enhancements and optimizing their implementation.

One case study focuses on the use of nanofluids, which are suspensions of nano-sized particles in conventional heat transfer fluids. CFD simulations and optimization algorithms have been employed to study the heat transfer characteristics of nanofluids in various heat exchanger configurations (Rao and Krishnamoorthy 2022). These studies aimed to identify the optimal nanoparticle type, concentration, and operating conditions to maximize heat transfer enhancement while minimizing pressure drop penalties and potential issues such as particle deposition or erosion.

Another example involves the optimization of surface modifications, such as micro/nanostructured surfaces or coatings, to enhance heat transfer through mechanisms like increased surface area, improved wettability, or nucleation site promotion. Numerical simulations, coupled with optimization techniques, have been used to explore various surface patterns, feature sizes, and material properties to achieve optimal heat transfer performance for specific applications, such as boiling or condensation heat transfer.

14.8.3 Design of Compact Heat Exchangers for Specific Applications

Compact heat exchangers are widely used in various industries, including aerospace, automotive, and electronics cooling, where space and weight constraints are critical design considerations. Numerical simulations and optimization techniques have played a vital role in developing compact and efficient heat exchanger designs tailored to specific applications. One case study involves the design of compact heat exchangers for aircraft environmental control systems (ECS). CFD simulations and multi-objective optimization algorithms have been employed to explore various fin geometries, flow channel configurations, and material selections to achieve high heat transfer rates while minimizing weight, volume, and pressure drop. These optimized designs have led to significant improvements in aircraft ECS performance and fuel efficiency. Another example is the design of compact heat exchangers for electronics cooling applications, such as data center cooling or high-performance computing systems. Numerical simulations have been used to investigate the thermal performance of various heat exchanger configurations, including microchannel heat sinks, vapor chambers, and two-phase cooling systems. Optimization techniques have been employed to identify optimal designs that can effectively dissipate high heat loads while meeting stringent space and weight constraints.

14.8.4 Integration of Heat Exchangers in Energy Systems

Heat exchangers play a crucial role in various energy systems, including power generation, HVAC, and refrigeration systems. Numerical simulations and optimization techniques have been instrumental in integrating and optimizing heat exchanger performance within these larger energy systems. One case study focuses on the optimization of heat recovery steam generators (HRSGs) in combined cycle power plants. CFD simulations and optimization algorithms have been used to analyze the complex flow patterns, heat transfer mechanisms, and thermal stresses within HRSGs. These studies have led to improved designs that maximize heat recovery from the gas turbine exhaust, enhancing overall plant efficiency and reducing emissions.

Another example is the optimization of heat exchanger networks in process industries or chemical plants. Process integration techniques, coupled with numerical simulations and optimization algorithms, have been employed to design and optimize heat exchanger networks that minimize energy consumption, reduce utility costs, and improve overall system efficiency. In the field of building energy systems, numerical simulations and optimization techniques have been used to integrate and optimize heat exchangers for HVAC applications, such as air-to-air heat recovery systems or ground-source heat pump systems. These studies have focused on enhancing energy efficiency, improving indoor air quality, and reducing operating costs through optimal heat exchanger design and system integration.

These case studies and practical applications demonstrate the versatility and effectiveness of numerical simulations and optimization techniques in addressing diverse challenges in heat exchanger design and analysis. By leveraging these advanced computational methods, researchers and engineers can develop innovative solutions, enhance thermal performance, optimize energy efficiency, and meet specific application requirements, ultimately contributing to the advancement of heat exchanger technologies across various industries.

14.9 AI/ML-BASED OPTIMIZATIONS

The integration of AI and ML techniques with numerical simulations and optimization algorithms has emerged as a promising avenue for enhancing the design and optimization processes of heat exchangers (Ma et al. 2021, Mozaffar et al. 2022, Ahmed et al. 2023). This section explores the applications of AI/ML techniques in heat exchanger optimization, including surrogate modeling, reinforcement learning, neural network-based optimization, and the challenges and future prospects of these approaches.

14.9.1 Machine Learning for Surrogate Modeling

Surrogate modeling, also known as metamodeling or response surface modeling, involves the construction of computationally efficient approximations of complex simulation models or experimental data (Gosselin et al. 2009, Misener and Biegler 2023). Machine learning techniques, such as ANNs (Diaz et al. 1999, Yang 2008, Tan et al. 2009, Mohanraj et al. 2015, Tahseen et al. 2019), Gaussian processes (Campet et al. 2020, Hosseini et al. 2022), or support vector machines (Al-Dhaifallah et al. 2017, Muthukrishnan et al. 2020), have been widely employed for building accurate surrogate models in heat exchanger optimization. One common approach is to use high-fidelity numerical simulations or experimental data to generate a training dataset, which is then used to train a machine learning model, such as an ANN or Gaussian process, to approximate the relationship between input parameters (e.g., geometrical variables, flow conditions) and output quantities of interest (e.g., heat transfer rates, pressure drops).

These surrogate models can then be used in optimization algorithms as computationally inexpensive replacements for the expensive high-fidelity simulations or experimental evaluations, enabling efficient exploration of the design space and identification of optimal solutions. For example, researchers have employed ANN-based surrogate models in conjunction with genetic algorithms or particle swarm optimization to optimize the fin geometries and tube configurations of plate-fin heat exchangers (Yousefi et al. 2012, Patel and Rao 2010, Peng et al. 2010, Sharma et al. 2023). By replacing the computationally expensive CFD simulations with the ANN surrogate model during the optimization process, significant computational savings can be achieved while maintaining reasonable accuracy in the predicted performance metrics.

14.9.2 Reinforcement Learning for Optimization

Reinforcement learning (RL) is a branch of machine learning that focuses on decision-making and control problems, where an agent learns to take optimal actions in an environment to maximize a cumulative reward signal. RL techniques have been applied to heat exchanger optimization problems, enabling the exploration of complex design spaces and the identification of optimal solutions through iterative learning and feedback. In the context of heat exchanger optimization, the RL agent can be trained to optimize various design variables or operating conditions by interacting with a simulation environment or a surrogate model. The agent takes actions by adjusting the design variables, and the simulation or surrogate model provides feedback in the form of a reward signal, such as the heat transfer performance or a multi-objective objective function.

One example is the use of deep reinforcement learning (DRL) algorithms, such as deep Q-networks (DQN) or policy gradient methods, for optimizing the fin geometries and tube configurations of compact heat exchangers. The DRL agent learns to navigate the design space by adjusting geometrical parameters and observing the resulting heat transfer performance and pressure drop from CFD simulations or surrogate models. Through iterative learning and exploration, the agent can converge toward optimal designs that maximize heat transfer while minimizing pressure drop. RL techniques can also be combined with other optimization algorithms or surrogate models, forming hybrid approaches that leverage the strengths of different methods. For instance, a surrogate model can be used to initially guide the exploration of the design space, and an RL agent can then refine the solution by interacting with the high-fidelity simulation model in promising regions of the design space.

14.9.3 Neural Network-Based Optimization Approaches

Neural networks, particularly deep neural networks (DNNs), have demonstrated remarkable performance in various machine learning tasks, such as image recognition, natural language processing, and decision-making. Researchers have explored the application of DNNs in heat exchanger optimization, either as surrogate models or as direct optimization algorithms. In the context of surrogate modeling, DNNs can be trained on simulation or experimental data to approximate the complex relationships between input design variables and output performance metrics. These DNN-based surrogate models can then be used in optimization algorithms, similar to the approaches discussed in Section 14.9.1.

Additionally, DNNs have been employed as direct optimization algorithms, where the neural network architecture is designed to learn the optimal design variables or operating conditions for a given heat exchanger configuration or performance objective. These approaches often involve training the DNN on a dataset of known optimal solutions or using reinforcement learning techniques to iteratively refine the network's predictions through interactions with the simulation environment. For example, researchers have developed DNN-based optimization frameworks for optimizing the fin geometries and tube configurations of plate-fin heat exchangers. The DNN is trained on a dataset of CFD simulation results or experimental data, learning to map the input design variables to the corresponding heat transfer performance and pressure drop. During the optimization process, the DNN generates candidate designs, which are evaluated using the simulation model or experimental setup, and the resulting performance metrics are used to update the network weights and refine the predictions.

Advances in deep learning techniques, such as convolutional neural networks (CNNs) and generative adversarial networks (GANs), have also opened up new avenues for heat exchanger optimization. CNNs have been employed for pattern recognition and feature extraction in heat exchanger designs, enabling the identification of optimal geometrical patterns or features that enhance heat transfer performance. GANs, on the other hand, have been explored for generating novel heat exchanger geometries or designs based on learned patterns from existing data.

14.9.4 Challenges and Future Prospects

While AI/ML-based optimization techniques hold great promise for advancing heat exchanger design and optimization, several challenges and limitations remain:

- *Data availability and quality.* The performance of machine learning models heavily relies on the availability of high-quality training data, which can be costly and time-consuming to obtain through numerical simulations or experiments. Techniques for efficient data generation, such as active learning or transfer learning, are needed to address this challenge.
- *Generalization and extrapolation.* Machine learning models may struggle to generalize or extrapolate beyond the training data distribution, limiting their applicability to new design spaces or operating conditions. Robust model architectures and regularization techniques are required to improve generalization capabilities.
- *Interpretability and trust.* Despite their impressive performance, deep neural networks often lack interpretability, making it challenging to understand the underlying decision-making process or gained insights. Developing explainable AI techniques can help build trust in optimized solutions and facilitate knowledge transfer.
- *Computational resources.* Training and deploying complex AI/ML models can be computationally intensive, especially for high-dimensional design spaces or large datasets. Efficient computational strategies, such as distributed training or model compression, are needed to address this challenge.
- *Integration with existing workflows.* Seamlessly integrating AI/ML techniques with existing numerical simulation and optimization workflows can be challenging, requiring robust interfaces, data pipelines, and workflow management systems.

Typical challenges for AI/ML-based optimization in heat exchanger design, as seen in Figure 14.5, include data quality, generalizability, and computational resources. Despite these challenges, the integration of AI/ML techniques with numerical simulations and optimization algorithms holds immense potential for advancing heat exchanger design and optimization. Future research efforts should focus on addressing the challenges mentioned above, as well as exploring new frontiers such as multi-fidelity modeling, transfer learning, and hybrid AI/physics-based approaches. Interdisciplinary collaborations between heat transfer experts, computer scientists, and AI researchers will be crucial in unlocking the full potential of AI/ML for heat exchanger optimization and pushing the boundaries of thermal performance and energy efficiency.

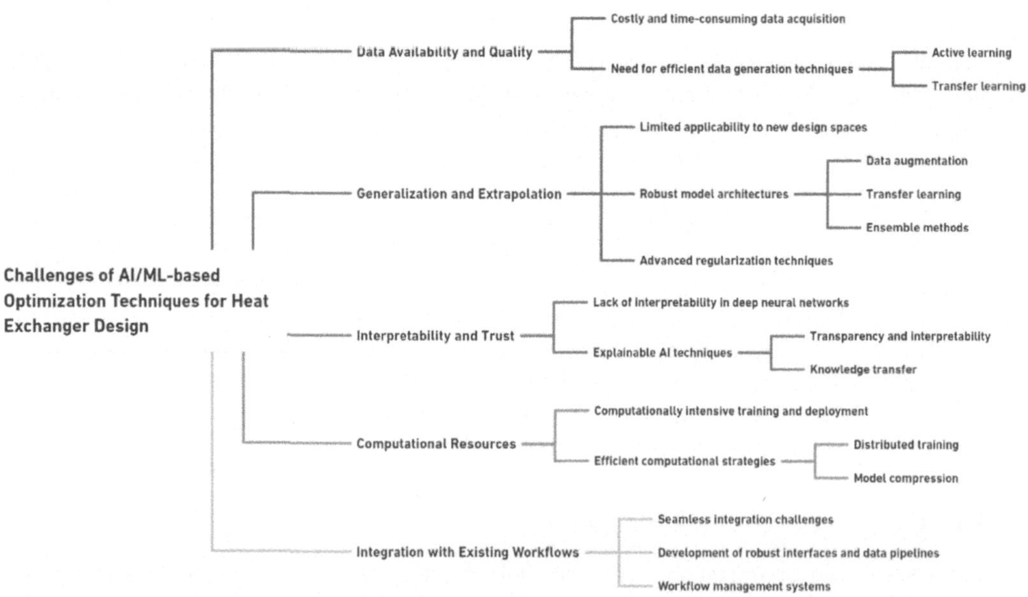

FIGURE 14.5 Typical challenges of AI/ML-based optimization techniques for heat exchanger design.

14.10 CONCLUSIONS AND FUTURE OUTLOOK

The field of heat exchanger simulations and optimization has witnessed significant advancements in recent years, driven by the development of powerful numerical techniques, advanced optimization algorithms, and the integration of artificial intelligence and machine learning methods. This chapter has provided a comprehensive overview of these state-of-the-art approaches, highlighting their applications, challenges, and future prospects.

Summary of key findings. Throughout this chapter, we have explored various numerical methods for simulating heat transfer phenomena in heat exchangers, including finite difference, finite volume, finite element, and lattice Boltzmann methods. These techniques have enabled accurate modeling and analysis of complex flow patterns, turbulence, phase change processes, radiation heat transfer, and multiscale phenomena. Optimization algorithms, such as gradient-based methods, derivative-free techniques, evolutionary algorithms, and hybrid approaches, have been discussed, highlighting their strengths and limitations in addressing the challenges of heat exchanger design and optimization. These algorithms have been instrumental in identifying optimal geometries, operating conditions, and control strategies that maximize heat transfer performance while minimizing pressure drop, material usage, and energy consumption.

The importance of verification and uncertainty quantification techniques has been emphasized, including code verification, solution verification, sensitivity analysis, and model validation and calibration. These methods ensure the accuracy and reliability of numerical simulations and optimization results, building confidence in the predicted performance and enabling informed decision-making. Multi-objective optimization and trade-off analysis techniques have been presented, highlighting the concept of Pareto optimality and the need to balance competing objectives in heat exchanger design. Various algorithms, such as classical methods and evolutionary multi-objective optimization algorithms, have been discussed, along with decision-making strategies for trade-off analysis.

The chapter has also explored practical applications and case studies, demonstrating the successful implementation of these advanced computational techniques in optimizing heat exchanger geometries, enhancing heat transfer through additives or surface modifications, designing compact heat exchangers for specific applications, and integrating heat exchangers into larger energy systems. Finally, the integration of artificial intelligence and machine learning techniques, such as surrogate modeling, reinforcement learning, and neural network-based optimization, has been discussed, showcasing the potential of these methods to further enhance the efficiency and accuracy of heat exchanger optimization processes.

Emerging trends and future research directions. As the field of heat exchanger simulations and optimization continues to evolve, several emerging trends and future research directions can be identified:

- *Multiscale and multiphysics modeling.* Capturing the intricate interactions between different physical phenomena and spatial scales remains a significant challenge. Future research efforts should focus on developing advanced multiscale and multiphysics modeling approaches that can accurately represent the complex interplay between fluid flow, heat transfer, phase change processes, and material behavior across various length and time scales.
- *Uncertainty quantification and robust optimization.* Accounting for uncertainties in input parameters, model forms, and operating conditions is crucial for ensuring reliable and robust heat exchanger designs. Future research should explore advanced uncertainty quantification techniques, such as polynomial chaos expansions, stochastic collocation methods, and Bayesian calibration approaches, coupled with robust optimization strategies to identify designs that are insensitive to uncertainties.
- *High-performance computing and exascale simulations.* As computational power continues to increase, the ability to perform high-fidelity simulations and explore larger design

spaces will become more feasible. Leveraging high-performance computing resources, such as exascale systems, and developing efficient parallel algorithms and scalable numerical methods will be essential for tackling computationally intensive simulations and optimizations.

- *Additive manufacturing and topology optimization.* The advent of additive manufacturing technologies has opened new possibilities for designing and fabricating complex and optimized heat exchanger geometries. Future research should focus on developing topology optimization techniques tailored for additive manufacturing, enabling the exploration of novel designs and unconventional flow paths that can further enhance heat transfer performance.
- *Digital twins and data-driven modeling.* The integration of numerical simulations with real-time sensor data and machine learning techniques can enable the development of digital twins for heat exchangers. These virtual representations can be used for predictive maintenance, online monitoring, and real-time optimization, leading to improved operational efficiency and reduced downtime.
- *Explainable AI and physics-informed machine learning.* While machine learning techniques have shown great promise in heat exchanger optimization, there is a need for developing explainable AI methods that can provide insights into the underlying physical mechanisms and decision-making processes. Physics-informed machine learning approaches, which incorporate physical laws and constraints into the learning process, can further enhance the accuracy and interpretability of optimized solutions.
- *Interdisciplinary collaborations.* Tackling the challenges of heat exchanger simulations and optimization requires a collaborative effort among researchers from various disciplines, including heat transfer experts, computational scientists, mathematicians, material scientists, and AI/ML specialists. Fostering interdisciplinary collaborations and encouraging knowledge exchange will be crucial for driving innovation and addressing the complex challenges in this field.

Interdisciplinary collaborations and industry-academia partnerships. The advancement of heat exchanger simulations and optimization techniques relies heavily on interdisciplinary collaborations and strong partnerships between academia and industry. Researchers from diverse fields, including heat transfer, computational fluid dynamics, applied mathematics, computer science, and materials science, must collaborate to develop comprehensive models, efficient algorithms, and innovative optimization strategies. Industry-academia partnerships play a vital role in translating theoretical advancements into practical applications and addressing real-world challenges. Industry partners can provide valuable insights into specific application requirements, operational constraints, and performance targets, guiding academic research toward practical and impactful solutions.

Furthermore, industry-academia collaborations can facilitate the exchange of knowledge, expertise, and resources. Academic institutions can leverage their expertise in numerical simulations, optimization techniques, and cutting-edge research, while industry partners can provide access to experimental facilities, real-world data, and manufacturing capabilities. By fostering interdisciplinary collaborations and strengthening industry-academia partnerships, the field of heat exchanger simulations and optimization can accelerate its progress, leading to the development of highly efficient and tailored heat exchanger solutions that address the growing energy demands and environmental challenges of various industries.

Thus, this chapter has provided a comprehensive overview of the state-of-the-art numerical simulations, optimization techniques, and AI/ML methods for heat exchanger design and optimization. While significant progress has been made, there are still numerous challenges and opportunities for future research. By embracing emerging trends, fostering interdisciplinary collaborations, and forging strong industry-academia partnerships, the field can continue to push the boundaries of thermal performance, energy efficiency, and sustainable heat exchanger technologies.

REFERENCES

Abeykoon, C. (2020). Compact heat exchangers–Design and optimization with CFD. *International Journal of Heat and Mass Transfer*, 146, 118766.

Afzal, A., Ansari, Z., Faizabadi, A. R., & Ramis, M. K. (2017). Parallelization strategies for computational fluid dynamics software: state of the art review. *Archives of Computational Methods in Engineering*, 24(2), 337–363.

Ahmed, N., Assadi, M., Ahmed, A. A., & Banihabib, R. (2023). Optimal design, operational controls, and data-driven machine learning in sustainable borehole heat exchanger coupled heat pumps: key implementation challenges and advancement opportunities. *Energy for Sustainable Development*, 74, 231–257.

Al-Dhaifallah, M., Nisar, K. S., Agarwal, P., & Elsayyad, A. (2017). Modeling and identification of heat exchanger process using least squares support vector machines. *Thermal Science*, 21(6 Part B), 2859–2869.

Aly, W. I. (2014). Computational fluid dynamics and optimization of flow and heat transfer in coiled tube-in-tube heat exchangers under turbulent flow conditions. *Journal of Thermal Science and Engineering Applications*, 6(3), 031001.

ANSYS Inc. (2023). *ANSYS Fluent* (Release 2023 R1) [Computer software]. Canonsburg, PA.

Baik, S. J., Kim, S. G., Bae, S. J., Ahn, Y. H., & Lee, J. I. (2014). Verification of heat exchanger design code KAIST_HXD by experiment. In *The Korean Nuclear Society Autumn Meeting*. The Korean Nuclear Society.

Bhutta, M. M. A., Hayat, N., Bashir, M. H., Khan, A. R., Ahmad, K. N., & Khan, S. (2012). CFD applications in various heat exchangers design: a review. *Applied Thermal Engineering*, 32, 1–12.

Campet, R., Roy, P. T., Cuenot, B., Riber, E., & Jouhaud, J. C. (2020). Design optimization of an heat exchanger using Gaussian process. *International Journal of Heat and Mass Transfer*, 150, 119264.

Cavallini, A. (2017). Heat transfer and heat exchangers. In *Organic Rankine Cycle (ORC) Power Systems*. Woodhead Publishing, 397–470.

De Bellis, F., & Catalano, L. A. (2012). CFD optimization of an immersed particle heat exchanger. *Applied Energy*, 97, 841–848.

Debtera, B., Neme, I., & Prabhu, V. (2018). CFD simulation of a double pipe heat exchanger: analysis conduction and convection heat transfer. *International Journal of Scientific Research and Review*, 7, 12.

Diaz, G., Sen, M., Yang, K. T., & McClain, R. L. (1999). Simulation of heat exchanger performance by artificial neural networks. *Hvac&R Research*, 5(3), 195–208.

Dixon, J. A., Valencia, A. G., Ireland, P., Ridland, P., & Hills, N. (2010). A coupled CFD finite element analysis methodology in a bifurcation pipe in a nuclear plant heat exchanger. In *Workshop Proceedings*. Vol. 3. CFD4NRS.

Dogan, S., Darici, S., & Ozgoren, M. (2019). Numerical comparison of thermal and hydraulic performances for heat exchangers having circular and elliptic cross-section. *International Journal of Heat and Mass Transfer*, 145, 118731.

Elsheikh, A. H., Panchal, H. N., Sengottain, S., Alsaleh, N., & Ahmadein, M. (2022). Applications of heat exchanger in solar desalination: current issues and future challenges. *Water*, 14(6), 852.

Firouzfar, E., Soltanieh, M., Noie, S. H., & Saidi, M. H. (2011). Application of heat pipe heat exchangers in heating, ventilation and air conditioning (HVAC) systems. *Scientific Research and Essays*, 6(9), 1900–1908.

Galvis, E., Jubran, B. A., Behdinan, F. X. K., & Fawaz, Z. (2008). Numerical modeling of pin–fin micro heat exchangers. *Heat and Mass Transfer*, 44, 659–666.

Garcia-Gasulla, M., Banchelli, F., Peiro, K., Ramirez-Gargallo, G., Houzeaux, G., Ben Hassan Saïdi, I., & Mantovani, F. (2020). A generic performance analysis technique applied to different CFD methods for HPC. *International Journal of Computational Fluid Dynamics*, 34(7–8), 508–528.

Gholap, A. K., & Khan, J. A. (2007). Design and multi-objective optimization of heat exchangers for refrigerators. *Applied Energy*, 84(12), 1226–1239.

Gilmour, C. H. (1960). Application of heat exchangers in chemical plants. *Industrial & Engineering Chemistry*, 52(6), 465–467.

Gomez-Garcia, F., Gauthier, D., & Flamant, G. (2017). Design and performance of a multistage fluidised bed heat exchanger for particle-receiver solar power plants with storage. *Applied Energy*, 190, 510–523.

Gosselin, L., Tye-Gingras, M., & Mathieu-Potvin, F. (2009). Review of utilization of genetic algorithms in heat transfer problems. *International Journal of Heat and Mass Transfer*, 52(9–10), 2169–2188.

Hatami, M., Ganji, D. D., & Gorji-Bandpy, M. (2014). A review of different heat exchangers designs for increasing the diesel exhaust waste heat recovery. *Renewable and Sustainable Energy Reviews*, 37, 168–181.

Hesselgreaves, J. E., Law, R., & Reay, D. (2016). *Compact Heat Exchangers: Selection, Design and Operation*. Butterworth-Heinemann.

Hosseini, S., Khandakar, A., Chowdhury, M. E., Ayari, M. A., Rahman, T., Chowdhury, M. H., & Vaferi, B. (2022). Novel and robust machine learning approach for estimating the fouling factor in heat exchangers. *Energy Reports*, 8, 8767–8776.

Huang, R., Mahvi, A., Odukomaiya, W., Goyal, A., & Woods, J. (2022). Reduced-order modeling method for phase-change thermal energy storage heat exchangers. *Energy Conversion and Management*, 263, 115692.

Hundy, G. F. (2016). *Refrigeration, Air Conditioning and Heat Pumps*. Butterworth-Heinemann.

Kishan, R., Singh, D., & Sharma, A. K. (2020). CFD Analysis of heat exchanger models design using ansys fluent. *International Journal of Mechanical Engineering and Technology*, 11(2): 1–9.

Končar, B., & Klobučar, L. (2018). Use of CFD codes for calculation of radiation heat transfer: from validation to application. In *Heat Transfer-Models, Methods and Applications*. IntechOpen.

Kumar, S., & Banerjee, D. (2023). PCMs for sustainability applications by leveraging machine learning. In *Phase Change Materials (PCMs) - Recent Advances, New Perspectives, and Applications*. IntechOpen.

Kumar, S., & King, M. D. (2022). Numerical investigation on indoor environment decontamination after sneezing. *Environmental Research*, 213, 113665.

Kumar, S., & Rathore, K. (2023). Renewable energy for sustainable development goal of clean and affordable energy. *International Journal of Materials, Manufacturing and Sustainable Technologies*, 2(1), 1–15.

Kumar, S., Vijayan, P. K., Sharma, M., Pilkhwal, D. S., Kannan, U., & Nayak, A. K. (2015). *Experimental and Numerical Analysis of Thermal Stratification in Large Water Pool*. Technical report.

Kumar, S., Vijayan, P. K., Kannan, U., Sharma, M., & Pilkhwal, D. S. (2017a). Experimental and computational simulation of thermal stratification in large pools with immersed condenser. *Applied Thermal Engineering*, 113, 345–361.

Kumar, S., Grover, R. B., Vijayan, P. K., & Kannan, U. (2017b). Numerical investigation on the effect of shrouds around an immersed isolation condenser on the thermal stratification in large pools. *Annals of Nuclear Energy*, 110, 109–125.

Kumar, S., Grover, R. B., Yadav, H., Vijayan, P. K., Kannan, U., & Agrawal, A. (2018). Experimental and numerical investigation on suppression of thermal stratification in a water-pool: PIV measurements and CFD simulations. *Applied Thermal Engineering*, 138, 686–704.

Kumar, S., Grover, R. B., Yadav, H., Vijayan, P. K., Kannan, U., & Agrawal, A. (2020a). Experimental investigations on thermal stratification in a large pool of water with immersed isolation condenser. In *Proceedings of the International Conference on Nuclear Engineering*. ICONE2020.

Kumar, S., Thyagarajan, A., & Banerjee, D. (2020b). An investigation on chevron plate heat exchanger filled with organic phase change material. In *Proceedings of the ASME 2020 Heat Transfer Summer Conference*. HT2020.

Kumar, S., Thyagarajan, A., & Banerjee, D. (2022). Experimental investigation of thermal energy storage (TES) platform leveraging phase change materials in a chevron plate heat exchanger. In *Proceedings of the ASME International Mechanical Engineering Congress and Exposition*. IMECE2022-96226.

Kumar, S., Klassen, M., Kalassen, D., Hardin, R., & King, M. (2023a). Dispersion of sneeze droplets in a meat facility indoor environment—without partitions. *Environmental Research*, 236, 116603.

Kumar, S., Hardin, R., & King, M. (2023b). Modeling of flows through porous media. In *Computational Fluid Flow and Heat Transfer: Advances, Design, Control and Applications*. CRC Press.

Liu, C., Bu, W., & Xu, D. (2017). Multi-objective shape optimization of a plate-fin heat exchanger using CFD and multi-objective genetic algorithm. *International Journal of Heat and Mass Transfer*, 111, 65–82.

Longatte, E., Baj, F., Hoarau, Y., Braza, M., Ruiz, D., & Canteneur, C. (2013). Advanced numerical methods for uncertainty reduction when predicting heat exchanger dynamic stability limits: review and perspectives. *Nuclear Engineering and Design*, 258, 164–175.

Luo, L., Wen, F., Wang, L., Sundén, B., & Wang, S. (2016). Thermal enhancement by using grooves and ribs combined with delta-winglet vortex generator in a solar receiver heat exchanger. *Applied Energy*, 183, 1317–1332.

Ma, T., Guo, Z., Lin, M., & Wang, Q. (2021). Recent trends on nanofluid heat transfer machine learning research applied to renewable energy. *Renewable and Sustainable Energy Reviews*, 138, 110494.

Menni, Y., Ahmad, H., Ameur, H., Askar, S., Botmart, T., Bayram, M., & Lorenzini, G. (2022). Effects of two-equation turbulence models on the convective instability in finned channel heat exchangers. *Case Studies in Thermal Engineering*, 31, 101824.

Misener, R., & Biegler, L. (2023). Formulating data-driven surrogate models for process optimization. *Computers & Chemical Engineering*, 179, 108411.

Mohanraj, M., Jayaraj, S., & Muraleedharan, C. (2015). Applications of artificial neural networks for thermal analysis of heat exchangers–a review. *International Journal of Thermal Sciences*, 90, 150–172.

Mozaffar, M., Liao, S., Xie, X., Saha, T., & Anitescu, C. (2022). Efficient modeling of multiphase CFD for high-dimensional design optimization via physics-informed neural networks. *Computers & Chemical Engineering*, 166, 107977.

Muthukrishnan, S., Krishnaswamy, H., Thanikodi, S., Sundaresan, D., & Venkatraman, V. (2020). Support vector machine for modelling and simulation of heat exchangers. *Thermal Science*, 24(1 Part B), 499–503.

Noie-Baghban, S. H., & Majideian, G. R. (2000). Waste heat recovery using heat pipe heat exchanger (HPHE) for surgery rooms in hospitals. *Applied Thermal Engineering*, 20(14), 1271–1282.

Oprițoiu, P. (2016). Prediction of turbulent flow using upwind discretization scheme and k-ε turbulence model for porous heat exchanger. *Hidraulica*, 2, 88–97.

Padoin, N., Dal'Toé, A. T., Rangel, L. P., Ropelato, K., & Soares, C. (2014). Heat and mass transfer modeling for multicomponent multiphase flow with CFD. *International Journal of Heat and Mass Transfer*, 73, 239–249.

Pandey, N., Murugesan, K., & Thomas, H. R. (2017). Optimization of ground heat exchangers for space heating and cooling applications using Taguchi method and utility concept. *Applied Energy*, 190, 421–438.

Papukchiev, A., Grishchenko, D., & Kudinov, P. (2020). On the need for conjugate heat transfer modeling in transient CFD simulations. *Nuclear Engineering and Design*, 367, 110796.

Patel, D. S., Parmar, R. R., & Prajapati, V. M. (2015). CFD analysis of shell and tube heat exchangers–a review. *International Research Journal of Engineering and Technology*, 17, 2231–2235.

Patel, V. K., & Rao, R. V. (2010). Design optimization of shell-and-tube heat exchanger using particle swarm optimization technique. *Applied Thermal Engineering*, 30(11–12), 1417–1425.

Peng, H., Ling, X., & Wu, E. (2010). An improved particle swarm algorithm for optimal design of plate-fin heat exchangers. *Industrial & Engineering Chemistry Research*, 49(13), 6144–6149.

Popov, D., Fikiin, K., Stankov, B., Alvarez, G., Youbi-Idrissi, M., Damas, A., Evans, J., & Brown, T. (2019). Cryogenic heat exchangers for process cooling and renewable energy storage: A review. *Applied Thermal Engineering*, 153, 275–290.

Prashant, S., & Pravin, S. (2018). Design, analysis & performance of heat exchanger using CFD. *International Journal for Research in Applied Science and Engineering Technology*, 6(4), 1261–1267.

Rao, A. V., & Krishnamoorthy, G. (2022). CFD analysis of heat transfer in a helical coil heat exchanger using nanofluids. *Journal of Thermal Analysis and Calorimetry*, 147(6), 4517–4530.

Rao, R. V., Saroj, A., Ocłoń, P., & Taler, J. (2020). Design optimization of heat exchangers with advanced optimization techniques: a review. *Archives of Computational Methods in Engineering*, 27, 517–548.

Rathore, K., & Kumar, S. (2023). Net zero emissions for our future generations through renewables: a brief review. *International Journal of Energy Resources Applications*, 2(1), 22–43.

Selamat, S., Miyara, A., & Kariya, K. (2016). Numerical study of horizontal ground heat exchangers for design optimization. *Renewable Energy*, 95, 561–573.

Seo, Y. B., Paramanantham, S. S., Bak, J. Y., Yun, B., & Park, W. G. (2020). CFD analysis of subcooled flow boiling in 4× 4 rod bundle. *Applied Sciences*, 10(13), 4559.

Shadloo, M. (2021). Application of support vector machines for accurate prediction of convection heat transfer coefficient of nanofluids through circular pipes. *International Journal of Numerical Methods for Heat & Fluid Flow*, 31(8), 2660–2679.

Shah, R. K., & Mueller, A. C. (1998). Heat Exchange. In *Ullmann's Encyclopedia of Industrial Chemistry*. Wiley-VCH, 309–419.

Shah, R. K., & Sekulic, D. P. (1998). Heat exchangers. In *Handbook of Heat Transfer*, 3rd edition. McGraw-Hill, 12–47.

Shabgard, H., Allen, M. J., Sharifi, N., Benn, S. P., Faghri, A., & Bergman, T. L. (2015). Heat pipe heat exchangers and heat sinks: Opportunities, challenges, applications, analysis, and state of the art. *International Journal of Heat and Mass Transfer*, 89, 138–158.

Sharadga, H., Dawahdeh, A., & Kumar, S. (2024). Accelerated MPPT controller for photovoltaic systems under rapid solar radiation changes using neural network. *Sustainability*, 16(3), 1021.

Sharma, R., Mishra, D. P., Wasilewski, M., & Brar, L. S. (2023). Application of response surface methodology and artificial neural network to optimize the curved trapezoidal winglet geometry for enhancing the performance of a fin-and-tube heat exchanger. *Energies*, 16(10), 4209.

Siemens Digital Industries Software. (2023). *STAR-CCM+* (Version 2023.1) [Computer software]. Plano, TX.

Skočilas, J., & Palaziuk, I. (2015). CFD simulation of the heat transfer process in a chevron plate heat exchanger using the SST turbulence model. *Acta Polytechnica*, 55(4), 267–274.

Sohrabi, N., Hammoodi, K. A., Hammoud, A., Jasim, D. J., Karouei, S. H. H., Kheyri, J., & Nabi, H. (2024). Using different geometries on the amount of heat transfer in a shell and tube heat exchanger using the finite volume method. *Case Studies in Thermal Engineering*, 55, 104037.

Sundén, B. (2007). Computational fluid dynamics in research and design of heat exchangers. *Heat Transfer Engineering*, 28(11), 898–910.

Tahseen, A., Yalçın, N., & Koçak, H. (2019). Use of artificial neural network in estimating heat transfer of helically corrugated heat exchangers. *Applied Thermal Engineering*, 153, 51–60.

Tan, Y. K., Wan, N. Y., & Yew, H. H. (2009). Application of CFD analysis on the shell-side of shell-and-tube heat exchanger. In 2009 *International Conference on Advances in Computational Tools for Engineering Applications*. IEEE, 70–74.

Thulukkanam, K. (2024). *Heat Exchangers: Operation, Performance, and Maintenance*. CRC Press.

Tian, E., He, Y. L., & Tao, W. Q. (2017). Research on a new type waste heat recovery gravity heat pipe exchanger. *Applied Energy*, 188, 586–594.

Timmerhaus, K. D., & Schoenhals, R. J. (1974). Design and selection of cryogenic heat exchangers. In *Advances in Cryogenic Engineering*. Springer, 445–462.

Verma, D., & Iyer, K. N. (2023). Dynamics and bifurcation characteristics of a boiling channel with forced circulation. *Sādhanā*, 48(2), 46.

Verma, D., Kalra, M. S., & Wahi, P. (2016). Dynamics of a BWR with inclusion of boiling nonlinearity, clad temperature and void-dependent core power removal: Stability and bifurcation characteristics of advanced heavy water reactor (AHWR). *Nuclear Engineering and Design*, 308, 297–311.

Verma, D., Kalra, M. S., & Wahi, P. (2017a). Effect of nonlinear void reactivity on bifurcation characteristics of a lumped-parameter model of a BWR: a study relevant to RBMK. *Nuclear Engineering and Design*, 315, 179–193.

Verma, D., Kalra, M. S., & Wahi, P. (2017b). A comparative study of bifurcation characteristics in single and two-zone fuel models of a BWR: a case study of the advanced heavy water reactor (AHWR). *Nuclear Engineering and Design*, 324, 158–170.

Verma, D., Paul, S., & Wahi, P. (2018). Stability and bifurcation characteristics of a forced circulation BWR using a nuclear-coupled homogeneous thermal-hydraulic model. *Nuclear Science and Engineering*, 190(1), 73–92.

Webb, R. L. (2000). *Heat Exchanger Design Handbook*. CRC Press.

Xie, T., Liu, S., & Peng, S. (2021). Dynamic simulation and optimization of a heat exchanger network using a multi-scale model. *Energy*, 223, 120085.

Yang, K. (2008). Artificial Neural Networks (ANNs): A new paradigm for thermal science and engineering. ASME. *Journal of Heat Transfer*, 130(9), 093001.

Yeşilyurt, S. (1995). Surrogates for numerical simulations; optimization of eddy-promoter heat exchangers. *Computer Methods in Applied Mechanics and Engineering*, 121(1–4), 231–257.

Yousefi, F., Karimi, H., & Papari, M. M. (2012). Modeling viscosity of nanofluids using diffusional neural networks. *Journal of Molecular Liquids*, 175, 85–90.

Zhang, L. W., Tafti, D. K., Najjar, F. M., & Balachandar, S. (1997). Computations of flow and heat transfer in parallel-plate fin heat exchangers on the CM-5: effects of flow unsteadiness and three-dimensionality. *International Journal of Heat and Mass Transfer*, 40(6), 1325–1341.

Zheng, N., Liu, P., Wang, X., Shan, F., Liu, Z., & Liu, W. (2017). Numerical simulation and optimization of heat transfer enhancement in a heat exchanger tube fitted with vortex rod inserts. *Applied Thermal Engineering*, 123, 471–484.

15 Heat Exchangers for Hostile Environments

Prashant Singh

15.1 FUNDAMENTALS OF HEAT EXCHANGERS

A heat exchanger is a device that facilitates the exchange of thermal energy from a source at a higher temperature to a sink at a relatively lower temperature. In such devices, theoretically, both hot and cold sides can be in motion, one side (hot or cold) can be in motion while the other at rest, or both hot and cold sides can be at rest. Heat exchangers find their applications in a wide range of industries, including power plants, waste heat recovery, refrigeration and air conditioning, and food processing, among others [1]. Traditionally, heat exchangers are classified based on the arrangement of the flow of the two streams at different temperatures and the type of construction to facilitate the relative motion between the two fluid streams [2].

Figure 15.1 shows two common types of heat exchangers based on the relative direction of the flow of two streams. If the two streams move in the same direction, the configuration is called "parallel flow" and when the two fluids move in the opposite directions, the configuration is called "counterflow." In this representation (Figure 15.1), the heat transfer (q) between the hot and cold channels occurs via the plate that separates the two channels. Through a simple control volume analysis, one can show that the net heat transfer (q) over a certain flow length can be determined by calculating the enthalpy change of either of the fluid streams. The thermal energy transfer between the hot and cold streams is provided by Equation 15.1.

$$q = \dot{m}_h c_{p,h} \left(T_{h,in} - T_{h,out}\right) = -\dot{m}_c c_{p,c} \left(T_{c,in} - T_{h,out}\right) \tag{15.1}$$

The heat transfer (q) occurs through the separator plate with a heat transfer area (A). Ideally, three major thermal resistances are present between the hot and cold streams. These are the thermal resistance due to convection on the hot side $\left(1/hA\right)_h$, conduction resistance due to the separator plate $\left(t_w / k_w A\right)$, and the convection resistance on the cold fluid side $\left(1/hA\right)_c$. Since these resistances are in series, the net thermal resistance between the two fluid streams is given as

$$R_{th,\ total} = \left(\frac{1}{hA}\right)_h + \frac{t_w}{k_w A} + \left(\frac{1}{hA}\right)_c \tag{15.2}$$

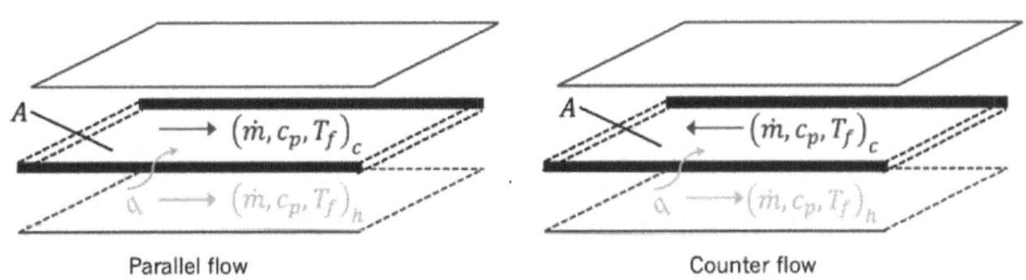

FIGURE 15.1 Parallel and counterflow heat exchangers.

DOI: 10.1201/9781003509837-15

The local heat transfer (q_x) at a general streamwise location (x) can be given as

$$q_x = \frac{T_h(x) - T_c(x)}{R_{th,total}} = \frac{T_h(x) - T_c(x)}{\left(\frac{1}{hA}\right)_h + \frac{t_w}{k_w A} + \left(\frac{1}{hA}\right)_c} \tag{15.3}$$

The local heat transfer as given in Equation 15.3 can be increased by increasing the temperature difference between the hot and cold sides and by decreasing the total thermal resistance $R_{th,\,total}$. For a given temperature difference between the hot and cold sides, which may be fixed due to operating conditions determined by the applications, heat transfer between hot and cold fluid streams can only be increased by decreasing the $R_{th,\,total}$. For the simple case shown in Figure 15.1, the heat transfer area "A" is fixed in all three individual thermal resistances, therefore, the thermal resistance can be simplified to

$$R_{th,\,total} = \frac{1}{A}\left[\frac{1}{h_h} + \frac{t_w}{k_w} + \frac{1}{h_c}\right] \tag{15.4}$$

For a given baseline heat transfer surface area (A), $R_{th,\,total}$ can be reduced by configurations that have high convective heat transfer coefficients on the hot and cold sides, and by keeping the separator plate thickness small, while utilizing highly conductive material for heat transfer. The local heat transfer (q_x) as given in Equation 15.3, can be represented as

$$q_x = (UA)\left(T_h(x) - T_c(x)\right) \tag{15.5}$$

where "U" is the overall heat transfer coefficient given as

$$U = \left[\frac{1}{h_h} + \frac{t_w}{k_w} + \frac{1}{h_c}\right]^{-1} \tag{15.6}$$

The heat transfer rate between the two fluid streams also depends on the nature of the variation in $\{T_h(x) - T_c(x)\}$, which in turn depends on their configuration and flow paths. For instance, the temperature variation of the hot and cold fluid streams with a streamwise location for the two heat exchangers configurations shown in Figure 15.1 is shown in Figure 15.2. As shown, the potential for heat transfer $(\Delta T = T_h - T_c)$ monotonically decreases in a parallel flow configuration, while such a trend is not observed in a counterflow configuration. The potential for heat transfer at the two ends of the heat exchanger is presented as ΔT_1 and ΔT_2. The overall heat transfer (q) in both these configurations can be given in terms of "UA" and a representative temperature difference between the hot and cold fluid streams.

$$q = UA\Delta T_{LM} = UA\frac{\Delta T_1 - \Delta T_2}{\ln\left(\Delta T_1 / \Delta T_2\right)} \tag{15.7}$$

where, ΔT_{LM} is the log mean temperature difference. Note that from Figure 15.2, ΔT_1 and ΔT_2 are defined differently for the two configurations. For a more detailed derivation of Equation 15.7, the readers are referred to [2].

In crossflow heat exchangers, the motion of the two fluid streams is orthogonal to each, where different arrangements can be designed. For instance, the tubes can accommodate sheet metal plates that can serve as fins or an extended surface for enhanced heat transfer, while also separating one fluid stream into different smaller sub-channels. Another possibility includes one fluid stream contained in tubes while the other is allowed to mix. In this case, heat transfer primarily occurs through the total curved surface area of the tubes. Note that these heat exchangers can be fabricated

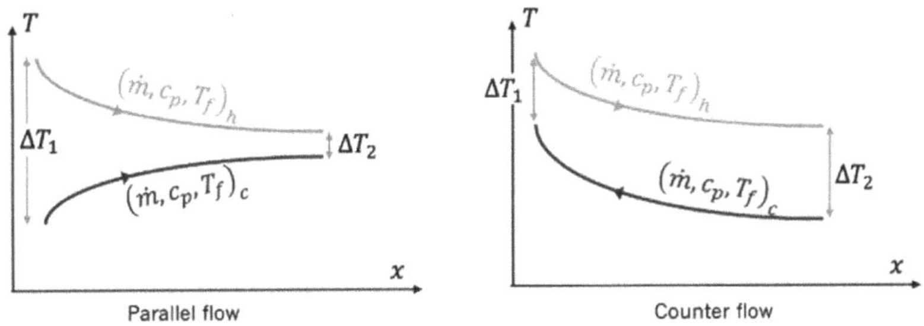

FIGURE 15.2 Temperature variation with a streamwise location for parallel and counterflow heat exchanger configurations.

using standard sheet metal plates and tubes, thus the cost of manufacturing is generally low, and mostly dependent on the choice of material.

Similarly, another configuration that is different from crossflow, parallel, or counterflow, is the cross-counterflow arrangement of the two fluid streams in a shell-and-tube heat exchanger. In this case, one fluid stream moves through the shell, which contains a bank of tubes carrying the secondary fluid. The shell-side fluid can be routed in a serpentine arrangement by placing baffles that partially block the shell. These baffles can also provide structural support to the tubes. Other crossflow configurations are also shown, such as fin tube and plate-fin heat exchangers. These heat exchangers are relatively more complex; however, their fabrication is still possible with the folding of sheet metal plates and brazing/welding them together to achieve heat exchangers with high heat duty.

As shown earlier, the overall heat transfer coefficient determines the net rate at which heat transfer occurs between the two fluid streams of a typical heat exchanger. The performance of large and complex heat exchangers can be characterized by evaluating their effectiveness ($\varepsilon = q / q_{max}$), which is the ratio of actual heat transfer (q) and the maximum possible heat transfer (q_{max}). The maximum possible heat transfer depends on the mass flow rates and the specific heat capacities of the two fluids.

$$q_{max} = C_{min}(T_{h,i} - T_{c,i}) \tag{15.8}$$

where, $C_{min} = C_h = \dot{m}_h c_{p,h}; C_h < C_c$ and $C_{min} = C_c = \dot{m}_c c_{p,c}; C_c < C_h$. The actual heat transfer for a heat exchanger can be determined by using

$$q = \varepsilon C_{min} \left(T_{h,i} - T_{c,i}\right) \tag{15.9}$$

where the heat exchanger effectiveness (ε) is a function of the number of transfer units ($NTU = UA / C_{min}$) and the ratio C_{min} / C_{max}. For instance, in a typical parallel flow heat exchanger, the effectiveness is given as

$$\varepsilon = \frac{1 - \exp\left\{-NTU\left[1 + \left(C_{min} / C_{max}\right)\right]\right\}}{1 + \left(C_{min} / C_{max}\right)} \tag{15.10}$$

For detailed expressions of ε for various heat exchanger types and configurations, readers are referred to [2].

15.1.1 Factors Involved in the Design of Heat Exchangers

Heat exchanger design and development is a complex process that requires the simultaneous consideration of several factors. This sub-section provides a brief account of various factors involved in conventional heat exchanger design. For detailed guidelines on heat exchanger design, the readers are referred to Shah and Sekulic [3]. The following factors are mentioned briefly here, and will be discussed in detail, in context with different applications and types of heat exchangers, at a later stage of this chapter.

15.1.1.1 Application-Specific Demands and Operating Conditions

Application-specific demands and operating conditions include operating temperature and pressure, working fluid types, heat duty requirements, available pumping power on the hot and cold fluid sides, external conditions to which the heat exchanger (HX) would be exposed, operational duration, the life of the HX, and the operational budget. Depending on the application, the maximum weight and volume occupied by the heat exchanger may also be constrained. The type of working fluid and the physical state in which it will be operated in the heat exchanger determine the corrosion and/or erosion rates and fouling properties, which need to be accounted for in the context of the minimum expected life of the heat exchanger. The unsteadiness in the operating conditions and off-design operating conditions must also be considered in the design of the heat exchanger.

15.1.1.2 Selection of Heat Exchanger Type and Configuration

With knowledge of the application-specific demands in terms of expected thermal-hydraulic performance, working fluid types, operating conditions, and manufacturing, as well as operational budget, the heat exchanger type and fluid flow configuration can be selected. Heat exchanger type selection depends on the nature of the working fluids. For instance, a simple heat exchanger with low pressure/low temperature working fluids can be designed in different ways. Conversely, high-pressure/high-temperature working fluids necessitate their containment in vessels/tubes with stringent mechanical constraints, which limits the number of choices in heat exchanger types. As mentioned earlier, the performance of heat exchangers can be quantified via heat exchanger effectiveness (ε), which varies with the type of heat exchanger, along with the flow configuration. Within the above constraints and expected performance, the design value of ε can be evaluated, and based on that, the appropriate heat exchanger type and flow configuration can be determined.

15.1.1.3 Thermal-Hydraulic Performance Quantification and Measures for Its Enhancement in Reference to Associated Pumping Power Cost

The heat duty (q) depends on the heat exchanger effectiveness (ε) and maximum possible heat transfer (q_{max}), once the heat exchanger type and flow configuration is decided (step 15.1.1.2). For a given heat carrying capacities of the two fluids (C_h, C_c), the ε can be increased by enhancing the "UA" of the heat exchanger. Note that C_h and C_c are dependent on the available pumping power for both fluid streams, which in turn contributes to the operational cost. For a given "U" value of heat exchanger, its size (A) can be decided to achieve a certain heat duty (q), or for a given size (A), appropriate enhanced heat transfer concepts can be applied on the hot and cold sides, along with the selection of heat transfer material, to achieve the required "U" value corresponding to a certain heat duty (q). Large-sized heat exchangers have their own manufacturing and operational challenges, and are not preferred in comparison with compact, high power-density heat exchangers, which can yield a desired heat duty in a compact volume and mass, where such heat exchanger designs involve high performance concepts that have high "U." However, there are innovative methods to enhance "A" in a compact volume, such as folded fins from sheet metal, which can also satisfy the requirements. The "U" can be enhanced by enhancing the convective heat transfer coefficients on the hot and cold sides by employing enhanced heat transfer concepts on the walls separating the two fluid streams. Some common examples include the employment of fins, and other types of roughness elements, which could result in local enhancement in heat transfer coefficient as well as heat transfer area. Note that in pursuit of enhancing U, A, or both U and A, the associated pumping power also increases. Hence, the heat exchanger design problem is to enhance the thermal performance corresponding to a given pumping power budget. Similarly, the conduction resistance between the hot and cold fluid streams can be reduced by choosing a highly conductive material or minimizing the sheet/tube wall thickness. This design choice does not impact the pumping power budget, however, certain highly conductive materials can be expensive and may be cost prohibitive, while on other occasions, such materials may not be a suitable choice due to the heat exchanger operational conditions in terms of temperature, pressure, and the corrosion/erosion nature of the working fluids and their reaction with the solid material. In summary, the heat duty or the outputs of heat exchanger can be expressed in terms of independent variables and parameters as

$$q = f\left(T_{h,i}, T_{c,i}, C_h, C_c, U, A, HX\ type\,\&\,conf.\right) \tag{15.11}$$

15.1.1.4 Consideration of Fouling and Its Effect on HX Performance

Figure 15.1 showcases an idealized scenario of heat exchange between hot and cold streams through a finite-wall thickness. In practical applications, with increasing operational time, fouling occurs on both hot and cold sides, which in turn adds thermal resistance. Fouling can occur as result of rust formation, chemical reactions between working fluids, or deposits from working fluids themselves, erosion, and so on. The layer formed on the hot and cold sides due to fouling has low thermal resistance. Fouling is an undesirable phenomenon, and at the same time, an unavoidable attribute of heat exchangers that are subjected to hostile environment. Heat exchanger lifetimes depend on fouling factors on hot and cold streams. Further, the design of heat exchangers must account for accruing fouling factors over time. With the accounting of fouling, the net thermal resistance between the hot and cold fluid streams as given in Equation 5.4 can be revised as

$$R_{th,\ total} = \left[\frac{1}{h_h A} + R''_{f,h} + \frac{t_w}{k_w A} + R''_{f,c} + \frac{1}{h_c A}\right] \tag{15.12}$$

where $R''_{f,h}$ and $R''_{f,c}$ are the fouling resistances per unit area on the hot and cold sides, respectively. From the heat exchanger design perspective, standard correlations on fouling-based thermal resistances can be incorporated in thermal-hydraulic performance calculations.

15.1.1.5 Mechanical Strength and Design Aspects

Depending on the application, heat exchangers are tasked to carry a wide variety of working fluids, which impose different types of mechanical constraints that must be accounted for in the design process. High-pressure/high-temperature working fluids dictate the minimum sheet/tube/shell thickness, and similar constraints are imposed on the supply/collection (headers, manifolds) sections. The working fluid's operating pressure and resultant thermal stress calculations are performed to ensure compliance with the ASME pressure vessel codes. Fabrication strategies, including critical joints, flanges, and sheet bonding methods, are also affected by mechanical constraints.

15.1.1.6 Manufacturing and Cost Aspects

The aforementioned steps in the heat exchanger design guide the manufacturing methods and hence the overall cost of heat exchanger fabrication. Manufacturing aspects include the direct availability of components such as flanges, tubes, pipe fittings, and plates, tools required for fabrication such as dies, the necessity to fabricate custom components for specific applications, the role of metal additive manufacturing, CNC machines, welding equipment, diffusion bonding, associated raw materials, brazing/soldering/welding of different sub-components, leak/pressure testing, and structural support and its testing, among others. The cost aspects in heat exchanger design are not limited to the manufacturing stage, and also include installation, operational, and disposal costs. The lifecycle cost of a heat exchanger is evaluated in reference to heat transfer parameters such as $q/\Delta T_{LM}$ or UA, resulting in metrics such as $\$/\{W/K\}$. The objectives expected to be accomplished by heat exchangers, whether it is power generation, process heat production, thermal management, or waste heat recovery, requires justification in terms of economic viability, which prioritizes the design and optimization of heat exchangers in terms of the ratio of lifetime cost to thermal performance ($q/\Delta T_{LM}$ or UA).

15.1.1.7 Multi-Objective Optimization

The above factors that influence or guide heat exchanger design and development indicate that it is a complex process with several parameters at play. To this end, a comprehensive optimization effort is imperative to design component and system-level concepts and strategies that could satisfy multiple objectives. Sole optimization in terms of heat transfer quantities is not sufficient, and other aspects such as pumping power, fouling probability, heat exchanger life, and overall fabrication, as well as operational cost, must be included as parameters.

15.1.2 Challenges in Modern Heat Exchangers Operating in Hostile Environments

The primary challenge in the development of modern heat exchangers comes from the demand for operations at high temperatures and high pressures. Such operating conditions are favorable for the thermodynamic efficiencies of power cycles and extraction of high-grade heat (generally associated with high temperatures) to accomplish a variety of process heat applications. In the growing push for increased energy efficiency, sustained and long-life operations of heat exchangers at such operating conditions are imperative. High temperature operations impose severe challenges in terms of materials, which undergo significant strength reduction for temperatures exceeding 500°C. Additionally, high temperature and high-pressure operations subject the heat exchangers to high localized thermal stresses, which need to be accounted for, in the design process itself. For high temperature operations, understanding material behavior when subjected to creep-fatigue loading is imperative [4]. Common modes of heat exchanger failure include fatigue and creep [5]. Thermal fatigue can occur due to repeated temperature fluctuations. Repeated thermal-mechanical stresses can lead to cracking. Different components or junctions involved in heat exchangers can undergo creep rupture due to hostile environments [6], which may include high temperature, high localized stress, and oxidation-related changes in tubing.

Further, working fluids are selected based on their high figure-of-merit depending on the applications. This choice, in reference to the heat exchanger material, can sometimes lead to corrosion over time, which reduces the life of the heat exchanger along with adding fouling resistance during regular operations. For instance, in co-combustion systems involving biomass, degradation is accelerated due to corrosion caused by hot gases and molten salt [7]. The corrosion problem arising from molten salt is also found in solar-thermal applications where molten salt is used, both as thermal energy storage as well as heat transfer fluid (on the hot side). Typically, molten salt-to-supercritical carbon dioxide (sCO_2) heat exchangers involve chloride salt and stainless steel, where high rates of corrosion are expected [8]. Corrosion-fatigue can result in crack development, which may potentially reduce the lifetime of heat exchangers [9]. In particle-to-sCO_2 heat exchangers, silica sand particles receive thermal energy in the receiver section of the solar tower and move downward toward the HX under the influence of gravity. The moving packed bed nature of particle movement through typical plate heat exchangers results in abrasion over time, which depends on the particle properties, operating temperatures, and the plate material [10, 11]. Apart from the interactions of working fluid with the material of a heat exchanger, the machinery required to transport the working fluid plays a critical role in the design of heat exchangers. Such devices also undergo degradation, blockages, and corrosion, and the lifecycle cost of these devices is also considered, along with the basic test of their feasibility to support operations at certain flow rates and at desired operating temperatures and pressure conditions.

Oxidation of heat exchangers and associated components is also a major challenge, which is caused by the unavoidable oxidizing environment, material affinity, and high temperatures [12]. The affinity of Heat exchanger material toward oxygen causes the formation of an oxide layer, which is further aggravated at high temperatures. For instance, in a typical fluidized bed combustor, carbon steel used as superheater tubes undergo oxidation that eventually leads to erosion [12].

In special circumstances, continued operations in hostile environments may result in material removal that may also lead to flow passage blockage and cause unwanted localized roughness scales. A classic example of a high temperature heat exchanger is a gas turbine blade in the high-pressure stages, which is subjected to a hostile environment on its outer end, while being cooled through internal passages. With accrued operational time, the external skin of the blades gets roughened due to wear and interactions with combustion products or ingestion of foreign particles. These roughness scales negatively impact the blades' life as they increase the net heat flux into the blade skin. Further, the ingestion or entry of foreign elements in large-scale heat exchangers can result in significant damage to heat exchangers over time.

15.2 HIGH TEMPERATURE HEAT EXCHANGER APPLICATIONS

As discussed in the previous section, heat exchangers for hostile environments are highly complex devices that face a variety of challenges. In this chapter, we have focused on the challenge of designing heat exchangers for high temperature and high-pressure conditions. This section presents a brief account of the wide range of applications that involve heat exchanger operations at high temperatures (>500°C).

15.2.1 High Temperature Gas-Cooled Reactors

In high temperature gas-cooled reactors (HTGRs), steam generators play a crucial role. Shell-and-tube heat exchangers involving helium and water/steam as working fluids have been used [13]. In such heat exchangers, helium is carried in the shell while the tubes carry the steam. The heat exchanger supplies superheated steam at 540°C and 1550 psig, as a result of helium supplied at 740°C and 329 psig [13].

Intermediate heat exchangers (IHX) in HTGRs facilitate process heat applications, where helically coiled heat exchangers and plate-fin type heat exchangers [14] have been employed in the past. IHXs are tasked to reliably operate at temperatures exceeding 900°C.

15.2.2 Recuperators for Gas Turbine and Microturbine Systems

Recuperators have the potential to yield efficiencies exceeding 60% for combined gas turbine cycles [15]. Printed circuit heat exchangers and plate-fin heat exchangers are considered to be suitable candidates that can serve as recuperators where the inlet gas temperatures in a typical combined gas turbine cycle exceed 825°C. In microturbine systems, the compressed air can be preheated in the recuperator by recovering thermal energy from the low-pressure stages of the turbine section. High effectiveness and low pressure drop recuperators can potentially increase the typical system efficiency of microturbines from 16–20% to over 30%. Some examples for recuperators in microturbine systems include plate heat exchangers [16].

15.2.3 Recuperator for sCO_2 Power Cycles

Compared with the conventional steam- [17, 18] or helium-based [19, 20] power cycles, the sCO_2-based power cycles have the potential to achieve higher efficiency when turbine inlet temperatures exceed 550°C. The recuperators are critical for sCO_2 cycles, and are expected to yield high NTUs, which necessitate the development of compact, high power-density heat exchangers, which can survive high temperature and extremely high-pressure conditions. Printed circuit heat exchangers (PCHEs) with different flow configurations have been considered as potential recuperators. PCHEs are several times more power-dense compared with typical shell-and-tube heat exchangers and are suitable for this application [21]. Wire-mesh and plate-fin heat exchangers have also been explored for recuperators in sCO_2 power cycles [21].

15.2.4 Waste Heat Recovery

In many applications, recovery of waste heat is desired to increase the process efficiency. This motivation is particularly high in the case when waste heat is high grade, i.e., at high temperatures (>500°C). In steel rolling furnaces, shell-and-tube heat exchangers are used to extract thermal energy from the exhaust flue gases at temperatures exceeding 600°C, where the shell carries the flue gases and the finned tubes carry the secondary fluid (air) used for waste heat recovery [22]. The heat transfer in the tubes carrying the tubes was enhanced by inserting twisted tubes inside the tubes carrying air to impart swirl and enhance the near-wall shear stress [22]. Shell-and-tube heat exchangers are also used in extracting waste heat from particles in solar-thermal applications [23], where the particles serve as thermal energy storage media as well as heat transfer fluids with temperatures > 700°C.

15.2.5 Particle-to-sCO_2 Heat Exchanger in CSPs

In a typical tower configuration of a concentrated solar power (CSP) plant, thermal energy is directed toward the receiver section, which contains moving particles, where the particle temperature is elevated to >700°C. To achieve 50% thermal-to-electric efficiency through recompression in the sCO_2 Brayton cycle, the turbine inlet temperatures and pressure should exceed 700°C and 20 MPa, respectively [24–26]. A particle-to-sCO_2 heat exchanger is the most critical component to achieve the sCO_2 conditions mentioned above. Different heat exchanger types such as fluidized beds with serpentine tubes carrying sCO_2 [27], shell-and-tube heat exchangers with shells carrying the moving packed bed of particles and tubes carrying sCO_2 [10], and shell-and-plate heat exchangers with alternate parallel plates carrying moving packed bed of particles, and minichannel arrays carrying sCO_2 [28] have been employed.

15.2.6 Hydrogen Production

Hydrogen is an attractive clean energy source and is gaining significant traction to diversify the clean energy portfolio. Hydrogen can be produced thermochemically using the sulfur-iodine cycle [29], where helium at 975°C heats the sulfur-iodine feed components to facilitate the decomposition reaction that occurs at temperatures exceeding 850°C [29]. Shell-and-plate heat exchangers and decomposer modules have been used where silicon carbide (SiC) was used as the material for the fabrication [29].

15.2.7 Gas Turbine Airfoils

High-pressure stage turbine airfoils are subjected to a hostile environment of temperatures of hot combustion gases exceeding 1,700°C, which is well above the material handling capabilities [30]. The turbine airfoils are cooled by drawing air from the compressor section where the "coolant" temperature is at 700°C. Gas turbine airfoils are highly sophisticated heat exchangers where the hot and cold streams are separated from each other by a thin skin of the airfoil. The airfoil is made hollow to accommodate the internal cooling channels. The internal passages feature enhanced heat transfer concepts such as leading-edge jet impingement, rib turbulators in serpentine passages for the mid-chord region of the airfoil, and an array of pin-fins for the trailing edge [31]. The internally routed air through these complex flow paths is purged into the hot gas path via film cooling holes arranged on the blade skin [32]. The research in this field is focused on the economic use of the coolant, which is bled off from the compressor as it imposes a penalty on the cost of operations. For energy-efficient power generation, both for aircraft propulsion and land-based electricity production, high-efficiency heat exchanger concepts need to be developed and employed in the high-pressure stages of the turbine section.

15.3. MATERIAL CANDIDATES AND SELECTION CRITERIA FOR HIGH TEMPERATURE HEAT EXCHANGERS

As identified in Section 15.1, the major challenge facing the design and development of high temperature heat exchangers is the selection of materials that can not only handle the thermal-mechanical load but can also provide efficient heat transfer from the hot to the cold side, in addition to being resistant to oxidation, corrosion, and creep-fatigue. In addition to the multi-functional attributes expected from the selected material, the choice of the heat exchanger type, working fluid flow configuration, header design, tubings, pipe fittings, and so on, also influence the selection of materials. This section presents an overview of the material candidates for high temperature heat exchangers. The materials for heat exchangers can be classified into four major categories: polymer-based, metallic, ceramics, and carbon-based [33]. Relevant to high temperature heat exchanger applications, metallic and ceramic-based materials are briefly discussed here.

15.3.1 Metals and Alloys

Metals are a preferred choice for heat exchanger fabrication due to the ease of manufacturing and superior thermal properties. However, at high temperatures, metals tend to lose their favorable attributes observed at low temperatures. The yield strength (MPa) versus temperature for various materials is shown in Figure 15.3.

For high temperature (>500°C) and high-pressure applications, iron- and nickel-based alloys can be used as the heat exchanger material, however, their employment has a limited range. Stainless steel is a popular choice for heat exchanger applications limited to 700°C due to its strength, thermal conductivity, manufacturability, availability of pipe and tube fittings, ease in welding, diffusion bonding, and its availability in different forms, such as plates of varied thickness and tubes of different sizes. Some common materials include SS316, SS347, SS304, and SS321, where SS316 has been used in a wide range of applications including intermediate heat exchangers (PCHE) in sodium-cooled fast reactors in nuclear applications [34] and particle-to-sCO_2 heat exchangers in

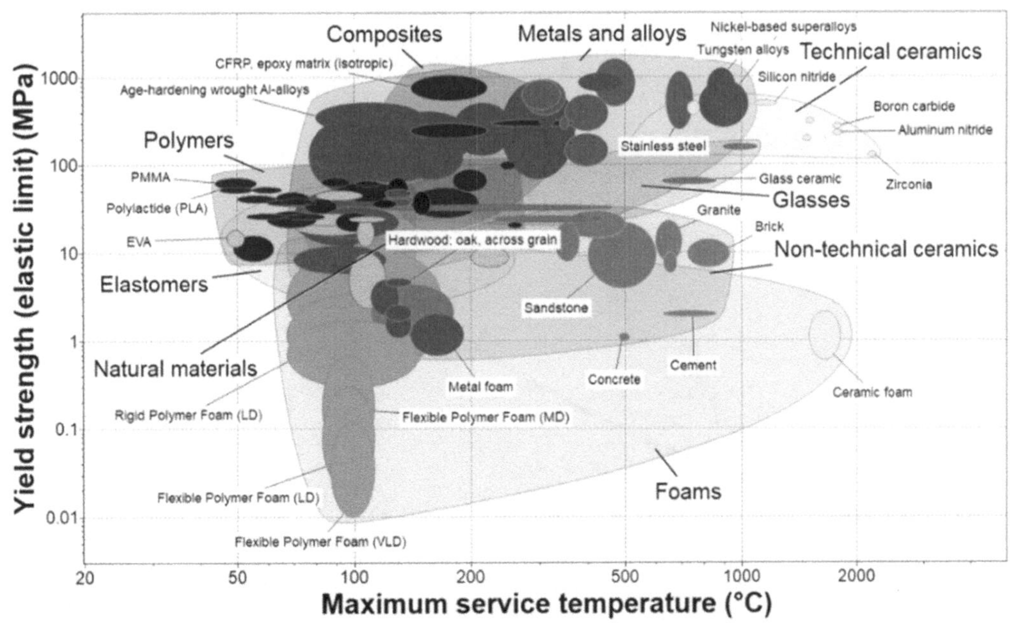

FIGURE 15.3 Yield strength (MPa) versus maximum service temperature (°C) (source: ANSYS; open-source figure).

CSP [35, 36], among others. SS347 usage has been limited to 600°C and is a popular choice for gas turbine recuperators [15]. Beyond this temperature, SS347 is susceptible toward oxidation and creep deformation [37]. HR120, AL120-25+Nb materials have been proven to be robust candidates for operations limited to 750°C and are explored as an alternative to SS347 [37]. Figure 15.4 shows the creep-strain versus time plot for tests conducted at 750°C and 100 MPa for SS347, AL 120-25+Nb, and SS625.

In contrast with iron-based alloys, nickel-based alloys exhibit superior mechanical strength (Figure 15.3), although they are significantly more expensive. Figure 15.5 shows a comparison between nickel- and iron-based alloys' maximum allowable stress. Haynes 282 and IN 740H have superior strength in the temperature band 700–800°C. Haynes 230 has better creep-rupture resistance in reference to iron-based alloys [39]. Nickel-based alloys are significantly more expensive than iron-based alloys. Figure 15.6 shows a comparison of normalized cost in reference to SS347. A summary of important thermo-physical properties of iron- and nickel-based alloys, in context with heat exchanger design, is presented in Table 15.1.

From Figure 15.5, it can be inferred that the mechanical strength of iron- and nickel-based alloys deteriorates rapidly for temperatures exceeding 700°C. Although several high temperature heat exchangers operate within this temperature limit, there exists a strong demand for materials that can perform reliably at much higher temperatures. To this end, ceramics offer a solution for operations far exceeding 1,000°C (Figure 15.3), while, at the same time, being resistant to corrosion, creep-fatigue [40].

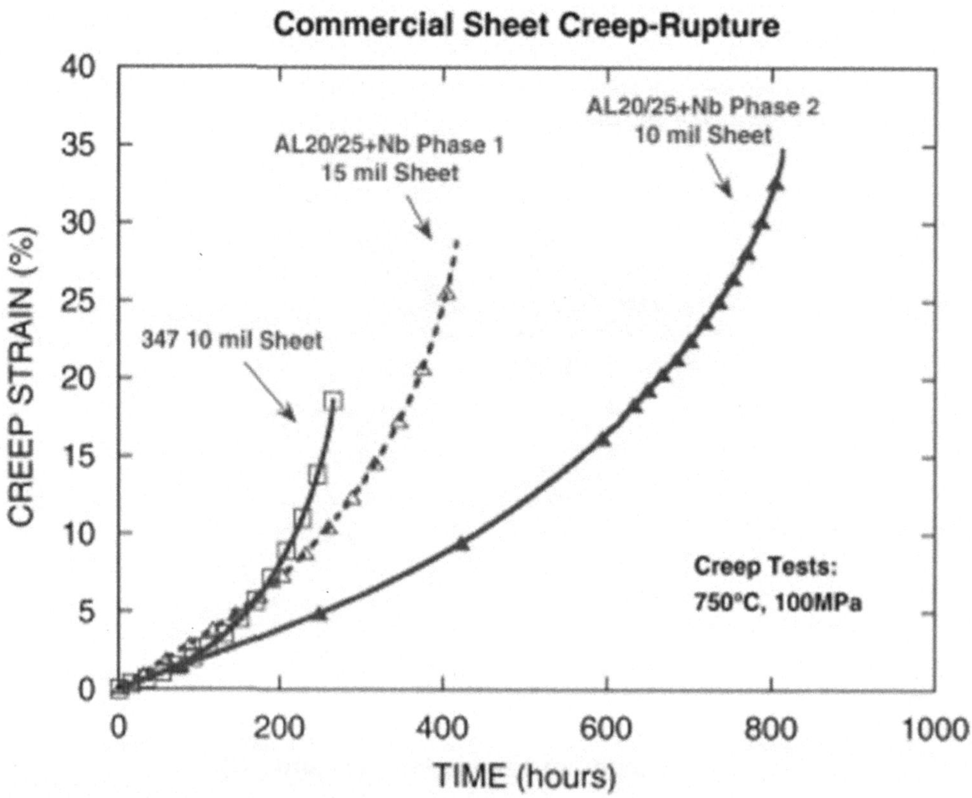

FIGURE 15.4 Creep-strain versus time for SS347, Al20-25+Nb [37] (with permission to reuse content).

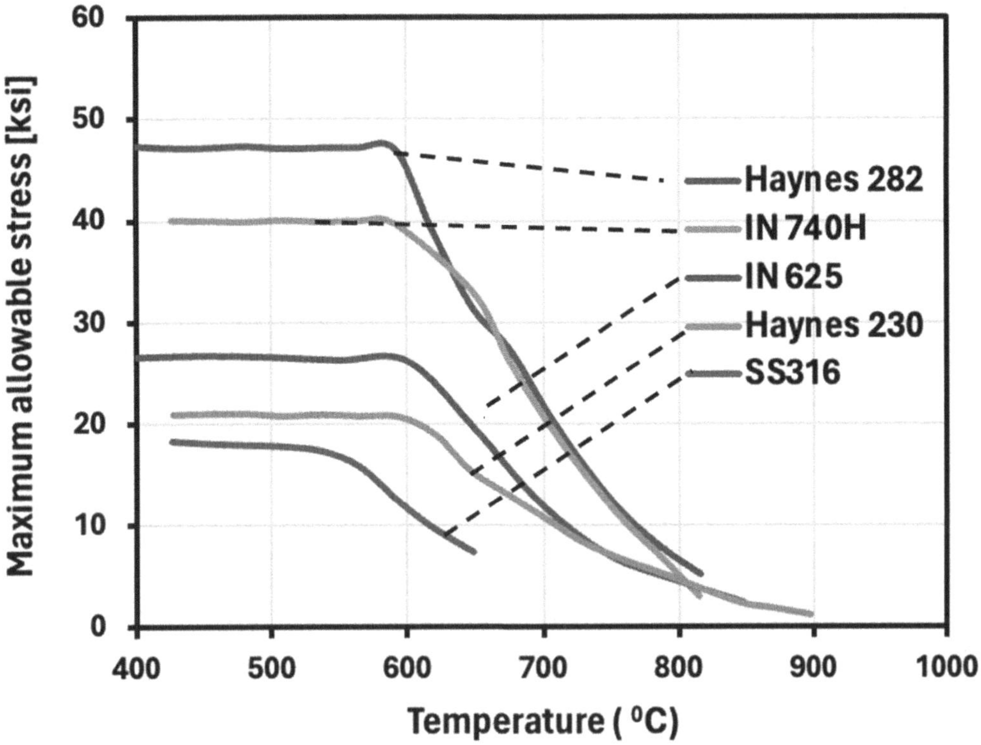

FIGURE 15.5 Maximum allowable stress versus temperature; comparison between iron- and nickel-based alloys (figure adapted from Chordia et al. [38]).

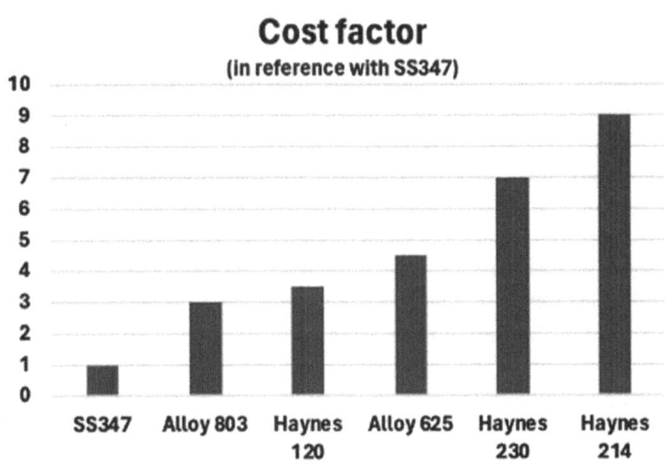

FIGURE 15.6 Normalized cost in reference to SS347 (adapted from [39]).

15.3.2 Ceramics

Due to the high temperature handling capability of ceramics and resistance toward oxidation, ceramic-based heat exchangers provide several advantages over metallic heat exchangers. Engineering ceramics can be divided into three categories: (1) ceramic oxides, (2) glass ceramics,

TABLE 15.1
Properties of Iron- and Nickel-Based Alloys

Material	Density (g/cm³)	Tensile strength, Yield (MPa)	CTE (sm/m-°C)	Specific heat capacity (J/g-°C)	Thermal conductivity (W/mK)
SS316	8	250	17.5 @ 540°C	0.5 @ 100°C	16.3 @ 100°C
SS347	8	240	18.4 @ 540°C	0.5 @ 100°C	16.3 @ 100°C
SS304	8	330	18.4 @ 538°C	0.5 @ 100°C	16.2 @ 100°C
SS321	8	240	18.5 @ 540°C	0.5 @ 100°C	16.1 @ 100°C
Haynes 230	8.97	405	14 @ 500°C 16.1 @ 1,000°C	0.419 @ 100°C 0.617 @ 1,000°C	10.4 @ 100°C 28.4 @ 1,000°C
Haynes Hastelloy X alloy	8.22	245 @ 540°C 91 @ 980°C	16.6 @ 1,000°C		28.7 @ 1,000°C
Haynes 214	8.05	545 @ 540°C 54 @ 980°C	18.6 @ 1,000°C	0.742 @ 1,000°C	34.7 @ 1,000°C
IN 718	8.19	980 @ 650°C	13 @ 100°C	0.435	11.4
IN 625	8.44	290 @ 650°C	12.8 @ 100°C	0.410	9.8
IN 740	8.05	818	12.38 @ 100°C	0.476 @ 100°C 0.635 @ 1,000°C	11.7 @ 100°C 23.8 @ 900°C

(Source: Matweb)

and (3) ceramic carbides and nitrides [41]. In terms of resistance to oxidation, silicon carbide and silicon nitride are the preferred choices.

Ceramic-based heat exchangers can be developed in both tubular and plate-fin-based configurations, with applications including waste heat recovery in metal processing industries, IHX in nuclear processes, gas turbine recuperators, and high-temperature gas-to-gas heat exchangers [41]. Some of the challenges for ceramic heat exchangers are their response to thermal shock [42], thermal fatigue [43], through-plane leakage [44], limited fabrication options, and limited bonding/joining techniques [45] with the same or dissimilar materials.

Ceramic material candidates for heat exchanger applications include silicon carbide (SiC), silicon nitride (Si_3N_4), alumina (Al_2O_3), zirconia (ZeO_2), and so on. [33]. SiC offers good resistance to thermal shock and is highly conductive in comparison with iron-based alloys. Silicon nitride offers good creep resistance; however, its usage is limited to 1,000°C due to the vulnerability toward oxidation. Compared with SiC and Si_3N_4, alumina is weaker in response to thermal shock, however, it is resistant toward oxidation. A comprehensive review of different ceramic materials in the context of heat exchanger development can be found in [33]. Thermal conductivities of the above candidate ceramic materials can be found in [46].

15.4 HIGH TEMPERATURE HEAT EXCHANGERS: TYPES, CONFIGURATIONS, AND APPLICATIONS

In this section, different heat exchanger types and flow configurations are discussed, covering a wide range of high temperature applications. Conventional heat exchanger types are presented in this section, while more recent, novel advanced heat exchanger concepts facilitated by additive manufacturing are discussed in the subsequent section. The important role of heat transfer fluid will be discussed in the context of the application and chosen material for the heat exchanger.

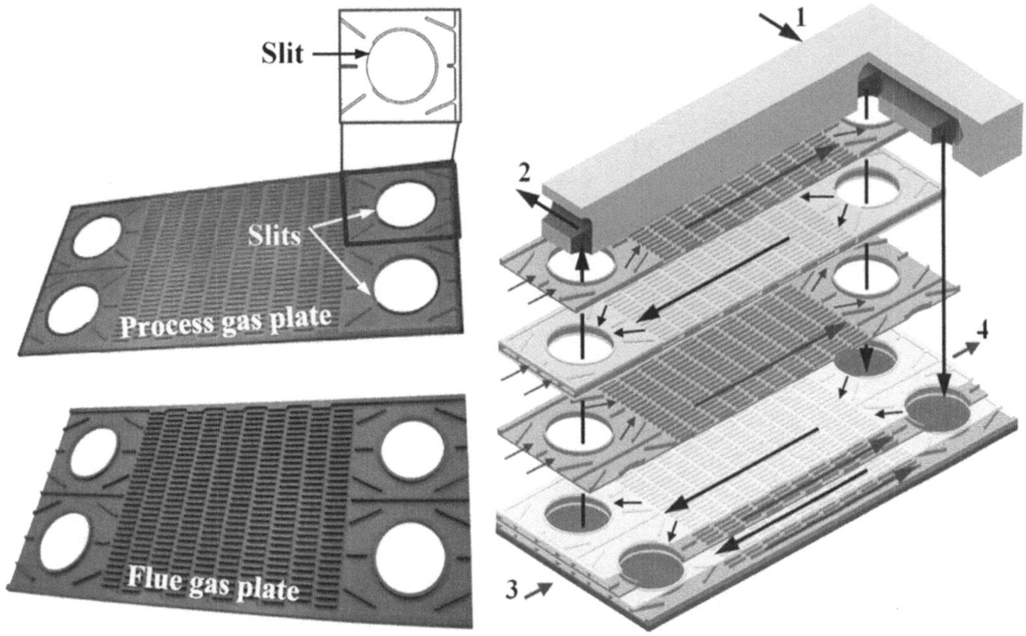

FIGURE 15.7 Ceramic plate-finned heat exchanger [47] (with permission to reuse content).

15.4.1 Plate-Fin Heat Exchangers

A sintered silicon carbide (SSiC) heat exchanger with offset strip fin (Figure 15.7) was experimentally tested by Haunstetter et al. [47]. The SSiC plates were sintered with each other, resulting in a monolith heat exchanger and eliminating the need for gasket material. The SSiC plates also featured offset strip fins to enhance the heat transfer coefficient due to the flow process and flue gas. The heat exchanger effectiveness was demonstrated by using air on both sides. On the hot side, air was heated up to 800°C, while the cold side temperature was varied up to 500°C. For the flow configuration where hot side air inlet temperature was 800°C and the cold side inlet air temperature was 450°C, heat exchanger effectiveness of over 95% was achieved. The SSiC plate provided a robust, leak-free, high-performance heat exchanger for high temperatures. A numerical study on flow and heat transfer for air through the offset strip fin flow channel revealed that high heat transfer zones exist at the leading edge of the fins where the flow stagnates, along with the nearby regions where the flow accelerates [48].

A ceramic heat exchanger in plate-finned configuration was demonstrated by Mello et al. [49], where alumina was used as the material. Air was used as the working fluid to demonstrate the performance at 800°C. It was shown that the heat exchanger effectiveness varied between 0.62 and 0.9. An offset strip fin with counterflow configuration was studied by Schulte-Fishedick et al. [50] where SSiC was used for temperatures exceeding 1,200°C. The flue gas inlet temperature was 1,215°C while the process gas inlet temperature was 700°C. The pressure ratio between the process gas and flue gas channel was ~14. The SSiC heat exchanger was developed for a heat duty of 50 kW_{th}.

A diffusion-welded intermediate heat exchanger for HTGRs was developed by Takeda et al. [14], where a compact plate-type heat exchanger was used (Figure 15.8). The plates were made from a nickel-based superalloy Hastelloy for operations at 950°C. It was determined that the diffusion-welded heat exchanger in Hastelloy could deliver 75 MW per unit, at a 1/16th volume of conventional helically coiled intermediate heat exchangers [14].

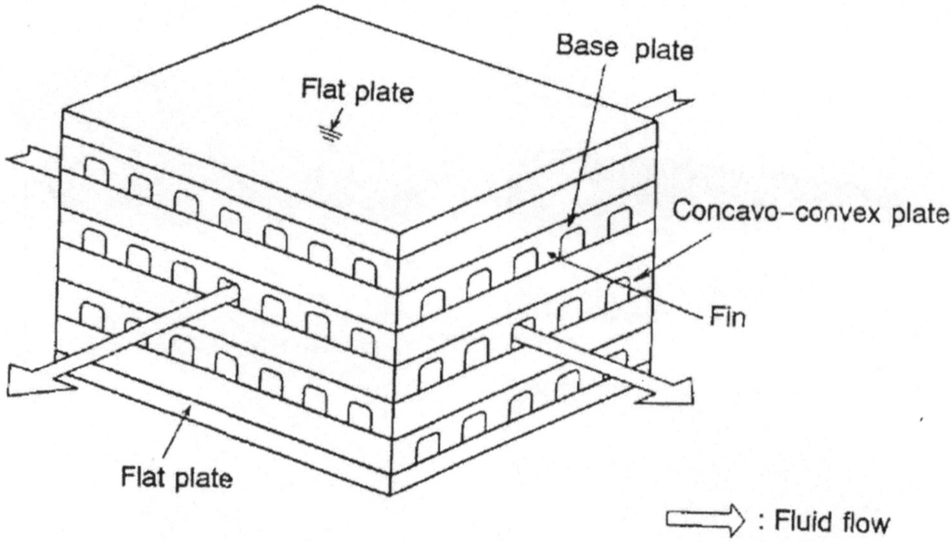

FIGURE 15.8 Compact plate-fin heat exchanger for HTGRs [14] (with permission to reuse content).

15.4.2 Shell-and-Tube Heat Exchangers

A ceramic matrix composite (CMC)-based shell-and-tube heat exchanger was tested for gas (combustion products) inlet temperatures ~1,200°C and secondary fluid (air) inlet temperatures of ~400°C [51]. The material choice of CMC was made for a high temperature application, which also involved a chemically aggressive environment and expected high degree of fouling. A cross-flow configuration was achieved by placing baffles. A low overall heat transfer coefficient (<30 W/m^{2}K) was achieved, resulting in heat exchanger effectiveness of less than 0.6. Recently, a shell-and-tube heat exchanger design and optimization study was conducted for waste heat recovery from flue gases, where the tubes were modified to achieve high heat exchanger effectiveness [22]. The recovered thermal energy preheated the ambient air entering the shell-and-tube heat exchanger, which was eventually fed back into the combustion chamber. Figure 15.9 shows the heat exchanger design, along with three different types of tubes, with a smooth tube as the baseline. The twisted tape inserted in a smooth tube imparts swirl to the moving air, increasing the near-wall shear and enhancing flow mixing, resulting in increased heat transfer. The flue gas inlet temperature was 800°C while the air entered at 20°C. Although modified tubes resulted in enhanced heat transfer, however, at a large cost of pressure drop penalty. For instance, for a two-fold increase in overall heat transfer, a ten-fold increase in pressure drop was observed, resulting in a 40% increase in heat exchanger effectiveness. The study was eventually optimized to strike a balance between an increase in effectiveness and the penalty on the pressure drop side—a classic problem in thermal-fluid science.

Similar to waste heat recovery from flue gases from combustion systems, in concentrating solar power tower configurations, waste heat from particles can be extracted to accomplish process heat objectives. Particles in CSP plants can be heated to temperatures exceeding 700°C—this high-grade thermal energy can be used to facilitate on-site process heat tasks. Based on novel tube arrangements in shell-and-tube heat exchanger configuration, Jiang et al. [23] explored heat extraction pathways from heated particles on the shell-side and gas contained in the tubes. Figure 15.10 shows different arrangements of serpentine tubes within a shell carrying particles, with an inlet temperature of 750°C. The dual-stage tube arrangement can be used to facilitate process heat applications in two distinctly different temperature bands.

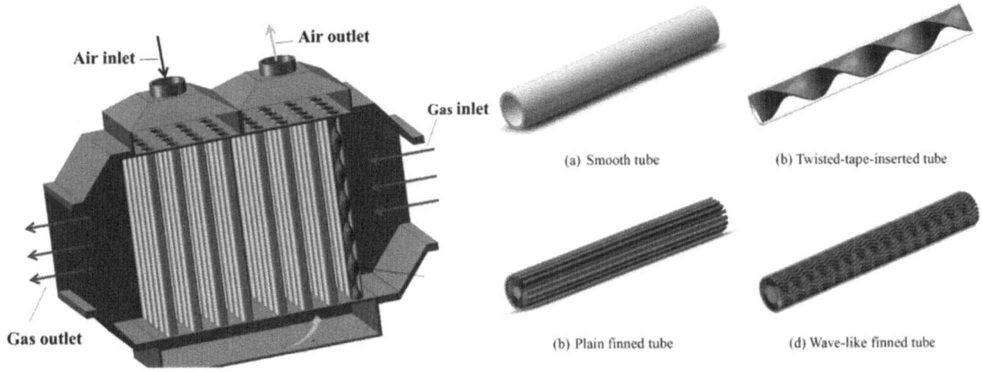

FIGURE 15.9 Shell-and-tube heat exchanger with s variety of tube configurations [22] (with permission to reuse content).

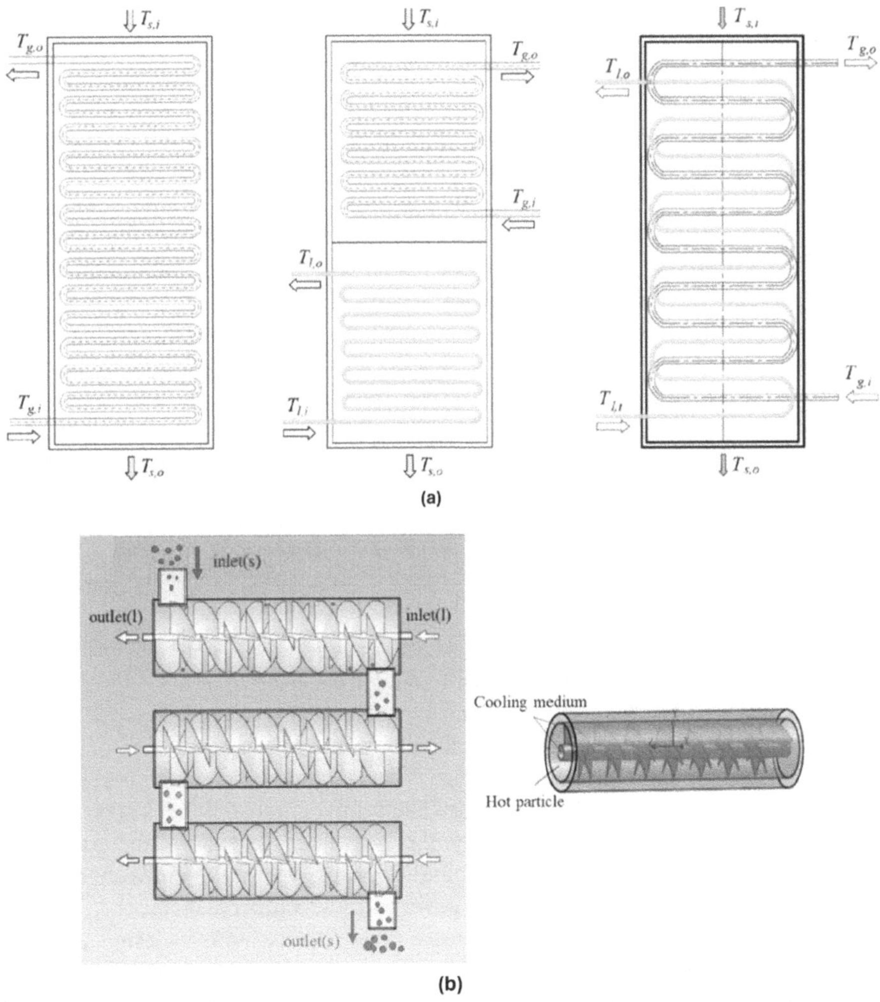

FIGURE 15.10 Shell-type and tube-type heat exchanger used for heat recovery process from vertically moving particles. (a) Different arrangements of tubes in a shell-and-tube heat exchanger for heat recovery from particles moving vertically on the shell-side; (b) horizontal movement of particles with novel screw-type fins added on the tubes [23] (with permission to reuse content).

Further, novel screw-type finned tubes serving the dual purpose of providing extended surface while also facilitating particle movement in horizontal direction was designed. The vertical motion of particles requires large tower height, which also adds to the cost of operations, since the particles have to be transported back to the receiver section located at the top of the tower [52]. A horizontal configuration can be used in low ground clearance situations, where the rotary motion of the tubes with screws can facilitate particle transport.

15.4.3 Fluidized Bed Heat Exchangers

Fluidized bed heat exchangers offer high rates of heat transfer between participating fluids and exhibit anti-fouling characteristics [53]. They find their applications in waste heat recovery from flue gas, carbon capture, geothermal applications, and desalination. Such heat exchangers are found in both direct and indirect mode configurations, where in direct mode, the hot and cold streams are allowed to mix with each other, for instance, gas-solid fluidized beds. In direct mode, the solid phase can be fluidized via a gas, while serpentine tubes carrying the secondary fluid exchanges heat via the tube walls. The indirect mode of heat exchange is typically accomplished in a shell-and-tube configuration. Fluidized beds can also be used to transfer stored energy in the solid-phase to the walls containing them, where fluidization helps increase the solid-phase and wall heat transfer rates by facilitating counter-rotating vortex pairs and bubbles [54]. Such a configuration is a direct heat exchange where the two fluids are mixed with each other, and as a result, a net heat transfer occurs between the walls and gas-solid mixture. A circulating fluidized bed configuration was used to transfer thermal energy from ash to steam carried in serpentine tubes in [55]. In concentrating solar power applications, quartz sand is considered as a thermal energy storage medium. Warerkar et al. [56] studied an air-to-sand heat exchanger configuration, where thermal energy from heated air was directly transferred to sand moving in a vertical direction, resulting in a crossflow, direct heat exchange configuration. The air temperature at the inlet of the heat exchanger was 800°C, while the sand entered at 200°C, resulting in heat exchanger effectiveness between 0.85 and 0.9. Ma and Martinek [27] studied a counterflow heat exchanger fluidized bed configuration for the transfer of thermal energy from particles heated to 775°C to sCO_2 entering at 550°C. The solid particles moved on the shell side of the shell-and-tube heat exchanger, while sCO_2 was carried in the tubes. To enhance the heat transfer between the hot and cold sides, the shell carrying the particles was fluidized. Fluidized bed heat transfer involving particles can yield heat transfer coefficients between 300 and 1,000 W/m^2K [57].

15.4.4 Parallel Plate Heat Exchangers

As mentioned earlier, a particle-to-sCO_2 heat exchanger is a critical component for sCO_2-based power cycle operations. In such heat exchangers, moving a packed bed of heated particles is routed between parallel plates, to which an array of minichannels carrying sCO_2 is attached (Figure 15.11). The thermal resistance between the hot and cold sides includes the convection resistance on the particle side, contact resistance between the particle and the plate, conduction resistance of the plate, and convection resistance on the sCO_2 side. The primary research focus is on reducing thermal resistance on the particle side, since that is the bottleneck. In state-of-the-art parallel plate heat exchangers, particles move between two parallel plates, where the convective heat transfer coefficient drops rapidly as it moves vertically downward, which necessitates the creation of different banks of heat exchangers. This is an expensive route that also adds to design and operational complexity. In an experimental study, the particle side convective heat transfer coefficients were found to be ~200 W/m^2K [36]. In a theoretical study, it was shown that the overall heat transfer coefficient for such particle-to-sCO_2 flows depends strongly on the particle diameter, with smaller particle diameters yielding high heat transfer [58]. Theoretically, the heat exchanger effectiveness for the flow configuration shown in Figure 15.11 is ~0.9, which could be improved further with the

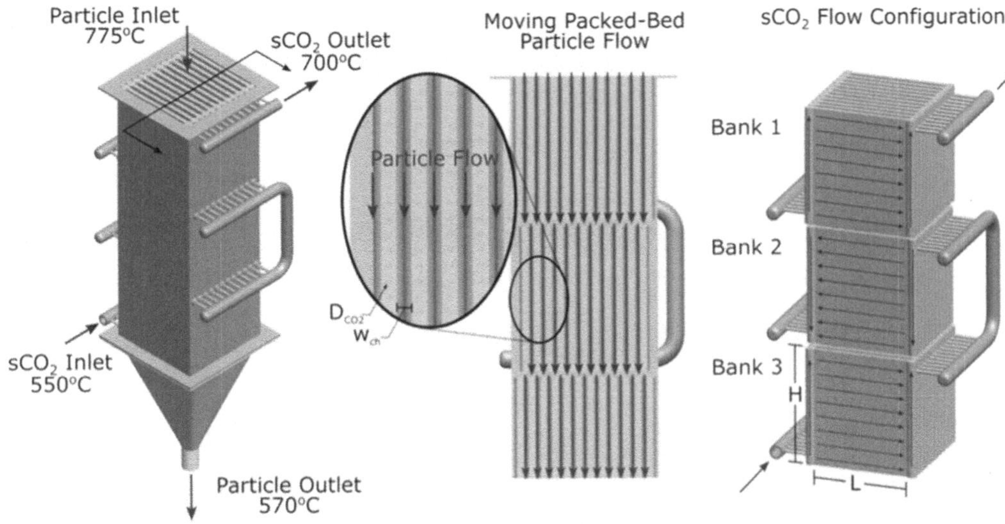

FIGURE 15.11 An illustration of a particle-to-sCO_2 heat exchanger, moving packed bed of particles flow between parallel plates, while sCO_2 flows through an array of minichannels attached to those plates [28] (with permission to reuse content).

incorporation of advanced concepts on the particle side. Recently, the parallel plate heat exchanger concept shown in Figure 15.11 was modified by Arthur-Arhin [59] through inclusion of fluidization in the particle flow channel. The heat exchanger provided heat duty of 35 kW, where only 2.5% of the total heat lost by the hot side was not transferred to the sCO_2 channels. For a wide range of fluidization velocities, the overall heat transfer coefficient of the heat exchanger was found to be lower than 200 W/m^2K [59]. Further, the heat exchanger effectiveness was <0.7 [59]. Particle-to-sCO_2 heat exchangers are complex devices, and their overall performance is evaluated considering a variety of factors that include scalability, manufacturability, heat transfer potential, erosion and corrosion, transient operations characteristics, and manufacturing, as well as operational costs and parasitic heat losses [10]. Among the three heat exchanger concepts, fluidized bed, shell-and-tube, and parallel plate heat exchangers, the parallel plate configuration is considered to be the most feasible route when all the above factors are accounted for, with proper weights are assigned to each [10].

15.4.5 Printed Circuit Heat Exchangers

PCHEs are built by diffusion-bonding thin metal plates that have photochemically etched flow passages. The cross-sectional shape of these flow passages is typically semi-circular, with diameters ranging from 1–1.5 mm [60]. The PCHEs are compact, high-power density heat exchangers that are particularly well-suited for high-pressure and high temperature applications [61]. The photochemically etched minichannels are suitable for carrying high-pressure fluids, such as helium and sCO_2. Helium-to-sCO_2 PCHE (made in SS316) was experimentally studied by Katz et al. [60] for nuclear engineering applications where helium could be used to cool graphite fuel elements. This heat exchanger, in addition to high temperature conditions, operates at high pressures on both hot (helium) and cold sides (sCO_2). To this end, PCHEs are a preferred choice to contain both pressurized streams and facilitate efficient heat exchange. A helium-to-helium (He-He) PCHE was experimentally tested by Chen et al. [62] for nuclear engineering applications involving an intermediate heat exchanger for employment in very-high temperature gas-cooled reactors (Figure 15.12). The inlet conditions of the hot and cold sides were 2.7 MPa/802°C and 2.7 MPa/464°C, respectively. In such heat exchangers, the overall heat transfer coefficient depends on the convective heat transfer resistances on the two fluids' sides and the conduction heat transfer resistance due to the material

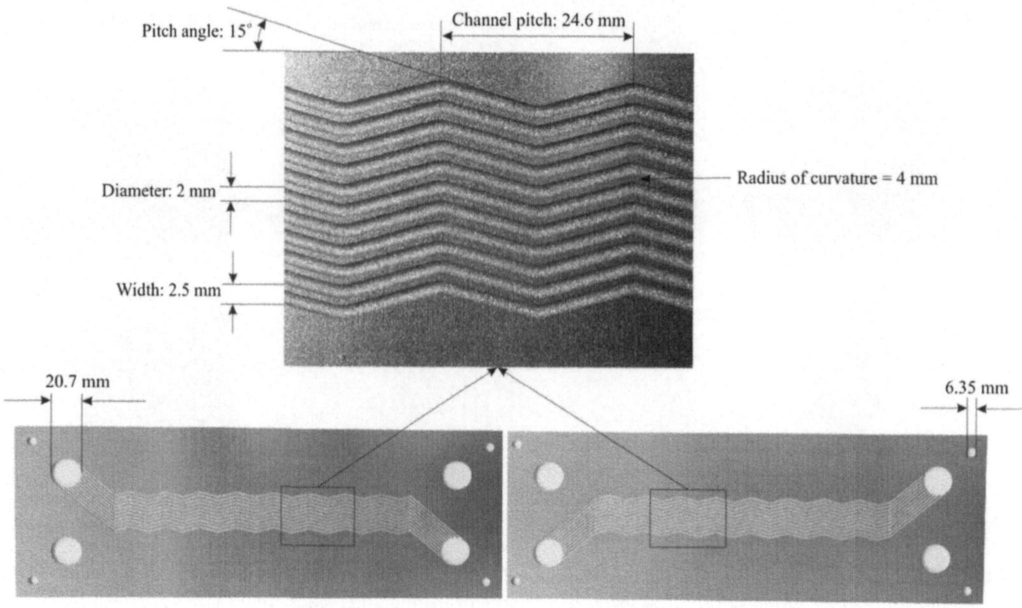

FIGURE 15.12 Printed circuit heat exchanger plates with chemically etched flow channels in zig-zag patterns for helium-helium heat transfer; plates made of Alloy 800H [62].

separating the two channels. The He-He PCHE tested in [62] was made in Alloy 800H, and the overall heat transfer coefficients exceeded 1,000 W/m^{2}K. For similar applications, Alloy 617 was used in [63, 64] for temperatures and pressure exceeding 900°C and 3 MPa, respectively. In a similar study, Alloy 617 PCHE (helium-helium) was tested with hot side inlet conditions (helium) of up to 800°C, where PCHE had ten hot and ten cold plates, with 12 flow channels with a semi-circular cross-section for each plate. For the case with an NTU of 3, the counterflow heat exchanger effectiveness was found to be nearly 0.75. PCHEs have also been used for solar salt-to-sCO_2 thermal energy transfer for concentrating solar-thermal-based power generation applications. The hot side included the flow of solar salt with inlet conditions of 560°C and 1 MPa [65]. For a heat exchanger with a design point heat duty of 27 kW, an overall heat transfer coefficient of 2,275 W/m^{2}K was reported.

15.4.6 Some Other Types of High Temperature Heat Exchangers

The prior sub-sections presented popular configurations of heat exchangers that are widely adopted in different industry sectors. In this sub-section, we have focused on some unique heat exchanger concepts, prior to discussing the additively manufactured heat exchangers. A screw heat exchanger was experimentally studied for concentrating solar-thermal application, where phase change material (eutectic mixture of $NaNO_3$ and KNO_3) could be transported during the phase change process [66]. A highly compact (50×100 mm^2 footprint) SiC-based counterflow microchannel heat exchanger was proposed by Kee et al. [67]. Figure 15.13 shows the concept, with typical flow along with conduction within the heat exchanger material, and an SEM view of the highly accurate microchannels with sharp corners. The air-air heat exchanger was tested for an inlet temperature of 750°C on the hot side and an inlet temperature of 30°C on the cold side. The all-ceramic heat exchanger was fabricated by stacking multiple layers with microchannels through the pressure laminated integrated structures (PLIS) technique, which is a low-cost manufacturing method. The heat exchanger facilitated high temperature operations due to the employment of SiC as the material, and the PLIS

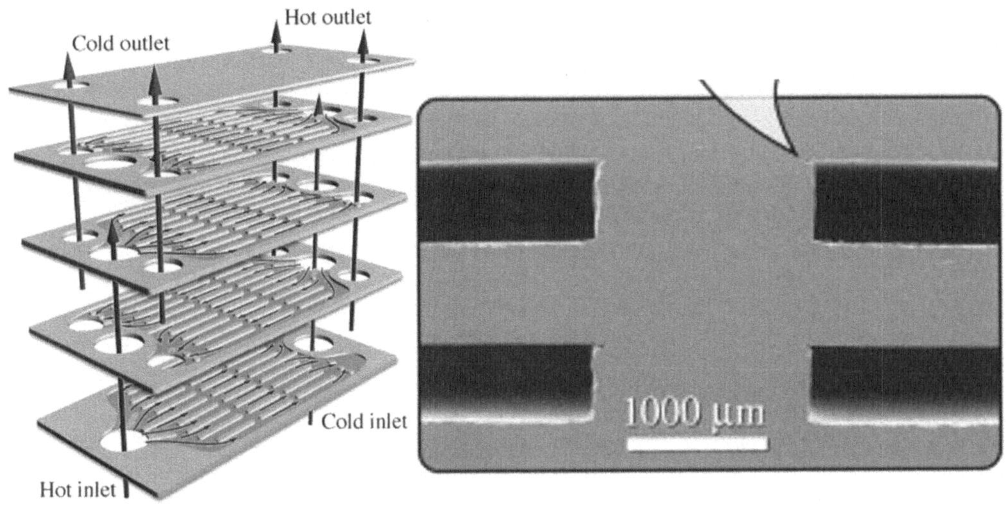

FIGURE 15.13 Highly compact high power-density, high temperature SiC-based PLIS-enabled counterflow microchannel heat exchanger [67].

manufacturing method enabled a highly compact and cost-effective heat exchanger. Heat exchanger effectiveness varied between 0.6 and 0.8 for the test conditions explored.

Robey et al. [68] explored a plate pin-fin heat exchanger for an sCO_2 recuperator, where the sCO_2 inlet conditions on the hot and cold ends were 854°C and 467°C, respectively (Figure 15.14).

In a typical heat exchanger involving two fluids and a wall separating the two, convective heat transfer resistances on the flow sides can be reduced by packing the flow channels with high porosity metal foams. Metal foam-based heat exchangers were evaluated by Muley et al. [69], where heat exchanger effectiveness between 0.7 and 0.95 was demonstrated. Ni-Cr and stainless-steel foams were employed for applications requiring operations up to 650°C. In a recent study by Banerjee et al. [70], alumina foams were employed in both the tube and the annular region of a tube-in-tube

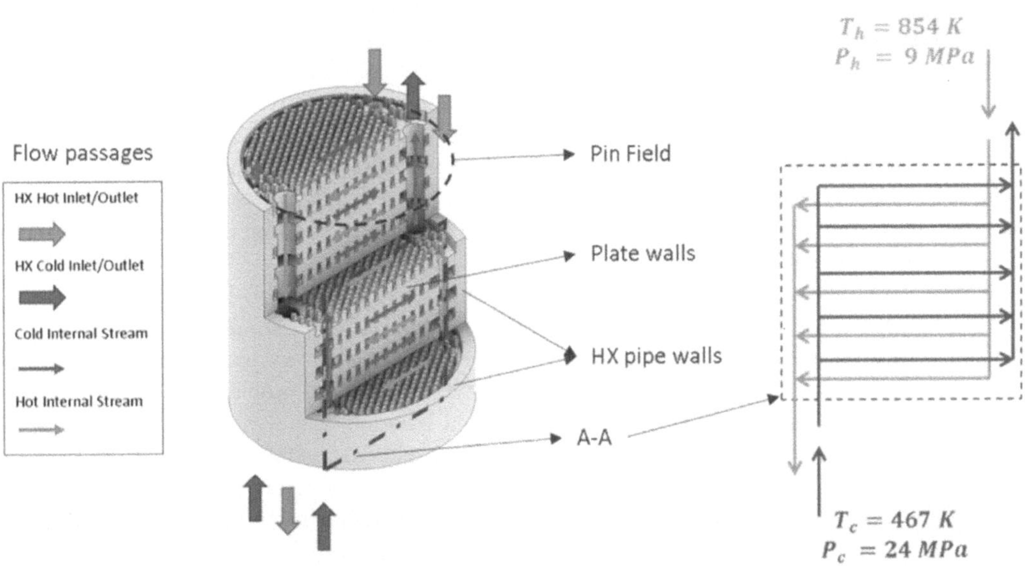

FIGURE 15.14 A plate pin-fin heat exchanger for a sCO_2 recuperator [68] (with permission to reuse content).

type heat exchanger for high temperature (~1,000°C) applications. Metal foams can be promising in high temperature heat exchangers as they offer a high surface-area-to-volume ratio and enable a lightweight and compact heat exchanger, resulting in high power density. Flow through metal foams' interconnected pores interact with the struts/ligaments, resulting in increased heat transfer and thermal dispersion [71–73]. However, metal foams enabled by the foaming process are commercially available only in fixed pore densities and typically at high porosity levels (>0.9). Some recent innovations in additively manufactured architectured materials [74–81] can be leveraged in the concepts discussed earlier to achieve a high degree of control of heat transfer and pressure drop, resulting in superior thermal-hydraulic performance. The next sub-section provides an overview of recent innovations on the AM front, facilitating the next generation of high-performance heat exchangers.

15.4.7 Additively Manufactured Heat Exchangers

The heat exchanger concepts for high-temperature/high-pressure applications discussed thus far were realized through conventional manufacturing processes (except Section 15.4.6). Such heat exchangers typically employ straight plates, serpentine tubes, chemically etched microchannels on straight plates, and so on. With the advent of metal and ceramic-based additive manufacturing in the past decade or so, the design domain of the heat exchangers has rapidly expanded, where unconventional concepts are demonstrated with high accuracy and high-quality prints. The research emphasis for AM-based heat exchangers is to develop concepts that are high performance, exhibit high-power density, monolith to the extent possible, and reduce surface roughness. Integrated monolith heat exchangers through AM have the potential to cut major costs that accrue in the assembly of conventional heat exchangers through welding, diffusion bonding, and so on. Innovations in material science and manufacturing have enabled high-performance compact heat exchangers for high-temperature/high-pressure applications. This sub-section presents some of the latest studies where AM-based heat exchangers were tested in hostile environments. For a detailed review of AM-based heat exchangers for a wider range of applications (not limited to high temperatures), see [82].

As discussed earlier, a molten salt-to-sCO_2 heat exchanger is an option that could facilitate solar-thermal energy to power generation through the efficient transfer of thermal energy from molten salt-to-sCO_2. Tano et al. [83] developed an additively manufactured molten salt-to-sCO_2 heat exchanger where molten salt was routed through the rectangular channels and the sCO_2 was routed through channels featuring cylindrical pin-fins. The overall concept is shown in Figure 15.15, and it can be seen that such a complex integrated heat exchanger design is only possible with AM. The design effort also included the header carrying sCO_2 at high pressure. A heat exchanger effectiveness of ~0.9 was demonstrated in [83]. An optimization study was carried out by Singh et al. [84], where a ceramic-based heat exchanger configuration was considered for thermal energy transfer from molten salt-to-sCO_2. In a follow-up study from the same research group [84], the SiC-based heat exchanger for molten salt-to-sCO_2 thermal exchange was printed using the binder jetting technique, where the Innovent+ machine (ExOne Inc.) was used [85].

Recently, the Lawrence Livermore National Laboratory (LLNL, USA) [86] explored additive manufacturing of heat exchangers based on triply periodic minimal surface (TPMS) concepts, which offer exceptionally high surface area-to-volume ratio. The heat exchanger development was focused on a chloride salt-to-sCO_2 heat exchanger, where the critical problem of corrosion due to molten salt was addressed, in addition to operability at high temperatures and high pressures on both hot and cold sides. A wide range of high-temperature material candidates were explored, where binder jet enabled ZrB_2-$MoSi_2$ was considered to be promising in terms of corrosion resistance. A parallel effort at LLNL [87] explored binder jetting additive manufacturing of silicon carbide (SiC), zirconium boride (ZrB_2), and tungsten carbide (WC). Material extrusion-based additive manufacturing of alumina was explored by Hadian et al. [88], where intertwined hot and cold passages (unmixed) were printed, including the flow supply ducts.

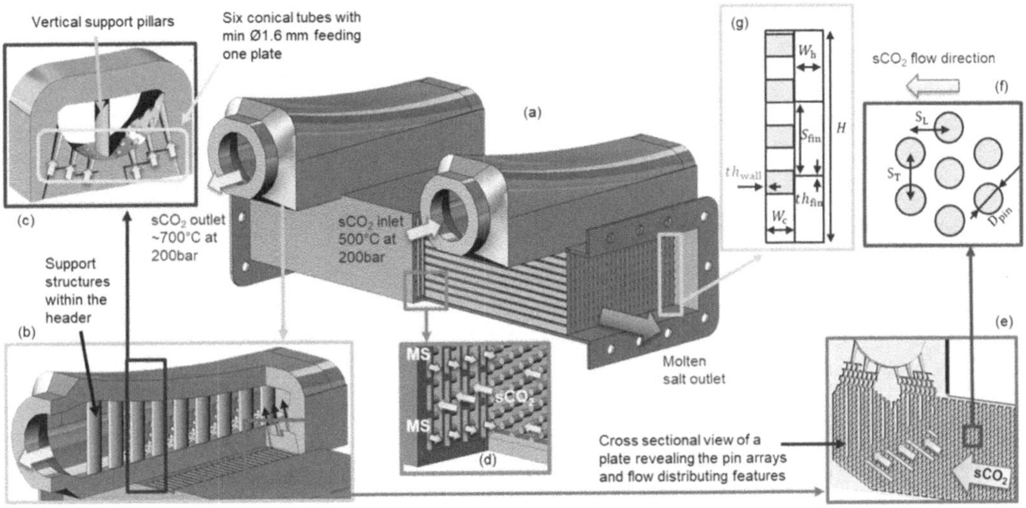

FIGURE 15.15 The design of an additively manufactured molten salt-to-sCO_2 heat exchanger [83] (open access).

Recently, an additively manufactured recuperator was designed and tested for sCO_2 inlet conditions at hot and cold sides, 800°C/250 bar and 300°C/80 bar, respectively [89]. The laser powder bed fusion (LPBF) technique was used with Haynes 282 as the material. For a sCO_2 recuperator, the nickel-based superalloy Mar-M247 was explored for inlet temperatures of 800°C, where the LPBF method was used [90].

In addition to novel heat exchanger designs, conventional heat exchanger configurations can also be potentially improved through the incorporation of AM-based enhanced heat transfer concepts. A parallel plate heat exchanger design was improved by incorporating additively manufactured 3D reticulated Kelvin unit cells in Si-SiC, into the space between parallel plates for both hot and cold sides [91]. Such concepts can be used for gas-to-gas heat exchange. A crossflow heat exchanger in compact form (Figure 15.16) was investigated by da Silva et al. [92], as an alternative to PCHEs, which require extensive diffusion bonding between the plates on which microchannels are etched. AM removes the expensive intermediate fabrication steps which help in the curtailment of the cost and at the same time increase the heat exchanger's reliability.

Additive manufacturing also enables intelligent manifold designs that could feed the hot and cold channels. A series of experimental investigations was conducted by Ohadi and co-researchers, where AM-based manifold microchannel heat exchangers were tested [93, 94].

15.4.7.1 Some Challenges in the Adoption of AM-Based Heat Exchangers

The overview of the AM-based heat exchanger concept demonstrates its potential to replace conventional heat exchangers by allowing complex topologies, enhanced heat transfer concepts, and different manifold schemes, along with a wide range of choices of high-temperature materials. AM-based designs tend to have higher power density due to their compact nature, along with potentially lower-cost manufacturing due to the elimination of intermediate bonding steps that are present in conventional manufacturing. The benefits of additive manufacturing should also be weighed and evaluated against existing challenges. AM-based parts are inherently rough [81], which could cause undesired hot or cold spots, increase pumping power penalty, and create issues with bonding two parts together. In addition to roughness, AM parts tend to have microstructural porosity, which is an undesirable attribute for high-pressure applications [95, 96]. Applications where the pressure differential between the hot and cold sides requires additional precautions if the AM route is considered for the integrated heat exchanger fabrication. Such application includes the particle-to-sCO_2

FIGURE 15.16 Crossflow compact heat exchanger printed in 316L stainless steel [92].

heat exchanger, where particle channels are not pressurized while the sCO_2 channels are at typical pressures of ~20 MPa. For such scenarios, a combination of additive manufacturing and conventional manufacturing can be adopted [52]. Furthermore, the quality of end AM parts is dependent on the process parameter, the AM process, and the chosen material. Some other types of defects include cracks, lack of fusion, surface-connected porosity and so on, which are undesirable attributes for hostile environment and cyclic thermo-mechanical load applications [97, 98].

The design and development of heat exchangers involves a wide range of factors, as discussed in Section 15.1.1. One such aspect is the cost of heat exchanger fabrication and its life. Although AM-based concepts typically eliminate intermediate bonding steps due to their integral nature, the AM processes, particularly the ones required for high-temperature materials, tend to be expensive. The benefits offered by AM-based solutions should also be evaluated in terms of their cost through the presentation of \$/UA or \$/kW_{th}.

15.5 THE ROLE OF MACHINE LEARNING IN NEXT-GENERATION HEAT EXCHANGER DEVELOPMENT

Machine learning (ML)-based methods are rapidly gaining prominence in thermal fluid science. Heat exchangers are complex devices with several parameters required to define their geometry, operating conditions, and thermal-fluid performance. In the design of heat exchangers, the thermal system under consideration is large, with coupled parameters determining performance. The use of computational fluid dynamics to simulate the entire thermal-fluidic system can be cost prohibitive [99], as well as limited in its predictive capabilities. When heat exchangers are in operation, typically, point-based measurements are monitored to determine the operating conditions and system performance, such as heat capacity ratio on the hot and cold sides, inlet/outlet temperatures of the hot and cold streams, local pressures, and so on. The health monitoring of heat exchangers is highly complex and point-based measurements are not sufficient to capture the true picture. The employment of sophisticated intrusive or non-intrusive measurement devices to monitor the health of these devices is costly and impractical. To this end, an ML-based approach can play a crucial role in both the design/development and the health monitoring of heat exchangers in operation. Furthermore, ML-based approaches can be used to predict the system life, based on datasets on the corrosion behavior of particular solid material against a particular working fluid and operating conditions.

ML techniques are not new to the world of heat transfer: early evidences of the employment of artificial neural networks (ANNs) can be found in the 1990s for a wide range of applications, including two-dimensional convective heat transfer coefficient determination [100, 101], thermodynamics of regenerator beds [102], and heat exchanger performance [103], to name a few.

Recently, physics-informed neural networks (PINNs) have been employed to classic heat transfer problems by Cai et al. [104], and have been extended to a variety of flow and heat transfer problems [105, 106]. A physics-informed deep learning and transfer learning approach has been employed for heat exchanger design by Wu et al. [107]. A standard PINN architecture is shown in Figure 15.17(a).

PINN and PINN-TL (transfer learning) architecture for optimization is shown in Figure 15.17(b) [107]. An integrated deep learning and reinforcement learning with heat transfer solver was implemented for shape optimization for reduced pressure drop and enhanced heat transfer, where the starting geometry was a simple rectangle [108]. ANN was recently employed to design heat exchangers with micro-finned tubes [109]. In addition to advancing and rapidly accelerating the design and development of heat exchangers, ML-based methods are also used in predicting the

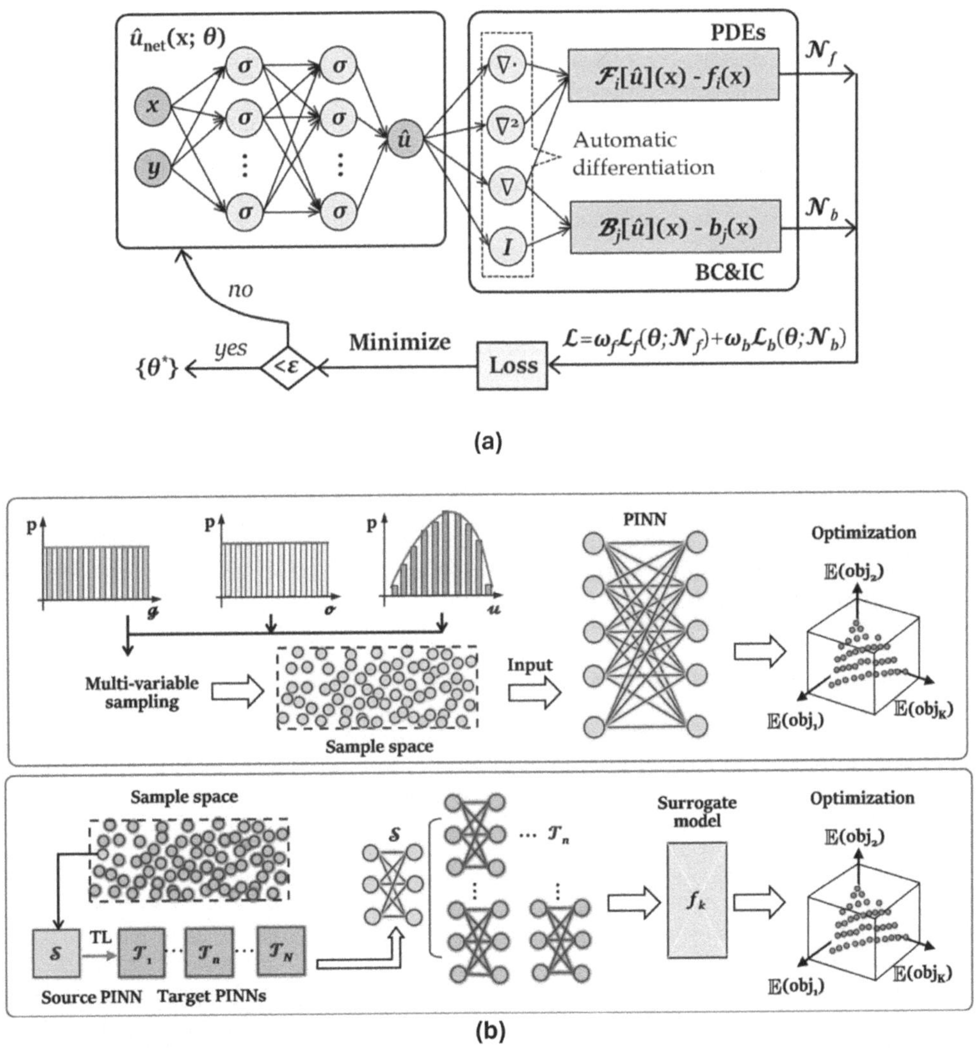

FIGURE 15.17 Standard architecture, optimization architecture and transfer learning method for PINNs. (a) Standard PINN architecture; (b) PINN architecture for optimization and the PINN-TL method for optimization [107].

fouling factor in heat exchangers. Recently, the ML approach was used to determine the fouling factor of heat exchangers based on the operating environment and constructing variables [110]. The prior knowledge of fouling can prevent significant downtime through the timely scheduling of maintenance when required.

15.6. CONCLUDING REMARKS

This chapter deals with modern high temperature heat exchangers with an emphasis on the rapidly evolving approach taken in the design, development, performance, and health monitoring of heat exchangers operating in hostile environments. The conventional manufacturing methods are mature technologies that result in the highly accurate fabrication of heat exchangers at lower costs of scale. Such heat exchangers involve relatively simpler enhanced heat transfer concepts, along with straightforward interactions between the hot and cold streams, such as counterflow, crossflow, and so on. With the advent of metal and ceramic-based additive manufacturing, highly complex and sophisticated enhanced heat transfer concepts can be employed to fabricate highly compact heat exchangers, resulting in high power density. Further, the AM-based design approach is targeted toward integrated concepts, where intermediate bonding steps between the sub-components of heat exchangers could be eliminated, resulting in direct cost savings in manufacturing and an increase in reliability during operations at high temperatures/high pressures. With the evolving and growing AM market, the costs of fabricated heat exchanger devices are expected to reduce in the near future with technology maturation. Ceramic-based AM technologies are expected to play a crucial role in future high temperature heat exchangers due to the strong demand for high-temperature operations, which helps to achieve high overall system efficiency. Machine learning is poised to dominate the design, development, and performance monitoring of heat exchangers. Next-generation heat exchangers demand a highly interdisciplinary approach between concentrated synergistic efforts from disciplines such as enhanced heat transfers, fluid mechanics, computational fluid dynamics, material science, mechanics of materials, advanced manufacturing, techno-economics, and ML-artificial intelligence. This chapter is expected to serve as a reference tool for such efforts in the near future.

REFERENCES

1. Heat Exchanger - An Overview I ScienceDirect Topics, (n.d.). https://www.sciencedirect.com/topics/engineering/heat-exchanger (accessed June 4, 2024).
2. T.L. Bergman, *Fundamentals of Heat and Mass Transfer*, John Wiley & Sons, 2011.
3. R.K. Shah, D.P. Sekulic, *Fundamentals of Heat Exchanger Design*, John Wiley & Sons, 2003.
4. M. Kolluri, P. ten Pierick, T. Bakker, Characterization of high temperature tensile and creep–fatigue properties of Alloy 800H for intermediate heat exchanger components of (V)HTRs, *Nuclear Engineering and Design* 284 (2015): 38–49. https://doi.org/10.1016/j.nucengdes.2014.12.017.
5. M. Ali, A. Ul-Hamid, L.M. Alhems, A. Saeed, Review of common failures in heat exchangers – Part I: Mechanical and elevated temperature failures, *Engineering Failure Analysis* 109 (2020): 104396. https://doi.org/10.1016/j.engfailanal.2020.104396.
6. M. Mazaheri, F. Djavanroodi, K.M. Nikbin, Creep life assessment of an overheated 9Cr-1Mo steel tube, *International Journal of Pressure Vessels and Piping* 87 (2010): 746–752. https://doi.org/10.1016/j.ijpvp.2010.08.013.
7. K. Li, Y. Zeng, Corrosion of heat exchanger materials in co-combustion thermal power plants, *Renewable and Sustainable Energy Reviews* 161 (2022): 112328. https://doi.org/10.1016/j.rser.2022.112328.
8. H. Sun, J. Wang, Z. Li, P. Zhang, X. Su, Corrosion behavior of 316SS and Ni-based alloys in a ternary NaCl-KCl-MgCl2 molten salt, *Solar Energy* 171 (2018): 320–329. https://doi.org/10.1016/j.solener.2018.06.094.
9. K. Ravindranath, N. Tanoli, H. Gopal, Failure investigation of brass heat exchanger tube, *Engineering Failure Analysis* 26 (2012): 332–336. https://doi.org/10.1016/j.engfailanal.2012.07.018.

10. C.K. Ho, M. Carlson, K.J. Albrecht, Z. Ma, S. Jeter, C.M. Nguyen, Evaluation of alternative designs for a high temperature particle-to-sCO2 heat exchanger, *Journal of Solar Energy Engineering* 141 (2019). https://doi.org/10.1115/1.4042225.
11. M.D. Carlson, K.J. Albrecht, C.K. Ho, H.F. Laubscher, F. Alvarez, High-temperature particle heat exchanger for sCO2 power cycles [Award 30342], Sandia National Lab. (SNL-NM), Albuquerque, NM (United States), 2020. https://doi.org/10.2172/1817287.
12. V. Kain, K. Chandra, B.P. Sharma, Failure of carbon steel tubes in a fluidized bed combustor, *Engineering Failure Analysis* 15 (2008): 182–187. https://doi.org/10.1016/j.engfailanal.2006.12.009.
13. W.G. Schuetzenduebel, *Steam generators for high-temperature gas-cooled reactor plants in the U.S.A.*, Gulf General Atomic, Inc., San Diego, CA, 1971. https://doi.org/10.2172/4053938.
14. T. Takeda, K. Kunitomi, T. Horie, K. Iwata, Feasibility study on the applicability of a diffusion-welded compact intermediate heat exchanger to next-generation high temperature gas-cooled reactor, *Nuclear Engineering and Design* 168 (1997): 11–21. https://doi.org/10.1016/S0029-5493(96)01361-1.
15. D. Aquaro, M. Pieve, High temperature heat exchangers for power plants: Performance of advanced metallic recuperators, *Applied Thermal Engineering* 27 (2007): 389–400. https://doi.org/10.1016/j.applthermaleng.2006.07.030.
16. G. Xiao, T. Yang, H. Liu, D. Ni, M.L. Ferrari, M. Li, Z. Luo, K. Cen, M. Ni, Recuperators for micro gas turbines: A review, *Applied Energy* 197 (2017): 83–99. https://doi.org/10.1016/j.apenergy.2017.03.095.
17. A. Ohji, M. Haraguchi, 2 - Steam turbine cycles and cycle design optimization: The Rankine cycle, thermal power cycles, and IGCC power plants, in: T. Tanuma (Ed.), *Advances in steam turbines for modern power plants*, Woodhead Publishing, 2017, 11–40. https://doi.org/10.1016/B978-0-08-100314-5.00002-6.
18. A. Sharma, A.K. Shukla, O. Singh, M. Sharma, Recent advances in gas/steam power cycles for concentrating solar power, *International Journal of Ambient Energy* 43 (2022): 4716–4727. https://doi.org/10.1080/01430750.2021.1919552.
19. C. Sánchez, R. Juárez, J. Sanz, M. Perlado, Design and analysis of helium Brayton power cycles for HiPER reactor, *Fusion Engineering and Design* 88 (2013): 2679–2683. https://doi.org/10.1016/j.fusengdes.2012.12.007.
20. L.-G. Chen, J.-L. Zheng, F.-R. Sun, C. Wu, Power, power density and efficiency optimization for a closed cycle helium turbine nuclear power plant, *Energy Conversion and Management* 44 (2003): 2393–2401. https://doi.org/10.1016/S0196-8904(03)00008-6.
21. J.S. Kwon, S. Son, J.Y. Heo, J.I. Lee, Compact heat exchangers for supercritical CO2 power cycle application, *Energy Conversion and Management* 209 (2020): 112666. https://doi.org/10.1016/j.enconman.2020.112666.
22. P. Zhang, T. Ma, W.-D. Li, G.-Y. Ma, Q.-W. Wang, Design and optimization of a novel high temperature heat exchanger for waste heat cascade recovery from exhaust flue gases, *Energy* 160 (2018): 3–18. https://doi.org/10.1016/j.energy.2018.06.216.
23. B. Jiang, D. Xia, H. Guo, L. Xiao, H. Qu, X. Liu, Efficient waste heat recovery system for high-temperature solid particles based on heat transfer enhancement, *Applied Thermal Engineering* 155 (2019): 166–174.
24. T. Neises, C. Turchi, A comparison of supercritical carbon dioxide power cycle configurations with an emphasis on CSP applications, *Energy Procedia* 49 (2014): 1187–1196. https://doi.org/10.1016/j.egypro.2014.03.128.
25. C.S. Turchi, Z. Ma, T.W. Neises, M.J. Wagner, Thermodynamic study of advanced supercritical carbon dioxide power cycles for concentrating solar power systems, *Journal of Solar Energy Engineering* 135 (2013). https://doi.org/10.1115/1.4024030.
26. S.A. Wright, P.S. Pickard, R. Fuller, R.F. Radel, M.E. Vernon, Supercritical CO2 Brayton cycle power generation development program and initial test results, in: *American Society of Mechanical Engineers Digital Collection*, 2010, 573–583. https://doi.org/10.1115/POWER2009-81081.
27. Z. Ma, J. Martinek, Fluidized-bed heat transfer modeling for the development of particle/supercritical-CO2 heat exchanger, *Energy Sustainability* 5 (2017): 1–2.
28. K.J. Albrecht, C.K. Ho, Design and operating considerations for a shell-and-plate, moving packed-bed, particle-to-sCO2 heat exchanger, *Solar Energy* 178 (2019): 331–340. https://doi.org/10.1016/j.solener.2018.11.065.
29. V. Ponyavin, Y. Chen, T. Mohamed, M. Trabia, A.E. Hechanova, M. Wilson, Design of a compact ceramic high-temperature heat exchanger and chemical decomposer for hydrogen production, *Heat Transfer Engineering* 33 (2012): 853–870. https://doi.org/10.1080/01457632.2012.654446.

30. J.-C. Han, Fundamental gas turbine heat transfer, *Journal of Thermal Science and Engineering Applications* 5 (2013). https://doi.org/10.1115/1.4023826.
31. J.-C. Han, Advanced cooling in gas turbines 2016 Max Jakob Memorial Award Paper, *Journal of Heat Transfer* 140 (2018). https://doi.org/10.1115/1.4039644.
32. R.J. Goldstein, Film cooling, in: T.F. Irvine, J.P. Hartnett (Eds.), *Advances in heat transfer*, Elsevier, 1971, 321–379. https://doi.org/10.1016/S0065-2717(08)70020-0.
33. A. Sommers, Q. Wang, X. Han, C. T'Joen, Y. Park, A. Jacobi, Ceramics and ceramic matrix composites for heat exchangers in advanced thermal systems—A review, *Applied Thermal Engineering* 30 (2010): 1277–1291. https://doi.org/10.1016/j.applthermaleng.2010.02.018.
34. Y. Lee, J.I. Lee, Structural assessment of intermediate printed circuit heat exchanger for sodium-cooled fast reactor with supercritical CO2 cycle, *Annals of Nuclear Energy* 73 (2014): 84–95. https://doi.org/10.1016/j.anucene.2014.06.022.
35. K.J. Albrecht, M.D. Carlson, C.K. Ho, Integration, control, and testing of a high-temperature particle-to-sCO2 heat exchanger, *AIP Conference Proceedings* 2126 (2019): 030001. https://doi.org/10.1063/1.5117513.
36. K.J. Albrecht, C.K. Ho, High-temperature flow testing and heat transfer for a moving packed-bed particle/sCO2 heat exchanger, *AIP Conference Proceedings* 2033 (2018): 040003. https://doi.org/10.1063/1.5067039.
37. P.J. Maziasz, B.A. Pint, J.P. Shingledecker, N.D. Evans, Y. Yamamoto, K.L. More, E. Lara-Curzio, Advanced alloys for compact, high-efficiency, high-temperature heat-exchangers, *International Journal of Hydrogen Energy* 32 (2007): 3622–3630. https://doi.org/10.1016/j.ijhydene.2006.08.018.
38. L. Chordia, M.A. Portnoff, E. Green, *High temperature heat exchanger design and fabrication for systems with large pressure differentials*, Thar Energy, LLC, Pittsburgh, PA, 2017. https://doi.org/10.2172/1349235.
39. C. Zhang, V. Gümmer, High temperature heat exchangers for recuperated rotorcraft powerplants, *Applied Thermal Engineering* 154 (2019): 548–561. https://doi.org/10.1016/j.applthermaleng.2019.03.119.
40. S. Mahmoudinezhad, M. Sadi, H. Ghiasirad, A. Arabkoohsar, A comprehensive review on the current technologies and recent developments in high-temperature heat exchangers, *Renewable and Sustainable Energy Reviews* 183 (2023): 113467. https://doi.org/10.1016/j.rser.2023.113467.
41. C.F. McDonald, The role of the ceramic heat exchanger in energy and resource conservation, *Journal of Engineering for Power* 102 (1980): 303–315. https://doi.org/10.1115/1.3230253.
42. H. Wang, R.N. Singh, Thermal shock behaviour of ceramics and ceramic composites, *International Materials Reviews* 39 (1994): 228–244. https://doi.org/10.1179/imr.1994.39.6.228.
43. Thermal Shock and Fatigue Behavior of Ceramic Tubes - ProQuest, (n.d.). https://www.proquest.com/docview/304019755?pq-origsite=gscholar&fromopenview=true&sourcetype=Dissertations%20&%20Theses (accessed June 9, 2024).
44. R. Bayaniahangar, I. Okoh, K. Nawaz, J. Cesarano, S. Bigham, Toward extreme high-temperature supercritical CO2 power cycles: Leakage characterization of ceramic 3D-printed heat exchangers, *Additive Manufacturing* 54 (2022): 102783. https://doi.org/10.1016/j.addma.2022.102783.
45. Z. Wang, Y. Liu, H. Zhang, J. Jiang, T. Lin, X. Liu, Z. Huang, Joining of SiC ceramics using the Ni-Mo filler alloy for heat exchanger applications, *Journal of the European Ceramic Society* 41 (2021): 7533–7542. https://doi.org/10.1016/j.jeurceramsoc.2021.07.056.
46. X. Zhang, H. Keramati, M. Arie, F. Singer, R. Tiwari, A. Shooshtari, M. Ohadi, Recent developments in high temperature heat exchangers: A review, *Frontiers in Heat and Mass Transfer* 11 (2018). https://doi.org/10.5098/hmt.11.18.
47. J. Haunstetter, V. Dreißigacker, S. Zunft, Ceramic high temperature plate fin heat exchanger: Experimental investigation under high temperatures and pressures, *Applied Thermal Engineering* 151 (2019): 364–372. https://doi.org/10.1016/j.applthermaleng.2019.02.015.
48. H.H.S. Villanueva, P.E.B. de Mello, Heat transfer and pressure drop correlations for finned plate ceramic heat exchangers, *Energy* 88 (2015): 118–125. https://doi.org/10.1016/j.energy.2015.04.017.
49. P.E.B. de Mello, H.H.S. Villanueva, S. Scuotto, G.H.B. Donato, F. dos S. Ortega, Heat transfer, pressure drop and structural analysis of a finned plate ceramic heat exchanger, *Energy* 120 (2017): 597–607. https://doi.org/10.1016/j.energy.2016.11.113.
50. J. Schulte-Fischedick, V. Dreißigacker, R. Tamme, An innovative ceramic high temperature plate-fin heat exchanger for EFCC processes, *Applied Thermal Engineering* 27 (2007): 1285–1294. https://doi.org/10.1016/j.applthermaleng.2006.11.007.

51. C. Luzzatto, A. Morgana, S. Chaudourne, T. O'Doherty, G. Sorbie, A new concept composite heat exchanger to be applied in high-temperature industrial processes, *Applied Thermal Engineering* 17 (1997): 789–797. https://doi.org/10.1016/S1359-4311(96)00060-9.
52. Y. Aider, I. Kaur, A. Mishra, L. Li, H. Cho, J. Martinek, Z. Ma, P. Singh, Heat transfer characteristics of particle and air flow through additively manufactured lattice frame material based on octet-shape topology, *Journal of Solar Energy Engineering* 145 (2023). https://doi.org/10.1115/1.4062196.
53. C. Wu, H. Yang, X. He, C. Hu, L. Yang, H. Li, Principle, development, application design and prospect of fluidized bed heat exchange technology: Comprehensive review, *Renewable and Sustainable Energy Reviews* 157 (2022): 112023. https://doi.org/10.1016/j.rser.2021.112023.
54. P. Basu, Heat transfer in high temperature fast fluidized beds, *Chemical Engineering Science* 45 (1990): 3123–3136. https://doi.org/10.1016/0009-2509(90)80058-M.
55. H. Wang, X. Lu, W. Zhang, Q. Wang, J. Chen, Y. Kang, W. Kuang, S. Xin, C. Liu, X. Xie, Study on heat transfer characteristics of the high temperature reheater tube panel in a 300 MW CFB boiler with fluidized bed heat exchanger, *Applied Thermal Engineering* 81 (2015): 262–270. https://doi.org/10.1016/j.applthermaleng.2015.02.025.
56. S. Warerkar, S. Schmitz, J. Goettsche, B. Hoffschmidt, M. Reißel, R. Tamme, Air-sand heat exchanger for high-temperature storage, *Journal of Solar Energy Engineering* 133 (2011). https://doi.org/10.1115/1.4003583.
57. D.C. Miller, C.J. Pfutzner, G.S. Jackson, Heat transfer in counterflow fluidized bed of oxide particles for thermal energy storage, *International Journal of Heat and Mass Transfer* 126 (2018): 730–745. https://doi.org/10.1016/j.ijheatmasstransfer.2018.05.165.
58. K.J. Albrecht, C.K. Ho, Heat transfer models of moving packed-bed particle-to-sCO2 heat exchangers, *Journal of Solar Energy Engineering* 141 (2018). https://doi.org/10.1115/1.4041546.
59. W.J. Arthur-Arhin, J.R. Fosheim, K.J. Brewster, D.A. Madden, K.J. Albrecht, G.S. Jackson, Demonstration of a multi-channel fluidized bed particle–supercritical carbon dioxide heat exchanger for concentrating solar applications, *Applied Thermal Engineering* 248 (2024): 123242. https://doi.org/10.1016/j.applthermaleng.2024.123242.
60. A. Katz, S.R. Aakre, M.H. Anderson, D. Ranjan, Experimental investigation of pressure drop and heat transfer in high temperature supercritical CO2 and helium in a printed-circuit heat exchanger, *International Journal of Heat and Mass Transfer* 171 (2021): 121089. https://doi.org/10.1016/j.ijheatmasstransfer.2021.121089.
61. R. de la Torre, J.-L. François, C.-X. Lin, Optimization and heat transfer correlations development of zigzag channel printed circuit heat exchangers with helium fluids at high temperature, *International Journal of Thermal Sciences* 160 (2021): 106645. https://doi.org/10.1016/j.ijthermalsci.2020.106645.
62. M. Chen, X. Sun, R.N. Christensen, I. Skavdahl, V. Utgikar, P. Sabharwall, Pressure drop and heat transfer characteristics of a high-temperature printed circuit heat exchanger, *Applied Thermal Engineering* 108 (2016): 1409–1417. https://doi.org/10.1016/j.applthermaleng.2016.07.149.
63. S.K. Mylavarapu, X. Sun, R.N. Christensen, R.R. Unocic, R.E. Glosup, M.W. Patterson, Fabrication and design aspects of high-temperature compact diffusion bonded heat exchangers, *Nuclear Engineering and Design* 249 (2012): 49–56. https://doi.org/10.1016/j.nucengdes.2011.08.043.
64. S. Mylavarapu, X. Sun, J. Figley, N. Needler, R. Christensen, Investigation of high-temperature printed circuit heat exchangers for very high temperature reactors, *Journal of Engineering for Gas Turbines and Power* 131 (2009). https://doi.org/10.1115/1.3098425.
65. S.R. Aakre, M.H. Anderson, Pressure drop and heat transfer characteristics of nitrate salt and supercritical CO2 in a diffusion-bonded heat exchanger, *International Journal of Heat and Mass Transfer* 189 (2022): 122691. https://doi.org/10.1016/j.ijheatmasstransfer.2022.122691.
66. V. Zipf, A. Neuhäuser, D. Willert, P. Nitz, S. Gschwander, W. Platzer, High temperature latent heat storage with a screw heat exchanger: Design of prototype, *Applied Energy* 109 (2013): 462–469. https://doi.org/10.1016/j.apenergy.2012.11.044.
67. R.J. Kee, B.B. Almand, J.M. Blasi, B.L. Rosen, M. Hartmann, N.P. Sullivan, H. Zhu, A.R. Manerbino, S. Menzer, W.G. Coors, J.L. Martin, The design, fabrication, and evaluation of a ceramic counter-flow microchannel heat exchanger, *Applied Thermal Engineering* 31 (2011): 2004–2012. https://doi.org/10.1016/j.applthermaleng.2011.03.009.
68. E. Robey, S. Ramesh, A.S. Sabau, A. Abdoli, J. Black, D. Straub, J. Yip, Design optimization of an additively manufactured prototype recuperator for supercritical CO2 power cycles, *Energy* 251 (2022): 123961. https://doi.org/10.1016/j.energy.2022.123961.
69. A. Muley, C. Kiser, B. Sundén, R.K. Shah, Foam heat exchangers: A technology assessment, *Heat Transfer Engineering* 33 (2012): 42–51. https://doi.org/10.1080/01457632.2011.584817.

70. A. Banerjee, R. Bala Chandran, J.H. Davidson, Experimental investigation of a reticulated porous alumina heat exchanger for high temperature gas heat recovery, *Applied Thermal Engineering* 75 (2015): 889–895. https://doi.org/10.1016/j.applthermaleng.2014.10.033.
71. P. Singh, K. Nithyanandam, R.L. Mahajan, An experimental and numerical investigation of forced convection in high porosity aluminum foams subjected to jet array impingement in channel-flow, *International Journal of Heat and Mass Transfer* 149 (2020): 119107. https://doi.org/10.1016/j.ijheatmasstransfer.2019.119107.
72. P. Singh, K. Nithyanandam, M. Zhang, R.L. Mahajan, The effect of metal foam thickness on jet array impingement heat transfer in high-porosity aluminum foams, *Journal of Heat Transfer* 142 (2020). https://doi.org/10.1115/1.4045640.
73. K. Nithyanandam, P. Singh, Enhanced forced convection through thin metal foams placed in rectangular ducts, *Heat Transfer Engineering* 44 (2023): 837–852. https://doi.org/10.1080/01457632.2022.2102960.
74. I. Kaur, P. Singh, Flow and thermal transport through unit cell topologies of cubic and octahedron families, *International Journal of Heat and Mass Transfer* 158 (2020): 119784. https://doi.org/10.1016/j.ijheatmasstransfer.2020.119784.
75. Y. Aider, I. Kaur, H. Cho, P. Singh, Periodic heat transfer characteristics of additively manufactured lattices, *International Journal of Heat and Mass Transfer* 189 (2022): 122692. https://doi.org/10.1016/j.ijheatmasstransfer.2022.122692.
76. I. Kaur, Y. Aider, K. Nithyanandam, P. Singh, Thermal-hydraulic performance of additively manufactured lattices for gas turbine blade trailing edge cooling, *Applied Thermal Engineering* 211 (2022): 118461. https://doi.org/10.1016/j.applthermaleng.2022.118461.
77. I. Kaur, P. Singh, Endwall heat transfer characteristics of octahedron family lattice-frame materials, *International Communications in Heat and Mass Transfer* 127 (2021): 105522. https://doi.org/10.1016/j.icheatmasstransfer.2021.105522.
78. I. Kaur, P. Singh, Direct pore-scale simulations of fully periodic unit cells of different regular lattices, *Journal of Heat Transfer* 144 (2022). https://doi.org/10.1115/1.4053204.
79. I. Kaur, Y. Aider, H. Cho, P. Singh, Granular flow in novel octet shape–based lattice frame material, *Journal of Solar Energy Engineering* 146 (2023). https://doi.org/10.1115/1.4064018.
80. I. Kaur, P. Singh, Critical evaluation of additively manufactured metal lattices for viability in advanced heat exchangers, *International Journal of Heat and Mass Transfer* 168 (2021): 120858. https://doi.org/10.1016/j.ijheatmasstransfer.2020.120858.
81. I. Kaur, P. Singh, Effects of inherent surface roughness of additively manufactured lattice frame material on flow and thermal transport, *International Journal of Heat and Mass Transfer* 209 (2023): 124077. https://doi.org/10.1016/j.ijheatmasstransfer.2023.124077.
82. I. Kaur, P. Singh, State-of-the-art in heat exchanger additive manufacturing, *International Journal of Heat and Mass Transfer* 178 (2021): 121600. https://doi.org/10.1016/j.ijheatmasstransfer.2021.121600.
83. I.-N. Tano, E. Rasouli, T. Ziev, Z. Wu, N. Lamprinakos, J. Seo, L. Schulze Balhorn, P. Vaishnav, A. Rollett, V. Narayanan, An additively-manufactured molten salt-to-supercritical carbon di-oxide primary heat exchanger for solar thermal power generation – Design and techno-economic performance, *Solar Energy* 234 (2022): 152–169. https://doi.org/10.1016/j.solener.2022.01.056.
84. D. Singh, W. Yu, D.M. France, T.P. Allred, I.-H. Liu, W. Du, B. Barua, M.C. Messner, One piece ceramic heat exchanger for concentrating solar power electric plants, *Renewable Energy* 160 (2020): 1308–1315. https://doi.org/10.1016/j.renene.2020.07.070.
85. W. Du, W. Yu, D.M. France, M. Singh, D. Singh, Additive manufacturing and testing of a ceramic heat exchanger for high-temperature and high-pressure applications for concentrating solar power, *Solar Energy* 236 (2022): 654–665. https://doi.org/10.1016/j.solener.2022.03.046.
86. J. Kelly, J. Haslam, L. Finkenauer, P. Roy, J. Stolaroff, D. Nguyen, M. Ross, A. Hoff, S. Raiman, D. Solejmanovic, B. Pint, Additive manufacturing of corrosion resistant UHTC materials for chloride salt-to-sCO2 Brayton cycle heat exchangers, 2021. https://doi.org/10.2172/1787194.
87. B. Carpman, Compositional exploration of additively manufactured ultra-high-temperature ceramic composites for use in concentrated solar power plant heat exchangers, 2022. https://doi.org/10.2172/1891216.
88. A. Hadian, B. Morath, M. Biedermann, M. Meboldt, F. Clemens, Selected design rules for material extrusion-based additive manufacturing of alumina based nozzles and heat exchangers considering limitations in printing, debinding, and sintering, *Additive Manufacturing* 75 (2023): 103719. https://doi.org/10.1016/j.addma.2023.103719.

89. S. Das, E. Rasouli, T. Ziev, N. Lamprinakos, J. Seo, A. Rollett, P. Vaishnav, V. Narayanan, Design and techno economic optimization of an additively manufactured compact heat exchanger for high temperature and high pressure applications, *Applied Thermal Engineering* 245 (2024): 122778. https://doi.org/10.1016/j.applthermaleng.2024.122778.
90. M. Marshall, M. Akbari, J.-C. Zhao, K. Hoopes, Design of an additively manufactured recuperator with 800 °C inlet temperature for sCO2 power cycle application, *Journal of Engineering for Gas Turbines and Power* 145 (2022). https://doi.org/10.1115/1.4055723.
91. M. Pelanconi, S. Zavattoni, L. Cornolti, R. Puragliesi, E. Arrivabeni, L. Ferrari, S. Gianella, M. Barbato, A. Ortona, Application of ceramic lattice structures to design compact, high temperature heat exchangers: Material and architecture selection, *Materials* 14 (2021): 3225. https://doi.org/10.3390/ma14123225.
92. R.P.P. da Silva, M.V.V. Mortean, K.V. de Paiva, L.E. Beckedorff, J.L.G. Oliveira, F.G. Brandão, A.S. Monteiro, C.S. Carvalho, H.R. Oliveira, D.G. Borges, V.L. Chastinet, Thermal and hydrodynamic analysis of a compact heat exchanger produced by additive manufacturing, *Applied Thermal Engineering* 193 (2021): 116973. https://doi.org/10.1016/j.applthermaleng.2021.116973.
93. F. Battaglia, X. Zhang, M.A. Arie, A. Shooshtari, A.P. Sarmiento, M. Ohadi, Additive manufacturing of a high temperature, Ni-based superalloy compact heat exchanger: A study on the role of select key printing parameters, *ASME Journal of Heat and Mass Transfer* 145 (2023). https://doi.org/10.1115/1.4056484.
94. X. Zhang, R. Tiwari, A.H. Shooshtari, M.M. Ohadi, An additively manufactured metallic manifold-microchannel heat exchanger for high temperature applications, *Applied Thermal Engineering* 143 (2018): 899–908. https://doi.org/10.1016/j.applthermaleng.2018.08.032.
95. A. Sola, A. Nouri, Microstructural porosity in additive manufacturing: The formation and detection of pores in metal parts fabricated by powder bed fusion, *Journal of Advanced Manufacturing and Processing* 1 (2019): e10021. https://doi.org/10.1002/amp2.10021.
96. B. Zhang, S. Liu, Y.C. Shin, In-process monitoring of porosity during laser additive manufacturing process, *Additive Manufacturing* 28 (2019): 497–505. https://doi.org/10.1016/j.addma.2019.05.030.
97. M.C. Brennan, J.S. Keist, T.A. Palmer, Defects in metal additive manufacturing processes, *The Journal of Materials Engineering and Performance* 30 (2021): 4808–4818. https://doi.org/10.1007/s11665-021-05919-6.
98. Z. Zhou, L. Huang, Y. Shang, Y. Li, L. Jiang, Q. Lei, Causes analysis on cracks in nickel-based single crystal superalloy fabricated by laser powder deposition additive manufacturing, *Materials & Design* 160 (2018): 1238–1249. https://doi.org/10.1016/j.matdes.2018.10.042.
99. M.T. Hughes, G. Kini, S. Garimella, Status, challenges, and potential for machine learning in understanding and applying heat transfer phenomena, *Journal of Heat Transfer* 143 (2021). https://doi.org/10.1115/1.4052510.
100. K. Jambunathan, S.L. Hartle, S. Ashforth-Frost, V.N. Fontama, Evaluating convective heat transfer coefficients using neural networks, *International Journal of Heat and Mass Transfer* 39 (1996): 2329–2332. https://doi.org/10.1016/0017-9310(95)00332-0.
101. S. Sreekanth, H.S. Ramaswamy, S.S. Sablani, S.O. Prasher, A neural network approach for evaluation of surface heat transfer coefficient, *Journal of Food Processing and Preservation* 23 (1999): 329–348. https://doi.org/10.1111/j.1745-4549.1999.tb00389.x.
102. S. Kumar Das, P. Nanda, Use of artificial neural network and leveque analogy for the exergy analysis of regenerator beds, *Chemical Engineering and Processing: Process Intensification* 39 (2000): 113–120. https://doi.org/10.1016/S0255-2701(99)00003-3.
103. G. Díaz, M. Sen, K.T. Yang, R.L. McClain, Simulation of heat exchanger performance by artificial neural networks, *HVAC&R Research* 5 (1999): 195–208. https://doi.org/10.1080/10789669.1999.10391233.
104. S. Cai, Z. Wang, S. Wang, P. Perdikaris, G.E. Karniadakis, Physics-informed neural networks for heat transfer problems, *Journal of Heat Transfer* 143 (2021). https://doi.org/10.1115/1.4050542.
105. D. Jalili, S. Jang, M. Jadidi, G. Giustini, A. Keshmiri, Y. Mahmoudi, Physics-informed neural networks for heat transfer prediction in two-phase flows, *International Journal of Heat and Mass Transfer* 221 (2024): 125089. https://doi.org/10.1016/j.ijheatmasstransfer.2023.125089.
106. R. Laubscher, Simulation of multi-species flow and heat transfer using physics-informed neural networks, *Physics of Fluids* 33 (2021): 087101. https://doi.org/10.1063/5.0058529.
107. Z. Wu, B. Zhang, H. Yu, J. Ren, M. Pan, C. He, Q. Chen, Accelerating heat exchanger design by combining physics-informed deep learning and transfer learning, *Chemical Engineering Science* 282 (2023): 119285. https://doi.org/10.1016/j.ces.2023.119285.

108. H. Keramati, F. Hamdullahpur, M. Barzegari, Deep reinforcement learning for heat exchanger shape optimization, *International Journal of Heat and Mass Transfer* 194 (2022): 123112. https://doi.org/10.1016/j.ijheatmasstransfer.2022.123112.
109. E. Efatinasab, N. Irannezhad, M. Rampazzo, A. Diani, Machine and deep learning driven models for the design of heat exchangers with micro-finned tubes, *Energy and AI* 16 (2024): 100370. https://doi.org/10.1016/j.egyai.2024.100370.
110. S. Hosseini, A. Khandakar, M.E.H. Chowdhury, M.A. Ayari, T. Rahman, M.H. Chowdhury, B. Vaferi, Novel and robust machine learning approach for estimating the fouling factor in heat exchangers, *Energy Reports* 8 (2022): 8767–8776. https://doi.org/10.1016/j.egyr.2022.06.123.

16 Future Trends and Emerging Technologies of Heat Exchangers

Subrata Jana, Dinkar Verma, and Shashikant Gupta

16.1 INTRODUCTION

This chapter explains the advancement of heat exchanger technologies and investigates emerging trends in various industrial applications. Initially with the brief of basic design advancements, focusing on the development of novel materials with superior thermal properties and corrosion resistance. We discuss the integration of smart technologies, including Internet of Things (IoT) and machine learning (ML), which open new avenues for real-time monitoring, control, and predictive maintenance of heat exchangers. It summarizes the benefits of artificial intelligence (AI) and ML integration in revolutionizing heat exchanger design and operation. The chapter then explores the realm of micro-scale and nanoscale heat exchangers, highlighting the benefits of miniaturization and the application of advanced fluids. Renewable energy integration and thermoelectric heat exchangers offer sustainable solutions for energy generation and utilization. Furthermore, the chapter explores biomimicry-inspired designs, drawing inspiration from nature to enhance heat exchange efficiency. Through these discussions, this chapter provides insights into the future of heat exchangers, emphasizing the importance of innovation in addressing the evolving needs of industries and promoting sustainability.[7] For example, the development of graphene-based materials with unparalleled thermal conductivity and robustness exemplifies the paradigm shift in materials science, offering transformative potential for heat exchanger applications across various sectors. Furthermore, the utilization of micro-channel heat exchangers in electronics cooling systems showcases the efficacy of miniaturization in achieving enhanced heat transfer[1] rates and compact system designs. Additionally, the advent of 3D printed heat exchangers, exemplified by intricate lattice structures optimized for specific thermal performance metrics, epitomizes the transformative capability of additive manufacturing in revolutionizing conventional fabrication methodologies.

The transformative impact of integrating AI and machine learning in heat exchanger technology is also considered for discussion. The chapter begins with an introduction to the significance of heat exchangers across industries and the challenges in design and optimization, highlighting AI's potential to address these issues. The discussion extends to heat exchanger design principles, emphasizing the importance of thermal efficiency, pressure drop, and material selection. It then examines how AI enhances design optimization by automating processes, improving thermal efficiency, and optimizing flow configurations.[37] The narrative covers machine learning's role in predictive maintenance and operation, showcasing real-time monitoring and anomaly detection capabilities. It discusses data-driven approaches to heat exchanger development, emphasizing the integration of simulation, historical, and real-time operational data for informed design improvements. The synergy between computational fluid dynamics (CFD) and AI is examined for its ability to enhance fluid flow and heat transfer understanding, with AI interpreting complex CFD results to guide design modifications. The chapter addresses challenges in applying AI, offering solutions like interdisciplinary collaboration and computational advances. It summarizes the benefits of AI

DOI: 10.1201/9781003509837-16

and machine learning integration in revolutionizing heat exchanger design and operation, focusing on enhanced efficiency, customization, and predictive capabilities.

16.1.1 Introduction to the Concept of Heat Exchanger Design

Heat exchanger design involves the complex engineering of systems that efficiently transfer heat from one fluid to another while keeping them physically separated to prevent mixing. This process plays a pivotal role in various industries, including power generation, heating, ventilation, and air conditioning (HVAC) systems, and chemical processing.[23] The design entails considerations such as maximizing heat transfer rates, minimizing pressure drops, ensuring uniform flow distribution, and optimizing material selection to withstand the operating conditions. Engineers must account for factors like fluid properties, operating temperatures and pressures, fouling potential, and safety requirements. The ultimate goal is to create a heat exchanger system that achieves optimal thermal performance, energy efficiency, and operational reliability within the given constraints and objectives. The scope and nature of these advancements are critical for evaluating the feasibility of design, analysis, and measurement optimization. The development specifications of the current chapter have been updated to describe the mechanisms through which these changes occur. Fundamentally, a heat exchanger facilitates heat transfer between a source and a working fluid. This model can be integrated into both cooling and heating processes. In the heat exchanger, the working fluid can be isolated by a solid barrier to prevent direct contact. This chapter concentrates on system design, emphasizing an in-depth understanding of the fundamental heat exchanger flow processes.

16.1.2 Types of Heat Exchangers

There are various types of heat exchangers, classified based on their design, flow arrangement, and application.

Types of heat exchangers are shown in Figure 16.1 as per the mechanisms of heat transfer, flow arrangement, and by purpose and construction. Table 16.1 provides an overview of various types of heat exchangers, along with brief specifications and corresponding schematic diagrams for each. Each type of heat exchanger is tailored to specific applications based on factors like temperature, pressure, and the nature of the fluids involved. In Figure 16.2, parallel and counterflow heat transfer have been described. Choosing the right heat exchanger depends on the system's thermal needs, available space, and operational environment. Table 16.2 presents a detailed comparison of various heat exchanger types based on their specific applications and utilization. Different types of heat exchangers are classified in the following sections.

16.1.3 Advancement for Heat Exchangers in Recent Years

Recent advancements in the design, analysis, and optimization of heat exchangers are known for their flexibility, scalability, efficiency, and efficacy. Innovations include the development of superior materials, the application of advanced mathematical models to refine existing designs, and the exploration of new applications through CAD design systems.[6] Recent advances in modern design, analysis, and optimization technologies have significantly increased heat exchanger efficiency. Simulations[41–50] are increasingly used to accurately predict the performance of various designs, utilizing new materials such as nanofluids,[8] which have superior thermal properties compared with traditional heat exchangers. To determine optimal design parameters for specific operating conditions and optimization techniques, genetic algorithms are being utilized.[21] Current research and development efforts are focused on advanced heat exchanger design, including sophisticated fin structure, compact heat exchanger design, tubular systems, double-pipe based systems, and the use of novel materials.[14] For instance, significant design improvements have been achieved through the increased use of finned tubes and micro-fin tubes, which enhance surface area and improve heat

FIGURE 16.1 Different types of heat exchangers.

transfer performance. Additionally, advancements in improved airflow on low-temperature finned tubes and modified header design are significant. In terms of analysis, notable advancements, including transient simulation, that capture the consequence of fouling and optimization techniques are being used to identify the optimal heat exchanger design.[22–38] These advances have collectively increased the dependability, robustness, and user-friendliness of heat exchangers, making them more effective and efficient in various applications. Significant progress has also been made in the optimization of heat exchangers, particularly in the development of improved manufacturing processes, the application of more efficient active management methods, and the design of optimal system architectures to reduce pressure drop.[3] In recent years, significant advances have been made in the design, analysis, and optimization of heat exchangers. These advancements have led to higher energy efficiency, lower operational and capital costs, improved thermal performance, and reduced environmental impact.[36] These developments have resulted in more efficient and cost-effective systems. Heat exchangers have evolved significantly since their inception, becoming more efficient and adaptable to a wide range of applications. Advances in design, analysis, and optimization have enhanced their ability to maximize thermal energy transfer, resulting in energy and resource savings. As metallurgy, materials engineering, and computational engineering progress, the potential for further energy optimization in heat exchangers increases. Finned surfaces and multi-positioned heat exchangers are examples of innovations that have been developed to increase heat transfer efficiency. Furthermore, advanced heat transfer methods have been developed to improve performance, such as metal foam technology and nanofluids. [6, 11, 13, 33, 34, 60] The use of CFD[39–51] analysis has facilitated the design and analysis of complex heat exchanger configurations. Optimization techniques, such as AI and nonlinear optimization, have further improved process design and

TABLE 16.1
Different Types of Heat Exchangers with Brief Specifications

Type of Heat Exchanger	Description	Figure
Shell-and-Tube heat exchanger[65, 66]	The shell-and-tube heat exchanger is one of the most widely utilized designs in industrial applications. It comprises a series of tubes enclosed within a cylindrical shell, with one fluid flowing through the tubes and the other fluid circulating over the tube surfaces, enabling heat transfer across the tube walls. This configuration involves one set of tubes carrying the hot fluid and another set carrying the cold fluid, facilitating the exchange of heat between the two fluids as they pass through the exchanger. These heat exchangers are commonly employed in industry due to their versatility, durability, and capability to withstand high pressures and temperatures.	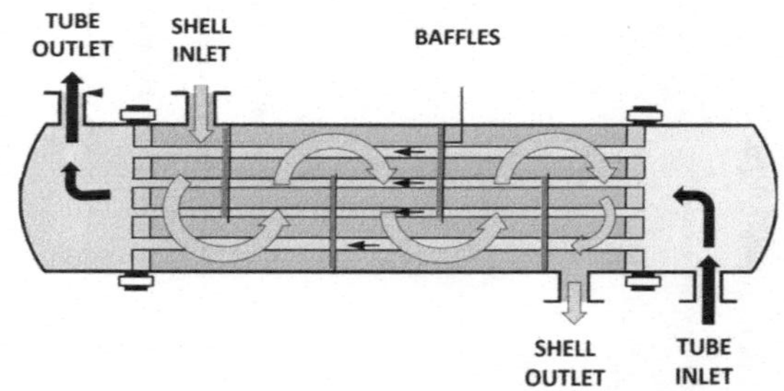 The schematic generally shows a large cylindrical shell with multiple tubes inside. Arrows indicate the direction of fluid flow (one through the tubes and one around the outside in the shell). In a counter-flow configuration, the fluids move in opposite directions; in parallel flow, they move in the same direction.
Plate-type heat exchanger[3, 43–45, 67–68]	Plate-type heat exchangers are a compact and efficient type of heat exchanger, consisting of a series of thin, parallel metal plates. Fluids flow alternatively between these plates, allowing for a large surface area that facilitates effective heat exchange and results in high heat transfer coefficients. This design makes plate heat exchangers cost-effective and suitable for applications requiring close temperature approaches, such as in heating, ventilation, and air conditioning systems. Additionally, their modular construction enables easy maintenance. The compact size and high thermal performance of plate heat exchangers make them a popular choice in a wide range of industrial and commercial applications where efficient heat transfer is critical.	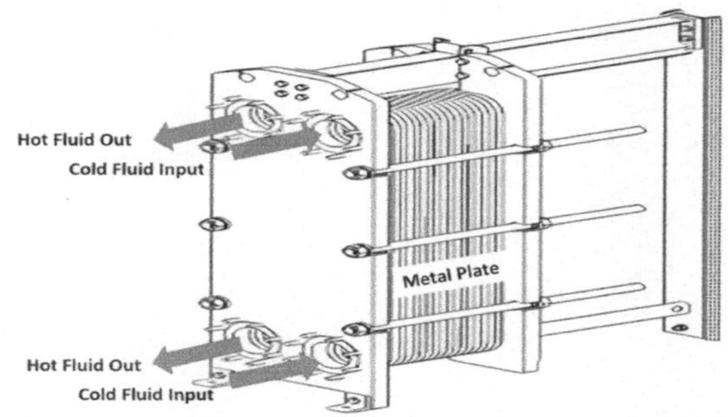 The drawing shows multiple parallel plates stacked together. Fluids flow in opposite directions (counter-flow) through alternate spaces between the plates. The plates may be drawn with wavy or corrugated lines to indicate the surface geometry that enhances heat transfer. This uses thin, corrugated plates stacked together to increase surface area for heat transfer. The hot and cold fluids flow between alternate plates.

(Continued)

TABLE 16.1 (CONTINUED)
Different Types of Heat Exchangers with Brief Specifications

Type of Heat Exchanger	Description	Figure
Double-Pipe heat exchanger[69, 70]	Double-pipe heat exchangers feature a simple design consisting of two concentric tubes. One fluid flows through the inner tube, while the other fluid circulates through the annular space between the inner and outer tubes. This straightforward configuration enables effective heat transfer as the fluids pass through the exchanger. Double-pipe heat exchangers are often utilized in smaller-scale applications due to their uncomplicated construction, ease of cleaning, and suitability for low to moderate pressure and temperature requirements. They are commonly found in laboratories, small-scale industrial processes, and heat recovery systems where efficient heat exchange is necessary but complex designs are not required. The compact and robust nature of double-pipe heat exchangers makes them a practical choice for applications where space and cost constraints are important considerations.	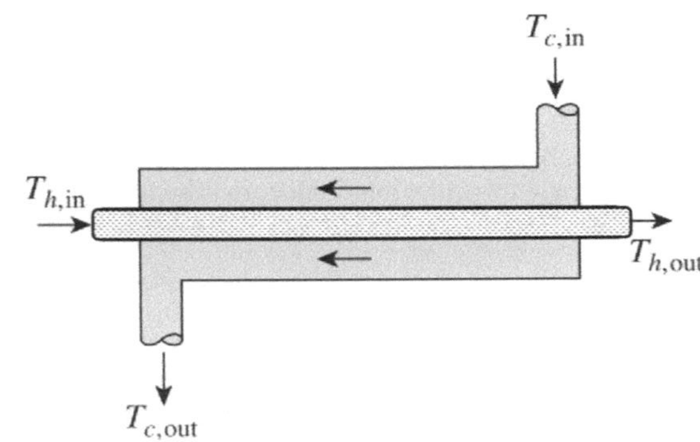 In this configuration, one fluid flows inside a pipe, and the other fluid flows in the annular space between the first pipe and a larger surrounding pipe. The drawing shows two concentric pipes with arrows showing the flow of fluids. One in the inner pipe and the other in the annulus between the pipes. Often represented with one fluid entering on one side and the other fluid entering on the opposite side to show counter-flow.
Air-cooled heat exchanger[7, 71]	Air-cooled heat exchangers rely on the transfer of heat from the fluid flowing through the tubes to the surrounding air. By using large fans to blow air across the finned tubes, these exchangers enhance heat dissipation and increase the overall heat transfer rate. Due to their simple design, ease of maintenance, and effective heat dissipation capabilities, air-cooled heat exchangers are widely utilized in various industrial processes, power generation facilities, and HVAC systems. These exchangers are particularly advantageous in applications where water is scarce or unavailable for liquid-based cooling systems, making them a popular choice for installations in remote or arid regions.	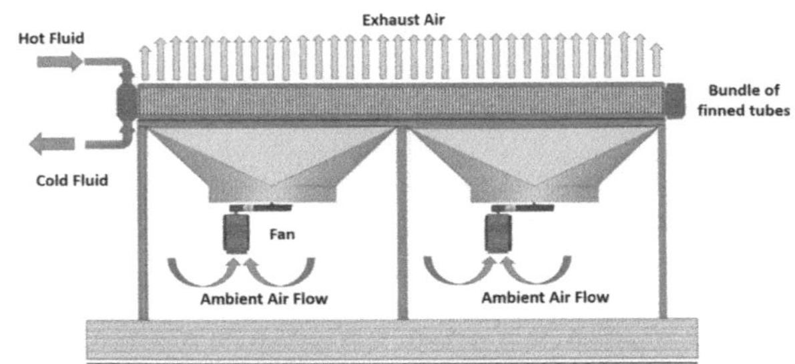 This schematic typically shows a bundle of finned tubes (often horizontally aligned) with arrows showing airflow passing over them. Fans may be depicted on one side to indicate forced air circulation.

(Continued)

TABLE 16.1 (CONTINUED)
Different Types of Heat Exchangers with Brief Specifications

Type of Heat Exchanger	Description	Figure
Regenerative heat exchanger[72, 73]	Regenerative heat exchangers are a specialized type of heat exchanger where the same fluid circulates through both sides of the device. This allows the fluid to store heat in one part of the cycle and then release that heat in another part, enabling the fluid to recycle the thermal energy. This design is particularly common in gas turbines and similar power generation applications, where the hot exhaust gases from the turbine are used to preheat the incoming combustion air, improving the overall efficiency of the system. The regenerative heat exchange process helps to capture and recycle waste heat, leading to significant improvements in the overall energy efficiency and performance of the system.	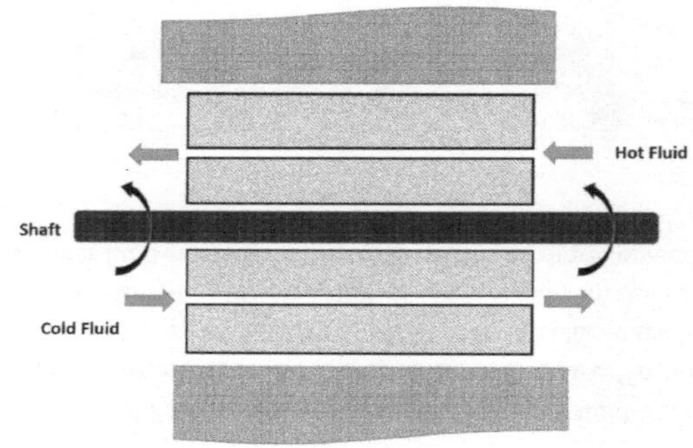 Often shown as a single fluid moving through a matrix that stores heat and then releases it when the fluid direction changes. Arrows indicate the flow path of the gas through the system.
Finned tube heat exchanger[74, 75]	This type of heat exchanger employs finned tubes to dramatically increase the surface area available for heat transfer, thereby enhancing its thermal efficiency and effectiveness. The fins can be either integral, meaning they are physically part of the tube, or external, where they are attached to the outside of the tubes. This finned tube design is extensively utilized in a wide range of air-cooled applications, such as HVAC systems, refrigeration equipment, and automotive radiators, where efficient heat dissipation is crucial for optimal performance and energy savings.	 This type adds fins to the tubes to increase the surface area and improve heat transfer between fluid inside the tubes and air (or another fluid) outside the tubes. The schematic shows tubes with fins around them, typically with air or gas passing over the finned area.

(Continued)

TABLE 16.1 (CONTINUED)
Different Types of Heat Exchangers with Brief Specifications

Type of Heat Exchanger	Description	Figure
Spiral heat exchanger[76]	Spiral heat exchangers are a specialized type of heat exchanger that consist of two long, spiral channels wrapped around a central core. This unique configuration allows one fluid to flow through each of the spiral channels, making the heat exchanger compact and efficient. Spiral heat exchangers are capable of achieving high rates of heat transfer due to the counter-flow arrangement of the fluids, which improves the temperature differences across the exchanger. Additionally, the spiral geometry increases the surface area available for heat transfer, further enhancing the thermal performance of these devices. These spiral heat exchangers are often employed in chemical and petrochemical processes, geothermal applications, and heat recovery systems where limited space is available but high thermal efficiency is a critical requirement.	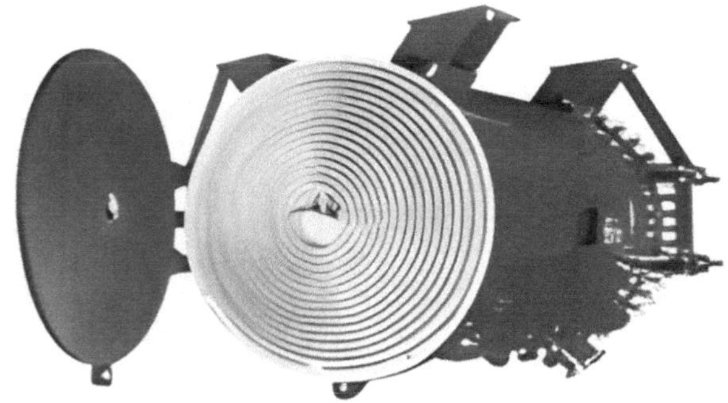 A schematic diagram of a spiral heat exchanger, showing two fluid flow paths coiled in a spiral configuration. The two fluids move through separate, parallel channels, with one fluid flowing from the center outward and the other from the outer edge inward, enabling efficient heat transfer across the walls of the channels.
Condensers and evaporators[77]	Condensation is the process whereby a gaseous substance is converted into a liquid state through the removal of heat. Conversely, evaporators perform the opposite function, vaporizing liquids into a gaseous form by applying heat. Heat exchangers known as condensers facilitate the transformation of vapor into liquid by cooling. Evaporation entails the conversion of a liquid into a vapor or gas phase, occurring when the heat supplied enables the molecules to acquire sufficient kinetic energy to escape the liquid's surface.	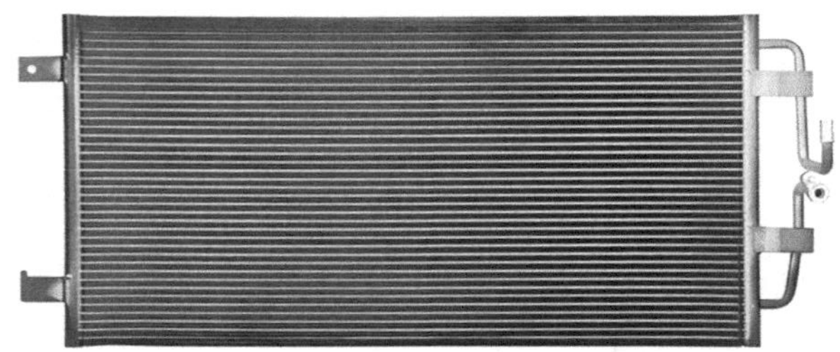 A condenser typically shows a coil with a vapor entering and liquid exiting, with heat being removed (air or water used to cool it). An evaporator shows a coil with liquid entering and vapor exiting, indicating that heat is absorbed.

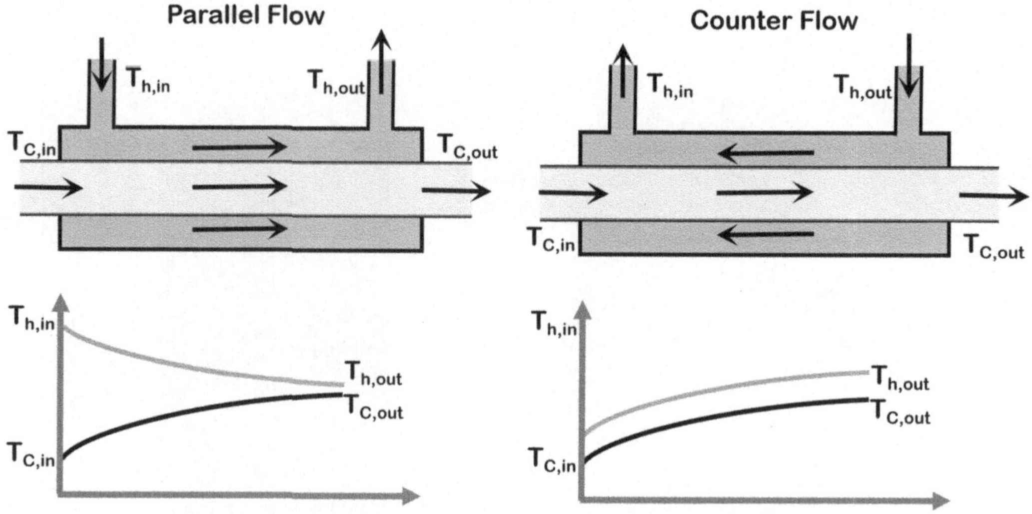

FIGURE 16.2 Schematic view of parallel and counter-flow heat transfer.

performance. The observed advancements in the field of heat exchangers demonstrate a reliance on technology for external distribution. For example, phase-changing material-induced cooling technology significantly impacts temperature distribution.[16–18] Heat transfer and pressure drop performance have been improved in systems that use horizontal cross-flow, coaxial roll-bond, and variable curvature fin-and-tube designs. Numerical techniques such as CFD, structural analysis, and nonlinear finite element methods have allowed for more accurate assessments of thermal-hydraulic performance and a better understanding of the underlying physical phenomena.[39–54] Another notable advancement in heat exchanger technology is the microplate heat exchanger, which enhances flow performance while reducing environmental impact. Understanding fundamental aspects, such as a heat exchanger's energy-storing capacity, is critical when evaluating these advancements. The use of latent heat in phase change material[15–17] (PCM) technology improves both functionality and accuracy. Existing thermal energy equations enable better management of PCM surroundings.[18, 49, 82] These considerations are essential for the controlled development of heat exchanger processes. The presence of various surrounding complexities and favorable initial conditions aids in solving numerical equations. Ghalambaz et al.[9] emphasize the importance of understanding storage phenomena in solar energy[17] conversion, specifically the need to maintain a controlled cooling temperature. To improve investigation efficiency, it is critical to focus on fusion configuration, which is the basic process of a tubular exchanger with five horizontal fins. The twisted heat exchanger represents a significant advancement in heat exchanger design as it combines tube assemblies without using baffles. This design employs overlapping heat tubes, allowing for greater positioning flexibility and distinguishing it from traditional tube shell heat exchangers. The number of twists per unit length can be increased depending on the design, which requires consideration of wall thickness and material yield limits. The twisting pattern is typically an "S" shape, maintaining a 360-degree twist with six main points of contact. The swirl intensity is determined by the cross-sectional shape and diameter ratio of the twisted pitch.[5–6] Assessing feasibility is critical for determining the suitability of new heat exchanger designs. Efforts to reduce manufacturing costs have resulted in simpler, more efficient designs. Analytical tools such as CFD, finite element analysis (FEA), artificial neural networks (ANNs), and particle tracking are being utilized to aid in the creation and optimization of new designs.[10] These tools enable designers to quickly evaluate system performance and optimize it prior to producing the final component. New optimization techniques are being developed to determine the best size, shape, and material for heat exchangers, ensuring that they meet customer needs without requiring excessive refrigeration or heat transfer. These advancements

TABLE 16.2
Comparison of Shell-and-Tube, Plate-Type, Finned Tube, Double-Pipe, Spiral, and Plate-Fin Heat Exchangers with Various Considerations[4, 12, 20, 28, 36, 64, 65]

Type	Construction	Heat Transfer Efficiency	Applications	Advantages	Disadvantages
Shell-and-tube	Tubes enclosed in a cylindrical shell	High	Industrial applications	Versatile, robust, handles high pressures and temperatures	Large size, potential for fouling
Plate-type	Multiple plates arranged in parallel	Very high	Compact, cost-effective, close temperature approaches	High heat transfer coefficients, easy maintenance	Can leak, not suitable for high-pressure applications
Finned tube	Finned tubes to increase surface area	High	Air-cooled applications (HVAC, refrigeration, automotive)	Enhanced thermal efficiency	More complex design, potential for fin fouling
Double-pipe	Concentric tubes (inner and outer)	Moderate	Laboratories, small-scale processes, heat recovery systems	Simple design, easy to clean	Limited to low/moderate pressure and temperature
Spiral	Concentric spiral channels	Very high	Viscous fluids, high fouling potential, limited space	Compact, self-cleaning design	More complex construction, difficult to clean if fouling occurs
Plate-fin	Stacked plates with alternating layers of fins	Very high	Aerospace, cryogenics, high-performance cooling	Lightweight, high thermal efficiency, withstands high pressures	Expensive, potential for fin fouling

enhance heat exchanger reliability, efficiency, and cost-effectiveness.[2] The use of twisted heat exchanger tubes provides several advantages. The increased thermal efficiency enhances heat storage and exchange functionality while reducing pressure drop, fouling, and vibration rates. This advancement demonstrates the potential for improved performance in heat exchanger design. These advancements offer comparative advantages such as higher heat transfer rates and economic benefits. Traditional heat exchangers have lower heat transfer rates and are less cost-effective. Eco-friendly configurations remain in high demand, and new developments meet these requirements. Advanced configurations also require less space and maintenance compared with traditional designs.[11] Recent developments in heat exchangers have resulted in significant improvements in design, analysis, and optimization. These advancements have reduced power requirements, increased efficiency, and enhanced the overall performance of heat exchanger systems. Advancements in heat exchanger design have included the use of computational fluid dynamics tools[40–65] and the development of new models that account for various heat transfer and convection processes. These advancements have enabled the enhancement of existing heat exchanger designs through the predictive optimization technique, refining the parameters that control heat transfer efficiency. Recent advances in heat exchanger design, analysis, and optimization have allowed researchers and engineers to improve performance while conserving energy and resources. The introduction of high-performance technologies capable of quickly adapting to changing conditions, such as plate-fin heat exchangers, rotary heat exchangers, and compact counter-flow exchangers, represents significant progress over traditional finned tube designs and methodologies. For example, due to their high surface area-to-volume ratio, plate-fin heat exchangers have been utilized in aerospace applications to improve thermal management efficiency.[2] Computational methodologies, techniques, and tools have been developed to simulate and optimize heat exchanger designs, resulting in improved performance, efficiency, and reduced energy waste. For example, FEA and CFD can simulate complex thermal loads and operating conditions within a single system, resulting in more precise and efficient heat exchanger designs. These advancements enable heat exchangers to simulate different thermal loads and operate under a variety of operating conditions within a single system. This capability has important implications for industries like power generation and HVAC, where adaptable and efficient heat exchangers can result in significant energy savings and increased system reliability.

16.1.4 A Brief Overview of the Importance of Heat Exchangers in Various Industries

Heat exchanger design involves the complex engineering of systems that efficiently transfer heat from one fluid to another while keeping them physically separated to prevent mixing. This process plays a pivotal role in various industries, including power generation, HVAC systems, and chemical processing.[45] The design entails considerations such as maximizing heat transfer rates, minimizing pressure drops, ensuring uniform flow distribution, and optimizing material selection to withstand the operating conditions. Engineers must account for factors like fluid properties, operating temperatures and pressures, fouling potential, and safety requirements. The ultimate goal is to create a heat exchanger system that achieves optimal thermal performance, energy efficiency, and operational reliability within the given constraints and objectives. Heat exchangers are indispensable components across industries, serving as the backbone of thermal management systems. In power generation, they play a critical role in both conventional and renewable energy sectors by facilitating the exchange of heat between coolant and working fluid in boilers, condensers, and heat recovery systems. Similarly, in the realm of HVAC, heat exchangers regulate indoor temperatures by transferring heat between indoor and outdoor air streams, ensuring comfort and energy efficiency in buildings. Moreover, in the chemical and petrochemical industries, heat exchangers are vital for processes such as distillation, evaporation, and chemical reaction cooling, where precise control of temperature is crucial for product quality and process efficiency.

High temperature heat exchangers[46–53] are critical components in many industrial applications like gas turbines, chemical plants, and nuclear[39–53] power plants. These heat exchangers must operate in severe environments and exhibit special characteristics to withstand high temperatures and other extreme conditions. Unlike typical heat exchangers, high temperature heat exchangers need to account for significant radiative heat transfer, which can play a major role in the overall heat transfer process. To handle the high temperatures, the tube diameters and pitch should be larger compared with conventional heat exchangers to reduce pressure drop and ensure efficient heat transfer. Additionally, the materials used in the construction of these heat exchangers must be carefully selected to provide the necessary strength, corrosion resistance, and thermal capabilities to operate reliably in the demanding conditions encountered in these industrial applications.

Beyond these industries, heat exchangers find application in transportation sectors, such as automotive and aerospace, where they aid in engine cooling, cabin climate control, and aircraft air conditioning systems. Additionally, in food and beverage processing, heat exchangers are instrumental for pasteurization, sterilization, and maintaining product consistency during production. The pharmaceutical sector relies on heat exchangers for drug synthesis, purification, and sterile processing. Overall, the importance of heat exchangers across various industries cannot be overstated, as they enable efficient heat transfer, energy conservation, process optimization, and product quality assurance. In Table 16.3, applications of heat exchangers in different industries with specific design considerations are explored.

16.2 POTENTIAL TECHNOLOGIES AND METHODOLOGIES FOR FUTURE ADVANCEMENTS IN HEAT EXCHANGER DESIGN

Future advancements in heat exchanger design are expected to be driven by a blend of innovative technologies and methodologies aimed at enhancing efficiency, sustainability, and adaptability. One area of exploration involves integrating advanced materials with superior thermal properties, such as nanomaterials and metamaterials, to optimize heat transfer rates and minimize energy consumption. Additionally, the advancement of additive manufacturing techniques, such as 3D printing, holds potential for creating intricate geometries and customized designs. This capability enables the development of highly efficient and compact heat exchangers tailored to specific applications. Moreover, advancements in CFD and AI offer promising avenues for optimizing heat exchanger performance. Through advanced simulations and machine learning algorithms, engineers can enhance their ability to predict and control fluid flow patterns, temperature distributions, and heat transfer characteristics. This capability leads to the development of more efficient and cost-effective heat exchanger systems. Furthermore, the integration of IoT sensors and data analytics platforms enables real-time monitoring and optimization of heat exchanger operations. This facilitates proactive maintenance, energy savings, and improved system reliability. By leveraging these emerging technologies and methodologies, future advancements in heat exchanger design have the potential to revolutionize thermal management across industries. They can effectively meet the increasing demands for sustainability and efficiency in thermal processes.

16.3 EMERGING NEW HEAT EXCHANGER TYPES

Heat exchangers can also be classified by the materials used in their construction, which is important because the material affects factors like thermal conductivity, corrosion resistance, weight, and cost. The choice of material depends on the application, the fluids being used, and the operating conditions (temperature, pressure, and corrosiveness). In Figure 16.3, heat exchangers as per material used are categorized.

TABLE 16.3
Applications of Heat Exchangers with Specific Design Considerations[8, 40–49, 65–77, 82–85]

Industry	Description	Design Considerations
Power generation[39–63]	Steam condenser in thermal and nuclear power plants[44]	High heat transfer efficiency, corrosion resistance due to untreated water, large surface area for rapid condensation.
HVAC systems[57, 81]	Air-to-air heat exchanger in building ventilation	Compact design to fit within buildings, high efficiency, corrosion-resistant materials.
Chemical processing[69–73]	Shell-and-tube heat exchanger in chemical plant	High pressure and temperature resistance, compatibility with chemical fluids, fouling minimization.
Oil and gas industry[74]	Plate heat exchanger in oil refineries	High thermal efficiency, compact design, easy maintenance, corrosion-resistant materials.
Food and beverage[21]	Pasteurization heat exchanger	Hygienic design, easy to clean, food-grade and corrosion-resistant materials, precise temperature control.
Aerospace industry[4, 20]	Heat exchanger in aircraft environmental control systems (ECS)	Lightweight materials, high reliability, resistance to high altitudes and pressures, compact design.
Automotive industry[69, 70]	Radiator in internal combustion engines	High heat dissipation, durability under varying conditions, compact design.
Renewable energy[18, 22, 30, 34, 40, 41, 59, 64]	Heat exchanger in geothermal power plants	Corrosion resistance, high temperature and pressure tolerance, efficient heat transfer.
Pharmaceutical industry[69–73]	Heat exchanger for sterilization processes	Hygienic standards, precise temperature control, materials suitable for frequent sterilization.
Metallurgy[10]	Heat exchanger in blast furnace systems	High heat transfer efficiency, robust materials to withstand extreme temperatures and corrosive environments.
Plastics industry[64–66]	Heat exchanger in extrusion processes	Precise temperature control, efficient heat transfer, durable materials to withstand high pressure.
Pulp and paper[65, 66]	Heat exchanger in paper drying machines	High efficiency to rapidly remove moisture, corrosion resistance, durable materials to handle high temperatures.
Textile industry[65, 66]	Heat exchanger in dyeing machines	Uniform temperature control, corrosion resistance, efficient heat transfer.
Manufacturing[20, 65]	Heat exchanger in injection molding machines	Efficient cooling to maintain precise temperatures, durable materials to withstand high pressure and temperature.
Electronics[80–82]	Heat exchanger in data centers, component cooling	Efficient heat dissipation, compact design, high reliability under continuous operation.
Mining industry[64]	Heat exchanger in underground ventilation systems	Robust design to withstand harsh conditions, efficient heat transfer, corrosion resistance.
Data center[83–85]	Data center industry for thermal management	Design a heat exchanger for data centers with optimized thermal management, ensuring efficient heat dissipation from high-density server racks. The design should prioritize energy efficiency, compactness, and reliability, while maintaining precise temperature control to prevent server overheating.

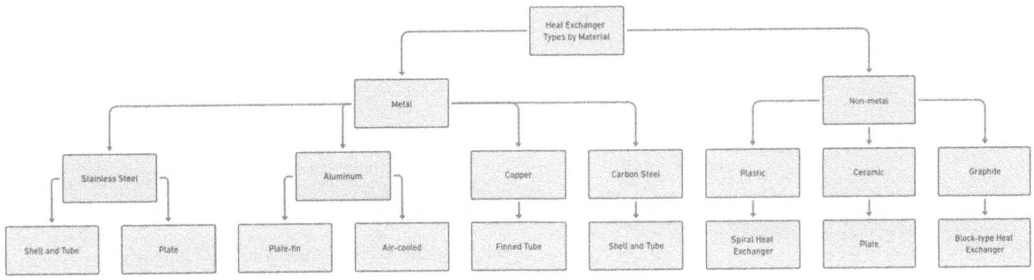

FIGURE 16.3 Types of heat exchangers as per materials used.

These are some of the most common types of heat exchangers, each offering unique advantages and suitability for specific applications based on factors such as fluid properties, operating conditions, space constraints, and efficiency requirements.

16.3.1 Innovative Approaches and Technologies Shaping the Future of Heat Exchange

The future of heat exchange is being shaped by a convergence of innovative approaches and technologies aimed at enhancing efficiency, sustainability, and adaptability. Nanotechnology offers the promise of nanostructured materials with superior thermal properties, enabling significant improvements in heat transfer rates and efficiency. Additive manufacturing, particularly 3D printing, enables the customization and optimization of heat exchanger designs, allowing for intricate geometries tailored to specific heat transfer requirements. Advanced computational modeling and simulation techniques,[26–52] including CFD and machine learning, empower engineers to optimize heat exchanger performance, predict behavior, and rapidly iterate designs. Together, these advancements herald a new era of heat exchange technology, characterized by increased efficiency, reduced environmental impact, and enhanced versatility across industries. Several innovative approaches and technologies are shaping the future of heat exchangers. In Figure 16.4, several types of innovative approaches and technology commonly utilized for heat exchanger design are shown.

16.3.1.1 Advanced Computational Modeling and Simulation

Advanced computational modeling and simulation[41–53] techniques, including CFD and FEA, are transforming the design and optimization of heat exchangers. By simulating fluid flow, heat transfer, and thermodynamic behavior within heat exchanger systems, engineers can gain insights into complex phenomena (such as mixing and instability development), optimize design parameters, and predict performance under various operating conditions.[42] Furthermore, the integration of machine learning algorithms with simulation tools enables the development of predictive models that can autonomously optimize heat exchanger designs based on real-time data and user-defined objectives. A flowchart of computational modeling with necessary criteria is represented in Figure 16.5. This convergence of simulation and artificial intelligence empowers engineers to rapidly iterate designs, explore innovative concepts, and achieve unprecedented levels of efficiency and performance in heat exchange applications across industries.

16.3.1.2 Nanotechnology

Nanotechnology is poised to revolutionize heat exchange through the development of nanostructured materials with enhanced thermal conductivity and surface properties. Nanomaterials such as carbon nanotubes, graphene, and nanoporous materials offer the potential to significantly improve heat transfer rates by providing large surface areas for interaction between fluids and solid surfaces. These materials can be incorporated into heat exchanger components to reduce thermal resistance, enhance heat transfer efficiency, and enable the design of more compact and lightweight systems,

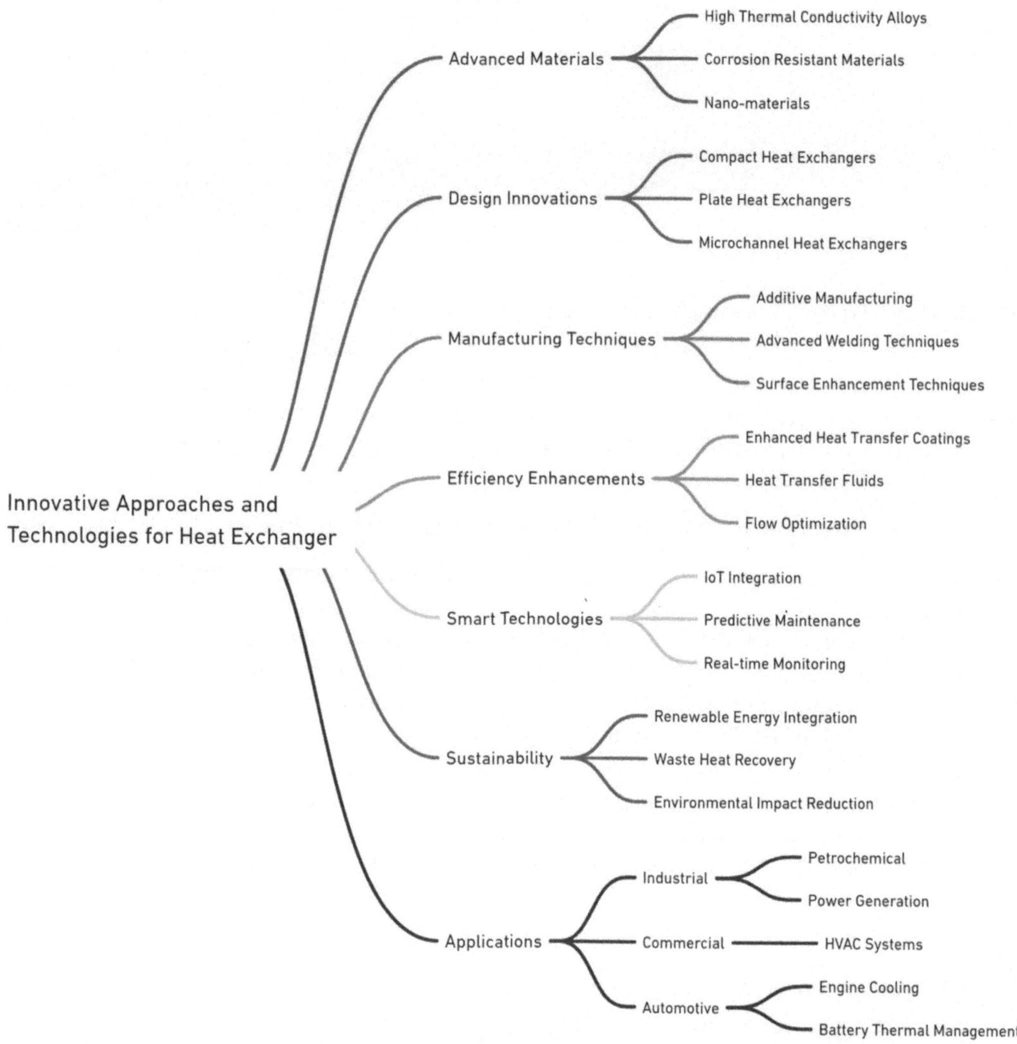

FIGURE 16.4 Types of innovative approaches and technology used for heat exchanger design.

making them particularly promising for applications in aerospace, electronics cooling, and renewable energy technologies.

16.3.1.3 Additive Manufacturing (3D Printing)

Additive manufacturing, commonly known as 3D printing, presents exciting opportunities for the customization and optimization of heat exchanger designs. By layering materials to build complex geometries, 3D printing enables the creation of heat exchanger components with intricate internal structures tailored to specific heat transfer requirements. This flexibility allows engineers to optimize flow paths, minimize pressure drops, and maximize heat transfer surfaces, leading to the development of highly efficient and compact heat exchangers. Furthermore, additive manufacturing facilitates rapid prototyping and on-demand production, reducing lead times and enabling the creation of bespoke heat exchanger solutions for diverse applications in automotive, medical devices, and consumer electronics.

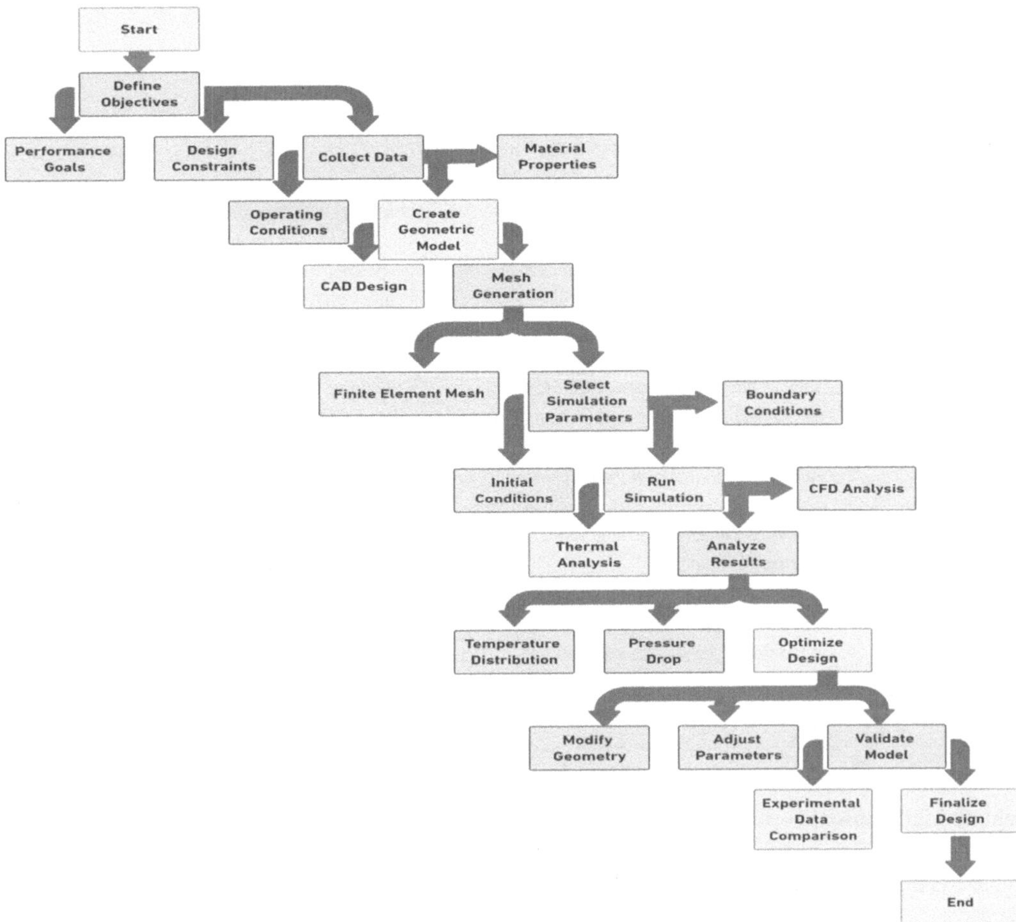

FIGURE 16.5 A Flowchart of computational modeling with necessary criteria.

16.4 MICRO-SCALE AND NANOSCALE HEAT EXCHANGERS

Micro-scale and nanoscale heat exchanger design principles revolve around maximizing surface area-to-volume ratios and exploiting unique heat transfer mechanisms at these small length scales. In micro-scale heat exchangers, which typically operate at sizes ranging from micrometers to millimeters, the focus is on miniaturization to achieve high heat transfer rates within compact geometries. Key principles include micro-fabrication techniques such as photolithography and etching to create intricate micro-channel networks, enhancing convective heat transfer and reducing thermal resistance. Additionally, micro-scale heat exchangers often leverage forced convection and enhanced fluid mixing to optimize thermal performance.

In nanoscale heat exchangers, which operate at dimensions on the order of nanometers, novel phenomena such as ballistic and phonon-mediated heat transfer come into play. Nanoscale heat exchanger design principles involve engineering surfaces with tailored nanostructures to manipulate phonon scattering and enhance thermal conductivity. Furthermore, nanoscale heat exchangers exploit phenomena like the Knudsen effect, where gas molecules behave differently in confined spaces, to achieve efficient heat transfer in micro/nanoscale channels. Overall, micro-scale and nanoscale heat exchanger design principles emphasize the utilization of advanced fabrication techniques, exploitation of unique heat transfer mechanisms, and optimization of fluid flow dynamics to

TABLE 16.4
Comparison of Micro-Scale and Nanoscale Heat Exchangers

Aspect	Micro-Scale Heat Exchangers	Nanoscale Heat Exchangers
Operating scale	Typically operates at sizes ranging from micrometers to millimeters.	Operates at dimensions on the order of nanometers.
Heat transfer mechanisms	Relies on convection and forced flow mechanisms, with an emphasis on miniaturization to achieve high heat transfer rates within compact geometries.	Exploits novel heat transfer mechanisms,[46–51, 58, 60] such as ballistic and phonon-mediated heat transfer, and phenomena like the Knudsen effect, where gas molecules behave differently in confined spaces.
Fabrication techniques	Utilizes micro-fabrication techniques such as photolithography, etching, and micromachining to create intricate micro-channel networks.	Relies on advanced nanofabrication techniques,[8, 22, 23, 27, 31] including self-assembly, atomic layer deposition, and nanoimprint lithography to engineer surfaces with tailored nanostructures.
Thermal performance	Offers high thermal performance within compact geometries, with enhanced convective heat transfer and reduced thermal resistance.	Capitalizes on unique heat transfer phenomena and surface effects to achieve efficient thermal performance[11, 42–53, 69, 80] even at extremely small scales, potentially surpassing the performance of micro-scale heat exchangers.
Applications	Commonly used in electronics cooling, microfluidics, and microreactors.	Finding applications in nanoelectronics, thermal management of nanoscale devices, and emerging fields such as nanomedicine and nanomanufacturing.[8, 12, 20, 21, 28, 36, 57, 69, 74]

achieve high thermal performance in extremely small devices. In Table 16.4, a suitable comparison of micro-scale and nanoscale heat exchangers is presented considering different aspects, and in Table 16.5, the advantages of miniaturization in heat exchangers are presented considering different aspects.

16.4.1 Advantages of Miniaturization in Heat Exchanger Technology

16.4.2 Application of Micro-Channels and Nanofluids for Enhanced Heat Transfer

Micro-channels and nanofluids offer several advantages for enhanced heat transfer in various applications. These applications demonstrate the versatility and effectiveness of utilizing micro-channels and nanofluids for enhanced heat transfer across diverse industries, ranging from electronics and energy to chemical processing and transportation. In Table 16.6, a comparison of different cooling techniques for heat exchangers is presented considering different aspects.

16.5 ADVANCED MATERIALS IN HEAT EXCHANGER DESIGN

16.5.1 Exploration of Novel Materials with Enhanced Thermal Conductivity and Corrosion Resistance

Exploring novel materials with enhanced thermal conductivity and corrosion resistance is crucial for advancing various industries, including electronics, energy, transportation, power plants, and manufacturing. The following sections are different avenues of exploration for such materials.

TABLE 16.5
Different Advantages of Miniaturization in Heat Exchangers

Advantages	Descriptions
Increased heat transfer rate	Miniaturization allows higher surface area-to-volume ratios, enhancing convective heat transfer.
Reduced thermal resistance	Smaller dimensions minimize the distance over which heat must be transferred, reducing thermal resistance.
Enhanced heat transfer uniformity	Miniaturization facilitates more uniform temperature distributions across the heat exchanger.
Improved energy efficiency	Higher heat transfer rates and reduced thermal resistance contribute to improved energy efficiency.
Compact and lightweight design	Miniaturized heat exchangers are smaller and lighter, making them suitable for space-constrained applications.
Faster response time	Smaller volumes result in quicker response times to changes in heat load or fluid flow conditions.
Lower material requirements	Reduced size allows for less material usage, potentially lowering costs and environmental impact.
Potential for integration with microsystems	Miniaturized heat exchangers can be integrated with microfluidic systems or micro-electromechanical system devices for compact and multifunctional designs.

TABLE 16.6
Comparisons of Different Cooling Techniques[12]

Cooling Type	Descriptions
Electronics cooling	Micro-channels embedded within electronic devices can efficiently remove heat generated by components such as CPUs, GPUs, and power electronics. Nanofluids, which are engineered suspensions of nanoparticles in a base fluid, can be circulated through these micro-channels to enhance heat transfer. The high thermal conductivity of nanofluids improves the cooling efficiency of micro-channels, leading to better thermal management and prolonged device lifespan in electronics.
Solar thermal collectors[11]	Micro-channel heat exchangers integrated into solar thermal collectors can significantly enhance heat transfer efficiency in capturing solar energy for heating applications. Nanofluids used as heat transfer fluids within these micro-channels exhibit superior thermal properties, such as higher thermal conductivity and better stability at elevated temperatures, leading to improved overall performance and energy conversion efficiency of solar thermal systems.
Micro-reactors for chemical synthesis	Micro-reactors featuring intricate micro-channel networks offer precise control over reactant mixing, residence time, and temperature distribution, enabling efficient and rapid chemical synthesis processes. Nanofluids circulated through these micro-channels facilitate enhanced heat transfer rates, enabling faster reaction kinetics and improved yields in various chemical synthesis applications, including pharmaceuticals, fine chemicals, and catalysis.
Microfluidic cooling systems	Microfluidic cooling systems, used in applications such as biomedical devices, lab-on-a-chip platforms, and micro-electromechanical systems, rely on micro-channels to dissipate heat generated by micro-scale components. Incorporating nanofluids as the cooling medium enhances heat transfer capabilities due to the superior thermal conductivity and stability of nanofluids, enabling efficient thermal management and ensuring reliable operation of microfluidic devices.
Aerospace and automotive cooling	Micro-channel heat exchangers integrated into aerospace and automotive cooling systems offer lightweight and compact solutions for dissipating heat from engines, power electronics, and other critical components. Nanofluids employed as the coolant medium enhance heat transfer efficiency, enabling more efficient thermal management and contributing to improved performance, reliability, and fuel efficiency of aerospace and automotive systems.

16.5.1.1 Nanostructured Materials

Researchers are investigating nanostructured materials such as graphene, carbon nanotubes, and nanocomposites for their exceptional thermal conductivity and mechanical properties. These materials offer high surface area-to-volume ratios and unique transport properties, making them promising candidates for applications requiring efficient heat transfer and corrosion resistance, such as heat exchangers, electronic devices, and aerospace components.

16.5.1.2 Metal Matrix Composites

Metal matrix composites (MMCs) incorporate high thermal conductivity reinforcements, such as ceramics or carbon fibers, into metal matrices to enhance both thermal and mechanical properties. By carefully tailoring the composition and microstructure of MMCs, researchers can achieve improved thermal conductivity and corrosion resistance, making them suitable for heat exchangers, automotive components, and marine applications.

16.5.1.3 Advanced Ceramics

Advanced ceramics, including silicon carbide (SiC) and aluminum nitride (AlN), exhibit high thermal conductivity and excellent corrosion resistance, particularly at elevated temperatures. These materials are being explored for use in harsh environments such as chemical processing, power generation, and electronic packaging, where thermal management and corrosion resistance are critical for long-term performance and reliability.

16.5.1.4 Metal-Organic Frameworks

Metal-organic frameworks (MOFs) are a class of porous materials composed of metal ions or clusters connected by organic ligands. Researchers are investigating MOFs for applications such as gas storage, catalysis, and thermal management due to their tunable pore structures, high surface areas, and potential for functionalization. By incorporating metals with high thermal conductivity and corrosion resistance into MOF frameworks, novel materials with enhanced thermal properties can be developed for diverse applications.

16.5.1.5 Biomimetic Materials

Inspired by natural materials, researchers are exploring biomimetic approaches to design materials with enhanced thermal conductivity and corrosion resistance. For example, bio-inspired coatings mimicking the structure and properties of natural materials like nacre (mother-of-pearl) are being developed for corrosion protection in marine environments. Similarly, hierarchical structures found in biological systems are being replicated to enhance thermal transport in synthetic materials.

16.5.2 Application of Nanotechnology in Heat Exchanger Material Design

Nanotechnology has revolutionized heat exchanger material design by offering innovative solutions to enhance thermal conductivity, corrosion resistance, and efficiency. One prominent application is the incorporation of nanoparticles, such as graphene or carbon nanotubes, into heat exchanger materials to significantly improve thermal conductivity. These nanoparticles create pathways for heat transfer within the material, resulting in enhanced heat transfer rates and reduced thermal resistance. Additionally, surface modifications at the nanoscale can improve corrosion resistance, extending the lifespan of heat exchanger materials in harsh operating environments. For instance, nanocoatings composed of corrosion-resistant nanoparticles can protect heat exchanger surfaces from degradation caused by corrosive fluids. Furthermore, the precise control over material properties enabled by nanotechnology allows for the design of lightweight and compact heat exchangers with optimized performance, making them suitable for applications in aerospace, the automotive industry, electronics cooling, and renewable energy systems. Composite materials offer significant potential for improving heat transfer efficiency in various applications by combining different materials to exploit their individual strengths. In Table 16.7, a comparison of different nanomaterial types for heat exchanger material design is shown with advantages and challenges.

TABLE 16.7
Comparison of Different Nanomaterial Types for Heat Exchanger Material Design

Nanomaterial Type	Key Properties	Applications	Advantages	Challenges
Carbon nanotubes (CNTs)	High thermal conductivity, strength, and stability	Coatings, composites for heat exchangers	Enhanced heat transfer, lightweight, increased mechanical strength	High cost, integration with conventional materials
Graphene	Extremely high thermal conductivity, flexibility	Coatings, conductive films, composites	Exceptional heat transfer efficiency, thin and flexible	Production scalability, cost, integration with existing systems
Nanofluids[13]	Fluids with suspended nanoparticles (e.g., Al2O3, CuO)	Heat transfer fluids in HVAC, automotive, electronics cooling systems	Increased thermal conductivity, improved heat transfer rates	Stability of nanoparticles, potential clogging, cost
Ceramic nanoparticles	High thermal stability, corrosion resistance	Coatings, heat exchanger surfaces	Improved thermal stability, reduced corrosion, enhanced durability	Cost, production complexity, material brittleness
Metallic nanoparticles	High thermal conductivity, chemical stability	Coatings, additives in heat transfer fluids	Enhanced heat transfer, improved surface properties, anti-fouling	Cost, potential environmental and health impacts
Nanocomposites	Combined properties of constituent nanomaterials	Structural components, coatings	Tailored thermal properties, enhanced mechanical strength, multifunctional	Complexity in manufacturing, cost, potential material incompatibilities
Nanostructured coatings	Nanoscale surface features, high surface area	Surface coatings for heat exchangers	Increased heat transfer area, reduced fouling, improved efficiency	Application uniformity, long-term stability, cost
Phase change nanomaterials[16–22]	High latent heat, enhanced thermal properties	Thermal energy storage, heat exchanger materials	Improved energy storage capacity, enhanced heat transfer during phase change	Complexity in design, cost, long-term stability
Aerogels with nanostructures	Extremely low thermal conductivity, lightweight	Insulating materials for heat exchangers	Superior insulation properties, lightweight, high porosity	Mechanical fragility, cost, production scalability
Quantum dots	Tunable thermal properties, nanoscale dimensions	Innovative heat transfer fluids, sensors	Precise control of thermal properties, potential for advanced thermal management	Cost, integration with conventional systems, long-term stability

One notable example is the incorporation of high thermal conductivity fillers, such as carbon fibers or metallic particles, into polymer matrices to create thermally conductive composites. These composites can be used in heat exchanger components such as fins, tubes, or plates to enhance heat transfer rates and reduce thermal resistance. Furthermore, composite materials offer the flexibility to tailor mechanical properties, corrosion resistance, and other factors to meet the specific requirements of different heat transfer applications, making them versatile solutions for improving thermal performance across industries.

16.6 AI AND MACHINE LEARNING INTEGRATION WITH HEAT EXCHANGERS

The role of AI and ML in future innovative heat exchanger design is poised to revolutionize the field by enabling data-driven optimization, predictive modeling, and autonomous design generation. AI and ML algorithms [78, 79] can analyze vast datasets of engineering systems (such as energy systems, including heat exchangers), performance parameters, fluid properties, and operational conditions to identify patterns and correlations that human designers may overlook.[39] This data-driven approach allows for the creation of predictive models that can accurately simulate heat exchanger behavior in various scenarios, enabling engineers to optimize designs for maximum efficiency, energy savings, and reliability. Moreover, AI and ML techniques can facilitate automated design generation, where algorithms iteratively explore and refine heat exchanger configurations to meet specified performance objectives. For example, AI-driven optimization algorithms can be used to generate novel micro-scale heat exchanger geometries with optimized fluid flow patterns, leading to enhanced heat transfer efficiency and compact designs. Overall, AI/ML holds immense potential for advancing heat exchanger design by leveraging data analytics, predictive modeling, and automated optimization to unlock new levels of performance and innovation.

The incorporation of AI and machine learning into heat exchanger design and operation represents a significant shift toward data-driven optimization and automation. AI algorithms can analyze massive datasets of heat exchanger performance, fluid dynamics, and operational parameters to find patterns and correlations that lead to more efficient designs. Machine learning models can predict heat exchanger behavior in various conditions, enabling engineers to optimize designs for maximum performance and energy efficiency. Additionally, AI-driven control systems can dynamically adjust operating parameters in real-time to adapt to changing conditions and optimize heat exchanger performance. For example, AI-based predictive maintenance algorithms can forecast potential issues and recommend proactive interventions to prevent costly downtime. Overall, the integration of AI and machine learning in heat exchangers enhances design efficiency, operational reliability, and energy savings, driving innovation in thermal management across industries. Machine learning models can anticipate possible equipment failures and suggest preventive maintenance measures to reduce downtime and avoid expensive disruptions by evaluating historical data on equipment performance, maintenance activities, and environmental conditions. For example, in manufacturing plants, ML algorithms can analyze sensor data from machinery to detect early signs of equipment degradation or malfunction, enabling proactive maintenance interventions before failures occur. Furthermore, ML-driven optimization techniques can continuously monitor and adjust operational parameters to maximize efficiency and minimize energy consumption. For instance, in smart grid systems, ML algorithms can optimize energy distribution and demand response strategies based on real-time data, improving system reliability and reducing energy costs. Overall, the application of ML algorithms for predictive maintenance and optimization empowers organizations to streamline operations, enhance asset performance, and achieve significant cost savings in the long run.

16.7 SMART HEAT EXCHANGERS AND IOT INTEGRATION

The concept of smart heat exchangers equipped with sensors and actuators represents a paradigm shift toward real-time monitoring, control, and optimization of thermal management systems. By integrating sensors throughout the heat exchanger, including temperature, pressure, flow rate, and fouling sensors, operators can continuously monitor key performance parameters and detect anomalies or inefficiencies. Actuators, such as variable flow valves or variable speed fans, enable dynamic adjustments to operating conditions based on real-time sensor data, optimizing heat transfer efficiency and energy consumption. For instance, in HVAC systems, smart heat exchangers can automatically adjust airflow rates or fluid flow velocities to maintain desired temperature levels while minimizing energy usage. Furthermore, predictive maintenance algorithms can analyze sensor data to forecast potential issues and trigger proactive maintenance actions, ensuring optimal performance and reliability. Overall, smart heat exchangers equipped with sensors and actuators offer enhanced functionality, efficiency, and reliability in thermal management applications across industries. The integration of IoT technology for real-time heat exchanger monitoring and control applications revolutionizes thermal management by enabling remote monitoring, data analytics, and automated control. IoT sensors embedded within the heat exchanger continuously collect data on temperature, pressure, flow rates, and other relevant parameters, transmitting this information to a central monitoring system via wireless connectivity. Through cloud-based platforms, operators can remotely access real-time performance data, detect anomalies, and identify potential issues before they escalate. Moreover, IoT-enabled heat exchangers can leverage machine learning algorithms to analyze data trends, predict equipment failures, and optimize operational parameters for energy efficiency and performance. For example, in industrial processes, IoT-enabled heat exchangers can dynamically adjust cooling rates or fluid flow rates based on real-time demand and environmental conditions, optimizing process efficiency while minimizing energy consumption. Overall, the integration of IoT technology in heat exchanger applications offers unprecedented levels of monitoring, control, and optimization, leading to improved reliability, efficiency, and cost savings across industries.

16.8 RENEWABLE ENERGY INTEGRATION IN HEAT EXCHANGE SYSTEMS

Heat exchangers play integral roles in renewable energy systems, including solar thermal and geothermal applications, facilitating the efficient transfer of thermal energy for power generation and heating. In solar thermal systems, heat exchangers are used to capture and transfer solar energy from sunlight to a working fluid, such as water or thermal oil, which is then circulated through a heat exchanger to generate steam for electricity production or for heating purposes. The heat exchanger enables the transfer of heat from the solar collector to the working fluid while maintaining the necessary separation between the two. Similarly, in geothermal systems, heat exchangers are employed to extract heat from underground reservoirs of hot water or steam and transfer it to a secondary fluid for electricity generation or heating. Heat exchangers in geothermal systems help maximize the efficiency of heat extraction and minimize heat loss during the energy conversion process. Overall, heat exchangers are essential components in renewable energy systems, enabling the efficient harnessing and utilization of renewable thermal energy sources for sustainable power generation and heating applications. Heat exchangers are poised to play a pivotal role in the transition toward sustainable energy sources by enabling the efficient utilization of renewable thermal energy for power generation, heating, and cooling applications. As the world shifts toward renewable energy sources such as solar, geothermal, and biomass, heat exchangers facilitate the conversion of thermal energy into usable forms, such as electricity or hot water, while minimizing energy losses and environmental impact.[40, 41] For instance, in concentrated solar power (CSP) plants, heat exchangers capture and transfer solar energy to generate steam for electricity production, contributing to the decarbonization of the energy sector. Additionally, heat exchangers play crucial roles in district heating systems, where they optimize the distribution of heat generated from renewable sources to residential and commercial buildings, reducing the reliance on fossil fuels for space heating and hot water. Overall, the prospects of heat exchangers in the transition toward sustainable energy sources are promising, as they enable the efficient harnessing and utilization of renewable thermal energy, driving global transition to a more sustainable and low-carbon future.

16.9 THERMOELECTRIC HEAT EXCHANGERS

Thermoelectric materials and the Seebeck effect are fundamental concepts in thermoelectricity, a phenomenon where a temperature gradient across a material leads to the generation of an electric voltage. The Seebeck effect describes the generation of a voltage when a temperature difference exists between two points in a conductor or semiconductor material. This effect occurs due to the movement of charge carriers (electrons or holes) in response to the temperature gradient, resulting in a potential difference across the material. Thermoelectric materials, such as certain semiconductors and alloys, exhibit a high Seebeck coefficient, indicating a strong response to temperature gradients. These materials can be used in thermoelectric generators to convert waste heat from sources, such as industrial processes, automotive engines, or even the human body, into useful electrical power. Conversely, thermoelectric coolers utilize the Peltier effect, the reverse of the Seebeck effect, to create a temperature difference when an electric current is passed through a thermoelectric material, enabling efficient solid-state cooling without the need for refrigerants. Continued research into thermoelectric materials and the optimization of the Seebeck effect hold promise for applications in energy harvesting, waste heat recovery, and thermal management.

The utilization of thermoelectric modules in heat exchangers for electricity generation represents a promising approach to harnessing waste heat and converting it into useful electrical power. Thermoelectric modules, composed of thermoelectric materials with high Seebeck coefficients, are integrated into heat exchangers to exploit temperature differentials between hot and cold streams. As heat flows through the heat exchanger, the temperature gradient across the thermoelectric modules induces a voltage difference, generating electricity through the Seebeck effect. This electricity can then be used to power sensors, actuators, or other electronic devices, offsetting energy consumption and reducing overall operating costs. For example, thermoelectric modules integrated into industrial heat exchangers can capture waste heat from manufacturing processes and convert it into electricity, improving energy efficiency and sustainability. Similarly, automotive exhaust systems equipped with thermoelectric heat exchangers can recover waste heat from engine exhaust gases to generate electrical power for onboard systems, enhancing vehicle efficiency and reducing emissions. Overall, the integration of thermoelectric modules in heat exchangers offers a novel approach to electricity generation from waste heat, contributing to the advancement of sustainable energy solutions. In Figure 16.6, the challenges of thermoelectric heat exchangers are shown, and Table 16.8 shows possible solutions for future upgrades.

These challenges and opportunities highlight the multifaceted nature of developing thermoelectric heat exchangers and the need for interdisciplinary approaches involving materials science, engineering, manufacturing, and sustainability considerations.

16.10 BIOMIMICRY AND NATURE-INSPIRED HEAT EXCHANGER DESIGNS

16.10.1 Exploration of Heat Exchanger Designs Inspired by Natural Phenomena with Examples

Exploring heat exchanger designs inspired by natural phenomena can lead to innovative solutions that enhance efficiency and sustainability. One example is biomimicry-inspired heat exchangers based on the principles of counter-current heat exchange found in the circulatory systems of marine mammals like whales and dolphins. These animals possess specialized anatomical structures, such as the rete mirabile, which enable efficient exchange of heat between arterial and venous blood vessels, minimizing heat loss during prolonged dives in cold waters. Biomimetic heat exchanger designs emulate this mechanism by incorporating intricate networks of micro-channels or fin arrangements that enable the counter-current flow of hot and cold fluids, maximizing heat transfer efficiency. Such biomimetic heat exchangers have applications in HVAC systems, refrigeration, and industrial processes, where minimizing energy consumption and optimizing thermal performance

FIGURE 16.6 Challenges of thermoelectric heat exchangers.

TABLE 16.8
Challenges of Thermoelectric Heat Exchangers in Specific Parameters

Challenges	Possible Solutions
Low efficiency	Improving thermoelectric material performance through advancements in material synthesis and processing techniques, such as nanostructuring and doping.
Cost-effectiveness	Reducing production costs by optimizing manufacturing processes, scaling up production, and exploring novel materials or hybrid approaches.
Temperature compatibility	Developing thermoelectric materials that can operate efficiently across a wide range of temperatures, including high-temperature and low-temperature differentials.
Durability and reliability	Enhancing the durability and reliability of thermoelectric modules through materials engineering, encapsulation techniques, and rigorous testing and validation protocols.
Scale-up and integration	Scaling up thermoelectric heat exchanger designs for practical applications, including integration into existing systems and compatibility with standard heat transfer equipment.
Heat transfer enhancement	Enhancing heat transfer within thermoelectric modules through innovative heat exchanger designs, fluid flow optimization, and thermal interface materials.
Environmental impact	Mitigating environmental concerns associated with thermoelectric materials, such as toxic elements or manufacturing processes, through sustainable practices and recycling initiatives.

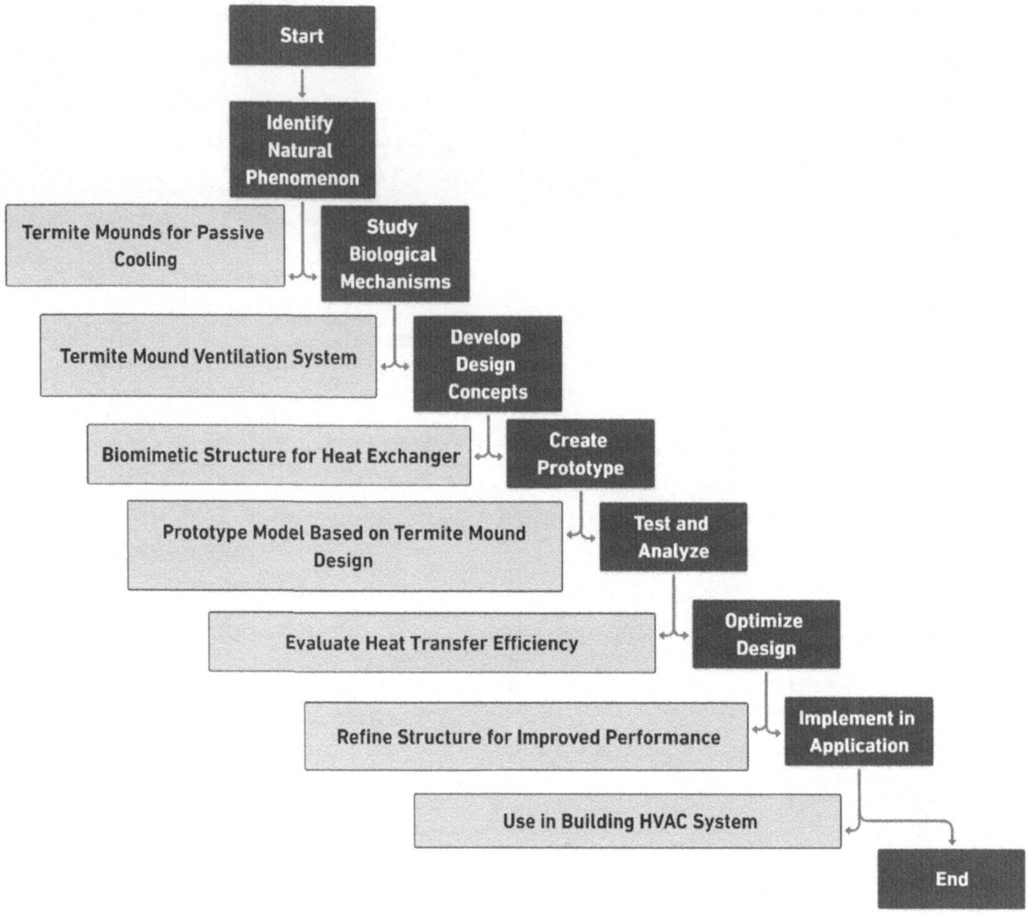

FIGURE 16.7 Flowchart of heat exchanger designs inspired by natural phenomena.

are critical. By drawing inspiration from nature's designs, engineers can develop heat exchangers that not only enhance efficiency but also contribute to sustainability by reducing energy consumption and environmental impact. A flowchart of heat exchanger designs inspired by natural phenomena is represented in Figure 16.7.

16.10.2 Potential Benefits and Applications of Nature-Inspired Heat Exchangers in Various Industries

Nature-inspired heat exchangers offer several potential benefits, including enhanced efficiency, minimized energy consumption, sustainable design, and adaptability to diverse environments. These benefits make them suitable for a wide range of applications across various industries, including HVAC systems, refrigeration, industrial processes, power generation, automotive, aerospace, renewable energy systems, bioengineering, environmental control, marine engineering, geothermal energy, biomedical devices, and food processing. By drawing inspiration from nature's designs, heat exchangers can be optimized for specific applications, leading to improved performance, energy savings, and environmental sustainability. In Figure 16.8, different potential benefits and applications of nature-inspired heat exchangers are elaborated.

FIGURE 16.8 Potential benefits and applications of nature-inspired heat exchangers.

16.11 CONCLUSION

Recent improvements in designing, analyzing, and optimizing heat exchangers have brought many benefits such as better performance, energy savings, cutting costs, and safety. These advancements offer a range of designs that suit different business needs. Engineers now use advanced computer simulations to improve designs effectively. They also study heat exchanger behavior closely using tools like CFD and FEA. These advances pave the way for even better heat exchangers in the future. They help tackle challenges in global energy use by boosting overall performance and energy efficiency. By using advanced designs, thorough analysis, and optimization techniques, businesses can significantly reduce emissions and increase energy efficiency. Innovation in heat exchanger design is paramount for achieving future sustainability and efficiency goals across industries. As global demand for energy continues to rise and concerns about climate change intensify, the development of more efficient and sustainable thermal management solutions becomes increasingly critical. Heat exchangers play a central role in numerous applications, from HVAC systems and industrial processes to renewable energy generation and transportation. By embracing innovation in heat exchanger design, we can unlock significant opportunities to enhance energy efficiency, reduce greenhouse gas emissions, and minimize resource consumption. Novel materials, advanced manufacturing techniques, and biomimetic design principles offer avenues for optimizing heat transfer performance, improving durability, and reducing environmental impact. Moreover, integrating emerging technologies such as AI, ML, and IoT enables real-time monitoring, predictive maintenance, and autonomous control, further enhancing the sustainability and efficiency of heat exchanger systems. Ultimately, investing in innovation in heat exchanger design not only drives technological advancement but also fosters a more sustainable and resilient future for generations to come.

16.12 FUTURE CHALLENGES AND OPPORTUNITIES IN THE FIELD

The field of heat exchangers faces several challenges and opportunities as it progresses into the future.

16.12.1 Challenges

- *Materials innovation.* Developing high-performance materials with improved thermal conductivity, corrosion resistance, and durability is crucial for advancing heat exchanger technology and meeting evolving industry demands.
- *Complex operating conditions.* Heat exchangers must operate effectively under diverse and often extreme conditions, including high temperatures, corrosive environments, and variable flow rates, posing challenges for design, materials selection, and maintenance.

- *Integration with emerging technologies.* Integrating heat exchangers with emerging technologies such as artificial intelligence, additive manufacturing, and nanotechnology requires overcoming technical barriers and ensuring compatibility with existing systems.
- *Compact and lightweight designs.* With industries like data centers and electric vehicles needing more compact solutions, designing smaller, lightweight, yet highly efficient heat exchangers is a challenge.
- *Increasing energy demand.* With growing energy demands globally, heat exchangers must continuously improve efficiency to meet increasing energy requirements while minimizing environmental impact.
- *Climate change mitigation.* Addressing climate change necessitates reducing greenhouse gas emissions, requiring heat exchangers to play a pivotal role in enhancing energy efficiency and promoting the adoption of renewable energy sources.
- *Sustainability and environmental impact.* The shift to renewable energy and stricter environmental regulations requires the development of eco-friendly materials and more sustainable manufacturing processes for heat exchangers.

16.12.2 Opportunities

- *Efficiency optimization.* Opportunities exist for optimizing heat exchanger efficiency through innovative design strategies, advanced materials, and improved manufacturing techniques, leading to energy savings and reduced operational costs.
- *Renewable energy integration.* Heat exchangers can facilitate the integration of renewable energy sources such as solar, geothermal, and waste heat recovery into existing infrastructure, contributing to decarbonization efforts and enhancing energy sustainability.
- *Smart and autonomous systems.* Leveraging IoT, AI, and machine learning technologies enables the development of smart heat exchanger systems capable of real-time monitoring, predictive maintenance, and autonomous operation, improving reliability and performance.
- *Cross-industry collaboration.* Collaborative efforts between industries, academia, and research institutions facilitate knowledge exchange, innovation, and the development of interdisciplinary solutions to complex heat exchanger challenges.
- *Circular economy principles.* Embracing circular economy principles in heat exchanger design promotes resource efficiency, recycling, and waste reduction, creating opportunities for sustainable materials sourcing, life-cycle management, and end-of-life recycling.

These evolving demands will drive innovation and create opportunities for new, more advanced heat exchanger technologies[40–67] in the coming years. Addressing these challenges and seizing opportunities will require interdisciplinary collaboration, investment in research and development, and a commitment to sustainability and innovation in the field of heat exchangers. By overcoming these challenges and embracing emerging opportunities, the future of heat exchanger technology holds immense potential for advancing energy efficiency, sustainability, and global resilience.

16.12.3 Recommendations

Recent advancements in heat exchanger design, analysis, and optimization have greatly helped businesses across different industries. Engineers now use better tools for calculating multi-phase flows, mathematical modeling, and simulations. These improvements have enabled the creation of more efficient and cost-effective heat exchangers.

- *Use of advanced materials.* Utilize high-performance alloys and sophisticated coatings to reduce size and weight while enhancing thermal efficiency. Incorporate modern materials such as nanofluids, micro-channel technology, and polymer-based materials for superior performance.

- *Adopt cutting-edge design and analysis tools.* Employ advanced numerical and simulation techniques for precise design, analysis, and optimization. Use new computational tools to achieve accurate control and improved efficiency.
- *Optimize thermal efficiency.* Use enhanced surface geometries such as fins, micro-channels, or lattice structures to maximize surface area and improve heat transfer. Incorporate counter-flow or cross-flow configurations for higher thermal efficiency.
- *Use of modern manufacturing methods.* The application of contemporary manufacturing techniques, such as 3D printing, computer numerical control machining, robotics, and smart manufacturing, enables the fabrication of complex and intricate heat exchanger geometries. This offers substantial advantages for the thermal management of data centers. Key benefits include enhanced design flexibility, accelerated prototyping, and improved thermal performance efficiency. These modern manufacturing approaches are transforming the design and optimization of heat exchanger systems.
- *Implement innovative technologies.* Design micro-channel heat exchangers and other creative configurations to enhance performance. Integrate artificial intelligence with existing optimization methodologies to optimize and improve performance.
- *Focus on long-term cost efficiency.* Aim for significant long-term cost reductions by adopting the latest technological advancements and improvements in materials.
- *Maximize energy efficiency and operational value.* Continuously update designs to incorporate the latest advancements for achieving the highest potential energy efficiency and operational value.
- *Prioritize compact and lightweight designs.* Focus on miniaturization for applications like electric vehicles, aerospace, or data centers, where space and weight are critical. Incorporate thin-walled, high-strength materials to reduce weight without compromising durability.
- *Consider modular and scalable designs.* Create modular heat exchanger systems that allow for easy scaling or replacement of individual components. Facilitate customization to adapt to varying industry needs or thermal requirements.
- *Utilize smart monitoring and IoT.* Integrate smart sensors and IoT technologies for real-time monitoring of temperature, pressure, and performance. Implement predictive maintenance algorithms using AI to reduce downtime and improve system longevity.

REFERENCES

1. Abouelregal, A.E. and Alesemi, M., 2022. Vibrational analysis of viscous thin beams stressed by laser mechanical load using a heat transfer model with a fractional Atangana-Baleanu operator. *Case Studies in Thermal Engineering*, 34, p.102028.
2. Auld, D.S., Coassin, P.A., Coussens, N.P., Hensley, P., Klumpp-Thomas, C., Michael, S., Sittampalam, G.S., Trask, Baroutaji, A., Arjunan, A., Ramadan, M., Robinson, J., Alaswad, A., Abdelkareem, M.A. and Olabi, A.G., 2021. Advancements and prospects of thermal management and waste heat recovery of PEMFC. *International Journal of Thermofluids*, 9, p.100064.
3. Berce, J., Zupančič, M., Može, M. and Golobič, I., 2023. Infrared thermography observations of crystallization fouling in a plate heat exchanger. *Applied Thermal Engineering*, 224, p.120116.
4. Diwania, S., Agrawal, S., Siddiqui, A.S. and Singh, S., 2020. Photovoltaic–thermal (PV/T) technology: a comprehensive review on applications and its advancement. *International Journal of Energy and Environmental Engineering*,11, pp.33–54.
5. Gao, J., Hu, Z., Yang, Q., Liang, X. and Wu, H., 2022. Fluid flow and heat transfer in microchannel heat sinks: modelling review and recent progress. *Thermal Science and Engineering Progress*, p.101203.
6. Ghachem, K., Aich, W. and Kolsi, L., 2021. Computational analysis of hybrid nanofluid enhanced heat transfer in cross flow micro heat exchanger with rectangular wavy channels. *Case Studies in Thermal Engineering*, 24, p.100822.
7. Ghalambaz, M., Mahdi, J.M., Shafaghat, A., Eisapour, A.H., Younis, O., Talebizadeh Sardari, P. and Yaïci, W., 2021. Effect of twisted fin array in a triple-tube latent heat storage system during the charging mode. *Sustainability*, 13(5), p.2685.

8. Hong, J., Su, M., Zhao, K., Zhou, Y., Wang, J., Zhou, S.F. and Lin, X., 2023. A minireview for recent development of nanomaterial-based detection of antibiotics. *Biosensors*, 13(3), p.327.
9. Kumar, L., Hasanuzzaman, M. and Rahim, N.A., 2019. Global advancement of solar thermal energy technologies for industrial process heat and its future prospects: a review. *Energy Conversion and Management*, 195, pp.885–908.
10. Liu, J., Zhou, Z., Zhang, J., Feng, W. and Zuo, J., 2019. Advances and challenges in commercializing radiative cooling. *Materials Today Physics*, 11, p.100161.
11. Maghrabie, H.M., Elsaid, K., Sayed, E.T., Abdelkareem, M.A., Wilberforce, T., Ramadan, M. and Olabi, A.G., 2021 Intensification of heat exchanger performance utilizing nanofluids. *International Journal of Thermofluids*, 10, p.100071.
12. Malinowski, R., Parkin, I.P. and Volpe, G., 2020. Advances towards programmable droplet transport on solid surfaces and its applications. *Chemical Society Reviews*, 49(22), pp.7879–7892.
13. Menni, Y., Chamkha, A.J. and Ameur, H., 2020. Advances of nanofluids in heat exchangers—A review. *Eat Transfer*, 49(8), pp.4321–4349.
14. Nafis, B.M., Whitt, R., Iradukunda, A.C. and Huitink, D., 2021. Additive manufacturing for enhancing thermal dissipation in heat sink implementation: a review. *Heat Transfer Engineering*, 42(12), pp.967–984.
15. Santosh, R., Arunkumar, T., Velraj, R. and Kumaresan, G., 2019. Technological advancements in solar energy driven humidification-dehumidification desalination systems-A review. *Journal of Cleaner Production*, 207, pp.826–845.
16. Tao, P., Chang, C., Tong, Z., Bao, H., Song, C., Wu, J., Shang, W. and Deng, T., 2019. Magnetically-accelerated large-capacity solar-thermal energy storage within hightemperature phase-change materials. *Energy & Environmental Science*, 12(5), pp.1613–1621.
17. Yuan, K., Shi, J., Aftab, W., Qin, M., Usman, A., Zhou, F., Lv, Y., Gao, S. and Zou, R., 2020. Engineering the thermal conductivity of functional phase-change materials for heat energy conversion, storage, and utilization. *Advanced Functional Materials*, 30(8), p.1904228.
18. Zhang, S., Feng, D., Shi, L., Wang, L., Jin, Y., Tian, L., Li, Z., Wang, G., Zhao, L. and Yan, Y., 2021. A review of phase change heat transfer in shape-stabilized phase change materials (ss-PCMs) based on porous supports for thermal energy storage. *Renewable and Sustainable Energy Reviews*, 135, p.110127.
19. Zhuang, S., Qi, H., Wang, X., Li, X., Liu, K., Liu, J. and Zhang, H., 2021. Advances in solar-driven hygroscopic water harvesting. *Global Challenges*, 5(1), p.2000085.
20. Elsheikh, A.H., Panchal, H.N., Sengottain, S., A Alsaleh, N. and Ahmadein, M., 2022. Applications of heat exchanger in solar desalination: current issues and future challenges. *Water*, 14(6), p.852.
21. Gilmour, C.H., 1960. Application of heat exchangers in chemical plants. *Industrial & Engineering Chemistry*, 52(6), pp.465–467.
22. Hatami, M., Ganji, D.D. and Gorji-Bandpy, M., 2014. A review of different heat exchangers designs for increasing the diesel exhaust waste heat recovery. *Renewable and Sustainable Energy Reviews*, 37, pp.168–181.
23. Hesselgreaves, J.E., Law, R. and Reay, D., 2016. *Compact heat exchangers: selection, design and operation*. Butterworth-Heinemann.
24. Hosseini, S., Khandakar, A., Chowdhury, M.E., Ayari, M.A., Rahman, T., Chowdhury, M.H. and Vaferi, B., 2022. Novel and robust machine learning approach for estimating the fouling factor in heat exchangers. *Energy Reports*, 8, pp.8767–8776.
25. Huang, R., Mahvi, A., Odukomaiya, W., Goyal, A. and Woods, J., 2022. Reduced-order modeling method for phase-change thermal energy storage heat exchangers. *Energy Conversion and Management*, 263, p.115692.
26. Hundy, G.F., 2016. *Refrigeration, air conditioning and heat pumps*. Butterworth-Heinemann.
27. Kishan, R., Singh, D. and Sharma, A.K., 2020. CFD Analysis of heat exchanger models design using ansys fluent. *International Journal of Mechanical Engineering and Technology*, 11(2). https://doi.org/10.34218/IJMET.11.2.2020.001.
28. Končar, B. and Klobučar, L., 2018. Use of CFD codes for calculation of radiation heat transfer: from validation to application. In *Heat transfer-models, methods and applications*. IntechOpen.
29. Liu, C., Bu, W. and Xu, D., 2017. Multi-objective shape optimization of a plate-fin heat exchanger using CFD and multi-objective genetic algorithm. *International Journal of Heat and Mass Transfer*, 111, pp.65–82.
30. Liu, S. and Sakr, M., 2013. A comprehensive review on passive heat transfer enhancements in pipe exchangers. *Renewable and Sustainable Energy Reviews*, 19, pp.64–81.

31. Longatte, E., Baj, F., Hoarau, Y., Braza, M., Ruiz, D. and Canteneur, C., 2013. Advanced numerical methods for uncertainty reduction when predicting heat exchanger dynamic stability limits: review and perspectives. *Nuclear Engineering and Design*, 258, pp.164–175.
32. Padalkar, A.S., Patil, A.S. and Sapali, S.N., 2011. Review of researchers on heat transfer augmentation techniques. In *Proceedings of the 6th national conference on advances in energy conversion technologies*. Manipal University, India, pp.2–8.
33. Rao, A.V. and Krishnamoorthy, G., 2022. CFD analysis of heat transfer in a helical coil heat exchanger using nanofluids. *Journal of Thermal Analysis and Calorimetry*, 147(6), pp.4517–4530.
34. Said, Z., Saidur, R., Mekhilef, S. and Rahim, N.A., 2014. Environmental impact of nanofluid in heat transfer. *Renewable and Sustainable Energy Reviews*, 31, pp.492–499.
35. Sheng, Q., Tao, W. and Zhang, W., 2022. Heat transfer and flow characteristics in helical tube heat exchanger with constant and variable helix angles. *Applied Thermal Engineering*, 209, p.118191.
36. Tan, Y.K., Wan, N.Y. and Yew, H.H., 2009. Application of CFD analysis on the shell-side of shell-and-tube heat exchanger. In *2009 International conference on advances in computational tools for engineering applications*. IEEE, pp.70–74.
37. Tapia, A., Salazar, A. and Palma, J.L., 2022. Computational fluid dynamics simulation to predict nucleation sites in low-pressure heat exchangers: a case study. *Energies*, 15(3), p.944.
38. Valero, M. and Cascales, J., 2022. Optimization of heat exchanger design for geothermal systems using computational fluid dynamics and response surface methodology. *Energies*, 15(4), p.1021.
39. Sharadga, H., Dawahdeh, A. and Kumar, S., 2024. Accelerated MPPT controller for photovoltaic systems under rapid solar radiation changes using neural network. *Journal of Sustainibility*, MDPI.
40. Kumar, S. and Rathore, K., 2023. Renewable energy for sustainable development goal of clean and affordable energy. *International Journal of Materials, Manufacturing and Sustainable Technologies*, 2(1), pp.1–15.
41. Rathore, K. and Kumar, S., 2023. Net zero emissions for our future generations through renewables: a brief review. *International Journal of Energy Resources Applications (IJERA)*, 2(1), pp.22–43.
42. Awasthi, M.K., Kumar, A., Dutt, N. and Kumar, S., 2023. Electrohydrodynamic capillary instability of Rivlin-Ericksen viscoelastic fluid film with mass and heat transfer. *Heat Transfer*, 53(1). https://doi.org/10.1002/htj.22944.
43. Kumar, S., Thyagarajan, A. and Banerjee, D., 2022. Experimental investigation of thermal energy storage (TES) platform leveraging phase change materials in a chevron plate heat exchanger. *International Mechanical Engineering Congress and Exposition*, pp.1–7. https://doi.org/10.1115/IMECE2022-96226.
44. Kumar, S., Grover, R.B., Yadav, H., Vijayan, P.K., Kannan, U., Agrawal, A., Grover, R.B., Kannan, U., Yadav, H. and Agrawal, A., 2020. Experimental investigations on thermal stratification in a large pool of water with immersed isolation condenser. *International Conference on Nuclear Engineering*, 3, pp.1–7.
45. Kumar, S., Thyagarajan, A. and Banerjee, D., 2020. An investigation on chevron plate heat exchanger filled with organic phase change material. In *Proceedings of the american society of mechanical engineers 2020 heat transfer summer conference HT2020 July 13-15*. Virtual, online.
46. Kumar, S., Grover, R.B., Yadav, H., Vijayan, P.K., Kannan, U. and Agrawal, A., 2018. Experimental and numerical investigation on suppression of thermal strati fi cation in a water-pool: PIV measurements and CFD simulations. *Applied Thermal Engineering*, 138, pp.686–704.
47. Kumar, S., Vijayan, P.K., Kannan, U., Sharma, M. and Pilkhwal, D.S., 2017. Experimental and computational simulation of thermal stratification in large pools with immersed condenser by. *Applied Thermal Engineering*, 113, pp.345–361.
48. Kumar, S., Grover, R.B., Vijayan, P.K. and Kannan, U., 2017. Numerical investigation on the effect of shrouds around an immersed isolation condenser on the thermal stratification in large pools. *Annals of Nuclear Energy*, 110, pp.109–125.
49. Kumar, S. and Banerjee, D., 2024. PCMs for sustainability applications by leveraging machine learning. In *Phase change materials (PCMs) - recent advances, new perspectives, and applications*. Intech Open Publication.
50. Verma, D., Kalra, M.S. and Wahi, P., 2017. A comparative study of bifurcation characteristics in single and two-zone fuel models of a BWR: a case study of the advanced heavy water reactor (AHWR). *Nuclear Engineering and Design*, 324, pp.158–170.
51. Verma, D., Kalra, M.S. and Wahi, P., 2017. Effect of nonlinear void reactivity on bifurcation characteristics of a lumped-parameter model of a BWR: a study relevant to RBMK. *Nuclear Engineering and Design*, 315, pp.179–193.

52. Verma, D., Paul, S. and Wahi, P., 2018. Stability and bifurcation characteristics of a forced circulation BWR using a nuclear-coupled homogeneous thermal-hydraulic model. *Nuclear Science and Engineering*, 190(1), pp.73–92.
53. Verma, D., Kalra, M.S. and Wahi, P., 2016. Dynamics of a BWR with inclusion of boiling non-linearity, clad temperature and void-dependent core power removal: stability and bifurcation characteristics of advanced heavy. *Nuclear Engineering and Design*, 308, pp.297–311.
54. Webb, R.L., 2000. *Heat exchanger design handbook*. CRC Press.
55. Zhang, G.F., Zhang, Y., Lin, Y. and Zhang, Y., 2018. Research on shell-and-tube heat exchanger with heat transfer enhancement by combining twisted tape inserts and multi-tube pass. *Applied Thermal Engineering*, 141, pp.400–410.
56. Debtera, B., Neme, I. and Prabhu, V., 2018. CFD simulation of a double pipe heat exchanger: analysis conduction and convection heat transfer. *International Journal of Scientific Research and Review*, 7, p.12.
57. Firouzfar, E., Soltanieh, M., Noie, S.H. and Saidi, M.H., 2011. Application of heat pipe heat exchangers in heating, ventilation and air conditioning (HVAC) systems. *Scientific Research and Essays*, 6(9), pp.1900–1908.
58. Gilmour, C.H., 1960. Application of heat exchangers in chemical plants. *Industrial & Engineering Chemistry*, 52(6), pp.465–467.
59. Hatami, M., Ganji, D.D. and Gorji-Bandpy, M., 2014. A review of different heat exchangers designs for increasing the diesel exhaust waste heat recovery. *Renewable and Sustainable Energy Reviews*, 37, pp.168–181.
60. Hesselgreaves, J.E., Law, R. and Reay, D., 2016. *Compact heat exchangers: selection, design and operation*. Butterworth-Heinemann.
61. Kishan, R., Singh, D. and Sharma, A.K., 2020. CFD analysis of heat exchanger models design using ansys fluent. *International Journal of Mechanical Engineering and Technology*, 11(2), pp.1–9.
62. Liu, S. and Sakr, M., 2013. A comprehensive review on passive heat transfer enhancements in pipe exchangers. *Renewable and Sustainable Energy Reviews*, 19, pp.64–81.
63. Luo, L., Wen, F., Wang, L., Sundén, B. and Wang, S., 2016. Thermal enhancement by using grooves and ribs combined with delta-winglet vortex generator in a solar receiver heat exchanger. *Applied Energy*, 183, pp.1317–1332.
64. Ma, T., Guo, Z., Lin, M. and Wang, Q., 2021. Recent trends on nanofluid heat transfer machine learning research applied to renewable energy. *Renewable and Sustainable Energy Reviews*, 138, p.110494.
65. Shah, R.K. and Sekulić, D.P., 2003. *Fundamentals of heat exchanger design.*, July 25. https://doi.org/10.1002/9780470172605.
66. Sheer, T., Klerk, G.B.D., Jawurek, H. and Lander, M., 2006. A versatile computer simulation model for rotary regenerative heat exchangers. *Taylor & Francis*, 27(5), pp.68–79.
67. Edreis, E. and Petrov, A., 2020. Types of heat exchangers in industry, their advantages and disadvantages, and the study of their parameters. *IOP Publishing*, 963(1), pp.012027–012027.
68. Master, B.I., Chunangad, K.S., Boxma, A.J., Kral, D. and Stehlík, P., 2006. Most frequently used heat exchangers from pioneering research to worldwide applications. *Taylor & Francis*, 27(6), pp.4–11.
69. Reppich, M., 1999. Use of high performance plate heat exchangers in chemical and process industries. *Elsevier BV*, 38(11), pp.999–1008.
70. Chen, K.J., 2014. Design of the plate heat exchanger. *Trans Tech Publications*, 912–914, pp.624–627.
71. Omidi, M., Farhadi, M. and Jafari, M., 2016. A comprehensive review on double pipe heat exchangers. *Elsevier BV*, 110, pp.1075–1090.
72. Serth, R.W. and Lestina, T.G., 2014. *Design of double-pipe heat exchangers*. November 5.
73. Hesselgreaves, J.E., Law, R. and Reay, D., 2016. Industrial compact exchangers. *Elsevier BV*, pp.35–89.
74. Dellenback, P.A., 2002 Improved gas turbine efficiency through alternative regenerator configuration. *ASM International*, 124(3), pp.441–446.
75. Andriani, R., Gamma, F. and Ghezzi, U., 2008. Regeneration and intercooling in gas turbine engines for propulsion systems. https://doi.org/10.2514/6.2008-4899.
76. Hassani, V., Dickens, J. and Bell, K.J., 2005. The Fin-on-plate heat exchanger: a new configuration for air-cooled power plants. *Taylor & Francis*, 26(6), pp.7–15.
77. Carnavos, T.C., 1979. Cooling air in turbulent flow with internally finned tubes. *Taylor & Francis*, 1(2), pp.41–46.
78. Ejigu, D. A., Tuo, Y. and Liu, X., 2024. Application of artificial intelligence technologies and big data computing for nuclear power plants control: a review. https://doi.org/10.3389/fnuen.2024.1355630.
79. Tanasawa, I., 1991. Advances in condensation heat transfer. *Elsevier BV*, 21, pp.55–139.

80. Wu, D., Wang, Z., Lu, G. and Peng, X., 2010. High-performance air cooling condenser with liquid–vapor separation. *Taylor & Francis*, 31(12), pp.973–980.
81. Matos, R.S., Vargas, J.V.C., Rossetim, M., Pereira, M.V.A., Pitz, D.B. and Ordóñez, J.C., 2020. Performance comparison of tube and plate-fin circular and elliptic heat exchangers for HVAC-R systems. *Elsevier BV*, 184, pp.116288–116288.
82. Promoppatum, P., Yao, S., Hultz, T. and Agee, D., 2016. Experimental and numerical investigation of the cross-flow PCM heat exchanger for the energy saving of building HVAC. *Elsevier BV*, 138, pp.468–478.
83. Jouhara, H. and Meskimmon, R., 2014. Heat pipe based thermal management systems for energy-efficient data centres. *Elsevier BV*, 77, pp.265–270.
84. Tang, G., Han, Y., Chen, H. and Zhang, X., 2018. Design, fabrication and characterization of a mini heat exchanger for data centre cooling application. pp.485–490. https://doi.org/10.1109/EPTC.2018.8654392.
85. Wang, Y., Zheng, W., Jiang, Y. and Sun, C., 2022. The effect of layout of finned tube heat exchangers in a novel cooling system on the thermal environment in data centers. *EDP Sciences*, 356, pp.01010–01010.

Index

A

B

D

I

J

K

L

M

R

S

T